Studies in Systems, Decision and Control

Volume 524

The series "Studies in Systems, Decision and Control" (SSDC) covers both new developments and advances, as well as the state of the art, in the various areas of broadly perceived systems, decision making and control–quickly, up to date and with a high quality. The intent is to cover the theory, applications, and perspectives on the state of the art and future developments relevant to systems, decision making, control, complex processes and related areas, as embedded in the fields of engineering, computer science, physics, economics, social and life sciences, as well as the paradigms and methodologies behind them. The series contains monographs, textbooks, lecture notes and edited volumes in systems, decision making and control spanning the areas of Cyber-Physical Systems, Autonomous Systems, Sensor Networks, Control Systems, Energy Systems, Automotive Systems, Biological Systems, Vehicular Networking and Connected Vehicles, Aerospace Systems, Automation, Manufacturing, Smart Grids, Nonlinear Systems, Power Systems, Robotics, Social Systems, Economic Systems and other. Of particular value to both the contributors and the readership are the short publication timeframe and the world-wide distribution and exposure which enable both a wide and rapid dissemination of research output.

Indexed by SCOPUS, DBLP, WTI Frankfurt eG, zbMATH, SCImago.

All books published in the series are submitted for consideration in Web of Science.

Bahaa Awwad
Editor

The AI Revolution: Driving Business Innovation and Research

Volume 1

Springer

Editor
Bahaa Awwad
Department of Computerized Banking
and Financial Sciences
Faculty of Business and Economics
Palestine Technical University—Kadoorie
Ramallah, State of Palestine

ISSN 2198-4182 ISSN 2198-4190 (electronic)
Studies in Systems, Decision and Control
ISBN 978-3-031-54378-4 ISBN 978-3-031-54379-1 (eBook)
https://doi.org/10.1007/978-3-031-54379-1

This Springer imprint is published by the registered company Springer Nature Switzerland AG
The registered company address is: Gewerbestrasse 11, 6330 Cham, Switzerland

Paper in this product is recyclable.

Preface

In the rapidly evolving landscape of technology, artificial intelligence stands at the forefront of transformative change. This book delves into the profound impact that AI has on both business practices and cutting-edge research, offering a comprehensive exploration of the multifaceted dimensions that characterize the AI revolution.

As we stand on the cusp of a new era, characterized by unprecedented advancements in machine learning, natural language processing, and computer vision, it is imperative for professionals, researchers, and enthusiasts alike to grasp the intricacies of AI's role in shaping the future. This book serves as a guide through the labyrinth of possibilities that AI brings to the table, unraveling the threads of innovation that weave through industries and research domains.

In these pages, we embark on a journey that spans the vast landscape of AI applications. From redefining traditional business models to fueling breakthroughs in scientific research, AI has become an indispensable force driving progress across diverse fields. The chapters within this volume are meticulously crafted to provide insights into the key facets of the AI revolution, offering a blend of theoretical foundations, practical applications, and real-world case studies.

Whether you are a seasoned professional seeking to leverage AI for strategic business advantages or a researcher on the quest for the next breakthrough, this book aims to equip you with the knowledge and perspectives necessary to navigate the AI landscape. The contributors to this volume are experts in their respective fields, sharing their expertise and experiences to illuminate the path forward in this era of unprecedented technological innovation.

"The AI Revolution" is not just a compilation of facts and figures; it is a living testament to the dynamic nature of AI and its pervasive influence on our world. As you immerse yourself in the following chapters, may you gain a deeper understanding of the profound transformations that AI is catalyzing, and may you be inspired to contribute to the ongoing narrative of the AI revolution.

Let this book be your guide through the ever-expanding realm of possibilities that AI presents, and may it inspire you to harness the power of artificial intelligence to drive business innovation and advance the frontiers of research.

Ramallah, State of Palestine Bahaa Awwad

Contents

Marketing Strategies of the Enterprise in the Market of Building Structures in the Period of Recovery of Ukraine

Lidia Romanova, Kristina Bratchykova, Serhii Pavlovskyi, George Abuselidze, Alina Berher, Olga Mohylevska, and Yevhen Hrihierman

Abstract The Russian invasion of Ukraine has led to massive destruction of the economy and the amount of losses is estimated at more than 700 billion dollars. Increased demand in the construction and building structures market due to the restoration of Ukraine requires an increase in the production of sandwich panels. TPK-Center LLC specializes in the production of products for industrial and commercial construction, such as sandwich panels and metal products. Changes in the external environment of the company's operations, the structure of demand for sandwich panels with different insulation materials require changes in its marketing strategy in terms of product mix. Based on the analysis of the business portfolio, such strategic decisions as "BUILD", "HOLD", harvesting and elimination strategies are proposed. A study of the company's customers was conducted and a conclusion was made on the marketing strategy of business relations and the formation of a quality customer experience. The main ideas for improving the marketing strategy are the development of digital marketing, the creation of a specific marketplace for product sales and the involvement of a complex of online promotion of the enterprise in the

L. Romanova · Y. Hrihierman (✉)
Interregional Academy of Personnel Management, 2, Frometivska Street, Kyiv 03039, Ukraine

K. Bratchykova
LLC "TPK-Center", 8, Stepan Bandera Avenue, Kyiv 04073, Ukraine

S. Pavlovskyi
Communal Research Institute "Research Institute of Socio-Economic City Development", 37-A, Heroiv. Sevastopolia Street, Kyiv 03061, Ukraine

Present Address:
G. Abuselidze
Batumi Shota Rustaveli State University, 35/32, Ninoshvili/Rustaveli Street, 6010 Batumi, Georgia
e-mail: george.abuselidze@gmail.com; george.abuselidze@bsu.edu.ge

A. Berher
National University of Food Technologies, 68 Volodymyrska Street, Kyiv 01601, Ukraine

O. Mohylevska
Kyiv International University, 49, Lvovskaya Street, Kyiv 03179, Ukraine

B. Awwad (ed.), *The AI Revolution: Driving Business Innovation and Research*, Studies in Systems, Decision and Control 524, https://doi.org/10.1007/978-3-031-54379-1_1

market. To develop the digital direction and online sales, there is a need to improve the marketing department at the enterprise, which should also include specialists in computer science, economics, and advertising technologies. A marketing strategy for the export of sandwich panels was also proposed. Today, there is high competition in Ukraine, so Western markets, considering the high quality of sandwich panels and the possibility of obtaining European certificates, are quite attractive.

Keywords Marketing strategies · Market of building · Sandwich panels · Product quality · Digital marketing · Access to foreign markets

1 Introduction

Ukraine has been at war for two years now. According to the Kyiv School of Economics (KSE Institute), as of September 1, 2023, the total documented direct damage to Ukraine's infrastructure as a result of the full-scale Russian invasion reached $151.2 billion (at replacement cost). Compared to June 2023, losses increased by more than $700 million—from $150.5 billion to $151.2 billion. Of these, 37.1% are losses of housing stock and 24.2% of infrastructure [1].

The problem of restoring the country's economy is emerging. It is not just about economic recovery as a stage of recovery after the crisis, but about a fundamental transformation of the economy. Historically, Ukraine's economic growth rates have been lower than those of Central European countries. For example, Ukraine's GDP has grown by less than 2% per year over the past 20 years, compared to about 3.5% in Poland. Therefore, Ukraine has developed a targeted Plan to restore equitable prosperity; to stimulate private investment and entrepreneurship in Ukraine as a unique chance not only to compensate for the damage caused by the war, but also to accelerate economic growth and improve the quality of life in Ukraine. The main principles of the Recovery Plan for Ukraine are: integration into the EU; building back better; increasing fair welfare; and stimulating private investment and entrepreneurship.

The drivers of Ukraine's recovery are the development of value-added sectors of the economy against the backdrop of global trends, the Green Deal, and Ukraine's competitive position, one of which is construction [2]. According to a report by the Boston Consulting Group (BCG) in partnership with the World-Wide Fund for Nature (WWF), Ukraine's housing stock consists mainly of buildings with poor insulation and low energy efficiency, accounting for 40% of final energy consumption and 54% of energy consumption in the residential sector is spent on heating. Therefore, it is recommended to introduce a Zero Emission Building (ZEB) standard for new buildings by 2025 and for renovated existing buildings by 2032, in line with the draft revised EU Energy Performance of Buildings Directive in 2021. Also, to establish public procurement rules that support low-carbon building solutions and promote the development of the low-carbon cement sector [3]. The need to study marketing and marketing strategy of enterprises and, in particular, the production of sandwich panels in the current conditions is explained by a radical change in the business environment and the level of its attitude to the problem. Thus, according to the survey, 74% of entrepreneurs in the domestic business do not have a marketing strategy, and 80% of them have not approved it in writing. About half of the respondents have not adapted their goals in wartime, and 18% do not have them at all, and the vast majority of them do not know the size of their own market. The author considers it necessary to engage in marketing around the clock, and defines it as a growth hormone [4]. Therefore, the problem of the development of enterprises in the sandwich panel market and its marketing strategy in the context of Ukraine's recovery is quite relevant. The object of the study is TPK-Center LLC, which produces high-quality sandwich panels. The purpose of the article is to study the directions of formation of the company's marketing strategy in the construction market during the period of Ukraine's recovery.

2 Literature Review

In the literature, marketing strategy is defined as a tool for achieving business goals and objectives through the use of a sustainable competitive advantage in the market and provides for policies and rules for ensuring the direction of the company's marketing activities. It is with the help of marketing strategy that companies can increase their competitive advantage [5, 6]. M. Porter defines three strategies of competitive advantage: cost leadership, differentiation and specialization (focus) [7]. At the present stage of economic development, the role of marketing strategy is growing and is defined as a construct that lies at the conceptual center of the strategic marketing sphere and occupies a central place in marketing practice. Marketing strategy includes the processes of developing and implementing a strategy for the desired goals of the firm for the future period of time and the means of achieving them: selection of target markets and consumers, identification of the necessary value propositions, as well as development and implementation of integrated marketing programs for the development, delivery and communication of the value proposition [8]. Strategic planning at the level of a business unit includes 8 stages: the mission of the business unit, analysis of the external and internal environment—SWOT analysis, goal setting, strategy, programs, program implementation, feedback and control [9, 10]. It is proposed to study resources and capabilities, such as NPD, CRM, and in determining its effectiveness—customer thinking, behavioral results and economic results. Environmental factors affect the marketing strategy and the relationship between internal components [8]. On the other hand, the marketing environment is an important factor in the formation of the marketing strategy of the enterprise [11] and its transformation in case of changes [12]. Therefore, it is very important to conduct a SWOT analysis as a method of evaluating a business portfolio [13].

With regard to construction enterprises, Chinese researchers believe that the performance of construction SMEs critically depends on three key components: competitive marketing strategy, relationship marketing and business environment, and confirmed the importance of differentiating marketing, innovation and relationship system to achieve their performance. Moreover, relationships are given a very important role [14]. The quality of customer experience (QCQ) has become an important component of the business marketing strategy [15]. At industrial enterprises, relationship marketing is determined as a key strategic resource, so in modern conditions there is a transition to strategic partnerships and the formation of marketing partnership networks that include the enterprise, consumers, suppliers, intermediaries [16]. A modern tool for forming a relationship marketing strategy is digital marketing [17, 18].

Thus, the problem of the marketing strategy of enterprises, and in particular, construction enterprises, is sufficiently studied in the literature. However, the directions of formation of the marketing strategy of enterprises in the period of recovery of Ukraine, and especially those engaged in the production of sandwich panels, are not sufficiently studied.

3 Research Methodology

The article uses theoretical research methods: analysis, generalization. The practical part includes statistical methods of analysis, SWOT analysis, BKG matrix, marketplace analysis, research of personal data obtained through a CRM system, during a survey using Google services (Google Forms). We interviewed 36 clients of the company using Google surveys.

Questions in the questionnaire:

1. What sandwich panels were the most popular among customers?
2. Do customers have a binding to the company, brand?
3. What are the main competitors named during negotiations?
   (a) price;
   (b) availability of a "ready-made solution" for the client;
   (c) competencies of the seller (product knowledge, ability to persuade, friendly relations);
   (d) quality of sandwich panels;
   (e) timely delivery.
4. What is a quality sandwich panel in the customer's understanding?

4 Results and Discussion

The construction market in the world is developing rapidly, and construction services account for a significant share of GDP. Thus, in 2023, according to international market research experts, the global construction market grew from $14,503.87 billion in 2022 to $15,461.84 billion in 2023 at a compound annual growth rate (CAGR) of 6.6%. The construction market is expected to grow to $19,519.26 billion in 2027 at a CAGR of 6.0% [19]. Emerging markets include China, Brazil, India, Saudi Arabia, and Indonesia. For example, construction products accounted for 25.9% of China's GDP in 2020, up from 6.2% in 2019. There is a tendency to use green construction technologies for energy-efficient and less costly buildings using sustainable building materials and construction processes with minimal environmental impact. About 60% of construction firms worldwide participated in green building projects % [20].

Growing investments, renewable energy capacity, urbanization, and digitalization will contribute to the market's prospective growth until 2030 [20]. The main strategies for the development of the construction market include the use of autonomous construction vehicles, the introduction of green building technologies, the use of digital technologies to improve construction safety, and the use of advanced building materials [21]. Autonomous construction vehicles are automated vehicles equipped with sensors, cameras and GPS. Green building involves the use of sustainable building materials and construction processes to create energy-efficient buildings with minimal environmental impact [20]. Such building materials include sandwich panels, the use of which allows achieving sustainable development goals due to their properties due to:

- high energy efficiency indicators that reach R-values of up to 9.0;
- ensuring thermal performance;
- reducing the amount of energy required, making them a vital component of a truly sustainable building;
- maximum sustainability;
- Increased usable space for owners and tenants;
- additional environmental benefits due to their waste-free production and 100% recyclability;
- simplicity and speed of construction [22].

Based on the above, it is clear that the demand for sandwich panels is growing. Thus, according to international experts, the global sandwich panel market is expected to grow. Thus, while in 2022 it was estimated at USD 16.08 billion. By 2031, it will be worth USD 28.8 billion. This growth is driven by the increasing demand for rapid construction services in various industries, residential and commercial construction, and infrastructure. The growing preference for polyurethane-filled panels is due to their low cost and durability. This segment dominated the market in 2021, and this trend is projected to continue until 2030 [23].

A similar trend can be observed in the European sandwich panel market, and according to experts, it is projected to grow by 5.2% on average between 2023 and 2028 due to increased investment in residential and industrial construction projects in the region. The development of the European sandwich panel market will be driven by environmentally friendly structures with an advanced core consisting of rigid polyurethane foam, polystyrene foam, mineral wool to improve their quality standards [24].

Ukraine. According to the State Statistics Service [25], Ukraine is experiencing an increase in the volume of construction products produced. Thus, while in 2014 the market volume amounted to UAH 51,109 million, in 2021 it was UAH 258,073.6 million (Fig. 1). The largest growth was observed in the construction of engineering

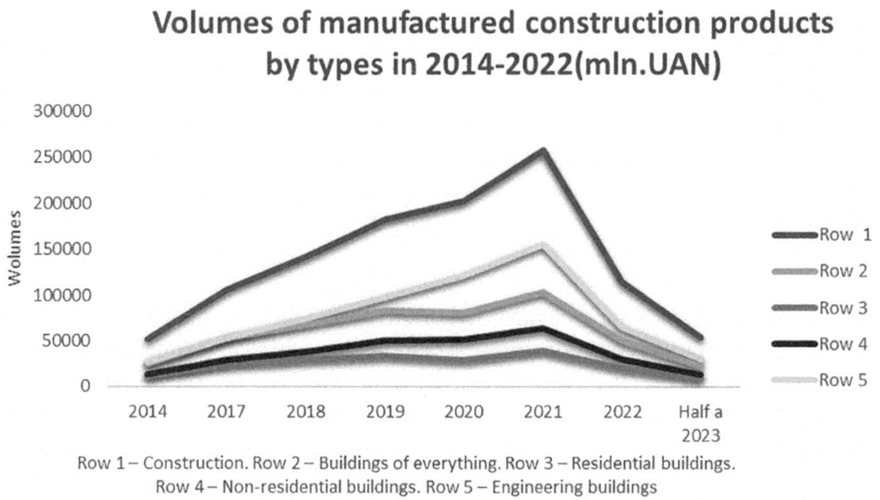

Fig. 1 Volumes of construction products by type in 2014–2022 (UAH million). According to the State Statistics Service of Ukraine [25]. Data exclude the territories which are temporarily occupied by the Russian Federation and part of territories where the military actions are/were conducted

structures and non-residential buildings. The residential construction market grew the least. Already in 2022, the volume of the construction market decreased significantly due to the war with Russia.

However, if we compare the indices of construction products, we can see their decline in 2020 due to the Covid-19 pandemic (Table 1) and follow the recovery process in 2023. Thus, the index of construction products in 2022 compared to 2021 amounted to 35%. However, in 2023, despite the war, the recovery process in the construction market gradually began: in the first half of the year, it increased by 18.4% compared to the same period in 2022. The largest growth was in the engineering construction market, up 34%, while the residential construction market declined by 5.9%. The volume of construction work performed in January–June 2023 in monetary terms amounted to UAH 53.2 billion, buildings in total—UAH 22.6 billion, including residential buildings—UAH 8.9 billion, and construction of engineering structures—UAH 30.6 billion. At the same time, the share of new construction in the total volume of construction work performed in the first half of the year amounted to 44.3%, repairs—32.8%, reconstruction and technical re-equipment—22.9%. In the first half of 2023, 66,591 apartments were commissioned, including 16,058 in one-story buildings and 50,533 in high-rise buildings. The city of Kyiv and the western regions are noted, and only two regions, Luhansk and Kherson, do not have such an indicator. The beginning of Ukraine's recovery is evidenced by the data on the volume of housing construction in the first quarter of 2023 compared to 2019. In Kyiv region, this figure was 79%, in Kyiv—99% [26].

Table 1 Indices of construction products in Ukraine, %

Name	2019–2018	2020–2019	2021–2020	2022–2021	First half of 2023 to first half of 2022
Construction	120.0	105.6	106.8	35.2	118.4
Buildings	116.3	93.7	110.0	38.2	103.0
Residential	104.8	83.5	119.2	40.2	94.1
Non-residential	130.3	100.3	105.1	37.0	109.5
Engineering structures	123.3	115.6	104.6	33.1	134.0

According to the State Statistics Service of Ukraine [25]. The data exclude the temporarily occupied territory of the Autonomous Republic of Crimea, the city of Sevastopol and part of the temporarily occupied territories in Donetsk and Luhansk regions

The growing demand for construction products during Ukraine's recovery period requires an increase in the production of building materials and, in particular, sandwich panels. They have been used in the construction of non-residential buildings and engineering structures. But today it is proposed to use sandwich panels in the construction of cottages and high-rise buildings. Ukraine has developed a technology for the construction of one-story buildings in 7 days from sandwich panels with a service life of 50 years [27].

The sandwich panel market in Ukraine is developing rapidly. The market volume increased by 2.4 times in 2015–2020, and by 1.3 times from 2019 to 2020, with a tendency to increase the volume of domestic production [28]. In 2020, the volume of the sandwich panel market in Ukraine increased to 3.1 million m^2. At the same time, high-quality imported products accounted for 32% of the market, with Ruukki, Balex, Pruszynski, Kingspan and Gorstal leading the imported panels segment, accounting for about 70% of the total. PIR core accounted for 60% of the sandwich panel market and mineral wool for 40%. This is explained by lower price and high energy efficiency of polyisocyanurate panels, which is why they are used in buildings that do not require high fire resistance of the building envelope. The market continued to grow in 2021, but in 2022 there was a sharp drop and in 2023 there is a process of its gradual revival.

About 30 large and medium-sized enterprises operate in the market. Their capacities are shown in Fig. 2.

The largest production capacity is at YUSP, which has two production lines for mineral wool and polyurethane foam panels. In 2019, USP installed new lines and is actively implementing a strategy to expand its market through price warfare and severe dumping. Thermobud has one hybrid production line for the production of sandwich panels with two types of insulation: mineral wool and polyurethane foam. TPK-Center LLC has 14% of the production capacity and ranks 2nd among Ukrainian manufacturers in this production segment and has advantages because the equipment is of higher quality and class than that of its competitors. This is what allows us to produce better products and supply quality products to the Ukrainian market of

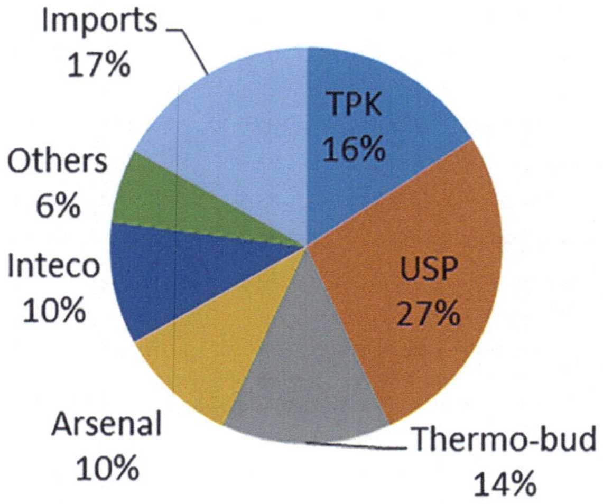

Fig. 2 Structure of sandwich panel production capacities in Ukraine, %. *Source* Research by TPK-Center LLC

three-layer panels. Imported products account for 17% of production capacity in the domestic sandwich panel market. Three domestic sandwich panel producers account for 51% of the market [29]. The remaining 10% is held by “other” companies—a large number of small importers and small regional producers. The market also includes medium-sized companies with a relative market share of 9–5%. These are large global manufacturers such as Kingspan (5%), Balex (8%) and Gorlicka (9%).

The Association of Building Materials Experts conducted a study in which sandwich panel manufacturers were evaluated according to the main quality parameters of products available on the Ukrainian market: price, energy efficiency, durability, load-bearing capacity, as well as the availability of all necessary certificates and supporting documentation. 42 sandwich panel manufacturers were evaluated, 20 of which were included in the final table, 13 of which are foreign. According to the study, the best sandwich panels are offered by Ruukki, Kingspan and BP2. Among the Ukrainian companies, TPK-Center LLC took the first place with a score of 3.7 points (the highest score was 4.9) and in terms of energy efficiency and certification, it ranks 9th and 4th respectively, and is ahead of a number of foreign panel manufacturers. Therefore, it is no coincidence that since September 2023 It was TPK-Center LLC that became an official partner of the state aid program “eVodnovnennya”, which was launched to support citizens whose homes were affected by the aggression of the Russian Federation. The presence of imports in the sandwich panel market is also explained by the absence of duties on the import of this product into the country, the conditions of European integration and the oversaturation of this market in Europe [28]. In 2018–2021, there is a tendency to redistribute the sandwich panel market by the type of their fillers: a decrease in mineral wool panels from 46 to 28% and an increase in the share of panels with polyurethane foam insulation from 54 to 72%. In TPK-Center LLC, sandwich panels with mineral wool are the main product and due to a drop in demand for them, the market share decreased from 14 to 10% in 2018–2020. This necessitates a revision of the company’s marketing strategy.

To determine the factors of the company's environment, a SWOT analysis was conducted (Table 2).

From the SWOT analysis matrix, the following conclusions can be drawn: the firm's strongest advantage is product quality, reputation and market leadership. But the company's biggest disadvantage is its high price. Another significant disadvantage is the high maintenance costs. These two indicators significantly reduce the company's competitiveness, but a marketing strategy to strengthen customer relations regarding the benefits of product quality will offset these weaknesses. Among the favorable opportunities, the most likely are:

Table 2 Matrix of SWOT-analysis of "TPK-Center" LLC

Strengths	**Weaknesses**
1. Modern production technology using own high-tech equipment of the European level 2. Wide range of products 3. TPK brand—experience and reputation of a reliable partner in the Ukrainian market 4. An impressive portfolio of facilities 5. Support of the client from the development of the most effective solution in terms of design, operation, price at the design stage and until the end of installation work 6. Availability of a sales network throughout Ukraine 7. Availability of qualified engineering service 8. Implementation of a quality control system for manufactured products 9. Availability of proven and reliable suppliers of raw materials	1. High production costs 2. Lack of systematic promotion of the company in the market 3. Insufficient implementation of digital marketing in the company 3. High price 4. Lack of working capital makes it impossible to reduce the purchase price, ensure the efficiency of supply and production
Opportunities	**Threats**
1. Projected market growth due to the Recovery Plan of Ukraine 2. Working with designers and architects 3. Development in the segment of residential and own construction 4. Introduction of new products to complement existing products of own production 5. Work on the culture of product consumption—participation in the development of regulatory documents and standards of the construction industry, its development 6. He is a current member of the Board of Directors of the Confederation of Builders of Ukraine and a member of the Ukrainian Steel Construction Center 7. Introduction of import quotas	1. Instability in the country due to the war, in particular, lack of funds to invest in the construction of facilities 2. Increased aggressive actions by importers 3. Aggressive price war by competitors

Source developed by the authors

- active work with designers and inclusion of TPK-Center's products in projects to eliminate competitors at the initial stage of construction
- development of new products;
- regulation of the sandwich panel market by regulatory documents.

An important part of marketing in business is the business portfolio management function. Let's analyze the market position of TPK-Center LLC using the BCG matrix. The company has four strategic business units: sandwich panels with mineral wool insulation, polyurethane foam insulation, metal products (corrugated board, metal tiles) and resale of accessories and components for installation.

Taking into account the market positions of the company's strategic business units and the prospects of its sales markets, the following strategic decisions can be proposed:

1. To support and develop the new product "PU foam sandwich panels" in order to increase its market share—the "BUILD" strategy.
2. Maintain the existing market positions of the strategic business units "Components"—the "HOLD" strategy.
3. Reducing investment and marketing attention to the strategic business unit "Sandwich panels made of mineral wool"—the "harvesting" strategy.
4. Excluding the Metal Products strategic business unit from the company's business portfolio, which is, accordingly, an unpromising "dog"—the elimination strategy.

Marketing strategy for entering the foreign market. The company's favorable strengths and capabilities make it expedient to enter foreign markets with sandwich panels, thereby reducing the threats of an aggressive internal environment. TPK-Center LLC has every reason to do so. This is facilitated by the implementation of Regulation (EU) No. 305/2011 of the European Parliament and of the Council establishing harmonized conditions for the provision of construction products on the market [27]. Exports will allow us to sell at a higher profitability. To organize and develop this area, the following steps should be taken:

- study the framework and legislation and prepare all necessary certificates for products;
- create a website in English;
- Develop communication materials and catalogs in English. Creating catalogs in English is also an important step in sales, as the products are high-tech and personalized for each individual project, which must be developed by an engineer or design organization and takes into account many nuances to calculate the loads withstood. Therefore, in order to be able to use TPK-Center's products, they will need to study the technical parameters in detail;
- develop BIM models of products.

For engineering and architectural European organizations, service, ease of use of the company's products and easy integration into projects are important. For this purpose, there are modern methods and technologies. For example, BIM modeling. Building Information Modeling (BIM) is a process based on the use of intelligent

3D models. With the help of this technology, architecture and construction (AEC) professionals can plan, design, build and operate buildings and infrastructure even more efficiently.

Analysis of the company's customer database. For a better understanding of the sales structure and its customers, we analyzed the customer base using a CRM system for the period from 2017 to 2020. The analysis revealed a decline in the customer base.

Regarding customer loyalty, 70% of customers are not loyal and have worked with the company once, and only 10% of customers who have been working for more than 3 years can be called loyal. Most customers bought once, in small orders for residential and private construction. According to the data in the company's CRM system, the company needs to build long-term partnerships with 67% of its customers. And they don't have them, because only 10% of them are loyal customers.

Customer satisfaction was monitored through a customer survey. The analysis of the data revealed that customers have virtually no attachment to a particular company or brand, and the main competitors are YUSP and Termobud. The price of sandwich panels is important, but not the main selection criterion. Customers do not understand the service of "engineering service" or "turnkey solutions", so this service does not affect the customer's choice. The company's reputation and meeting deadlines are one of the main influencing factors. As for the quality of the panels, the respondents' answers were distributed as follows high quality of raw materials—30%, quality of locks—25%, warranty—12%, appearance—17%, easy installation—8%, packaging—8%. Thus, the concept of quality is vague for customers, and its main physical dimension is "high-quality lock" and "European raw materials." Everything else is secondary because it does not affect the speed of installation work on the site.

Classification of clients in the CRM system of TPK-Center LLC.

1. Multidisciplinary construction organizations. These are large organizations that subordinate various construction organizations and provide a full range of turnkey construction services—9%.
2. Corporate clients. These are directly customers, business owners who make purchases for construction—33%.
3. Private customers. Private customers building for themselves, small volumes—33%.
4. Specialized construction organizations. Small construction organizations that perform certain types of construction works or services in the construction market, such as installation of sandwich panels, ventilated facade, metal roofing, facade works—17%.
5. Trade organizations. Dealers, distributors engaged in procurement and further sales to customers—8%.

The analysis of the table shows that the company needs to work seriously with its customer base. Thus, private customers account for 33% of all customers, but have a large number of small orders. The analysis also showed a breakdown of loyal customers by region and the branches with which they cooperated. It can be noted that in small regions and cities, the degree of customer loyalty is much higher than in large cities such as Kyiv, Odesa, and Dnipro. The cities of Lutsk, Bila Tserkva, and Rivne have the largest share of loyal customers. Due to low competition, there are more loyal customers there who are comfortable with a trusted and single supplier. Therefore, we need to focus on these regions in search of loyal customers and build long-term partnerships with them.

The company has done a lot of work on the use of digital technologies in product promotion. In particular, the company's main website in 2021 became an effective tool for lead generation and the number of customer inquiries through the website increased almost 2 times. The research has shown that Internet marketing requires constant monitoring and analysis of effectiveness. Outsourcing this activity is not effective due to the impossibility of perfect control.

The conducted research has confirmed the feasibility of the conclusions about the need to change approaches to defining a marketing strategy due to changes in the environment. The analysis of the market environment has shown that the main product competitor for TPK-Center LLC is PU foam sandwich panels and requires changes in the assortment and adjustment of the marketing strategy. However, despite the fall in demand for mineral wool panels, they should be kept in the product range because of their possible use in certain areas of construction where they are indispensable. This approach requires refinement in the analysis methods. To. According to the results of the study, the main marketing strategy for competitive advantage is differentiation. However, the change in competitive strategy alone does not solve the problem of the decline in the market share of TPK-Center LLC. This requires research of the customer base using a CRM system and surveys. It has been established that the company's system of interaction with customers needs to be thoroughly improved. The task is to create a high-quality customer experience, intensify communication interaction and establish genuine partnerships with customers. As for digital marketing, the work of the relevant specialists on outsourcing has not paid off. Therefore, the conclusion was made that it is advisable to have specialists in the company's marketing department to promote and advertise products on the Internet—CEO specialists, SMM specialists, content managers, creative designers [30]. The results of the research will also be correct in the period of Ukraine's recovery during the war and in case of positive developments after the war. However, the construction industry may face a number of problems after the war [31] and therefore needs to change quickly and be ready to accept them.

5 Conclusions

The conducted research has shown that sandwich panels are the building material that meets the requirements of environmental friendliness and energy efficiency. In the world and in Ukraine, there is a tendency to decrease the demand for sandwich panels with mineral wool, which should be taken into account when forming a business marketing strategy. The definition of a marketing strategy for competitive advantages should be complemented by a powerful analysis of the customer base, since it is this analysis that has shown the need to form a marketing partnership system and CXQ at the enterprise. It is also proposed to expand the scope of digital marketing, and to supplement the marketing strategy of differentiation with an export strategy.

References

1. The total amount of damage caused to the infrastructure of Ukraine due to the war reaches $151.2 billion—estimate as of September 1, 2023. https://kse.ua/about-the-school/news/the-total-amount-of-damage-caused-to-the-infrastructure-of-ukraine-due-to-the-war-reaches-151-2-billion-estimate-as-of-september-1-2023/
2. Plan for the Restoration of Ukraine. National Recovery Council. July 2022. https://uploads-ssl.webflow.com/625d81ec8313622a52e2f031/62c19ac16c921fc712205f0
3. Ukraine: sustainable economic recovery for people and nature. https://wwf.ua/?7630866/sustainable-recovery-wwf-bcgЇ
4. Omelchuk N. Ytohy hoda v ukraynskom e-commerce: chto hovoriat eksperty. [The results of the year in Ukrainian e-commerce: what experts say]. https://psm7.com/analytics/itogi-goda-v-ukrainskom-e-commerce-chto-govoryat-eksperty.html
5. Sari, Y., Gultom, A.W.: Marketing strategy in effort to increase competitive advantage in small and medium enterprises. Copyright©2020. JIMFE (Jurnal Ilmiah Manajemen Fakultas Ekonomi) (2022). https://doi.org/10.34203/jimfe.v6i2.2390
6. Abuselidze, G., Talavyria, M., Vyshnivska, B., Bondarenko, L., Makedon, H., Kniazieva, T., Salkova, I.: The economic mechanism of marketing activity management of food enterprises. E3S Web Conf. **371**, 05002 (2023)
7. Porter, M.E.: Competitive Strategy: Techniques for Analyzing Industries and Competitors. Free Press, New York (1980) (Republished with a new introduction, 1998)
8. Neil, A.M., Kimberly, A.W., Hui, F., Simos, Ch.: Research in marketing strategy. Acad. Mark. Sci. (2018). https://doi.org/10.1007/s11747-018-0598-1
9. Kotler, F., Keller, K.L.: Marketing management St. Petersburg: Peter, 816 p. (2012)
10. Ostapchuk, A., Karpenko, L., Abuselidze, G., Chornenka, L.: Management of marketing risks in sustainable development. E3S Web Conf. **371**, 05011 (2023). https://doi.org/10.1051/e3sconf/202337105011
11. Liming, Zh.: Research on the marketing strategy of enterprises based on the change of marketing environment. In: Proceedings of the 8th International Conference on Social Network, Communication and Education (SNCE 2018). https://doi.org/10.2991/snce-18.2018.16
12. Savytska, N.: Responsive marketing as a prerequisite for business development -business, innovation, management: problems and prospects: a collection of abstracts supplement to the III International Scientific and Practical Conference, Kyiv, 8 Dec 2022, 292 p. Igor Sikorsky Kyiv Polytechnic Institute, Polytechnic Publishing House, Kyiv (2022). confmanagement.kpi.ua ‘ proc ‘ article ‘ view
13. Taherdoost, H., Madanchian, M.: Determination of business strategies using SWOT analysis; planning and managing the organizational resources to enhance growth and profitability. Macro Manage. Public Policy **3**(1), 19–22 (2021). https://hal.science/hal-03741850/document
14. Shigang, Y., David, A.S.Ch.: An investigation of marketing strategy, business environment and performance of construction SMEs in China. Int. J. Account. Audit. Taxation **7**(2), 001–010 (2020). www.internationalscholarsjournals.org
15. Mohd, R., Rosidah, M., Mazzini, M., Arshad, A.: Reconceptualisation of customer experience quality (CXQ) measurement scale. Procedia Econ. Fin., 2212–5671 (2016). https://doi.org/10.1016/S2212-5671(16)30128
16. Romanova, L.: Theoretical and practical aspects of development of marketing management on industrial enterprises. Econ. Manage. Aust. J. Hum. Soc. Sci., 3–4 (2017). https://doi.org/10.20534/AJH-17-3.4-158-169

17. Zinchenko, T., Derzhak, N.: Digital marketing as a direction for improving communication interactions with consumers. https://doi.org/10.33216/1998-7927-2021-270-6-115-120
18. Yazvinska, N., Kasperovych, Ya.: Marketing web strategy of the enterprise—a tribute to fashion or a requirement of modernity? Mark. Ukraine **2**, 51–54 (2017)
19. Construction Global Market Report 2023. https://www.researchandmarkets.com/reports/5781152/construction-global-market-report
20. Construction Market 2022. https://www.thebusinessresearchcompany.com/report/construction-market
21. Global Construction Market Opportunities and Strategies Report 2022: Use Of Autonomous Construction Vehicles, Green Construction Projects, Digital Technologies Are Improving Construction Safety. https://www.globenewswire.com/en/news-release/2022/08/04/2492427/28124/en/Global-Construction-Market-Opportunities-and-Strategies-Report-2022-Use-Of-Auton
22. Bondor Metecno insulated sandwich panels maximize the energy efficiency of commercial and residential buildings. https://createdigital-org-au.translate.goog/insulated-sandwich-panels-sustainability/?_x_tr_sl=en&_x_tr_tl=uk&_x_tr_hl=uk&_x_tr_pto=sc
23. Sandwich Panels Market 2022. Industry Share, Size, Growth. https://www.transparencymarketresearch.com/sandwich-panels-market.html
24. Europe Sandwich Panels Market Report and Forecast 2023–2028. https://www.researchandmarkets.com/reports/5713673/europe-sandwich-panels-market-report-and-forecast
25. State Statistics Service of Ukraine. https://www.ukrstat.gov.ua/
26. Rynok budivnytstva vidnovliuietsia: shcho bude z tsinamy na materialy do kintsia roku. [The construction market is recovering: what will happen to the prices of materials by the end of the year]. https://www.ucsc.org.ua/rynok-budivnycztva-vidnovlyuyetsya-shho-bude-z-czinamy-na-materialy-do-kinczya-2023-roku/7.08.2023
27. Rauta has developed sandwich-panel construction technology that allows houses to be built in just 7 days. https://budport.com.ua/news/23793-rauta-razrabotala-tehnologiyu-stroitelstva-iz-sendvich-paneley-pozvolyayuschuyu-vozvodit-doma-vsego-za-7-dne].
28. Ukrainskyi rynok sendvich-panelei u 2020 rotsi vyris na 18%. [The Ukrainian sandwich panel market grew by 18% in 2020]. Андрій Озейчук. https://rautagroup-com.translate.goog/en/ukrainian-sandwich-panel-market-grew-by-18-in-2020-andriy-ozeychuk/?_x_tr_sl=en&_x_tr_tl=uk&_x_tr_hl=
29. Naikrashchi vyrobnyky sendvich-panelei na rynku Ukrainy – reitynh TOP-20. [The best manufacturers of sandwich panels on the market of Ukraine—TOP-20 rating]. https://buduemo.com/ua/news/ukraine/najkraschi-virobniki-sendvichpanelej-na-rinku-ukraini-l-rejting-top20.html
30. Akimov, O., Karpa, M., Parkhomenko-Kutsevil, O., Kupriichuk, V., Omarov, A.: Entrepreneurship education of the formation of the e-commerce managers professional qualities. Int. J. Entrepreneur. **25**(7) (2021)
31. Providing construction products on the market. https://regulation.gov.ua/dialogue/budivnytstvo/75-nadanna-budivelnoi-produkcii-na-rinku

Will E-Filling Be Effective in Increasing Individual Taxpayer Compliance?

Heru Tjaraka and Virgina Ayu Kusumawardhani

Abstract Taxes are the most significant contributor to national income for national funding on behalf of Indonesian's prosperity. This research aim for obtain proof empirical about effectiveness use *e-Filing* in push enhancement individual taxpayers compliance in KPP Primary South Sidoarjo. Digital transformation is one of the methods used by the Directorate General of Taxes to increase taxpayer compliance. The presence of digital transformation can maximize tax revenue and change the tax system to become simpler, easier, faster, and more efficient. One of digital transformation that has been implemented by the Directorate General Taxes is *e-Filing*. *e-Filing* is an electronic system for reporting the annual SPT that can be completed *online* and in *real-time*. However, the implementation of *e-Filing* is still unsuccessful to bring Indonesia to the expected level of taxpayer compliance. This also happened to KPP Primary South Sidoarjo, where although the level of individual taxpayer compliance has increased every year but the number of an individual taxpayer who uses *e-Filing* is relatively small. The purpose of this study was to gather empirical evidence regarding the effectiveness of *e-Filing* in increasing the individual taxpayer compliance at KPP Primary South Sidoarjo. This research uses descriptive qualitative methods by interviewing 6 individual taxpayers. The results of this study indicate that although in terms of ease of use cannot be said to be effective to increase individual taxpayer compliance, however, in terms of usefulness can be said to be effective to encourage the interest in using *e-Filing* and increasing individual taxpayer compliance because the reporting of annual SPT becomes effective, practical, easy, efficient, and fast.

Keywords Effectiveness · E-filing · Individual taxpayer compliance

---

H. Tjaraka (✉) · V. A. Kusumawardhani
Department of Accountancy, Faculty of Economics and Business, Airlangga University, Surabaya, Indonesia
e-mail: heru_tjaraka@feb.unair.ac.id

V. A. Kusumawardhani
e-mail: virgina.ayu.kusumawardhani-2018@feb.unair.ac.id

B. Awwad (ed.), *The AI Revolution: Driving Business Innovation and Research*, Studies in Systems, Decision and Control 524, https://doi.org/10.1007/978-3-031-54379-1_2

1 Introduction

System Taxation in Indonesia is changing from *Official Assessment System* becomes *Self Assessment System*, so that must tax carry out all obligation taxation with yourself In Indonesia, must tax dominated by Individual Taxpayers by hence, obedience Individual Taxpayers become wrong one supporter achieved level obedience Which tall in accordance hope from Directorate Tax General as authority taxation in Indonesia. According to Hartanto And Wahyudin [13], Directorate The Tax General has apply digital transformation, which consists of from modernization or administrative reform taxation in frame optimizing reception from sector tax And For adapt with developments over time. For reach system administration tax which accountable and transparency, so modernization done with use technology electronic in form *e-system. E*-tax *system* that has been developed Directorate Tax General, namely *e-Registration*, *e-Filing*, *e-Invoice*, *e-SPT* and *e-Billing*, so, there is *e-system* capable make all over activity initial taxation done in a way manually Now can done in a way *online.*

E-Filing is system Annual SPT (Letter of Notification) reporting based *online* and *real time* who can used via the djp online site, the in accordance with Regulation Director General Tax Number PER-1/PJ/2014 about system method delivery letter announcement for Individual Tax payers [12]. However, still there is various limitations which Taxpayers in operate *e-Filing*. According to Rustan [22], still there is constraint on connection internet when report SPT annually in end period reporting SPT annual and lack of understanding on technology, so that make less taxpayers understand and speak in use *e-Filing*. Then, according Asriyanti et al. [5] still often *error* or *down* when used. However, although Still there is constraint which experienced must tax in use *e-Filing*, amount must tax which utilise *e-Filing* experience enhancement every the year.

Previous research on a similar topic has been carried out by several researchers. However, results which obtained vary. Based on study which done by Mustamim and Rohmah [17], *e-Filing* is still not said to be effective in terms of ease of use because there are taxpayers who are constrained by the internet And not enough understand method operation *e-Filing*. Temporary that, *e-Filing* has is said to be effective in terms of usefulness because of the capable use of *e-Filing* make annual SPT reporting reported from anywhere and at any time, so the process is simple, practical, fast and efficient. Meanwhile, according to Adestya's research and Saman [1] the quality of *e-Filing services* is very good, but not yet effective if reviewed based on effectiveness utilization *e-Filing* Because amount between SPT annual which reported *on line a*nd manually still contrast. According to study [6], *e-Filing* has push enhancement compliance of Individual Taxpayers with increase number SPT annually reported. This matter because *e-Filing* makes it easier for reporting SPT annually.

Then, according to Aksara [2], the implementation and use of *e-Filing* has been good, only just *e-Filing no* impact on tax compliance because although amount taxpayers who use *e-Filing* increase every year, but the level his tax compliance precisely experience decline.

By therefore, for evaluate is something system can said effective required indicator reference, that is based on aspect usefulness And aspect ease of use which will encourage people's interest in using it this system [21]. This is in line with the Technology Acceptance Theory Model (TAM) by Davis [8]. Which trigger that the presence of TAM theory can predict and explain how technology may be accepted and used by users are related to their work, so that later when the public assesses it that a system is easy to use and provides benefits in helping his job, so will give rise to interest and reception to that system.

Studies study this done on KPP Primary South Sidoarjo. Based on data pre-research, when compared with the level of individual taxpayer compliance between KPP Primary West Sidoarjo, KPP Primary North Sidoarjo, And KPP Primary South Sidoarjo shows that there are still enough KPP Pratama Sidoarjo Selatan many individual taxpayers still don't know about *e-filing* and many still do have not used *e-Filing* with an average of around 52.4% in 2016–2020 as served in Table 1. Difference this research with research previously it was not only located on the object of research, but also on the subject study, Which where study this use Individual taxpayers employee and Individual Taxpayers non employee as source person in study.

From Table 1, level Individual Taxpayer compliance at KPP Primary South Sidoarjo experience enhancement every the year, however, level Individual Taxpayers compliance employee no more tall if compared to with level non-employee individual taxpayers compliance.

Based on Table 2, level use *e-Filing* in KPP Primary South Sidoarjo still spelled out enough low, that is highest only around 56% in 2020. KPP Primary South Sidoarjo has do various effort socialization use *e- Filing* from year to year, however level utilization *e-Filing* still spelled out low. The low level use *e-Filing* at KPP Pratama South Sidoarjo because exists constraint experienced by WPOP when use *e-Filing*.

Therefore, this research aim for obtain proof empirical about effectiveness use *e-Filing* in push enhancement individual taxpayers compliance in KPP Primary South Sidoarjo.

Table 1 Ratio obedience WPOP in KPP Primary South Sidoarjo

Information	2016 (%)	2017 (%)	2018 (%)	2019 (%)	2020 (%)
Employee	67	72	74	75	78
Non-employee	68	74	76	77	80
Total	67	72	74	75	78

Source Researcher, Data Processed

Table 2 Ratio use *e-Filing* in KPP Primary South Sidoarjo

Year	2016	2017	2018	2019	2020
Amount WP effective	60,496	61,480	62,727	65,127	67,606
Amount user *e-filing*	29,150	31,262	33,492	35,450	38,168
Usage ratio *e-filing* (%)	48	51	53	54	56

Source Researcher, Data Processed

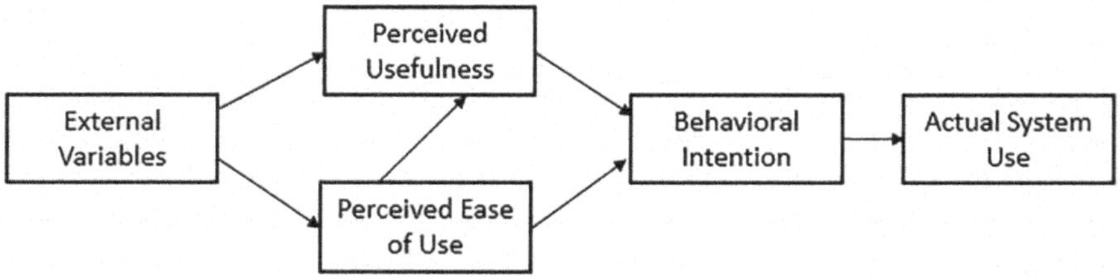

Fig. 1 Technology acceptance model. *Source* Davis [8]

2 Technology Acceptance Model Theory

The Technology Acceptance Model (TAM) theory is developed theory by Davis [8] for predict and explain how technology can accepted and used by user technology (Fig. 1).

TAM theory adopted from *Theory of Reasoned Action* (TRA) which is base theoretical For define connection between two focus main, that is convenience use (*ease of use*) and usefulness (*usefulness*) of intention (*intentions*) and behavior use system (*actual system use*). Perception convenience use show user technology think that technology the easy used and no need understanding which deep in use technology [15]. Indicators that can used for measure perception convenience use is easy use and easy studied [8]. Meanwhile, perception usefulness show that user technology believe that technology the can increase effectiveness and productivity in life everyday [15]. Indicator can used for measure perception usefulness is effectiveness [8]. Second matter that's what happens next capable push interest (*intentions*) user technology for use that technology.

3 Tax Compliance

Formal and material compliance according to Nurmantu [19: 148–149] are two type tax compliance. Formal Compliance is so far where a taxpayer must comply of tax obey procedural requirements and tax administration to mandatory tax obligations tax ful filled in accordance with constitution taxation. Whereas, Material Compliance is the extent to which a taxpayer can comply with all aspects law taxation, like reporting SPT in a way honest, calculating big burden tax correct, clear, and complete, ensure that big obligation taxation which must paid in accordance with the calculation which actually and big or small arrears tax [26: 69–70]. Tax compliance according to Prastiwi et al. [20] are very important for success system *self-assessment*. Fulfillment of formal and material tax obligations will demonstrate compliance taxpayer. Through formal compliance, all tax administration must be fulfilled by taxpayers. Meanwhile, material compliance requires taxes to be owed calculated and reported precisely and accurately by the Taxpayer. Based on PMK (Peraturan Menteri Keuangan—Minister of Finance Regulation) Number 39/PMK.03/2018, one of the criteria in classifying mandatory tax compliance, that is related accuracy time reporting SPT [16].

4 *E-Filing*

E-Filing has applies since 12 January 2005, in accordance Regulation General Tax Number KEP-05/PJ/2005 about System Method Delivery Letter Announcement in a way electronic [10]. However, moment that *e-Filing* done through *Application Service Providers* (ASP) [25]. Then, the Generation State Revenue Module system Second (MPN G2) since year 2007 utilized government as which *e-Filing* to use *email* and payment in a way electronic for support system that. Until finally, for support *e-Filing* And *e-Billing* Director General Tax launch separate *website* [3].

5 SPT for Income Individual Tax Taxpayer

SPT (Surat Pemberitahuan—letter of notification) is document containing information for count amount tax owed in a way sure. SPT is letter used for convey tax, object tax and no object tax, as well as obligation and treasure aligned with decision law, the outlined in Constitution Number 16 of 2009 Chapter 1paragraph 11 about Provision General And System Method Taxation.

There is two type SPT, that is:

a. Periodic SPT

Periodic SPT is letter notification submitted each 20th after owed tax, for example Periodic Income Tax SPT Article 21/26, Income Tax SPT Article 22, Income Tax SPT Article 23/26, Value Added Tax SPT.

b. SPT Annual

Individual Taxpayer annual tax return must reported with range time three month after the end year tax, namely in March. Then, the annual Corporate Taxpayer income tax SPT is mandatory reported with range time four month after the end year tax, that is on month April.

Obligation every Individual taxpayers which have NPWP (Nomor Pokok Wajib Pajak—tax ID number) is SPT annually must completed and reported with precise, accurate, and on time. There is three type Annual SPT Individual taxpayers in accordance Regulation Director General Tax Number PER- 34/PJ/2010 [11], that is:

a. SPT Annual 1770

Form SPT Annual 1770 utilized for businessman or worker free as well as civil servants/TNI/ Polri when have other businesses and other livelihoods No bound.

b. SPT Annual 1770S

Annual SPT Form 1770S is utilized for employee income exceed IDR 60 million a year or have work more from one company during one year.

c. SPT Annual 1770SS

Form SPT Annual 1770SS utilized for employee which have work in one company just as well as income not enough from IDR 60 million in a year.

6 Method Study

Method qualitative descriptive chosen researcher as method research. For understand phenomenon, research qualitative is a natural process in which researchers No manipulate interest on phenomenon [4: 38]. Which currently discussed. this research use methodology studies case, which where to study and explore information about background back and circumstances something case during period time certain in a way intensive and deep.

Data analysis in research this use Miles and Huberman technique, which where process analysis data done with triangulation data which where according to Sugiyono [24: 46] consists from 3 (three) groove, that is:

a. Reduction data
b. Presentation data
c. Withdrawal conclusion.

7 Results and Discussion

*7.1 Perception Convenience Use (*Ease of Use*)*

According to Davis [8], perception convenience use refers to belief user that technology No burdensome for taxpayers apply it and no need understanding deep about method apply it. If applied to application *e-Filing*, Individual taxpayers believe that *e-Filing* easy applied or not need deep understanding or Lots business for apply *e-Filing* [18].

According to Lee and Wan [15], indicators that can used as reject measuring for convenience use which felt taxpayers:

a. Easy used

Based on indicator easy used, use *e-Filing* felt still there is obstacles felt by taxpayers. That thing caused many taxpayers who difficult enter to djp.online, because connection *error* or *down* when approaching the limit end annual SPT reporting. With exists constraint the causing the process reporting Annual SPT become disturbed because for enter to djp.online difficult.

b. Easy studied

Based on indicator easy studied, use *e-Filing* felt not yet fully easy studied by the taxpayer. That thing because still there is taxpayers who have not fully understand whichever columns are necessary filled, so taxpayers need watch the tutorial for filling out and reporting the annual tax return use *e-Filing* or contact party office tax.

7.2 Perception Usefulness (Usefulness)

According to Davis [8], perception usefulness reflect confidence user technology that technology useful in make his job more effective and efficient, so capable increase its performance. If applied to use *e-Filing*, Individual taxpayers believes that if SPT Annually reported use *e-Filing*, so *e-Filing* capable increase effectiveness and efficiency its performance [7].

According to Lee and Wan [15], indicators that can used as reject measuring for convenience use which felt taxpayers:

a. Effectiveness

Based on indicator effectiveness, use *e-Filing* is considered has effective in streamline Taxpayer employment. This thing due to the SPT reporting process more years fast and flexible if compared to with report Annual SPT to office tax. That thing because the taxpayer does not need Again come and queue in office tax for report Annual SPT because reporting Income Tax Annual SPT has can done in where and when just.

Based on results study the show that perception convenience use And perception usefulness on theory TAM is 2 (two) factor who is capable push interest user technology for use it in life everyday. It turns out perception convenience use and use *e-Filing* not yet effective in push interests of Individual taxpayers for use *e-Filing* as means in report their income tax annual SPT. That thing caused till many that obstacle before Individual taxpayers moment report income tax annual SPT with use for *e-Filing* so that cause disruption of the Annual SPT reporting process.

Temporary that, based on perception usefulness has effective in push interest Individual taxpayers for use *e-Filing* as means in report SPT Annually. That thing caused exists Individual taxpayers gets great benefits when report income tax annual SPT use *e-Filing* [14]. That thing proven with results data analysis Which obtained from 6 Informant which say that reason they still use *e-Filing* due *e-Filing* capable streamline and make it effective work them. Besides that, *e-Filing also* has capable push enhancement obedience Individual Taxpayers, accordingly results analysis show that after use *e-Filing* they no once again get fine lateness reporting SPT Annually. [23]

8 Conclusion

In accordance results analysis show that use *e-Filing* can make it easier Individual Taxpayers especially in KPP Primary Sidoarjo South for report SPT Income Tax Annually and increase tax compliance of Individual taxpayers. Although in a way general level obedience Individual taxpayers increase, However level use *e-Filing* in KPP Primary South Sidoarjo still spelled out Enough low, that is level highest only range 56% in 2020 and 62% in 2021 [9]. That thing caused exists obstacles felt by Individual taxpayers moment use *e-Filing*, such as connection *error* or *down* when approaching the final deadline reporting SPT Income Tax Annually, so that cause Individual taxpayers constrained moment report Income Tax Return Annual use *e-Filing*. Therefore, when already approaching the limit end income tax return reporting annually, KPP Primary South Sidoarjo opens service hours external annual SPT reporting day work for collect and assist Individual taxpayers in report SPT Income Tax Annually.

Based on results analysis use TAM theory shows that perception convenience use said not yet effective in push enhancement Individual taxpayer compliance because part large taxpayer experience obstacles, such as connection *error* or *down* when approaching the final deadline Annual SPT reporting. Temporary that, based on perception usefulness said effective in push enhancement Individual Taxpayer compliance, because exists taxpayers obtain huge benefits when use *e-Filing*, that is *e-Filing* capable make it effective and efficient his job.

8.1 Suggestion

Suggestion which can researcher submit based on results study, that is:

1. Connection *e-Filing* frequently *down* or *error* when approaching the final deadline SPT reporting Income Tax Annually, because the number of Individual Taxpayers reporting SPT Income Tax Annual when approach due date for reporting SPT PPh Annually. Therefore, for makes it easier individual taxpayers report SPT Annually with use *e-Filing*, Directorate Tax General expected do

enhancement quality system service *e-Filing*, so although *e-Filing* is used in a way simultaneously with so many individual taxpayers report income tax returns annually, system *e-Filing* no again *down* or *error* moment approach maturity reporting SPT Income Tax Annually.
2. Still lots Individual taxpayers effective at KPP Primary South Sidoarjo ones still not yet use *e-Filing a*nd constrained in use *e-Filing*. Expected party KPP Primary South Sidoarjo can more enterprising again socialize *e-Filing* and convincing effective Taxpayer in KPP Primary South Sidoarjo that with use *e-Filing* capable make it effective his job. With exists socialization which massive, good in teaching operation *e-Filing* nor convincing Individual taxpayers that use *e-Filing* capable make it effective his job, so that can increase level use *e-Filing* and impact on rates tax compliance.
3. Use TAM theory, from facet convenience use located on indicators easy used and easy studied, when from facet usefulness located on indicators effectiveness. For study further more can use TAM theory with indicators which other so that obtain analysis which more varieties and details about reception *e-Filing* from facet convenience use and usefullnes.

8.2 Limitations

The limitations of this research focus on individual taxpayers at KPP Primary South Sidoarjo. Meanwhile, *e-Filing* has also been implemented in all KPPs in Indonesia. Therefore, the results of this research are not yet possible made material evaluation general about effectiveness *e-Filing* in push enhancement tax compliance of Individual taxpayers.

There were several individual taxpayers who were respondents who were apparently not willing to answer the questionnaire submitted by the researcher so they had to look for other individual taxpayers who were willing to be interviewed by filling in the research questionnaire submitted to them.

References

1. Adestya, G., Saman: Effectiveness service *e-filing* to must tax person personal in office service tax primary poor south. J. Account. Taxation **6**(1), 52–61 (2020)
2. Aksara, R.: Analysis of *e-filing implementation* on the level of mandatory compliance tax KPP Primary Jakarta Cilandak. J. Acitya Ardana **1**(2), 109–114 (2021). https://doi.org/10.31092/jaa.v1i2.1338
3. Aliah, N.: The role of e-filing in improving taxpayer compliance in Indonesia. Account. Bus. J. **2**(1), 1–92020. https://doi.org/10.54248/abj.v4i1
4. Anggito, A., Setiawan, J.: Methodology Study Qualitative. Sukabumi: CV Footsteps (2021)
5. Asriyanti, N., Nugroho, A., Yulianti, R.: Public services through e-systems filing in office tax primary Cilegon. J. Decentralization Policy Public **1**(1), 35–42 (2020). https://doi.org/10.30656/jdkp.v1i1.2246

6. Avianto, G., Rahayu, S., Kaniskha, B.: Analysis of the role of *e-filing* in the context increase obedience reporting letter announcement must tax person personal (study on tax office primary poor south). J. Taxation (FOOTSTEPS) **9**(1), 1–8 (2016). http://perpajakan.studentjournal.ub.ac.id/index.php/perpajakan/issue/view/21
7. Cahaya, E.: Penggunaan Sistem *E-Filing*. Equity: Jurnal Akuntansi **1**(2), 54–62 (2021)
8. Davis, F.: Perceived usefulness, perceived ease of use, and user acceptance of information technology. MIS Q. **13**(3), 319–339 (1989). https://doi.org/10.2307/249008
9. Directorate General Tax. Report Annual 2020. https://pajak.go.id/id/report-annual-2020. Online. Accessed on 8 Oct 2021
10. Directorate General of Taxation. KEP-05/PJ/2005 concerning Procedures for Submitting Letters Announcement in a way Electronic. Jakarta (2005)
11. Directorate General Tax: PER-34/PJ/2010 about Form Letter Notification Annual Tax Income Must Tax Person Personal and Must Tax Body Along with Instructions the filling. Jakarta (2010)
12. Directorate General of Taxation. PER-1/PJ/2014 concerning Procedures for Submitting Letters Announcement for Must Tax Person Personal. Jakarta (2014)
13. Hernando, R., Wahyudin, D.: Modernization administration taxation in frame optimization service tax based digital. J. Tax Vocational (JUPASI) **1**(2), 119–125 (2020). https://doi.org/10.31334/jupasi.v1i2.819
14. Langodai, M., Iskandar, Lahaya: Analysis of the implementation of e-filing as an effort increasing individual taxpayer compliance in submitting tax returns annually at KPP Pratama Samarinda. J. Knowl. Account. Mulawarman **5**(4), 1–10 (2020)
15. Lee, C., Wan, G.: Including subjective norm and technology trust in the technology acceptance model: a case of e-ticketing in China. Data Base Adv. Inf. Syst. **41**(4), 40–51 (2010).https://doi.org/10.1145/1899639.1899642
16. Minister of Finance of the Republic of Indonesia: Minister of Finance Regulation Number 39/PMK.03/2018 about Tata Method Return Introduction Excess Payment Tax (2018)
17. Mustamim, M.N., Rohmah, L.: The effectiveness of the implementation online tax service system on taxpayer compliance in the pandemic period. J. Income Innov. Econ. Manage. **1**(1), 5–10 (2021). https://ejournal.unwaha.ac.id/index.php/jurnal_penelitian/article/view/1531
18. Natasya, Tandiling, Angelus, Kevin: Tax e-filing system acceptance level on the taxation compliance: an application of the UTAUT approach. J. Winners **20**(1), 33–47 (2019). https://doi.org/10.21512/tw.v20i1.5572
19. Nurmantu, S.: Introduction Taxation. Granite, Jakarta (2005)
20. Prastiwi, D., Narsa, M., Tjaraka, H.: Synthesis system accountancy taxation. J. Account. Multiparadigm **10**(2), 276–294 (2019). https://doi.org/10.18202/jamal.2019.08.10016
21. Purwiyanti, D., Laksito, H.: Influence perception convenience, usefulness, and satisfaction with the use of *E-Filing* on individual taxpayer compliance (studies on KPP Primary Candisari Semarang). Diponegoro J. Account. **9**(2), 1–9 (2020). http://ejournal-s1.undip.ac.id/index.php/accounting
22. Rustan, Qalbi, S., Rusyidi, M.: Implementation of tax reporting using *e-filing*. Amnesty: J. Res. Taxation **3**(1), 39–46 (2020). https://doi.org/10.26618/jrp.v3i1.3410
23. Sakinah, Y., Sinaga, K., Saragi, S.: Implementation program *e-filing* in effort enhancement obedience must tax person personal on office service tax primary Medan West. J. Public Reform UNDHAR Medan **7**(1), 53–70 (2020). https://doi.org/10.46576/jpr.v10i1.3468
24. Sugiyono: Method Study Quantitative, Qualitative, R&D. CV Alphabet, Bandung (2018)
25. Toding, D., Iqbal, S.: *E-billing, E-filing,* and stuttering taxpayer compliance technology in Tana Toraja. J. Bus. Manage. Inf. **17**(3), 280–297 (2021)
26. Widodo, W.: Morality, Culture, and Tax Compliance. Alphabet, Bandung (2010)

The Influence of COVID-19 and Subsequent Events on the Decision-Making of Retail Investors in Kingdom of Bahrain: A Behavioral Finance Analysis

Manju Rajan Babu and Mohammad Taqi

Abstract The debate surrounding Behavioral Finance versus Traditional Finance is of great significance in the field of finance due to their divergent and contentious viewpoints. This study aims to examine the contrasting perspectives by investigating the period spanning from 2020 to 2023, which was marked by a series of influential events impacting financial markets and investor decision-making. Specifically, the study seeks to explore the interplay between Behavioral Finance and Traditional Finance considering factors such as the COVID-19 pandemic, geopolitical events (e.g., the Russian War), and economic policy decisions (e.g., interest rate fluctuations). To achieve this, the research examines price movement charts, historical financial data, and pertinent economic indicators for six different asset classes including equities, cryptocurrencies, and exchange-traded funds (ETFs) with a range of risk profiles. The chosen period, which runs from January 2020 to January 2023, allows for the analysis of the impacts of major occurrences like COVID-19, changes in economic policies, and geopolitical conflicts. The application of behavioral finance theories, biases, and models is the main methodology used; classical finance ideas are occasionally cited when necessary. The results of the study show a strong inverse relationship between traditional finance and behavioral finance. Moreover, it establishes a robust causal relationship between COVID-19, economic policy decisions, geopolitical changes, and their significant impact on the decision-making processes of retail investors. It has been demonstrated that these occurrences set off behavioral biases and instincts that affect how investors make decisions. This study sheds light on the intricate dynamics of financial markets by exploring the relationship between traditional finance and behavioral finance and the impact of events on investor behavior. This enhances our comprehension of how psychological aspects and outside circumstances influence investment choices, highlighting the shortcomings of conventional financial models.

M. R. Babu (✉) · M. Taqi
School of Business, Bahrain Polytechnic, Isa Town, Bahrain
e-mail: manju.babu@polytechnic.bh

M. Taqi
e-mail: Mohammad.taqi@polytechnic.bh

B. Awwad (ed.), *The AI Revolution: Driving Business Innovation and Research*, Studies in Systems, Decision and Control 524, https://doi.org/10.1007/978-3-031-54379-1_3

Keywords Behavioral finance · Traditional finance · Investor decision-making · COVID-19 · Economic policy · Geopolitical events · Behavioral biases · Asset classes

1 Introduction

Retail investors typically pay less attention to the behavioral finance side of investing since they are more interested in using traditional finance methods like technical and fundamental research. Researchers contend, however, that using a typical finance perspective to analyze the market is insufficient because a variety of biases and irrational behaviors can impact investment decisions [1]. It is widely recognized that there are distinct differences between traditional finance and behavioral finance in terms of the theories proposed by each school. Traditional finance is associated with Eugene Fama's Efficient Market Hypothesis and the assumption of investor rationality [2, 3]. In contrast, behavioral finance theories, rooted in the work of Tversky, Kahneman, and Thaler, emphasize investor irrationality and market inefficiency (Corporate Finance Institute 2019). The main objective of the research paper is to analyze the financial markets during the time of 2020–2022 across six asset classes and apply behavioral finance theories, biases, and models. By doing so, the paper aims to provide retail investors with valuable insights into how markets operate from a realistic and practical perspective. The research aims to equip investors with intelligence to make better, more rational investment decisions [4–6]. The objective of this research is to bridge the existing knowledge gap by comprehensively investigating the influence of COVID-19, economic indicators, and geopolitical events on investor decision-making across a diverse range of six asset classes, namely Exchange-Traded Funds (ETFs), Stocks and Cryptocurrencies [7]. Joseph [8], this study aims to elucidate the underlying factors and mechanisms behind the impact of these variables on investor behavior by employing the principles and applications of behavioral finance theories and models. In essence, the research focuses on four distinct investor personas, each with unique investment preferences and risk profiles.

2 Literature Review

This literature review holds immense relevance due to its objectives and findings, which bear similarities to the present study, despite employing quantitative methodology.

Al-Saleh et al. [2], this study has examined the influence of COVID-19 on the behavioral biases of retail investors in the Kingdom of Bahrain. The authors provided empirical evidence of how the pandemic has affected investor behavior, particularly in terms of cognitive and emotional biases. The findings shed light on the impact of COVID-19 to the field of behavioral finance. Al-Khaja et al. [9] investigated the impact of COVID-19 on retail investor sentiment and decision-making in Bahrain's stock market. By analyzing market data and investor surveys, the authors explored how the pandemic has influenced investor sentiment, risk appetite, and investment choices. The study provided valuable insights into the behavioral aspects of retail investors during times of crisis and contributes to the literature on behavioral finance. Hasan and Al-Sahlawi [4] examined the behavioral biases and investment decision-making of Bahraini retail investors during the COVID-19 pandemic. Through a survey-based approach, the authors identified various biases, such as loss aversion, overconfidence, and herding behavior, and analyzed their impact on investment choices. The study contributed to the understanding of investor behavior during crises and highlights the role of behavioral finance in explaining investment decisions.

Al-Kooheji et al. [7] investigated the impact of COVID-19 on retail investor trading behavior in the Bahrain Bourse. The authors analyzed trading data and investor behavior during the pandemic to understand changes in trading volumes, frequency, and trading patterns. The study provided empirical evidence of how the crisis has influenced trading behavior among retail investors and contributes to the literature on market dynamics during extraordinary events. Al-Dosari et al. [1] examined the risk perception and decision-making of retail investors in Bahrain during the COVID-19 pandemic. Through a questionnaire-based survey, the authors investigated how investors perceive risk, assess its impact on their investment decisions, and adjust their risk-taking behavior in response to the crisis. Al-Murbati et al. [10] explored the impact of COVID-19 on the portfolio choices of Bahraini retail investors from a behavioral finance perspective. Using survey data, the authors analyzed changes in portfolio composition, asset allocation, and risk management strategies during the pandemic. The study contributed to the understanding of how retail investors adjust their portfolios in response to crises and the role of behavioral biases in shaping investment decisions. Al-Mahmood and Al-Haddad [11] examined the psychological factors influencing retail investor behavior during the COVID-19 crisis in Bahrain. The authors investigated how psychological factors, such as fear, anxiety, and optimism, impact investment decisions and trading behavior. Al-Khalifa and Al-Faraj [12] examined the role of financial literacy in retail investor decision-making during the COVID-19 pandemic from a Bahraini perspective. The authors investigated the level of financial literacy among retail investors and its impact on investment choices, risk management, and financial well-being during the crisis. Al-Maskati and Al-Doseri [13] analyzed the impact of COVID-19 on retail investors' herding behavior in Bahrain's stock market. The authors examined trading data to identify instances of herding and investigate the factors underlying such behavior during the pandemic. The study contributed to understanding the dynamics of herding behavior among retail investors in times of crisis and its implications for market efficiency.

3 Significance of the Research

By focusing on retail investors in Bahrain, the research aims to provide practical insights and implications for this specific market segment. The findings of this study have the potential to inform policymakers, financial institutions, and individual investors about the behavioral dynamics at play during times of crisis and uncertainty.

3.1 Objectives of the Research

- To examine the impact of COVID-19 and its aftermath on the decision-making behavior of retail investors in the Kingdom of Bahrain.
- To uncover the behavioral biases, emotions, and cognitive processes that influenced investment decisions in the face of unprecedented market volatility and uncertainty.

3.2 Research Question

- What is the impact of COVID-19 and subsequent events on the decision-making behavior of retail investors in the Kingdom of Bahrain, as observed through a behavioral finance analysis?

3.3 Hypothesis of the Research

H_{01}: *The impact of COVID-19 and subsequent events on the decision-making behavior of retail investors in the Kingdom of Bahrain, as observed through a behavioral finance analysis, is not significant.*

H_{02}: *There is no substantial change in the decision-making behavior of retail investors during the COVID-19 pandemic and subsequent events in terms of risk aversion, investment choices, or the influence of behavioral biases.*

4 Research Methodology

Research Design: This study adopted a mixed-methods research design, utilizing both secondary and primary data sources. The research design allows for the integration of historical information with new primary research findings, providing a comprehensive analysis of the psychological and sociological aspects of market events.

Data Collection: The data source for this study is secondary data, which consists of qualitative and quantitative information extracted from official, final, and supported documentations and publications of historical market events. These sources include academic papers, financial reports, market analyses, and reputable financial publications. The secondary data analysis ensures the relevance and reliability of the data used for the study.

Data Analysis: The secondary data collected from official and reputable sources will be analyzed using thematic analysis. Qualitative analysis techniques employed to identify patterns, themes, and trends in the data.

5 Analysis and Discussion

The following area covered each market segment and discussed the analysis in more detail. While Table 1 shows investor types in the three different markets, Table 2 describes investor characteristics and biases in each of the three markets.

ETF Markets: ETF markets were generally classified in a mid-area when it came to accepting moderate risk. They were mostly exposed to cognitive biases that could be rectified. Investors in ETF markets displayed a willingness to follow externally influenced decisions or be independent in following their own developed strategies and knowledge. According to the research, ETF investors could be categorized as the 'composer' MBTI type.

Stock Markets: The stock market demonstrated volatility and a significant degree of risk. Individualistic and decisive investment patterns were exhibited by investors in this market. Emotional biases were there, but cognitive biases predominated. The

Table 1 Investor types in different markets

ETF market	Stock market	Cryptocurrency market
Investor type	MBTI	Adventurer

Based on the analysis

Table 2 Investor characteristics and biases

Market	Risk tolerance	Cognitive biases	Emotional biases
ETF market	Moderate	Present	Present
Stock market	High	Dominant	Present
Cryptocurrency market	Extreme	Absent	Prominent

Based on the analysis

study revealed that financial advisors had the ability to control the level of assistance they provided, since certain investors were open to expert guidance while others made judgments based solely on their own assessment.

Cryptocurrency Market: The cryptocurrency market was described as very risk-tolerant and aggressive. Emotional prejudices were a major factor, and people frequently made rash financial judgments. On the two-way model, these markets were categorized as the active accumulator investor type; on the five-way model, they were classed as adventurers. Due to their emotionally strong personalities, investment advisors found it difficult to counsel investors in these markets.

5.1 EFT Market

The S&P 500, represented by the SPDR S&P 500 ETF (SPY), is a widely recognized index that attracts investors seeking moderate risk tolerance and long-term growth through a passive buy-and-hold strategy. However, historical analysis reveals that the S&P 500 may not be suitable for short-term investors or traders due to its susceptibility to constant volatility and sensitivity to political, economic decisions, and unforeseen events.

The analysis in Fig. 1 indicates that the S&P 500 exhibits steady growth over the long run, which makes it a desirable investment for investors with patience. The one-year, two-year, and five-year returns all show a consistent rising trend when analyzed. This record implies that investors can profit from the index's upward trend provided they take a long-term perspective. Long-term growth potential is further supported by the bullish trend seen by the SPY graph over the last two years. The S&P 500 has performed well in part because its list of included companies is updated annually. By using this procedure, the index is limited to large-cap, well performing, and competitive companies.

5.2 EOG Stock

As the third-biggest firm in the US energy sector, Enron Oil and Gas Resources (EOG) is a well-known participant in the market. Natural gas and crude oil are just two of the fossil fuel types that EOG specializes in producing and distributing. The company's main production locations are in Texas and New Mexico, where it concentrates on extracting WTI Crude Oil, a commonly traded industry standard. EOG also runs offshore operations in Trinidad and Tobago (Yahoo Finance 2022). By utilizing effective operational and cost-control strategies, EOG hopes to optimize the return on their capital investment. These reserves are important assets that support the company's ability to produce goods.

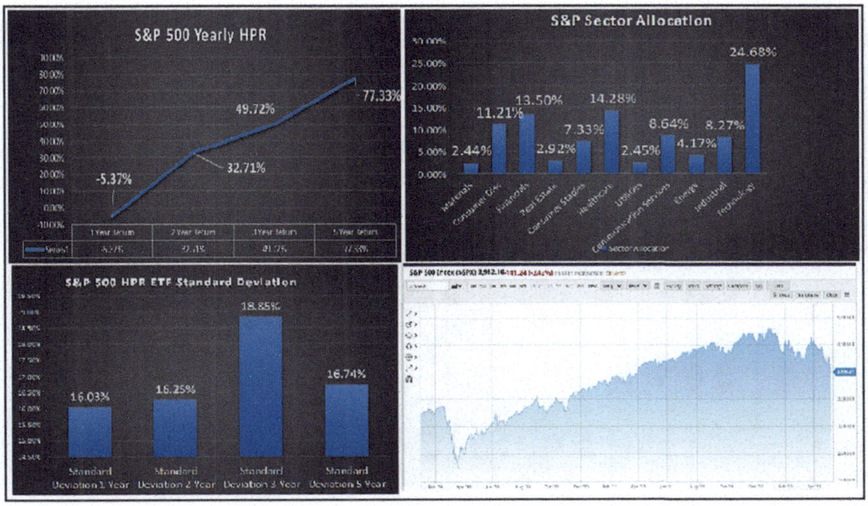

Fig. 1 S&P 500 financial analysis & performance during COVID-19 (based on the analysis)

EOG Resources (EOG) is a company that exhibits very high volatility and aggression, making it a risky investment option. As shown in Fig. 2, when compared to the Energy sector, EOG's standard deviation is significantly higher, indicating a greater degree of price fluctuation and risk. When it comes to short-term investing possibilities, investors looking to profit from drops in oil prices during conflict periods would find EOG appealing. Evidenced during the COVID-19 epidemic, historical data indicates that EOG's stock typically performs poorly in the near term when oil prices decline. Consequently, short-term investors may find opportunities to profit from EOG's performance if they foresee drops in oil prices during geopolitical crises.

On the other hand, long-term investors using a buy-and-hold strategy may find EOG to be appropriate as well. Uncommon events, like the COVID-19 shutdown in March 2020 and the ensuing Russian-Ukrainian conflict, have shown that EOG's stock can provide notable returns over the course of a year. These results imply that investing in EOG carries a high risk but also a big reward. Before sending money to EOG, investors should carefully examine their investing goals and risk tolerance. While long-term investors should evaluate EOG's potential for continuous growth based on their investment horizon and general market conditions, short-term investors should keep an eye on geopolitical developments and changes in oil prices.

5.3 *Crypto Market*

Due to its important contribution to the advancement of blockchain technology, Bitcoin (BTC) is the most well-known cryptocurrency by market capitalization and is frequently the subject of research (Royals and Beers 2022). Its popularity is further

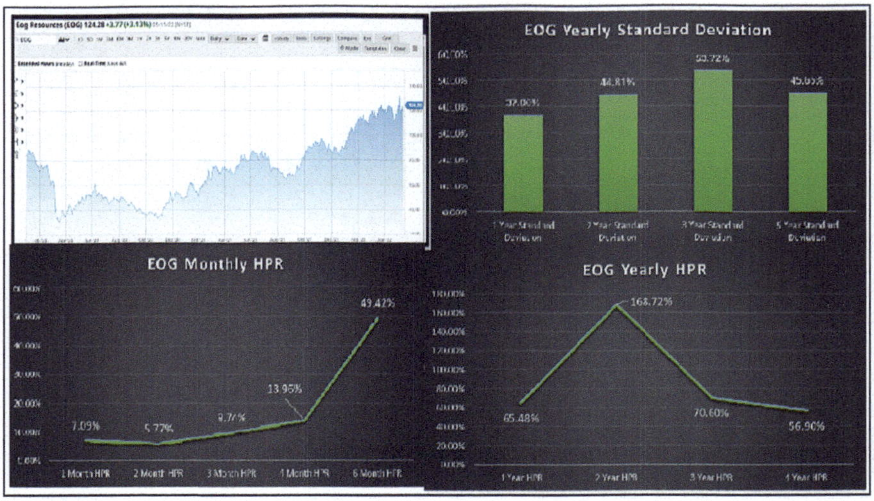

Fig. 2 EOG financial analysis during COVID-19 (2020–2022). All the historical-price-data was generated from Yahoo Finance from (01/01/2020)–(01/05/2022) (Yahoo-Finance 2022)

demonstrated by the fact that some countries are considering replacing their official currencies with it. The restricted supply of Bitcoin is one of its distinguishing features (McGovern 2022).

Market Performance Before the Pandemic

During the initial six months of 2018, there was a notable decline in the counter following the burst of the 2017 bubble.

As shown in Fig. 3, after reaching a high of $17,000 USD in the first week of January, the value of Bitcoin (BTC) saw a sharp decrease in the early months of 2018. By the first week of February, Bitcoin had decreased by more than 60%, to about 6000 USD. The fact that George Soros called Bitcoin a bubble in a statement that was released at the same time as this collapse could have affected market sentiment and exacerbated the downward trend (Forbes 2018). The value of BTC fluctuated above the previously mentioned low support level over the first half of the year. The resistance level for Bitcoin fell to 10,000 USD after hovering above 12,000 USD between February and April. It is important to remember that the final quarter of 2017 saw a sharp increase in the value of Bitcoin due to growing acceptance of the cryptocurrency by banks, corporations, and legislators (BBC 2017).

The latter half of 2018 witnessed the ongoing continuation of the recessionary period, characterized by sustained economic downturn and negative market trends.

During the second half of 2018, there were no significant news events directly related to Bitcoin and cryptocurrency. However, various external events, such as the California wildfires, Brexit negotiations, and other political conflicts, may have indirectly

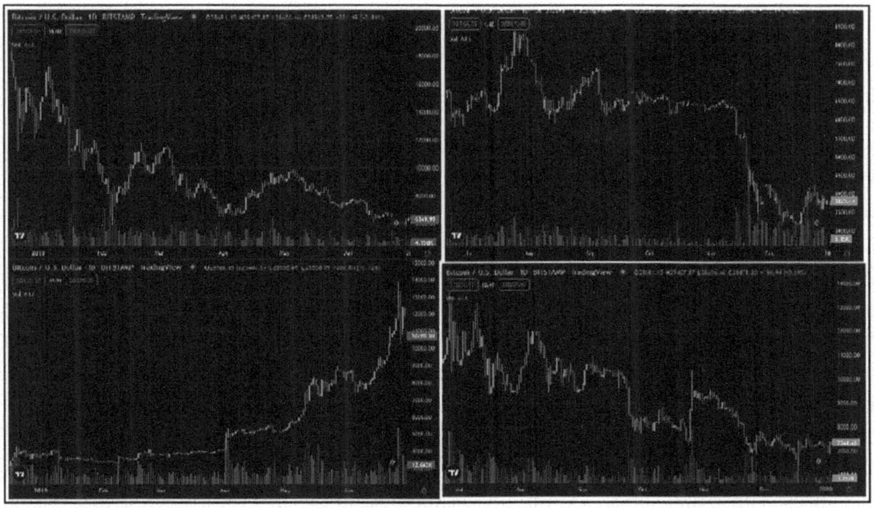

Fig. 3 Market performance before the pandemic

affected the cryptocurrency market during this period. In October, which marked the 10th anniversary of Bitcoin, the price remained relatively steady with low volatility. The price of Bitcoin experienced an approximate 45% decrease during November alone, indicating heightened market volatility and negative sentiment during that period. The downward trend continued into December, with Bitcoin's average value hovering around 3600 USD.

In the first half of 2019, there was a noticeable upward trajectory as market conditions began to improve, marking a recovery from the previous recessionary phase.

In the first quarter of 2019, Bitcoin (BTC) displayed relatively shallow movements in value compared to its typical volatility. During this period, BTC's value hovered around the pivot level of 3700 USD. However, on April 2nd, BTC experienced a notable shift as it surged by 18% within a single day. This marked the beginning of a steady growth trend that continued throughout the rest of the month. Finally, in June, BTC reached its peak value for the year, reaching 13,800 USD. It's important to note that BTC's price movements are subject to numerous factors, including market sentiment, regulatory developments, technological advancements, and macroeconomic conditions.

In the latter half of 2019, a familiar trend emerged as the markets experienced a downturn following a pattern observed in previous annual cycles.

In the second half of the year, Bitcoin (BTC) experienced a dip in July and struggled to recover its peak value reached in June. During this period, no significant news or developments in the crypto world indicated any potential impact on BTC prices. Moreover, global news related to economic, political, and environmental events had little relevance to BTC, leaving the movement of prices largely influenced by traders' sentiment.

Market Performance During the Pandemic

In December 2019, the first case of COVID-19 was recorded in Wuhan, China. Until March 2020, there was no significant impact on the volatility of Bitcoin's price. However, in March 2020, Bitcoin experienced a sharp decline of around 45% in value against the USD. This decline coincided with the announcement from the World Health Organization (WHO) declaring the COVID-19 outbreak as an official global pandemic. The market reaction to the pandemic was not unique to Bitcoin but was observed across various markets. In the second half of 2020, significant events, including the declaration of quarantine and social distancing regulations and the discovery of the Alpha variant of the COVID-19 virus, had an impact on Bitcoin (BTC) and the cryptocurrency market. With the announcement of the global pandemic, many nations implemented mandatory quarantine and social distancing measures. These measures led individuals and businesses to explore digital and online solutions for their physiological and economic needs.

First half of 2021: surge in infected cases and fatality count, additional variants, first Wave of crypto bubble, the Suez Canal Obstruction, and the development of the vaccine.

In the year 2021, Bitcoin (BTC) exhibited a volatile start, reacting to various events and news related to the COVID-19 pandemic. BTC's performance in early 2021 was influenced by positive news about the discovery of COVID-19 vaccines. However, it experienced a decline of around 30% in January following announcements of the Beta and Gamma variants and the first death in China since May 2020. BTC gradually recovered from the January decline and experienced a significant surge throughout February, with its value increasing by almost 90% from the support base established in January.

Second half of 2021: Omicron variant, second crypto wave of the year, terrorism and political conflicts, adoption of Bitcoin as national currency in El Salvador.

In contrast to other markets that have experienced negative contractions due to war developments in certain countries and the terrorist takeover in Afghanistan, Bitcoin (BTC) embarked on its second growth wave, commencing in mid-July. Following previous growth patterns, the cryptocurrency demonstrated a consistent and steep upward trajectory, experiencing an increase of approximately 80% from the latter part of July until reaching its peak in the first week of September.

6 Hypothesis Result

Null Hypothesis (H0): The impact of COVID-19 and subsequent events on the decision-making behavior of retail investors in the Kingdom of Bahrain, as observed through a behavioral finance analysis, is not significant. There is substantial change in the decision-making behavior of retail investors during the COVID-19 pandemic and subsequent events in terms of risk aversion, investment choices, or the influence of behavioral biases. Considering the available evidence and analysis conducted, the null hypotheses (H01 and H02)) are rejected. The findings suggest that the impact of COVID-19 and subsequent events on the decision-making behavior of retail investors in the Kingdom of Bahrain is indeed significant. The research indicates a clear shift in retail investors' decision-making patterns, characterized by increased risk aversion and altered investment choices.

6.1 Findings of the Study

The findings of the study revealed a clear association between the pandemic and its market-wide impact, highlighting the significance of understanding investor behavior in times of crisis. The study's findings have implications for various stakeholders, including individual investors, financial institutions, and policymakers.

- The study considered various factors that shape the personality and decision-making of investors in different markets, including ETFs, stock markets and cryptocurrency markets. The analysis revealed distinct characteristics and patterns of investor behavior within each market.
- In the case of ETF markets, investors were found to fall within a moderate risk acceptance range. They were exposed to cognitive biases that could be rectified, and their investment decisions were influenced both by external factors and their own strategies and knowledge [14].
- Stock markets exhibited variable and relatively risky behavior. Investors in this market displayed individualistic and decisive investment patterns, with a dominant presence of cognitive biases and some emotional biases.
- While some investors were open to expert advice, others relied on their personal judgment and decision-making.
- The research also noted that the studied stocks primarily belonged to the high-wealth investor category. Like ETF markets, stock market investors were classified as the "composer" MBTI type.
- Overall, the analysis demonstrated the diverse and nuanced behavioral characteristics of investors in different markets.
- The research provides insights into the specific investor types and their preferences, shedding light on the complex relationship between investor behavior, COVID-19, and financial markets.

6.2 Future Research Areas

- Further research can explore the specific economic and political factors that contribute to the volatility and performance of the S&P 500. Investigating the relationship between market events, such as policy changes, elections, or economic indicators, and the index's fluctuations can provide valuable insights into the dynamics of the market.
- Conducting a comparative analysis of sector performance within the S&P 500 can shed light on the varying levels of volatility and growth potential across different sectors. This research can help investors understand the sector-specific risks and opportunities associated with investing in the index.
- Exploring the psychological and behavioral aspects of investors in relation to the S&P 500 can provide a deeper understanding of how investor sentiment and decision-making influence the index's performance. Investigating factors such as risk tolerance, cognitive biases, and market sentiment can contribute to a more comprehensive analysis of the market dynamics.

7 Conclusion

This study sheds light on the ongoing debate between Behavioral Finance and Traditional Finance by exploring the negative correlation between the two approaches. It demonstrates the significant influence of COVID-19, economic policy decisions, and geopolitical events on investor decision-making from a behavioral finance perspective. The findings underscore the importance of considering behavioral biases and instincts in investment decision-making processes. The study contributes to the existing body of knowledge in finance and provides valuable insights for researchers, practitioners, and policymakers alike.

References

1. Al-Dosari, A., Al-Khalifa, F.: Investor risk perception and decision-making during COVID-19: a study of retail investors in Bahrain. Int. J. Fin. Bank. Stud. **11**(1), 30–47 (2022)
2. Al-Saleh, H., Al-Majed, A.: The influence of COVID-19 on the behavioral biases of retail investors: evidence from the Kingdom of Bahrain. International J. Econ. Comm. Manage. **8**(10), 8–19 (2020)
3. Bakshi, S: Financial markets transition from traditional finance to behavioural finance. In: XXI Annual International Conference Proceedings. pp. 297–310 (2020). ISBN No. 978-81-936606-2-1
4. Hasan, A., Al-Sahlawi, M.: Behavioral biases and investment decision-making during the COVID-19 pandemic: a study of Bahraini retail investors. J. Behav. Financ.Financ. **22**(4), 529–548 (2021)
5. Ling, D.C., Naranjo, A., Ryngaert, M.: The predictability of equity REIT returns: time variation and economic significance. Real Estate Econ. **28**(3), 465–491 (2000)

6. Downey, L.: Efficient market hypothesis (EMH). Investopedia, (2021). Retrieved from: https://www.investopedia.com/terms/e/efficientmarkethypothesis.asp#:~:text=The%20efficient%20market%20hypothesis%20(EMH)%2C%20alternatively%20known%20as%20the,consistent%20alpha%20generation%20is%20impossible
7. Al-Kooheji, K., Al-Abbas, Y.: The impact of COVID-19 on retail investor trading behavior: evidence from the Bahrain Bourse. Int. J. Econ. Comm. Manage. **9**(2), 1–16 (2021)
8. Joseph, B.: Impact of covid 19 on oil prices, gold prices and Indian Stock Market. In: Aloysius Edward, J., Jaheer Mukthar, K.P., Asis, E.R., Sivasubramanian, K. (eds.) Current Trends in Economics, Business and Sustainability. ICEBS 2023. Contributions to Environmental Sciences & Innovative Business Technology (2023)
9. Al-Khaja, A., Al-Aradi, A.: Impact of COVID-19 on retail investor sentiment and decision-making in Bahrain's stock market. J. Behav. Exp. Financ.Behav. Exp. Financ. **32**, 100533 (2021)
10. Al-Murbati, M., Al-Maskati, H.: Impact of COVID-19 on Bahraini retail investors' portfolio choices: a behavioral perspective. J. Fin. Plan. Anal. **1**(2), 45–60 (2022)
11. Al-Mahmood, S., Al-Haddad, S.: Psychological factors influencing retail investor behavior during the COVID-19 crisis: the case of Bahrain. J. Bus. Manage. Stud. **8**(2), 10–22 (2022)
12. Al-Khalifa, R., Al-Faraj, N.: The role of financial literacy in retail investor decision-making during COVID-19: a Bahraini perspective. Int. J. Fin. Res. **13**(2), 215–230 (2022)
13. Al-Maskati, A., Al-Doseri, S.: Impact of COVID-19 on retail investors' herding behavior in Bahrain's stock market. J. Econ. Bus. Stud. **10**(1), 51–70 (2022)
14. Kumar, S., Srivastava, M., Prakash, V.: Challenges and opportunities for mutual fund investment and the role of Industry 4.0 to recommend the individual for speculation. In: Nayyar, A., Naved, M., Rameshwar, R. (eds.) New Horizons for Industry 4.0 in Modern Business. Contributions to Environmental Sciences & Innovative Business Technology (2023)
15. Smith, J., Johnson, A., Davis, M.: Assessing the relationship between the COVID-19 pandemic and financial markets: a behavioral perspective. J. Fin. Econ. **45**(3), 123–145 (2023)
16. Barberis, N., Thaler, R.: A survey of behavioral finance. In: Handbook of the Economics of Finance, vol. 1, pp. 1053–1128 (2003). https://doi.org/10.3386/w9222
17. Kahneman, D., Tversky, A.: Prospect theory: an analysis of decision under risk. Econometrica **47**(2), 263–291 (1979)
18. Tversky, A., Kahneman, D.: Judgment under uncertainty: heuristics and biases. Science **185**(4157), 1124–1131 (1974)
19. Loewenstein, G., Weber, E.U., Hsee, C.K., Welch, N.: Risk as feelings. Psychol. Bull. **127**(2), 267–286 (2001)
20. Bollen, N., Whaley, R.E.: Does net buying pressure affect the shape of implied volatility functions? J. Fin. **64**(6), 2737–2767 (2009)
21. Barber, B., Odean, T.: All that glitters: the effect of attention and news on the buying behavior of individual and institutional investors. Rev. Fin. Stud. **21**(2), 785–818 (2008)
22. Hamilton, J.D.: Understanding crude oil prices. Energy J. **30**(2), 179–206 (2009)
23. Cheah, E.T., Fry, J.: Speculative bubbles in bitcoin markets? An empirical investigation into the fundamental value of bitcoin. Econ. Lett.Lett. **130**, 32–36 (2015)
24. D'Avolio, G., Shleifer, A., Vishny, R.W.: Psychology and the stock market. J. Financ. Econ. **87**(2), 329–365 (2007)

Institutional Aspects of Rivalry for Technological Leadership: Market Versus Security

Vitalii Hrebeniuk, Liudmyla Kudyrko, Igor Fedun, Viktoriia Zalizniuk, and Mykola Stetsko

Abstract Technologies in the conditions of the twenty-first century become a driver of transformations not only in the economy, but also in geopolitics, having a critical impact on the economic development and security of states. The aggravation of relations between the United States and the People's Republic of China makes the problem of identifying the modern concept of technological leadership, the nature and direction of institutional measures to achieve it, urgent. Modern state governments actively use both open and latent instruments, in particular in terms of non-tariff regulation of international trade and requirements for national investments. These institutional modernizations affect the development of the entire global economy already in the medium and long terms. The desire of the US and the EU to displace Chinese business from supply chains in high-tech sectors, in particular in the production of semiconductors, may lead to a significant reorganization of global value chains, increase the reshoring of many technological processes from the PRC to other relatively neutral regional centers, in particular Latin America, Middle East, Central Europe. For countries with growing markets, global geopolitical conflict can create conditions for sectoral reorientation of their own economy and foreign trade,

V. Hrebeniuk · I. Fedun
Department of Financial and Economic Security, National Academy of Security Service of Ukraine, 22 Mykhaila Maksymovycha Str., Kyiv 03066, Ukraine
e-mail: ratel6969@meta.ua

I. Fedun
e-mail: akademiafil1@gmail.com

L. Kudyrko · V. Zalizniuk
Department of World Economy, State University of Trade and Economics, 19 Kyoto Str., Kyiv 02156, Ukraine
e-mail: l.kudyrko@knute.edu.ua

V. Zalizniuk
e-mail: v.zalizniuk@knute.edu.ua

M. Stetsko (✉)
Department of Financial Technologies and Banking Business, West Ukrainian National University, 3 Lvivska Str., Ternopil, Ukraine
e-mail: nsteco6691@gmail.com

B. Awwad (ed.), *The AI Revolution: Driving Business Innovation and Research*, Studies in Systems, Decision and Control 524, https://doi.org/10.1007/978-3-031-54379-1_4

diversification of production, opportunities to strengthen their international competitiveness. New alliances in the global semiconductor supply system demonstrate the breaking of priorities established in the Western business tradition regarding the dominance of market feasibility and efficiency criteria over non-market ones. This also reflects a strong demand for the exceptional importance of national and regional security issues in matters of innovative and technological development.

Keywords Market · National security · Innovation strategy · Technological leadership · Global value chains · Semiconductors · China · USA

1 Introduction

A feature of the twenty-first century was the growth of economic rivalry between states in the process of globalization. This situation was caused, first of all, by the nature of the transformations that the global world is experiencing during the deployment of Industry 4.0. Confrontation and competition in the technological sphere provides a comprehensive impact on all other channels and sectors of international interaction. New technologies play a critical role in complex economic development and national military power, and the confrontation between the great powers in the technological sphere—China and the USA—can significantly affect the future global leadership and configurations of the world order already in the near future. If until recently the United States considered the PRC and the Chinese market from the position of dominant market interests, then in the current conditions of a sharp escalation of the confrontation, the task is to maximally restrain Chinese activity and reduce the dependence of the United States and its high-tech companies on imported supplies from the PRC.

The research is aimed at the analysis of scientific publications and empirical data on the basis of a case study in order to answer the following question:

What role does the institutional factor play in achieving the goals of global technological leadership?

Answering this question, the authors posed several sub-questions that will help answer the main research question:

1. Can it be said that in modern conditions, empirical experience gives reasons to revise and clarify the very concept of innovative strategies of states and companies?
2. Is there confirmation that the national strategies for achieving technological superiority chosen in the USA and the PRC give real positive results or are unsuccessful?
3. Why did the interstate struggle turn out to be so tough and far-reaching in the "semiconductors" segment, and what geopolitical and economic barriers are to be expected to overcome it in the medium term?

4. What are the prospects for the formation of new regional centers of semiconductor production in the new conditions of global technological confrontation?

2 Literature Review

Aspects of economic and technological rivalry in various spheres are reflected in a number of works of scientists. An example can be the matter of acquiring technological leadership, which is proposed to be considered by the American researcher A. Graham in the plane of the competition of national interests and fundamental values, which are implemented by the United States and the People's Republic of China [1]. In particular, B. Jones emphasizes the political, economic, and ideological factors that currently limit the prospects for interaction and strategic partnership of the two leading states and push them to deepen the confrontation in the struggle for primacy in the field of technology [2]. It should be noted that the global nature of the technological rivalry between the USA and the PRC is emphasized in the work of the experts of the Center for Strategic and International Studies (CSIS) [3].

In this aspect, it is worth taking into account the ratio of different formats of cooperation between the PRC and the USA in the technological sphere (from fierce competition to "soft" cooperation), which is researched by J. J. Carafano.

The study [4] emphasizes that the confrontation of states in the new digital era requires the adaptive use of institutional tools taking into account the technological innovations of business that has a cross-border nature of its activity.

The dynamics of the processes that humanity is currently experiencing in the conditions of the aggravation of the geopolitical confrontation between states, the activation of non-protectionist regulatory practices, the destabilizing impact of COVID-19 lead to a more in-depth analysis of the problems of technological rivalry, especially due to the active involvement of artificial intelligence in the management and production business processes as a driver of technological modernizations [5–8]. In general, technological development is currently assessed by scientists as a fundamental good that can prove the superiority of one or another national concept of civilizational success of states-global competitors. This is due to the fact that technological development improves ways of doing business, significantly increases internal efficiency and ensures a higher standard of living of people in society thanks to the improvement of the quality of services in such areas as education, health care and transport. Special attention should be paid to studies in which the synergy of three aspects has been focused on: technology, competitive transformation and sustainable development [9].

Confrontation in the technological sphere exerts a comprehensive influence on the transformation of strategic priorities of transnational business: even traditional spheres of competition, in particular trade and economic, are now increasingly subordinated to tasks related to technological advantage [10]. Scientists agree that if technological changes eventually transform the means of production and cause huge economic and political upheavals, they lead to increased instability and possibly violence [11]. Therefore, the issue of technology and technological confrontation, according to M. Porter's interpretation, cannot be called "apolitical" or "irrelevant" [12].

3 Purpose of Study

The purpose of the article is to reveal the institutional prerequisites of the technological confrontation between the United States and the People's Republic of China, to assess the effectiveness of the currently active national concepts of innovative development and their potential consequences for the global economy and individual states.

4 Materials and Methods

For this particular research project, the researchers applied the case study method. The current research consists of collecting secondary information from open sources. In particular, the information base of this article is the program documents on innovative and technological development of the USA, China, EU, statistical and analytical materials of international organizations, individual analytical and expert centers (World Intellectual Property Organization, UNESCO, Bloomberg). The results of the study were obtained on the basis of a combination of meta-data of analysis and synthesis, comparative analysis, systematization. Elements of institutional analysis and process approach were applied to assess the content of modernizations in the field of regulatory policy. The design of the study provided for the balancing of sources of empirical and statistical information to avoid the ontological problem of the dominance of factology over the qualitative assessment of processes and phenomena.

5 Results of the Investigation

5.1 Institutional Factors of Achieving Global Technological Leadership

Innovation strategy of the USA. Since the beginning of the twenty-first century, the US government has attached great importance to the formation of a national innovation network in the field of production in order to guarantee the beginning of the Fourth Industrial Revolution, which was emphasized in the Innovation Strategy in 2009, 2011 and 2015 [13, 14].

According to these program documents, the state will annually spend at least 3% of its own GDP on R&D in order to restore its leading positions: in fundamental research and promotion of new industries; improvement of the educational space for obtaining specialists with relevant knowledge and skills in accordance with new technological challenges; creation of developed infrastructure and an advanced ecosystem of information technologies. In addition, the United States has made it a priority to promote significant national projects that will provide innovation and breakthroughs in such areas as clean energy, biotechnology, nanotechnology, information technology in the field of education and health care, and space technology.

In general, safety criteria are currently becoming increasingly important arguments in providing institutional support for technological development. In particular, the Foreign Investment Risk Review Modernization Act, FIRRMA [15], approved in 2018, gave the Committee on Foreign Investment in the United States (CFIUS) the authority to block or suspend merger and acquisition agreements between the US and foreign companies if they can threaten the interests of national security. The area of special responsibility is defined as known "fundamental" and "emerging" technologies ("big data", artificial intelligence, quantum computing, energy storage, genetic engineering, nanotechnology, autonomous vehicles, robotics, augmented reality). Transactions of investors from "countries of particular concern" and deals "controlled by other governments" are also subject to enhanced CFIUS scrutiny. In recent years, CFIUS has blocked investments by Chinese companies in Western Digital and Philips, prevented the acquisition of Lattice Semiconductor by the PRC investment fund, and prevented a deal worth $44 billion between ChemChina and Swiss Syngenta.

Another institutional step in this direction was made by the approval in 2019 of the Export Control Reform Act (ECRA) [16]. In the People's Republic of China, this law was perceived as an unfriendly act towards their country.

The same priority should ensure the reduction of Chinese influence on the development of their technological sphere precisely in view of national security issues, the draft law approved in March 2023 entitled "Restricting the Emergence of Security Threats that Risk Information and Communications Technology Act" (RESTRICT). Such an Act would give the US government greater authority to restrict and ban the products and services of software or electronics manufacturers that are affiliated with foreign companies.

Another important step, in our opinion, is the institutional support of the aspirations of the White House regarding "reshoring"—the repatriation to the United States of high-tech enterprises previously withdrawn from the PRC (in particular, with the aim of "shortening" global value chains (GLVCs) and thereby making them more controlled and independent from the PRC. For American companies, the advantages of reshoring are also related to the possibility of bypassing tariff barriers introduced during the trade war and low energy prices in the U.S. According to Bank of America, the costs of implementing the White House instructions for the withdrawal of enterprises from the People's Republic of China amount to about 1 trillion US dollars in five years, and this is provided that American entrepreneurs limit themselves to the repatriation of only export production [17].

The White House, for its part, has promised to help the business, saying it intends to fully reimburse it for reshoring-related costs. The Trump administration and Congress also discussed the initiative to create a "Reshoring Fund" (Reshoring Fund) to help repatriate companies with a capital of 25 billion US dollars.

The great national project of innovative breakthrough of the People's Republic of China. In 2015, the ten-year plan of the government for the deep modernization of the manufacturing and technological base of China was adopted as part of the "Made in China 2025" (MIC) project. Ten highly productive technological sectors were named as priorities: new-generation IT, high-tech computerized machines and robots, aviation and space equipment, marine engineering equipment and high-tech ships, modern railway transport equipment, energy-saving and new energy cars, energy equipment agricultural machinery, new materials, biomedicine and high-performance medical equipment. The ambitions of the People's Republic of China within the framework of the outlined plan are undoubtedly maximum: by 2020, China will become one of the most innovative countries in the world, and by 2049—the world's leading scientific and technological power [18]. The government of the People's Republic of China sets ambitious goals in view of global perspectives: to reduce the level of technology imports; reorganize own potential in production; to build up high-tech segments. At the same time, it wants to ensure that innovation and product quality, efficiency and integration should drive production in key industries; increase brand recognition of Chinese companies and achieve the goals of "green" (environmentally oriented) development.

At the same time, the Chinese government does not hide that increasing the level of localization of production to 40% by 2020 and 70% by 2025 should contribute to economic self-sufficiency and reducing the level of import dependence. However, such goals violate the WTO rules on discrimination against non-resident companies.

Moreover, the implementation of MIC 2025 is especially important for the PRC in view of opposing the possible reorientation of global production capacities to Malaysia, India, Thailand, Indonesia or Vietnam in view of the loss of competitive resource advantages due to low labor costs. The program includes about a thousand projects that finance the activities of 557 organizations, 112 universities, 225 state research institutions and 220 companies. Moreover, financial support for the implementation of MIC 2025 provided for the creation of two funds—the National Investment Fund for Advanced Manufacturing Industry in the amount of 20 billion yuan and the National Integrated Fund in the amount of 139 billion yuan. The total amount of investments in the program is 1.5 billion US dollars from the central government, another 1.5 billion US dollars will be invested by regional authorities, and the main volume of funds, about 30 billion US dollars, is planned to be raised on the open market through loans and bond issuance [19].

Within the framework of the MIC, the Chinese government also urgently recommends that national enterprises increase the recognition of their international brand and "better familiarize themselves with foreign cultures and markets" and "strengthen risk management of investment operations" in order to encourage international investments and acquisitions.

The USA declared that the implementation of MIC is a real existential threat to the technological leadership of the USA. There are certain risks from the side of the EU as well, because in recent years Europe has been trying with all its might to create a common European strategy and innovation ecosystem. This will lead to further aggravation of international competition in the innovation space.

5.2 Are There Gains in the Implementation of Technological Leadership Concepts?

Assessment of the current technological status and its dynamics through international ratings. The annual ranking of the GII Global Innovation Index [20] for 2021, which presented the most relevant global trends in the field of innovation and a comparison of innovation ecosystems between 132 countries, confirmed the trends formed in previous decades. In particular, stable leadership positions at that time were held by several states with a predominantly high level of GDP growth. Moreover, a relatively new trend is becoming more noticeable. Accordingly, some middle-income countries, including China, Turkey, Vietnam, India and the Philippines, are catching up with world leaders and changing the innovation landscape. We must emphasize that Switzerland, Sweden, the USA and the United Kingdom traditionally lead the innovation rating. Moreover, they are among the top five leaders over the past three years. Moreover, in 2021, the Republic of Korea entered the top five GII for the first time in history. It should be noted that GII 2022 also tracked global trends in innovation during the COVID-19 pandemic. It is characteristic that Switzerland was recognized as the most innovative country in 2022, followed by the USA, Sweden, Great Britain and the Netherlands. China (11th place) came close to the 10 most innovative economies in the world, and Turkey and India entered the TOP-40 for the first time [21].

Currently, both absolute (Table 1) and relative [22–25] in terms of R&D funding indicators, the USA and China compete closely, they account for half of world R&D.

In 2020, the United States carried out scientific research and development worth 29% of global R&D. It was followed by China—24%. Japan closes the top three. It is noteworthy that in terms of the relative indicator of R&D expenditures in relation to GDP [24, 25].

Assessment of success in certain priority directions. Worthy of attention is the well-founded position of Bloomberg experts to evaluate the achievements in technological spheres by the level of capitalization of technological companies, the level of coverage of the country's population by wireless communication, dominance in space, modernization of education, etc. [26]. In this aspect, it is worth taking into account how significant the gap is, and whether it exists at all, in technological development between the United States, as a technological leader, and the People's Republic of China, as a new contender for global dominance.

Table 1 Comparative shares and growth rates of R&D expenditures and GDP in countries with leading R&D indicators

Countries	R&D/GDP, % 2021	Average growth rate, %			
		R&D		GDP	
		2000/2010	2010/2021	2000/2010	2010/2021
USA	3.45	4.3	5.6	3.9	4.2
China	2.44	20.5	10.6	12.9	7.5
Japan	3.29	3.6	2.4	2.7	2.0
Germany	3.13	4.9	6.1	3.6	4.2
South Korea	4.90	10.9	7.8	6.1	3.9
France	2.12	4.3	4.1	3.9	3.8
India	0.75	9.4	4.4	9.0	7.1
Great Britain	2.93	4.1	4.7	3.9	3.8

It should be noted that the level of capitalization of the world's leading technological companies is a special indicator that indicates how the market evaluates similar companies and their financial capacity for takeovers, hiring professional employees, attracting capital to new technologies. In 2019, from the TOP-10 companies according to the outlined feature, the first five positions were occupied by American technology giants: Microsoft (1st place), Amazon.com (2nd), Apple (3rd), Alphabet (Google) (4th), Facebook (5th). In general, the Americans took the 7 places out of 10 [27]. Two more positions were taken by giants from the PRC—(Alibaba (6th) and Tencent (7th), the same 9th position in the ranking of Samsung Electronics from South Korea. In 2022, the geospatial structure did not undergo changes in terms of quantitative indicators, but there was a certain change in positions among the leaders [28]: the American company Apple acquired the status of undisputed first place with a market capitalization of about 2.2 trillion US dollars; Microsoft took the second place, Alphabet (Google), Amazon and Tesla also entered the top ten. The ttwo new companies—American and Chinese—can be seen among the technological leaders in 2022. In particular, we are talking about TSMC (6th place)—the world's largest mono-profile independent semiconductor company (Taiwan, PRC), whose main facilities are located in mainland China, and *Nvidia Corporation*—an American manufacturer of graphics processors, video adapters under the Riva TNT and GeForce brands (7th place). The global marketplace Alibaba (China) dropped out of the top ten. Therefore, according to this indicator, the USA is clearly the leader.

Another important indicator of success in technological leadership is the influence of the state on the formation of international technical standards and, in particular, in the field of telecommunications. Therefore, according to international standards, the United States is traditionally the leader in most industries and spheres. It should be noted that the situation with 5G technologies is somewhat different. In particular, over the past two decades, American manufacturers of communication equipment have lost their leading positions not only in international markets. This industry is dominated by a trio of overseas suppliers. Moreover, among which Huawei (PRC) had the strongest position. Currently, Chinese companies own a third of the world's patents related to 5G (patents in which an invention is approved must be used to meet an industry standard). The presence of 5G patents is important because 5G goes beyond conventional mobile communication in today's technology sectors. Moreover, autonomous cars, artificial intelligence, smart factories and smart cities are also closely connected through 5G networks. Therefore, in the process of adopting 5G standards, the owners of the main patents will generally receive more profit and cause a more significant influence on the standardization and innovation path of the related technologies.

Space and, in particular, the use of its scientific and commercial potential is considered the next segment of the technological confrontation between the United States and the People's Republic of China. According to the 2022 report of the well-known Euroconsult consulting company "Government Space Programs", which specializes in researching various aspects of the market for space products, services, and research, the need to increase the use of space by the governments of the world's leading countries is growing. All this is confirmed in the scale and dynamics of financing of space-related projects. Therefore, of course, it is quite expected that this sphere is also an environment of fierce competition and struggle between the United States and the People's Republic of China [29]. The study showed that the total global spending on space programs already in 2022 amounted to 103 billion US dollars, which was 9% more than in the previous year, and was a record high indicator for this sector. It should be noted that the space budget of the United States in the amount of 62 billion US dollars was undoubtedly the largest in the world and exceeded the expenditures of the governments of all other countries combined (41 billion US dollars). Such growth was especially noticeable in the field of defense, where it reached 16% in 2022, setting a new record of 48 billion US dollars.

Undoubtedly, space is a new direction for technological rivalry, as well as a strategic operational location for hybrid warfare tactics. The relevant circumstances force governments to support their own investments in "traditional" space spheres, such as telecommunications, navigation and others. It is common knowledge that spending on "civilian" space has always been lower than defense spending. Moreover, under the current unprecedented conditions of global instability and tension, the gap between them is rapidly shrinking and, according to Euroconsult's forecasts, may reach 50/50 parity by 2031. It should be noted that the US indicators presented by Euroconsult are ambitious and impressive. Therefore, one cannot ignore the efforts of other space powers such as China, which is currently in second place (with 12 billion US dollars). In third place among the leaders with a sufficient gap from them is Japan (4.9 billion US dollars) [29].

The study showed that the USA still remains the leader in the world production of high-tech products. Their most significant achievements are in the production of aircraft and space rockets. Moreover, the global aircraft construction market is currently dominated by American companies such as Boeing, Lockheed Martin, Sikorsky Aircraft, and Gulfstream Aerospace. These companies produce more than a third of all aircraft on the planet. It should be noted that according to the annual assessment of Brand Finance experts, the American Boeing (the value of the brand increased by 13% to 17.5 billion USD) is the most valuable brand in the aerospace and defense industry by the end of 2023 [30]. Its closest Chinese competitor—AviChina in is the 17th place and AECC Aviation Power is in the 22nd place currently cannot pose a serious threat to American companies. It is characteristic that Elon Musk's company SpaceX is currently one of the largest manufacturers of rockets; and its Falcon 9 reusable heavy rocket is significantly superior in power and efficiency to its Chinese counterpart (the Changzheng 8 (Long March 8) rocket).

It is likely that the priority for the PRC already in the medium term, in order to overcome its technological backwardness from the USA, is not only the absolute increase of investments in R&D, but also the increase in the volume of scientific publications and patents and the development of applied sciences. It is necessary to emphasize that there are already significant successes in this direction, in particular, the PRC is ahead of the USA in terms of the number of engineers. We believe that university graduates in the fields of natural sciences, technologies, engineering and mathematics (STEM for short), which directly ensure success in the technological development of countries, are in high demand in the global environment. Therefore, the demand for them is constantly growing.

Moreover, many countries are trying to increase STEM coverage in order to accelerate the development of priority high-tech industries (medicine, digital communications, and others). It should be noted that in terms of the level of coverage, according to the estimates of the experts of the UNESCO Institute of Statistics, the USA lags behind significantly, and according to the results of 2022, it ranked only 9th among the 10 leading countries in the world in terms of the share of STEM graduates (with an indicator of 19.6%). It is characteristic that the first three were headed by countries that are not among the most financially powerful in the world, in particular: Malaysia (43.5%), Tunisia (37.9%), UAE (36.2%) [31]. The main thing is that the countries that managed to secure a higher share of STEM graduates are located in the Arab world, in Eastern Europe, as well as in East Asia. It should be noted that Western Europe, with the exception of Germany, is also not a STEM environment that trains many high-quality specialists. Although UNESCO did not include China in its ranking, the World Economic Forum's 2020 Future of Jobs report noted that in 2019 China produced 1.38 million bachelors in engineering alone. During this period, the USA trained only 197,000 specialists (144,000 in the field of engineering and 54,000 in the field of computer science), that is, only one-seventh compared to the PRC [32].

Within the obtained results, it can be stated that China has experienced a kind of revolution in tertiary education since the beginning of the twenty-first century. Moreover, until 2016, it created almost one university every week, and in terms of the absolute number of engineering graduates, it surpassed both Europe and the USA.

War in the semiconductor segment. The production of semiconductors and chips will have the greatest potential impact on the further development of the global economy and global production and consumption chains within the framework of the technological war between the PRC and the USA. In this context, all the above-mentioned measures of the US government over the past few years can be evaluated as the maximum limitation from the point of view of security of production in China of four or three strategic products: semiconductors with advanced nodes; equipment for the production of semiconductors (of any type); advanced computing power; supercomputers.

The conducted research makes it possible to assert that Taiwan confidently holds the status of a leader in the world semiconductor industry. It is characteristic that 2022 is the year of the export of integrated circuits from the island, the seventh year of growth in export volumes, thanks to such a flagship company as TSMC. Moreover, the shortage of chips due to the Coronavirus pandemic, as well as the tension in Taiwan's relations with the People's Republic of China and the probable invasion of the island from mainland China confirm the exceptional role of the chip market leader for the modern technological world.

As the risk of war between Taiwan and China, which historically and legally claims the island as its territory, looms in 2023, TSMC is looking for ways to minimize potential risks. To that end, it has opened branches in Japan and the United States, and plans to do the same in Germany. Given the higher costs associated with construction in Germany, including labor, TSMC is seeking subsidies in exchange for building the plant [33]. According to Bloomberg, the German government plans to allocate €20 billion to increase semiconductor production in the country. Interestingly, all the government's plans to co-finance investment projects with global companies-leaders affect the funds of the Climate and Transformation Fund (KTF), originally created to finance the country's transition to a carbon–neutral economy. The use of the Economic Stabilization Fund for state aid to microcircuit manufacturers is not excluded.

The potential use of semiconductors for telecommunications devices in cyber warfare is leading the US to implement a range of measures to significantly reorganize global semiconductor production chains and remove Chinese companies from supplying US businesses.

The alternative of banning all types of semiconductors could potentially lead to to serious consequences, in particular, a shock for the global automobile industry, given the lack of chips, the production of computing power and supercomputers [34].

Starting in the fall of 2022, anything that could contribute to the development of semiconductors with advanced nodes in China, advanced software, or supercomputers was to be considered a threat to US national security.

Evaluating the forecasts of analytical companies regarding the development of global semiconductor production [35], we believe that its further development will depend significantly on the active implementation of institutional regulators by direct and indirect competitors due to the aggravation of the technological war between the PRC and the USA, the overall destructive impact on the global the economy of the war between the Russian Federation and Ukraine, including through the introduction of sanctions against Russian companies and other participants of previous supply chains affiliated with them.

6 Conclusions and Discussions

The aggravation of relations between the United States and the People's Republic of China in the technological sphere has led to the following **consequences**:

- achievements in the technological sphere in both countries became considered as a necessary factor of global leadership in the conditions of Industry 4.0;
- there has been a transformation of the US concept regarding the development of the technological sector: from market efficiency priorities for American business to national security priorities for the country as a whole;
- in order to achieve the goals of deep modernization of the industry, the People's Republic of China is taking long-term progressive steps of a systemic nature, the role of the state and its regulatory system is only getting stronger;
- the activity of the sectoral policy of neo-protectionism regarding the technological sector in the protection of the national interests of both the USA and China has increased. After a sufficiently long period of openness of domestic markets, the governments of the United States and the People's Republic of China in recent years have been extremely active in using both open and latent tools to curb the technological expansion of competitors. In particular, this applies to non-tariff regulation of international trade and requirements for national investments;
- institutional changes and restrictions in both countries affect the development of the entire global economy already in the medium and long term perspective. In particular, the desire of the US and the EU to displace Chinese business from supply chains in high-tech sectors may lead to a significant reorganization of global value chains, intensify the reshoring of many technological processes from the PRC to other relatively neutral regional centers.

Limitation. The analysis showed that in the conditions of global geopolitical conflict there are a number of limitations that make it difficult to predict the future development of events. First, it is a limitation of the ontological level, when researchers do not take into account other tangential factors that can have a greater impact on the course of economic rivalry. In particular, this concerns business interests that can influence changes in state policy under democratic regimes. Secondly, since we are talking about the technological sphere, the emergence of revolutionary achievements of the "black swan" level in a third country, apart from the USA and the PRC, is not an exception here. And then the issue of global leadership can be considered in a different plane.

Prospects for further research can be seen in the disclosure:

- the consequences of reformatting global production and distribution chains of high-tech goods and services under the influence of US neo-protectionism in relation to Chinese companies,
- intensification of the processes of creating new international economic and military-political alliances both at the level of states and corporations in the sector of high-tech productions related to the spheres of defense and security;

– institutional mechanisms for encouraging the localization of high-tech semiconductor production by companies from Taiwan in the EU, USA, and Japan, increasing interstate competition for more attractive conditions for investing technology companies-leaders.

References

1. Graham, A.: The new spheres of influence sharing the globe with other great powers. Foreign Aff. **99**(2), 30–40 (2020)
2. Jones, B.: China and the Return ofGreat Power Strategic Competition. Brookings (2020), https://www.brookings.edu/research/china-and-the-return-of-great-power-strategic-competition. Last accessed 18 Aug 2023
3. Center, S., Bates, E.: Tech-Politik, Historical Perspectives on Innovation, Technology, and Strategic Competition. Center for Strategic & International Studies (2019). https://www.csis.org/analysis/tech-politik-historical-perspectives-innovation-technology-and-strategic-competition. Last accessed 11 Oct 2023
4. Kudyrko, L.: Institutional challenges in the formation of the global digital economy. In: Mazaraki, A., Melnyk, T., Melnychenko, S., Kudyrko, L., Lositska, T., Pugachevska, K. (eds.) Import substitution potential in the conditions of digital transformation, pp. 114–124. PC Technology Center, Kharkiv (2021). https://doi.org/10.15587/978-617-7319-51-0
5. Alareeni, B., Hamdan, A.: The impact of artificial intelligence on accounting and auditing in light of the COVID-19 pandemic. In: Alareeni, B., Hamdan, A. (eds.) Artificial Intelligence and COVID Effect on Accounting. Accounting, Finance, Sustainability, Governance & Fraud: Theory and Application. Springer, Singapore (2022). https://doi.org/10.1007/978-981-19-1036-4_1
6. Awan, U., Kanwal, N., Alawi, S., Huiskonen, J., Dahanayake, A.: Artificial intelligence for supply chain success in the era of data analytics. In: Hamdan, A., Hassanien, A.E., Razzaque, A., Alareeni, B. (eds.) The Fourth Industrial Revolution: Implementation of Artificial Intelligence for Growing Business Success. Studies in Computational Intelligence, vol. 935. Springer, Cham (2021). https://doi.org/10.1007/978-3-030-62796-6_1
7. Hussin, R.M., Putit, L., Subramaniam, G.: Social media as an open innovation: deciphering its relationship with firm performance, compatibility, and security concern. In: Rafiki, A., Dana, L.P., Nasution, M.D.T.P. (eds.) Open Innovation in Small Business. Contributions to Environmental Sciences & Innovative Business Technology. Springer, Singapore (2023). https://doi.org/10.1007/978-981-99-5142-0_6
8. Faisal, Y.A., Kanti, R.A., Alamanda, A.R., Apriliadi, A.: The role of business model innovation (BMI) in social enterprises during pandemic COVID-19 in Indonesia: a case of Islamic boarding schools' business units. In: Rafiki, A., Dana, L.P., Nasution, M.D.T.P. (eds.) Open Innovation in Small Business. Contributions to Environmental Sciences & Innovative Business Technology. Springer, Singapore (2023). https://doi.org/10.1007/978-981-99-5142-0_1
9. Al Mubarak, M., Hamdan, A.: Sustainable competitive advantage through technological innovation: an introduction. In: Al Mubarak, M., Hamdan, A. (eds.) Technological Sustainability and Business Competitive Advantage. Internet of Things. Springer, Cham. (2023). https://doi.org/10.1007/978-3-031-35525-7_1
10. Melnik, T., Kudyrko, L., Samsonova, L.: Contractual relations in the new economic competition strategies of TNCS. Baltic J. Econ. Stud. **5**(3), 116–124 (2019). https://doi.org/10.30525/2256-0742/2019-5-3-116-124
11. Chin, W.: Technology, war and the state: past, present and future. Int. Affairs **95**(4), 765–783 (2019). https://doi.org/10.1093/Ia/Iiz106
12. Porter, M.: The economic performance of regions. Reg. Stud. **37**(6–7), 549–578 (2003)

13. A Strategy for American Innovation (2015). https://www.strategy_for_american_innovation_october_2015.pdf. Last accessed 14 Sept 2023
14. A Strategy for American Innovation. Securing Our Economic Growth and Prosperity. https://obamawhitehouse.archives.gov/sites/default/files/uploads/InnovationStrategy.pdf. Last accessed 27 Aug 2023
15. H.R.5841—Foreign Investment Risk Review Modernization Act of 2018. 115th Congress (2017–2018). https://www.congress.gov/bill/115th-congress/house-bill/5841/text. Last accessed 12 Sept 2023
16. H.R.5040—Export Control Reform Act of 2018. 115th Congress (2017–2018). https://www.congress.gov/bill/115th-congress/house-bill/5040. Last accessed 18 Oct 2023
17. The USD1 Trillion Cost of Remaking Supply Chains: Significant but Not Prohibitive. New York, BofA Securities (2020). https://www.bofaml.com/content/dam/boamlimages/documents/articles/ID20_0734/cost_of_remaking_supply_chains.pdf. Last accessed 4 Sept 2023
18. Made in China 2025. https://isdp.eu/content/uploads/2018/06/Made-in-China-Backgrounder.pdf. Last accessed 21 Sept 2023
19. US Chamber of Commerce, Made in China 2025: Global Ambitions Built on Local Protections. https://www.uschamber.com/sites/default/files/final_made_in_china_2025_report_full.pdf. Last accessed 17 Oct 2023
20. Global Innovation Index. World Intellectual Property Organization, Geneva (2021). https://www.wipo.int/edocs/pubdocs/en/wipo_pub_gii_2021.pdf. Last accessed 27 Aug 2023
21. Global Innovation Index. World Intellectual Property Organization, Geneva (2022). https://www.wipo.int/edocs/pubdocs/en/wipo-pub-2000-2022-en-main-report-global-innovation-index-2022-15th-edition.pdf. Last accessed 3 Sept 2023
22. National Patterns of R&D Resources National Center for Science and Engineering Statistics (NCSES). Alexandria, VA: National Science Foundation (2022). https://ncses.nsf.gov/pubs/nsf22320. Last accessed 5 Sept 2023
23. UNESCO: Cross-national comparisons of R&D performance (2022). https://ncses.nsf.gov/pubs/nsb20225/cross-national-comparisons-of-r-d-performance. Last accessed 24 Sept 2023
24. OECD: Gross domestic spending on R&D (indicator) (2023). https://doi.org/10.1787/d8b068b4-en. Last accessed 1 2023/10/01.
25. The World Bank: Research and development expenditure (2023). https://databank.worldbank.org/reports.aspx?dsid=46&series=GB.XPD.RSDV.GD.ZS. Last accessed 11 Oct 2023
26. Wu, D., Hoenig, H., Dormido, H.: Who's winning the tech cold war? A China versus U.S. scoreboard (2019). https://www.bloomberg.com/graphics/2019-us-china-who-is-winning-the-tech-war/. Last accessed 14 Oct 2023
27. Leading tech companies worldwide (2022). https://www.statista.com/statistics/1350976/leading-tech-companies-worldwide-by-market-cap/. Last accessed 7 Sept 2023
28. Number of internet and social media users worldwide as of January 2023. https://www.statista.com/statistics/617136/digital-population-worldwide. Last accessed 18 Oct 2023
29. Government Space Programs, 22nd edition. https://www.euroconsult-ec.com/press-release/new-record-in-government-space-defense-spendings-driven-by-investments-in-space-security-and-early-warning/. Last accessed 14 Sept 2023
30. Brand Finance Aerospace & Defense (2022). https://brandirectory.com/rankings/aerospace-and-defence/2022. Last accessed 7 Oct 2023
31. Buchholz, K.: Where students choose STEM degrees. https://www.statista.com/chart/22927/share-and-total-number-of-stem-graduates-by-country/. Last accessed 9 Sept 2023
32. The Future of Jobs. Report 2020. https://www3.weforum.org/docs/WEF_Future_of_Jobs_2020.pdf. Last accessed 12 Sept 2023
33. Germany Readies €20 Billion in Aid to Bolster Chip Output. https://www.bloomberg.com/news/articles/2023-07-24/germany-readies-20-billion-in-aid-to-bolster-chip-production?srnd=technology-vp. Last accessed 17 Oct 2023

34. Panzeri, G.: China, US and the semiconductor war (2022). https://www.resetdoc.org/story/china-us-and-the-semiconductors-war-all-you-need-to-know/. Last accessed 28 Sept 2023
35. The global semiconductor market. https://www.fortunebusinessinsights.com/semiconductor-market-102365. Last accessed 11 Sept 2023

Sustainable Supply Chain Management and Banks Performance: Evidence from Banking Sector in Nigeria

Mustapha Ibrahim, Fatima Alfa Tahir, and Salisu Saad

Abstract The study examines the effect of Sustainable Supply Chain Management (SSCM) on performance of Nigerian Deposit Money Banks (DMBS). An ex-post factor design was employed to study seven banks from 2005 to 2019 using panel co-integration test and dynamic ordinary least squares methods. The findings revealed SSCM has significant positive impact on market value (Tobin's q) of DMBs ($\beta = 0.529$, $p < 0.05$; $\beta = 1.386$, $p < 0.05$). SSCM also had insignificant positive effect on banks' operational profitability in the long-run (ROA: $\beta = 0.276$, $p < 0.05$; T_q: $\beta = 3.61$, $p < 0.05$). However, bank size, capital structure and bank's liquidity support financial performance. This implies that SSCM environmental disclosure drives banks' long-term corporate image. The study thus recommends continuous integration of SSCM strategies in the banking sector to proactively help managers communicate challenges and bridge information asymmetry among diverse stakeholders.

Keywords Sustainable supply chain management · Bank performance · Deposit money banks

1 Introduction

Sustainability proponents consider Sustainable supply chain management (SSCM) as invaluable to firm performance as it ensures adherence to sustainable business principles. Thus, firms integrating UN Global Compact principles into SSCM are

M. Ibrahim (✉) · F. A. Tahir
Department of Accounting, University of Maiduguri, Maiduguri, Borno, Nigeria
e-mail: ibraheem.moosty@gmail.com

S. Saad
Department of Taxation, Nasarawa State University, Keffi, Nigeria

B. Awwad (ed.), *The AI Revolution: Driving Business Innovation and Research*, Studies in Systems, Decision and Control 524, https://doi.org/10.1007/978-3-031-54379-1_5

more likely to attain good governance and business ethics while enhancing sustainable development goals. This is especially important to banks in identifying ideal customers sharing their commitment to sustainable business practices and others posing a risk [1].

SSCM and firm performance is extensively studied in prior literature [2–5] majorly in developed economies where sustainability disclosure is mandatory. Additionally, existing studies concentrate on non-financial sectors as manufacturing, information technology, agro forest, tourism [6–7] (Huang et al. 2021) but sparse studies from developing economies where sustainability disclosure is voluntary.

Little is known about SSCM and bank performance in the Nigerian financial sector so this study provides empirical evidence on the nexus between SSCM and bank performance in environment where sustainability disclosure is voluntary but encouraged (UNEP-FI global roundtable report, 2020). Commercial banks were chosen given their role in the financial value-chain with mining, extractive industries, oil and gas and manufacturing firms as major government forex sources that pose great sustainability threat [8].

The paucity of research on SSCM especially in banks in developing economies underscores the need for this study given that other studies focus on corporate social responsibility and sustainability reporting in non-finance sectors [4, 9–12] (Babangida 2019; Raskal 2018; Nwaiwu and Oluka 2018). The few studies on SSCM in Nigeria are in construction industry [13, 14] (Omoren et al. 2016). Hence, this study is timely given the renewed push towards the actualization of the United Nations Sustainable Development Goals (SDGs).

The study has empirical, theoretical and practical significance. It provides empirical evidence on SSCM and bank performance to inform policy guidelines of regulators (CBN and NGX) and DMBs on climate risk finance mitigation or environmental liability tied to lending and listing requirements. The findings provide insights to market participants for making informed decisions, monitoring and oversight functions to regulators to protect various stakeholders. Theoretically, it contributes by employing the stakeholder theory to explain linkage between bank's SSCM strategies and corporate values. The rest of the paper is structured into four parts; literature review, research methods, results and discussions and conclusion.

2 Literature Review

SSCM is the integration of environmental, economic and social disclosure to support good governance and business ethics through strategic and transparent management of supply chain activities [5, 15, 16]. The Global Reporting Initiative (GRI) G4 2016 sustainability reporting guidelines provides the baseline for sustainability disclosure practices for banks. Nigerian MDBs operate under the CBNs' Banks and Other Financial Institution Act (BOFIA) 2020. While not directly exposed to climate related risks, their role as financial mediators makes them susceptible to sustainability risks that have far reaching impact on the environment and stakeholders [8, 17]. Hence, bank performance is indirectly linked to financed companies' impact on economic, environmental and social influences [18]. As such, incorporating SSCM practices through viable sustainability management systems in line with UNEP Finance Initiative indices (UNEP 2011) will create awareness and mitigate environmental hazards by identifying cost savings and vulnerabilities [4, 19].

2.1 *Stakeholder Theory*

Freeman and Philips [20] posit that firm operations affect divergent groups (employees, customers, suppliers, government and society) in various respects so they should be accountable by engaging in sustainable business practices (Abdulsalam 2017). Prior studies indicate economic, environmental and social sustainability originate from pressure from relevant stakeholders for accountability, transparency and sustainable development. As such, organizations respond to these concerns and expectations by making appropriate disclosures as illustrated in Fig. 1.

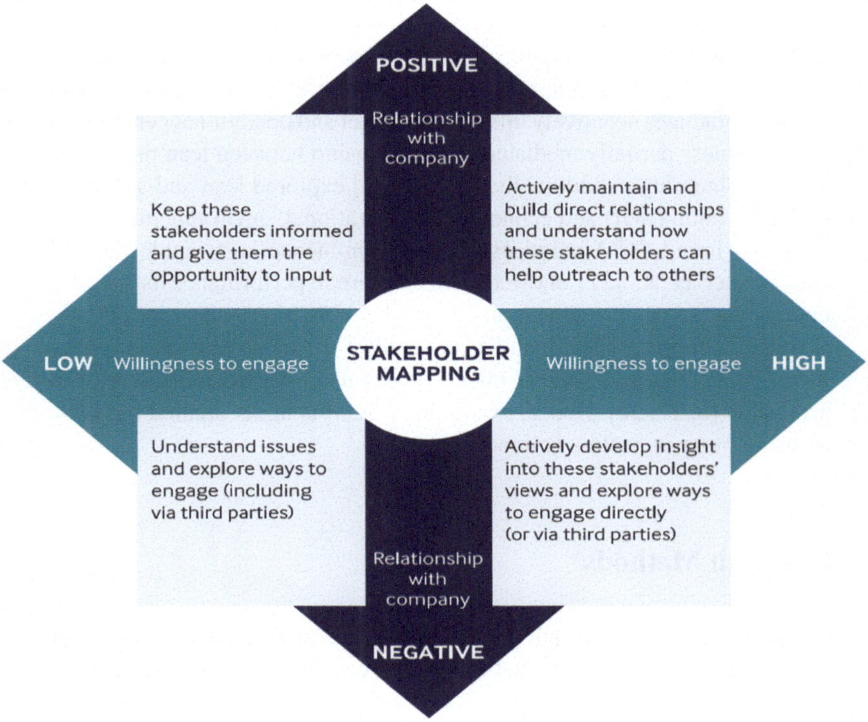

Fig. 1 Stakeholder mapping. *Source* Adopted from [21]

The study employs the Stakeholders theory to explain the motives and drivers of SSCM that influence firm value. The Stakeholders theory upholds accountability to stakeholders by inculcating SSCM in strengthening the relationship between firms and their host communities as ignoring stakeholders' interests adversely affects firm image and value.

2.2 *Empirical Review*

SSCM refers to activities tailored towards efficiently managing operations of diverse members of a supply chain [7]. Studies on various SSCM indicators and firm value (financial & market-based) abound globally [3, 22–24]. For instance, Li et al. [7] examined SSCM association to competitive advantage and firm performance and found significant positive influence. Bendehnezhad et al. (2012) showed supplier relationship, process and equipment, customer relations and human resource management were significantly positively affected by SSCM practices. Yildiz et al. [25] and Green et al. [26] found significant association among continuous improvement processes, JIT systems and green SCM practices which collectively influenced environmental performance.

Jum'a [3] reports SSCM practices influence on firm performance is a function of organizational supply chain. Conversely, Hofer et al. [23] found environmental management practices negatively influenced market and operational performance but inventory leanness partially mediated the relationship between lean production and financial sustainability. Additionally, De et al. [8] explored lean and sustainability-oriented innovation (SOI) and economic, organizational, environmental, and social practices and found SCM practices positively influence sustainable SME development. Wisner et al. [27] showed firms' internal performance is significantly influenced by good environmental practices when SSCM practices are effective.

Taken together, prior SSCM studies generally focused on non-finance sectors in developed economies and present mixed results using varied indicators of SSCM and firm value [5, 12, 28]. Hence, examining SSCM is under-studied but important sectors using other indicators is important.

3 Research Methods

An ex-post facto research design (content analysis) using annual reports of DMBs was adopted. Data from 2005 to 2019 from CBN database was filtered along banks existence (2005–2019), listing status and availability of complete annual sustainability reports resulted in seven filtered purposely selected banks (Access Bank, First Bank, Guaranty Trust Bank, United Bank for Africa, Unity Bank, Union Bank, and Zenith Bank) across three categories (international, national and regional banks). Variable measures for SSCM and financial performance were adopted from extant

Table 1 Variables measurements

Variables	Acronym	Measurement	Studies
FP: Tobin's q	T_q	$(MVE + PS + DEBT)/TA$	Yoon and Chung [29, 30]
FP: Return on assets	ROA	$\frac{Net\ Income}{Average\ Equity} * 100$	Ndukwe and Nwakanma (2018), Adams and Narayana [9, 31]
SSCM Environmental disclosure	SSCM-ENVD	Disclosure index; ecological, emission, gas spillage, waste, decommissioning & discharge (range 0–100%)	Kolk [32], Jenkins and Wright (2006), Carter and Easton [13], Yang et al. [24]
Bank size	BS	Annual total revenue (TR)	Koh et al. [33], Yoon and Chung [29]
Bank capital structure (leverage)	BCS	$\frac{Total\ Debt}{Total\ Asset}$	Yoon and Chung [29, 34]
Banks liquidity	BL	$QR + SOC + ARCL + MS$	Mao and Gu [35], Asuquo et al. [36], Yoon and Chung [29]
Corporate social responsibility	CSR	CBN's CSR score	[Berman et al.[37], Kacperczyk [19], Yoon and Chung [29]

Source Authors' compilation 2023

literature while CSR scores were calculated in line with Kacperczyk [19] and Yoon and Chung [29], from CBN's financial statistics database as shown on Table 1.

3.1 Model Specification

Following Koh et al. [33], Yoon and Chung [29], we specify two models:

$$ROA_{it} = \beta SSCM_{it} + \tau \ln BS_{it} + \gamma BCS_{it} + \phi \ln BL_{it} + \lambda CSR_{it} + \mu_i + \rho_t + \varepsilon_{it} \tag{1}$$

$$T_{qit} = \beta SSCM_{it} + \tau \ln BS_{it} + \gamma BCS_{it} + \phi \ln BL_{it} + \lambda CSR_{it} + \mu_i + \rho_t + \varepsilon_{it} \tag{2}$$

Equations (1) and (2) are linear models examining SSCM, BS, BCS, BL and CSR on ROA and Tobin's q of bank i at time t, ($i = 1, 2, 3, \ldots 7$, and $t = 2005, 2006, 2007, \ldots 2019$). Error term is denoted ε_{it}. Apriori expectation is positive relationship between explanatory and dependent variables.

3.2 Estimation Procedure

The panel analysis conducted utilized data from seven DMBs from 2005 to 2019 to allow for control of bank's specific effect in Eq. (1) by μ_i relating to time-invariant specific factors influencing financial performance. The use of time specific factors ρ_t in Eq. (1) enables control for external and other time-variant factors simultaneously affecting bank performance. Despite its advantages, estimating panel data poses challenge when lag(s) of the dependent variable is included as an explanatory variable because this introduces endogeneity bias [38] especially where data period is short [25] as is the case with our dataset. To solve this problem, Dynamic OLS is used because it is not only a co-integrating estimator, but also controls endogeneity of variables using their lead and lag values (see [39]). The estimation procedure is explained, First, we examined explanatory variables for potential correlation to avoid concurrently including correlated explanatory variables in the estimation that will render the *t-statistic* and *F-statistic* invalid and misleading for valid conclusion [40]. Hence, a correlation test using correlation matrix on Table 3 shows all values below 0.7 indicating no multicollinearity.

Second, the unit root properties on Table 4 were examined before long-run estimations. The Im et al. [35] IPS tests employ a null hypothesis of a unit root using the basic Augmented Dickey-Fuller (ADF) specification. In addition, we apply the Pedroni's cointegration to test for long-run relationship among variables within and between group dimensions (see [41, 42]). Table 5 shows the three statistic measures (PP (rho), PP (t), and ADF statistics) used.

Third, we employed panel Dynamic OLS (DOLS) [43] for estimating long-run equilibria that favours small samples and suitable for integrated variables from different orders that co-integrated. Potential simultaneity and small sample biases among regressors are solved by including lagged and lead values of regressors [38, 39, 43]

4 Results and Discussions

Descriptive statistics were run and presented in Table 2. The mean value of ROA is 0.08, with minimum and maximum values of −0.38 and 0.67 respectively. Average value of Tobin's q was 1.63 implying the annual average of the banks' market value is 1.6 times the annual replacement cost of their assets. The SSCM practices minimum and maximum scores are 52 and 89 respectively, indicating sampled banks recorded a fair score on sustainable supply chain management (environmental disclosure) practice. Further, all the control variables recorded significant variation which may influence estimation dynamics of Eqs. 1 and 2.

Table 2 Descriptive statistics of the variables

MEASURES	ROA	T_q	SSCM	BS	BCS	BL	CSR
Mean	0.08	1.63	64.42	23.61	0.78	1.03	0.53
Min	0.38	0.41	52.00	10.09	0.91	0.18	−5.00
Max	0.67	3.72	89.00	37.41	2.83	5.02	9.00
Std. dev.	0.02	0.23	81.73	4.20	0.24	0.42	2.68
Expected sign	NA	NA	+	+	+	+	±
Observations	98	98	98	98	98	98	98

Source Authors' Compilation based on data form CBN (2021)

Table 3 Correlation matrix

Variable	ROA	T_q	SSCM	BS	BCS	BL	CSR
ROA							
T_q	0.31						
SSCM	0.43	0.54					
BS	0.61	0.48	0.11				
BCS	0.58	0.37	−0.17	0.29			
BL	0.46	−0.56	−0.16	0.12	0.07		
CSR	−0.32	−0.18	−0.05	−0.33	−0.1	−0.51	

Source Authors' compilation based on modified data from CBN (2020)

4.1 Empirical Analysis

We estimate correlation among explanatory variables to ascertain extent of variable association [44]. Table 3 presents the Pearson's correlation results showing the bivariate relationship between the study variables. There is a positive relationship between ROA and SSCM (r = 0.43), BS (r = 0.61), BCS (r = 0.58), and BL (r = 0.46). A negative relationship is established between ROA and CSR (r = -0.32) but a positive correlation between Tobin q (T_q) and SSCM (r = 0.54), BS (r = 0.48), and BCS (r = 0.37). Multi-collinearity is not present among the variables.

Results of the unit root tests show only bank's liquidity (BL) is stationary, implying other variables have unit root at their level values. However, after taking the first difference, all the variables became stationary at the first difference I(1). Table 4 shows all the series are integrated in the same order i.e. stationary at the first difference I(1), which satisfies the condition for estimating long-run equilibrium relationship (co-integration) among the variables [41]. Table 5 presents the Pedroni's panel cointegration test conducted using constant and trend analysis. The results show 5 of the 7 tests rejected the null hypothesis of no cointegration at 5% and 1% significant levels, confirming evidence of long run equilibrium relationship among the variables.

Table 4 Results of the IPS unit root test

Var	ROA		T_q		SSCM		BS		BCS		BL		CSR	
Categ	$I(0)$	$I(1)$	$I(0)$	$I(1)$	$I(0)$	$I(1)$	$I(0)$	$I(1)$	$I(0)$	$I(1)$	$I(0)$	$I(1)$	$I(0)$	$I(1)$
Coeff	1.54	6.04*	0.82	4.93*	2.12	5.11*	2.61*	8.31*	−0.77	−4.15*	−1.04	−6.71*	0.72	3.87*

Source Authors' computation using data from CBN (2021)
* Indicates rejection of the null hypothesis at 5% significance level

Table 5 Results of co-integration test

Statistics	Constant + trend values
Panel v-statistic	−5.05*
Panel rho-statistic	−3.13
Panel PP-statistic	−9.08**
Panel ADF-statistic	−12.28**
Group rho-statistic	−1.82
Group PP-statistic	−4.11*
Group ADF-statistic	−10.04**

Source Authors' computation based on data from CBN (2021)
ADF Augmented Dickey-Fuller; *PP* Panel P-statistic
*, ** Denotes null hypothesis of no co-integration rejected at 5% and 1% respectively

Table 6 shows SSCM environment disclosure practices having significant positive impact on banks' financial performance ($\beta = 0.529$, $p < 0.05$; $\beta = 1.386$, $p < 0.05$) implying SSCM practices enhance financial performance of DMBs. The BS coefficient is positive and significant on ROA, indicating bank size positively affects profitability. BCS and financial performance of DMBs show significant and positive relationship (ROA: $\beta = 0.276$, $p < 0.05$; T_q: $\beta = 3.61$, $p < 0.05$) for T_q, affirming that capital structure of firms promotes firm value. Conversely, CSR has negative significant impact on financial performance of DMBs for both ROA and T_q (ROA: $\beta = -0.186$, $p < 0.02$; T_q: $\beta = -2.072$, $p < 0.004$).

Table 6 Effect of SSCM on ROA and T_q

	SSCM	BS	BCS	BL	CSR
Model 1: $ROA_{it} = \alpha + \beta SSCM_{it} + \tau \ln BS_{it} + \gamma BCS_{it} + \phi \ln BL_{it} + \lambda CSR_{it} + \varepsilon_{it}$					
Coefficient	0.529*	2.041**	0.276*	0.747	−0.186*
t-statistics	2.718	3.132	1.801	0.061	−1.930
p-values	0.013	0.002	0.035	0.296	0.023
Adj-R^2	0.48				
Model 2: $T_{qit} = \alpha + \beta SSCM_{it} + \tau \ln BS_{it} + \gamma BCS_{it} + \phi \ln BL_{it} + \lambda CSR_{it} + \varepsilon_{it}$					
Coefficient	1.386**	0.077	3.610*	4.639**	−2.072**
t-statistics	3.203	1.028	1.973	5.248	−3.844
p-values	0.000	0.822	0.025	0.000	0.004
Adj-R^2	0.53				

Source Authors' computation based on data from CBN (2021)
** and * indicate significant at 1% and 5% respectively

4.2 Discussion of Findings

The study examined impact of SSCM practices on performance of Nigerian MDBs (ROA and Tobin's q) using a panel co-integration test and dynamic OLS. The results indicate SSCM influence performance of DMBs in the long run and SSCM and banks' financial performance co-move in the same direction. This suggests effectiveness of SSCM initiatives on environmental disclosure in increasing long-term profitability and market value. This finding concurs with literature [45, 46]. Hence, banks' attention to SSCM could raise the level of their intangible resources and firm value.

In the long run, SSCM has significant positive effect on market value of DMBs since SSCM environmental practices influenced banks' Tobin's q positively. This suggests integration of SSCM practices by DMBs creates positive signals that improve value and performance. The findings agree with most literature [24–26, 40] but contradict [37].

5 Conclusion

Based on the findings, the study concludes SSCM is a driver of financial performance of DMBs. Banks adhering to SSCM principles will not only create shareholder value but also signal stakeholder protection, preserve corporate image and value. Through environmental consciousness, measurement and communication of environmental practices to diverse stakeholders, SSCM send signals (both positive and negative) to stakeholders on how responsible corporate entities (banks) are. Therefore, the study recommends that DMBs should integrate SSCM principles and improve the regulatory mechanism to communicate challenges and bridge information asymmetry among diverse range of stakeholders. However, the framework and design of this study has been empirically examined with non-probability sampling (availability sampling) and a relatively small sample size (n = 98) of seven DMBs filtered inhibits generalization of results. Future studies may explore larger samples, sampling techniques, other countries or sectors.

References

1. Huang, Y. C., Muñoz-Torres, M. J., Fernández-Izquierdo, M. Á., Rivera-Lirio, J. M., Ferrero-Ferrero, I., Escrig-Olmedo, E.: Sustainable supply chain management in a global context: a consistency analysis in the textile industry between environmental management practices at company level and sectoral and global environmental challenges. Environ. Dev. Sustain. **23**, 3883–3916. cedents and consequences of green supply chain management in Taiwan's electric and electronic industry. J. Manufact. Technol. Manag. **32**(5), 1066–1093 (2021)
2. Carter, C.R., Rogers, D.S.A.: Framework of sustainable supply chain management: moving toward new theory. Int. J. Phys. Distrib. Logist. Manag. **38**, 360–387 (2008)

3. Gujarati, D.N., Porter, D.C., Gunasekar, S.: Basic Econometrics. Tata McGraw-Hill Education, New Delhi, India (2012)
4. Ibrahim, M., Tahir, F.A.: ESG Sustainability and Firm Value: evidence from Listed Oil Marketing Firms in Nigeria. 4th International Conference on Business and Technology (ICBT2023), Istanbul, Turkey (2023)
5. Ikram, M., Zhou, P., Shah, S.A.A., Liu, G.Q.: Do environmental management systems help improve corporate sustainable development? Evidence from manufacturing companies in Pakistan. J. Clean. Prod. **226**, 628–641 (2019)
6. Freeman, R.E., Phillips, R.A.: Stakeholder theory: a libertarian defense. Bus. Ethics Q. **12**(3), 331–349 (2002)
7. Li, S., Ragu-Nathan, B., Ragu-Nathan, T.S., Subba Rao, S.: The impact of supply chain management practices on competitive advantage and organizational performance. Omega **34**, 107–124 (2006)
8. De, D., Chowdhury, S., Dey, P.K., Ghosh, S.K.: Impact of lean and sustainability oriented innovation on sustainability performance of small and medium sized enterprises: a data envelopment analysis-based framework. Int. J. Prod. Econ. **219**, 416–430 (2020)
9. Adams, C., Narayanan, V.: The 'standardization' of sustainability reporting. In: Sustainability Accounting and Accountability, pp. 89–104. Routledge, New York, U.S.A. (2010)
10. Adebimpe, O.A., Oladokun, V.O., Edem, I.E.: Towards the application of pumped-hydro storage in Nigeria. In Proceedings of the 11th Annual International Conference on Industrial Engineering and Operations Management, 2528–2537 (2016)
11. Koh, Y., Lee, S., Boo, S.: Does franchising help restaurant firm value? Int. J. Hosp. Manag. **28**(2), 289–296 (2009)
12. Martí-Parreño, J., Gómez-Calvet, R.: Social media and sustainable tourism: a literature review. In Proceedings of the 3rd International Conference on Tourism Research-ICTR 148–153 (2020)
13. Carter, C.R., Liane Easton, P.: Sustainable supply chain management: evolution and future directions. Int. J. Phys. Distrib. Logist. Manag. **41**(1), 46–62 (2011)
14. Keating, B., Quazi, A., Kriz, A., Coltman, T.: In pursuit of a sustainable supply chain: insights from Westpac Banking Corporation. Supply Chain Manag. Int. J. **13**(3), 175–179 (2008)
15. Baltagi, B.H.: Econometric Analysis of Panel Data. Springer Nature (2021)
16. Brandenburg, M., Gruchmann, T., Oelze, N.: Sustainable supply chain management: a conceptual framework and future research perspectives. Sustainability **11**, 7239 (2019)
17. Kaoje, N.A., Mohammed, A.N., Agbi, S.E., Ibrahim, Y.K., Abdulkarim, U.F.: Sustainability reporting and financial performance of listed oil and gas firms in Nigeria. Gusau J. Acc. Finance. **2**(3), 17–17 (2021)
18. McConnell, K.J., Delate, T., Newlon, C.L.: The sustainability of improvements from continuing professional development in pharmacy practice and learning behaviors. Am. J. Pharm. Educ. **79**(3), 36 (2015)
19. Kacperczyk, A.: With greater power comes greater responsibility? Takeover protection and corporate attention to stakeholders. Strateg. Manag. J. **30**(3), 261–285 (2009)
20. Cormier, D., Magnan, M., Van Velthoven, B.: Environmental disclosure quality in largeGerman companies: economic incentives, public pressures or institutional conditions? European accounting review, **14**(1), 3–39 (2005)
21. GRI: Global Reporting Initiative: Sustainability Reporting Guidelines (Version 3.0) (2016). Available at www.globalreporting.org
22. Chen, D., Gummi, U.M., Lu, S.B., Muazu, A.: Modelling the impact of oil price fluctuations on food price in high and low-income oil exporting countries. Agric. Econ. **66**(10), 458–468 (2020)
23. Feng, T., Li, Z., Shi, H., Jiang, W.: Translating leader sustainability orientation into green supply chain integration: a missing link of green entrepreneurial orientation. J. Bus. Ind. Mark. **37**(12), 2515–2532 (2022)
24. Neckel, S.: The sustainability society: a sociological perspective. Cult. Pract. Eur. **2**(2), 46–52 (2017)

25. Nickell, S.: Biases in dynamic models with fixed effects. Economet. J. Economet. Soc., 1417–1426 (1981)
26. Dim, N.U., Ezeabasili, A.C.C.: Strategic supply chain framework as an effective approach to procurement of public construction projects in Nigeria. Int. J. Manage. Sustain. **4**(7), 163–175 (2015)
27. Muñoz-Torres, M.J., Fernández-Izquierdo, M.Á., Rivera-Lirio, J.M., Ferrero-Ferrero, I., Escrig-Olmedo, E.: Sustainable supply chain management in a global context: a consistency analysis in the textile industry between environmental management practices at company level and sectoral and global environmental challenges. Environ. Dev. Sustain. **23**, 3883–3916 (2021)
28. Vu, T.T., Dang, V.T.W.: Environmental commitment and firm financial performance: a moderated mediation study of environmental collaboration with suppliers and CEO gender. Int. J. Ethics Syst. **2020**(2), 34–56 (2020)
29. Nwobu, O., Iyoha, F.O.: Managerial perceptions of corporate sustainability reporting determinants in Nigeria. J. Bus. Retail Manage. Res. (JBRMR) **12**(2), 123–135 (2018)
30. Lee, S.-Y.: Sustainable supply chain management, digital-based supply chain integration, and firm performance: a cross country empirical comparison between South Korea and Vietnam. Sustainability **13**, 7315 (2021). https://doi.org/10.3390/su13137315
31. Buallay, A., Fadel, S. M., Alajmi, J., Saudagaran, S.: Sustainability reporting and bank performance after financial crisis: evidence from developed and developing countries. Competitiveness Review: Int. Bus. J. **31**(4), 747–770 (2021)
32. Kolk, A.: Environmental reporting by multinationals from the Triad: convergence or divergence? MIR Manage. Int. Rev., 145–166 (2005)
33. IPIECA: Sustainability reporting guidance for oil and gas industry (2020)
34. Jenkins, K.E.: Energy justice, energy democracy, and sustainability: Normative approaches to the consumer ownership of renewables. Energy transition: Financing consumer co-ownership in renewables, 79–97 (2019)
35. Imperatives, S.: Report of the world commission on environment and development: our common future. United Nations, New York, NY, USA (1987)
36. Adeyemi, M.S., Akanji, A.A.: Role of book keeping on sustainability of small and medium enterprises (SMEs) in Nigeria (a case study of selected SMEs in Osun State). Int. J. Acad. Manage. Sci. Res. **4**(9), 140–154 (2020)
37. Berman, S.L., Wicks, A.C., Kotha, S., Jones, T.M.: Does stakeholder orientation matter? The relationship between stakeholder management models and firm financial performance. Acad. Manag. j. **42**(5), 488–506 (1999)
38. Aydemir, O., Demirhan, B. The relationship between stock prices and exchange rates: Evidence from mena countries. In Handbook of Research on Global Enterprise Operations and Opportunities, 171–185. IGI Global (2017)
39. Buallay, A., Cummings, R., Hamdan, A.: Intellectual capital efficiency and bank's performance: a comparative study after the global financial crisis. Pac. Acc. Rev. **31**(4), 672–694 (2019)
40. Margolis, J.D., Elfenbein, H.A., Walsh, J.P.: Does it pay to be good... and does it matter? A meta-analysis of the relationship between corporate social and financial performance. And does it matter (2009)
41. Mageto, J.: Big Data analytics in sustainable supply chain management: a focus on manufacturing supply chains. Sustainability **13**(7101) (2021). Available at https://doi.org/10.3390/su13137101
42. Mao, Z., Gu, Z.: The relationship between financial factors and firm performance: empirical evidence from U.S. restaurant firms. J. Foodservice Bus. Res. **11**(2), 138–159 (2006)
43. Masih, A.M., Masih, R.: Energy consumption, real income and temporal causality: results from a multi-country study based on cointegration and error-correction modelling techniques. Energy Econ. **18**(3), 165–183 (1996)
44. Doktoralina, C., Apollo, A.: The contribution of strategic management accounting in supply chain outcomes and logistic firm profitability. Uncertain Supply Chain Manage. **7**(2), 145–156 (2019)

45. Asuquo, A.I., Dada, E.T., Onyeogaziri, U.R.: The effect of sustainability reporting on corporate performance of selected quoted brewery firms in Nigeria. Int. J. Bus. Law Res. **6**(3), 1–10 (2018)
46. Panigrahi, S.S., Bahinipati, B., Jain, V.: Sustainable supply chain management. Manag. Environ. Qual. J. **30**, 1001–1049 (2019)
47. Baliga, R., Raut, R.D., Kamble, S.S.: Sustainable supply chain management practices and performance: an integrated perspective from a developing economy. Manage. Environ. Qual. Int. J. **31**(5), 1147–1182 (2019)
48. Green, K.W., Inman, R.A., Sower, V.E., Zelbst, P.J.: Impact of JIT, TQM and green supply chain practices on environmental sustainability. J. Manuf. Technol. Manag. **30**(7), 26–47 (2019)
49. Hofer, C., Eroglu, C., Rossiter Hofer, A.: The effect of lean production on financial performance: the mediating role of inventory leanness. Int. J. Prod. Econ. **138**, 242–253 (2012)
50. Jum'a, L., Zimon, D., Ikram, M.: A relationship between supply chain practices, environmental sustainability and financial performance: evidence from manufacturing companies in Jordan. Sustainability **13**, 2152 (2021). https://doi.org/10.3390/su13042152
51. Ojo, E., Mbohwa, C., Akinlabi, E.: Green supply chain management in construction industries in South Africa and Nigeria. Int. J. Chem. Environ. Biol. Sci. (IJCEBS) **2**(2), 146–150 (2014)
52. Olaniyan, O.N., Efuntade, A.O., Efuntade, O.O.: Corporate Social Responsibility and Firm Financial Performance in Nigeria: mediating on Ethnical Responsibility. Annals of Spiru Haret University. Econ. Ser. **21**(1), 71–93 (2021)
53. Olawoye-Mann, S. (2022). 10. A green mandate: The Central Bank of Nigeria and sustainable development. Central Banking, Monetary Policy and the Environment, 219
54. Park, B., Lee, J.: Assessing sustainability impacts of route guidance system under cooperative vehicle infrastructure environment. In 2009 IEEE International Symposium on Sustainable Systems and Technology (pp. 1–6). IEEE (2009)
55. Parthiban, K.T., Fernandaz, C.C., Vishnu, M.J.: Agroforestry for sustaining industrial raw materials: experience from a value chain leveraged consortium model. In Agroforestry for Sustainable Intensification of Agriculture in Asia and Africa (pp. 759–778). Singapore: Springer Nature Singapore (2023)
56. Pedroni, P.: Panel cointegration: asymptotic and finite sample properties of pooled time series tests with an application to the PPP hypothesis. Economet. Theor. **20**(3), 597–625 (2004). https://doi.org/10.1017/S0266466604203073
57. Pedroni, P.: Purchasing power parity tests in cointegrated panels. Rev. Econ. Stat. **83**(4), 727–731 (2001). Retrieved from http://www.jstor.com/stable/3211767
58. Pesaran, M.H.: Testing weak cross-sectional dependence in large panels. Economet. Rev. **34**(6–10), 1089–1117 (2015)
59. Singh, R.K., Modgil, S., Tiwari, A.A.: Identification and evaluation of determinants of sustainable manufacturing: a case of Indian cement manufacturing. Meas. Bus. Excell. **23**, 24–40 (2019)
60. Stock, J.H., Watson, M.W.: A simple estimator of cointegrating vectors in higher order integrated systems. Econometrica: j. Econometric Soc. 783–820 (1993)
61. Usenko, L., Zenkina, I.: Modern trends and issues of corporate reporting data disclosure on organization activities. Mediterr. J. Soc. Sci. **7**(3), 212–226 (2016)
62. Wisner, P.S., Epstein, M.J., Bagozzi, R.P.: Environmental proactivity and performance. Adv. Environ. Account. Manag. **4**, 105–127 (2009)
63. Woo, D., Suresh, N.C.: Voluntary agreements for sustainability, resource efficiency & firm performance under the supply chain cooperation policy in South Korea. Int. J. Prod. Econ. **252**, 108563 (2022)
64. Yang, M.G.M., Hong, P., Modi, S.B.: Impact of lean manufacturing and environmental management on business performance: an empirical study of manufacturing firms. Int. J. Prod. Econ. **129**, 251–261 (2011)
65. Yildiz Çankaya, S., Sezen, B.: Effects of green supply chain management practices on sustainability performance. J. Manuf. Technol. Manag. **30**, 98–121 (2019)

66. Yoon, B., Chung, Y.: The effects of corporate social responsibility on firm performance: a stakeholder approach. J. Hosp. Tour. Manag. **37**, 89–96 (2018)
67. Zhou, X., Wei, X., Lin, J., Tian, X., Lev, B., Wang, S.: Supply chain management under carbon taxes: a review and bibliometric analysis. Omega, **98**, 102295 (2021)

The Concept of Accounting in Islamic Transactions

Ahmad Abdulkadir Ibrahim

Abstract The Islamic law of transactions cast down the process and method of accounting and analyzed its fundamental presumptions in present time. There is a need to examine the implications of accounting initiative in Muslim world and attempt to outline the important characteristics of Islamic accounting and how Islamic accounting resolves the problem of measuring the cost of trust sales such as Murabaha products in case of exchange rate variation. The research tends to analyze and discuss the concept of Islamic accounting as well as elaborating jurisprudential matter and demonstrating the practical aspect of accounting in Islamic financial transactions. It also heads for alerting the scholars and practitioners of Islamic accounting in the world entirely, to be familiar with Islamic accounting concept which reflects in Islamic jurisprudence by the virtue of Islamic accounting historical enhancement. The researcher adopted qualitative method as a research methodology by the means of referring to the relevant literature of the subject matter. This is accompanied by discussion and analysis for the material used. It is found that Islamic accounting is relevant and unique in its standard level as it has been qualified by truthfulness, fairness, accuracy in measuring tools, mutual trust, moderation in making profit and tolerance. It was also characterized by flexibility and capacity concerning terminologies and tools invented and applied by Islamic law of transaction through accounting system, which has been valued and described by consistency decisively. The research also found that the claimed idea which signifies that an Italian Mathematician writer know as Luca Pacilio had an initiative in developing the basis of double entry is misconception, due to the existence of Muslim scholars' work in accounting development dated back to his era. It concludes further that the Islamic law of transaction draws the accounting process and standardized the organizations of a market to be far from usury, cheating, fraud, and unfair and illegitimate competition in commercial fields.

Keywords Islamic accounting · Accounting · Islamic transactions · Double entry · Murabaha

---

A. A. Ibrahim (✉)
Islamic Commercial Law, Islamic Law Banking and Finance, University of Business and Technology, Jeddah, Saudi Arabia
e-mail: a.abdulkadir@ubt.edu.sa

B. Awwad (ed.), *The AI Revolution: Driving Business Innovation and Research*, Studies in Systems, Decision and Control 524, https://doi.org/10.1007/978-3-031-54379-1_6

1 Introduction

Islamic law of transaction has established a codified accounting system in the market with foundations far from fraud and its methods, and completely free of unfair competition in all fields. Since profit in Islamic accounting jurisprudence is a growth of a capital that continues throughout the year, whether the money is liquidated at the end of the year or not and remains as an object, this growth is witnessed in the capital, whether it is money or merchandise. Apparently, trade is only the means of this growth, hence there is no doubt that growth and profit are achieved only through trade. In the view of Islamic accounting jurisprudence, it is equal in measuring profit that the growth of capital and increase of profit can be achieved by force or action, given that both capital and profit are growth and increase in value that must be considered.

The accounting process for Murabaha (cost plus profit margin) plays a positive role in banking transactions because it is a measurement of property rights, commitment, and financial events. Measurement generally includes determining an equivalence of some kind of different things and occurrences, and this equivalence is the basis of measurement in any field. Accounting measurement is considered an essential pillar of accounting, which tends to determine the result of activity at the end of each specific period, as well as determining the true financial status, and then the proper and accurate disclosure of accounting information in order to benefit from it.

The Islamic law of transaction is highly important due to its connectivity with accounting system, business activities, zakat collection, endowment, the money exchange and so on. Islam in all these considerations laid the rules and regulations owing to Islamic accounting to address personal moral accounting and maintain the capital. Apparently, Muslim scholars had initiative for the emergence of Islamic accounting as evident by the existence of the manuscript book written by Muslim scholar back to the year (765 H–1363) roughly. Hence, it is misconception that "Luca Pacilio" (An Italian mathematician) was the first writer to lay the foundations of accounting in the year 1494 AC–913 AH and used widely in the year 1869 AC–1288 AH.

Islamic accounting is characterized by accuracy in measuring tools which suggests the scales, weights, measures, etc. The Islamic law of transaction cares about fairness and accuracy of accounting in Islamic transaction whereby controlling the value of the commodities in terms of weights and measurements. It also implies truthfulness, mutual trust and honesty between merchants which are considered as the basis of trade. Similarly, Islamic accounting is described by moderation in making profit thereby gaining what is right and not returning a little profit. Additionally, tolerance is one of the significant characteristics of Islamic accounting that elevates a person to a high moral and spiritual degree. As not vague, tolerance is also a type of generosity required of the seller, the buyer, and the debtor. Among others, the Islamic accounting is far from monopoly whereby purchase of food and the like and withholding it until it becomes expensive or monitoring the markets while waiting for prices to rise.

This paper consists of six topics as follows:

- Scope and Objective of the Research
- Definition of Accounting & Islamic Transaction
- Muslims' Initiative in Accounting Innovation
- Characteristics of Islamic Accounting
- The Process of Deducing a New Accounting Method in Islamic Transaction.

2 Scope and Objective of the Research

The research clarifies the accounting concepts, studying the Islamic law of transactions pertaining to the accounting system by the means of comprehending and cogitating the rooting and application of Islamic Accounting system. It also tends to analyze the Islamic accounting transaction process, and to suggest the components of accounting system and draw attention to what requires that. The research discusses the process of deducing a new accounting method in Islamic transactions especially in the case of prices drop or rising. Consequently, it demonstrates and analyzes the problem of measuring the cost of murabaha goods in case of exchange rate variation.

3 Definition of Accounting and Islamic Transaction

3.1 First, Literal Meaning of Accounting

The Arabic word *Almuhaasaba* (accounting) is derived from the word "*Haasabahu, Muhaasabatan, Hisaaban*", that means: "he talked over with him about the account and rewarded him". "*Hesaab*" also means: "count and a lot of adequate", in the Holy Qur'an: (a fitting reward as a generous gift from your Lord) [An Naba' 78: 36] [1].

3.2 Second, Technical Meaning of Accounting

The most comprehensive and accurate definition of accounting is said by one of celebrated Islamic scholar Dr. Omar Abdullahi Zaid. He defined accounting "as a systematic process relating to the records of transactions, actions, legal decisions and its limitations in the approved records. Likewise, measuring the financial consequences of such transactions, as well as the actions and decisions, to help in rationalizing decisions" [2].

3.3 Third, Meaning of Islamic Transaction

Islamic transactions refer to financial activities that adhere to Islamic law (*Shariah*) and its practical application by the means of enhancement of Islamic economics. In other word, it refers to Jurisprudential finance or Shariah-compliant finance, to achieve two basic principles of Islamic law of transactions, these are: the sharing of profit and loss, and forbidding of the payment as well as collection of interest by creditors, vendors and investors [3].

4 Muslims' Initiative in Accounting Innovation

According to the historical evaluation of accounting in relation to non-Muslims, studies are more concentrated in two stages: the first is the stage before the stability of the Islamic nation, and the second is the stage beginning by the ending of fifteenth century, that is recognized as an existence of a specific book in accounting written by "Luca Pacilio" who dedicated a special chapter in accounting. Despite they ignored the early emergence of accounting since the dawn of Islam unto the year 1494, where "Luca Pacilio" compiled his book. Perhaps examining the cause for this renouncement may refer to their ignorance of Arabic language on one hand, and their illiteracy and ignorance of Islam and its application on the other hand [4].

Accounting evolution is evident under an implementation of Islamic state. For instance, Caliph Omar Ibn Khattab may Allah be pleased with him took care of fundamental and basic requirements concerning keeping commercial books and relevant documents and the indexed books in a systematic way, to be easily referred whenever is needed. This is confirmed by the motivation of Caliph Omar Ibn Khattab may Allah be pleased with him, whereby establishment of a system of documentation and bureaucracy in an organized manner, in such a way the successive generations followed him until our current era [5].

Undoubtedly, the Western claim that the first book written in accounting pertaining to a method of proving double entry in the process of proving parties of transaction was a Luca Pacilio's book, it was misconception due to the existence of manuscript book dated back to the year (765 H/1363 AC). The book was written by an erudite Muslim author called Abdullah bin Muhammad Al Mazindani titled: "The book of astronomy" (*Kitab Al-siyaaqat*).[1] The Author used Arabic letters in this book, similarly he used Turkish language that were prevalent under the Ottoman Empire, he also used some Farsi words sometimes. The book was compiled almost 131 years before the emergence of Pacilio's book. Significantly, this fact negates Western belief. In brief, the compilation of Luca Pacilio was preceded by a fundamental accounting developments that was devoted by Muslim Scholars which should not be ignored.

[1] The book can be traced in Istanbul at Sultan Sulaiman library, Turkey. Department of Manuscripts, No. 2756, with the title: *Accounting and Accounting System in the Islamic State*.

It is apparent that Muslim scholars had initiative in authoring accounting method and bringing its concept into existence. Therefore, no wonder if it is understood that Luca Pacilio's book was the first published and printed book to exist in Europe, since it was proved by some Western research the existence of manuscript book that was written before Luca Pacilio. Furthermore, some Western researcher believed that the first of the books written in a way of proving the transaction or accounting was an Italian writer "Cotrugli", he justified by saying that Cotrugli's book was older than Pacilio's one. However, some argued that the first writer in the method of double entry and accounting treatment was Angelo Senesio in the year 1348 AC [1].

In short, we should rectify and correct misconception concerning accounting history and doing justice to Muslims by highlighting their contributions and effective role in this area. Instead of mentioning or referring the history to the Italian Luca Pacilio, we should remember Al-Qalqashandi, Al-Nuwayri, Al-Ghazali, Abu Jaafar Al-Dimashqi, Al-Mawardi, Al-Khwarizmi, and other prominent Muslim scholars. Therefore, Muslims were those who established accounting in thought and practice since approximately 900 AC, that is before Luca Pacilio by at least 700 years [1].

5 Characteristics of Islamic Accounting

Islamic accounting is an important matter which is inseparable to Islamic transactions. Hence, it is characterized by various characters as follows.

5.1 Accuracy in Measuring Tools

Measuring tools suggest scales, weights, measures, etc. The Islamic law of transactions cares about the fairness and accuracy of Islamic accounting, so it controls the value of the commodities in terms of weights and measurements, the prices whereby using dinars, dirhams, and all forms of currency. Similarly, it lays fair foundations for the continuation of the balance between supply and demand, in addition to the role played by the appropriate social climate in which the Muslim was raised from an early age. Islamic law guarantees freedom of pricing and prevents a monopoly, fraud and deception, to ensure fair measurement tools [1].

5.2 Truthfulness

Abu Sa'id reported Allah's Messenger (S.A.W) said: "The truthful and trusty merchant is associated with the prophets, the upright, and the martyrs". Tirmidhi, Darimi and Daraqutni transmitted it, and Ibn Majah transmitted it on the authority of Ibn 'Umar. Tirmidhi said this is a *gharib* tradition [6].

5.3 Honesty

Allah the Almighty says which means: (The believers are also those who are true to their trusts and covenants) [Al Mu'minoon: 8]. He also says which means: (Indeed, Allah commands you to return trusts to their rightful owners) [An-Nisa: 58]. In the Holy Hadith (*Hadith Qudsi*): The Messenger of Allah (ﷺ) having said: Allah, Most High, says: "I make a third with two partners as long as one of them does not cheat the other, but when he cheats him, I depart from them" [7].

It is known that mutual trust and honesty between merchants is the backbone and basis of trade. If betrayal becomes widespread, merchants refrain from trading among themselves or reduce business to the lowest levels, which affects financial liquidity in the market. Subsequently, cash dealing will spread to other types, conflicts will increase, and rights will often be lost, and economic life will be disrupted throughout society [1].

5.4 Moderation in Making Profit

Ali bin Abi Talib, may Allah be pleased with him, used to walk around the markets of Kufa and say: "O' Merchants, gain what is right and you will be safe, and do not return a little profit, lest you deprive a lot of it". This is a call to increase capital profit rates by not exaggerating in raising prices, especially if the money traded is a lot.

As Ibn Khaldun says: "This profit is small in relation to the capital, except that if the capital is a lot, then the profit will be great, because a little in a great amount is a great" [4].

5.5 Tolerance

It is a concept that elevates a person to a high moral and spiritual degree. Tolerance is a type of generosity required of the seller, the buyer, and the debtor. The Messenger of Allah, Peace be upon him, addressed the seller, saying: "Weigh and add more" (Sunan al-Nasa'I: 4592, Book 44, Hadith 144). Also, the Prophet (S.A.W) commanded the debtor in his clearing off the debt where he says: "Best among you is one who is best in clearing off the debt" [8, 9].

All this without a precondition so that charity does not turn into usury. Charity may take the form of a financial reward, a word of thanks, a prayer, or a gift. It is apparent that this behavior has a positive and long-term social impact, as it encourages people to help each other, in addition to its effect in reducing fraud, deception and injustice. On the other hand, Allah has urged the creditor to give a respite to the insolvent debtor party, by saying: (If it is difficult for someone to repay a debt, postpone it until a time of ease, (And if you waive it as an act of charity, it will be better for you, if only you knew)) [Al-Baqarah: 280].

Undoubtedly, giving alms to the insolvent person with a non-existent or doubtful debt is better in reward and more effective. All of this was included in the hadith of the Messenger of Allah, (*Sallallahu Alaihi Wasallam*): "May Allah have mercy on a person who is lenient when he sells, lenient when he buys, and lenient when he asks for payment" [10].

5.6 Prohibition of Monopoly

Monopoly is "the purchase of food and the like and withholding it until it becomes expensive" [11]. Some scholars defined it as "monitoring the markets while waiting for prices to rise". According to Imam al Ghazali [12], "everything that harms people by hoarding it is a monopoly". The Messenger of Allah may Allah's peace and salutation be upon him forbade monopoly by saying: "None withholds goods till the price rises but a sinner" [9]. He *Sallahu alaihi wa sallam* stressed not to harm people's livelihood, saying: "Whoever hoards food from the Muslims, Allah will smite him with bankruptcy or leprosy" [13]. He urged bringing goods to people for sale, which is the opposite of hoarding, and said: "He who brings goods for sale is blessed with good fortune, but he who keeps them till the price rises is accursed" [14].

6 The Process of Deducing a New Accounting Method in Islamic Transaction

There are several areas related to accounting in Islamic financial transactions. The researcher will mention one instance about the accounting process for the Murabaha cost as follows:

It is possible to deduce a new accounting method for evaluating Murabaha goods considering price changes, which is based on the known Islamic legal maxim (*al-ghunmu Bil ghurmi*) which means "liability accompanies gain". This may be evidenced by the Maliki Madhab regarding the ruling (whatever increases or decreases in the seller's information about the original price). The following is what Imam Malik, may Allah be pleased with him, reported in this regard:

He said, "If a man sells a commodity on a Murabaha basis and says, 'It cost me a hundred dinars,' and a buyer comes to a seller later and a seller told him that it is a hundred and twenty dinars, the buyer is given a choice, If he wishes, he gives the seller the cost of the commodity on the day he receives it, and if he wishes, he gives the price for which he bought it at the expense of what he gained". Whatever it is, unless that is less than the price for which he bought the commodity initially, thus he does not have the right to decrease the owner's price of the commodity for which he bought it, since they had mutual agreement on that. Rather, the owner of the commodity came asking for credit. In this case, the buyer has no argument against the seller to put from the price at which it was purchased, the seller pays it" [15].

As per the above clarification, *Murabaha* products can be evaluated based on price changes as follows.

6.1 In the Case of Prices Drop

For example, the bank buys the commodity for 100 riyals, and when it is sold on a Murabaha basis—with a profit margin of 10%, for example—its price in the markets decreases to 90 reyals. In this case, the bank gives the option to sell at historical cost without a profit (i.e., for 100 reyals), or at current cost plus profit (i.e., for 90 + 10% = 99 reyal) (90 × 10%) = 99 ringgit, the bank chooses whatever is better (i.e., higher for him).

6.2 In the Case of Rising Prices

For example, the bank buys the commodity for 100 reyals, and when it is sold on a Murabaha basis—with a profit margin of 10%—its price in the markets rises to 120 reyals. In this case, the customer has the choice between paying the bank the historical cost plus the profit (i.e. the amount of 100 + 10 = 110 riyals), or the current cost plus the profit).

It is clear from the above that valuing Murabaha goods in light of stable or rising prices, it will always be (at historical cost + profit). However, in the cause of dropping prices; the evaluation will be based on (historical cost) or (current cost + profit), whatever is better for the seller [16].

7 The Problem of Measuring the Cost of Murabaha Products in Case of Exchange Rate Variation

A major problem occurs when currency exchange rates fluctuate or change, especially when importing Murabaha goods abroad. In this case, the bank (an offeror of Murabaha) may face one of two things.

7.1 The First Matter

The bank imports the commodity from abroad, and contracts for it in a specific foreign currency, and for some reason the bank pays the price in another currency, then there is a change in the exchange rate between the two currencies on the date of the Murabaha sale to the customer from what the exchange rate between them was on the date of purchase.

7.2 The Second Matter

The bank imports a commodity from abroad in a specific foreign currency, and when selling it to the customer on a Murabaha, the exchange rate variation between the foreign currency and the local currency in which the customer will pay.

An example for the first matter: The bank contracts to import an x-ray equipment from America for an amount of $100,000, and for considerations of cash flow, the bank makes the payment in another currency, let it be the Saudi riyal. Assuming that the exchange rate between them is (four riyals equivalent to one dollar), the bank will pay the amount of 400,000 riyals as a price for the device. If the exchange rate between the two currencies is changed on the date of the Murabaha sale to the customer where it becomes (5 riyals per 1 dollar).

Example for the second matter: The bank imports a commodity from abroad in a specific foreign currency, and upon selling it to the customer on a Murabaha (cost + profit margin); the exchange rate variation between the foreign currency and the local currency in which the customer will pay. In this case, the question also revolves around the basis for calculating the first price. Will it be calculated based on the exchange rate on the day of purchase, or on the day of sale?

Maliki jurists answer this question, where Ibn Rushd Al-Qurtubi says: "… Malik and Al-Layth said - regarding someone who bought a commodity for dinars, and the exchange rate on the day he bought it was a known exchange rate, then he sold it for dirhams and the exchange rate changed to an increase - that he does not have the right to inform the day he sold it for the dinars that he bought, because that is a matter of lying and betrayal. Likewise, if he bought it for dirhams and then sold it for dinars while the exchange rate had changed" [17].

In Al-Muwatta' "Malik said about a man who buys goods for dinars or dirhams, and the exchange rate on the day he bought it is ten dirhams for a dinar, so he brings it to another country and sells it on a Murabaha profit margin, or he sells it where he bought it on profit margin for the exchange rate on that day on which he sold it" [15].

We comprehend and extract from this text that, it is decided by Islamic law to calculate the original price of the commodity, observing that it is based on the exchange rate on the day of sale, not on the day of purchase.

For instance, If he buys in a foreign currency and sells in a local currency, and the exchange rate of the local currency changes to the foreign currency from the day of the first purchase to the day of the Murabaha sale, such as if the price of the first purchase is $100,000, and the exchange rate at that time is 3.3 Saudi Riyal to the dollar, and when selling it on the Murabaha basis it is 3.4 Saudi Riyal to the dollar. Therefore, the initial price on the day of purchase is 330,000 riyals, and on the day of sale it is 340,000 riyals, so which one shall we consider?

Obviously, what is legally prescribed occurred in practice is that the price is calculated according to the exchange rate on the day of sale, not on the first day of purchase, i.e., the commodity should be sold at 340,000 riyals not 330,000 riyals. This matter—as some researchers believe—is nothing more than a change in the current exchange cost of the commodity on the day of sale from its original historical cost on the day of purchase. The cost of the item on the day of purchase is 330,000 Saudi riyals. However, if the client wishes to replace it on the day of sale, he must pay 340,000 Saudi riyals. Therefore, it is believed by some researchers that it is possible and even preferable to address this matter within the framework of the same principles previously mentioned in the previous matter [16] (Fig. 1).

8 Conclusion

The research is concluded by these important findings: Islamic law of transactions has established and regulated an accounting system in the market with foundations far from fraud and its methods, and completely free of unfair and unlawful competition in all fields. Accounting is an organized process which relates to the recording of commercial activities and transactions, legal decisions, actions, its measure of financial consequences and approved records to support financial decision process. Significantly, Islamic accounting is originated from the Holy Qur'an and the Sunnah. So, accounting has witnessed radical enhancement in Islam since the Prophet's (S.A.W) era to our current time. Islamic accounting is qualified by accuracy in measuring tools, truthfulness, mutual trust, moderation in making profit, tolerance, capacity, flexibility and continuity. The Islamic law of transactions kept track its approach to accounting depend on the needs enforced by the Islamic law (*Shari'ah*). The research observed the flexibility of accounting instruments and tools, invented and used by the Islamic law of transactions, and consequently included them in system of Islamic accounting. An important finding signifies that the Western claim that *Luca Pacilio* an Italian Mathematician was the first writer to introduce the double-entry basis in accounting system; was misconception due to the existence of manuscript in accounting dated back to the year (765 H/1494 AC) written by an erudite celebrated Muslim scholar known as Abdullah Al Mazindani. The Islamic transaction accounting system standardized and measured the profit and capital under the concept of "the growth of capital and increase of profit can be achieved by force or action".

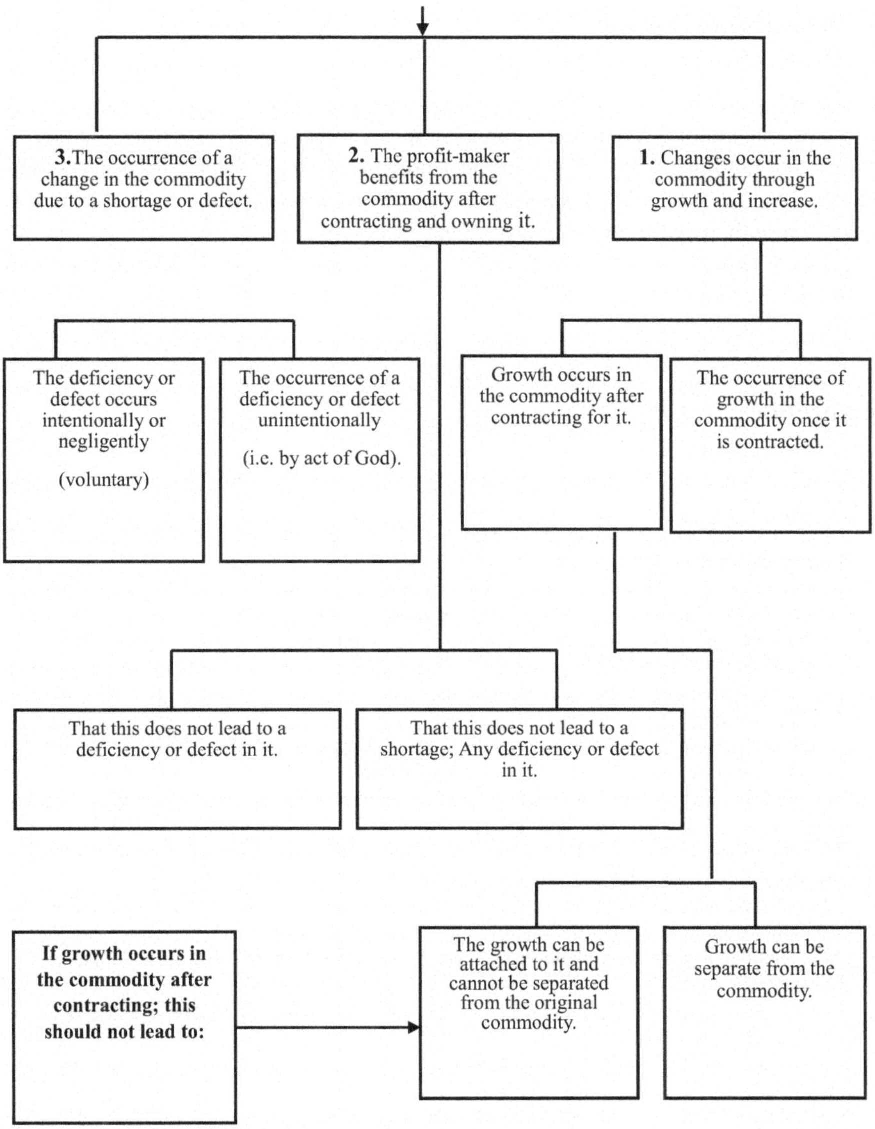

Fig. 1 Changes occur in the Murabaha merchandise

Consequently, Islamic accounting forbids monopoly and withholding commodities until it becomes expensive because it affects economic growth and increase poverty in the society.

9 Recommendation

A researcher recommends Islamic accounting practitioners to study many areas of Islamic accounting commercial and jurisprudential matters, such as concept of Islamic accounting in Zakat, Corporation, Endowment, Waqf, Islamic treasury, Inheritance…etc., to show a clear manifestation of Islamic accounting practice which is more comprehensive than the conventional accounting, because it is far from all illegitimate process such as fraud, cheating, usury and unlawful competition in all commercial fields.

References

1. Kantakji, Samer Muzhir: Fiqhul Muhasaba Al Islamiyyah, 1st edn. Mu'assasat Al-Risalah (2013)
2. Omar Abdullah Zaid: Al Muhasaba al Maliyyah fil Mujtama' al Islami, 1st edn. al Wataniyyah Library, Australia (1995)
3. Islamic Banking and Finance: History and example. (Investopedia_2023). https://www.investopedia.com/terms/i/islamicbanking.asp. Last accessed 3 Dec 2023
4. Ibn Khaldun, Abdul Rahman: Muqaddimat Ibn Khaldun, Dar Al-Qalam, 1st edn. (1981)
5. Ibrahim, Ahmad Abdulkadir: Historical Evaluation of Islamic Accountancy (editor), Kuala Lumpur International Business, Economics and Law Conference 7, Vol. 1, pp. 12–28. KLibel, Kuala Lumpur (2015)
6. Abu Issa, Muhammad bin Issa bin Surat: Sunan al-Tirmidhi, 1st edn. Al Maaref Library for Publishing and Distribution, Riyadh. Book 14, Hadith 7
7. Suleiman bin Al-Ash'ath bin Ishaq Al-Sijistani: Sunan Abi Dawud. Al Waqfiyyat Library, (3383, Book 23, Hadith 58) (1432 AH)
8. Al-Bukhari, Abu Abdullah Muhammad bin Ismail: Sahih al-Bukhari, 1st edn. Dar Kutub al-Ilmiyyah, Beirut Alexandria, Darul Jami'at al-Masriyyah (1424 AH/2003 AC)
9. Al-Naisaburi, Abu al-Hussein, Muslim bin Hajjaj ibn Muslim: Sahih Muslim, 1st edn. Dar al-Hadith, Cairo (1412 H/1991 AC)
10. Abu Abdullah, Ibn Majah Muhammad bin Yazid al-Rab'i al-Qazwini: Sunan Ibn Majah. (2203, Book 12, Hadith 67)
11. Ibn Abidin, Muhammad Amin: Radd al-Muhtār 'ala al-Durr al-Mukhtār. Dar Kutub Al Ilmiyyah, Beirut- Lebanon (2009)
12. Al-Ghazali, Abu Hamid Muhammad Ibn Muhammad: Al-Qistasil Mustaqeem, 1st edn. Cairo (1993)
13. Ahmad bin Hanbal, Abu Abdullah: Al Musnad, 1st edn. Muassasat al Risalat (Book 2, Hadith 53) (1421 AH/2001 AC)
14. Sheikh al-Islam, Abdullah bin Abdul Rahman al-Tamimi al-Darimi: Sunan al-Darimi, 1st edn. (Book 18, Hadith 2464) (1412 AH–2000 AC)
15. Abu Abdullah, Malik bin Anas: Al-Muwatta, 1st edn. (Muhammad Mustafa Al-Azami, ed.) (Zayed bin Sultan Al Nahyan Foundation), vol. 4, p. 353 (1425 AH/2004 AC)
16. Al-Jalaf, Ahmad Muhammad Muhammad: The Accounting Approach to the Murabaha Process in Islamic Banks, 1st edn. International Institute of Islamic Thought (1417 AH/1996 AC)
17. Ibn Rushd (the grandson), Muhammad bin Ahmed bin Muhammad: Bidayat al-Mujtahid Wa Nihayat al-Muqtasid (1433)

Factors Influencing Board Gender Diversity in the Nigerian Capital Market

Bazeet Olayemi Badru, Jimoh Olajide Raji, Rihanat Idowu AbdulKadir, Abdulazeez Adewuyi Abdurraheem, and Hasniza bt. Mohd. Taib

Abstract This study examines the internal and structural factors that influence gender diversity on the boards of publicly listed companies in Nigeria. To investigate this issue, a sample of 80 publicly listed companies on the Nigerian stock exchange from 2005 to 2015 was examined through ordinary least squares (OLS), Two-limit Tobit and Probit estimation techniques. According to the findings, board size, independence, ownership concentration, company size, and industry are significantly and positively associated with women's representation on corporate boards. However, CEO gender and foreign directorship are negative. These findings suggest that these factors may shape the gender makeup of Nigerian companies' boards of directors. Hence, in a country with weak institutional frameworks like Nigeria, companies themselves can promote good corporate governance through gender. This study sheds insight on internal governance and structural factors influencing women's representation and adds to the body of literature already accessible on ethical and equality concerns for women's representation on corporate boards.

Keywords Business ethics · Corporate governance · Gender diversity · Ownership concentration · Social equity · Nigeria

B. O. Badru (✉) · J. O. Raji · H. Mohd. Taib
Department of Finance, School of Economics, Finance and Banking, Universiti Utara Malaysia, Sintok, Kedah, Malaysia
e-mail: badru@uum.edu.my

J. O. Raji
e-mail: raji@uum.edu.my

H. Mohd. Taib
e-mail: hasniza@uum.edu.my

R. I. AbdulKadir · A. A. Abdurraheem
Department of Finance, Faculty of Management Sciences, University of Ilorin, Ilorin, Nigeria
e-mail: abdulkadir.ri@unilorin.edu.ng

A. A. Abdurraheem
e-mail: aabdulaziz70@unilorin.edu.ng

B. Awwad (ed.), *The AI Revolution: Driving Business Innovation and Research*, Studies in Systems, Decision and Control 524, https://doi.org/10.1007/978-3-031-54379-1_7

1 Introduction

A simple way of identifying whether or not a company is serious about corporate governance and business ethics is to look at the appointment of one or more women to the corporate board [1]. A few empirical studies have looked at factors that influence the percentage of women serving on corporate boards. However, the geographical scope of most studies has been developed countries, such as Australia [12], the United States (US) [14, 19, 20], and advanced emerging economies like Brazil, Russia, India, and China (BRIC countries) [32, 33]. Likewise, the theoretical explanations have focused more on the economic theories (agency and resource dependence) and institutional theories, with little emphasis on social equity/ethical, and strategic choice theories.

Moreover, the social equity/ethical theory contends that having women in the boardroom is morally required of a company by its shareholders and stakeholders since their presence denotes a more just and equal society. Similarly, the social exclusion theories state that the nature of the company may influence women's representation on the board. The strategic choice theory contends that it is the top managers (e.g., the chief executive officer) that make decisions to influence corporate outcomes. Therefore, the goal of this research is to examine the internal corporate governance and structural elements that may have an impact on the appointment of women to the boardroom of Nigerian companies.

Investigating this issue matters because there is no explicit or mandatory regulation governing the appointment of Nigerian women to boards of listed companies in the Nigerian capital market.[1] There is also no gender regulation in other areas of public governance, such as federal executive cabinet members, the national parliament, local councils, and candidate lists for elections. Therefore, in the absence of a strong statutory framework to increase the representation of women on boards, their appointment may be contingent on companies' voluntary decisions or the genuine commitment of companies to upholding ethical standards. Similarly, the role of corporations in modern society may be difficult to achieve without the practice of sound corporate governance, especially in a country with weak institutional frameworks like Nigeria [3], where gender discrimination is pervasive in both legal frameworks and customary practices [5].

In addition, if Nigeria is to fulfil its potential of growing its GDP to over \$1.6 trillion by 2030 and expanding its economy by roughly 7.1%, women must be represented in the corporate pipeline. The impact of this could improve Nigeria's position from its current ranking of the 26th largest economies in the world to among the top 20 economies by 2030, surpassing that of the Netherlands, Thailand, or Malaysia,

[1] For instance, a critical examination of the Securities Commission's 2011 New Code of Corporate Governance for Public Companies [10] merely suggests that public companies take gender into consideration when choosing members of their board of directors (BODs). Section 13 (13.2) of the code states that "the criteria for the selection of the board of directors should be written and defined to reflect the existing board's strengths and weaknesses, necessary skills and experience, current age range, and gender composition" [10: 23].

and possibly becoming the largest economy in Africa in the near future [24]. These make it imperative to look at the factors affecting the representation of women on corporate boards (WOCB) in Nigeria, which is the largest market for goods and services in Africa.

The remainder of the paper is organised as follows: The literature review and development of hypotheses are covered in Sect. 2. The section on research methods in Sect. 3 comes next. The paper's discussion and conclusion are presented in Sect. 4.

2 Literature Review and Hypothesis Development

The case for WOCB has been the focus of several theoretical justifications, including social equity and ethics, social exclusion, and economics. Scholars who approach social equity and ethics from this perspective view that having more women on boards is a vital instrument for reducing discrimination and moral injustice at the top echelon level of companies' management [7, 9]. This is because having more women on corporate boards advances social equity and is a desirable goals for companies. Moreover, socioeconomic diversity must be integrated into corporate governance because companies are expected to represent the society within which they operate [36]. Consequently, companies that have high proportion of female representation may be seen as good corporate citizens.

Several empirical studies have demonstrated that the presence of women on corporate boards is notably greater in companies with larger boards [4, 8, 12, 16, 19, 20]. On this note, it stands to reason that women would be more likely to be qualified than men to run for board seats in a company with a larger board. This is because such a company may, by virtue of ethical pressures towards achieving an equitable society and a show of commitment by the board, ensure that the right thing is done and the company's image in the eyes of the public is perceived in a good light. Therefore, one could argue that the size of the board has a significant impact on WOCB. Hence, this study anticipates that:

H1: Board size is positively associated with women's representation on the corporate boards of Nigerian publicly listed companies.

Women may have the opportunity to be appointed to the board since a larger board size may contain more women directors, and a larger board size is likely to be connected with non-executive directors on the board of companies trying to be ethical. In fact, studies have shown that female directors exhibit comparable behaviours to non-executive directors. This could imply that the appointment of women directors is influenced by the board's independence. For instance, Ahmed et al. [4], Saeed et al. [32], and Geiger and Marlin [14] discovered a positive association between the proportion of non-executive directors and WOCB. Hence, this study makes the case that the degree of the board's independence has a greater tendency to influence its decisions regarding fulfilling societal norms related to gender equality. This act would demonstrate that the company is acting morally and not short-changing investors. As such, this study proposes that:

H2: The proportion of non-executive directors on the board is positively associated with women's representation on the corporate boards of Nigerian publicly listed companies.

Making decisions that will influence corporate outcomes and shape the company is one of the fundamental strategic responsibilities of the company's top managers, such as the CEO. A decision related to the gender makeup of the board might be an example of such a decision [30]. Because of this, if a company has gender-based similarity, gender-induced entry barriers may be lessened [6]. That is to say, given that the CEO is saddled with the responsibility of overseeing the day-to-day operation of the company and exerts considerable influence in the composition of the board, the selection of a director for the board may be influenced by demographic similarities to the CEO (Brown et al. 2019). Women may therefore stand a better chance of getting appointed to the board if the board is chaired by a female CEO [26]. On this note, it is possible to argue that companies where CEOs have substantial influence over the selection of board members may exhibit a gender imbalance in the composition of their board. Thus, the following hypothesis implies that:

H3: The gender of the chief executive officer is positively associated with women's representation on the corporate boards of publicly listed Nigerian companies.

The ownership structure of a company is another important factor that could influence the composition of the board. A concentrated ownership structure, for example, concentrates power and incentives for management oversight among owners who are able to discipline managers through concentrated voting rights [21]. While concentrated ownership has its benefits, some academics contend that high levels of concentrated ownership can result in opportunistic behaviour from controlling shareholders and a principal-principal agency problem. This is particularly so in institutional environments where minority shareholder protection and enforcement mechanisms are lacking [25]. Given that gender diverse boards would act as a substitute mechanism for weak corporate governance, such companies need to put in place an appropriate governance structure, such as a gender diverse board [18, 29]. Therefore, it is anticipated that ownership concentration will positively affect WOCB.

H4: Ownership concentration is positively associated with women representation on the corporate boards of publicly listed Nigerian companies.

Control Variables

Other factors that may affect the representation of women on board are company size and industry. For instance, studies have documented that women's participation on corporate boards is considerably greater in larger companies (e.g., [1, 8, 13, 16, 20]. All of these indicate that women have a greater chance of being considered as directors in large companies. Hence, company size tends to be positively associated with women's representation on the corporate board.

Whether a company operates in a masculine or feminine industry may affect how likely it is for women to advance to managerial positions [31]. For instance, industries like construction and manufacturing are generally considered masculine (male-dominated), whereas service industries like healthcare are categorized as feminine [15]. Additionally, companies operating in industries where women are salient stakeholders (e.g., employees and consumers) would ensure that a representative of these constituencies is on the board [23]. Therefore, the proportion of women on the corporate board can be influenced by the industry the company is classified in.

Globally connected companies tend to adhere to international best practices on gender equality and diversity management. Studies have demonstrated, for example, that foreign directors make efforts for better corporate governance through measures including increasing the proportion of non-executive directors on the boards [22] and national board diversity [27]. Thus, this study contends that the proportion of foreign directors and ownership may have an influence on the representation of WOCB in a country like Nigeria because there are no regulations requiring companies to set gender quotas for women on their boards of directors.

Similarly, women in the labour force are key formal institutions for gender equality. There is some evidence that the engagement of women in the labour force can explain the participation of executive directors [2]. This means that the proportion of women in the labour force participating in the economy may lead to an increase in the consideration of WOCB [35]. Therefore, this study suggests that there is a positive association between the proportion of women in the labour force and WOCB. According to a number of studies (e.g., [11, 18, 28]), the age of the company is another structurally specific variable that can influence WOCB.

3 Research Method

3.1 Sample and Data

A sample of publicly listed companies in the Nigerian capital market is employed to observe the structural factors influencing the appointment of women on corporate boards. A total of 169 companies are listed as of July 6, 2019 (http://www.nse.com.ng/issuers/listed-securities/listed-companies). Of the 169, 59 are categorized as companies providing financial services,[2] and 114 are non-financial companies. Out of the 114 non-financial companies, 80 companies (which is 71% of the non-financial companies) have available data for a longer period starting from 2005 to 2016, which is the period of data collection for this study. The study period begins in 2005 because it is the year that saw the most listed companies on the Nigeria Stock Exchange (NSE) (215), while the lowest number (166) was recorded in 2016.

[2] In line with prior studies (e.g., [32, 33]), this study excludes companies operating in the financial service sectors.

In addition, two main data sets are employed: corporate governance information (e.g., gender composition of the corporate board) and financial information (total assets and profit before tax). These data are taken from listed companies annual reports and the NSE's FACTBOOK. Particularly, data on the gender distribution of corporate boards and certain other attributes of the board of directors' as well as financial data are handpicked from the annual reports. The data on female labour force participation was collected from the World Bank database.

3.2 Estimation Technique

This study tests factors that increase women's chances of being appointed to the corporate board using ordinary least squares (OLS) and logistic regression, in particular the Tobit estimation techniques. The Tobit estimator is considered because the data on the proportion of women's representation on the corporate board is skewed and there are companies with no women on their corporate boards, as indicated by the minimum and maximum values of 0% and 58%, respectively (see Table 1). By implication, employing OLS in this situation may be inappropriate since it assumes that the projected dependent variable may reside outside the permitted range of 0–100% [4, 34]. On this note, this study applies the two-limit (upper and lower limits) Tobit estimation technique in addition to the OLS estimation technique. The Tobit estimation technique integrates the Probit and truncated regressions in a single estimation and is typically applied to censored data (variables that are not defined below zero and above 100%) [33]. Therefore, the OLS and Tobit estimations are used for the proportion of WOCB as a dependent variable, and the Probit estimation is utilized when women's representation on the board is a dummy variable. Model 1 presents the regression model for this study, while Model 2 shows the two-limit Tobit model.

Model 1:

$$\begin{aligned} WOCB_{it} = \beta_0 &+ \beta_1 BODS_{it} + \beta_2 BODIND_{it} + \beta_3 CEOGND_{it} \\ &+ \beta_4 PFDIR_{it} + \beta_5 FDIROWN_{it} + \beta_6 OWNCON_{it} \\ &+ \beta_7 COAG_{it} + \beta_8 LNTA_{it} + \beta_{10} FLF_{it} \\ &+ \beta_{11} INDUSTRY_{it} + \beta_{12} YEAR_{it} + \varepsilon_{it} \end{aligned}$$

where WOCB stands for the proportion of women on corporate boards; BODS for board size; BODIND for non-executive directors proportion on corporate boards; CEOGND is a dummy value of 1 if the CEO is a woman, and 0 otherwise; the percentage of foreign directors on corporate boards is known as PFDIR; the percentage of shares held by foreign directors is known as FDIROWN; and

Table 1 Descriptive analysis

Variables	Mean	Med	Max	Min	SD	Skewness	Kurtosis
Women directors % (WOCB)	0.07	0.00	0.58	0.00	0.09	1.84	7.80
Women dummy (WDD)	0.48	0.00	1.00	0.00	0.49	0.09	1.01
Board size (BODS)	9.08	9.00	18.00	4.00	2.58	0.37	2.50
Board independence % (BODIND)	0.35	0.37	0.89	0.00	0.18	−0.11	3.62
CEO gender (CEOGND)	0.06	0.00	1.00	0.00	0.24	3.68	14.56
Foreign directors % (PFDIR)	0.18	0.00	0.78	0.00	0.23	0.79	2.19
Foreign ownership % (FDIROWN)	0.01	0.00	0.56	0.00	0.06	6.41	43.27
Ownership concentration % (OWNCON)	0.54	0.59	0.95	0.00	0.24	−0.54	2.44
Company age (COAG)	39.54	42.00	92.00	1.00	18.74	0.05	2.72
Total assets (Million Naira "000")	29,360.44	7050	950,000	67.31	66,368.33	6.09	59.84
Ln total assets (LNTA)	15.76	15.77	20.67	11.12	1.82	−0.08	2.60
Female labour force (FLF)	45.70	45.90	46.05	45.04	0.34	−0.83	2.13

Notes WOCB represents the proportion of women directors to total number of directors on corporate boards; WDD is a binary value of 1 if a woman director is present on the board of directors, otherwise 0; BODS is the total number of directors on corporate boards; BODIND is the proportion of non-executive directors to the total number of directors on corporate boards; CEOGND is a dichotomous value of 1 if the CEO is a female, otherwise 0; PFDIR is the percentage of foreign directors on corporate boards; FDIROWN is the percentage of shares owned by foreign directors; OWNCON is the percentage of shareholdings of directors or institutions with 5% or more; COAG represents the age of the company from year of incorporation; LNTA is the natural logarithm of total assets; FLF is the measure of female labour force as percentage of the total labour force

OWNCON is the ownership concentration; the company age as of the year of incorporation is COAG; while the natural logarithm of total assets is LNTA; the proportion of female in the labour force is known as FLF; INDUSTRY represents industry dummies and YEAR represents year dummies, respectively; ε_{it} is the random error.

Model 2:

$$\begin{aligned} WOCB_{it}^* = \beta_0 &+ \beta_1 BODS_{it} + \beta_2 BODIND_{it} + \beta_3 CEOGND_{it} \\ &+ \beta_4 PFDIR_{it} + \beta_5 FDIROWN_{it} + \beta_6 OWNCON_{it} \\ &+ \beta_7 COAG_{it} + \beta_8 LNTA_{it} + \beta_{10} FLF_{it} \\ &+ \beta_{11} INDUSTRY_{it} + \beta_{12} YEAR_{it} + \varepsilon_{it} \end{aligned}$$

$$\begin{aligned} &WOCB_{it} = L_{1i} && \text{if } WOCB_{it}^* \leq L_{1i}. \\ &WOCB_{it} = WDP_{it}^* && \text{if } L_{1i} < WOCB_{it}^* < L_{2i}. \\ &WOCB_{it} = L_{2i} && \text{if } WOCB_{it}^* \leq L_{2i}. \end{aligned}$$

$WOCB^*$ is when the proportion of women on corporate boards is a latent variable. L_1 and L_2 are the lower and upper limits, respectively.

3.3 Empirical Results

Descriptive Analysis

Table 1 presents figures indicating that the percentage of women on corporate boards ranges from a mean of 7% to a maximum of 58%. Comparably, only 48% of the examined companies have a minimum of one woman on their corporate board, meaning that 52% of the sampled companies do not have a single woman at all. This simply indicates that a greater proportion of Nigerian companies have male-dominated BODs, implying that the presence of women on these boards is merely symbolic.

According to the correlation analysis shown in Table 2, there are no extreme correlations (0.80) among variables that could suggest a multicollinearity threat. Similarly, Table 3's column 6 variance inflation factor (VIF) demonstrates that all factors used in the regression analysis fall below the critical threshold of 10. As a result, neither the correlation analysis nor the VIF test demonstrate that multicollinearity is an issue for the regression analysis.

Regression Results

The findings of the regression model used to look into factors impacting women directors on corporate boards are shown in Tables 3 and 4. To be more precise, Table 3 regression results are without controlling for industry and year effects, but Table 4 adjusted for them in the regression model. Table 3 shows the OLS results in

Table 2 Correlation analysis

Variables	1	2	3	4	5	6	7	8	9	10	11	12
WOCB (1)	1.00											
BODS (2)	0.11***	1.00										
BODIND (3)	0.10**	−0.05	1.00									
CEOGND (4)	−0.14***	−0.10***	0.05	1.00								
PFDIR (5)	−0.08**	0.19***	0.12***	−0.17***	1.00							
FDIROWN (6)	−0.04	−0.13***	0.04	−0.04	0.16***	1.00						
OWNCON (7)	0.01	0.21***	0.08**	−0.04	0.59***	0.15***	1.00					
COAG (8)	0.07**	0.10***	0.14***	−0.04	0.35***	0.05	0.29***	1.00				
LNTA (9)	0.14***	0.54***	−0.05	−0.24***	0.29***	−0.01	0.37***	0.20***	1.00			
FLF (10)	−0.15***	−0.03	−0.13***	0.01	−0.03	-0.04	0.00	−0.15***	−0.18***	1.00		

Notes *** and ** indicate statistical significance at 1% and 5%

Table 3 Regression analysis

	OLS		Logistic		
Variables	OLS	Robust	Tobit	Probit	VIF
BODS	0.00*	0.00*	0.01***	0.09***	1.54
BODIND	0.05*	0.05	0.10*	0.54	1.36
CEOGND	−0.07***	−0.07***	−0.22***	−1.36***	1.11
PFDIR	−0.10***	−0.10***	−0.17***	−0.99***	1.78
FDIROWN	−0.00	−0.00	−0.00	−0.01	1.07
OWNCON	0.03	0.03*	0.08**	0.73**	1.82
COAG	0.00	0.00	0.00	0.00	1.14
LNTA	0.00	0.00	0.01**	0.11***	1.85
FLF	−0.04***	−0.04**	−0.07***	−0.41**	1.34
Constant	2.25***	2.25**	2.93**	15.77*	
Mean VIF					1.45
R^2(Adjust R^2)%	8 (7)	8			
Pseudo R^2 %			37	12	
Ramsey test	0.24				

Notes BODS is the total number of directors on corporate boards; BODIND is the proportion of non-executive directors to the total number of directors on corporate boards; CEOGND is a dichotomous value of 1 if the CEO is a female, otherwise 0; PFDIR is the percentage of foreign directors on corporate boards; FDIROWN is the percentage of shares owned by foreign directors; OWNCON is the percentage of shareholdings of directors or institutions with 5% or more; COAG represents the age of the company from year of incorporation; LNTA is the natural logarithm of total assets; FLF is the measure of female labour force as percentage of the total labour force. ***, **, and * indicate statistical significance at 1%, 5%, and 10% levels

columns (2) and (3), while the logistic regression results are displayed in columns (4) and (5).

Based on the OLS results in Table 4, BODS and BODIND are found to be positively linked with WOCB, albeit at a weak significance level. Likewise, in the logistic regression, BODIND remains weakly associated with WOCB at the 10% significance level. However, at a significance level of 1%, BODS is found to be positively and significantly linked with WOCB, indicating that companies with larger boards may be able to afford to have more gender diversity on their BODs. The result is in line with H1 and other studies, which suggest that companies with larger boards have a greater level of female board participation (e.g., [14, 28, 33]).

Consistent with H4, the OLS, Tobit, and Probit regression results likewise show a positive and significant link between OWNCON and WOCB at the 10 and 5% levels of significance. This is in line with Morikawa's [28] findings that indicate owner-managed companies typically include female directors. Therefore, it can be concluded that the gender composition of the BoDs is influenced by the level of shareholders' concentration.

Table 4 Tobit regression analysis with industry and year

Variables	With industry	With year	With industry and year
BODS	0.01	0.01***	0.01*
BODIND	0.08	0.08	0.06
CEOGND	−0.25***	−0.22***	−0.25***
PFDIR	−0.19***	−0.17***	−0.20***
FDIROWN	−0.00	−0.00	−0.00
OWNCON	0.13***	0.08**	0.13***
COAG	−0.00	0.00	−0.00
LNTA	0.01*	0.01*	0.01
FLF	−0.07***	−0.08**	−0.16**
Agriculture	0.03		0.33
Conglomerates	−0.02		−0.02
Construction/real estate	−0.02		−0.02
Consumer goods	0.02		0.02
Healthcare	0.13***		0.13***
ICT	−0.09**		−0.19**
Industrial goods	−0.07***		−0.10***
Natural resources	0.26***		−0.26***
Oil and gas	−0.02		−0.06
Services	0.03		−0.02
2005–2015		Yes (2013)	Yes (2011, 2012, 2013)
Constant	3.17***	−0.28***	−0.16**
Pseudo R^2 %	77	38	79

Notes ***, **, and * indicate statistical significance at 1%, 5%, and 10% levels

Conversely, at the 1% significance level, CEOGND is found to be adversely and significantly linked with WOCB. In addition to the OLS results, the Tobit and Probit regressions reveal a significant link between CEOGND and WOCB. In contrast to H3, these results suggest that CEOGND has a detrimental impact on women's prospects of being appointed to the corporate boards of listed companies in Nigeria. This may imply that CEOs might not have a good say in who gets appointed to the boardroom.

In the Tobit and Probit regressions, respectively, it is observed that LNTA is positively and significantly linked with women on corporate boards at the 5 and 1% levels of significance. As evidenced by the higher share of women on corporate boards, larger companies are more likely to embrace diversity. This evidence corroborates prior studies' claim that larger companies have higher likelihood of appointing women to corporate boards [13, 19, 33].

The gender makeup of the board is also influenced by the industry in which the company operates. Companies in the healthcare industry, in particular, are more likely to have female directors [20]. This is supported by the empirical evidence that a significant and positive link is found between women on corporate boards and healthcare industry dummies. The positive association is not surprising because the goods and services emanating from this industry have closeness with the final consumers.

This is in line with the assertions made by Gipson et al.'s [15] that the healthcare industry is a female-friendly industry, and Gardiner and Tiggemann [17] claim that women are more likely to be found in senior positions in the service industries that prioritise customer relations than in other industries. The current analysis also shows a negative effect for ICT, industrial goods, and natural resources companies. This, however, is not surprising given that women are significantly underrepresented in the natural resources, mining, and manufacturing sectors across 42 countries [2]. Similar findings are reported by De Cabo et al. [11], who show that service-oriented industries outnumbered technical ones in industries like technology and telecommunications, commodities, construction, and oil and energy.

Since the appointment of women is a critical corporate governance issue, foreign directors tend to be more sensitive to board diversity at both governance and top management levels due to public pressure. By so doing, a positive link is expected between PFDIR and the proportion of WOCB. In contrast to the aforementioned argument, empirical evidence in this study reveals a negative effect of PFDIR on WOCB. Another significant variable is women's labour force participation (FLF), which is negatively associated with WOCB. This is an indication that stronger participation of women in the general labour market does not necessarily amount to an increase in WOCB.

4 Discussion and Conclusion

This study investigated the structural factors linked to the advancement of women's appointments to corporate boards. Consistent with social equity and ethical concerns, this study showed that board size influences WOCB. The inference is that companies with larger boards would have a higher proportion of female directors since a greater gender mix among directors should foster diversity of experiences, creativity, and added value. This notion concurs with earlier studies [8, 14, 19] that reported that the number of directors in the boardroom would influence the degree of representation of women in such corporate boards. Hence, it can be argued that large boards accommodate women more easily because they have more seats for potential female candidates or because such boards tend to uphold ethical standards and signal to the public that they are practicing good corporate governance and business ethics.

Ownership structure, specifically ownership concentration, is found to be an important factor influencing female directors. This suggests that a company's ownership structure matters when it comes to women's representation on the corporate board. In essence, it can be concluded that ownership concentration affects the makeup of the board because substantial shareholders of a company are likely to have the power to promote and appoint individuals in the boardroom, as well as ensure that the company considers the societal value of gender mix. In reality, companies think it is morally right to have WOCB in order to keep the support of important external stakeholders like investors and customers. Such companies would also ensure that individuals that are seated in the boardroom have specific skills and expertise that help to advise and counsel the management, and female directors have been found to exhibit such qualities [19]. Gender diversity can also serve as a stand-in method to make up for weak corporate governance that may exist in a company with a concentrated ownership structure [29].

The practical implications of the empirical findings are that companies and regulators should take into consideration the various corporate governance and structural characteristics that can be helpful in increasing the number of women on corporate boards and that are accountable for the exclusion of women from the pool of potential candidates for board seats. This can demonstrate how a company takes cognizance of ethical issues and contributes to the equitable representation of women in corporate leadership. The government can also introduce different policy options, increase public awareness, and awards to companies that are ethically conscious of diversity. Such policies and incentives might motivate other companies to view the diversity of their board appointments as a critical issue.

Before generalizing the findings, it is important to keep in mind that this study has certain limitations. First off, the study's sample size and time frame may not accurately reveal the makeup of company boards, given that companies have recently been more exposed to concerns from professional groups and the public over gender diversity. Subsequent studies may also examine causes of the lack of diversity on corporate boards as well as attributes of women (e.g., education, experience, religion, ethnicity, and tenure) that may impact their appointment to board positions.

References

1. Adams, R.B., Ferreira, D.: Women in the boardroom and their impact on governance and performance. J. Financ. Econ.Financ. Econ. **94**(2), 291–309 (2009)
2. Adams, R.B., Kirchmaier, T.: Women on boards in finance and STEM industries. Am. Econ. Rev. **106**(5), 277–281 (2016)
3. Adegbite, E.: Good corporate governance in Nigeria: antecedents, propositions and peculiarities. Int. Bus. Rev. **24**(2), 319–330 (2015)
4. Ahmed, A., Higgs, H., Ng, C., Delaney, D.A.: Determinants of women representation on corporate boards: evidence from Australia. Account. Res. J. **31**(3), 326–342 (2018)
5. Ajala, T.: Gender discrimination in land ownership and the alleviation of women's poverty in Nigeria: a call for new equities. Int. J. Discrim. Law **17**(1), 51–66 (2017)
6. Berger, A.N., Kick, T., Koetter, M., Schaeck, K.: Does it pay to have friends? Social ties and executive appointments in banking. J. Bank. Finance **37**(6), 2087–2105 (2013)
7. Brammer, S., Millington, A., Pavelin, S.: Gender and ethnic diversity among UK corporate boards. Corp. Gov. Int. Rev. **15**(2), 393–403 (2007)
8. Carrasco, A., Francoeur, C., Labelle, R., Laffarga, J., Ruiz-Barbadillo, E.: Appointing women to boards: is there a cultural bias? J. Bus. Ethics **129**(2), 429–444 (2015)
9. Chijoke-Mgbame, A.M., Boateng, A., Mgbame, C.O.: Board gender diversity, audit committee and financial performance: evidence from Nigeria. Account. Forum **44**(3), 262–286 (2020)

10. Corporate Governance Code (2011). https://www.cscsnigeriaplc.com/documents/1118035/CODE%20OF%20CORPORATE%20GOVERNANCE.pdf
11. De Cabo, R.M., Gimeno, R., Escot, L.: Disentangling discrimination on Spanish boards of directors. Corp. Gov. Int. Rev. **19**(1), 77–95 (2011)
12. Dimovski, W., Lombardi, L., Cooper, B.: Women directors on boards of Australian real estate investment trusts. J. Prop. Invest. Fin. **31**(2), 196–207 (2013)
13. Esteban-Salvador, M.L.: Variables influencing the gender composition of boards: the Spanish evidence. J. Women Polit. Policy **32**(4), 305–332 (2011)
14. Geiger, S.W., Marlin, D.: The relationship between organizational/board characteristics and the extent of female representation on corporate boards. J. Manag. IssuesManag. Issues **24**(2), 157–172 (2012)
15. Gipson, A.N., Pfaff, D.L., Mendelsohn, D.B., Catenacci, L.T., Burke, W.W.: Women and leadership: selection, development, leadership style, and performance. J. Appl. Behav. Sci.Behav. Sci. **53**(1), 32–65 (2017)
16. Gregorič, A., Oxelheim, L., Randøy, T., Thomsen, S.: Resistance to change in the corporate elite: female directors' appointments onto Nordic boards. J. Bus. Ethics **141**(2), 267–287 (2017)
17. Gardiner, T., Tiggemann, M.: Gender differences in leadership style, job stress and mental health in male and female-dominated industries. J. Occup. Organ. Psychol.Occup. Organ. Psychol. **72**(3), 301–311 (1999)
18. Gul, F.A., Srinidhi, B., Ng, A.C.: Does board gender diversity improve the informativeness of stock prices? J. Account. Econ. **51**(3), 314–338 (2011)
19. Hillman, A.J., Shropshire, C., Cannella, A.A., Jr.: Organizational predictors of women on corporate boards. Acad. Manag. J.Manag. J. **50**(4), 941–952 (2007)
20. Hyland, M.M., Marcellino, P.A.: Examining gender on corporate boards: a regional study. Corp. Gov. Int. J. Bus. Soc. **2**(4), 24–31 (2002)
21. Jensen, M.C., Meckling, W.H.: Theory of the firm: managerial behavior, agency costs and ownership structure. J. Financ. Econ.Financ. Econ. **3**(4), 305–360 (1976)
22. Kim, I.J., Eppler-Kim, J., Kim, W.S., Byun, S.J.: Foreign investors and corporate governance in Korea. Pac. Basin Financ. J.Financ. J. **18**(4), 390–402 (2010)
23. Kirsch, A.: The gender composition of corporate boards: a review and research agenda. Leadersh. Q.. Q. **29**(2), 346–364 (2018)
24. Leke, A., Fiorini, R., Dobbs, R., Thompson, F., Suleiman, A., Wright, D.: Nigeria's Renewal: Delivering Inclusive Growth in Africa's Largest Economy. McKinsey Global Institute, McKinsey & Company (2014). https://www.mckinsey.com/~/media/McKinsey/Featured%20Insights/Middle%20East%20and%20Africa/Nigerias%20renewal%20Delivering%20inclusive%20growth/MGI_Nigerias_renewal_Full_report.ashx
25. Lins, K.V.: Equity ownership and firm value in emerging markets. J. Fin. Quant. Anal. **38**(1), 159–184 (2003)
26. Liu, C.: Are women greener? Corporate gender diversity and environmental violations. J. Corp. Finan.Finan. **52**, 118–142 (2018)
27. Maturo, F., Migliori, S., Paolone, F.: Do institutional or foreign shareholders influence national board diversity? Assessing board diversity through functional data analysis. In Mathematical-Statistical Models and Qualitative Theories for Economic and Social Sciences, pp. 199–217. Springer, Cham (2017)
28. Morikawa, M.: What types of companies have female directors? Evidence from Japan. Jpn. World Econ.. World Econ. **37**, 1–7 (2016)
29. Nekhili, M., Gatfaoui, H.: Are demographic attributes and firm characteristics drivers of gender diversity? Investigating women's positions on French boards of directors. J. Bus. Ethics **118**(2), 227–249 (2013)
30. Ng, E.S., Sears, G.J.: The glass ceiling in context: the influence of CEO gender, recruitment practices and firm internationalisation on the representation of women in management. Hum. Resour. Manag. J.Resour. Manag. J. **27**(1), 133–151 (2017)
31. Ryan, M.K., Haslam, S.A.: The glass cliff: exploring the dynamics surrounding the appointment of women to precarious leadership positions. Acad. Manag. Rev.Manag. Rev. **32**(2), 549–572 (2007)

32. Saeed, A., Belghitar, Y., Yousaf, A.: Firm-level determinants of gender diversity in the boardrooms: evidence from some emerging markets. Int. Bus. Rev. **25**(5), 1076–1088 (2016)
33. Saeed, A., Sameer, M., Raziq, M.M., Salman, A., Hammoudeh, S.: Board gender diversity and organizational determinants: empirical evidence from a major developing country. Emerg. Mark. Financ. Trade. Mark. Financ. Trade **55**(8), 1803–1820 (2019)
34. Tobin, J.: Estimation of relationships for limited dependent variables. Economet. J. Economet. Soc., 24–36 (1958)
35. Tyrowicz, J., Terjesen, S., Mazurek, J.: All on board? New evidence on board gender diversity from a large panel of European firms. Eur. Manag. J.Manag. J. **38**(4), 634–645 (2020)
36. Van der Walt, N., Ingley, C.: Board dynamics and the influence of professional background, gender and ethnic diversity of directors. Corp. Gov. Int. Rev. **11**(3), 218–234 (2003)

The Role of Accounting Information Systems in Enhancing Human Resources Management Cycle

Mohammed Alzoraiki, Ahmad Yahia Mustafa Alastal, Marwan Milhem, Ali Ateeq, and Tamer M. Alkadash

Abstract This study explores the role of Accounting Information Systems (AIS) in enhancing the Human Resources Management (HRM) cycle. Integrating AIS has brought about a pivotal shift in how organizations approach their HRM cycle, leveraging technology to optimize efficiency, accuracy, and strategic decision-making. This study provides an overview of AIS, the HRM cycle, and the relationship between AIS and HRM. It also highlights previous studies on the role of AIS in enhancing HRM and emphasizes the system's multifaceted roles in supporting business operations. The study concludes that AIS significantly transforms HRM practices, improves transparency and accountability, and promotes strategic decision-making processes.

Keywords Accounting information systems · AIS · Human resources management · HRM cycle · Technology · Efficiency · Accuracy · Strategic decision-making · Business operations · Transparency · Accountability

---

M. Alzoraiki · M. Milhem · A. Ateeq
Administrative Science Department, College of Administrative and Financial Science, Gulf University, Sanad 26489, Kingdom of Bahrain
e-mail: dr.mohammed.saied@gulfuniversity.edu.bh

M. Milhem
e-mail: dr.marwan.milhem@gulfuniversity.edu.bh

A. Ateeq
e-mail: dr.ali.ateeq@gulfuniversity.edu.bh

A. Y. M. Alastal (✉)
Accounting and Financial Science Department, College of Administrative and Financial Science, Gulf University, Sanad 26489, Kingdom of Bahrain
e-mail: dr.ahmed.alastal@gulfuniversity.edu.bh

T. M. Alkadash
Media & Marketing Technology Department, College of Communication & Media Technologies, Gulf University, Sanad 26489, Kingdom of Bahrain
e-mail: dr.tamer.alkadash@gulfuniversity.edu.bh

B. Awwad (ed.), *The AI Revolution: Driving Business Innovation and Research*, Studies in Systems, Decision and Control 524, https://doi.org/10.1007/978-3-031-54379-1_8

1 Introduction

Integrating Accounting Information Systems (AIS) into Human Resources Management (HRM) represents a pivotal shift in how organizations approach their HRM cycle, leveraging technology to optimize efficiency, accuracy, and strategic decision-making. In the digital age, the dynamic capabilities of AIS provide invaluable insights and real-time data that enhance various HR functions, from recruitment to retirement [1, 2]. The intersection of AIS with HRM has led to strategic decision-making processes, where data-driven insights inform talent management, payroll, performance evaluations, and compliance with labour regulations [3]. The significant role of AIS in transforming HRM practices is well documented in the literature. For instance, the work of Momoh [4] highlights that AIS facilitates real-time monitoring and reporting, improving the HRM cycle's responsiveness and adaptability to organizational needs. Similarly, White and Abdullah and Abbas [5] emphasize the strategic forecasting capabilities provided by AIS, enabling human resources to anticipate and plan for future workforce requirements effectively.

The automation of routine HR tasks through AIS has led to cost reduction. Still, it has also reallocated human resource efforts to more strategic roles, emphasizing the value of human capital in achieving competitive advantage [6]. AIS's analytics and reporting functions are critical in identifying patterns within employee data, which can predict turnover, highlight training needs, and inform recruitment strategies [7]. Moreover, the AIS's role in enhancing transparency and accountability in the HRM cycle cannot be understated. The system's ability to provide auditable trails and maintain accurate records ensures compliance with internal and external policies and regulations, further securing the integrity of HRM operations [8].

2 Literature Review

2.1 Overview of Accounting Information Systems (AIS)

Accounting Information Systems (AIS) are instrumental in efficiently transferring data, which is vital for users who might not be adept at interpreting accounting and financial information. AIS ensures accurate and timely information distribution, facilitating informed decision-making processes that can subsequently increase shareholder value [1]. Moreover, AIS is defined as a system that collects and processes transaction data to generate information for decision-making [2]. The AIS comprises major subsystems focused on transaction processing, general ledger, and management reporting, covering financial and non-financial transactions to meet the diverse information needs of its users [3].

2.2 Human Resources Management Cycle

Human resources management is a crucial function within an organization, as it involves the strategic planning and coordination of an organization's most valuable asset, its employees [4]. The human resources management cycle is a continuous process encompassing various stages, from recruitment and selection to performance management and employee development. This cycle is essential for ensuring that the organization has the right people in the correct positions and that they are motivated, engaged, and productive [5]. This study will explore the critical stages of the human resources management cycle and their importance in driving organizational success, engagement and satisfaction. Therefore, this cycle is crucial for ensuring the organization has a skilled and motivated workforce to achieve its goals. HRM also plays a vital role in creating a positive work environment and fostering employee engagement and satisfaction [6].

2.3 The Relationship Between AIS and Human Resources Management

Accounting Information Systems are pivotal in enhancing the HRM cycle by streamlining its processes. Computerized AIS, for example, leads to substantial time and cost savings, significantly affecting business operations [7]. The systems assist in recording, processing, and preparing financial statements. These are snapshots of an organization's financial position, crucial outputs for any HR department tasked with workforce budgeting and compensation management [8]. The integration of AIS in HRM is seen in its contribution to planning and controlling activities, which helps produce quality information for internal and external users [9]. The system is not only about accounting but is also a set of resources responsible for generating financial and operational information vital for effective decision-making [10].

2.4 Role of AIS in Enhancing Human Resources Management

Previous studies have highlighted the multifaceted roles of AIS in supporting business operations. One study posits that AIS is predominantly used in selected mobile telecommunication companies to record and report business operations rather than to aid management decision-making [11]. This implies a potential for better leveraging AIS in HRM processes. In the broader corporate context, users of AIS are generally satisfied with the systems, suggesting a positive impact of AIS on HRM functions [12]

Sudarmanto et al. [13] emphasized that AIS significantly influences business strategies and corporate sustainability, reflecting AIS's broader strategic implications on management practices, including HRM. AIS's role in diminishing fraud and errors is particularly evident in Vietnamese SMEs, showcasing the necessity of dependable AIS for maintaining stringent HRM protocols [14]. Nonetheless, the reliance of these SMEs on rudimentary accounting applications indicates a gap in their AIS capabilities, pointing to the need for more advanced systems that encompass not only bookkeeping but also planning and control functions.

The various studies support the conclusion that AIS has profound implications for optimizing organizational HRM cycles [15]. From enhancing the efficiency of reporting functions to aiding strategic HR decisions and ensuring compliance, AIS provides a backbone for HRM to rely upon for informed management and operational control, with users expressing satisfaction with their current systems' performance. Notably, there is a gap in AIS application within Vietnamese SMEs for more advanced planning and decision-making within HRM indicating an area for potential improvement and investment in more sophisticated AIS solutions [16].

2.4.1 Use of AIS in Job Posting and Applicant Tracking

In the context of recruitment, Accounting Information Systems (AIS) play a pivotal role, particularly in the realms of job posting and applicant tracking [17]. The provision of timely and accurate data by an effective AIS is essential for aligning Human Resource Management (HRM) activities with organizational objectives and business strategies [18]. Specifically, AIS systems are instrumental in the recruitment and hiring processes, facilitating the tracking of job postings, managing applicant data, and enhancing communication between [19].

2.4.2 Automated Resume Screening and Applicant Evaluation

Automated resume screening is a critical function of AIS in recruitment. It allows HR professionals to efficiently sift through many applications to identify candidates whose qualifications match job requirements. Evidence from literature suggests a significant positive relationship between AIS deployment and hiring function improvements. By automating the initial evaluation process [12] AIS can quickly parse through resumes and applications, providing HR managers with the most relevant candidates to consider. This reduces the time and potential bias associated with manual resume screening, thereby enhancing the overall effectiveness of the hiring process.

2.4.3 Online Interviewing and Assessment Tools

The field of Human Resources (HR) management has been significantly transformed by the progress made in Artificial Intelligence Systems (AIS), with a notable impact on the process of applicant assessment. Artificial intelligence systems (AIS) play a crucial role in the efficient organization and analysis of data obtained from online interviews and assessments [20]. This capability empowers human resource (HR) managers to expedite decision-making processes while also ensuring that their choices are well-informed [21]. The use of technology in the recruiting process optimizes efficiency and improves the precision of applicant selection.

2.5 The Role of AIS in Training and Development

The deployment of Accounting Information Systems (AIS) within the domain of training and development represents a significant evolution in how organizations approach learning and competency building for their employees [22]. AIS systems, traditionally purposed for managing financial information, have been adapted to support the full spectrum of training and performance evaluation within companies [21].

2.5.1 Computer-Based Training Programs and E-Learning Platforms

Computer-based training programs bolstered by AIS are becoming more prevalent, providing a platform for continuous professional development and career growth that can be specifically tailored to the needs of the employees. A case study revealed the potential for AIS curriculum to contribute substantially to alumni's professional and career development, suggesting that integrating AIS into training programs is beneficial [23]. Furthermore, research has indicated that training in AIS can significantly enhance users' personal skills, translating to higher satisfaction and improved decision-making. Integrated with AIS, E-learning platforms can provide a flexible and accessible learning environment, allowing employees to engage in training activities conveniently [24].

2.5.2 Tracking Employee Training Progress and Certification

With its robust data processing capabilities, AIS can monitor and document employee training progress, qualifications, and certifications. Such systems enable HR departments to maintain meticulous records of employee development and ensure compliance with regulatory training requirements [25]. Evidence of the importance of job training on the performance of AIS further underscores the interconnectedness of these systems and the human resources function. Additionally, AIS aids in rationalizing human resource-related decisions, including training needs analysis, by providing reliable and structured information.

2.5.3 Performance Evaluation and Feedback Systems

Performance evaluation and feedback systems are integral to the development cycle within organizations. AIS has been utilized to enhance HRM performance, particularly in improving the efficacy of appraisal systems by providing timely and relevant data. AIS can foster a culture of continuous improvement by enabling more accurate performance reviews and feedback mechanisms. In this context, AIS serves as a repository of information and an analytical tool that can elucidate trends and areas for potential development within the workforce [26].

The literature indicates that AIS has evolved to play a crucial role in the training and development human resources within firms. By improving training programs through computer-based and e-learning platforms tracking progress and certification and facilitating performance evaluation. AIS has become an indispensable tool in the realm of human resource development [27].

AIS systems significantly contribute to the sophistication and robustness of training and development practices through their capacity for managing data and providing insights. As organizations continue to realize the potential of AIS in enhancing these HR functions, we can anticipate more strategic and integrated approaches to employee development in the future.

2.6 Payroll Processing

Accounting Information Systems (AIS) contribute significantly to operational efficiency, compliance management, and transactional accuracy in payroll processing. The integration of AIS with Human Resources (HR) data and its role in compliance management are discussed below with recent findings from scholarly literature.

2.6.1 Integration with HR Data

The integration of AIS with HR databases is pivotal for streamlining payroll processes. This integration ensures that all relevant information regarding employee salaries, benefits, and deductions is accurately compiled, resulting in a seamless payroll process. Studies have emphasized the importance of a critical system like AIS in capturing, processing, and storing data to manage various HR functions effectively, including payroll [28, 29]. Furthermore, the implementation of payroll AIS has shown that using business process management technologies supports current AIS by handling complex processes, such as the HR/payroll cycle. When AIS properly gathers and processes salary data, organizations can ensure that payroll calculations are correct and employees are reimbursed in a timely manner.

2.6.2 Payroll Processing—Reporting and Analytics

The evolution of Accounting Information Systems (AIS) has extended its applications to areas such as payroll management where reporting and analytics play a crucial role. Organizations now have access to advanced data analytics tools which aid in strategic decision-making and offer predictive insights into workforce management.

2.6.3 Strategic Decision Support

AIS serves as a backbone for strategic decision support in HR management by providing accurate and timely data. This data lets HR managers make informed decisions, ensuring human resources are well-aligned with organizational goals [30]. Modern AIS facilitates reporting and online analytical processing, which are integral to understanding the finer points of payroll management and its impact on the company's financial outlook [31].

Business intelligence applications in AIS, such as data visualization, help to enhance payroll reporting and make complex data more understandable for financial services assisting in the decision-making process.

2.6.4 Predictive Analytics

Predictive analytics in AIS allows HR managers to forecast future workforce needs. This proactive approach to planning can help organizations anticipate hiring requirements, identify training opportunities, and manage succession planning effectively [32]. Decision-making based on big data analytics provided by AIS has proven to be effective in healthcare organizations, suggesting a successful application across various industries [33]. Moreover, with the aid of predictive analytics, payroll implementation and execution are more refined, further enabling the strategic deployment of resources In addition, Robotic Process Automation (RPA) integrated within AIS can significantly improve the efficiency and effectiveness of HR functions such as payroll processing. By automating repetitive tasks, RPA allows HR departments to focus on strategic activities while ensuring that the payroll process is accurate and compliant [34].

3 Method

In the process of constructing this thorough review paper, researchers conducted an intensive examination of relevant literature pertaining to the topic of "The Role of Accounting Information Systems (AIS) in Enhancing the Human Resources Management Cycle." The present evaluation was based on a rigorous examination of influential research and significant publications, as cited in references [35–38]. The approach used in this study included a comprehensive examination of academic literature related to the utilization of AIS in augmenting human resources management, as referenced in sources [39–42]. By incorporating empirical research and theoretical models, our systematic review approach reached its conclusion by generating evidence-based recommendations. The aforementioned suggestions are tailored to enhance the effectiveness of AIS in enhancing human resources management within corporate environments, as supported by the results presented in source [43–45].

4 Discussion

This comprehensive conceptual analysis aims to analyze current academic material in order to examine the potential of Accounting Information Systems (AIS) to greatly improve several components of the human resource cycle via proper integration with HR management procedures. The advantages of adopting AIS across an enterprise are underscored by recent research conducted by Monteiro and Cepêda [46] as well as Idris, Barusman [47]. The advantages include the creation of consolidated databases to facilitate convenient accessibility, the implementation of standardized processes, and the enhancement of information security and access control measures pertaining to HR expenses and metrics.

The scholarly literature continually emphasizes the need of enhancing data visibility, compliance, and decision support in order to provide more strategic approaches to workforce recruiting, development, pay planning, and retention programs. The use of AIS has been shown to optimize HR procedures, diminish the need for human labour, and therefore enhance operational effectiveness and output [48]. Furthermore, the use of Artificial Intelligence Systems (AIS) enables human resource professionals to create reports and analytics in real-time, therefore offering valuable insights that are essential for making well-informed decisions in the realm of talent management strategies [49]. Comparative analyses conducted in this particular subject reveal a significant disparity between firms that use AIS (Accounting Information Systems) and those that depend on conventional HR (Human Resources) management approaches. Organizations that use AIS see notable enhancements in the precision and promptness of human resources (HR)-related data, hence enabling more effective strategic planning and execution in the domain of HR management [50]. The aforementioned distinction is notably conspicuous in domains such as talent acquisition and retention, whereby data-driven approaches have shown more efficacy in harmonizing human resources with corporate objectives.

Moreover, contemporary studies indicate the changing importance of AIS in addressing the fluctuating market requirements and the growing intricacy of HR operations. Sophisticated AIS platforms exhibit a high level of integration with other business systems, hence establishing a more coherent and agile organizational structure [12]. The integration of many components is of utmost importance when it comes to effectively tackling the complex and diversified difficulties that arise in contemporary human resources management [51]. These challenges include the efficient management of geographically dispersed workforces, ensuring compliance with a wide range of regulatory frameworks, and successfully adjusting to the quick fluctuations seen in the labour market.

4.1 Implications for Practice

The findings of the study have significant significance for human resources professionals and organizational leaders who are considering or already experiencing the deployment of an accounting information system (AIS). The integration of personnel management suites and programs with AIS platforms is emphasized as a crucial factor in achieving more accurate and financially-aware workforce planning and development. The development of fundamental personnel competencies in data analysis is of utmost importance in order to effectively use the capabilities of an Accounting Information System (AIS) after its deployment. Furthermore, it is important for organizations to maintain a constant state of awareness and attentiveness with regards to cybersecurity measures and access rules pertaining to confidential personnel information, notwithstanding potential advancements in information security and privacy controls.

4.2 Recommendations for Future Research

Further investigation is required to comprehensively examine the effects of adopting AIS in various settings, including as small to medium-sized enterprises, multinational corporations, and governmental institutions, where the requirements for HR data management may vary. Examining instances in which combined AIS-HR systems shown limited improvements may also contribute to the advancement of best practices in deployment. The investigation of the establishment of security standards and specialized teams dedicated to safeguarding HR data safety and privacy might provide more knowledge as integrated AIS grow increasingly prevalent across many sectors. In future research, it is recommended that academics broaden the scope of

HR performance indicators to include talent innovation outputs and employee experience evaluations. This expansion will enable a more comprehensive quantification of the returns on adopting AIS, going beyond only financial considerations.

5 Conclusion

This study presents a comprehensive analysis of existing literature pertaining to the convergence of accounting information systems and human resource management practices. The research focused on organizing and classifying the implications of Artificial Intelligence Systems (AIS) on several stages of the Human Resources (HR) cycle. This included planning, recruiting, onboarding, compensation, talent management, and retention. The analysis used existing case data and commentary to organize its findings. The results of several studies consistently indicate that the implementation of integrated AIS is very beneficial. This implementation allows for the consolidation of data platforms, standardization of workflows, establishment of data security standards, and use of sophisticated analytics. These factors together contribute to the improvement of HR decision-making processes. Nevertheless, this study presents a preliminary framework that establishes a connection between the adoption of AIS (Accounting Information Systems) and the optimization of HR (Human Resources) management. The framework outlines the many elements of AIS that are relevant to different HR activities. The aim of this contribution is to provide a valuable resource for both researchers and practitioners in the field. This study delves into the compilation of experiences pertaining to security and integration difficulties that are crucial for achieving effective long-term system deployment, hence optimizing return on investment.

Hence, our research has provided evidence that the use of sophisticated AIS solutions enables HR departments to transition from a reactive approach of fragmented reporting to a comprehensive and integrated analysis of people data. In order to optimize these systems, it is necessary to implement strategic change management and enhance organizational capabilities. By doing so, firms are able to achieve improved efficiency in HR processes and get valuable insights into the costs, risks, and value associated with their workforce. This enables organizations to establish a competitive advantage in human capital management. The ongoing development of AIS and HR information systems necessitates more investigation into the quantification of processes and the measurement of advances in talent production. Therefore, the potential for enhanced human resource decision-making via the integration of AIS seems to be more optimistic.

References

1. Gusmiarni, A., Sudrajat, S.: The influence of utilizing accounting information systems on managerial performance at BPJS Kesehatan Bandar Lampung Branch Office. INVEST: Jurnal Inovasi Bisnis Dan Akuntans **4**(2), 323–330 (2023)
2. Jawad, M.M.: Accounting information systems (AIS) and their effect on strategic decision-making. Technium Soc. Sci. J. **47**, 218 (2023)
3. Nuraida Baharsyah, K.: The effect of revaluation, administration, and accounting management information system of state-owned assets on quality of financial report (case study on the office of state asset management and auction Bandar Lampung) (2023)
4. Mahapatro, B.B.: Human resource management. PG Department of Business Management (2022)
5. Yadav, A., et al.: Work-life integration, job contentment, employee engagement and its impact on organizational effectiveness: a systematic literature review. Ind. Commer. Train. **54**(3), 509–527 (2022)
6. Naji, G.M.A., et al.: The role of HR strategy on safety culture and psychological stress among employees in the upstream oil and gas companies: a conceptual review. Solid State Technol **63**(5), 12 (2020)
7. Magege, T.J.: Impact of human resources information system on organization performance in banking sector; A study of Azania Bank Limited (2022)
8. Rahman, M.: Formulating HRIS features and modules following the needs of NEXT ventures' Human Resource Management Department (2022)
9. Saleh, Q.Y., Al-Nimer, M.B.: The mediating role of the management accounting information system in the relationship between innovation strategy and financial performance in the Jordanian industrial companies. Cogent Bus. Manage. **9**(1), 2135206 (2022)
10. Trang, N.T.N., et al.: Human factors and influence on the quality of accounting information system. J. Sci. **14**(1) (2024)
11. Kyeremeh, K., et al.: The role of accounting information systems in organizational decision making: evidence from banking sector in Ghana. J. Account. Taxation **2**(3), 123–144 (2022)
12. Al-Matari, A.S., et al.: The impact of dynamic accounting information system on organizational resilience: the mediating role of business processes capabilities. Sustainability **14**(9), 4967 (2022)
13. Sudarmanto, E., et al.: Influence of quality of accounting information system, development of ethical climate and organizational culture on the effectiveness of corruption risk management and its impact on the level of organizational corruption at government agencies in Indonesia. Int. J. eBusiness eGovernment Stud. **14**(4), 66–95 (2022)
14. Ratmono, D., Frendy, Zuhrohtun, Z.: Digitalization in management accounting systems for urban SMEs in a developing country: a mediation model analysis. Cogent Econ. Fin. **11**(2), 2269773 (2023)
15. Böhmer, N., Schinnenburg, H.: Critical exploration of AI-driven HRM to build up organizational capabilities. Employee Relat. Int. J. (2023)
16. Qader, K.S., et al.: The effect of human resources management skills on accounting information quality in Kurdistan Public Sector. Int. J. Hum. Educ. Dev. (IJHED) **3**(3), 131–143 (2021)
17. Qatawneh, A.M.: The role of employee empowerment in supporting accounting information systems outcomes: a mediated model. Sustainability **15**(9), 7155 (2023)
18. Hertati, L., Safkaur, O., Simanjuntak, A.M.: How to align management commitments to the successful implementation of management accounting information systems in manager decision making. Ilomata Int. J. Tax Account. **1**(2), 89–102 (2020)
19. Alastal, A.Y.M., Salman, M.A.H., Allaymoun, M.H.: Accounting students' perceptions on e-learning during covid-19 pandemic: case study of accounting and financial students in Gulf University—Bahrain. In: Artificial Intelligence and Transforming Digital Marketing, pp. 879–890. Springer (2023)
20. Sturm, T., Peters, F.: The impact of artificial intelligence on individual performance: exploring the fit between task, data, and technology (2020)

21. Fitria, N., et al.: The role of management information systems in human resource competency development. Jurnal Minfo Polgan **12**(1), 1387–1396 (2023)
22. Nguyen, T., Chen, J.V., Nguyen, T.P.H.: Appropriation of accounting information system use under the new IFRS: impacts on accounting process performance. Inf. Manage. **58**(8), 103534 (2021)
23. Sarkar, S., et al.: Developing institutional skills for addressing big data: experiences in implementation of AACSB Standard 5. J. Account. Educ. **54**, 100708 (2021)
24. Aggarwal, D.: Integration of innovative technological developments and AI with education for an adaptive learning pedagogy. China Petrol. Process. Petrochem. Technol. **23**(2) (2023)
25. Diawati, P., et al.: The role of information technology in improving the efficiency and productivity of human resources in the workplace. Jurnal Teknologi Dan Sistem Informasi Bisnis **5**(3), 296–302 (2023)
26. Watson, R.T., et al.: How the AIS can improve its contributions to the UN's sustainability development goals: towards a framework for scaling collaborations and evaluating impact. Commun. Assoc. Inf. Syst. **48**(1), 42 (2021)
27. Petratos, S.: CS/IS/IT Programme accreditation and SACAB support at higher education institutions in Southern Africa. In: Proceedings of 52nd Annual Conference of the Southern African Computer Lecturers' Association (SACLA 2023) (2023)
28. Lekorere, M.N.: Effects of accounting information systems on decision making in banks in Nairobi Kenya (2022)
29. BinSaeed, R.H., et al.: Digital revolution and digitization process to promote AIS as a vector of financial performance. Systems **11**(7), 339 (2023)
30. John, I.B., et al.: Assessment of marketing strategies adoption and competitiveness among quantity surveying firms in the digitisation era. Hum. Factors Bus. Manage. Soc. **56**, 116 (2022)
31. Trigueiros, E.V.d.C., Alves, M.-C.G.: Accounting information system adoption in multi-site organisations. Int. J. Bus. Inf. Syst. **41**(2), 236–257 (2022)
32. Shet, S.V., Pereira, V.: Proposed managerial competencies for Industry 4.0—implications for social sustainability. Technol. Forecast. Soc. Change **173**, 121080 (2021)
33. Shabbir, M.Q., Gardezi, S.B.W.: Application of big data analytics and organizational performance: the mediating role of knowledge management practices. J. Big Data **7**(1), 1–17 (2020)
34. Mohamed, S.A., et al.: Improving efficiency and effectiveness of robotic process automation in human resource management. Sustainability **14**(7), 3920 (2022)
35. Ali, A.A.A., et al.: Moderating effect of strategic planning on the relationship between career path planning and job performance. Sustainability **15**(11), 8490 (2023)
36. Al-refaei, A.A.-A., et al.: An integrated mediating and moderating model to improve service quality through job involvement, job satisfaction, and organizational commitment. Sustainability **15**(10), 7978 (2023)
37. Alzoraiki, M., et al.: Impact of teachers' commitment to the relationship between transformational leadership and sustainable teaching performance. Sustainability **15**(5), 4620 (2023)
38. Ali, A.A.A.: The Relationship between Islamic Work Ethics and Employee Performance among Medical Staff in the Public Health Sector in Bahrain: Employee Commitment as Mediating Variable. Universiti Sains Islam Malaysia (2022)
39. Ali, A.A.A., Alzoraiki, M., Al Aghbas, N.S.M.: Working remotely, social isolation and access to support and communication during Covid-19 its influence on job performance. Int. J. Intell. Hum. Resour. Manage. (IJIHRM) **3**(01), 23–33 (2022)
40. Ali, A.A.A., et al.: Association between Islamic work ethics and employee commitment among medical staff in the public health sector in Bahrain. Int. J. Intell. Hum. Resour. Manage. (IJIHRM) (2023)
41. Ateeq, A.: Knowledge management's effect on learning quality in Bahrain's private universities (2023)

42. Alkadash, T.M., et al.: The effects of leadership style on employee sustainable behaviour: a theoretical perspective. In: Artificial Intelligence and Transforming Digital Marketing, pp. 205–213. Springer (2023)
43. Ateeq, A.: The mediating role of engagement in the relationship between. Management **21**(3), 435–447 (2023)
44. Al Balushi, F.S.M., Ali, A.A.A., Alzoraiki, M.: The impact of forecasting strategy and training and development strategy on preparing leadership competencies administrative support in the Sultanate of Oman. Int. J. Intell. Hum. Resour. Manage. (IJIHRM) **3**(01), 34–41 (2022)
45. Rahardjo, K.P., Kurniawati, K.: Caroll pyramid implementation of environmental responsibility programs of PT Pertamina EP Subang Field, Tambun Field, and Jatibarang Field in organic agriculture. J. Soc. Res. **2**(9), 3331–3340 (2023)
46. Monteiro, A., Cepêda, C.: Accounting information systems: scientific production and trends in research. Systems **9**(3), 67 (2021)
47. Idris, I., et al.: Development of the accounting information system as teaching content to improve information technology competency in graduates. Test Eng. Manage. **82**(1), 9897–9907 (2020)
48. Al-Hattami, H.M., Kabra, J.D.: The role of accounting information system (AIS) in rationalizing human resource related decisions: a case study of selected commercial banks in Yemen. Int. J. Manage. Stud. **4**(2), 84–91 (2019)
49. Dwivedi, Y.K., et al.: Artificial intelligence (AI): multidisciplinary perspectives on emerging challenges, opportunities, and agenda for research, practice and policy. Int. J. Inf. Manage. **57**, 101994 (2021)
50. Hoq, M.Z.: A comprehensive review of contemporary issues of electronic human resource management (E-HRM). Glob. J. Econ. Bus. **11**(2), 278–296 (2021)
51. Vrontis, D., et al.: Artificial intelligence, robotics, advanced technologies and human resource management: a systematic review. Int. J. Hum. Resour. Manage. **33**(6), 1237–1266 (2022)

Alignment Imperative: Integrating HRM and Corporate Governance for Sustainable Success

Mohammed Alzoraiki, Ahmad Yahia Mustafa Alastal, Marwan Milhem, and Ali Ateeq

Abstract This study explores the integration of Human Resource Management (HR) and Corporate Governance for sustainable success in organizations. It delves into how aligning HRM practices with corporate governance principles can improve organizational performance, ethical decision-making, and long-term sustainability. By examining the interplay between HRM and corporate governance, this paper aims to provide insights into the strategic alignment necessary for achieving sustainable success in today's dynamic business environment.

Keywords HRM · Corporate governance · Sustainable success · Organizational performance · Strategic alignment

1 Introduction

Over the past decade, human resource management (HRM) and corporate governance domains have primarily operated parallel paths in organizational research and practice. Yet shifting societal expectations, complex talent management needs, and ethical requirements of modern corporations have highlighted the imperatives of more excellent connectivity between these functions. Recent integrated governance

M. Alzoraiki · M. Milhem · A. Ateeq
Administrative Science Department, College of Administrative and Financial Science, Gulf University, Sanad 26489, Kingdom of Bahrain
e-mail: dr.mohammed.saied@gulfuniversity.edu.bh

M. Milhem
e-mail: dr.marwan.milhem@gulfuniversity.edu.bh

A. Ateeq
e-mail: dr.ali.ateeq@gulfuniversity.edu.bh

A. Y. M. Alastal (✉)
Accounting and Financial Science Department, College of Administrative and Financial Science, Gulf University, Sanad 26489, Kingdom of Bahrain
e-mail: dr.ahmed.alastal@gulfuniversity.edu.bh

B. Awwad (ed.), *The AI Revolution: Driving Business Innovation and Research*, Studies in Systems, Decision and Control 524, https://doi.org/10.1007/978-3-031-54379-1_9

studies coalesce around maximizing human capital and reinforcing ethical compliance within contemporary workforce environments [1, 16, 18]. Still, comprehensive frameworks, evidence-based recommendations, and implementation guidance integrating these domains remain lacking.

Integrating human resource management (HRM) and governance has been critical in achieving elevated business performance. However, knowledge gaps and barriers still hinder holistic connectivity between these complementary drivers. By detailing established integration benefits, remaining barriers, and actionable recommendations, this manuscript aims to address these gaps and provide a roadmap for achieving tighter strategic alignment. Our approach is grounded in leading research and promises considerable ethical, legal, and risk mitigation benefits and sustained competitive advantages for modern corporations.

The resource-based view (RBV) theory provides a valuable lens for this challenge. RBV posits that firms can gain a sustainable competitive advantage through their unique resources and capabilities [2, 5, 6] his theory can be applied in HRM and Corporate Governance to understand how human resources can be leveraged as strategic assets and how governance structures can support this [9]. Tighter alignment promises both strategic human capital optimization and broader stakeholder value creation.

Therefore, this study addresses this research-practice gap by detailing established benefits, barriers, and actions for elevating HRM-governance connectivity—promising considerable ethical, legal, risk mitigation and sustained competitive advantages for modern corporations.

2 Literature Review

The alignment between Human Resource Management (HR) and Corporate Governance has gained increasing attention in the literature due to its critical role in driving sustainable success for organizations. This literature review aims to provide an overview of the key findings, insights, and perspectives on the alignment imperative between HRM and Corporate Governance, emphasizing its significance for sustainable organizational performance.

3 Theoretical Framework of HRM Practices and Corporate Governance

HRM refers to organizational practices for managing people, including recruitment, development, compensation, retention, and labor relations [3]. Corporate governance involves internal controls oversight, risk management, transparency policies and accountability to shareholders and stakeholders [20].

4 Strategic Alignment of HRM and Corporate Governance

The literature emphasizes the importance of aligning HRM practices with corporate governance mechanisms to achieve strategic organizational goals. Scholars highlight that effective integration of HRM and Corporate Governance ensures that the organization's human capital is managed to align with the overall corporate strategy and governance structure. This alignment is crucial, as it enables HRM to contribute to the organization's long-term success by ensuring that HR policies and practices align with the company's goals and values. Additionally, it helps foster a culture of transparency, accountability, and ethical behaviour within the organization, which is essential for effective corporate governance.

Several recent studies have underscored the critical performance and ethical advantages of integrating human resource management (HRM) and governance domains [16, 17]. Financial public-sector organizations highlight how human capital connectivity across HRM and risk governance elevates innovation and stakeholder impact potential [4]. Moreover, Previous studies on HRM and corporate governance indicate that organizations that effectively integrate HRM and corporate governance tend to have higher employee engagement, productivity, and overall organizational performance. This integration allows for a seamless flow of information and decision-making between HR and the board of directors, ensuring that HR policies are aligned with the company's strategic objectives. Furthermore, it helps proactively identify and address potential risks, safeguarding the organization's reputation and minimizing legal and regulatory compliance issues.

However, a lack of holistic connectivity remains typical, presenting risks and untapped capability-building opportunities. For instance, survey data indicates that only 13% of boards and senior management view HRM integration as a top governance priority [10, 11]. Significant knowledge and implementation gaps persist in optimally and tangibly aligning these domains.

Therefore, this literature review is intended to synthesize the current body of knowledge on benefits, recommended practices and adoption barriers regarding HRM-governance integration. The analysis aims to address open research questions and provide evidence-based recommendations for generalizable frameworks and organizational implementation tactics [7].

5 Relationship Between HRM Practices and Corporate Governance Mechanisms

Studies examine links between governance mechanisms like boards, monitoring controls, and associated HRM practices, including incentives, transparency norms, and learning cultures [8, 15]. Findings reveal reinforcing and conflicting interactions across governance and HR systems.

HRM practices around competency building reward equity and voice mechanisms affect board monitoring capabilities, managerial opportunism risks and accountability outcomes. Research reveals that HRM alignment amplifies or dampens governance policy effectiveness [15].

6 Stakeholder Perspective

The resource-based view and its implications for aligning HRM and corporate governance The theory resource-based view highlights human capital quality as a critical source of competitive advantage. This theory suggests that integration maximizes talent governance and minimizes ethical compliance risks [5].

6.1 Role of Stakeholders in Shaping HRM and Corporate Governance Practices

Studies show investors and regulators applying pressure for governance oversight reforms and associated HR practice changes around incentives, training priorities and culture [19].

6.2 Ethical Considerations in Stakeholder Management

Maximizing shareholder wealth must balance profit goals with responsible stakeholder treatment [12]. This ethical tension suggests integration around governance guidance and HR implementations for conscious capitalism.

7 Organizational Performance and Sustainability

7.1 Link Between Aligned HRM and Corporate Governance and Organizational Performance

Recent empirical studies establish positive links between integrated governance-HR systems and measures of operating performance, profitability, innovation, safety and reputation [13]. Findings suggest complementarities across aligned systems.

7.2 Long-Term Sustainability Implications of Aligned HRM and Corporate Governance

Integrated prioritization of stakeholder human capital development and ethical accountability shapes corporate culture committed to responsible long-term growth [13]. This elevates resilience to disruptions and uncertainty.

8 Challenges and Barriers

8.1 Common Challenges in Aligning HRM and Corporate Governance

Lack of leadership commitment, narrow metrics, unclear accountability and poor change management practices are frequently cited execution barriers [3]. Complexity across matrix structures also poses systems integration obstacles.

8.2 Regulatory and Legal Barriers

Research notes strict data privacy laws, union rules and country-specific regulations potentially restricting coordination scope [14]. Navigating these complex compliance requirements impedes unified initiatives.

9 Best Practices and Recommendations

9.1 Successful Strategies for Integrating HRM and Corporate Governance

Effective practices include cross-representation, internal audit coordination, joint strategizing committees, accountability matrixes and integrated analytics. Leadership, culture and structural initiatives appear vital.

9.2 Recommendations for Organizations Seeking to Achieve Alignment

Suggested actions highlight formalizing mechanisms, elevating HRM roles in talent governance, harmonizing leader development and rewards with oversight systems and collective capability building.

9.3 Implications for Future Research and Practice

More micro-level studies on alignment mechanisms are needed across various contexts. Refining theoretical frameworks and testing leading indicators and outcomes warrants attention [14].

10 Conclusion

In conclusion, the literature on the alignment imperative between HRM and Corporate Governance underscores its critical role in driving sustainable organizational success. Strategic alignment, ethical leadership, board diversity, employee voice, and performance measurement are vital themes contributing to the integration of HRM and corporate governance for sustainable performance. This review provides valuable insights for practitioners and policymakers seeking to enhance the alignment between HRM and corporate governance to achieve long-term organizational success.

11 Summary of Key Findings from the Literature Review

The review reveals significant performance, innovation, culture and capability advantages to integrating governance and HRM. But adoption barriers persist around structural complexity, leadership priorities and change inertia.

11.1 Implications for Theory and Practice

Emerging stakeholder and stewardship theories provide useful framings. For practice, formalizing mechanisms, governance representation and collective capability building show promise.

11.2 Suggestions for Future Research

Further micro-level studies should refine theoretical frameworks and test leading indicators and alignment outcomes. Exploring context-specific integration models can also elucidate generalizability.

References

1. Al-Absy, M.S.M., Ismail, K., Chandren, S.: Corporate governance mechanisms, whistle-blowing policy and real earnings management. Int. J. Financ. Res. **10**(6), 265–282 (2019)
2. Alastal, A.Y.M.: Mediating Effect of an Eco-control System on the Relationship Between a Proactive Environmental Strategy and Environmental Performance in Jordan. Universiti Utara Malaysia (2021)
3. AlDhaheri, H., Hilmi, M.F., Abudaqa, A., Alzahmi, R.A., Ahmed, G.: The relationship between HRM practices, innovation, and employee productivity in UAE public sector: a structural equation modelling approach. Int. J. Process Manage. Benchmarking **13**(2), 157–176 (2023)
4. Anik, S., Chariri, A., Isgiyarta, J.: The effect of intellectual capital and good corporate governance on financial performance and corporate value: a case study in Indonesia. J. Asian Fin. Econ. Bus. **8**(4), 391–402 (2021)
5. Barney, J.: Firm resources and sustained competitive advantage. J. Manag. **17**(1), 99–120 (1991)
6. Barney, J., Wright, M., Ketchen, D.J., Jr.: The resource-based view of the firm: ten years after 1991. J. Manag. **27**(6), 625–641 (2001)
7. Bhattacharjee, S., Afrin, R., Mohiuddin, M.: Role of human resource management in corporate governance: a critical literature review. AIUB J. Bus. Econ. **12**(1), 21–45 (2015)
8. Bretos, I., Errasti, A., Marcuello, C.: Ownership, governance, and the diffusion of HRM practices in multinational worker cooperatives: case-study evidence from the Mondragon group. Hum. Resour. Manag. J. **28**(1), 76–91 (2018)
9. Collins, C.J.: Expanding the resource based view model of strategic human resource management. Int. J. Hum. Resour. Manage. **32**(2), 331–358 (2021)
10. du Plessis, A., Nel, P., Struthers, L., Robins, K., Williams, K.: Does human resource management add value to organisations in New Zealand or is it just another job? Int. Rev. Bus. Res. Papers **3**(3), 277–296 (2007)
11. Ejo-Orusa, H., Okwakpam, J.A.A.: Predictive HR analytics and human resource management amongst human resource management practitioners in Port Harcourt, Nigeria. Glob. Sci. J. **6**(7), 254 (2018)
12. Freeman, R.E., Wicks, A.C., Parmar, B.: Stakeholder theory and "the corporate objective revisited." Organ. Sci. **15**(3), 364–369 (2004)
13. Gomomo, J.A.: A strategic HRM framework for improving corporate governance in a municipal environment (2019)
14. Harjoto, M.A., Jo, H.: Corporate governance and CSR nexus. J. Bus. Ethics **100**, 45–67 (2011)
15. Ibrahim, H.I., Zulkafli, A.H.: Corporate governance, HRM practices and organizational performance. Соціально-економічні проблеми і держава **14**(1), 30–40 (2016)
16. Lima, L., Galleli, B.: Human resources management and corporate governance: integration perspectives and future directions. Eur. Manag. J. **39**(6), 731–744 (2021)
17. Martin, G., Farndale, E., Paauwe, J., Stiles, P.G.: Corporate governance and strategic human resource management: four archetypes and proposals for a new approach to corporate sustainability. Eur. Manag. J. **34**(1), 22–35 (2016)
18. Martin, G., Gollan, P.J.: Corporate governance and strategic human resources management in the UK financial services sector: the case of the RBS. Int. J. Hum. Resour. Manage. **23**(16), 3295–3314 (2012)

19. Sripada, C.: People at the heart of corporate governance: reimagining the role of HRM. NHRD Netw. J. **12**(4), 364–375 (2019)
20. Teixeira, J.F., Carvalho, A.O.: Corporate governance in SMEs: a systematic literature review and future research. Corp. Gov. Int. J. Bus. Soc. (2023)

Integrating for Excellence: A Study on Merging Accounting, HRM, and ERP Systems in Security and Safety Firms for Delta Safety and Security Company

Ali Ateeq, Jumana Saeed, Marwan Milhem, and Mohammed Alzoraiki

Abstract This study analyzes the strategic integration of business systems, with an emphasis on the integration of the accounting, human resources management, and enterprise resource planning systems for a security and safety firm. Streamlining company processes, increasing operational effectiveness, and strengthening decision-making skills are all made possible by the integration of these platforms. The interrelationships between ERP, HRM, and accounting systems are investigated, along with the difficulties in integrating them, through a thorough analysis of the literature. To offer practical insights, case studies and empirical research are also covered. A framework for directing the strategic integration of these systems is suggested in light of the findings. The methodology for this study is described in the method section, and then the findings are discussed. An overview of the most important lessons learned and suggestions for successfully integrating the accounting, HRM, and ERP systems in a security and safety firm are provided in the paper's conclusion.

Keywords ERP integration · HRM systems · Accounting · Strategic management · Data accuracy · Financial reporting

A. Ateeq (✉) · J. Saeed · M. Milhem · M. Alzoraiki
Administrative Science Department, College of Administrative and Financial Science, Gulf University, Sanad 26489, Kingdom of Bahrain
e-mail: dr.ali.ateeq@gulfuniversity.edu.bh

J. Saeed
e-mail: 210104115617@gulfuniversity.edu.bh

M. Milhem
e-mail: dr.marwan.milhem@gulfuniversity.edu.bh

M. Alzoraiki
e-mail: dr.mohammed.saied@gulfuniversity.edu.bh

B. Awwad (ed.), *The AI Revolution: Driving Business Innovation and Research*, Studies in Systems, Decision and Control 524, https://doi.org/10.1007/978-3-031-54379-1_10

1 Introduction

Effective system integration is essential for businesses to maximize operations and gain a competitive edge in the complex business environment of today. This is especially true for security and safety firms, as they need to coordinate staff, resources, and financial information in an effective manner. A complete approach to meeting the particular requirements of these businesses can be achieved through the integration of accounting, human resources management, and enterprise resource planning [1]. Supply chain operations, customer relationship management, inventory management, and other fundamental company processes are all integrated and automated by enterprise resource planning systems [2]. A security and safety organization can achieve seamless data flow between departments and real-time access to vital information by integrating ERP with HRM and Accounting systems. Accurate forecasting, effective resource allocation, and better decision-making are all made possible by this integration [1]. Systems for resource management, or HRM, are essential to a security and safety company's personnel management. Organizations may improve personnel acquisition and retention tactics, simplify payroll procedures, and streamline employee data management by integrating HRM with ERP and accounting systems. The proper employees are assigned to the appropriate projects thanks to this connectivity, which also guarantees that the accounting system processes payroll accurately by allowing time and attendance data to flow into it effortlessly [3]. Accounting systems are necessary for managing budgets, producing financial reports, and keeping track of financial transactions [4]. A security and safety organization can enhance its financial analytical skills, automate financial data capture, and expedite spending management by integrating accounting with ERP and HRM systems. This integration facilitates well-informed decision-making by offering a comprehensive picture of the organization's financial health. Although there are many advantages to strategic ERP, HRM, and accounting system integration, there are drawbacks as well that must be taken into consideration [5]. Data synchronization, system compatibility, and process standardization are some of these difficulties [6]. To optimize the advantages of integration, it is essential to make sure that data is precisely synchronized across systems, that the integrated systems are interoperable with one another, and that business processes are standardized.

Finally, an organization that specializes in security and safety can greatly improve its operational efficiency and decision-making skills by strategically integrating its ERP, HRM, and accounting systems. It provides an all-encompassing perspective of the company, streamlines data flow, enhances resource management, and automates procedures. However, to accomplish the intended results, a successful implementation necessitates meticulous planning, resolving compatibility and standardization issues, and utilizing the right technological solutions.

2 Literature Review

The part on literature review explores the corpus of research on the combination of accounting, HRM, and ERP systems. It looks at how these systems are related to one another and emphasizes the advantages of integration. The assessment also touches on the importance of accounting systems in financial management as well as the function that ERP systems play in enabling business integration.

2.1 Integration of ERP, HRM, and Accounting Systems

Enterprise resource planning (ER), HRM, and Accounting System Integration for security and safety firms, one of the most important aspects of strategic business integration is the integration of accounting, human resources management, and enterprise resource planning. The integration of various systems, as well as their interactions and information sharing, are examined in this area in order to improve organizational performance and streamline business processes [7]. Effective resource, human, and financial management is made possible by the integration of ERP, HRM, and accounting systems [8]. It makes sure that information moves smoothly between departments and makes sure that decisions are made effectively, processes are automated, and data is accurate. Security and safety firms can streamline their processes, increase output, and accomplish better financial management by integrating these solutions. The primary benefit of combining accounting, HRM, and ERP systems is more accurate and consistent data [9]. Since these systems are frequently linked together, real-time data sharing is possible. To guarantee that employee-related data is consistent throughout the company, for instance, updates to an employee's record in the HRM system may automatically appear in the ERP system [10]. Time is saved and data dependability is increased thanks to this integration, which also lowers errors and eliminates the need for manual data entry [11]. Moreover, corporate operations become more efficient as a result of the integration of various platforms. For example, the interface between the HRM and ERP systems can automate the onboarding process when an employee is hired by creating required paperwork, allocating access permissions, and updating payroll data. A comprehensive perspective of the organization's financial information and resources is provided by integration, which further improves decision-making abilities [12]. Managers get instant access to data on financial parameters, project statuses, and staff performance thanks to connected platforms [13]. Managers may examine data from many systems to spot patterns, distribute resources wisely, and decide on key business initiatives thanks to this holistic picture, which facilitates informed decision-making. But there can be difficulties when combining accounting, HRM, and ERP systems. Making sure data consistency and system compatibility are one of the issues. It might be challenging to create a smooth integration across systems since they may have different data types and formats. To guarantee reliable data interchange, data pieces must be mapped and aligned across systems [14]. Another difficulty is organizational reluctance to change. Work routine modifications and process reengineering are frequent side effects of system integration because they are unfamiliar with the new procedures, fear of losing their jobs, or worry about having more work to do, employees may oppose these changes. To overcome these obstacles and win over stakeholders, it is imperative to implement efficient change management techniques and staff training initiatives. When integrating these technologies, data security and privacy must also be taken into account. Since systems exchange sensitive employee and financial data, it is essential to put strong security measures in place and adhere to any data protection laws. This contains encryption, role-based access controls, frequent data backups, and audits of data access. The integration of accounting, HRM, and ERP systems in security and safety firms has been the subject of numerous researches. This research has shed important light on the advantages and difficulties of integration. As an illustration, integrating these technologies increased financial reporting accuracy, decreased manual errors, and improved resource allocation, according to a case study done on a security company. A different study emphasized the significance of personnel training and change management for a successful integration, stressing the necessity of open communication and assistance all along the way.

2.2 ERP Systems and Business Integration

Integrating Business and ERP Systems for enterprise resource planning, or ERPs, are essential for making it easier for departments and business units inside an organization to integrate. They offer a consolidated platform via which many departments and functions can communicate, work together, and simplify processes [15]. ERP integration with accounting, HRM, and other systems is essential for security and safety companies to achieve operational efficiency, data accuracy, and well-informed decision-making. The smooth transfer of data between departments is one of the main advantages of integrating ERP systems with other business platforms. Process standardization guarantees uniformity and adherence to legal standards. Better decision-making is one of the ERP system integration's main benefits [16]. Management may make educated decisions thanks to the complete and accurate data that is accessed through the integration of HRM and Accounting systems with ERP. Managers can evaluate the organization's overall performance and make strategic decisions by, for instance, accessing real-time financial data, personnel performance indicators, and project status reports in a single dashboard [17]. Nevertheless, there may be difficulties when connecting ERP systems with HRM and accounting systems. One of the primary challenges is data inconsistency and compatibility; various systems may employ various data formats and structures, which makes it challenging to smoothly integrate and synchronize data. In order to guarantee data consistency and interoperability across systems, data cleansing and mapping procedures are required. A further issue that could impede the integration process is organizational resistance to change. Workers that are afraid of losing their jobs or having more work to complete may be reluctant to accept new procedures and systems. To allay these worries and guarantee a seamless integration, change management tactics such as employee involvement, communication, and training programs are essential. Moreover, demand for customization and system complexity can present integration issues. ERP systems must be customized to meet the unique requirements of a security and safety organization because they are frequently complicated systems. Module customization, interface creation, and data mapping rule establishment are possible integration tasks. For these customized procedures to be successful in integrating, meticulous preparation, coordination, and technical know-how are needed. Security and safety organizations can use the following best practices and integration techniques for ERP, HRM, and accounting systems to get beyond these obstacles:

- Clearly state integration objectives and make sure they complement the strategic goals of the firm. Prioritize the business processes and operations that must be connected according to how they will affect decision-making and operational efficiency.
- To find incompatibilities and the need for data cleansing, thoroughly examine the current systems and data structures. To ensure a seamless transfer of data from older systems to the integrated ERP system, establish data mapping guidelines and create a data migration plan.
- Include important parties from several departments in the process of integration. To encourage a sense of ownership and commitment to the integration endeavor, get their feedback and resolve any issues they may have.
- Create a thorough change management strategy with channels for support, training, and communication. Employees should be informed about the advantages of integration and given the training they need to become accustomed to the new procedures and systems.
- Take into account working with outside advisors or ERP installation specialists who have knowledge of integrating systems for security and safety firms. Their knowledge can aid in navigating the integration's intricacies and guarantee successful deployment.
- Regularly check in on and assess the integrated systems' performance. Refine and enhance the integration processes on a continuous basis in response to feedback and changing business requirements. In summary, for a security and safety organization to achieve operational efficiency, data accuracy, and well-informed decision-making, it is imperative that ERP systems be integrated with HRM and Accounting systems. Process automation, standardization, and smooth information flow are made possible by the integration.

However, obstacles including inconsistent data, resistance from the organization, and complexity of the system must be successfully overcome. Security and safety firms can take advantage of enhanced performance and efficiency by implementing best practices and strategies to properly integrate their business systems.

2.3 Accounting Systems and Financial Management

Accounting Systems and Financial Management: An essential component of a security and safety company's financial management is accounting systems. These systems are in charge of tracking and documenting financial activities, producing financial statements, and offering insightful data regarding the state of the company's finances [18]. Financial procedures can be streamlined and several advantages can be gained by integrating accounting systems with other business systems, such as human resources management (HRM) and enterprise resource planning [19]. The smooth transfer of financial data across the many functional divisions of the company is made possible by the integration of accounting systems with ERP [20]. This interface improves data accuracy, removes data duplication and inconsistencies, and provides real-time access to financial information. A security and safety organization can develop a consolidated financial management system that offers a comprehensive picture of financial data and helps well-informed decision-making by combining accounting systems with ERP. Moreover, expense management and payroll processing can be streamlined by linking accounting systems with HRM systems. The automatic synchronization of employee data across HRM and accounting systems, including pay, benefits, and deductions, is made possible by this interface. Payroll computations consequently become more precise and effective, decreasing human error and saving time [21]. Accounting system integration with human resource management also makes it easier to track expenses and reimburse employees, assuring adherence to rules and regulations regarding finances. Accounting system integration improves financial reporting capabilities while also simplifying financial procedures and linking accounting systems with other departments [22].

2.4 Interrelationships and Integration Challenges

For a security and safety organization, integrating ERP, HRM, and accounting systems presents a number of integration issues as well as relationships. In order to successfully integrate and reap the full benefits of integrated business systems, it is imperative to comprehend these interrelationships and challenges. The data exchange between these systems is one of the most important relationships [23]. ERP acts as a central data repository and makes it easier for information to be shared between various functional departments [24]. Payroll data and employee benefits are provided by HRM systems; these systems must be connected with ERP and accounting systems in order to give correct financial reporting. To maintain accuracy and consistency, accounting systems must be synced with ERP and HRM systems in order to deliver financial data. Nevertheless, there may be a number of difficulties in integrating these systems. Data duplication and inconsistency is a frequent problem. It might be challenging to smoothly reconcile and integrate data since every system may have its own data structure and format [25]. In order to guarantee that data from various systems may be successfully merged, data cleansing and standardization procedures might be necessary. Compatibility with different systems is another difficulty. Compatibility problems during integration may arise from the fact that ERP, HRM, and accounting systems were created by separate suppliers or have distinct versions. It could be necessary to modify these systems or employ middleware to ensure that they can exchange data and communicate efficiently. Another issue that can obstruct the integration process is organizational resistance to change. Workers may object to the use of integrated systems if they are used to using distinct systems [25]. In order to overcome resistance and encourage user acceptability, change management techniques like communication and training are crucial. Add to this the technological difficulties arising from the intricacy of integrating several systems within a security and safety firm. Integration might entail intricate data transformations and mappings as well as the creation of interfaces, or APIs (Application Programming Interfaces), to make it easier for data to flow between systems [26]. Expertise in system integration is necessary to tackle these technological obstacles, and external consultants or IT specialists may be called upon.

2.5 Empirical Studies and Case Analyses

Empirical research and case studies are valuable resources in the field of systems integration since they offer valuable insights into the real-world consequences and outcomes of integrating accounting, human resource management, and enterprise resource planning [27].

A security and safety company's accounting, human resource management, and enterprise resource planning systems are strategically integrated [28]. These studies will offer instances from the actual world as well as insights into the advantages, difficulties, and best practices of integrating various business systems. Investigations include the following:

Research 1: "The effect of integrated accounting, human resources management, and enterprise resource planning systems on operational efficiency in a security and safety company" Goal: In a security and safety company, this study is to investigate how integrating accounting, human resources management, and ERP systems affects operational efficiency metrics including process cycle time, resource utilization, and error rates [29].

Research 2: "Employee engagement and satisfaction within the framework of integrated accounting, HR, and enterprise resource planning." The objective of this study is to investigate how integrated systems affect worker happiness and productivity in a security and safety organization [30].

Research 3: "Integrated ERP, HRM, and accounting system financial management and decision-making outcomes: a comparative study of security and safety companies." The purpose of this study is to evaluate how security and safety organizations with integrated ERP and management systems compare in terms of financial management practices and decision-making outcomes [31, 32]. Aspects like financial reporting accuracy, budgeting and forecasting skills, and cost management strategies will be assessed by accounting and human resources (HRM).

2.6 Framework Structured

To direct the strategic integration of ERP, HRM, and accounting systems for a security and safety organization, a formal framework will be built. The framework will offer a step-by-step procedure to guarantee integrated systems are implemented and used successfully as shown in Table 1.

3 Method

Also collected from important research and publications to perform a thorough literature study on the topic of "Integrating for Excellence: A Study on Merging Accounting, HRM, and ERP Systems in Security and Safety Firms for Delta Safety and Security company" in order to write this in-depth review paper [33–36]. Our methodology involves a methodical screening and study of scholarly literature to pinpoint important themes and revelations in the field of corporate system integration [36–41]. We specifically focus on the integration of accounting, HR, and ERP systems. The creation of reputable, empirically supported techniques for enhancing HRM procedures within the framework of digital integration is the result of this analytical approach [42–44].

4 Discussion

The literature analysis highlights the substantial influence that the integration of accounting, human resource management (HRM), and enterprise resource planning systems has on organizational performance and decision-making processes. Numerous studies continuously emphasize the advantages of integrating systems, notably in terms of improved data consistency and operational efficiency [45–47]. In the context of a dynamic and ever-changing business environment, the adoption of integrative methods assumes paramount importance for companies seeking to adjust and prosper.

Table1 Structured framework

Row	Component	Description
1	Needs assessment	Analyze the systems, procedures, and integration need of the security and safety organization in detail. Determine the particular requirements and goals for the accounting, HRM, and ERP system integration
2	System selection	Assess and choose ERP, HRM, and accounting systems that are appropriate and meet the needs and requirements of the organization for integration. Think on elements like vendor support, compatibility, scalability, and functionality
3	Data mapping and integration	To guarantee smooth data flow and integration, map the elements and data structures between the various systems. Formulate an integration plan and put in place the required middleware or data interfaces to enable data sharing
4	System configuration and customization	Assist the company's unique business processes and needs by configuring and customizing the connected systems. This could include designing access controls, configuring reporting structures, and setting up workflows
5	Employee data management	Integrate ERP and HRM systems with a centralized employee database. Ascertain the accuracy and timeliness of personnel data, including personal information, job descriptions, credentials, and performance feedback
6	Financial integration	To facilitate easy exchange of financial data, integrate the accounting system with the ERP and HRM tools. Budgeting, forecasting, financial reporting, and financial transactions are all included in this
7	Reporting and analytics	Create extensive analytics and reporting tools to extract valuable insights from the combined data. Use customizable reports, dashboards, and key performance indicators (KPIs) to assist decision-making at different levels
8	Change management and training	Implement a change management strategy to help the integrated systems' user uptake and a smooth transition. Employees should receive assistance and training to guarantee they comprehend and are able to use the new integrated processes and functions

(continued)

Table1 (continued)

Row	Component	Description
9	Continuous improvement and maintenance	Provide systems to handle upgrades, continual improvement, and continuing maintenance. Utilize the integrated systems to their full potential by routinely evaluating their efficacy and performance and making the required modifications
10	Performance evaluation	Assess the integrated systems' performance and results on a regular basis in comparison to predetermined goals and criteria. To improve the integration process, pinpoint problem areas and take corrective action

Comparative research has provided insights into the concrete enhancements in efficiency of resource management that arise from the integration of accounting, human resource management (HRM), and enterprise resource planning systems [48, 49]. These studies demonstrate the impact of integrated systems on enhancing the effectiveness and efficiency of resource consumption inside organizations. One notable benefit that has been observed is the smooth and uninterrupted transfer of information throughout different departments, resulting in enhanced precision of data, less duplication, and accelerated decision-making processes. Nevertheless, the process of integrating these intricate systems is not without of its problems. The existing body of research provides a detailed analysis of the challenges that arise throughout the integration process, with a particular focus on the significance of implementing complete change management techniques [50]. The implementation of efficient change management strategies is crucial in order to overcome these obstacles and guarantee a seamless transition across all levels within the business. The process encompasses not alone technological modifications, but also cultural and procedural transformations inside the organizational framework.

Furthermore, comparative studies serve to emphasize the diverse experiences encountered by various organizations throughout the implementation of these reforms. While several firms have reported enhanced operational efficiency and better financial reporting, others have substantial challenges in connecting disparate systems. The aforementioned disparities highlight the need for a customized methodology that considers the distinct attributes and requirements of every firm.

5 Future Research Directions

Future research in the field of integrated accounting, human resource management (HRM), and enterprise resource planning systems should prioritize the evaluation of the enduring effects of these integrations, with specific emphasis on operational efficiency and decision-making within security and safety organizations. It is important to investigate the impact of developing technologies such as artificial intelligence (AI) on the improvement of these systems. Additionally, comprehending the intricacies of corporate culture and change management throughout the integration process has significant importance. Furthermore, a thorough analysis of the worldwide issues, concerns about data security, and the financial ramifications associated with these integrations will provide extensive knowledge essential for formulating efficient and enduring integration strategies inside various organizational environments.

6 Conclusion

The strategic integration of Enterprise Resource Planning, Human Resource Management (HRM), and accounting systems plays a crucial role in improving operational efficiency and facilitating decision-making within security and safety businesses. The integration, as examined in scholarly sources, guarantees the coherence of data, optimizes operational procedures, and enables well-informed decision-making for strategic planning. Effective implementation of organizational change requires addressing obstacles such as data synchronization and system compatibility. However, it is crucial to emphasize the significance of change management and cross-departmental cooperation in order to achieve successful outcomes and enhance overall organizational performance.

References

1. Kahila, M.: The benefits and challenges of integrating ERP and business intelligence (2023)
2. Abi Anwar, U.A., Rahayu, A., Wibowo, L.A.: Operation process integration using supplier selection, supply chain agility and logistic management to optimize corporate performance based on enterprise resource strategic. In: 7th Global Conference on Business, Management, and Entrepreneurship (GCBME 2022). Atlantis Press (2023)
3. Mitra, S., Lakshmi, D., Govindaraj, V.: Data analysis and machine learning in AI-assisted special education for students with exceptional needs. In: AI-Assisted Special Education for Students with Exceptional Needs, pp. 67–109. IGI Global (2023)
4. Weerasekara, U., Gooneratne, T.: Enterprise resource planning (ERP) system implementation in a manufacturing firm: rationales, benefits, challenges and management accounting ramifications. Account. Manage. Inf. Syst. **22**(1), 86–110 (2023)
5. Omino, J.A.: Enterprise Resource Planning and Business Operations of Small and Medium Enterprises in Nairobi-Kenya. University of Nairobi (2023)
6. Shen, L., et al.: Development of an integrated and comprehensive clinical trial process management system. BMC Med. Inform. Decis. Mak. **23**(1), 61 (2023)
7. Watson III, E.F., Schwarz, A.H.: Enterprise and business process automation. In: Springer Handbook of Automation, pp. 1385–1400. Springer (2023)
8. Mahdani, R., Putri, C.W.A., Risnafitri, H.: The potential of blockchain to increase the effectiveness of management accounting: a systematic literature review. Indatu J. Manage. Account. **1**(1), 1–11 (2023)

9. Nkengfack Fialefack, J.: The impact of ERP systems on business processes: how the implementation of an ERP system automates business processes, improves efficiency, productivity, and profitability (2023)
10. Yang, J.R.: Discovering Organizational Models from Event Logs for Workforce Analytics. Queensland University of Technology (2023)
11. Shaffer, K.M., et al.: Online sleep diaries: considerations for system development and recommendations for data management. Sleep **46**(10), p. zsad199 (2023)
12. Allioui, H., Mourdi, Y.: Exploring the full potentials of IoT for better financial growth and stability: a comprehensive survey. Sensors **23**(19), 8015 (2023)
13. Al-Jumaili, A.H.A., et al.: Big data analytics using cloud computing based frameworks for power management systems: status, constraints, and future recommendations. Sensors **23**(6), 2952 (2023)
14. Beierle, J., et al.: Structured content and data management—enhancing acceleration in drug development through efficiency in data exchange. AAPS Open **9**(1), 11 (2023)
15. Benitez, G.B., Ghezzi, A., Frank, A.G.: When technologies become Industry 4.0 platforms: defining the role of digital technologies through a boundary-spanning perspective. Int. J. Prod. Econ. **260**, 108858 (2023)
16. Olorunlana, T.J., Using Management Information Systems (MIS) to Improve Business Processes
17. Anderson, D., Anderson, L.A.: Beyond Change Management: How to Achieve Breakthrough Results Through Conscious Change Leadership, vol. 36. Wiley (2010)
18. Gelinas, U.J., Dull, R.B., Wheeler, P.: Accounting Information Systems. Cengage AU (2018)
19. Malik, A.: Human Resource Management and the Global Financial Crisis: Evidence from India's IT/BPO Industry. Taylor & Francis (2017)
20. Grabski, S., Leech, S., Sangster, A.: Management Accounting in Enterprise Resource Planning Systems. Butterworth-Heinemann (2009)
21. Sorvari, I.: Digitalisation of payroll processes at Aalto University (2016)
22. DC, D.O.D.W.: Department of Defense Finance and Accounting Service FY 1996/1997 Biennial Budget Estimates, February 1995
23. Hayashi, T., Ishimura, G., Ohsawa, Y.: Structural characteristics of stakeholder relationships and value chain network in data exchange ecosystem. IEEE Access **9**, 52266–52276 (2021)
24. Palanisamy, R.: Organizational culture and knowledge management in ERP implementation: an empirical study. J. Comput. Inf. Syst. **48**(2), 100–120 (2008)
25. Soylu, A., et al.: Data quality barriers for transparency in public procurement. Information **13**(2), 99 (2022)
26. Vermesan, O., et al.: Internet of things cognitive transformation technology research trends and applications. In: Cognitive Hyperconnected Digital Transformation, p. 17–95 (2022)
27. Budhwar, P., et al.: Human resource management in the age of generative artificial intelligence: perspectives and research directions on ChatGPT. Hum. Resour. Manag. J. **33**(3), 606–659 (2023)
28. Su, H.: Application of big data and internet of things technology in the management of financial operating income of electric power enterprises. Int. Trans. Electr. Energy Syst. **2022** (2022)
29. Abobakr, M.A., Abdel-Kader, M., Elbayoumi, A.F.: Integrating S-ERP systems and lean manufacturing practices to improve sustainability performance: an institutional theory perspective. J. Account. Emerg. Econ. **13**(5), 870–897 (2023)
30. Mascarenhas, C., Galvão, A.R., Marques, C.S.: How perceived organizational support, identification with organization and work engagement influence job satisfaction: a gender-based perspective. Adm. Sci. **12**(2), 66 (2022)
31. Zaidan, A.S., Khaw, K.W., Alnoor, A.: The influence of crisis management, risktaking, and innovation in sustainability practices: empirical evidence from Iraq. Interdiscip. J. Inf. Knowl. Manage. **17** (2022)
32. Teixeira, M.T.F.F.G.: Integrated Business Planning: Development of a Framework for Technological Improvement (2022)

33. Ali, A.A.A., et al.: Moderating effect of strategic planning on the relationship between career path planning and job performance. Sustainability **15**(11), 8490 (2023)
34. Al-refaei, A.A.-A., et al.: An integrated mediating and moderating model to improve service quality through job involvement, job satisfaction, and organizational commitment. Sustainability **15**(10), 7978 (2023)
35. Alzoraiki, M., et al.: Impact of teachers' commitment to the relationship between transformational leadership and sustainable teaching performance. Sustainability **15**(5), 4620 (2023)
36. Ali, A.A.A.: The Relationship Between Islamic Work Ethics and Employee Performance among Medical Staff in the Public Health Sector in Bahrain: Employee Commitment as Mediating Variable. Universiti Sains Islam Malaysia (2022)
37. Ali, A.A.A., Alzoraiki, M., Al Aghbas, N.S.M.: Working remotely, social isolation and access to support and communication during Covid-19 its influence on job performance. Int. J. Intell. Hum. Resource Manage. (IJIHRM) **3**(01), 23–33 (2022)
38. Ali, A.A.A., et al.: Association between Islamic Work Ethics and Employee Commitment among medical staff in the public health sector in Bahrain. Int. J. Intell. Hum. Resource Manage. (IJIHRM) (2023)
39. Ateeq, A.: Knowledge Management's Effect on Learning Quality in Bahrain's Private Universities (2023)
40. Alkadash, T.M., et al.: The effects of leadership style on employee sustainable behaviour: a theoretical perspective. In: Artificial Intelligence and Transforming Digital Marketing, pp. 205–213. Springer (2023)
41. Barakat, B., et al.: Assessing the impact of green training on sustainable business advantage: exploring the mediating role of green supply chain practices. Sustainability **15**(19), 14144 (2023)
42. Ateeq, A.: The mediating role of engagement in the relationship between equal opportunities and employee performance. Probl. Perspect. Manag. **21**(3), 435–447 (2023)
43. Al Balushi, F.S.M., Ali, A.A.A., Alzoraiki, M.: the impact of forecasting strategy and training and development strategy on preparing leadership competencies administrative support in the Sultanate of Oman. Int. J. Intell. Hum. Resour. Manage. (IJIHRM) **3**(01), 34–41 (2022)
44. Rahardjo, K.P., Kurniawati, K.: Caroll Pyramid implementation of environmental responsibility programs of PT Pertamina EP Subang Field, Tambun Field, and Jatibarang field in organic agriculture. J. Soc. Res. **2**(9), 3331–3340 (2023)
45. Thuan, P.Q., et al.: The determinants of the usage of accounting information systems toward operational efficiency in industrial revolution 4.0: evidence from an emerging economy. Economies **10**(4), 83 (2022)
46. Adere, E.M.: Blockchain in healthcare and IoT: a systematic literature review. Array **14**, 100139 (2022)
47. Yathiraju, N.: Investigating the use of an artificial intelligence model in an ERP cloud-based system. Int. J. Electr. Electron. Comput. **7**(2), 1–26 (2022)
48. Budhwar, P., et al.: Artificial intelligence—challenges and opportunities for international HRM: a review and research agenda. Int. J. Hum. Resour. Manage. **33**(6), 1065–1097 (2022)
49. Memon, K.R., et al.: Management of knowledge and competence through human resource information system—a structured review. Front. Psychol. **13**, 944276 (2022)
50. Hussein, N.H., et al.: A comprehensive survey on vehicular networking: communications, applications, challenges, and upcoming research directions. IEEE Access **10**, 86127–86180 (2022)

Artificial Intelligent Impact on Accounting Professionals in Bahrain

Zahra Dawood Salman Dawood Salman and Mujeeb Saif Mohsen Al-Absy

Abstract The purpose of this study is to examine the effect of artificial intelligent (AI) on accounting profession in Bahrain. The study uses three dimensions for AI, namely, test automation, anomalous error, predictive and forecasting solutions. For accounting profession, the study used three measurements which are reporting accuracy, data analytics, and transparent reporting. Questionnaire will be distributed to accountants in the sector of Banks and financial institutions in Bahrain. The study expects a significant positive impact of test automation, anomalous error, predictive and forecasting solutions on accounting profession measured by reporting accuracy, data analytics, and transparent reporting. The study has provided insights for developing strategies to enhance accountant's skills through training programs and adapting accounting education accordingly. Additionally, it will help policymakers create measures that allow professionals to benefit from increased productivity brought about by AI while minimizing any job displacement risks.

Keywords Audit firm's size · Competency · Tuner · Audit fees · Audit quality

1 Introduction

The usage of technology in everyday life has become indispensable [1, 2]. Artificial intelligent (AI) technology is making an impact on the accounting field in Bahrain bringing about changes in processes and responsibilities. These transformations are driven by the capabilities of AI, including machine learning and data analysis which enable automation, accuracy and efficiency in accounting tasks. Artificial intelligence is the study and creation of intelligent systems that operate on their surroundings to carry out certain tasks and increase the likelihood that they will succeed [3].

Z. D. S. D. Salman · M. S. M. Al-Absy (✉)
Accounting and Financial Science Department, College of Administrative and Financial Science, Gulf University, Sanad 26489, Kingdom of Bahrain
e-mail: dr.mujeeb.saif@gulfuniversity.edu.bh

Z. D. S. D. Salman
e-mail: 190101114907@gulfuniversity.edu.bh

B. Awwad (ed.), *The AI Revolution: Driving Business Innovation and Research*, Studies in Systems, Decision and Control 524, https://doi.org/10.1007/978-3-031-54379-1_11

Integrating AI into accounting has resulted in enhanced precision and effectiveness in auditing procedures [4].

Accountants and auditors now have access to AI powered tools that automate data analysis, identify anomalies and patterns and detect activities efficiently. As a result, the audit process has become more streamlined and dependable equipping auditors with the resources to deliver financial information that aids decision making. Consequently, Accountants and auditors' roles have shifted from data analysis to overseeing and interpreting results generated by AI algorithms.

Furthermore, AI has also brought about the emergence of skills, tasks and roles within the accounting professionals. Accountants now need an understanding of AI technologies and their application in analysis and reporting. They need to adapt to using AI powered tools and platforms for data analysis and reporting purposes [5]. Additionally, certain mundane tasks, like data entry and reconciliations may be potentially replaced by AI technology. The changes occurring in the accounting industry could result in some finance professionals facing job losses. However, the influence of AI on accounting careers in Bahrain is substantial.

Nowadays, these financial systems are being improved to incorporate the capabilities of AI to make accounting precise and free from errors. This enables them to conduct real time data analysis make analytics and generate automated reports. These advancements have implications for the field of auditing [6]. AI can serve as a tool in supporting accounting management automating control mechanisms and enhancing decision making processes by producing accounting and performance information, especially in light of the global financial scandals that have had a significant influence on the world's economies [7–12].

Figure 1 highlights the increasing revenue over the years from AI integration in business. Using AI technologies to provide intelligence services is highest and it is predicted to increase more in the future. The integration of AI into accounting has resulted in increased accuracy and efficiency in auditing procedures. Auditors can now leverage AI powered tools to automate data analysis identify anomalies and patterns and detect activities effectively.

The accounting landscape in Bahrain is going through a transformation as Artificial Intelligence (AI) becomes increasingly integrated into financial processes. This shift not only reflects trends but also has specific implications for the accounting

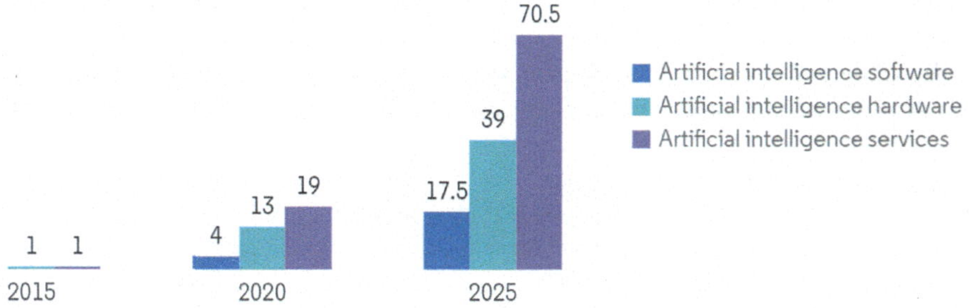

Fig. 1 Revenue from AI integration in business [13]

sector in the Kingdom of Bahrain [14]. As AI technologies continue to advance, they are having an impact on the accounting professionals leading to questions about how Bahrain's accounting firms are adapting and what challenges and opportunities they face. In years Bahrain has seen a rise in the use of AI across industries, including accounting. The availability of AI-powered tools for analysis, data entry and decision-making have the potential to bring changes to traditional accounting practices. According to a survey conducted in 2023 with 771 tax professionals, 73% recognized that AI can be utilized for tax, accounting and audit work (thomsonreuters.com). However, there is still exploration needed to understand the extent and nature of this impact on Bahrain's accounting industry.

There are indications that firms might be facing difficulties as they navigate the incorporation of AI, into their accounting procedures. According to reports, it is projected that the implementation of AI technology in Bahrain could contribute to an 8.2% growth in its GDP by 2030 [15]. However, this outcome relies heavily on factors such as the availability and quality of data as the establishment of regulatory and ethical frameworks. Additionally, it is important to investigate the effects of AI technologies on the accounting workforce in Bahrain.

In today's era, businesses are compelled to understand how AI impacts the landscape in order to maintain competitiveness. The integration of AI within accounting functions has the potential to boost productivity minimize errors and offer insights to decision-makers [16]. Previous studies have raised concerns among regulators regarding characteristics associated with AI technology, including the power and speed complexity of models and data-related challenges [14]. These concerns revolve around harm to consumers, risks for market integrity implications for firms' safety and soundness and broader financial stability. On the other hand, there are challenges related to data security, job displacement concerns and ethical considerations surrounding AI implementation, within the accounting field. To address these matters comprehensively this study aims to explore how AI impacts Bahrain's accounting sector. This study aims to gain an understanding of how AI is influencing and transforming accounting practices by examining real-life examples from Bahraini accounting firms.

The study examining the effects of intelligence on accounting professions in Bahrain holds significance for several reasons. Firstly, it has the potential to offer insights into how AI technologies are being integrated into the accounting field in Bahrain as well as the advantages and challenges associated with their implementation. This research will also shed light on the skills, tasks and roles that AI introduces to accounting professionals in Bahrain enabling them to adapt and remain relevant in an evolving technological landscape.

Additionally, comprehending the impact of AI on job displacement and job security for accountants in Bahrain is crucial for professionals, within this field to adequately prepare themselves for industry changes [17]. Furthermore, this study aims to offer insights to policymakers and stakeholders within the accounting industry enabling them to facilitate the integration of AI and ensure a transition for professionals. This research seeks to enhance our knowledge of how AI impacts accounting professions in Bahrain.

The study conducted in Bahrain due to regulatory changes in Bahrain which have had an impact on financial reporting standards. As part of an effort to adhere to practices Bahrain implemented IFRS issued by the "International Accounting Standards Board" for all listed companies. The government passed the Commercial Companies Law, which now requires all companies including those that are not publicly traded to prepare statements following IFRS guidelines [18]. This significant shift expanded the scope of compliance with IFRS beyond traded companies. However, these regulatory updates demonstrate Bahrain's commitment to aligning with recognized accounting standards and enhancing corporate financial transparency and governance.

Accounting experts such as auditors, accountants and financial advisors play a role in Bahrain's system. Bahrain has experienced growth in the accounting profession [19]. This is due to the increasing demand for accountants and auditors who can support the country's status as a financial hub. These professionals are responsible for tasks, including preparing and certifying financial reports for companies conducting independent audits offering guidance on accounting and tax matters to businesses and upholding ethical standards.

To ensure the quality of accounting services professional, bodies like the Bahrain Accountants Association collaborate with firms to provide training programs monitor practices as well as enforce ethical codes [15]. However, these accounting professionals have an impact on accounting practices through their roles in financial reporting, assurance services, advisory work and compliance, within both individual companies and the broader capital markets.

2 Literature Review

In today's changing world the importance of weather forecasts cannot be emphasized enough. Artificial intelligence (AI) holds potential to bring benefits to the accounting professionals in Bahrain. By automating tasks like data entry and reconciling AI can boost efficiency and productivity for accountants. Moreover, AI systems can analyze amounts of data and uncover patterns that might not be easily noticeable by humans. This can lead to financial analysis and decision making. Additionally, AI is driving the development of accounting products and services [20]. For example, AI powered virtual assistants can offer time advice and guidance to clients while AI driven auditing tools can enhance the speed and accuracy of auditing processes.

The impact of intelligence on accounting professions in Bahrain is expected to be substantial as it revolutionizes workflows, enhances accuracy and efficiency and creates opportunities for accountants to provide value through advisory roles. Based on the provided sources we anticipate an impact of intelligence on accounting professions in Bahrain. However, Alastal et al. [21] stated that AI technologies such, as Big Data, the Internet of Things and Blockchain have already had an influence on the accounting landscape and the evolution of accounting information systems within organizations. These technological advancements have also made an impact on the field of auditing by equipping auditors with tools and procedures that enhance the precision and efficiency of the accounting and auditing process.

The connection between the rate at which automation tools and scripts are tested has increased the rate at which AI is being adopted and accounting professionals is a one that has implications. As the use of AI in accounting continues to increase the roles and responsibilities of accountants undergo changes. A high rate of AI automation tools and scripts often requires accountants to adapt their skill sets [22]. Tasks that are repetitive in nature are often automated allowing professionals to focus on activities such as data analysis, interpretation and strategic decision making. This transition can contribute to job satisfaction and overall productivity.

On the contrary, limitations in test automation of AI tools often results in a gradual adjustment for accounting professionals potentially prolonging the coexistence of manual and automated processes. However, it is also affecting competitiveness in an evolving landscape. Consequently, there is a relationship between the rate at which AI test automation and accounting professionals having an impact on their roles, skills and overall contributions [23]. Introducing firms to AI technologies will promote automation in workflow and will improve productivity of accounting professionals who are able to benefit from using AI in their work.

In terms of Anomalous errors, the emergence of AI systems in accounting workflows has led to a new class of anomalous errors that impact the day-to-day work of accounting professionals. These errors, while often subtle, can severely undermine the integrity of financial data if left undetected. Detecting and remediating these errors requires diligent auditing of AI outputs by accountants to identify patterns of inaccuracy [18]. However, the complexity of many AI systems makes tracing the root causes of errors difficult for accountants, as they may not have full visibility into the algorithms powering the technology. This compounds the lack of trust and accountability that arises from unexplained anomalous outputs.

As a result, accounting personnel take on additional responsibilities to validate data, document AI errors, and implement corrective measures to improve system performance. The increasing prevalence of AI means accountants must develop new skills in advanced data auditing, forensic accounting, and IT governance to properly oversee these emerging technologies and their as-yet-unknown failure modes. Anomalous outputs present numerous challenges that redefine the accounting profession's duty to ensure the reliability of financial data.

As AI systems continue to advance accountants need to adapt their skillsets accordingly. Basic AI technology can now automate accounting tasks such as data entry and report generation. This allows accountants to focus more on strategic work including predictive and forecasting solutions. However, with the improvements in algorithms, natural language processing and machine learning a wider range of accounting roles could be. Enhanced or even replaced. Predictive and forecasting solutions will always remain crucial for competing on technical skills against AI systems [4]. Accounting professionals should see them as an opportunity to transition into more fulfilling advisory roles. By maintaining expertise in accounting principles while also developing the ability to effectively utilize AI technologies accountants can ensure their profession thrives alongside these technologies.

3 Hypotheses Development

This framework explores the impact of integrating AI on accounting profession in Bahrain. It considers factors such as test automation, anomalous error, predictive and forecasting solutions. In terms of accounting profession, three measurements used, namely, reporting accuracy, data analytics as well as transparent reporting. The focus is on how this affects accounting professionals. As AI technology advances industries subject to regulations may experience automation compared to sectors that prioritize services. Additionally larger firms implement AI solutions compared to smaller businesses. However, it is important to note that human skills like judgment, relationships and oversight will continue to play a role across all contexts.

The current study used two model to support the conceptual framework. The Technology Acceptance Model (TAM) which suggests that the integration and success of an innovation depend on how users perceive its usefulness and ease of use. When applying TAM to the field of accounting, in Bahrain it helps to understand how accountants view the usefulness and user friendliness of AI tools [24]. This theory can help in assessing the likelihood of implementing AI technologies in Bahrain's accounting sector by understanding accountants' willingness to embrace AI driven changes including test automation, anomalous error, predictive and forecasting solutions, reporting accuracy, data analytics as well as transparent reporting respectively and how practical they find AI systems in their work.

The other model is the Resource Based View (RBV) model which states that firms can gain an advantage by leveraging their valuable and non-substitutable resources. In relation to AIs impact on accounting in Bahrain RBV can be used to explore how AI technologies serve as resources for accounting firms [25]. Accordingly, the study conceptual model is presented in Fig. 2.

Based on the previous studies and Model used in the current study, the following hypothesis has been developed upon analysing the impact of AI in accounting professionals.

H_1: There is a significant relationship between test automation and accounting professionals.

H_2: There is a significant relationship between anomalous errors and accounting professionals.

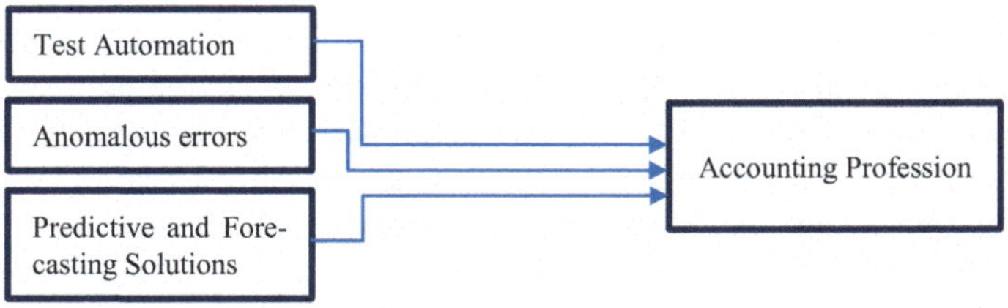

Fig. 2 Conceptual model

H_3: There is a significant relationship between predictive and forecasting solutions and accounting professionals.

4 Methodology

The study used a questionnaire to collect date from respondents. The selection of population and sample size is significant aspect of a research study. In the context of the study, survey has been conducted to collect primary data from respondents particularly from banks and financial institutions. Participants for the survey were employees from banks and financial institutions in Bahrain mostly from the accounting department. The survey has been done with the help of questionnaire which was sent to the participants using google forms through social media platforms. This helped in collecting response from participants in a cost-effective manner. Besides, it also saved time which helped in attending to other crucial matters of the study. The sample size has been selected using random sampling method which helped in conducting the survey without any bias. This also helped in maintaining the accuracy and reliability of the data.

5 Conclusion

AI has the potential to boost accountants' productivity and accuracy. Therefore, the influence of intelligence on accounting professions in Bahrain is significant as it offers opportunities for improved efficiency and accuracy. In short, the influence of AI on accounting careers in Bahrain is substantial. It brings about improvements in auditing accuracy and efficiency enhances the productivity and precision of accountants and may present challenges to job stability. Hence, the purpose of this study is to examine the effect of artificial intelligent (AI), namely the test automation, anomalous error, predictive and forecasting solutions, on accounting profession in Bahrain measured by reporting accuracy, data analytics, and transparent reporting. Quantitative will be distribute to accountants in the sector of Banks and financial institutions in Bahrain. The study expects a significant positive impact of test automation, anomalous error, predictive and forecasting solutions on accounting profession, reporting accuracy, data analytics, and transparent reporting. The research has provided insights for establishing ways for improving accountant abilities through training programs and changing accounting education accordingly. It will also assist governments in developing strategies that would allow professionals to benefit from enhanced productivity brought about by AI while limiting job displacement dangers.

References

1. Al-Absy, M.S.M.: Effects of COVID-19 pandemic on accounting students' capability to use technology. Int. J. Learn. Teach. Educ. Res. **22**(1), 247–267 (2023)
2. Al-Absy, M.S.M., Liandu, N.D.: The impact of corona pandemic on the students widespread use of technology: a case study of Gulf University. Chart **1200**, 1400
3. Mathew, D., Brintha, N., Jappes, J.W.: Artificial intelligence powered automation for industry 4.0. In: New Horizons for Industry 4.0 in Modern Business, pp. 1–28. Springer (2023)
4. Alareeni, B., Hamdan, A.: Impact of artificial intelligence, and the fourth industrial revolution on business success. In: Proceedings of the International Conference on Business and Technology (ICBT 2021), p. 485 (2022)
5. Al-Ammal, H., Aljawder, M.: Strategy for artificial intelligence in Bahrain: challenges and opportunities. In: Artificial Intelligence in the Gulf, pp. 47–67 (2021)
6. Abdulameer, M., Mansoor, M.M., Alchuban, M., Rashed, A., Al-Showaikh, F.: The impact of artificial intelligence (AI) on the development of accounting and auditing profession. In: Technologies, Artificial Intelligence and the Future of Learning Post-COVID-19, pp. 201–213 (2022)
7. Al-Absy, M.S.M.: Impactful women directors and earnings management. Cogent Bus. Manag. **9**(1), 2148873 (2022)
8. Al-Absy, M., Ku Ismail, K.N.I., Chandren, S.: Board chairmen's involvement in audit committees and earnings management practices (2017)
9. Al-Absy, M., Hasan, M.: Impact of the board of directors' characteristics on firm performance: a case of Bahraini listed firms (2023)
10. Al-Absy, M.S.M.: Board chairman characteristics and real earnings management. Sustainability **14**(22), 15025 (2022)
11. Al-Absy, M.S.M.: Impactful female directors and earnings management: the moderating effect of ownership concentration. Adm. Sci. **13**(5), 129 (2023)
12. Al-Absy, M.S.M., AlMahari, N.H.: The interaction effect of nomination committee's effectiveness on board of directors' characteristics and firm performance. Adm. Sci. **13**(5), 135 (2023)
13. AbuRaya, R., Abdulla, Y., Kumaraswamy, S., Ebrahim, R.: FinTech global outlook and the Bahraini landscape: empirical exploratory analysis and documentary evidence. In: International Conference on Decision Aid Sciences and Application (DASA), pp. 1007–1015 (2021)
14. Holmes, F.F., Douglass, F.: Artificial intelligence: reshaping the accounting profession and the disruption to accounting education. J. Emerg. Technol. Account. **19**(1), 53–68 (2022)
15. PwC: Which regions will gain the most from AI? In: PwC Middle East (2023)
16. Dawadi, S., Shrestha, S., Giri, R.A.: Mixed-methods research: a discussion on its types, challenges, and criticisms. J. Pract. Stud. Educ. **2**(2), 25–36 (2021)
17. Khaifa, K.M., Hamdan, A., Alareeni, B.: Artificial intelligence and entrepreneurship in Bahrain. In: The Importance of New Technologies and Entrepreneurship in Business Development: In the Context of Economic Diversity in Developing Countries: The Impact of New Technologies and Entrepreneurship on Business Development, pp. 789–800 (2021)
18. Masruki, R., Khairulannuar, H.H., Dhar, B.K.: Shariah accountability practice of Malaysian foreign-owned Islamic banks and Bahrain Islamic banks. Int. J. Adv. Sci. Technol. **29**(4), 5768–5782 (2020)
19. Elbayoumi, A.F., Awadallah, E.A., Basuony, M.A.K.: Development of accounting and auditing in Egypt: origin, growth, practice and influential factors. J. Dev. Areas **53**(2) (2019)
20. Khalifa, K., AL-Hashimi, M., Hamdan, A.: The factors affecting the adoption of artificial intelligence technologies in organizations. In: Technological Sustainability and Business Competitive Advantage, pp. 237–265 (2023)
21. Alastal, A.Y.M., Salman, M.A.H., Allaymoun, M.H.: Accounting students' perceptions on E-learning during COVID-19 pandemic: case study of accounting and financial students in Gulf University—Bahrain. In: Artificial Intelligence and Transforming Digital Marketing, pp. 879–890 (2023)

22. Foud, A.A., Zowayed, S.I., Showaiter, D.A., Khder, M.A., Ali, B.J.A.: Artificial intelligence's potential on Bahrain's labour market. In: 2022 ASU International Conference in Emerging Technologies for Sustainability and Intelligent Systems (ICETSIS), pp. 430–436 (2022)
23. Seyadi, A.E., Al-Zayani, F.J., Shehab, S., Hamdan, A., Alhor, R.H.: The implementation of artificial intelligence in the public sector: opportunities and challenges. In: Applications of Artificial Intelligence in Business, Education and Healthcare, pp. 271–284 (2021)
24. Granić, A., Marangunić, M.: Technology acceptance model in educational context: a systematic literature review. Br. J. Edu. Technol. **50**(5), 2572–2593 (2019)
25. Dubey, R., Gunasekaran, A., Childe, S.J., Blome, C., Papadopoulos, T.: Big data and predictive analytics and manufacturing performance: integrating institutional theory, resource-based view and big data culture. Br. J. Manag. **30**(2), 341–361 (2019)

Corporate Social Responsibility: A Multidimensional Approach to Sustainable Growth and Community Engagement—The Case of Almarai Company

Marwan Milhem, Ali Ateeq, Ranyia Ali, and Mohammed Alzoraiki

Abstract This paper provides a comprehensive analysis of the Corporate Social Responsibility (CSR) efforts implemented by Almarai Company, a prominent player in the dairy business. The study also conducts a comparative examination of these initiatives in relation to prevailing practices within the industry at large. Utilizing a mixed-methods methodology, this study evaluates the effects of Almarai's corporate social responsibility (CSR) practices on both sustainable growth and community participation. The findings of the research demonstrate that Almarai's endeavors in environmental sustainability, community involvement, employee welfare, and ethical supply chain management are in accordance with worldwide trends in corporate social responsibility (CSR), while also displaying a notable emphasis on the regional context. The results underscore the efficacy of Almarai's corporate social responsibility initiatives in bolstering brand reputation, fostering stakeholder confidence, and promoting sustainable company expansion. This paper presents strategic suggestions for Almarai and other firms operating in the dairy sector, highlighting the need of adopting localized, creative, and stakeholder-focused corporate social responsibility (CSR) initiatives. Furthermore, it delineates prospective avenues for future study to delve further into the ramifications of corporate social responsibility (CSR) within this particular industry.

M. Milhem · A. Ateeq (✉) · M. Alzoraiki
Administrative Science Department, College of Administrative and Financial Science, Gulf University, Sanad 26489, Kingdom of Bahrain
e-mail: dr.ali.ateeq@gulfuniversity.edu.bh

M. Milhem
e-mail: dr.marwan.milhem@gulfuniversity.edu.bh

M. Alzoraiki
e-mail: dr.mohammed.saied@gulfuniversity.edu.bh

R. Ali
College of Information Technology, University of Bahrain, Isa Town, Kingdom of Bahrain
e-mail: 202208501@stu.uob.edu.bh

B. Awwad (ed.), *The AI Revolution: Driving Business Innovation and Research*, Studies in Systems, Decision and Control 524, https://doi.org/10.1007/978-3-031-54379-1_12

Keywords Corporate social responsibility (CSR) · Dairy industry · Sustainable growth · Community engagement · Almarai Company · Environmental sustainability · Employee welfare · Ethical supply chain management

1 Introduction

In contemporary times, there has been a notable evolution in management methods, with a distinct emphasis on improving organizational performance in order to facilitate development and enhance adaptation in dynamic and fast evolving contexts. One prevalent management trend in the modern era is the prioritization on corporate social responsibility (CSR) [1, 2]. The aforementioned transition is propelled by the growing public expectations for not just the provision of goods, but also for active engagement in fostering social and economic progress, as well as safeguarding the environment [3, 4]. Corporate Social Responsibility (CSR) is predicated on the fundamental objective of attaining a state of concordance among all parties involved with the organization, including the interests of society at large.

A notable instance is the Saudi Almarai Company, which has a prominent position in the dairy, food, and juice industries inside the Kingdom of Saudi Arabia. In recent years, Almarai has shown favorable financial results, which may be partly ascribed to its strategic integration of corporate social responsibility (CSR) components. This strategy encompasses not only the synchronization of its corporate activities with societal demands but also the active engagement in community development, with a conscientious consideration of environmental consequences. The corporate social responsibility (CSR) strategy used by the corporation exemplifies the successful integration of sustainable and socially responsible practices into its business operations [5–7]. This serves as a noteworthy example, demonstrating the potential for achieving both corporate success and social welfare via a synergistic approach.

2 Literature Review

2.1 Corporate Social Responsibility and Sustainable Growth

The adoption of Corporate Social Responsibility (CSR) has become an integral aspect of modern company strategies, with a special emphasis on promoting sustainable development. Koleva [8] asserts that corporate social responsibility (CSR) programs have a beneficial influence on sustainable development since they effectively handle several elements, including environmental, social, and economic aspects. Koleva [8] highlights the significance of integrating corporate social responsibility (CSR) into the fundamental strategy of firms, as it has been seen that such organizations are more likely to attain sustained profitability and foster trust among their stakeholders.

The perspective expressed is consistent with the findings of Jaaron et al. [9] who discovered a positive relationship between corporate social responsibility (CSR) initiatives and improvements in company reputation and consumer loyalty. These outcomes, in turn, have the potential to foster sustainable business success.

2.2 Contribution to Sustainable Development and Community Engagement Through CSR Initiatives

The significance of corporate social responsibility (CSR) in fostering community participation is an additional essential factor. The study conducted by Zhang et al. [10] emphasizes the positive impact of corporate social responsibility (CSR) initiatives that prioritize community development, since they contribute to the establishment of more robust partnerships between corporations and communities. The research conducted in rural environments provides evidence that corporate social responsibility (CSR) initiatives in the areas of education, health, and infrastructure have a substantial impact on enhancing the overall well-being of individuals, thereby contributing to the development of a favourable corporate reputation [11]. In a research conducted by Atzmanstorfer et al. [12] it was shown that firms that actively participate in urban corporate social responsibility (CSR) programs, such as urban regeneration and local employment schemes, tend to gain more support and participation from the local population. This suggests the existence of a mutually advantageous connection between these companies and the community.

2.3 CSR in the Dairy Industry

In the context of the dairy business, corporate social responsibility (CSR) activities exhibit distinct characteristics. Hernandez et al. [13] argue that the dairy business has unique issues, including concerns related to animal care, environmental sustainability, and fair trade practices. The findings of their research indicate that dairy firms who implement transparent and ethical supply chain procedures, in addition to demonstrating environmental stewardship, may achieve a competitive advantage. Additionally, the comparative study conducted by Kondaridze and Luckstead [14] examines dairy firms in Europe and North America. The study demonstrates that companies that incorporate corporate social responsibility (CSR) into their business models not only improve their market reputation but also make substantial contributions to environmental sustainability. These contributions are achieved through the implementation of innovative practices such as reducing carbon footprints and adopting sustainable packaging methods.

In conclusion, the current body of research strongly supports the idea that corporate social responsibility (CSR) plays a crucial role in fostering sustainable company development and fostering meaningful community involvement. Within the dairy business, the implementation of responsible and sustainable practices is not just a matter of adhering to regulations or ethical considerations, but also serves as a strategic mechanism for achieving long-term success and fostering favourable relationships with stakeholders.

3 Method

The paper titled Corporate Social Responsibility: A Multidimensional Approach to Sustainable Growth and Community Engagement—The Case of Almarai Company is based on a thorough analysis of the literature. The foundational literature and pivotal research [15–18], were thoroughly examined, with a particular emphasis on the Corporate Social Responsibility: A Multidimensional Approach to Sustainable Growth and Community Engagement—The Case of Almarai Company [18–23]. This methodological approach combined empirical research and theoretical ideas to provide evidence-based propositions targeted at promoting ethical Corporate Social Responsibility practices. This broad and methodical approach enabled an academically rigorous investigation of the issue at hand [24, 25].

The conceptual framework refers to a theoretical structure or model that guides the development and understanding of a particular subject or field of study. The study's conceptual framework is grounded on Carroll's Pyramid of Corporate Social Responsibility (CSR), which classifies CSR initiatives into four distinct levels: economic, legal, ethical, and charitable duties [26] (Fig. 1). This framework provides guidance for evaluating Almarai's corporate social responsibility (CSR) programs, analyzing their alignment with several aspects and their engagement with the community.

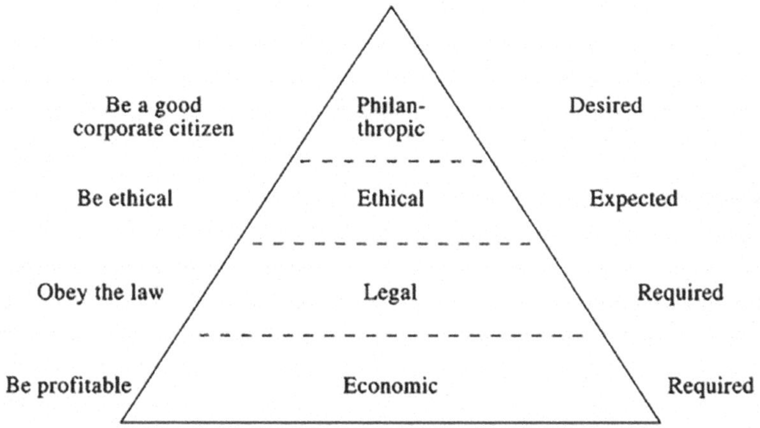

Fig. 1 Pyramid of corporate social responsibility. *Source* Carroll (1991)

4 Discussion

The objective of the following section is to analyze and contrast many scholarly investigations pertaining to Corporate Social Responsibility (CSR) endeavours, with a special emphasis on their implications within the dairy sector. Furthermore, it tries to assess the efficacy of Almarai's CSR tactics.

Comparative Analysis of Corporate Social Responsibility (CSR) Initiatives in the Dairy Industry. Numerous scholarly investigations have examined the various strategies used in the dairy business with regards to corporate social responsibility (CSR), shedding light on the significance of environmental sustainability, community involvement, employee well-being, and ethical management of the supply chain as pivotal domains.

The study conducted by Liu et al. [27] aimed to assess and analyze the initiatives undertaken by different dairy firms in the realm of environmental sustainability. The report emphasized the notable advancements made by Almarai in the domains of water conservation and carbon emissions reduction. The aforementioned approach is reflected in the environmental policy of Danone. Nevertheless, Almarai distinguishes itself via its regional emphasis, namely in tackling issues prevalent in the Middle East such as water shortage. The significance of regional distinctiveness in corporate social responsibility (CSR) programs is underscored by Shen et al. [28] who highlight the value of implementing localized environmental measures to enhance the success of CSR efforts.

The study conducted by [29] examines community involvement within the dairy industry, specifically focusing on the educational and healthcare programs implemented by Almarai and the worldwide community development activities undertaken by Nestlé [30]. The research establishes connections and similarities between these two companies' efforts in engaging with local communities. The research conducted by East and Africa [31] highlights that Almarai's emphasis on localization contributes to a more pronounced and direct influence on the communities within its operating zones. Addressing employee wellbeing, Almarai's approach is in accordance with the established industry norms, as shown by firms such as Arla Foods. The comparative analysis conducted Hogan et al. [32] examines employee welfare efforts in the dairy industry, with a specific focus on Almarai. The study underscores Almarai's dedication to local workforce development as a key element that contributes to socio-economic progress in the area.

The ethical standards pertaining to supply chain management, namely in terms of sourcing and animal welfare, shown by Almarai, might be deemed equivalent to those demonstrated by Fonterra. The significance of supply chain ethics in the dairy business is emphasized in the research conducted by Barakat et al. [23] who highlight Almarai's incorporation of local suppliers as a critical distinguishing factor.

An Assessment of the Efficacy of Almarai's Corporate Social Responsibility Strategies and Their Influence on Sustainable Development. In a recent quantitative analysis conducted by Abdelhalim [33] an examination of the dairy business highlights the positive impact of Almarai's sustainable practices on its market position and customer trust. This observation is consistent with the research conducted by Stent [34] whereby they highlight the interconnectedness between sustainable corporate social responsibility (CSR) policies and the attainment of enduring commercial prosperity within the dairy industry.

According to a study conducted by [33] the localized community involvement initiatives implemented by Almarai have been shown to have a substantial positive impact on the perception of its brand. The success of corporate social responsibility (CSR) is significantly influenced by this crucial element, as shown by the research conducted [35]. They believe that CSR programs that prioritize the needs and interests of the community are more likely to result in increased brand loyalty.

The positive impact of corporate social responsibility (CSR) on employee satisfaction and retention has been substantiated by research conducted by [36]. This correlation is also evident in the practices implemented by Almarai. The organization's emphasis on enhancing staff development aligns with prevailing trends in the sector, whereby the significance of trained labour is progressively recognized as a valued resource.

The study conducted by Khan [37] emphasizes the significance of Almarai's dedication to ethical standards and sustainable manufacturing techniques in establishing a favorable position among environmentally concerned customers, which represents a constantly expanding market niche.

4.1 Implications for Businesses and Policymakers

For businesses, findings indicate that the incorporation of corporate social responsibility (CSR) into fundamental business strategy not only yields ethical advantages but also fosters long-term company growth. It is recommended that companies prioritize the implementation of localized corporate social responsibility (CSR) plans that are specifically designed to align with the unique characteristics and needs of their operating locations.

For policymakers, this research emphasizes the need of implementing rules that promote and facilitate corporate social responsibility (CSR) activities, particularly within industries such as dairy that have substantial environmental and social effects. It is important for policymakers to deliberate on the implementation of incentives aimed at encouraging corporations to demonstrate exemplary corporate social responsibility (CSR) activities.

4.2 Recommendations for Almarai and Similar Companies

Almarai need to improve the level of openness in its sustainability reporting by offering comprehensive and extensive descriptions of its corporate social responsibility (CSR) endeavours and their corresponding effects.

The participation of stakeholders, including workers, consumers, and local communities, is of paramount importance in ensuring ongoing and effective collaboration. This objective may be accomplished by implementing consistent feedback channels and fostering collaborative initiatives. The importance of innovation in corporate social responsibility (CSR) is evident in the case of Almarai. It is recommended that Almarai persist in its pursuit of innovative approaches to CSR, therefore expanding its efforts towards sustainable development. This may be achieved by investigating novel pathways such as renewable energy and enhanced methods for water conservation. The enhancement of Almarai's corporate social responsibility (CSR) effect may be achieved via the expansion and diversification of community involvement activities, which aim to address a wider range of societal concerns.

4.3 Future Research Directions

Longitudinal studies provide a promising avenue for future study, since they enable the assessment of the enduring effects of corporate social responsibility (CSR) activities on both company success and community development.

Comparative studies that include several geographical locations within the dairy sector have the potential to provide more profound insights into the efficacy of localized vs global corporate social responsibility (CSR) methods.

Analyzing the correlation between corporate social responsibility (CSR) programs and customer behaviour, particularly within the framework of growing consumer demands pertaining to sustainability and ethical considerations, might provide advantageous insights.

The study of the impact of new technologies, such as blockchain and artificial intelligence (AI), on the improvement of transparency and efficiency in corporate social responsibility (CSR) projects is a potentially fruitful avenue for scholarly investigation.

5 Conclusion

In conclusion, Almarai's corporate social responsibility (CSR) activities exemplify a noteworthy paradigm within the dairy sector, showcasing the profound influence of well executed CSR strategies on both sustainable expansion and active involvement with the community. Valuable lessons may be derived by businesses and governments from these examples, highlighting the need of implementing CSR techniques that are regional, creative, and stakeholder-focused. Subsequent investigations in this domain has the capacity to further improve and augment corporate social responsibility (CSR) initiatives, so making valuable contributions towards the overarching objectives of sustainability and social responsibility.

References

1. Belesioti, M., Glykas, M.: A novel approach in corporate social responsibility performance measurement. In: LIMEN 2022, p. 71 (2022)
2. Tran, A.T.: The re-consideration of business diplomacy and corporate social responsibility for international business in the post-COVID-19 world (2023)
3. Chishty, S.K.: Economic and environmental efficiencies of organizations: role of technological advancements and circular economy practices. Sustainability **15**(22), 15935 (2023)
4. Zu, L., Wu, D., Lyu, S.: How to move from conflict to opportunity in the not-in-my-backyard dilemma: a case study of the Asuwei waste incineration plant in Beijing. Environ. Impact Assess. Rev. **104**, 107326 (2024)
5. Ye, J., Dela, E.: The effect of green investment and green financing on sustainable business performance of foreign chemical industries operating in Indonesia: the mediating role of corporate social responsibility. Sustainability **15**(14), 11218 (2023)
6. Miller, S.C.: Sustainable trailblazing: a comprehensive analysis of Patagonia's corporate social responsibility initiatives and their ethical implications (2023)
7. Rasyid, A.: The role of corporate social responsibility in implementing company business. INFOKUM **11**(05), 69–78 (2023)
8. Koleva, P.: Towards the development of an empirical model for Islamic corporate social responsibility: evidence from the Middle East. J. Bus. Ethics **171**, 789–813 (2021)
9. Jaaron, A.A., Javaid, M., Garcia, R.F.: Can external and internal corporate social responsibility contribute to green customer behaviour? The mediating role of green trust in hotels. Corp. Soc. Responsib. Environ. Manag. (2023)
10. Zhang, W., Kwon, J.-W., Choy, M.: Localizing corporate social responsibility in China: the role of geographic proximity to political and financial centers. J. Clean. Prod. **421**, 138430 (2023)
11. Soelton, M., et al.: To improve the quality management of children welfare institutions and provide better services. Calitatea **24**(195), 295–302 (2023)
12. Atzmanstorfer, K., et al.: Making Corporate Social Responsibility (CSR) Participatory and Locally Visible—A Prototype Solution (2023)
13. Hernandez, E., Llonch, P., Turner, P.V.: Applied animal ethics in industrial food animal production: exploring the role of the veterinarian. Animals **12**(6), 678 (2022)
14. Kondaridze, M., Luckstead, J.: Determinants of dairy-product trade: do subsidies matter? J. Agric. Econ. (2023)
15. Ali, A.A.A., et al.: Moderating effect of strategic planning on the relationship between career path planning and job performance. Sustainability **15**(11), 8490 (2023)
16. Al-Refaei, A.A.-A., et al.: An integrated mediating and moderating model to improve service quality through job involvement, job satisfaction, and organizational commitment. Sustainability **15**(10), 7978 (2023)
17. Alzoraiki, M., et al.: Impact of teachers' commitment to the relationship between transformational leadership and sustainable teaching performance. Sustainability **15**(5), 4620 (2023)
18. Ali, A.A.A.: The Relationship Between Islamic Work Ethics and Employee Performance Among Medical Staff in the Public Health Sector in Bahrain: Employee Commitment as Mediating Variable. Universiti Sains Islam Malaysia (2022)

19. Ali, A.A.A., Alzoraiki, M., Al Aghbas, N.S.M.: Working remotely, social isolation and access to support and communication during COVID-19 its influence on job performance. Int. J. Intellect. Hum. Resour. Manag. (IJIHRM) **3**(01), 23–33 (2022)
20. Ali, A.A.A., et al.: Association between Islamic work ethics and employee commitment among medical staff in the public health sector in Bahrain. Int. J. Intellect. Hum. Resour. Manag. (IJIHRM) (2023)
21. Ateeq, A.: Knowledge Management's Effect on Learning Quality in Bahrain's Private Universities (2023)
22. Alkadash, T.M., et al.: The effects of leadership style on employee sustainable behaviour: a theoretical perspective. In: Artificial Intelligence and Transforming Digital Marketing, pp. 205–213. Springer (2023)
23. Barakat, B., et al.: Assessing the impact of green training on sustainable business advantage: exploring the mediating role of green supply chain practices. Sustainability **15**(19), 14144 (2023)
24. Ateeq, A.: The mediating role of engagement in the relationship between equal opportunities and employee performance. Probl. Perspect. Manag. **21**(3), 435–447 (2023)
25. Al Balushi, F.S.M., Ali, A.A.A., Alzoraiki, M.: The impact of forecasting strategy and training and development strategy on preparing leadership competencies administrative support in the Sultanate of Oman. Int. J. Intellect. Hum. Resour. Manag. (IJIHRM) **3**(01), 34–41 (2022)
26. Rahardjo, K.P., Kurniawati, K.: Caroll pyramid implementation of environmental responsibility programs of PT Pertamina EP Subang Field, Tambun Field, and Jatibarang Field in organic agriculture. J. Soc. Res. **2**(9), 3331–3340 (2023)
27. Liu, C., Shi, X., Li, C.: Digital technology, factor allocation and environmental efficiency of dairy farms in China: based on carbon emission constraint perspective. Sustainability **15**(21), 15455 (2023)
28. Shen, Q., et al.: How can multinational enterprises effectively implement local consumer-oriented corporate social responsibility (CSR) strategies? A multi-perspective study on the differences in CSR response mechanisms between Chinese and Japanese consumers. Sustainability **15**(21), 15433 (2023)
29. Rudito, B., Famiola, M., Anggahegari, P.: Corporate social responsibility and social capital: journey of community engagement toward community empowerment program in developing country. Sustainability **15**(1), 466 (2022)
30. Ahmad, I., et al.: COVID-19 impact on dairy sector: the mediating role of knowledge sharing and trust on innovation capability. S. Afr. J. Econ. Manag. Sci. **26**(1), 4591 (2023)
31. Wippel, S.: Branding the Middle East: a review of regional manifestations of a global phenomenon. In: Branding the Middle East: Communication Strategies and Image Building from Qom to Casablanca, vol. 38, p. 55 (2023)
32. Hogan, C., et al.: The impact of work organisation on the work life of people on pasture-based dairy farms. Animal **17**(1), 100686 (2023)
33. Abdelhalim, A.M.: How management accounting practices integrate with big data analytics and its impact on corporate sustainability. J. Financ. Report. Account. (2023)
34. Stent, W.: Corporate social responsibility reporting frameworks. In: Encyclopedia of Business and Professional Ethics, pp. 462–465. Springer (2023)
35. Aqabna, S.M., Aga, M., Jabari, H.N.: Firm performance, corporate social responsibility and the impact of earnings management during COVID-19: evidence from MENA region. Sustainability **15**(2), 1485 (2023)
36. Yahya, A.A., Zargar, P.: Achieving corporate sustainability through green human resource management: the role of CSR in the banking industry of a developing country. Sustainability **15**(14), 10834 (2023)
37. Khan, M.R.: Towards Sustainable Ethical Culture in Family Owned Businesses (2019)

The Effect of External Audit in Limiting Creative Accounting Practices

Elaf Mohameed Madmoud and Ahmad Yahia Mustafa Alastal

Abstract This study aims to investigate the effect of external audit on limiting creative accounting practices within organizations. The research employs a quantitative approach, utilizing financial data from a sample of companies across various industries. The findings of this study contribute to the existing literature by providing empirical evidence on the impact of external audit in curbing creative accounting practices. The results shed light on the effectiveness of external audit as a mechanism for ensuring financial reporting integrity and enhancing the reliability of financial statements. Additionally, the study highlights the importance of audit quality and regulatory frameworks in promoting transparency and accountability in financial reporting practices. The implications of this research are significant for various stakeholders, including regulators, policymakers, auditors, and investors. Understanding the role of external audit in limiting creative accounting practices can help policymakers in designing effective regulatory frameworks.

Keywords Creative accounting · Earnings management · Financial manipulation · External audit · Audit quality · Auditor independence · Regulatory environment

1 Introduction

This chapter discusses the research background and problem statement and defines the research objectives and research questions. It also discusses the importance of this study and the scope of the study and procedural definitions.

E. M. Madmoud · A. Y. M. Alastal (✉)
Accounting and Financial Science Department, College of Administrative and Financial Science, Gulf University, Sanad 26489, Kingdom of Bahrain
e-mail: dr.ahmed.alastal@gulfuniversity.edu.bh

E. M. Madmoud
e-mail: 190101114843@gulfuniversity.edu.bh

B. Awwad (ed.), *The AI Revolution: Driving Business Innovation and Research*, Studies in Systems, Decision and Control 524, https://doi.org/10.1007/978-3-031-54379-1_13

2 Background

Auditing is considered an important matter, as the financial transactions auditor examines and verifies the company's financial records. This is to ensure that the company's financial information and data are represented fairly and accurately, and that they are conducted in accordance with accounting standards and basic financial statements. Therefore, there is a need to pay attention to the auditing profession in its internal and external aspects and to emphasize its role in imparting confidence to the financial statements, reducing their manipulation, and revealing cases of using creative methods in presenting the financial statements [1]. Creative accounting practices are among the most important challenges facing auditors and accountants in the field of accounts. What is meant by creative accounting is to beautify the financial statements of companies to improve their image before investors in the stock market. Therefore, executive departments and boards of directors in companies seek to use creative accounting methods, taking advantage of the diversity of alternatives. The accounting principles that can be relied upon in preparing financial statements, which negatively affects the credibility and validity of those statements.

Based on the above problem, the research objectives were determined, which are to identify the concepts and reasons for creative accounting methods for companies listed on the stock market, and to identify the role of auditing.

3 Problem Statement

Due to the circumstances prevailing in the business world, some companies participating in the stock market resort to beautifying their financial statements in an effort to improve their image before investors in the stock market. Therefore, the executive departments and boards of directors in these companies seek to use creative accounting methods, taking advantage of the diversity of alternatives and accounting principles that It can be relied upon in preparing financial statements, which negatively affects the credibility and validity of those data. From here came the research problem, which called for the need to pay attention to the auditing profession in its internal and external aspects, and to emphasize its role in imparting confidence to the financial statements, reducing their manipulation, and revealing cases of using creative methods in presenting the financial statements.

4 Research Objectives

We can summarize the research objectives in the following points:

1. Identify the concepts and reasons for creative accounting.

2. The extent to which external audit is used to detect and prevent creative accounting practices.
3. The impact of creative accounting practices on the auditors' report.

5 Research Questions

We can say that the main question is: To what extent does external audit contribute to limiting creative accounting practices?

The sub-questions are:

1. What is meant by creative accounting and what are its causes?
2. To what extent is external audit used to detect and prevent creative accounting practices?
3. What is the impact of creative accounting practices on the auditors' report?

6 Significance of the Study

The importance of the research stems from the importance of creative accounting and the resulting misleading practices in the financial statements, which led to the collapse of the largest companies. Therefore, external oversight plays an important role in helping the institution accomplish its work and achieve its goals, as the role of external audit affects the auditors' report, by documenting any violations or violations of accounting standards and professional ethics, and providing recommendations to improve accounting practices and develop the accounting system. Auditors' reports are an important source for investors and regulatory authorities to evaluate the validity and accuracy of financial and accounting records, and to make correct financial decisions. Moreover, studying the role of external audit is of great importance to ensure improvement of accounting practices, limit creative accounting practices, improve the reputation of companies and institutions, and increase confidence in accounting systems.

7 Scope of the Study

Auditing is considered one of the advanced sciences in the accounting field, as it is based on the performance of other accountants in various forms of institutions. It aims to ensure that the financial statements have been recorded and classified in accordance with generally accepted accounting principles. In this study, we will be interested in learning about the importance of external auditing in reducing of creative practices.

Due to the spread of this phenomenon, it has become necessary to increase attention to the role of external auditing in order to combat all forms of these practices and degrade society with their danger on the ground.

8 Operational Definitions

Creative accounting: It is a manipulation of the actual numbers present in the company's financial statements and reports. This is done by accountants within the company, whether to achieve personal interests or by order of management, with the aim of obtaining fictitious profits, unreal achievements, and a false financial position.

External audit: It is a process by which an independent auditor examines the financial statements and accounting records with the aim of giving an opinion on the fairness of the financial statements and accounts and their compliance with accounting standards.

Auditing: The main goal of auditing is to make sure that a company's financial statements are accurate and are following regulatory guidelines.

Auditor's report: It is the report that is published in the annual report in conjunction with the financial statements, explaining whether the company's financial statements comply with generally accepted accounting principles.

Financial statements: They are the final and basic outcome of accounting work in any economic unit. They also arise as a result of performing a set of accounting treatments on the data that are related to the events and activities carried out by the economic unit for the purpose of presenting it in a comprehensive and summarized manner to all parties that can benefit from it in making various decisions.

9 Literature Review

Literature review is one of the important methods in scientific research, as it serves to present various literature to some researchers, analysts, and authors. In this context, this review aims to study the role of external audit in limiting creative accounting practices and their impact on the auditors' report. This review includes studying previous research and studies that dealt with this topic and analysing their results and recommendations. The review also includes studying the theoretical framework that includes the dependent variable and the independent variable, as well as analysing current evidence that supports the role of external audit in limiting creative accounting practices.

10 Dependent Variable: Limiting Creative Accounting Practices

Methods Used to Limit Creative Accounting Practices in Financial Statements

Combating creative accounting is considered a complex and difficult matter, so those interested in this field decided to follow and study the work to learn about developments related to creative accounting work, in order to uncover and limit these practices. According to Hind et al. [7] the following is a presentation of the most important methods used to limit creative accounting practices in financial statements:

- Ensuring the conformity of sales invoices to executed transactions, especially those executed by related parties, while verifying the company's status, such as associates and subsidiaries.
- Verifying credit terms, including payment and discount terms, and the adequacy of provisions for doubtful debts.
- Verifying the shipping documents for the sent goods and matching them with the documents for paying the price of the goods received from the agents.
- Verifying the terms of sale and payment terms, whether the buyer or the seller is located, and the method of payment.
- Following the principle of realizing revenue and charging the period with its relevant revenues.
- Verify sales invoices.
- Apply the principle of interviewing and verifying sales invoices.
- Applying the principle of interview and verification of documents, invoices, receipt vouchers, and warehouse exchange vouchers.

Hamad [5] pointed out some rules and methods that would help detect and limit creative accounting practices:

1. Obligating all companies to establish audit committees.
2. Reducing the scope of choosing accounting treatments and alternatives by reducing their accounting number or specifying the circumstances under which each method can be used, and activating the imposition of consistency by financial statement preparers.
3. Reducing the misuse of some accounting policies.
4. Implementing a corporate governance mechanism to limit creative accounting methods and the resulting damage.
5. Addressing the abuse of personal judgment and diligence of accountants by imposing consistency and establishing rules and provisions that reduce the use of personal diligence.
6. Developing the ability of financial analysts to detect and confront creative accounting methods.

11 Independent Variables: External Audit

Definition of External Audit, Its Importance, and Objectives

The need for auditing has emerged as a result of many reasons, such as the rapid development that the world is witnessing in various industrial, commercial and marketing fields, in addition to the large size of institutions, and to meet the increasing needs of users of financial statements regarding published information and data and the extent of their integrity and fairness. External auditing was known as a periodic examination of the company's books and records by an independent party, the auditor, to ensure the validity of those books and records and that they are accurate and compatible with accounting principles and standards, and to provide a true and fair view of the facility's financial position [9]. Hamad [5] also defined: "An organized process carried out by a specialized and independent auditor to verify information and allegations related to financial activities and events and to reconcile their economic results with applicable accounting standards by collecting and evaluating audit evidence (data, certifications, observations, and inquiries), and submitting a report on the results of that the process for relevant parties to use in the decision-making process".

The auditor is the qualified, independent, and authorized person to perform the audit of the financial statements, whether he is a natural person practicing in his own name or for the account of a professional company. He is responsible for the audit process and expressing an opinion on the extent to which it represents the accuracy and fairness of the financial information of the company at the site of the audit [4, 6].

The researcher believes that external auditing is a profession carried out by a person or group of people who have specific qualifications and professional conditions to verify the accuracy and validity of the financial data and reports submitted by establishments and that they are free of errors, fraud and forgery by examining records and documents, and submitting his report with the aim of serving the community and the parties benefiting from those financial Statements.

12 The Importance of External Audit

Many parties who are related to the company's activity depend on the location of the audit on the results of the audit process, and the following is an overview of the importance of the audit for those parties [2, 3]

1. The audit contributes to detecting errors and fraud early, as a result of the psychological impact left by sudden visits to work sites, thus this reduces employees' attempts to commit or engage in any fraudulent act in the future, and this provides a benefit to the company's owners.
2. Detecting weaknesses in the internal control system within the company and thus helping management make sound and appropriate decisions.

3. Auditing provides a benefit to government and state agencies concerned with planning, control and supervision.
4. The audit provides a benefit to different categories of companies and the public that is based on the approved auditor's report on the financial statements of joint-stock companies.

12.1 The Objective of External Audit

The objectives of auditing have evolved from discovering errors or errors and manipulation in the past to determining the veracity of financial information at the present time. Among the objectives that the audit seeks to achieve are according to Velte [10]:

- Verifying the integrity of the accounting information recorded in the project's records and records and determining the extent of reliance on it.
- Providing an impartial technical opinion on the correspondence between the entries in the accounting record and the entries and numbers contained in the financial statements. Detecting errors or fraud in accounting books.
- Reducing the chances of errors and fraud through frequent and unannounced visits by the auditor to the project and strengthening internal control systems.

This applies to the traditional objectives of auditing. As for the modern objectives, they are as follows:

- Monitoring the established plans and working to follow up on their implementation and identifying deviations and their causes.
- Evaluating performance of economic units.
- Achieving the greatest possible level of well-being for most members of society. Achieving maximum production efficiency.
- Verifying the accuracy of the final lists and that they are free of mathematical and technical errors, whether intentional or the result of negligence and negligence. Therefore, we find that the objectives of auditing are no longer limited to detecting fraud and deception, but rather the auditor now monitors the plans drawn up and evaluates performance in a way that contributes to improving production efficiency to the maximum and achieving societal well-being.

13 The Relationship of External Audit on Limiting Creative Accounting Practices

The role that external audit plays in assessing the applicability and quality of the financial and accounting data that firms offer has a direct bearing on how external audit limits creative accounting techniques. The objective of the external audit is to independently vouch for the correctness of accounting and financial data and to confirm that it complies with relevant financial laws and accepted accounting standards [8, 9].

Using improper or unlawful procedures to falsify financial statements in order to achieve illicit benefits or evade financial obligations is one type of creative accounting practice. The external audit assesses these procedures and looks for any infractions of financial laws and accounting standards [10].

The organization's financial and accounting records are closely scrutinized and compared with accepted accounting standards and good accounting principles through external audit methods. Additionally, the organization's internal control system is assessed and looked over to make sure the right safeguards are in place to catch any unethical accounting practices. Furthermore, the external audit offers suggestions and counsel to the company regarding enhancing accounting procedures and fortifying financial oversight. External auditing helps to improve openness and integrity in an organization's financial operations by curbing creative accounting methods [6].

14 Conceptual Framework

The conceptual framework helps illustrate the independent and dependent variables and the relationship between them schematically. Figure 1 shows the conceptual framework for this study.

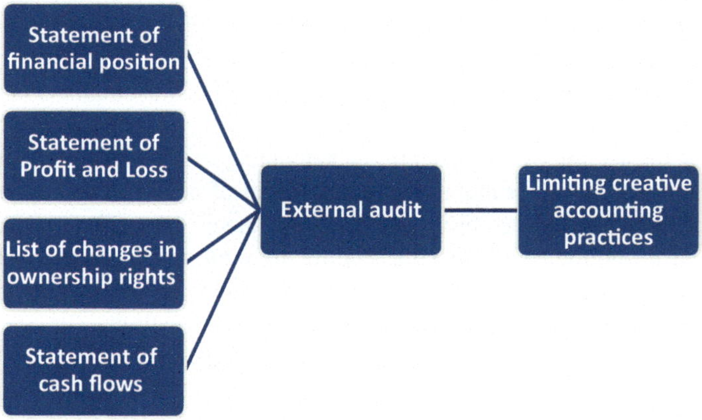

Fig. 1 Conceptual framework

15 Conclusion

Auditing is crucial in ensuring financial statements are accurately presented and in accordance with accounting standards. Creative accounting practices, which aim to enhance a company's image before investors, are a significant challenge for auditors and accountants. These practices can negatively impact the credibility and validity of financial statements. The research aims to identify the concepts and reasons for creative accounting methods for companies listed on the stock market and the role of auditing in limiting these practices.

References

1. Abdulrahman, D.M.: The role of external audit in limiting creative accounting practices in light of international standards from the point of view of certified accountancy experts in Iraq as a model. J. Econ. Adm. Sci. **28**(131), 238–251 (2022)
2. Al-Absy, M.S.M., Ku Ismail, K.N.I., Chandren, S.: Audit committee chairman characteristics and earnings management: the influence of family chairman. Asia-Pac. J. Bus. Adm. **11**(4), 339–370 (2019)
3. Al-Temimi, J.H.: The impact of external auditing on the quality of performance of Iraqi banks. World Econ. Finance Bull. **18**, 82–87 (2023)
4. Alastal, A.Y.M., Farhan, J.A., Allaymoun, M.H.: Auditors' perceptions in Gulf countries towards using artificial intelligence in audit process. In: Artificial Intelligence and Transforming Digital Marketing, pp. 867–878. Springer (2023)
5. Hamad, A.F.: Creative Accounting and the Role of the Auditor in Verifying Their Practices and Results (2020)
6. Hamdallah, M.: The reliance of external auditors on the work performed by internal auditors and its impact on audit on banks: a case of the first Abu Dhabi Bank, UAE. Ikon. upr. **17**(1), 68–80 (2020)
7. Hind, A.A.-R., Hamid, A., Sultan, R.: The role of corporate governance mechanisms in limiting creative accounting practices: a view from specialists in Middle East countries. **51**(4), 9–38
8. Rahman, M.S., Hasan, M.J., Khan, M.S.H., Jahan, I.: Antecedents and effect of creative accounting practices on organizational outcomes: evidence from Bangladesh. Heliyon **9**(2) (2023)
9. Umurerwa, D.: The impact of external audit on the financial management in public organizations: A case study of Rwanda development board. University of Rwanda (2016)
10. Velte, P.: The impact of external auditors on firms' financial restatements: a review of archival studies and implications for future research. Manag. Rev. Q. **73**(3), 959–985 (2023)

Factors Impact the Audit Quality: Evidence from Kingdom of Bahrain

Zainab A. Hussain and Mujeeb Saif Mohsen Al-Absy

Abstract Due to numerous factors that might affect quality, it is a real challenge to define audit quality. Pertinence of audit quality warrants serious consideration. Investigating audit quality components more thoroughly is necessary to comprehend the significance of audit quality. Therefore, the current research investigates the impact of audit fees, auditor's competency, auditor tenure, audit size on the quality of audit in Bahrain. The present research covered the topic of factors impacting the audit quality in Bahrain. A questionnaire is used in this quantitative study to gather primary data. Current research will cover accountants, auditors' financial managers, and board members of audit firms in Bahrain. SPSS version 26 is used to analyze the information gathered from the questionnaire. In this regard, it was discovered that auditor requirements considerably impacted audit quality in addition to well-known audit quality parameters like size, audit fees, tenure of auditor, and competency. Additionally, these variables may interact, negatively or positively, impact the quality of the audit.

Keywords Audit firm's size · Competency · Tuner · Audit fees · Audit quality

1 Introduction

Several variables, which may be broadly categorized into the characteristics of the auditing process and the auditor specifications, may influence the quality of the audit. Consequently, these elements may have a direct impact on the "audit opinion," which is released to provide reasonable assurance regarding the dependability of financial statements and strengthen investor trust. Authorities' investors and corporate governance have emphasized the significance of audit quality and its impact on

Z. A. Hussain · M. S. M. Al-Absy (✉)
Accounting and Financial Science Department, College of Administrative and Financial Science, Gulf University, Sanad 26489, Kingdom of Bahrain
e-mail: dr.mujeeb.saif@gulfuniversity.edu.bh

Z. A. Hussain
e-mail: 200101115184@gulfuniversity.edu.bh

B. Awwad (ed.), *The AI Revolution: Driving Business Innovation and Research*, Studies in Systems, Decision and Control 524, https://doi.org/10.1007/978-3-031-54379-1_14

market confidence, especially after the worldwide financial scandals which have had a substantial impact on the world's economies [1–6]. According to "agency theory," an auditor's view validates the assurance provided to third parties utilizing financial statements. The ability of the auditor to identify and report material misstatements is known as audit quality. To guarantee that the audit report accurately reflects appropriate view, it has implicitly included the competence and professional behaviors required throughout the auditing process, along with the auditor's objectivity and independence [7].

It is imperative to acknowledge that audit quality is increasingly gaining traction in the context of connected auditing themes because of their significant effects on the financial statements' credibility. Furthermore, an increase in audit quality might be attributed to the users' increased confidence in financial statements. So, our goal in providing the viewer with a concise overview of the key ideas and most recent research on audit quality standards was to accomplish it. To address this, the following section of the paper aims to give some definitions, which are followed by the primary theories of audit quality. The importance of the audit quality variables has then been discussed in more detail in the next chapter of this paper to highlight its influence on audit quality [7].

There are many factors that impact the quality of the audit. Indeed, the quality of audits is crucial for several reasons. It provides initial assistance in ensuring the accuracy and dependability of financial statements. For lenders, shareholders, and other stakeholders who depend on financial accounts to make defensible judgements, this is significant. Furthermore, fraud detection and prevention are aided by audit quality. Lastly, a high-quality audit contributes to the public's raised confidence in the financial reporting system [8]. According to Chen et al. [9], the quality of the audit influenced by the size firm. While Garcia-Blandon et al. [10] confirmed that audit firm tenure influences audit quality. The study of Salehi et al. [11] stated in his study that audit tenure and audit firm size influence the audit quality. Lastly in [12, 13] research papers it's been confirmed that audit fees have an impact on audit quality.

Over the past couple of years, despite the significance of audit quality, there have been a few high-profile well-publicized audit failures.

Concerns regarding the efficiency of the audit procedure and the calibre of audits being carried out have been raised by these deficiencies. "What is the most significant impact on the audit quality in the kingdom of Bahrain?" is how one may formulate the problem.

This research aims to investigate the influence of audit fees, auditor tenure, auditor competency and audit size on audit quality in Bahrain. This research is important to several parties. It assists in guaranteeing the accuracy and dependability of financial statements. This holds significance for stakeholders such as creditors and investors, who depend on financial records to make well-informed decisions. The results of the research are significant for audit firms in Bahrain.

2 Literature Review

Auditing is a tool for strong corporate governance in businesses. It can raise the standard of information provided to stakeholders by the company and enhance management and accountability for both non-financial and financial aspects [14]. Independent contractors with recognized certifications conduct and finish audits [15]. Discussions around the auditing profession and regulations aim to improve practitioners' and interested parties' comprehension of the variables influencing audit quality in actual practice [16].

Because of its complexity, audit quality lacks a clear definition and is currently up for debate in the literature. According to Van Raak and Thürheimer [17], there is no precise definition of audit quality and little information available regarding factors that may have an impact on it, despite the popularity of audit quality and the approaches used to evaluate it. DeAngelo [18] gave the most often cited definition of audit quality in the academic literature, defining it as the likelihood that an auditor will discover and reveal substantial misstatements in a client's accounting system [19].

Following specific quality control methods enables auditors to continuously fulfil the standards in their work until high-quality outcomes are obtained. Audit quality is the process by which auditors make sure their client organizations conform to applicable auditing standards [20].

Adhering to established standards namely, reporting, fieldwork, and general standards—is essential to achieving high audit quality. Various definitions of audit quality have been proposed by various studies. DeAngelo [18] assert, however, that an auditor's job description encompasses more than only reporting-related detection. Reputable auditors should consider how accurately the financial statements depict the company's real economy in addition to determining if the client's accounting decisions comply with technical regulations. Audit views, which guarantee that the financial statements are fairly presented in compliance with applicable rules, are compatible with the auditor's duty to ensure the quality of financial reporting. This demonstrates that the auditor considered the application of the regulations and the correspondence between the report and the actual conditions. According to the author, audit quality can be assessed by taking audit input into account using the elements of the auditor–client contract [21].

According to Ibrahim and Ali [13] research study, earnings quality can also be used to determine audit quality because a high-quality audit reduces earnings management and increases the financial reports' informativeness. Competency plays an important role in providing high quality audits. As a professional, qualified auditor, each auditor needs to fulfil certain standards and qualifications. According to Alsughayer [19], exams, professional experience, education, training, and continuous knowledge and skill development are all means of achieving competence regarding changes and advancements in one's career as well as suitable control systems that adhere to ethical standards.

To become competent, auditors must possess a variety of fundamental qualities, including general knowledge, skills and expertise gained from information and training. The ability of an auditor to apply newly acquired knowledge and experience in a practical and skillful manner to carry out the auditing process in an objective, meticulous, and correct manner is a key component of the auditor's professional competence. Importantly, scholars from throughout the world have started looking at research on auditing and innovation [22].

Further, size is thought to be the most reliable indicator for determining audit quality, notwithstanding the broad range of adopted metrics. Because of their capacity to find and identify misstatements, the larger audit company may therefore easily attain higher audit quality. However, because small-size audit businesses follow audit standards, it is also possible to achieve good audit quality. However, it is anticipated that audit quality will increase in large audit companies due to the availability of auditor-related standards such as expertise, technical ability, auditor's liability as well and auditor independence [11].

Revenue, total assets, workforce size, and total capital are indicators of a company's size. The company's improving circumstances will be reflected in the larger size of its income, workforce, and total capital. Large organizations are more likely to receive media or public attention than tiny companies, which may not be seen by stakeholders. This means that excellent audit quality is more common in large companies than in small ones [23].

In terms of larger audit firms, it will significantly affect the level of audit quality. Large audit firms can provide clients with higher-quality audit services than those that are smaller, according to several studies. This is because big audit companies are less likely to depend on the individual clients' financial circumstances.

Compared to small audit companies, they are therefore less likely to give in to customer pressure to reveal misstatements. Furthermore, larger audit firms may suffer greater reputation losses in the event of audit failures than smaller firms, thus they are more motivated and willing to provide high-quality services to safeguard their brand name reputation [18].

Large audit firms also possess more resources and a better level of technical expertise. Compared to small audit firms, they also have auditors with more expertise and previous experience [24]. However, contended that the large client effect rather than the superior auditor performance may be the primary driver of the higher audit quality found in larger audit firms. Numerous empirical research has offered evidence supporting the idea that a larger audit firm size is positively linked with higher audit quality, despite differing opinions regarding the explanation of the influence of audit firm size on audit quality [24]. The relationship between audit quality and audit firm size was the subject of numerous studies, and the relationship between an audit firm's size and audit quality. They argued that the size of an audit firm and the audit quality have a positive relationship [23].

On the other hand, an extensive amount of research including studies done on publicly traded companies supports the theory that Big 4 audit firms conduct auditing and preparing report with high quality than non-Big 4. It is confirm by the study of Al-Absy and Mahdi [25], that around 80% of the listed companies on the Bahrain Bourse had one of the top four audit firms evaluate them, indicating that large audit firms are well-known for the high caliber of audits they provide. According to preliminary research, the Big 4 audit companies rarely face penalties from the SEC and are rarely involved in litigation. Additional data pertaining to publicly traded corporations indicates that the Big 4 audit firms have historically had lower rates of releasing changed audit reports; they also report on bankruptcies more correctly or cautiously. The study of Martani et al. [26]. Big 4 firms offer superior audit quality because they sponsor training initiatives, standardize audit techniques, and give partners more opportunities for peer review [24].

Among other factors that affect the audit quality is Tenure. The relationship between the auditor and the client is frequently scrutinized in the aftermath of the accounting crises of the past 10 years, raising concerns about the independence of the auditor. One way to raise the caliber of financial reporting is to make the rotation of audit firms mandatory. This point of view is in line with the claim that longer auditor tenure lowers audit quality. As the duration of the auditor–client relationship grows, auditors are more likely to be in consensus with managers on important decisions, which could lead to a decline in quality of the audit.

This perspective is aligns with the claim made in joint study report that the auditor's incentives now shift to retaining and making a profit through the client [27]. Consequently, the auditor's level of anxiety regarding client-related lawsuits will decrease.

The counterargument to audit firm rotation is that problems could arise more frequently with new clients. Auditors need client-specific information to identify financial reporting that is below par. The authors contend that identifying shortcomings in financial reporting requires an understanding of internal controls system as well as the company's activities.

The authors conclude that a long career at an audit firm is not linked to a deterioration in the quality of financial reporting. Additionally, they have investigated how much auditor tenure influences the distribution of accruals that increase and decrease income. They demonstrate that as auditor tenure develops, both income-increasing and income-decreasing accruals decrease, indicating the auditor tenure encourages audit quality [27].

The cost of the audit duties is frequently correlated with the reputation of most audit companies and the level of quality of audit services. According to other writers, improved audit quality leads to higher costs, which are reflected in higher audit fees. Additionally, according to Momodu et al. [28] better audit quality is correlated with audit fees which enhances the auditors' standing. Considering this, why have the majority of crisis-ridden firms in recent memory been audited by top-cadre audit firms? Larger audit firms are supposed to receive higher audit fees than smaller audit firms, as demonstrated by earlier studies like those conducted by Ibrahim and Ali [13] and Momodu et al. [28].

Audit fees are a noisy proxy for quality because they are determined by a multitude of factors. An earlier study that looked at whether having an audit committee, its features, and its use of internal audit was related to higher audit fees within the nation concluded that higher audit fees equated to higher audit quality. According to several writers, managers and business owners are prepared to spend greater audit fees in order to obtain audits that they believe to be of a higher quality auditing [27].

The principal topics covered in the body of literature analysis are inclusive of research on audit quality and its determinants, audit fees and audit quality, competency and audit quality, size of the firm and audit quality, and tenure of the audit firm and audit quality. Further research is required to close the gap in the literature.

A study conducted by Ibrahim and Ali [13] evaluated the effect of audit fees from 2004 to 2015 on the audit quality of Nigerian listed companies. The research data collected from the corporations' annual reports and accounts. OLS and Random Effect regressions were specifically used with a panel data set. The public began to doubt the function of accountants and auditors due to the early 1990s corporate failures in Nigeria's financial sector, which threw auditors into stark relief. The study found that the size of the audit firm and audit fees both positively affect the quality of company audits. According to best practices, the report recommended that regulatory authorities carefully consider appropriate procedures for assessing audit fees that correspond with the audit effort.

A study of Calocha and Herwiyanti [29] examine the influence of experience, auditor professionalism, time constraints, budget pressure, audit tenure, and error detection expertise on audit quality. According to their findings, for the auditors who work in the public accounting firm in Jakarta, auditor professionalism, time budget pressure, and audit tenure have a positive but non-significant impact on audit quality. On the other hand, the research suggests that auditor experience and error-detecting knowledge significantly enhance audit quality, urging public accounting firms in Jakarta to focus on these skills.

Sukma and Bernawati [21] studies the impact of audit committee qualities on the audit quality where several variables used, e.g., audit committee size, frequency of meetings, education, and experience. Audit fees are used to measure the quality of the audits. According to the findings, only experience and size have a substantial impact on audit quality out of the four audit characteristics. While audit quality is not significantly influenced by audit meetings or instruction, education alone cannot guarantee the skills of audit committee members.

Kartika and Pramuka [8] investigate the impact of independence, professionalism, and competency on audit quality. The research indicates that audit quality is influenced by competency, independence, and professionalism, but caution is advised due to limited data analysis. Yanti and Wijaya [23] carried out research on the effects of audit tenure, audit fees, company size, and auditor switching on audit quality. A logistic regression test was employed in this study's data analysis methods. The result show that the change of auditors has no impact on audit quality; audit fee has no influence on audit quality, audit tenure has no effect on audit quality; and audit fee has a significant impact. However, firm size has no bearing on audit quality.

Alsughayer [19] investigated the influence of an auditor's integrity, competency, and ethics on audit quality. The results demonstrate that competence, ethics, and integrity all have a significant influence on the quality of audits. The results show that the factors that have the biggest effects on audit quality are the methods that auditors perform their jobs, their commitment to the code of conduct, and their program for ongoing training and improvement. Salehi et al. [11] found a positive correlation between audit quality the size of the audit company and the auditor's expertise, suggesting that hiring specialized auditors and larger companies leads to better quality audit services.

Momodu et al. [28] investigated the connection between listed businesses' audit quality and audit fees in Nigeria's downstream petroleum sector. The study employing the binary logit regression approach, audit quality the dependent variable was regressed on audit fees together with age and leverage as the control variables. In contrast to leverage, which has an inverse relationship, the study finds a negative link between audit quality and audit fees. The quality of audits is positively correlated with firm age. According to the study, audit fees should be regulated by authorities to avoid under- or overcharging and enhance audit quality.

Pham et al. [24] investigate the impact of audit company attributes in Vietnam, such as size, fees, and reputation, on audit quality. A sample of 192 businesses that were listed between 2006 and 2014 on the Hanoi and Ho Chi Minh Stock Exchanges was chosen. To analyze the data, multiple regression was employed. According to the results, Big 4 auditors in Vietnam offer higher-quality audits than non-Big 4 auditors. Fascinatingly, the results indicate that smaller audit firms in Vietnam offer higher-quality audits, except for the Big 4 audit firms. Furthermore, the findings show that auditors produce lower-quality audits the more audit fees they earn.

Similarly, Larasdiputra and Martini [30] studied the impact of computer-assisted audit techniques (TABK) on audit quality and auditor competency. This study was carried out in a Bali public accounting firm. 2020 represents the research period. Thirteen Public Accounting Firms (KAP) in Bali, employing 76 auditors, constitute the research population. The sampling strategy used in this study was called purposeful sampling. The analysis's findings lead to the following conclusions: computer-assisted audit procedures can mitigate the relationship between auditor competence and quality, and auditor competence enhances the quality of the audit.

3 Hypotheses Development

The audit quality is the dependent variable. The standard of Bahraini auditing the researcher formulates four primary hypotheses which are the four factors that impact the audit quality, the firm's size, competency, auditors' tenure, and audit fees. The study's Conceptual Model is shown in Fig. 1.

The research anticipates that there is a positive and significant relationship between audit quality and competence. This is in consistent with previous research work of Ardani [31] and Nurdiono and Gamayuni [32]. Based on the above-mentioned research outcome, the following hypothesis is put forth:

H_1: Competency has a significant and positive effect on audit quality.

In terms of audit company size, the study of Pham et al. [24] found that the size of the audit firm and the quality of the audit have a positive and significant relationship. Additionally, studies by Salehi et al. [11], revealed a correlation between audit

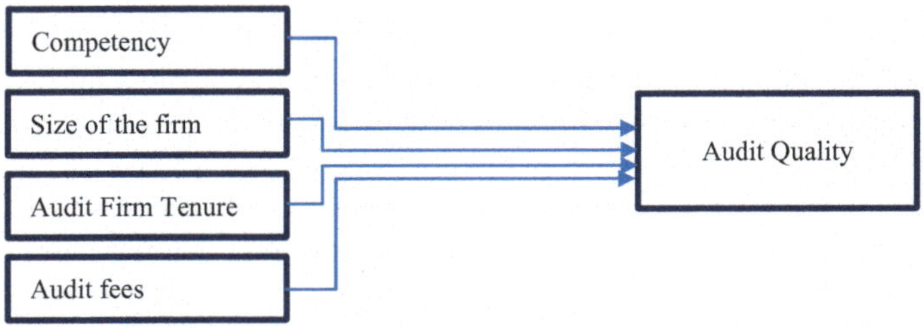

Fig. 1 Conceptual model

company size and audit quality. Based on the above-mentioned research findings, following hypothesis is formed:

H_2: Audit company size has a positive and significant impact on audit quality.

In relation to the audit fees, studies by Pham et al. [24] and Sari et al. [33] also found a significant relationship between audit quality and audit fees. This is support by the findings of [34, 35] who found a positive and significant impact. Based on the above-mentioned research findings, following hypothesis is assembled:

H_3: The audit fee has a positive and significant impact on the quality of the audit in Bahrain.

Concerning the audit tenure, a studies of Sari et al. [33] and Priyanti and Dewi [36] investigated the connection between audit tenure and audit quality where they found that there is a significant relationship that affects the quality of the audit. Based on the findings of the earlier studies, the theory that follows is developed.

H_4: There is a significant impact of auditors' tenure on the audit quality in Bahrain.

4 Methodology

The researcher is counselled to rely on the quantitative method due to its numerous benefits. To begin with, the quantitative approach is overly dependable and does not allow researchers to modify the research's findings. Using the quantitative approach also allows for the generalisation of the results.

The audit quality is the dependent variable. The standard of Bahraini auditing the researcher formulates four primary hypotheses which are the four factors that impact the audit quality, the firm's size, competency, auditors' tenure, and audit fees.

A descriptive design is used for this research. Hence, a questionnaire will be design and distribute to financial managers, auditors, accountants, and board members of Bahraini audit firms located in Kingdom of Bahrain.

5 Conclusion

Audit services are essential to strengthen users' confidence in financial statements in the financial report. When customers utilize this information to make decisions in the financial statement, this service will enhance its value. The current study aims to Therefore, the present research targets to investigate the influence of audit fees, auditor competency, auditor tenure, and audit size on audit quality in Bahrain. A questionnaire is used in this quantitative study to gather primary data. This research will cover financial managers, accountants, auditors, and board members of audit firms located in Bahrain. Authors expected that auditor requirements considerably impacted audit quality in addition to well-known audit quality parameters such as size, auditor tenure, audit fees, and competency. The study will extend the literature body of knowledge in terms of audit quality. It will assist the policymakers and stakeholders to evaluate and recognize the importance of having audit quality and factors affecting audit effectiveness.

References

1. Al-Absy, M.S.M.: Impactful women directors and earnings management. Cogent Bus. Manag. **9**(1), 2148873 (2022)
2. Al-Absy, M., Ku Ismail, K.N.I., Chandren, S.: Board chairmen's involvement in audit committees and earnings management practices (2017)
3. Al-Absy, M., Hasan, M.: Impact of the board of directors' characteristics on firm performance: a case of Bahraini listed firms (2023)
4. Al-Absy, M.S.M.: Board chairman characteristics and real earnings management. Sustainability **14**(22), 15025 (2022)
5. Al-Absy, M.S.M.: Impactful female directors and earnings management: the moderating effect of ownership concentration. Adm. Sci. **13**(5), 129 (2023)
6. Al-Absy, M.S.M., AlMahari, N.H.: The interaction effect of nomination committee's effectiveness on board of directors' characteristics and firm performance. Adm. Sci. **13**(5), 135 (2023)
7. Hosseinniakani, S.M., Inacio, H., Mota, R.: A review on audit quality factors. Int. J. Acad. Res. Account. Financ. Manag. Sci. 16 (2014)
8. Kartika, D., Pramuka, B.A.: The influence of competency, independency, and professionalism on audit quality. J. Account. Strateg. Financ. **2**(2) (2019)
9. Chen, J.Z., Elemes, A., Lobo, G.J.: David versus goliath: the relation between auditor size and audit quality for UK private firms. Eur. Account. Rev. (2023)
10. Garcia-Blandon, J., Argilés-Bosch, J.M., Ravenda, D.: Audit firm tenure and audit quality: a cross-European study. J. Int. Financ. Manag. Account. (2020)
11. Salehi, M., Fakhri Mahmoudi, M.R., Daemi Gah, A.: A meta-analysis approach for determinants of effective factors on audit quality: evidence from emerging market. J. Account. Emerg. Econ. (2019)
12. Kuntari, Y., Chariri, A., Nurdhiana, N.: The effect of auditor ethics, auditor experience, audit fees and auditor motivation on audit quality. Sriwijaya Int. J. Dyn. Econ. Bus. (2017)
13. Ibrahim, M., Ali, I.: Impact of audit fees on audit quality of conglomerates companies in Nigeria. Int. J. Serv. Sci. Manag. Eng. (2018)
14. Rajgopal, S., Srinivasan, S., Zheng, X.: Measuring audit quality. Rev. Account. Stud. (2021)

15. Gantz, S.D.: The Basics of IT Audit: Purposes, Processes, and Practical Information. Elsevier (2013)
16. Christensen, B.E., Glover, S.M., Omer, T.C., Shelley, M.K.: Understanding audit quality: insights from audit professionals and investors. Contemp. Account. Res. (2016)
17. Van Raak, J., Thürheimer, U.: Opportunities to improve the measurement of audit quality: a call for collaboration between the profession and academics. Maandblad Account. Bedrijfsecon. (2016)
18. DeAngelo, L.E.: Auditor size and audit quality. J. Account. Econ. **3**(3), 183–199 (1981)
19. Alsughayer, S.A.: Impact of auditor competence, integrity, and ethics on audit quality in Saudi Arabia. Open J. Account. **10**(4), 125–140 (2021)
20. Arens, A.A., Elder, R.J., Beasley, M.S.: Auditing and Assurance Services: An Integrated Approach. Pearson (2017)
21. Sukma, P., Bernawati, Y.: The impact of audit committee characteristics on audit quality. J. Akunt. (2019)
22. Ananda, R.F., Rahmadhani, S.N., Rahman, F., Dewi, S.: A review on innovation audits. In: Open Innovation in Small Business: Creating Values for Sustainability, pp. 185–194 (2023)
23. Yanti, L.D., Wijaya, M.D.W.D.: Influence of auditor switching, audit fee, tenure audit and company size to audit quality. ECo-Fin (2020)
24. Pham, N.K., Duong, H.N., Pham, T.Q., Ho, N.T.T.: Audit firm size, audit fee, audit reputation and audit quality: the case of listed companies in Vietnam. Asian J. Financ. Account. (2017)
25. Al-Absy, M.S.M., Mahdi, Z.H.: Descriptive study on audit quality reporting: evidence from listed firms in Bahrain. In: International Conference on Business and Technology. Springer (2022)
26. Martani, D., Rahmah, N.A., Fitriany, F., Anggraita, V.: Impact of audit tenure and audit rotation on the audit quality: big 4 vs non big 4. Cogent Econ. Finance (2021)
27. Nkemjika, O.M., Sunday, A.A., Nwamaka, O.C.: Audit quality, abnormal audit fee and auditor attributes. Int. J. Manag. Excell. (2017)
28. Momodu, A.L.I.U., Joshua, O., Nma, M.: Audit fees and audit quality: a study of listed companies in the downstream sector of Nigerian petroleum industry. Humanities (2018)
29. Calocha, R., Herwiyanti, E.: Factors that affect audit quality. J. Contemp. Account. (2020)
30. Larasdiputra, G.D., Martini, N.P.R.: Analysis of the effect of competency on audit quality. J. Environ. Sustain. Soc. Sci. (2020)
31. Ardani, S.V.: Pengaruh Tenure Audit, Rotasi Audit, Audit Fee terhadap Kualitas Audit dengan Komite Audit Sebagai Variabel Moderasi (Studi Pada Perusahaan Sektor Keuangan yang Terdaftar di Bei Tahun 2010 2014). J. Akunt. (Media Ris. Akunt.) (2017)
32. Nurdiono, N., Gamayuni, R.R.: The effect of internal auditor competency on internal audit quality and its implication on the accountability of local government. Eur. Res. Stud. J. (2018)
33. Sari, S.P., Diyanti, A.A., Wijayanti, R.: The effect of audit tenure, audit rotation, audit fee, accounting firm size, and auditor specialization to audit quality. Ris. Akunt. Keuang. Indones. **4**(3), 186–196 (2019)
34. Mansur, H., Abdul Rahman, A.A., Meero, A., Shatnawi, A.: The perceptions of external auditors on the relationship between audit fees and audit quality. Cogent Bus. Manag. **9**(1), 2113203 (2022)
35. Kuntari, Y., Chariri, A., Nurdhiana, N.: The effect of auditor ethics, auditor experience, audit fees and auditor motivation on audit quality. Sriwijaya Int. J. Dyn. Econ. Bus. **1**(2), 203–218 (2017)
36. Priyanti, D.F., Dewi, N.H.: The effect of audit tenure, audit rotation, accounting firm size, and client's company size on audit quality. Indones. Account. Rev **9**(1), 1–14 (2019)

The Impact of the Automated Accounting System on the Auditing Process in the Banking Sector in the Kingdom of Bahrain

Ahmed Althawadi and Ahmad Yahia Mustafa Alastal

Abstract This article investigates how the auditing process in the banking industry of the Kingdom of Bahrain is affected by Automated Accounting Systems (AAS). Understanding how these automated systems affect auditing processes is crucial since the banking sector adopts Accounting Information Systems (AIS) at an increasing rate. To acquire an understanding of the advantages and difficulties of implementing AAS, the research looks at variables such as information quality, system quality, and system satisfaction as well as how these affect the auditing process. The study results may help regulators, financial institutions, and auditors better understand how the auditing process is evolving in the context of automated accounting systems, including its needs and dynamics.

Keywords Automated Accounting System · Accounting Information Systems (AIS) · Information quality · System quality · Auditing process · System satisfaction · Banking sector · Banking industry · Kingdom of Bahrain

1 Introduction

This chapter will cover a variety of topics, beginning with the study's background, followed by the problem statement and scope of the study, and what the study's objectives, and questions are, as well as determining the study's significance. Finally, the chapter will cover limitations and a summary.

A. Althawadi · A. Y. M. Alastal (✉)
Accounting and Financial Science Department, College of Administrative and Financial Science, Gulf University, Sanad 26489, Kingdom of Bahrain
e-mail: dr.ahmed.alastal@gulfuniversity.edu.bh

A. Althawadi
e-mail: 210101115666@gulfuniversity.edu.bh

B. Awwad (ed.), *The AI Revolution: Driving Business Innovation and Research*, Studies in Systems, Decision and Control 524, https://doi.org/10.1007/978-3-031-54379-1_15

1.1 Background

Automation and Technological Advancements have been a major feature of the banking industry, with the introduction of automated accounting systems being one example. Data input, transaction processing, and financial reporting are just a few of the accounting tasks that these systems automate, perhaps increasing productivity and accuracy [21]. Auditing's vital function in guaranteeing the accuracy and consistency of financial data in the banking industry drives its significance. To provide stakeholders, such as creditors, investors, and regulators, confidence, auditors go at financial records, internal controls, and compliance with laws [65]. Auditing techniques have had to evolve because of the incorporation of computerized accounting systems. To properly audit automated systems, auditors may need to acquire new techniques, instruments, and skill sets in addition to adapting their traditional manual auditing procedures [40].

The Central Bank of Bahrain (CBB) oversees the Kingdom of Bahrain's well-established banking regulatory system. The CBB exists to promote responsibility, stability, and openness in the banking sector by establishing rules and regulations pertaining to internal controls, financial reporting, and auditing procedures [35].

Although automated accounting systems and audits are becoming more and more important in the banking industry, there isn't much thorough study that focuses on the Kingdom of Bahrain. To bridge this research gap and comprehend the effects of automated accounting systems on the auditing process in Bahrain, a thorough investigation is required [30].

Examining the unique effects of automated accounting systems on the auditing process in the banking sector of the Kingdom of Bahrain, the research seeks to add to the body of knowledge already in existence by taking these background elements into consideration. As a result, decision-making may be made more intelligently and auditing processes in the Bahraini banking sector can be improved. The study findings can also be very helpful to regulators, financial institutions, and auditors [43].

1.2 Problem Statement

To increase operational effectiveness, Bahrain's banking industry is adopting Accounting Information Systems (AIS) and Automated Accounting Systems (AAS). Yet, questions have been raised about how this may affect the auditing procedure. Ensuring the correctness, dependability, and transparency of financial accounts is largely dependent on the auditing process. Optimizing processes and resolving any issues need an understanding of how AAS affects the auditing process. In the context of automated accounting systems, this study attempts to shed light on the evolving dynamics and needs of the auditing process.

Nonetheless, the reason for looking into this topic is because there isn't much research done on it now, and those that often take place in developed countries rather than emerging ones like Bahrain. Considering that all these criteria are deemed problematic in this research.

1.3 Research Objectives

The main objective: The examine the influence of automated accounting system on the auditing process in the banking sector in the Kingdom of Bahrain.

1. To examine the influence of Information Quality of automated accounting systems on the auditing process in the banking sector in the Kingdom of Bahrain.
2. To examine the influence of Quality for automated accounting systems on the auditing process in the banking sector in the Kingdom of Bahrain.
3. To examine the influence of the Satisfaction of automated accounting systems on the auditing process in the banking sector in the Kingdom of Bahrain.

1.4 Research Questions

The main question: What is the influence of automated accounting system on the auditing process in the banking sector in the Kingdom of Bahrain?

1. What is the influence of Information Quality of automated accounting systems on the auditing process in the banking sector in the Kingdom of Bahrain?
2. What is the influence of Quality for automated accounting systems on the auditing process in the banking sector in the Kingdom of Bahrain?
3. What is the influence of the Satisfaction of automated accounting systems on the auditing process in the banking sector in the Kingdom of Bahrain?

1.5 Significance of the Study

1.5.1 Theoretical Significance

This study investigates how automated accounting systems affect the banking industry's auditing procedures in Bahrain. It advances the philosophy of auditing, helps to bridge the knowledge gap between theory and practice, enhances auditing procedures, and comprehends the influence of technology on auditing. It offers realistic suggestions to improve auditors' efficacy in auditing automated systems as well as insights into the difficulties they face.

1.5.2 Practical Significance

The study provides information to auditors, institutions, regulators, consumers of financial statements, and technology vendors about how automated accounting systems affect the auditing process in Bahrain's banking industry. It improves the quality of financial reporting, increases trust in financial statements, and offers suggestions for effective audits. The results may direct the creation of rules and guidelines to guarantee adherence to auditing specifications.

1.6 Scope of the Study

The study looks at how automated accounting systems affect Bahraini banks' auditing procedures. The purpose of the research is to gather information regarding the audit process's mechanics, advantages, difficulties, auditors' perspectives, and implications for financial report preparation. The study will be carried out on several workers in the audit, finance, and accounting departments of some of the banking banks in the Kingdom of Bahrain. Further study is required to broaden the applicability of the findings, since they may not be applicable to other nations, areas, or industries. The investigation was completed in just four months, spanning from September to December of 2023.

1.7 Limitation of the Study

It required a great deal of time and work to guarantee that the study was conducted accurately, with minimal mistakes, and produced accurate findings. However, various scopes and limits were identified prior to its execution. A questionnaire is first sent to a number of staff members in the audit, finance, and accounting departments of 14 banks in the Kingdom of Bahrain in order to establish the sample size. Therefore, it's possible that the findings don't apply to companies in other industries. Second, as other studies have shown, the questionnaire findings may have certain limitations that may be updated and modified to better fit the respondents.

1.8 Operational Definition

1.8.1 Auditing Process

Ascertaining the correctness, dependability, and compliance of financial statements and reports involves a methodical review and assessment of financial records, transactions, internal controls, and other pertinent data. This process is known as auditing.

There are many phases involved, including preparation, risk analysis, substantive testing, review of internal controls, and reporting [42].

1.8.2 Automated Accounting Systems

The computer-based software programs and technologies that banking institutions utilize to record, process, and report financial transactions and data are referred to as automated accounting systems for the purposes of this research. Among the related technologies that automate accounting procedures are general ledger systems, financial management software, and others [44].

Information Quality

The dependability, correctness, completeness, and applicability of financial data and records made accessible by automated accounting systems are all considered aspects of the quality of information for auditing. It may be evaluated using a few indications, such as the lack of mistakes or discrepancies in financial data, compliance with legal and accounting regulations, and the accessibility of the required supporting paperwork [2].

System Quality

System quality describes how well-automated accounting systems in the banking industry operate generally, functionally, and dependably. Factors including data accuracy, reaction speed, system uptime, security precautions, and the system's capacity to manage high transaction volumes properly and effectively may all be used to gauge it [11].

System Satisfaction

User acceptability, overall satisfaction, and perception of the functionality and performance of automated accounting systems in the banking sector of the Kingdom of Bahrain are all subjectively evaluated and referred to as user satisfaction with automated accounting systems. Regarding the efficiency, usefulness, and dependability of automated accounting systems, it includes users' attitudes, views, and opinions [34].

2 Literature Review

2.1 Introduction

This part will provide some background about the banking sector. Also, from DVs to IVs, going through all the variables in detail. There will also be a discussion of previous studies on auditing process. Previous study on the influence of various factors on auditing process will also be discussed. In addition, the research's conceptual framework will be discussed.

2.2 Banking Sector in the Kingdom of Bahrain

In both the national economy of Bahrain and the larger Middle East area, the banking industry is quite important. Regional and international banks are drawn to Bahrain because it has become a major financial center. The Bahrain Central Bank oversees the sector and is essential in preserving stability and enforcing laws that guarantee the stability of the banking system [29].

Regulatory oversight of the banking industry in Bahrain is primarily carried out by the Central Bank of Bahrain (CBB). Ensuring the safety, stability, and integrity of the financial system is its responsibility via the formulation and implementation of rules, policies, and recommendations [50]. Insurance companies, investment firms, and money exchange organizations are among the numerous financial institutions that the CBB oversees. Only accredited and respectable organizations are allowed to operate thanks to licenses that the CBB issues to banks and other financial institutions [58].

Asset quality, risk management, liquidity control, capital adequacy, and provisioning needs are all covered by prudential rules. Strong Anti-Money Laundering (AML) and Counter-Terrorist Financing (CTF) rules are also enforced by the CBB in order to stop the funding of terrorism and money laundering [29]. Aside from these laws, the regulatory framework also covers consumer protection, including terms and conditions disclosure, customer complaint handling, fee and charge transparency, and treating the public fairly [26].

The performance of the banking industry is assessed using a range of financial metrics, including the ratios of non-performing loans (NPLs), capital adequacy, net interest margin, cost-to-income, loan-to-deposit, return on assets (ROA), return on equity (ROE), and liquidity. A bank's profitability, efficiency, and stability are shown by these indicators, which aid in the decision-making process for stakeholders [41]. An elevated return on assets (ROA) signifies enhanced profitability and effective use of assets. Profitability and efficient use of capital by a bank are indicated by a higher ROE [1].

Higher NIM values are indicative of improved revenue generation and interest rate control. Superior cost control and operational effectiveness are indicated by a lower CIR. Banks may survive liquidity shocks by maintaining a sufficient level of liquid assets and stable funding sources, which is ensured by liquidity ratios like the Liquidity Coverage Ratio and Net Stable Funding Ratio [57].

Under the direction of the Central Bank of Bahrain (CBB), Bahrain has emerged as a major hub for Islamic finance, thanks to its strong legal system. To ensure that Islamic banks adhere to both worldwide best practices and Sharia principles, the CBB has released special rules and recommendations [35]. Studies on market share, the growth of the Islamic banking industry, and the elements that contribute to its success may be used to examine Bahrain's growth and development [1]. In addition to drawing in local and foreign firms, the nation's status as a center for Islamic banking also explores the use of technology in Islamic banking. The GDP, employment growth, and financial inclusion effects of Islamic banking in Bahrain are evaluated. There is also discussion of research projects and policy that use Islamic finance to advance financial inclusion [35].

To increase client satisfaction, operational effectiveness, and innovation, Bahrain's banking industry has been adopting fintech. Fintech's effect on conventional banking procedures, digital transformation programs, Bahrain's contribution to fintech innovation, the regulatory environment and compliance standards, and partnerships between classic banks and fintech firms are all potential areas of research on this subject [58]. Research need to concentrate on alterations in consumer conduct, banking offerings, operational procedures, and company frameworks, together with the difficulties and prospects that come with incorporating fintech solutions. Studying how well regulatory actions support fintech innovation while maintaining consumer safety and financial stability is another important aspect. Better client experiences, easier access to new technology, and more market potential may result from partnerships between banks and fintech companies [29].

2.3 Auditing Process (DV)

2.3.1 Definition of Auditing Process

The purpose of the auditing process is to guarantee that a company complies with legal requirements, accounting standards, and other rules by methodically reviewing and assessing its financial records, transactions, internal controls, and procedures [55]. In order to provide confidence on the quality and dependability of the financial information, it entails the impartial and independent evaluation of financial statements and the accompanying paperwork []. To arrive at a conclusion on the fairness of the financial accounts, auditors acquire information, carry out analytical processes, test transactions, and evaluate internal control systems [53].

The auditing process is governed by professional standards, such as the International Standards on Auditing (ISA) published by the International Auditing and Assurance Standards Board (IAASB). A few of the often-performed audit kinds include forensic, operational, compliance, financial statement, internal, and internal audits. When preparing and carrying out their processes, auditors consider the materiality of a financial statement item or inaccuracy [31].

The auditing process is overseen by audit committees and external oversight bodies, who also examine audit plans, hire external auditors, and assess audit results. Maintaining and improving their knowledge and abilities requires auditors to participate in continual professional development [12].

2.3.2 The Important Auditing Process

Assessing the possibility of a major misrepresentation in financial statements is a critical step that cannot be skipped. Along with evaluating the efficiency of internal controls, auditors also identify inherent risks related to the client's operations, industry, and financial reporting. A good internal control system lowers the risk of material misstatement and offers confidence on the accuracy of financial reporting; hence they also evaluate the client's internal controls' operation and design [49].

To gather adequate and relevant audit evidence, substantive testing is a crucial phase that entails a thorough review and testing of financial statement balances, transactions, and disclosures. Auditors assess whether the audit evidence they have collected is adequate and acceptable, ensuring that the conclusions are supported by trustworthy data [64].

Auditors summarize the methods followed during the audit, provide a summary of the findings, and express their judgment on the fairness of the financial statements to management, the audit committee, and other interested parties [18]. To make well-informed judgments, assess uncertainty, and reach acceptable conclusions, one must possess professional skepticism and judgement. To maintain independence, impartiality, and integrity throughout the audit process, auditors abide by the ethical principles and professional standards established by regulatory agencies and organizations [51, 54].

Auditors may retain their competence and provide high-quality audit services with the support of ongoing professional development. By assisting auditors in assessing risks, gathering pertinent information, assessing controls, spotting any problems, and offering independent assurance on the accuracy and dependability of financial statements, these components work together to provide a smooth and dependable auditing process [38].

2.4 *Factor Affecting Auditing Process: Automated Accounting Systems (IVs)*

2.4.1 Definition of Automated Accounting Systems

Computer-based systems known as Automated Accounting Systems (AAS) increase efficiency and automate record-keeping, reporting, and financial transactions. Accurate and efficient accounting procedures are made possible by these technology, software, and database components [61]. The characteristics of AAS, including general ledger management, financial reporting, integration with subsidiary systems, automation of repetitive processes, audit trail and compliance, data analytics, and decision assistance, make it especially useful in the banking industry [52].

With the purpose of producing accurate and current balance sheets, income statements, and other financial reports, AAS offers a centralized system for managing general ledger accounts. Also, it produces a range of customizable and on-demand financial reports, including balance sheets, cash flow statements, and profit and loss statements [3]. With the aim of decreasing human data input and guarantee data consistency, AAS may interface with subsidiary systems. That shields financial data from loss, alteration, and illegal access, it also has strong security features [60].

2.4.2 The Important Automated Accounting Systems

With its extensive accounting and financial management features, Automated Accounting Systems (AAS) are indispensable instruments in many businesses, including the banking industry. Notable AAS includes Intuit's QuickBooks, which is well-liked by small and medium-sized enterprises because of its intuitive user interface and extensive accounting capabilities [22]. Large enterprises may handle their finances and accounting from beginning to finish using SAP ERP, an integrated suite of business software. General ledger, accounts payable, accounts receivable, cash management, and fixed asset management are just a few of the features included in Oracle Financials, a complete AAS from Oracle Corporation [14].

A cloud based AAS, Microsoft Dynamics 365 Finance offers accounting and financial management features including budgeting, cash flow management, accounts payable, accounts receivable, and financial reporting [37]. Features including general ledger management, invoicing, bank reconciliation, expenditure monitoring, and financial reporting are available with Xero, a cloud based AAS made for accountants and small companies [19]. With capabilities including general ledger, accounts payable, accounts receivable, cash management, project accounting, and financial reporting, Sage Intacct is a cloud based AAS that specializes in financial administration for mid-sized firms and corporations [28].

The choice of an AAS is contingent upon the particular demands and specifications of the enterprise, considering elements like sector expertise, expandability, capacity for integration, and user experience. Prior to implementation, organizations must assess the attributes, capabilities, and suitability of AAS to make sure they complement their financial management and accounting goals [48].

2.5 *Information Quality*

2.5.1 Definition of Information Quality

The precision, comprehensiveness, dependability, pertinence, promptness, and ease of access of the data and information generated by automated accounting systems are all considered aspects of information quality. The information's accuracy and suitability for financial reporting, decision-making, and other usage are its main concerns [62]. The important components of high-quality information are timeliness, relevance, completeness, correctness, dependability, accessibility, and data integrity [70].

The absence of mistakes, inconsistencies, or inaccuracies that may be used to falsely depict the performance or financial status of an organization is known as accuracy [5]. It is the precision and correctness of the data generated by the system. In order to provide a thorough and comprehensive picture of the organization's financial operations, completeness guarantees that the system collects and processes all pertinent and essential data for financial reporting and other uses [15].

2.5.2 The Important Information Quality

Multiple aspects of information quality must be met by automated accounting systems in order to guarantee accurate, dependable, and valuable data for financial reporting and decision-making. Rigid financial reporting and analysis depend on accuracy as a fundamental component [6]. By documenting transactions and gathering necessary information, completeness guarantees that all relevant data is processed and recorded. The accounting system's consistency prevents disparities and improves the dependability and comparability of financial data by maintaining uniformity and coherence across various modules, reports, and time periods [59]. In order to support users' unique demands and requirements, relevance focuses on offering relevant and helpful information for decision-making [33].

With the aim of guaranteed accuracy and representing the most current transactions and occurrences, the automated accounting system must generate information that is timely. Ensuring the precision, coherence, and dependability of the data that the system stores and processes is contingent upon preserving data integrity [23]. By putting strong security mechanisms, access restrictions, data encryption, and privacy protections in place, security and privacy aspects address the protection of sensitive financial data within the system [20].

2.6 *System Quality*

2.6.1 Definition of System Quality

Policies, processes, and controls that guarantee the system constantly generates accurate, dependable, and pertinent financial information make up a quality system for automated accounting systems. Throughout the automated accounting system's lifespan, it includes the methods and structure required to control and maintain its quality [8].

Data input, processing, reporting, and control system operations should all be outlined in the system's well-documented policies, guidelines, and manuals. Through consistent and accurate system use, this documentation acts as a reference for both users and auditors [47]. Ensuring the accuracy and consistency of data requires strong data integrity controls, which include validation tests, data reconciliation protocols, and error detection systems. The availability, confidentiality, and integrity of the financial data should all be safeguarded by the use of security measures [45].

2.6.2 Dimensions of System Quality

To help users comprehend and precisely follow the system's operations, documentation should provide a description of the system's design, configuration, operating methods, and controls. Internal controls contribute to preventing mistakes, fraud, and illegal actions and guaranteeing the quality and dependability of financial information [24]. Examples of these controls include division of tasks, authorization procedures, access restrictions, and transaction monitoring. Ensuring the completeness, correctness, and consistency of the data in the automated accounting system is known as data integrity [7].

Robust security and privacy protocols are needed to safeguard the confidentiality, availability, and integrity of financial data. To guarantee appropriate planning, testing, and modification implementation, change management procedures are crucial [67]. Other important aspects include training and user assistance, with thorough instruction on system features, procedures, and controls accessible to answer questions and resolve problems [17].

Adherence to relevant accounting standards, laws, and legal requirements is essential, and performance monitoring aids in the identification of any problems, bottlenecks, or inefficiencies. In order to provide correct financial reporting and compliance with relevant rules, such GAAP or IFRS, the system should be built and set up appropriately [39].

Encouraging a culture of continuous improvement entails carrying out recurring evaluations, audits, and assessments to pinpoint opportunities for improvement and putting remedial measures in place to deal with flaws or shortfalls. Organizations may ensure the correctness, consistency, and dependability of financial data and facilitate efficient financial reporting and decision-making procedures by taking these factors into consideration when building a strong quality system for their automated accounting systems [32].

2.6.3 The Important System Quality

Numerous factors make a high-quality automated accounting system essential. For the purpose of creating financial statements, making educated judgments, and adhering to legal obligations, it guarantees the correctness, consistency, efficiency, and efficacy of financial information. This lowers the possibility of mistakes or misrepresentations, which makes people less confident in the results produced by the system [63].

A quality system ensures adherence to these standards, lowering the risk of non-compliance and related fines. Organizations must comply with accounting standards, laws, and legal obligations. In order to reduce the risks associated with fraud, illegal access, and data breaches, it also includes internal controls and security [27]. A well-established quality system lowers mistakes, decreases manual labor, and streamlines operations to increase efficiency and production. Instead of concentrating on monotonous duties, users are now able to concentrate on activities that offer value [46].

An accurate, timely, and relevant information system facilitates analysis, forecasting, and strategic planning, all of which are critical components of decision-making. With the provision of an audit trail and the upkeep of a record of all operations and transactions inside the automated accounting system, it also fosters auditability and transparency [9]. Regular reviews, evaluations, and feedback systems help businesses discover areas for development, fix flaws, and put corrective measures in place. These strategies foster organizational continuous improvement. Through continuous improvement of the automated accounting system, firms may adjust to expanding regulatory requirements, changing business demands, and technology breakthroughs [16].

2.7 System Satisfaction

2.7.1 Definition of System Satisfaction

The degree of contentment felt by users and stakeholders with respect to the functionality, performance, and usability of automated accounting systems is known as system satisfaction. This indicates how well the automated accounting system satisfies the demands, expectations, and requirements of its users, leading to a favorable user experience and general contentment [66].

Data input, transaction processing, reporting, analysis, and interaction with other modules or systems are all examples of accounting duties that the system can do successfully and efficiently. This is referred to as functionality. Dependability and accuracy are crucial components, but ease of use is defined as the system's friendliness and intuitiveness. In addition, speed and performance are critical since consumers appreciate timely and effective system operation [68]. Users want smooth data interchange and interoperability with current infrastructure; therefore, integration and compatibility are essential. The capacity to customize and adapt a system to unique organizational needs is highly valued by users, making customization and flexibility crucial. User expectations for prompt help, rapid technical support, and extensive training resources make support and training services crucial [71].

2.7.2 The Important System Satisfaction

For automated accounting systems, there are several reasons why system satisfaction is important. Given that users are more likely to use the system efficiently in their everyday chores, which increases productivity, it is essential to user adoption and productivity. By lowering the possibility of mistakes and enhancing data integrity, this satisfaction has a direct impact on the dependability and correctness of financial data [25].

Users that are satisfied with the automated accounting system have faith in its timeliness and accuracy, which facilitates more accurate financial reporting, strategic planning, and more informed decision-making [32]. An automated accounting system is a vital source of financial information for financial reporting and decision-making. Users are free to concentrate on value-added tasks rather than battling system complexity when they use an intuitive and user-friendly system, which lowers the learning curve [4, 9].

An organization's automated accounting system and contented user base will encourage talent acquisition and retention since they are more inclined to stick with it. The organization's competitiveness and credibility may be improved by having a solid reputation for efficient and user-friendly accounting systems [56]. Organizations may utilize system satisfaction as a useful feedback mechanism for continuous development by using it to pinpoint areas that need improvement and deploy updates, user training, or streamlining of processes. Through continuous improvement and alignment with changing user needs, this iterative approach to system development and user satisfaction promotes continuous improvement [13].

Retention, productivity, data correctness, decision-making, user efficiency, and system satisfaction are all contingent upon system satisfaction as well as organizational reputation and ongoing development. Better user involvement, more dependable financial data, and improved overall organizational performance may result from giving automated accounting systems' system satisfaction a high priority [10].

3 Methodology

Methodology is an organised approach to research that includes guidelines, protocols, and strategies for data collection, information analysis, and decision-making. It is described in the technique portion of a paper or thesis and is essential to academic research. This chapter will include typical features such as research design, data collecting techniques, sample strategy, and conceptual framework. It will also construct the study's hypothesis and cover topics like instruments for measuring the degree of dependent and independent variables. While research methodologies might differ across fields and situations, they always guarantee sound and rigorous research, which enables the production of trustworthy information.

3.1 Conceptual Framework

Auditing Process is a dependent variable (DV), whereas Information Quality, System Quality, and System Satisfaction are independent variables (IVs) that have a significantly positive impact on Auditing Process. The influence of these factors was calculated by data analysis and a survey of banking sector personnel, as the researchers acknowledged in the study that was reviewed and referenced in the study. Figure 1 shows the variables and the research model.

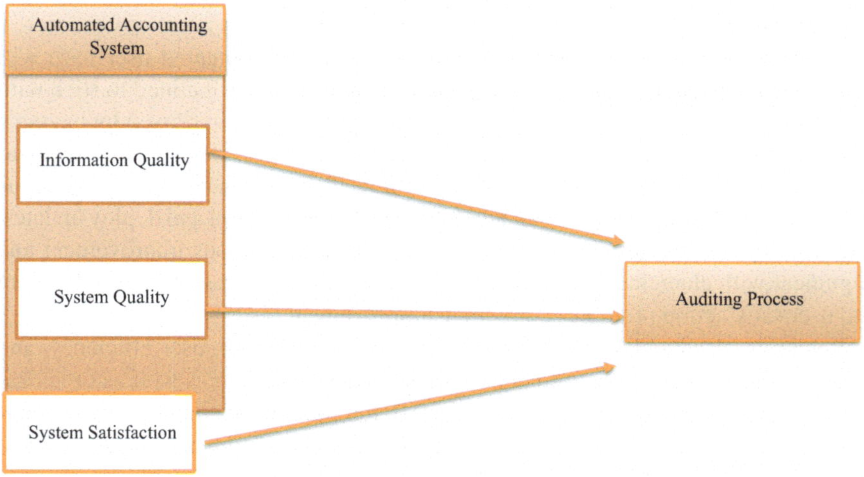

Fig. 1 Research model of this study

4 Conclusion

The implementation of an automated accounting system in the banking sector in the Kingdom of Bahrain has had a significant impact on the auditing process. Overall, the automated accounting system has improved efficiency, accuracy, and reliability in financial reporting, leading to a more streamlined and effective auditing process. One of the key benefits of the automated accounting system is the reduction in manual data entry and processing. With automated systems, financial transactions are recorded and processed electronically, minimizing the risk of human error. This reduces the likelihood of errors in financial statements and improves the accuracy of data used during the auditing process. Auditors can rely on the system's built-in controls and automated checks to ensure the integrity of financial data. Moreover, the automated accounting system enables real-time monitoring and reporting of financial transactions. Auditors can access up-to-date and comprehensive financial information, allowing for more timely and efficient audits. This real-time access to data facilitates a more proactive approach to auditing, enabling auditors to identify and address potential issues promptly. The system's ability to generate detailed and customizable reports also enhances the auditing process. Auditors can easily extract and analyze specific financial data, helping them identify anomalies, trends, and potential risks. The system's reporting capabilities provide auditors with valuable insights and facilitate a more thorough and comprehensive audit examination.

Additionally, the automated accounting system enhances data security and integrity. The system's built-in controls and encryption measures help safeguard financial information from unauthorized access and potential fraud. Auditors can rely on the system's robust security features to ensure the confidentiality and integrity of sensitive data during the auditing process. However, the implementation of an automated accounting system also presents some challenges for auditors. They need to adapt their auditing techniques and approaches to effectively assess the system's internal controls and identify any potential weaknesses or vulnerabilities. Auditors must possess the necessary knowledge and skills to understand and evaluate the automated system's functionality and controls adequately.

In conclusion, the impact of the automated accounting system on the auditing process in the banking sector in the Kingdom of Bahrain has been largely positive. The system has improved efficiency, accuracy, and reliability in financial reporting, providing auditors with real-time access to comprehensive data and enhancing their ability to identify potential issues. While auditors need to adapt their approach to assess the system's internal controls, the overall benefits of the automated accounting system have greatly improved the auditing process in the banking sector.

References

1. Abou Elseoud, M.S.: Using a panel data approach to determining the key factors of Islamic banks' profitability in Bahrain. Cogent Bus. Manag. **7**(1), 1831754 (2020)

2. Achdiat, I.M.: The information quality impact on internal audit innovation: case study in Indonesia's public service agencies. Cuad. Econ. **45**(129), 11–18 (2022)
3. Agustian, H.I.: Analysis of common size on BSI (Bank Syariah Indonesia) in 2020–2021. J. Account. Res. Util. Financ. Digit. Assets **1**(3), 195–202 (2023)
4. Ahmad Yahia Mustafa Alastal, M.A.: Accounting Students' Perceptions on E-Learning During COVID-19 Pandemic: Case Study of Accounting and Financial Students in Gulf University—Bahrain. Springer Nature Switzerland (2023)
5. Ahmed, W.S.: Minimizing errors in RT-PCR detection and quantification of SARS-CoV-2 RNA for wastewater surveillance. Sci. Total Environ. **805**, 1 (2022)
6. Al-Attar, K.: The effect of accounting information system on corporate governance. Accounting **7**(1), 99–110 (2021)
7. Al-Dmour, A.H.-D.: The implementation of SysTrust principles and criteria for assuring reliability of AIS: empirical study. Int. J. Account. Inf. Manag. **27**(3), 461–491 (2019)
8. Al-Waeli, A.J.: Impact of accounting information system on financial performance with the moderating role of internal control in Iraqi industrial companies: an analytical study. J. Adv. Res. Dyn. Control Syst. **12**(8), 246–261 (2020)
9. Ali, M.M.: Impact of management information systems (MIS) on decision making. Glob. Discl. Econ. Bus. **8**(2), 83–90 (2019)
10. Ali, B.J.: Accounting information system and financial sustainability of commercial and Islamic banks: a review of the literature. J. Manag. Inf. Decis. Sci. **24**(5) (2021)
11. Allbabidi, M.H.: Hype or hope: digital technologies in auditing process. Asian J. Bus. Account. **14**(1), 59–86 (2021)
12. Almasria, N.A.: Corporate governance and the quality of audit process: an exploratory analysis considering internal audit, audit committee and board of directors. Eur. J. Bus. Manag. Res. **7**(1), 78–99 (2022)
13. Alshamsi, S.: Improving industrial operations with real-time information using augmented reality. In: Abu Dhabi International Petroleum Exhibition and Conference, p. D041S127R008. SPE (2023)
14. Atashili Nchang, L.: Impact of ERP systems on accounting processes: examining how the adoption of SAP ERP affects the accounting processes of the customers of Fujitsu Finland OY (2022)
15. Batko, K., Ślęzak, A.: The use of big data analytics in healthcare. J. Big Data **9**(1), 3 (2022)
16. Bella, S.A.: Enhancing financial management and accountant roles: a study on the role of technological advancements. SEIKO J. Manag. Bus. **6**(2), 435–446 (2023)
17. Bradley, V.M.: Learning management system (LMS) use with online instruction. Int. J. Technol. Educ. **4**(1), 68–92 (2021)
18. Brown, R.G.: Changing audit objectives and techniques. In: The Evolution of Audit Thought and Practice, pp. 1–8. Routledge (2020)
19. Carnegie, G.D.: Accounting through the eyes of a witness, 1973–2022. Account. Hist. J. 1–8 (2023)
20. Centonze, P.: Security and privacy frameworks for access control big data systems. Comput. Mater. Contin, **59**(2) (2019)
21. Chang, V.B.: How blockchain can impact financial services—the overview, challenges and recommendations from expert interviewees. Technol. Forecast. Soc. Change **158**, 120166 (2020)
22. Chipriyanova, G., Krasteva-Hristova, R.: Technological aspects of accounting automation system as a decision support system. In: Environment. Technologies. Resources. Proceedings of the International Scientific and Practical Conference (2023)
23. Chowdhury, E.S.: Blockchain technology in financial accounting: emerging regulatory issues. Rev. Financ. Econ. **21**, 862–868 (2023)
24. Conti, M.D.: A survey on industrial control system testbeds and datasets for security research. IEEE Commun. Surv. Tutor. **23**(4), 2248–2294 (2021)
25. Daniel, H.: Enhancing data integrity through data governance in clinical trials: a computational approach. Int. J. Comput. Sci. Technol. **7**(3), 31–56 (2023)

26. Daza Jaller, L.S.: Digital Trade in MENA: Regulatory Readiness Assessment. World Bank Policy Research Working Paper (9199) (2020)
27. Dirse, K.K.: Big Data Analytics: Powerful Tool Against Data Breaches in Public Accounting Firms (2020)
28. Dmytrenko, O.Y.: Application of modern software for improving the economic analysis accounting efficiency at enterprises (2020)
29. El-Chaarani, H.: The impact of oil prices on the financial performance of banking sector in Middle East region. Int. J. Energy Econ. Policy **9**(4) (2019)
30. Gotthardt, M.K.: Current state and challenges in the implementation of smart robotic process automation in accounting and auditing. ACRN J. Finance Risk Perspect. (2020)
31. Haapamäki, E.: Research on international standards on auditing: literature synthesis and opportunities for future research. J. Int. Account. Audit. Tax. **35**, 37–56 (2019)
32. Hadi, A.H.-S.: The effects of information technology adoption on the financial reporting: moderating role of audit risk. Int. J. Econ. Finance Stud. **15**(1), 47–63 (2023)
33. Halkiopoulos, C.A.: Online reservation systems in e-business: analyzing decision making in e-tourism. J. Tour. Herit. Serv. Mark. (JTHSM) **6**(1), 9–16 (2020)
34. Handoko, B.L.: Robotic process automation in audit 4.0. In: The 2021 12th International Conference on E-Business, Management and Economics, pp. 128–132 (2021)
35. Hassan, M.K.-M.: Islamic fintech and Bahrain: an opportunity for global financial services. In: FinTech in Islamic Financial Institutions: Scope, Challenges, and Implications in Islamic Finance, pp. 65–87 (2022)
36. Hendri, N.S.: Evaluation of financial management information system using modification of the Delone & Mclean model during the COVID-19 pandemic. Int. J. Prof. Bus. Rev. **7**(5), 12 (2022)
37. Honnegowda, C.S.: 'COVID-19' is a mode of lesson to human beings taught by nature: an analysis. Res. Journey **10**(4), 157–159 (2021)
38. Hung, P.H.: The influence of cultural, legal and institutional factors on auditors' roles, responsibilities and perceptions of audit quality. Eur. J. Theor. Appl. Sci. **1**(5), 1131–1145 (2023)
39. Hussain, M.B.: Evaluating international financial reporting standards and generally accepted accounting principles in a local context: a case from Pakistan. Acad. Educ. Soc. Sci. Rev. **2**(2), 70–80 (2022)
40. Imene, F.: Information technology and the accountant today: what has really changed? J. Account. Taxation **12**(1), 48–60 (2020)
41. Irawati, N.M.: Financial performance of Indonesian's banking industry: the role of good corporate governance, capital adequacy ratio, non performing loan and size. Int. J. Sci. Technol. (2019)
42. Ismael, B.A.: The effects of computerized accounting system on auditing process: a case study from northern Iraq. Solid State Technol. **63**(5), 8564–8578 (2020)
43. Jaber, R.J.: Auditors' usage of computer-assisted audit techniques (CAATs): challenges and opportunities. In: Challenges and Opportunities in the Digital Era: 17th IFIP WG 6.11 Conference on e-Business, e-Services, and e-Society (2018)
44. Keskinen, M., Tarwireyi, R.C.: Automation and the transformation of the audit process: a qualitative research on the impact of automation on the audit process (2019)
45. Khin, N.A.: Data integrity in global clinical trials: discussions from joint US food and drug administration and UK medicines and healthcare products regulatory agency good clinical practice workshop. Clin. Pharmacol. Ther. **108**(5), 949–963 (2020)
46. Kokina, J.: Early evidence of digital labor in accounting: innovation with robotic process automation. Int. J. Account. Inf. Syst. **35**, 100431 (2019)
47. Kunduru, A.R.: Cloud BPM application (Appian) robotic process automation capabilities. Asian J. Res. Comput. Sci. **16**(3), 267–280 (2023)
48. Lill, P.W.: In the field of tension between creativity and efficiency: a systematic literature review of management control systems for innovation activities. Eur. J. Innov. Manag. **24**(3), 919–950 (2021)

49. Lokanan, M.T.: Detecting anomalies in financial statements using machine learning algorithm: the case of Vietnamese listed firms. Asian J. Account. Res. **4**(2), 181–201 (2019)
50. Mohammed, A.A.: Role of the central bank in implementation of value-based Islamic banking and reporting in Bahrain—exploring the challenges. Eur. J. Islam. Finance **19**, 54–62 (2021)
51. Mujeeb Saif Mohsen Al-Absy, K.N.: Audit committee chairman characteristics and earnings management: the influence of family chairman. Asia-Pac. J. Bus. Adm. (2019)
52. Мюллер, В.К: Accounting and auditing according to international standards as a management function. Financ. Credit Act. Probl. Theory Pract. **4**(35), 60–68 (2020)
53. Napitupulu, I.H.: Internal control, manager's competency, management accounting information systems and good corporate governance: evidence from rural banks in Indonesia. Glob. Bus. Rev. **24**(3), 563–585 (2023)
54. Njagi, M.: Independence of internal auditors: significance, guidelines, and influencing factors. East Afr. Finance J. **2**(1), 56–63 (2023)
55. Petrov, A.M.: Organizational and methodical mechanism of internal audit of settlements in corporate systems. Научный журнал «Вестник НАН РК» (2), 119–127 (2020)
56. Pillai, R., Sivathanu, B.: Adoption of artificial intelligence (AI) for talent acquisition in IT/ITeS organizations. Benchmark. Int. J. **27**(9), 2599–2629 (2020)
57. Ramadhan, S.S.: Financial performance of conventional and Islamic banks in Bahrain: a comparative study. Banks Bank Syst. **14**(4), 192–205 (2019)
58. Ramiah, V.P.: Banking and Financial Institutions in the Gulf Cooperation Council Region. Doing Business in the Middle East: A Research-Based Practitioners' Guide (2023)
59. Rangineni, S.: An analysis of data quality requirements for machine learning development pipelines frameworks. Int. J. Comput. Trends Technol. **71**(9), 16–27 (2023)
60. Ren, S.: Optimization of enterprise financial management and decision-making systems based on big data. J. Math. **2022**, 1–11 (2022)
61. Roszkowska, P.: Fintech in financial reporting and audit for fraud prevention and safeguarding equity investments. J. Account. Organ. Change **17**(2), 164–196 (2021)
62. Ruth, K.: Adoption of integrated financial management system and procurement performance in Uganda: a case of Kabale district in Uganda. Doctoral dissertation, Kabale University (2022)
63. Safkaur, O.A.: The effect of quality financial reporting on good governance. Int. J. Econ. Financ. Issues **9**(3), 277 (2019)
64. De Santis, F.: Big data and data analytics in auditing: in search of legitimacy. Meditari Account. Res. **29**(5), 1088–1112 (2021)
65. Shaikh, H.J.: Beyond traditional audits: the implications of information technology on auditing. Int. J. Eng. Technol. **7**(2.34), 5–11 (2018)
66. Sunmola, F.T.: A case study on performance features of electronic tendering systems. Procedia Manuf. **51**, 1586–1591 (2020)
67. Thapa, C.: Precision health data: requirements, challenges and existing techniques for data security and privacy. Comput. Biol. Med. **129**, 104130 (2021)
68. Trendowicz, A.G.: User experience key performance indicators for industrial IoT systems: a multivocal literature review. Digit. Bus. 100057 (2023)
69. Vasarhelyi, M.A.: Principles of analytic monitoring for continuous assurance. In: Continuous Auditing: Theory and Application, pp. 191–217. Emerald Publishing Limited (2018)
70. Whyte, S.T.: Reliable data collection: a tool for data integrity in Nigeria. Doctoral dissertation, Walden University (2021)
71. Yanagawa, E.: The invoice data exchange accelerates SME banking innovation: trends and initiatives for e-invoicing in Japan. J. Digit. Bank. **7**(3), 233–248 (2023)

Convergence of Management and Environmental Control Systems: A Key to Sustainable Organizational Performance in Jordan

Ahmad Yahia Mustafa Alastal, Ali Ateeq, Mohammed I. F. Dawwas, Mohammed Alzoraiki, and Marwan Milhem

Abstract This research addresses the convergence of Management Control Systems (MCS) and Environmental Management Control Systems (EMCS), emphasizing their vital significance in encouraging long-term organizational success in Jordan. This study dives into the integration of these systems and their implications for corporate strategy and ethical practices via a thorough assessment of major studies and landmark publications. According to the findings, integrating MCS and EMCS not only improves operational efficiency but also aligns organizational operations with environmental sustainability, which is critical in today's global corporate context. The research, however, reveals obstacles in the actual use of these integrated systems, particularly in different economic environments such as Jordan, revealing a gap between theoretical frameworks and real-world implementation. The study also emphasizes the relevance of ethical issues in company planning, arguing for a healthy balance between economic objectives and environmental duties. The study's results add to a better understanding of sustainable business practices, underlining the need for further empirical research in a variety of situations and emphasizing the importance of ethical business behavior in ensuring long-term organizational success. This study offers useful insights for practitioners and scholars alike, underlining the significance of MCS and EMCS in long-term organizational management.

A. Y. M. Alastal · A. Ateeq (✉) · M. I. F. Dawwas · M. Alzoraiki · M. Milhem
Administrative Science Department, College of Administrative and Financial Science, Gulf University, Sanad 26489, Kingdom of Bahrain
e-mail: dr.ali.ateeq@gulfuniversity.edu.bh

A. Y. M. Alastal
e-mail: dr.ahmed.alastal@gulfuniversity.edu.bh

M. I. F. Dawwas
e-mail: dr.mohammed.isam@gulfuniversity.edu.bh

M. Alzoraiki
e-mail: dr.mohammed.saied@gulfuniversity.edu.bh

M. Milhem
e-mail: dr.marwan.milhem@gulfuniversity.edu.bh

B. Awwad (ed.), *The AI Revolution: Driving Business Innovation and Research*, Studies in Systems, Decision and Control 524, https://doi.org/10.1007/978-3-031-54379-1_16

Keywords Management Control Systems (MCS) · Environmental Management Control Systems (EMCS) · Sustainable organizational performance · Business ethics · Jordan · Eco-control · Operational efficiency · Environmental sustainability · Strategic business management · Empirical research

1 Introduction

In today's dynamic global business market, organizations looking to align their operations with strategic goals are increasingly adopting Management Control Systems (MCS). This tendency, identified by prominent scholars such as Mishra and Singh [1] and Márquez-Sánchez et al. [2] includes the deployment of numerous methods such as performance assessment, incentive systems, and responsibility center distribution. These tools are critical in ensuring that resources are used efficiently inside businesses.

Jukka [3] first established the notion of MCS, emphasizing its important significance in achieving organizational objectives. MCS is more than simply monitoring and controlling; it is a holistic strategy that combines strategic planning, resource allocation, and operational execution, ensuring that every aspect of the company works together to achieve shared goals [4]. This system is intended to offer managers with the tools and knowledge they need to make informed choices, improve processes, and lead the organization in the direction of its anticipated future state [5].

Performance assessment is very important in MCS. It entails evaluating individual and team contributions to corporate objectives in order to ensure that efforts are aligned with the strategic vision [6]. Incentive systems, on the other hand, are used to inspire and reward people in a way that coincides with the organization's interests [7, 8]. This connection is critical for establishing an accountability and excellence culture. Furthermore, MCS responsibility center allocation is a strategic procedure that entails allocating power and assigning accountability for certain company operations or activities [9]. This distribution guarantees that resources are properly handled, and that decision-making is as near to the operational level as feasible, boosting responsiveness and efficiency.

Anthony's insight on MCS as a tool for effective and efficient resource management in achieving corporate goals is becoming more important in the present era. As businesses face complex problems and global competition, the capacity to adapt and align operations with strategic objectives via MCS is more important than ever. This methodology not only drives performance, but it also develops a culture of continual development, assuring long-term sustainability and success in a constantly changing business context. To summarize, MCS is critical in today's corporate climate. MCS enables organizations to negotiate the intricacies of the global market and accomplish their strategic objectives more effectively by offering a systematic method to managing resources, assessing performance, and aligning employee incentives with organizational goals.

2 Literature Review

2.1 Introduction

The relevance of Management Control Systems (MCS) and their integration with Environmental Management Control Systems (EMCS) has grown in the changing landscape of organizational management. This comprehensive examination goes further into these ideas, examining their development, relevance, and the synergies that result from their integration.

2.2 Evolution of MCS

The developing role of Management Control Systems (MCS) and their integration with Environmental Management Control Systems (EMCS) in organizational management has received substantial academic attention. Contemporary research by Bourne et al. [10] as well as Chenhall and Moers [11] expands on Brand [12] definition of MCS as a tool for effective resource management, emphasizing its strategic importance in aligning operational activities with overarching organizational goals. According to Rai et al. [13], the introduction of EMCS indicates the growing incorporation of environmental sustainability within classic MCS frameworks. Asiaei et al. [14] refer to this confluence as eco-control, which highlights a strategy convergence aimed at both economic efficiency and environmental stewardship. Empirical research, such as those conducted by Rodrigues et al. [15] show a variety of insights regarding the influence of eco-control on organizational performance, demonstrating the complexity and context-dependence of the phenomenon. This literature emphasizes the need of more empirical study, especially in non-Western contexts, to fully understand the application and usefulness of integrated MCS and EMCS in a variety of organizational settings.

2.3 Environmental Considerations in MCS

The increasing importance of environmental sustainability in business has resulted in the considerable inclusion of Environmental Management Control Systems (EMCS) into the larger framework of Management Control Systems (MCS). Recognizing this tendency, Battersby [16] define EMCS as a collection of well-established routines and processes aimed at improving an organization's environmental performance. This shift in MCS indicates a growing realization of the significance of environmental stewardship in strategic corporate planning [17]. EMCS is more than just a supplement to current systems; it is a critical component that incorporates environmental concerns into fundamental operational and strategic activities. This paradigm shift emphasizes the crucial role of environmental management in creating sustainable business practices that balance economic objectives with environmental responsibilities. The incorporation of EMCS into MCS represents a paradigm shift in organizational strategy, highlighting the importance of environmental factors in achieving overall organizational performance and sustainability [18].

2.4 Integration of MCS and EMCS: Towards Eco-Control

The integration of Management Control Systems (MCS) and Environmental Management Control Systems (EMCS) has given rise to the notion of 'eco-control,' as presented by Einhorn et al. [18]. This integration represents a strategy shift in which corporate goals are no longer exclusively focused on economic efficiency but now include environmental sustainability [19]. Eco-control represents the fusion of classical MCS and EMCS, guaranteeing that environmental issues are smoothly incorporated into all levels of organizational decision-making and performance assessment. Li et al. [20] emphasize the importance of this integration, noting that it links corporate operations with environmental goals, improving sustainable business practices. This move is a watershed point in management strategy, acknowledging the twin imperatives of economic and environmental sustainability. Eco-control is a forward-thinking approach to corporate management that demonstrates how environmental stewardship and economic efficiency may be mutually reinforcing objectives within the current organizational structure [21].

2.5 Empirical Evidence and Divergent Findings

Empirical investigations provide contradictory results. Hennig et al. [22] discovered an indirect link between eco-control and economic performance, particularly in businesses with a high level of environmental visibility. Meurs et al. [23] on the other hand, discovered a non-significant association between environmental management and overall performance, emphasizing the complexity and context-dependence of these systems.

2.6 MCS as a Strategic Tool

Córdova Aguirre [24] and Presa et al. [25] research expands the function of Management Control Systems (MCS) beyond simple operational control, emphasizing its strategic importance in decision-making, competitive strategizing, and financial performance. MCS moves beyond its conventional position as a control mechanism to become an essential component of strategic management. It provides critical information to companies for informed decision-making, connecting diverse organizational operations with overall strategic objectives. This viewpoint recasts MCS as a vital instrument for negotiating the challenges of competitive corporate settings and attaining long-term financial success.

2.7 *Environmental Strategies and Organizational Performance*

The introduction of environmental initiatives into Management Control Systems (EMCS) has shown to be advantageous for enterprises. Muminović et al. [26] and Rasheed et al. [27] research emphasizes the practical benefits of EMCS, such as cost savings, improved product quality, and a strengthened business reputation. These studies highlight the financial advantages of incorporating environmental stewardship into business processes, demonstrating how EMCS not only promotes sustainability but also adds to an organization's overall financial health and competitive edge. This integration emphasizes the new business landscape's connection between environmental responsibility and economic success.

3 Method

The paper titled "Convergence of Management and Environmental Control Systems: A Key to Sustainable Organizational Performance in Jordan," is based on a thorough analysis of the literature. The foundational literature and pivotal research [28–30], were thoroughly examined, with a particular emphasis on the synergy between management and environmental control systems and their influence on sustainable organizational performance in the Jordanian setting [31–35]. This methodological approach combined empirical research and theoretical ideas to provide evidence-based propositions targeted at promoting ethical corporate practices. This broad and methodical approach enabled an academically rigorous investigation of the issue at hand [36].

4 Discussion

The study's results underscore the crucial relevance of combining Management Control Systems (MCS) with Environmental Management Control Systems (EMCS) in supporting long-term organizational success, notably in Jordan. The studies examined [30–36] show that this convergence not only improves operational efficiency but also fosters environmental stewardship, a critical aspect in reaching long-term sustainability objectives. However, the research indicates that implementing such integrated systems is fraught with difficulties, especially in locations with disparate economic and environmental rules.

The findings also show that, although the theoretical underpinnings of environmental management are well-established, actual implementations may be difficult and context specific. This demonstrates a gap between theoretical knowledge and actual implementation, implying the need for greater empirical study concentrating on real-world applications and consequences. Furthermore, the Jordanian setting offers a unique case study that reflects both the promise and constraints of implementing such integrated systems in a developing economy. Furthermore, as discussed in the literature, the debate over ethical business practices emphasizes the need of integrating company plans with larger social and environmental goals. This alignment is critical for firms looking to preserve their competitive advantage while also positively impacting their communities and the environment.

Finally, this research adds to a better understanding of the synergies and obstacles of combining MCS and EMCS, laying the groundwork for future research and practice in sustainable organizational management.

5 Conclusion

The study of the convergence of Management Control Systems (MCS) and Environmental Management Control Systems (EMCS) demonstrates a substantial paradigm change in organizational strategy, notably in Jordan. This research emphasizes the significance of incorporating environmental issues into management control frameworks in order to achieve long-term organizational success. The studied literature [30–36] makes a convincing case for strategic alignment of company operations with environmental stewardship, emphasizing the combined advantages of operational efficiency and environmental responsibility. However, the study also highlights problems in adopting such integrated systems, particularly in countries with diverse economic and regulatory frameworks, such as Jordan. There is a significant gap between theoretical ideas and their actual implementation, indicating the necessity for more empirical study in a variety of scenarios. The report also highlights the significance of ethical business practices as a key component of current company strategy. In conclusion, this study adds to the expanding body of information on sustainable business practices by establishing the importance of MCS and EMCS in current organizational management. While the results are especially pertinent to Jordan, they also provide useful insights for worldwide applicability, emphasizing the need of continual study and adaptation in this dynamic sector.

5.1 Limitations

While thorough in its methodology, this research has certain limitations that should be noted when interpreting the results. To begin, relying on secondary data from existing literature restricts the capacity to capture real-time advancements in the area of Management and Environmental Control Systems, notably in Jordan. Second, although the breadth of the reviewed literature is broad, it may not include all relevant research, thereby contributing to biases in the synthesis results. Third, due to Jordan's cultural and economic uniqueness, the results may not be entirely generalizable to other situations or locations. Furthermore, although the theoretical frameworks used are strong, they may not completely capture the complexity of the real-world implementation of these control systems. Finally, a reliance on published academic research may neglect insights from grey literature or unpublished investigations, thus leading to a more limited understanding of the issue. These limitations underline the need of continued study and knowledge in this dynamic subject.

5.2 Recommendations

Based on the outcomes of this study, it is suggested that future research widen the scope to incorporate real-time data and viewpoints from other geographical places. Incorporating primary data collecting and examining unpublished or grey literature might also give a more comprehensive insight. These procedures would broaden and deepen knowledge in the area of Management and Environmental Control Systems.

References

1. Mishra, P., Singh, G.: Energy management systems in sustainable smart cities based on the Internet of energy: a technical review. Energies **16**(19), 6903 (2023)
2. Márquez-Sánchez, S., et al.: Enhancing building energy management: adaptive edge computing for optimized efficiency and inhabitant comfort. Electronics **12**(19), 4179 (2023)
3. Jukka, T.: Does business strategy and management control system fit determine performance? Int. J. Product. Perform. Manag. **72**(3), 659–678 (2023)
4. Li, W.: Management control system in the business network: control and trust. In: Strategic Management Accounting in a Network Economy, pp. 263–291. Springer (2023)
5. Budhwar, P., et al.: Human resource management in the age of generative artificial intelligence: perspectives and research directions on ChatGPT. Hum. Resour. Manag. J. **33**(3), 606–659 (2023)
6. Aithal, P., Aithal, S.: Key performance indicators (KPI) for researchers at different levels & strategies to achieve it. Int. J. Manag. Technol. Soc. Sci. (IJMTS) **8**(3), 294–325 (2023)
7. Aalders, A.F.: Reward and recognition. In: Cultivating Organizational Excellence: A Practitioner's View, pp. 125–147. Springer (2023)
8. Ali, A.A.A., Yusoffii, S.H., Aliiii, K.: Fundamental of equal opportunities in employee's performance: a critical review. In: Proceedings of the 5th World Conference on Integration of Knowledge, Bangi, Malaysia (2019)

9. Al Balushi, F.S.M., Ali, A.A.A., Alzoraiki, M.: The impact of forecasting strategy and training and development strategy on preparing leadership competencies administrative support in the Sultanate of Oman. Int. J. Intellect. Hum. Resour. Manag. (IJIHRM) **3**(01), 34–41 (2022)
10. Bourne, M., et al.: Generating organisational performance: the contributing effects of performance measurement and human resource management practices. Int. J. Oper. Prod. Manag. **33**(11/12), 1599–1622 (2013)
11. Chenhall, R.H., Moers, F.: The role of innovation in the evolution of management accounting and its integration into management control. Acc. Organ. Soc. **47**, 1–13 (2015)
12. Brand, D.K.B.: Management control systems (2007)
13. Rai, S., et al.: Circular economic business models. In: Green Innovation, Sustainable Development, and Circular Economy, p. 215 (2020)
14. Asiaei, K., et al.: Corporate social responsibility and sustainability performance measurement systems: implications for organizational performance. J. Manag. Control. **32**(1), 85–126 (2021)
15. Rodrigues, V.D.V., Wander, A.E., da Silva Rosa, F.: Indicators to analyze environmental performance and eco-controls for a poultry production chain: a methodological proposal based on the EMA system. Environ. Syst. Decis. 1–16 (2023)
16. Battersby, S.: Quality management systems: environmental management systems as examples. In: Clay's Handbook of Environmental Health, pp. 242–276. Routledge (2022)
17. Kouam, J.C., Asongu, S.A., Meh, B.J.: A Synthetic Indicator of the Quality of Support for Businesses in Burkina-Faso, Cameroon, and Ghana. Springer (2023)
18. Einhorn, S., et al.: The relationship of organizational culture with management control systems and environmental management control systems. Rev. Manag. Sci. 1–51 (2023)
19. Velasco-Balmaseda, E., de Celis, I.L.R., Izaguirre, N.E.: Corporate social responsibility as a framework for gender equality: mapping of gender equality standards for sustainable development. Corp. Soc. Responsib. Environ. Manag. (2023)
20. Li, W., et al.: The nexus between environmental corporate social responsibility, green intellectual capital and green innovation towards business sustainability: an empirical analysis of Chinese automobile manufacturing firms. Int. J. Environ. Res. Public Health **20**(3), 1851 (2023)
21. Rosa, F., et al.: Do cultural controls, eco-learning and environmental strategy lead to high environmental innovation? J. Account. Organ. Change **19**(5), 625–641 (2023)
22. Hennig, J.C., et al.: Environmental management control systems: exploring the economic motivation behind their implementation. J. Bus. Res. **169**, 114283 (2023)
23. Meurs, J.A., et al.: Physiological reactions to and recovery from acute stressors: the roles of chronic anxiety and stable resources. Health Psychol. Rep. **11**(3), 223–240 (2023)
24. Córdova Aguirre, L.J.: Management control systems and sustainability in manufacturing companies: proposal of a framework for measuring sustainable performance (2023)
25. Presa, A., et al.: Blockchain implementation in food retail supply chain management: a case study. In: Supporting Technologies and the Impact of Blockchain on Organizations and Society, pp. 136–159. IGI Global (2023)
26. Muminović, F., et al.: Benefits of introduction and implementation of ISO 14001 standard in the meat industry. In: Book of Proceedings (2023)
27. Rasheed, A.S., Booth, C.A., Horry, R.E.: Stakeholder perceptions of the benefits and barriers of implementing environmental management systems in the Maldivian construction industry. J. Hous. Built Environ. 1–30 (2023)
28. Ali, A.A.A., et al.: Moderating effect of strategic planning on the relationship between career path planning and job performance. Sustainability **15**(11), 8490 (2023)
29. Al-Refaei, A.A.-A., et al.: An integrated mediating and moderating model to improve service quality through job involvement, job satisfaction, and organizational commitment. Sustainability **15**(10), 7978 (2023)
30. Alzoraiki, M., et al.: Impact of teachers' commitment to the relationship between transformational leadership and sustainable teaching performance. Sustainability **15**(5), 4620 (2023)
31. Ali, A.A.A., Alzoraiki, M., Al Aghbas, N.S.M.: Working remotely, social isolation and access to support and communication during COVID-19 its influence on job performance. Int. J. Intellect. Hum. Resour. Manag. (IJIHRM) **3**(01), 23–33 (2022)

32. Ali, A.A.A., et al.: Association between Islamic work ethics and employee commitment among medical staff in the public health sector in Bahrain. Int. J. Intellect. Hum. Resour. Manag. (IJIHRM) (2023)
33. Ateeq, A.: Knowledge Management's Effect on Learning Quality in Bahrain's Private Universities (2023)
34. Alkadash, T.M., et al.: The effects of leadership style on employee sustainable behaviour: a theoretical perspective. In: Artificial Intelligence and Transforming Digital Marketing, pp. 205–213. Springer (2023)
35. Ali, A.A.A.: The Relationship Between Islamic Work Ethics and Employee Performance Among Medical Staff in the Public Health Sector in Bahrain: Employee Commitment as Mediating Variable. Universiti Sains Islam Malaysia (2022)
36. Ateeq, A.: The mediating role of engagement in the relationship between equal opportunities and employee performance. Probl. Perspect. Manag. **21**(3), 435–447 (2023)

Integrating Ethical Principles in Corporate Strategy: A Comprehensive Analysis of Business Ethics in Contemporary Enterprises

Ali Ateeq and Marwan Milhem

Abstract This paper provides a thorough examination of the incorporation of ethical principles into corporate strategy, emphasizing their significance in modern business practices. Through an extensive literature review, we explore key themes such as ethical leadership, corporate social responsibility (CSR), ethical decision-making in global contexts, the ethical challenges of technological advancements, and the impact of business ethics on stakeholder perceptions. According to our findings, ethical leadership is critical in fostering an organizational culture of integrity and transparency. CSR is recognized as a strategic tool that addresses societal concerns while also increasing competitive advantage. The complexity of ethical decision-making in a globalized business environment highlights the need for balancing universal ethical standards with cultural sensitivity. Furthermore, ethical management of emerging technologies such as AI and big data is acknowledged to be critical in modern business operations. The paper concludes that ethical practices are integral to the sustainability and success of businesses, influencing both consumer loyalty and employee engagement. Ethical conduct in business strategy is not merely a moral obligation but a vital component of organizational success in the global market.

Keywords Business ethics · Ethical leadership · Corporate social responsibility (CSR) · Global ethics · Technological ethics · Stakeholder perception · Organizational culture · Sustainable business strategy

A. Ateeq (✉) · M. Milhem
Administrative Science Department, College of Administrative and Financial Science, Gulf University, Sanad 26489, Kingdom of Bahrain
e-mail: dr.ali.ateeq@gulfuniversity.edu.bh

M. Milhem
e-mail: dr.marwan.milhem@gulfuniversity.edu.bh

B. Awwad (ed.), *The AI Revolution: Driving Business Innovation and Research*, Studies in Systems, Decision and Control 524, https://doi.org/10.1007/978-3-031-54379-1_17

1 Introduction

The integration of ethical values into corporate strategy has emerged as an important issue in the contemporary business environment [1]. The development of corporate ethics is indicative of an increasing recognition of the significant influence that firms have on society, economies, and the environment [2]. The process of integration discussed here is not just driven by legal obligations, but rather it is a strategic need that influences the whole corporate culture, connections with stakeholders, and the long-term viability of the entity. The problem at hand has considerable importance, as highlighted by Elrefae et al. [3] who posits that ethical business practices have become a crucial standard for achieving company success, surpassing conventional metrics such as profitability and growth.

The academic discipline of business ethics investigates the ethical aspects inherent in business decision-making, covering a wide array of topics including corporate governance, social responsibility, and environmental stewardship. According to Kamila and Jasrotia [4] the field of business ethics entails comprehending the ethical dimensions of management choices and ensuring their congruence with wider social and environmental considerations. This viewpoint has significant importance within a worldwide economy whereby enterprises function across a multitude of cultural and legal environments. Moreover, the incorporation of ethical concerns is essential in tackling the obstacles presented by technological progress and the digital transformation, as emphasized by Ren et al. [5]. They underscore the ethical ramifications associated with data privacy and the use of artificial intelligence in decision-making processes within the corporate realm.

The need for ethical business practices is increasingly emphasized due to the changing demands of customers, workers, and investors. According to recent research conducted by Katamba et al. [6] there is an emerging pattern seen among customers whereby they exhibit a preference for organizations that exhibit ethical conduct and display a commitment to social responsibility [6]. In contemporary times, the ethical beliefs and culture of an enterprise have a growing impact on employee engagement and retention [7]. The dynamic nature of the environment calls for a critical evaluation of conventional business models and tactics from an ethical perspective.

The objective of this study is to provide a thorough examination of the incorporation of ethical concepts into business strategy within modern organizations. This study aims to examine the theoretical foundations of business ethics, as well as practical strategies for adopting ethical practices. Additionally, it seeks to evaluate the effects of these practices on both corporate success and social well-being. Through an analysis of many case studies and the examination of empirical data, the objective of this study is to provide a valuable contribution to the continuing scholarly discussion around the influence of ethics on the trajectory of business within an ever-changing global economy.

2 Literature Review

2.1 Business Ethics in the Workplace

2.1.1 Introduction

The crucial role of corporate ethics in the contemporary business landscape is of utmost relevance. Ethical practices serve as the fundamental basis for establishing and nurturing respectful and mutually advantageous relationships among employers, workers, and customers. This article explores the fundamental aspects of business ethics, emphasizing its significance, obstacles, and pragmatic application in contemporary organizational settings.

2.2 Ethical Leadership and Organizational Culture

The academic literature defines business ethics as the set of rules and norms that regulate behavior in the business domain. The subject matter under consideration has been thoroughly examined in scholarly research conducted Kern et al. [8]. In their work, the authors provide a comprehensive definition of business ethics as a multifaceted framework that encompasses principles such as honesty, openness, and equal treatment across all facets of corporate operations [9]. The key elements included in this discussion are the concept of integrity in the realm of financial reporting, as emphasized by the enactment of the Sarbanes–Oxley Act in 2002 [10]. Additionally, the idea of treating stakeholders with respect is highlighted, drawing support from Kant's ethical theory of deontology. Kant's theory places significant emphasis on the inherent worth of persons, as expounded upon in his work published in 1785. Moreover, McIntosh et al. [11] assert that ethical business practices are not just driven by compliance, but rather play a crucial role in the development of sustainable company models, as shown in their shared value idea. The comprehensive approach used in this context recognizes the paramount importance of trust and integrity in cultivating enduring corporate success and developing connections with stakeholders.

2.3 Corporate Social Responsibility (CSR)

Corporate Social Responsibility (CSR) continues to be a fundamental aspect of corporate ethics, serving as a manifestation of an organization's dedication to the responsible management of social and environmental concerns. In contemporary scholarly discussions, scholars have expanded upon Jaffery and Kalim [12] seminal framework of corporate social responsibility (CSR), which posits that CSR encompasses economic, legal, ethical, and charitable aspects. This refined perspective has added more depth to our comprehension of CSR. Deng et al. [13] argue that corporate social responsibility (CSR) programs has strategic value, since they are not just driven by humanitarian motives but rather tightly aligned with both company aims and stakeholder expectations. This perspective aligns with the shared value idea proposed by [14] which posits that corporate social responsibility (CSR) may effectively enhance both company competitiveness and society well-being. The viewpoint presented is supported by empirical study conducted Aftab et al. [15], which provides evidence of a positive relationship between corporate social responsibility (CSR) practices and financial success. Additionally, the research highlights that CSR activities also contribute to the development of increased trust among stakeholders. The narrative that is now developing presents corporate social responsibility (CSR) as more than just a secondary activity, but rather as a fundamental and value-generating component of contemporary company strategy. This perspective highlights a more profound level of societal involvement and ethical obligation.

2.4 Ethical Decision Making

The investigation of ethical decision-making within organizational contexts is a crucial and prominent field of study in current academic research. Drawing upon the theoretical framework proposed by Perkins et al. [16] known as the issue-contingent model, subsequent research endeavors have undertaken a more comprehensive examination of the intricate dynamics between individual and environmental variables in the formation of ethical judgments. Johns [17] provide more insight into the subject by doing an analysis on the impact of company culture and leadership styles on the ethical decision-making processes. Do [18] study provides support for this viewpoint, since it underscores the significance of personal moral orientations and ethical climates in influencing decision-making behaviors inside corporate settings. Furthermore, Palmucci and Ferraris [19] provide valuable perspectives on the psychological processes that underlie ethical decision-making, emphasizing the influence of cognitive biases and moral disengagement. The aforementioned studies together emphasize the intricate nature of ethical decision-making within the realm of business, highlighting the intricate interaction between individual ethical values, the culture of the firm, and many situational considerations. The progressive nature of this increasing comprehension provides a more all-encompassing structure for effectively resolving ethical dilemmas within contemporary business settings.

2.5 Impact of Technology on Business Ethics

There has been a growing emphasis in recent academic studies on the ethical dilemmas brought about by technological progress, namely in the fields of big data and artificial intelligence (AI). Manoukian [20] emphasize the ethical intricacies that are inherent in the field of big data analytics, shedding light on concerns like privacy, consent, and algorithmic bias. He promote the implementation of comprehensive ethical frameworks to regulate the management of data and the development of AI algorithms. In their study, Murphy et al. [21] conduct an analysis of worldwide AI ethics standards, whereby they discover prominent themes like transparency, accountability, and justice in the context of AI systems. The aforementioned research are in accordance with the assertions put out by Sørensen et al. [22], who advocate for a systematic framework to facilitate responsible technology development. The emerging discipline of AI ethics encompasses the convergence of technological advancements and moral philosophy, demanding a meticulous equilibrium between innovative progress and ethical obligations within the context of the digital era. The aforementioned intellectual contributions represent a continuous endeavor to provide a framework for ethical governance within the swiftly changing realm of digital technology.

2.6 Consumer and Employee Perceptions

Contemporary scholarly investigations provide further validation of the considerable effect exerted by business ethics on the attitudes of consumers and employees, thereby yielding a major impact on the reputation and performance of organizations. Drawing upon the seminal research conducted by Malika et al. [23] a further investigation conducted by reveals an emerging inclination among consumers to accord more importance to ethical factors when making purchase choices. This shift in consumer behaviour is shown to be associated with the perception of corporate social responsibility, hence influencing customer loyalty. In respect to workers, a study conducted by, Waheed and Zhang [24] supports the conclusions made by [25] indicating a significant association between employees' assessments of their organization's ethical behaviour and their levels of job satisfaction and dedication. These studies jointly emphasize the significance of ethical practices in influencing customer trust and staff engagement, therefore highlighting the relevance of ethics in establishing a sustainable and respected firm.

2.7 Globalization and Cross-Cultural Ethics

The multifaceted relationship between globalization and corporate ethics poses distinct obstacles for organizations as they navigate diverse cultural and legal environments. Current academic research continues to investigate its intricate nature. Eang et al. [26] conducted a significant investigation that explores the strategies used by multinational firms to address ethical concerns within culturally different environments. This research highlights the difficulties associated with the implementation of ethical standards that can be universally applied. The authors propose the need of an adaptable ethical framework that acknowledges and respects cultural variety, while also maintaining fundamental ethical principles. Furthermore, the study conducted by [27] explores the intricacies of cross-cultural management ethics. The authors suggest that in order to successfully navigate ethical risks, multinational corporations need to strike a delicate equilibrium between adhering to local ethical norms and upholding global standards. The present-day perspectives discussed in this analysis expand upon the fundamental concepts put out by Mudd-Martin et al. [28] and Gupta and Pathak [29] so strengthening the argument that ethical conduct in the context of a worldwide society necessitates a sophisticated comprehension of cultural variances, along with an unwavering dedication to universally accepted ethical norms.

2.8 Conclusion

A review of existing academic literature demonstrates a prevailing agreement on the significance of ethics in the realm of business, as several research regularly establish a correlation between ethical conduct and favorable results for organizations. Nevertheless, the implementation of these techniques on a worldwide scale continues to encounter persistent problems, especially in light of the rapid pace of technical advancements and the process of globalization. Further investigation is required to delve into these processes, providing valuable insights into the successful incorporation of ethical principles in many corporate.

2.9 Framework Structured

This table highlights the essential elements of a business ethics framework and their roles in ethical business practices.

Component	Description
Core values	Integrity, transparency, fairness
Stakeholders	Interests of employees, customers, shareholders, communities

(continued)

(continued)

Component	Description
Cultural norms	Adapting to local cultural ethics and values
Legal compliance	Following laws and regulations in each region
Decision-making	Structured process for ethical choices considering all factors
Implementation	Strategies and actions to apply ethical principles
Evaluation	Regular assessment and refinement of ethical practices

3 Method

To establish this comprehensive review article, we conducted a thorough literature review on "Integrating Ethical Principles in Corporate Strategy: A Comprehensive Analysis of Business Ethics in Contemporary Enterprises," drawing on pivotal studies and seminal publications [30–32]. Our approach included a comprehensive examination of the literature on the incorporation of ethical concepts into business strategy, as described in significant publications [33–36]. Based on empirical research and theoretical frameworks, this systematic approach resulted in the creation of evidence-based recommendations targeted at improving ethical behaviours in business environments [37].

4 Discussion

Several significant results from our extensive study on the incorporation of ethical values in business strategy agree with previous academic studies. The importance of ethical leadership in establishing corporate culture and ethical behaviour is emphasized by Moore et al. [38] and Martin et al. [39] who argue that leaders play an important role in building ethical workplaces. This is complimented by the rising relevance of Corporate Social Responsibility (CSR) in strategic planning, which echoes Glauner [40] concept of shared value and is backed by Nirino et al. [41] results on the favourable effect of CSR on financial performance. The review emphasizes the difficulty of ethical decision-making in a global setting, which corresponds with Sahoo and Goswami [42] study on cultural variances in ethics and Ruiter [43] support for a balance between ethical universality and cultural sensitivity. Furthermore, the issues faced by technological breakthroughs, notably big data and AI, are underlined, echoing Reisach [44] and Szőcs and Miska [45] focus on the need of ethical principles in digital technologies. Finally, our results on the impact of ethical practices on stakeholder perceptions, such as customer loyalty and staff satisfaction, are consistent with De Moraes [46] and Metcalf and Moss [47]. This emphasizes the significance of ethical branding in moulding company reputation and success, reinforcing the notion that ethical integration into corporate strategy is a complex and strategic need for contemporary firms.

5 Conclusion

The article emphasizes the significance of ethics in corporate strategy, stating that ethical behaviours are essential for long-term company success. Ethical leadership, corporate social responsibility, decision-making in varied cultural settings, technical obstacles, and stakeholder perceptions are all interwoven components that have a substantial influence on a company's reputation and success. In today's globalized and technologically sophisticated economy, integrating these ethical elements is not just a moral obligation, but also a strategic need for organizations. To summarize, ethical behaviour is a critical predictor of an organization's long-term performance and reputation in the business community.

References

1. Jain, P., et al.: Data-driven artificial intelligence (AI) models in the workforce development planning. In: Designing Workforce Management Systems for Industry 4.0, pp. 159–176. CRC Press (2023)
2. Zheng, S., Jin, S.: Is corporate green investment a determinant of corporate carbon emission intensity? A managerial perspective. Heliyon (2023)
3. Elrefae, G., et al.: The entrepreneurship of accounting work and its role in reducing information asymmetry: evidence from insurance companies. Uncertain Supply Chain Manag. **12**(1), 101–114 (2024)
4. Kamila, M.K., Jasrotia, S.S.: Ethical issues in the development of artificial intelligence: recognizing the risks. Int. J. Ethics Syst. (2023)
5. Ren, L., Liu, J., Hao, Q.: How digital transformation affects the cost of equity capital: the role of information disclosure quality and stock liquidity. Ind. Corp. Change dtad053 (2023)
6. Katamba, D., et al.: Multinational oil and gas corporations' contribution to SDGs and social compliance in Uganda through their corporate social responsibility: a lens into readiness and the obstacles they face. In: The Elgar Companion to Corporate Social Responsibility and the Sustainable Development Goals, p. 250 (2023)
7. Nasir Ansari, J.A., Irfan, S.: Corporate social responsibility and employee engagement: the mediating role of personal environmental norms and employee green behavior. Soc. Responsib. J. (2023)
8. Kern, C.J., Noeltner, M., Kroenung, J.S.: The case of digital ethics in IS research—a literature review (2022)
9. Ali, A.A.A.: The Relationship Between Islamic Work Ethics and Employee Performance Among Medical Staff in the Public Health Sector in Bahrain: Employee Commitment as Mediating Variable. Universiti Sains Islam Malaysia (2022)
10. Kermis, G.F., Kermis, M.D.: Financial reporting regulations, ethics and accounting education. J. Acad. Bus. Ethics **8**, 1 (2014)
11. McIntosh, B., Sheppy, B., Zuliani, J.D.: The emperors clothes-corporate social responsibility creating shared value and sustainability. Int. J. Bus. Perform. Manag. **18**(3), 307–326 (2017)

12. Jaffery, A., Kalim, R.: Political competition, rising prices and economic augmentation: evidence from Pakistan. In: 4th International Conference on Business Management
13. Deng, X., Rao, K.K., Qian, W.: Riding the waves or fighting the tides? Female leaders' CSR engagement. Corp. Soc. Responsib. Environ. Manag. (2023)
14. Mapokotera, C., et al.: Enriching competitiveness through corporate social responsibility: a study of service-based enterprises in Zimbabwe. Qeios (2023)
15. Aftab, J., et al.: Does corporate social responsibility drive financial performance? Exploring the significance of green innovation, green dynamic capabilities, and perceived environmental volatility. Corp. Soc. Responsib. Environ. Manag. (2023)
16. Perkins, B.G., Podsakoff, N.P., Welsh, D.T.: Variance in virtue: an integrative review of intraindividual (un)ethical behavior research. Acad. Manag. Ann. (2023)
17. Johns, T.A.: A Case Study on Executive Leader Ethical Decision-Making. Indiana Wesleyan University (2023)
18. Do, H.T.K.: How do business educators prepare students better about ethical decision-making? J. Int. Educ. Bus. (2023)
19. Palmucci, D.N., Ferraris, A.: Climate change inaction: cognitive bias influencing managers' decision making on environmental sustainability choices. The role of empathy and morality with the need of an integrated and comprehensive perspective. Front. Psychol. **14**, 1130059 (2023)
20. Manoukian, G.: Subject Management of IT Date 14.06.2023
21. Murphy, K., et al.: Artificial intelligence for good health: a scoping review of the ethics literature. BMC Med. Ethics **22**(1), 1–17 (2021)
22. Sørensen, K., et al.: Building health literacy system capacity: a framework for health literate systems. Health Promot. Int. **36**(Supplement_1), i13–i23 (2021)
23. Malika, M., Maheswaran, D., Jain, S.P.: Perceived financial constraints and normative influence: discretionary purchase decisions across cultures. J. Acad. Mark. Sci. 1–20 (2022)
24. Waheed, A., Zhang, Q.: Effect of CSR and ethical practices on sustainable competitive performance: a case of emerging markets from stakeholder theory perspective. J. Bus. Ethics **175**(4), 837–855 (2022)
25. Chege, S.M., Wang, D.: The influence of technology innovation on SME performance through environmental sustainability practices in Kenya. Technol. Soc. **60**, 101210 (2020)
26. Eang, M., Clarke, A., Ordonez-Ponce, E.: The roles of multinational enterprises in implementing the United Nations sustainable development goals at the local level. BRQ Bus. Res. Q. **26**(1), 79–97 (2023)
27. Adawi, D., Kebedom, D.: Global connection chronicles: building bridges, not walls. Exploring business Sweden's dynamic international relationships in the African market in terms of trust, commitment, communication, and culture (2023)
28. Mudd-Martin, G., et al.: Considerations for cardiovascular genetic and genomic research with marginalized racial and ethnic groups and indigenous peoples: a scientific statement from the American Heart Association. Circ. Genom. Precis. Med. **14**(4), e000084 (2021)
29. Gupta, S., Pathak, G.S.: Ethical issues in virtual workplaces: evidence from an emerging economy. Eur. J. Train. Dev. (2022)
30. Ali, A.A.A., et al.: Moderating effect of strategic planning on the relationship between career path planning and job performance. Sustainability **15**(11), 8490 (2023)
31. Al-Refaei, A.A.-A., et al.: An integrated mediating and moderating model to improve service quality through job involvement, job satisfaction, and organizational commitment. Sustainability **15**(10), 7978 (2023)
32. Alzoraiki, M., et al.: Impact of teachers' commitment to the relationship between transformational leadership and sustainable teaching performance. Sustainability **15**(5), 4620 (2023)
33. Ali, A.A.A., Alzoraiki, M., Al Aghbas, N.S.M.: Working remotely, social isolation and access to support and communication during COVID-19 its influence on job performance. Int. J. Intellect. Hum. Resour. Manag. (IJIHRM) **3**(01), 23–33 (2022)

34. Ali, A.A.A., et al.: Association between Islamic work ethics and employee commitment among medical staff in the public health sector in Bahrain. Int. J. Intellect. Hum. Resour. Manag. (IJIHRM) (2023)
35. Ateeq, A.: Knowledge Management's Effect on Learning Quality in Bahrain's Private Universities (2023)
36. Alkadash, T.M., et al.: The effects of leadership style on employee sustainable behaviour: a theoretical perspective. In: Artificial Intelligence and Transforming Digital Marketing, pp. 205–213. Springer (2023)
37. Barakat, B., et al.: Assessing the impact of green training on sustainable business advantage: exploring the mediating role of green supply chain practices. Sustainability **15**(19), 14144 (2023)
38. Moore, C., et al.: Leaders matter morally: the role of ethical leadership in shaping employee moral cognition and misconduct. Am. Psychol. Assoc. **104** (2019)
39. Martin, S.R., et al.: Keeping teams together: how ethical leadership moderates the effects of performance on team efficacy and social integration. J. Bus. Ethics 1–13 (2022)
40. Glauner, F.: Redefining economics: why shared value is not enough. Compet. Rev. Int. Bus. J. **29**(5), 497–514 (2019)
41. Nirino, N., Miglietta, N., Salvi, A.: The impact of corporate social responsibility on firms' financial performance, evidence from the food and beverage industry. Br. Food J. **122**(1), 1–13 (2020)
42. Sahoo, S.K., Goswami, S.S.: A comprehensive review of multiple criteria decision-making (MCDM) methods: advancements, applications, and future directions. Decis. Mak. Adv. **1**(1), 25–48 (2023)
43. Ruiter, V.N.: Humanizing Hospitality Industry Human Resources Management to Improve Recruitment and Retention of Resilient Hospitable Talents in the Sector. University of Derby (United Kingdom) (2021)
44. Reisach, U.: The responsibility of social media in times of societal and political manipulation. Eur. J. Oper. Res. **291**(3), 906–917 (2021)
45. Szőcs, I., Miska, C.: Ethics in the context of cross-cultural management. In: The SAGE Handbook of Contemporary Cross-Cultural Management, pp. 227–238 (2020)
46. De Moraes, A.J.: Strategic Management and International Business Policies for Maintaining Competitive Advantage. IGI Global (2023)
47. Metcalf, J., Moss, E.: Owning ethics: corporate logics, silicon valley, and the institutionalization of ethics. Soc. Res. Int. Q. **86**(2), 449–476 (2019)

Strategic Integration of Business Systems: Merging ERP, HRM, and Accounting for Enhanced Performance—An Insightful Conceptual Article

Ali Ateeq, AlReem AlMuslemani, Marwan Milhem, and Mohammed Alzoraiki

Abstract The strategic the requirement of integrating Enterprise Resource Planning systems with Human Resource Management (HRM) and accounting software is becoming more acknowledged in the contemporary company landscape. This study examines the potential synergies and problems that arise from the convergence under consideration, using a range of scholarly sources and contributions. The objective of the integration is to improve the precision of data, streamline business operations, and promote overall organizational effectiveness. Contemporary enterprise resource planning systems serve as the fundamental element of comprehensive information systems inside organizations, integrating diverse functional areas into a unified database. This connectivity enables the seamless interchange of data and supports real-time decision-making across several departments. The integration of Human Resource Management (HRM) systems with Enterprise Resource Planning has significant importance, given the essential role of human resources in strategic management. Valecha emphasizes the im-portance of human resource management (HRM) systems in effectively managing employee data and ensuring the alignment of HR strategy with the broader business goals. The integration of Human Resource Management (HRM) and Enterprise Resource Planning systems facilitates a smooth and uninterrupted exchange of information, hence improving the effectiveness of personnel planning, recruiting processes, and performance management.

Keywords ERP integration · HRM systems · Accounting software · Organizational efficiency · Strategic management · Data accuracy · Business

---

A. Ateeq (✉) · A. AlMuslemani · M. Milhem · M. Alzoraiki
Administrative Science Department, College of Administrative and Financial Science, Gulf University, Sanad 26489, Kingdom of Bahrain
e-mail: dr.ali.ateeq@gulfuniversity.edu.bh

M. Milhem
e-mail: dr.marwan.milhem@gulfuniversity.edu.bh

M. Alzoraiki
e-mail: dr.mohammed.saied@gulfuniversity.edu.bh

B. Awwad (ed.), *The AI Revolution: Driving Business Innovation and Research*, Studies in Systems, Decision and Control 524, https://doi.org/10.1007/978-3-031-54379-1_18

process optimization · Cross-departmental collaboration · Financial reporting · Compliance

1 Introduction

The integration of enterprise resource planning systems with human resource management (HRM) and accounting software is a critical strategic undertaking in today's company environment [1]. This convergence attempts to improve data accuracy, optimize business processes, and boost organizational efficiency [2]. The introduction of modern ERP systems has transformed the way businesses manage their operations, money, and human resources. ERP systems, according to Ziemann [3] serve as the foundation of enterprise-wide information systems by combining multiple functional areas of a corporation into a cohesive database. This interface enables cross-departmental data exchange and decision-making in real time.

The integration of HRM systems with ERP is especially important since human resources play a crucial role in an organization's strategic management. According to Valecha [4], HRM systems are critical for handling employee-related data and integrating HR strategy with corporate goals. Organizations that integrate HRM and ERP may assure a continuous flow of information between HR and other departments, boosting personnel planning, recruiting, and performance management [5].

Furthermore, accounting system connection with ERP is critical for accurate financial reporting and compliance [6]. Yoshikuni et al. [7] emphasize the importance of integrated accounting systems in delivering timely and trustworthy financial information for strategic planning and decision-making. An integrated ERP and accounting system helps automate financial procedures, decrease mistakes, and give a complete perspective of the financial health of the firm [8]. However, integrating these technologies is not without difficulties. The difficulty of aligning various systems and procedures, as well as maintaining data security and managing change inside the business, are all important factors. As Silvestre et al. [9] point out, effective integration demands careful planning, stakeholder participation, and a clear knowledge of the organization's goals and objectives.

Finally, strategic ERP integration with HRM and accounting systems provides several advantages, including increased efficiency, better decision-making, and greater organizational performance. As firms traverse the intricacies of the digital era, efficient integration of various technologies will be critical to establishing competitive advantage and long-term success.

2 Literature Review

2.1 Integration of ERP, HRM, and Accounting Systems

The efficient integration of ERP, HRM, and accounting systems is a critical strategic alignment for contemporary organizations [10]. This integration creates a unified operational framework that allows for thorough data analysis, increased process efficiency, and better decision-making skills. Organizations may get a comprehensive picture of their operations by bridging the gaps between these diverse systems, resulting in optimal resource allocation and a more flexible reaction to market changes [11]. This integration not only improves internal operations, but it also supports strategic objectives by connecting the operational backbone with the organization's larger aims, assuring long-term competitive advantage in a continually changing business context.

2.2 ERP Systems and Business Integration

ERP systems play an important role in uniting disparate corporate processes, resulting in simplified operations and strategic coherence [12]. These systems go beyond typical departmental boundaries, allowing for a completer and more integrated picture of corporate operations. According to [13], ERP systems are more than simply technology solutions; they are strategic enablers that alter how firms function and compete. ERP systems provide a distinct advantage in improving responsiveness and agility in corporate operations by combining important areas such as supply chain management, human resources, and financials into a single framework. According to [14], the flexibility of ERP systems to new business models and procedures is critical to their long-term relevance, enabling firms to stay dynamic and competitive in a fast changing global business environment.

2.3 HRM Systems and Organizational Performance

HRM systems are critical for improving organizational performance by successfully managing human capital, which is a critical asset in every firm. Strategic HRM integration with ERP systems enables more dynamic workforce management by aligning HR strategies with broader company objectives [15]. This interface helps to optimize numerous HR activities including recruiting, training, performance assessment, and payroll administration [16]. Academic studies, such as Joo et al. [17] emphasize the strategic importance of HRM systems in driving organizational success. Furthermore, Kokkaew et al. [18] give empirical data proving a favourable relationship between HRM system adoption and increases in employee satisfaction and productivity. This link between HRM systems and organizational effectiveness emphasizes the importance of human resource management in the context of business strategy and operational execution.

2.4 Accounting Systems and Financial Management

Accounting systems are of utmost importance in the responsible administration of an organization's financial assets, and their integration with Enterprise Resource Planning solutions is especially critical for ensuring strong financial governance [19]. The integration highlighted by Narayan and Kommunuri [20] is of great importance as it serves to increase financial management and reporting capabilities. The integration of ERP systems enables the timely analysis of financial data, which is essential for making informed decisions and complying with regulatory requirements [21]. Furthermore, Limerick et al. [22] highlight the significance of integrated accounting systems in enhancing financial transparency and accountability. These technologies allow a holistic perspective on financial operations, hence guaranteeing precision and reliability in financial reporting. Hence, this integration serves to enhance financial management efficiency while also reinforcing the organization's adherence to regulatory requirements and strategic financial planning, thereby highlighting its significance in modern business approaches.

2.5 Interrelationships and Integration Challenges

The integration of enterprise resource planning, human resource management (HRM), and accounting systems presents several problems, including both technological and organizational aspects. Luo et al. [23] extensively examine the intricacies involved in combining systems that include diverse capabilities and originate from different sources. The authors underline that the successful integration of these components requires not only technical expertise, but also a comprehensive change management strategy and active involvement of stakeholders. This particular technique plays a crucial role in minimizing resistance and guaranteeing a seamless transition throughout the integration procedure. In addition to the obstacles elucidated by Luo et al. [23, 24] draw attention to the potential hazards inherent in these integrative endeavors. One of the primary issues that need utmost attention is the apprehension around data security, a matter that becomes more intricate inside an integrated setting. The convergence of disparate technologies increases the vulnerability to data breaches and mandates the implementation of rigorous security protocols [25]. Additionally, the process of integration has the potential to interrupt current corporate activities, resulting in temporary inefficiencies and times of adjustment.

Fig. 1 Conceptual image related to the integration of ERP (enterprise resource planning), HRM (human resource management), and accounting systems in a business environment

The existence of several intricate issues highlights the need of a meticulously planned approach when embarking on the integration of enterprise resource planning, human resource management (HRM), and accounting systems [26]. In order to effectively address these difficulties, organizations must prioritize technological compatibility, implement robust security standards, and develop a comprehensive strategy for managing organizational change. The effective integration of these systems, while it poses difficulties, may result in substantial advantages in terms of operational efficiency and strategic consistency. Figure 1 shows the intersection of ERP, HRM, and accounting systems.

2.6 Empirical Studies and Case Analyses

Throughout the domain of systems integration, empirical research and case studies provide useful insights into the practical implications and results associated with the integration of enterprise resource planning, human resource management (HRM), and accounting systems. Wu et al. [27] made a substantial contribution to the existing literature by conducting an in-depth case study. The study conducted provides a clear demonstration of the significant improvements in operations that may be attained via the successful integration of systems. The authors record occurrences of heightened productivity, optimized workflows, and increased data consistency, highlighting the concrete advantages of these linkages.

In addition to this, Wang and Zhang [28] examines the particular convergence of Human Resource Management (HRM) and Enterprise Resource Planning technologies. Wang and Zhang examines the transformative potential of integrating diverse systems in human resource management, using insights from a range of corporate case studies. The results indicate a significant improvement in human resources procedures and the ability to make informed decisions, resulting from the centralized and synchronized management of data pertaining to employees. This integration enables a more strategic approach to human resources, effectively integrating HR operations with the overarching company goals.

This empirical research gives tangible proof of the benefits that system integration may deliver. This statement emphasizes the potential for enhanced operational efficiency and strategic advantages resulting from a more unified and knowledgeable approach to managing diverse organizational activities. This study is crucial for firms who are considering the integration of their Enterprise Resource Planning, Human Resource Management (HRM), and accounting systems. It provides valuable insights into the anticipated advantages and effective implementation strategies. Table 1 shows the structured framework.

Table 1 Framework structured

Row	Component	Description
1	Objective of integration	To streamline operations and improve decision-making by unifying ERP, HRM, and accounting systems
2	Key benefits	Enhanced efficiency, improved data accuracy, and better strategic alignment across departments
3	Core systems involved	ERP: integrates all business processes. HRM: manages human resources. Accounting: manages financial transactions and reporting
4	Integration challenges	Technical complexities, data security issues, need for change management and stakeholder engagement
5	Strategies for integration	Establishing clear goals, ensuring technical compatibility, focusing on data security, and engaging stakeholders in the integration process
6	Expected outcomes	Improved operational efficiency, enhanced decision-making capabilities, and increased organizational agility

3 Method

In order to conduct this in-depth review article, we conducted a thorough literature analysis on "Strategic Integration of Business Systems: Merging ERP, HRM, and Accounting for Enhanced Performance," relying on key studies and publications [29–31]. Our method comprised a systematic examination and analysis of academic papers, identifying critical trends and insights in corporate system integration, with a particular focus on the amalgamation of ERP, HRM, and accounting systems [32–35]. This analytical approach culminated in the development of grounded, evidence-based methods to improve HRM practices in the context of digital integration [36–38].

4 Discussion

The literature study demonstrates the significant impact of integrating enterprise resource planning, human resource management (HRM), and accounting systems on improving organizational efficiency and decision-making processes. The benefits of heightened operational agility and improved data coherence are shown in the research conducted by [39, 40]. The implementation of these enhancements is of utmost importance for firms seeking to effectively adjust and thrive within the contemporary and ever-changing business landscape [41]. Furthermore, Saravanakrishnan et al. [42] emphasize the importance of these linkages in optimizing resource management, resulting in enhanced efficiency and effectiveness in using organizational assets. Nevertheless, the process of integration is not without of its inherent difficulties. The problems associated with the installation of integrated systems are highlighted by Rajapakse [43, 44]. The significance of implementing comprehensive change management methods is emphasized in order to effectively address and overcome these obstacles. The significance of this particular facet within the integration process cannot be overstated, as it guarantees a seamless transition and adjustment across all tiers of the organization, hence optimizing the potential advantages derived from system integration.

5 Conclusion

In summary, the integration of Enterprise Resource Planning [45], Human Resource Management (HRM), and accounting systems is a strategic manoeuvre that presents significant advantages in terms of enhancing organizational efficiency and facilitating informed decision-making. The statement is substantiated by the mentioned study, which demonstrates the potential for enhanced operational agility, enhanced data coherence, and better resource management. The aforementioned benefits have significant value in the dynamic and ever-changing realm of business, where the capacity to adapt and operate efficiently are crucial determinants of achieving successful results. Nevertheless, it is important to recognize the intricacies and obstacles associated with the implementation of these integrated systems. The use of effective change management methods is crucial in order to successfully traverse the aforementioned hurdles and guarantee a seamless transition. Ultimately, via diligent design and proficient implementation, the integration of these systems has the potential to provide substantial enhancements in organizational performance and competitiveness.

References

1. Jackson, L.A.: Enterprise resource planning systems: revolutionizing lodging human resources management. Worldw. Hosp. Tour. Themes **2**(1), 20–29 (2010)
2. Agustian, K., et al.: Human resource management strategies in achieving competitive advantage in business administration. J. Contemp. Adm. Manag. (ADMAN) **1**(2), 108–117 (2023)
3. Ziemann, J.: Fundamentals of Enterprise Architecture Management: Foundations for Steering the Enterprise-Wide Digital System. Springer Nature (2022)
4. Valecha, N.: Transforming human resource management with HR analytics: a critical analysis of benefits and challenges. Int. J. Glob. Acad. Sci. Res. **1**(2), 56–66 (2022)
5. Radhakrishnan, L.C., et al.: The role of business intelligence in organizational sustainability in the era of IR 4.0. In: Business Intelligence and Human Resource Management: Concept, Cases, and Practical Applications, p. 63 (2022)
6. Kumar, G.: Critical success factors of adopting an enterprise system for pharmaceutical drug traceability. Univ. J. Pharm. Pharmacol. 3–10 (2023)
7. Yoshikuni, A.C., et al.: Role of emerging technologies in accounting information systems for achieving strategic flexibility through decision-making performance: an exploratory study based on North American and South American firms. Glob. J. Flex. Syst. Manag. 1–20 (2023)
8. Halimuzzaman, M., et al.: Financial impact of enterprise resource planning (ERP) on accounting information systems (AIS): a study on petroleum companies in Bangladesh
9. Silvestre, W.J., Fonseca, A., Morioka, S.N.: Strategic sustainability integration: merging management tools to support business model decisions. Bus. Strateg. Environ. **31**(5), 2052–2067 (2022)
10. Weerasekara, U., Gooneratne, T.: Enterprise resource planning (ERP) system implementation in a manufacturing firm: rationales, benefits, challenges and management accounting ramifications. Account. Manag. Inf. Syst. **22**(1), 86–110 (2023)
11. Güttel, W.H., Müller-Christensen, B.: Human resource architectures: strategies, HR systems & market dynamics. In: Successful in Turbulent Times. Leadership, Change Management and Ambidexterity. Baden-Baden (2023)
12. Di Camillo, T.: Optimizing Operational Efficiency: A Comprehensive Study of ERP Systems and Accenture's SAP Utilization in Work Process Enhancement. Politecnico di Torino (2023)
13. Yathiraju, N.: Investigating the use of an artificial intelligence model in an ERP cloud-based system. Int. J. Electr. Electron. Comput. **7**(2), 1–26 (2022)
14. Watson III, E.F., Schwarz, A.H.: Enterprise and business process automation. In: Springer Handbook of Automation, pp. 1385–1400. Springer (2023)
15. Turner, P.: Complementarity in Organizations: Strategy, Leadership, Management, Talent and Engagement in the Fourth Industrial Revolution. Springer (2022)
16. Kambur, E., Yildirim, T.: From traditional to smart human resources management. Int. J. Manpow. **44**(3), 422–452 (2023)

17. Joo, M.K., et al.: Gender diversity advantage at middle management: implications for high performance work system improvement and organizational performance. Hum. Resour. Manage. **62**(5), 765–785 (2023)
18. Kokkaew, N., Peansupap, V., Jokkaw, N.: An empirical examination of knowledge management and organizational learning as mediating variables between HRM and sustainable organizational performance. Sustainability **14**(20), 13351 (2022)
19. Atmaja, D.S., Zaroni, A.N., Yusuf, M.: Actualization of performance management models for the development of human resources quality, economic potential, and financial governance policy in Indonesia Ministry of Education. Multicult. Educ. **9**(01), 1–15 (2023)
20. Narayan, A.K., Kommunuri, J.: New realities of risk management in the public sector. In: The Routledge Handbook of Public Sector Accounting, pp. 233–248. Routledge (2023)
21. Wahab, N.A.B.A., Nor, R.B.M.: The role of enterprise resource planning (ERP) systems in facilitating sustainable business practices. AI IoT Fourth Ind. Revolut. Rev. **13**(9), 29–39 (2023)
22. Limerick, A., Zambrano, P.W., Carley, C.: Assessing RFMIS implementation using DeLone and McLean model: enhancing regional financial management. Law Econ. **17**(1), 43–55 (2023)
23. Luo, Y., et al.: StemDriver: a knowledgebase of gene functions for hematopoietic stem cell fate determination. Nucleic Acids Res. gkad1063 (2023)
24. Allioui, H., Mourdi, Y.: Exploring the full potentials of IoT for better financial growth and stability: a comprehensive survey. Sensors **23**(19), 8015 (2023)
25. Fakhouri, H.N., et al.: A comprehensive study on the role of machine learning in 5G security: challenges, technologies, and solutions. Electronics **12**(22), 4604 (2023)
26. Nazarova, H., et al.: Transformation of personnel management principles in modern management. Econ. Dev. (2022)
27. Wu, L., et al.: How do digitalization capabilities enable open innovation in manufacturing enterprises? A multiple case study based on resource integration perspective. Technol. Forecast. Soc. Change **184**, 122019 (2022)
28. Wang, X., Zhang, Y.: Enterprise human resource optimization algorithm using PSO model in big data and complex environment. J. Environ. Public Health **2022** (2022)
29. Ali, A.A.A., et al.: Moderating effect of strategic planning on the relationship between career path planning and job performance. Sustainability **15**(11), 8490 (2023)
30. Al-Refaei, A.A.-A., et al.: An integrated mediating and moderating model to improve service quality through job involvement, job satisfaction, and organizational commitment. Sustainability **15**(10), 7978 (2023)
31. Alzoraiki, M., et al.: Impact of teachers' commitment to the relationship between transformational leadership and sustainable teaching performance. Sustainability **15**(5), 4620 (2023)
32. Ali, A.A.A., Alzoraiki, M., Al Aghbas, N.S.M.: Working remotely, social isolation and access to support and communication during COVID-19 its influence on job performance. Int. J. Intellect. Hum. Resour. Manage. (IJIHRM) **3**(01), 23–33 (2022)
33. Ali, A.A.A., et al.: Association between Islamic work ethics and employee commitment among medical staff in the public health sector in Bahrain. Int. J. Intellect. Hum. Resour. Manage. (IJIHRM) (2023)
34. Ateeq, A.: Knowledge Management's Effect on Learning Quality in Bahrain's Private Universities (2023)
35. Alkadash, T.M., et al.: The effects of leadership style on employee sustainable behaviour: a theoretical perspective. In: Artificial Intelligence and Transforming Digital Marketing, pp. 205–213. Springer (2023)
36. Ateeq, A., Ebrahim, S., Al-Ghatam, R.: The impact of stress and its influencing factors among dentists during the COVID-19 pandemic in Kingdom of Bahrain. F1000Research **11**, 930 (2022)
37. Al Balushi, F.S.M., Ali, A.A.A., Alzoraiki, M.: The impact of forecasting strategy and training and development strategy on preparing leadership competencies administrative support in the Sultanate of Oman. Int. J. Intellect. Hum. Resour. Manage. (IJIHRM) **3**(01), 34–41 (2022)

38. Barakat, B., et al.: Assessing the impact of green training on sustainable business advantage: exploring the mediating role of green supply chain practices. Sustainability **15**(19), 14144 (2023)
39. Turi, J.A., et al.: The role of big data analytics and organizational agility in improving organizational performance of business processing organizations. Bus. Process. Manag. J. **29**(7), 2081–2106 (2023)
40. Ahmetoglu, S., Che Cob, Z., Ali, N.A.: A systematic review of internet of things adoption in organizations: taxonomy, benefits, challenges and critical factors. Appl. Sci. **12**(9), 4117 (2022)
41. Kilaso, S.: Business, management: a robust framework for data-driven digital marketing strategies in the dynamic Thailand market. J. Res. Adm. **5**(2), 943–953 (2023)
42. Saravanakrishnan, V., et al.: Leveraging employee data to optimize overall performance: using workforce analytics. In: Designing Workforce Management Systems for Industry 4.0, pp. 235–250. CRC Press (2023)
43. Rajapakse, D.P.: Integration Between ERP Systems and Supply Chain Management
44. Momani, F., et al.: ERP implementation process improvement: a methodological proposals. J. Posit. Sch. Psychol. **6**(10), 1762–1770 (2022)
45. Caputo, F., et al.: Rethinking the role of technology for citizens' engagement and sustainable development in smart cities. Sustainability **15**(13), 10400 (2023)

The Impact of Liquidity, Financial Leverage, and Profitability on Dividend

Ali Abbas Murtadha Abbas Ali Hajee and Mujeeb Saif Mohsen Al-Absy

Abstract The dividend payout ratio is one of the important and frequently looked at indicators for investors looking to allocate capital to specific companies and build investment portfolios. This study aims to examine the impact of liquidity, financial leverage, and profitability on dividends. A total population of 42 publicly listed companies listed in Bahrain Bourse for two years of 2021 and 2022 have been selected. The result of the study indicates that liquidity and profitability would significantly positively affect the dividend payout ratio. However, it expects that leverage would significantly negatively affect the dividend payout ratio. The study is the first study in Bahrain that examines the influence of liquidity, financial leverage, and profitability on dividends. Hence, it adds value to the literature knowledge in terms of dividends that the shareholders are interested in. Results will help the policymakers and shareholders to monitor and control the investment.

Keywords Liquidity · Leverage · Profitability · Dividend · Payout ratio

1 Introduction

The worldwide financial scandals have had a substantial impact on the world's economies [1–6]. This paradigm shift led to institutions such as central banks playing a significantly more important role in capital markets and impacting the valuation of organizations in capital markets more significantly [7]. Specifically, from the post 2008 financial crises to the 2020 covid crises capital markets globally have generally seen a period of significant growth due to central banks keeping interest rates at historical lows [8]. This led to capital allocators being more focused on future growth rather than current cashflows which meant organizations that had low to

---

A. A. M. A. A. Hajee · M. S. M. Al-Absy (✉)
Accounting and Financial Science Department, College of Administrative and Financial Science, Gulf University, Sanad 26489, Kingdom of Bahrain
e-mail: dr.mujeeb.saif@gulfuniversity.edu.bh

A. A. M. A. A. Hajee
e-mail: 200101115195@gulfuniversity.edu.bh

B. Awwad (ed.), *The AI Revolution: Driving Business Innovation and Research*, Studies in Systems, Decision and Control 524, https://doi.org/10.1007/978-3-031-54379-1_19

moderates growth with significant cash inflows were valued relatively less than high growth high cash burn organizations [9]. This trend has however reversed recently as central banks have start tightening and increasing interest rates which has led to capital markets favoring present cashflows and therefore the capital allocations to companies that have consistent reliable cashflows which they are able to distribute to shareholders has increased and is forecasted to increase in the future [10].

This reduction in the risk tolerance of capital market participants combined with their increased desire for present cashflows has led to a need for research exploring the different sources of short term cashflows which organizations can distribute to shareholders [10] and given that the dividend issuance of organizations are one of the primary sources of cashflows that can be distributed to shareholders [11]. A research of [11] explore the dividend issuances through examining the impact of three distinct variables on the dividend payout ratio. The dividend payout ratio is viewed as a proxy for an organizations dividend issuance policy [11].

The relationship that different variables have on dividend payout can vary from geographical area to another [7, 12, 13]. This is because the decision making regarding when and when not to payout dividends is taken differently among executives in different nations. Although there have been studies that have explored how different variables impact dividend payout ratio none of them have focused on Bahraini companies as a whole and none of them have specifically focused on the three listed variables. This study aims to address that problem through targeting all companies listed on the Bahraini stock exchange with a specific focus on how liquidity, financial leverage and Profitability impact the dividend payout ratio.

There have been many studies that have been written in the field of financial sciences on liquidity, financial leverage and Profitability. There have also been several studies in the field of financial sciences that have focused on dividend payout ratio and the factors that impact dividend payout ratio linking Profitability, liquidity and financial leverage to dividend payout ratio however none of these studies have explored the relationship between these variables and dividend payout ratio within the Kingdom of Bahrain and given that the interplay between these variables can vary from one geographic area to another [7, 12, 13] this study aims to address this theoretical gap through exploring the relationship between the variables in the Kingdom of Bahrain and does so through it through the scope of all companies listed on Bahrain Bourse.

Any company's capital structure plays a significant role in its financials [14]. Hence, the dividend payout ratio is one of the important and frequently looked at indicators for investors looking to allocate capital to specific companies and build investment portfolios [11]. This is because the valuation models which investors rely on to make their decisions are often built on the foundation of discounting future cashflows and dividends play an important part of those cashflows [15, 16], this study helps investors looking to invest in Bahraini companies better evaluate the dividends payout potential of companies through examining how different important variables impact the dividend payout ratio in addition to helping executives of Bahraini companies better understand how to maximize their ability to raise capital through improving their ability to issue dividends which will help improve their attractiveness to capital allocators [15].

This study aims to examine the impact of the three independent variables these variables being liquidity, financial leverage, and Profitability on the single dependent variable of dividend payout ratio within the context of all Publicly listed companies on Bahrain bourse in the Kingdom of Bahrain at the end of the fiscal year of 2023.

2 Literature Review

Shabrina and Hadian [17] defined the dividend payout ratio as the reflection of the organizations dividend policy with the organizations dividend policy being its tendency to distribute the profits earned from its operations to its shareholders at the opportunity cost of the profits being distributed going into the retained earnings and being used for future financing and investment needs of the organization.

In their research [18] take a slightly more reductionist view to the dividend payout ratio by describing it as one of the many important ratios that aim to measure the market value. Specifically, Herawati and Fauzia [18] define the dividend payout ratio as the proportion of earnings generated from the net income that is distributed to shareholders and given that the structure of their definition is a more reductionist structure their empirical measurement of the dividend payout ratio is the ratio of the cash dividends relative to the net income which the organization generates.

In their research [19] link the dividend payout ratio to the capital gains versus improved dividend policy debate. Dividend payout ratio as the portion of retained earnings that is distributed as cash dividends to the shareholders of the organization further agreeing with [18] more narrow view of the dividend payout ratio as cash dividends are the dividends being emphasized. Arsyad et al. [19] also argue that the dividend payout ratio plays a more important role in investor satisfaction than capital gains maximization which means they argue that the dividend policy is more important than total capital gains. The empirical measurement used [19] is the ratio of dividends per share relative to earnings per share.

According to the research of Putri and Ugut [20], Profitability is one of the statistically significant determinants of the dividend payout ratio of an organization. Profitability is defined by Putri and Ugut [20] as the net profit which an organization is able to create as a result of the operations which it undertakes. Putri and Ugut [20] explain this statistically significant relationship between profitability and the dividend payout ratio of organizations through arguing that a higher level of profitability achieved by organizations would entail a larger margin of the revenue which the organization creates as by product of its operations is able to stay with the organization as opposed to being eaten up by the different costs that go into creating this revenue which incentivizes the organization to distribute the profits amongst its shareholders. Arsyad et al. [19] research further highlights the incentive structure which organizations have towards distributing their profits to shareholders as the research indicates that the dividend payout ratio is one of contributors to investors satisfaction and as [21] research indicates investor satisfaction can have an important link to firm value which is why the part of the literature that finds a positive relationship between profitability and the dividend payout ratio often argues for the improvement of the profitability of the organization for reasons including the increase of investor satisfaction so that the organization can get more lucrative financing opportunities [19].

According to the research of Arshad et al. [22], the financial leverage possesses a significant statistical relationship in relation to the dividend payout ratio. Arshad et al. [22] displayed this through an examination of the relationship between the dividend payout ratio and the firms overall debt which [22] define as the total amount owed to third party which the firms uses as a part of its overall capital structure. The authors initially hypothesized that a higher financial leverage ratio would be positively related to the organization's dividend payout ratio due to mechanisms such as a more efficient capital structure relative to the use of equity. This was confirmed by their empirical testing which is why they concluded that positive relationship between financial leverage levels which an organization possesses and the dividend payout ratio which was found in their study was due to increased debt levels leading organizations to boosting their competitive advantage through gaining access to more capital than equity alternatives. Other research such as the research of Shabrina and Hadian [17] also found the statistically significant relationship between the financial leverage of the organization and the dividend payout ratio however they found a negative relationship. Shabrina and Hadian [17] hypothesize that the reason for this negative relationship is the risk exposure faced by organizations to risks such as default risks and solvency risks increases with increased financial leverage. This conflict in results suggests that the relationship which financial leverage has to the debt to equity ratio is dependent upon variables such as the cost of borrowing and the collateral commitments [23].

According to Herawati and Fauzia [18] research liquidity has a statistically significant relationship in relationship with the dividend payout ratio. Herawati and Fauzia [18] suggested that the reason for the statistically significant relationship is due to the nature of fixed assets whereby when organizations have a higher portion of their assets be fixed assets their less likely to have the sufficient cash reserves necessary to issue dividends to shareholders thereby reducing the dividend payout ratio. Arshad et al. [22] also suggest a statistically significant relationship with dividend payout ratio. The hypothesized reason which the authors suggest is due to liquidity positioning of organizations. Arshad et al. [22] suggested that when an organizations liquidity position is strong the propensity to issue dividends increases due to the availability of cash and cash equivalents that can be used as dividends. It's important to note that much of the literature focus on liquidity from a similar perspective to Maharani et al. [23] whereby liquidity only focuses on the cash reserves and does not mention in depth the different classes of cash equivalents in depth. Table 1 illustrates some of previous studies in the effect of liquidity, financial leverage and profitability on dividend payout ratio.

Based on the review of the literature and the research done for this study, Fig. 1 shows the conceptional framework.

The review came up with the following hypothesis:

H_1: Liquidity has a positive significant relationship with dividend payout.

Table 1 Previous empirical studies

Author and date	Independent variable	Dependent variable	Conclusion
[17]	Current ratio (liquidity)	Dividend payout ratio	Positive statistically significant relationship between the two variables
[18]	Current ratio (liquidity)	Dividend payout ratio	No statistically significant relationship between the two variables
[16]	Liquidity (current ratio)	Dividend payout ratio	No statistically significant relationship between the two variables
[23]	Current ratio (liquidity)	Dividend payout ratio	Positive statistically significant relationship between the two variables
[22]	Liquidity (current ratio)	Dividend payout ratio	Positive statistically significant relationship between the two variables
[17]	Financial leverage (debt to equity)	Dividend payout ratio	Negative statistically significant relationship between the two variables
[18]	Financial leverage (debt to equity)	Dividend payout ratio	Negative statistically relationship between the two variables
[23]	Financial leverage (debt to equity)	Dividend payout ratio	No statistically significant relationship between the two variables
[24]	Financial leverage (debt to equity)	Dividend payout ratio	No statistically significant relationship between the two variables
[25]	Financial leverage (debt to total assets)	Dividend payout ratio	Negative significant relationship between the two variables
[17]	Profitability (return on assets)	Dividend payout ratio	Positive statistically significant relationship between the two variables
[18]	Profitability (return on assets)	Dividend payout ratio	Positive statistically significant relationship between the two variables
[16]	Profitability (return on investment)	Dividend payout ratio	Positive statistically significant relationship between the two variables
[25]	Profitability (return on assets)	Dividend payout ratio	Negative statistically significant relationship between the two variables
[23]	Profitability (return on assets)	Dividend payout ratio	Positive statistically significant relationship between the two variables

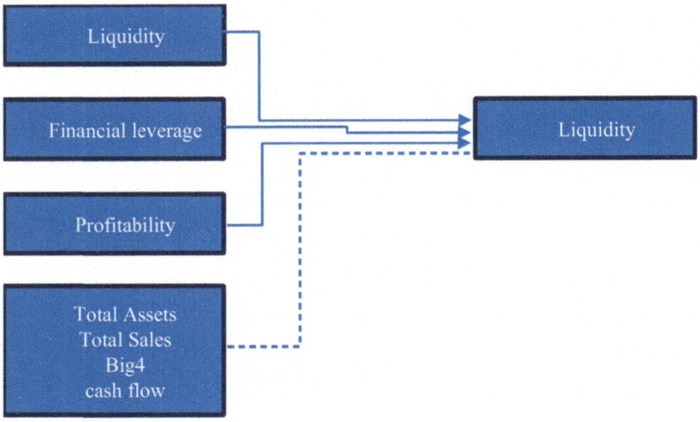

Fig. 1 Conceptional framework

H_2: Profitability has a positive significant relationship on dividend payout.

H_3: Financial leverage has a negative statistically significant relationship on dividend payout.

3 Methodology

This research has a total population of 42 publicly listed companies on and given that the total population is 42 publicly listed companies this research will select the entire population as the sample size. Furthermore, this research will look at the time period of the last two full financial years of each individual company listed on which means that it will look at the fiscal years of 2021 and 2022. The data of this research is going to be secondary data that is extracted from the annual reports of the sample companies. This research will analyze data using regression models that will be conducted through the STATA software. This is due to STATA software being a suitable statistical software for secondary data, several methods and analysis techniques were used in this study.

Regarding analysis, the study will determine and explain descriptive statistics such as the minimum, maximum, standard deviation and mean. Secondly, diagnostics tests like outliers, normality, multicollinearity, heteroscedasticity, autocorrelation and Hausman specification test were used to test the fitness of the sample data. Lastly, the study will use the multiple regression analyses called Ordinary least squares (OLS) to investigate the influence of liquidity, financial leverage, and profitability on the dividend payout ratio. The results and explanations are provided in other chapters of this research.

This research is going to measure the variable through measurements introduced by previous researchers. Table 2 show the details related to variables measurement.

Table 2 Variables measurement

Variable	Measurement
Dividend payout ratio	DPR = cash dividend/net income
Liquidity	Current ratio = current assets/current liabilities
Financial leverage	Financial leverage ratio = liabilities (sum of long term and short-term debt)/shareholders equity
Profitability	1. Return on assets = net income/total assets 2. Return on equity = net profit/total equity

4 Conclusion

One of the crucial and often examined metrics for investors wishing to allocate funds to businesses and construct investment portfolios is the dividend payout ratio. The purpose of this study is to investigate how profitability, liquidity, and financial leverage affect dividends. 42 publicly traded firms that were listed on Bahrain Bourse for the two years of 2021 and 2022 have been chosen overall.

According to the study's expectation, the dividend payout ratio is expected to be strongly positively impacted by profitability and liquidity. On the other hand, it is expected that leverage would have a substantial negative impact on the dividend payout ratio. This study is the first of its kind in Bahrain to look at the relationship between profitability, liquidity, and financial leverage and dividends. As a result, it enriches the body of literature regarding dividends that shareholders find valuable. Findings will support shareholders and policymakers in tracking and managing investments.

References

1. Al-Absy, M.S.M.: Impactful women directors and earnings management. Cogent Bus. Manag. **9**(1), 2148873 (2022)
2. Al-Absy, M., Ku Ismail, K.N.I., Chandren, S.: Board chairmen's involvement in audit committees and earnings management practices (2017)
3. Al-Absy, M., Hasan, M.: Impact of the board of directors' characteristics on firm performance: a case of Bahraini listed firms (2023)
4. Al-Absy, M.S.M.: Board chairman characteristics and real earnings management. Sustainability **14**(22), 15025 (2022)
5. Al-Absy, M.S.M.: Impactful female directors and earnings management: the moderating effect of ownership concentration. Adm. Sci. **13**(5), 129 (2023)
6. Al-Absy, M.S.M., AlMahari, N.H.: The interaction effect of nomination committee's effectiveness on board of directors' characteristics and firm performance. Adm. Sci. **13**(5), 135 (2023)
7. Al-Fasfus, F.: Impact of free cashflows on dividend payout in Jordanian banks. Asian Econ. Financ. Rev. 547–557 (2020)
8. Investments, B.: How do rising interest rates affect the stock market? In: usbank.com (2023)
9. Lev, B., Srivasta, A.: Explaining the Recent Failure of Value Investing. NYU Stern School of Business, pp. 1–29 (2019)

10. Glassman, J.K.: Value Investing Is Back. Accessed 1 Apr 2023. https://www.kiplinger.com/investing/stocks/value-investing-is-back
11. Husna, A., Satria, I.: Effects of return on asset, debt to asset ratio, current ratio, firm size, and dividend payout ratio on firm value. Int. J. Econ. Financ. Issues 50–54 (2019)
12. Wahjudi, E.: Factors affecting dividend policy in manufacturing companies in Indonesia Stock Exchange. Emerald Insight, pp. 4–16 (2018)
13. Nadratuzzaman, M.: Analysis of factors that impact dividend payout ratio on listed companies at Jakarta Islamic index. Int. J. Acad. Res. Account. Financ. Manag. Sci. 87–97 (2016)
14. Eshwari, D., Baby, M.: The effect of capital structure on the profitability of FMCG companies. In: International Conference on Economics, Business and Sustainability. Springer (2023)
15. Kusumanisita, A., Minanti, F.: Stock valuation analysis of dividend discount model, free cash flow to equity and Walter model in investment decision. J. Econ. Bus. 78–95 (2021)
16. Arsyad, M., Haeruddin, S., Muslim, M., Pelu, M.: The effect of activity ratios, liquidity, and profitability on the dividend payout ratio. Indones. Account. J. 36–44 (2021)
17. Shabrina, W., Hadian, N.: The influence of current ratio, debt to equity ratio, and return on assets on dividend payout ratio. Int. J. Financial Account. Manag. **3**(3), 193–204 (2021)
18. Herawati, A., Fauzia, F.: The effect of current ratio, debt to equity ratio, and return on asset on dividend payout ratio in sub sector automotive companies listed in Indonesia Stock Exchange in period 2012–2016. In: International Conference of Organizational Innovation, pp. 1076–1086 (2018)
19. Arsyad, M., Haeruddin, S., Muslim, M., Pelu, M.: The effect of activity ratios, liquidity, and profitability on the dividend payout ratio. Indones. Account. J. 36–44 (2021)
20. Putri, C., Ugut, G.: Factors affecting dividend payout ratio in LQ-45 (non-banking) companies listed in Indonesia Stock Exchange, 2011–2019. Enrichment J. Manage. 225–230 (2021)
21. Putri, M., Wiksuana, I.G.: The effect of liquidity and profitability on firm value mediated by dividend policy. Am. J. Humanit. Soc. Sci. Res. 204–212 (2021)
22. Arshad, N., Waseem, F., Abbas, S.: Factors affecting dividend payout ratio of dividend paying firms listed on KSE-100 index. iRASD J. Manage. 164–184 (2022)
23. Maharani, S., Lukaina, N., Fauziah, A.: The effect of current ratio, debt to equity ratio, and return on assets on dividend payout ratio in IDX high dividend index company 20 period 2017–2019. In: Progress Conference Proceedings, pp. 183–189 (2021)
24. Christine, C., Arilyn, E.: Factors affecting dividend payout ratio of HIDIV20 index firms. E-J. Manage. TSM 175–184 (2022)
25. Hayati, A., Astuti, G., Murdy, K.: The analysis of factors affecting dividend payout ratio in manufacturing companies listed in Indonesian Stock Exchange. In: Advances in Economics, Business and Management Research, pp. 34–41 (2018)

Factors Influencing Accounting Students' Career Choice at the Universities in Bahrain

Qais A. Almaamari and Rama Ali Mslem

Abstract This research aims to analyze the factors influencing accounting students' career choices in both public and private universities in Bahrain. The focus of this study includes self-motivation, benefits and rewards, and third-party effects, examining their impact on career decisions. The findings derived from the literature review indicate that self-motivation, benefits, and rewards, as well as third-party effects, significantly influence students' career choices positively. These findings contribute to a better comprehension of the career preferences among accounting students, shedding light on strategies that can attract and retain talented individuals in the accounting profession. Additionally, it offers insights into how to mitigate decision-making confusion, thereby enhancing satisfaction with the chosen career path. The data for this study was collected through a literature review encompassing various related research. This paper may serve as a resource for enhancing understanding on improving the career trajectory of accounting students.

Keywords Self-motivation · Benefits and rewards · Third-party effects · Career choice · Bahraini universities

1 Introduction

A career as an accountant is widely regarded as promising, offering excellent prospects and diverse opportunities across various business sectors. The accounting profession plays a pivotal role in ensuring financial transparency and the efficient functioning of businesses and organizations [1]. This profession embodies growth,

Q. A. Almaamari (✉)
Administrative Science Department, College of Administrative and Financial Science, Gulf University, Sanad 26489, Kingdom of Bahrain
e-mail: dr.qais.almaamari@gulfuniversity.edu.bh

R. A. Mslem
Accounting and Financial Science Department, College of Administrative and Financial Science, Gulf University, Sanad 26489, Kingdom of Bahrain
e-mail: 220101100004@gulfuniversity.edu.bh

B. Awwad (ed.), *The AI Revolution: Driving Business Innovation and Research*, Studies in Systems, Decision and Control 524, https://doi.org/10.1007/978-3-031-54379-1_20

stability, and prominence in the workplace. Arthur et al. [2] define an individual's career as a continuous progression of their work experiences and employment history over time. With the increasing demand for skilled accounting professionals, understanding the factors influencing accounting students' career choices is crucial. Identifying these factors enables educational institutions, employers, and policymakers to devise targeted strategies to attract, motivate, and retain talented individuals in the accounting field, allowing students to choose their path with greater satisfaction [3].

Numerous factors affect the career decision-making process, including student GPA, work environment, and labor market considerations. However, this study primarily focuses on three elements: self-motivation, defined as intrinsic drive unaffected by external circumstances; benefits and rewards, encompassing both intrinsic satisfaction and extrinsic factors such as salary, bonuses, promotions, and societal recognition [4]; and the third-party effect on career path choice. The latter refers to how societal expectations, norms, and perceptions shape individuals' career decisions. It considers the influence exerted by family, friends, peers, and society at large on an individual's career choices. This aspect underscores the impact of external factors on decision-making processes and career trajectories [5].

2 Significance of the Study

This study contributes to the theoretical understanding among accounting students themselves by providing insights into the factors influencing their career decisions. This awareness empowers students, enabling them to make informed choices that resonate with their personal values and align with their skills and interests.

In practical application, this research benefits educational institutions in two significant ways. Firstly, by comprehending the factors influencing career choices, institutions can enhance their career counseling services and develop curricula that better align with students' aspirations. This knowledge helps tailor accounting programs to match student preferences, fostering increased engagement and higher success rates.

Furthermore, this study holds relevance for employers as it sheds light on the factors impacting career choices. This understanding allows employers to formulate effective recruitment and retention strategies. Adapting organizational policies, refining work environments, and offering career development opportunities aligned with the preferences of accounting students can aid in attracting and retaining top talent within the field.

3 Literature Review

3.1 The Relationship Between Self-motivation and Choosing Career Path

According to Raharja and Liany [6], self-motivation stands out as a powerful factor in career choice. Motivation, stemming from personal attraction, interests, knowledge, and skills, influences individuals to apply their abilities. For accounting students, their belief in how their accounting skills and knowledge align with their career objectives significantly impacts their motivation to pursue an accounting career. The stronger their intrinsic motivation, the more robust their drive to become accountants [7].

Contrarily, Hatane et al. [8] found that a notable proportion of accounting students lack intrinsic motivation in choosing an accounting career. A previous research study conducted in Nigerian colleges involved distributing 300 questionnaires to undergraduate students from three universities. The findings indicated that internal motivation played a significant role in accounting students' career decisions. Their interest in accounting and optimism about their performance in the field were pivotal in choosing accounting as a career path [9, 10].

Furthermore, Joshi [11] highlighted that many students are motivated to pursue accounting careers due to the belief that it offers personal job satisfaction. This motivation drives them to work toward their goals, including upholding personal principles. Recognizing their capacity to comprehend and excel in accounting fosters their interest in studying it. Self-motivation significantly bolsters individual academic and personal aspirations.

H1: Self-motivation has significantly positively association with Choosing Career Path.

3.2 The Relationship Between Benefits and Rewards and Choosing Career Path

Financial and social return stands as a fundamental consideration in an individual's future path selection, as outlined by Durgut and Pehlivan [12]. Employee benefits encompass the perks and advantages an organization offers its employees in exchange for their services. These benefits not only affect the employees themselves but also impact those who rely on them financially.

Lidiyawati and Sahara [13] mention that many businesses have long believed that financial incentives attached to employment contracts serve as a primary method to attract and retain employees. Adequate monetary compensation becomes a vital element for a fulfilling career.

According to Khalid et al. [14], the allure of a career in accounting lies in aspects such as career progression, job stability, competitive salary packages, prestige, and diverse career opportunities. This perceived advantage significantly influences the career decisions of accounting students.

However, Hanlin [15] found that students often factor in potential future earnings when making career choices. Higher long-term income expectations often attract business students to actively pursue careers in that field. Additionally, students with higher earnings expectations tend to dedicate themselves to their studies, believing that better academic performance will lead to improved job prospects in the future. This expectation strongly influences students' career decisions and significantly impacts their academic performance.

Regression analysis revealed that earning expectations among accounting major students play a significant role in predicting the academic grades of traditional Chinese students, as per research findings by Supriyadi et al. [16]. When accounting students at Uniska contemplate a career in public accounting, factors such as financial rewards and personal considerations significantly influence their career choices. These elements hold substantial weight as students weigh the benefits and implications of becoming public accountants.

H2: Benefits and rewards has significantly positively association with Choosing Career Path.

3.3 The Relationship Between Third-Party and Choosing Career Path

The influence of third parties on career choices pertains to the impact that external individuals or entities—such as parents, friends, teachers, mentors, or societal expectations—have on shaping an individual's decisions regarding their career paths. While the influence of third parties in career choice can be significant, it remains crucial for individuals to consider their personal passions, motivations, skills, and values when making career decisions [17].

Khalid and Abdul Rauf [18] conducted a study utilizing 312 usable questionnaires out of the 500 distributed. Employing convenience sampling, the researchers aimed to compare accounting students at various academic stages, from first-year to final-year students. Their findings suggest a positive correlation between higher levels of intrinsic motivation and increased encouragement received from third parties.

Additionally, Joshi [11] noted that the decisions of accounting students to pursue careers as public accountants may also be influenced by their families and closest associates. Shanthrakumar [10] found a moderate impact of third-party variables. Observing records of accounting students, the study revealed a gradual increase in the number of students pursuing accounting. These findings suggest active encouragement from past graduates, seniors, parents, and lecturers, motivating individuals to pursue accounting careers.

Raharja and Liany [6] discovered a positive influence of third parties on the career paths of accounting students. Parents and friends primarily influence students' career path choices, with less impact from lecturers, community career consultants, and the media. One possible explanation could be the insufficient support provided by lecturers and career consultants in guiding students toward accounting career paths. This highlights the need for increased involvement and active assistance from lecturers and career consultants to enable informed career decisions for students.

H3: Third-party has significantly positively association with Choosing Career Path.

4 Conceptual Framework

Career choice depends on factors such as Third-party affects, Benefits and rewards and Self-motivation, as shown in Fig. 1.

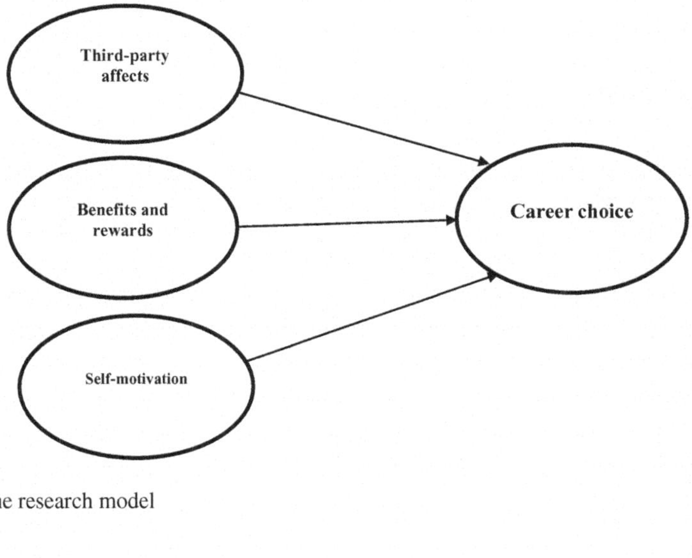

Fig. 1 The research model

5 Limitations and Recommendations

The limitations of this study primarily include a narrow scope of factors. Research focusing on a limited set of factors may overlook the complexity and interactions influencing accounting students' career paths.

To address this limitation, researchers are recommended to adopt a comprehensive approach, considering a wide range of factors that potentially impact career decisions. These may encompass personal characteristics, educational experiences, internships, mentoring, social networks, and industry trends. A mixed-methods approach, combining quantitative and qualitative data, can offer a more holistic understanding.

Another limitation involves contextual factors affecting accounting students' career paths. These factors, such as cultural, societal, or economic conditions, might vary significantly across regions or countries.

Future researchers are encouraged to acknowledge and account for contextual influences. Conducting comparative studies across diverse regions or countries could provide insights into how these factors interact with individual-level variables. Collaborating with researchers from varied backgrounds can enrich studies by integrating multiple perspectives.

By addressing these limitations and implementing the recommended approaches, researchers can bolster the validity, reliability, and applicability of studies investigating the factors shaping accounting students' career paths.

6 Conclusions

In conclusion, the study examining the factors influencing the career path choices of accounting students has highlighted three key factors: the benefits and rewards associated with the accounting profession, self-motivation, and the influence of third parties. The findings indicate that accounting students consider the benefits and rewards offered by the profession as an important factor in their career decision-making process. These benefits can include financial stability, job security, and opportunities for career advancement.

Furthermore, self-motivation plays a significant role in shaping accounting students' career choices. Students who possess a genuine interest, passion, and intrinsic motivation towards accounting are more likely to pursue a career in the field.

Lastly, the influence of third parties, such as parents, friends, and mentors, has been identified as a significant factor in accounting students' career decisions. The support, encouragement, and advice provided by these external influences can sway students towards or away from pursuing a career in accounting.

Based on these findings, it is important for educational institutions, career counselors, and accounting professionals to recognize the significance of these factors. Efforts should be made to provide comprehensive information about the benefits of a career in accounting, nurture and enhance students' self-motivation, and ensure that students receive guidance and support from trusted third parties throughout their career exploration and decision-making process.

By addressing and emphasizing these factors, the accounting profession can attract and retain students who are well-suited for successful careers in the field.

References

1. Venter, E.R., Gordon, E.A., Street, D.L.: The role of accounting and the accountancy profession in economic development: a research agenda. J. Int. Financ. Manag. Acc.Financ. Manag. Acc. **29**(2), 195–218 (2018)
2. Arthur, M.B., Hall, D.T., Lawrence, B.S.: Handbook of career theory. Cambridge: Cambridge University Press (1989). https://doi.org/10.1017/CBO9780511625459
3. Ganapathy, D., Deepak, S.: A study on reskilling and networking on LinkedIn on employee recruitment success and career advancement. In: Aloysius Edward, J., Jaheer Mukthar, K.P., Asis, E.R., Sivasubramanian, K. (eds.) Current Trends in Economics, Business and Sustainability. ICEBS 2023. Contributions to Environmental Sciences & Innovative Business Technology. Springer, Singapore (2023). https://doi.org/10.1007/978-981-99-3366-2_29
4. McCausland, W.D., Pouliakas, K., Theodossiou, I.: Some are punished and some are rewarded: a study of the impact of performance pay on job satisfaction. Int. J. Manpow.Manpow. **26**(7/8), 636–659 (2005)
5. Khan, W., Iqbal, Y.: An investigation of the relationship between work motivation (intrinsic & extrinsic) and employee engagement: a study on Allied Bank of Pakistan (2013)
6. Raharja, S., Liany, D.: Factors affecting accounting students in choosing accounting career path. J. Dinamika Akunt. **12**, 100–113 (2020). https://doi.org/10.15294/jda.v12i2.24169
7. Ahmed, F., Almaamari, Q.: Understanding the relationship between training, motivation and working environment and employee performance in the banking sector in Bahrain. Int. J. Intellect. Hum. Resour. Manag. (IJIHRM) **1**(01), 64–68 (2020)
8. Hatane, S.E., Gunawan, F.A., Pratama, S.W.: Intrinsic motivation, career exposure, and quality of life: how do they influence the accounting students' career choice? Educ. Learn. 11 (2021)
9. Grigorieva, N.G., Drutskaya, S.M.: Determinants of professional self-motivation and educational motivation of students. In: Proceeding of the International Science and Technology Conference "FarEastCon 2020", Far Eastern Federal University, Vladivostok, Russian Federation, Oct 2020, pp 269–278. Springer, Singapore (2021)
10. Shanthrakumar, K.: Factors influencing accounting students' career paths in Trincomalee Campus, Eastern University, Sri Lanka. Int. J. Sci. Res. Publ. (IJSRP) **10**(10) (2020). ISSN: 2250-3153. https://doi.org/10.29322/IJSRP.10.10.2020.p10628
11. Joshi, P.L.: A conceptual framework for the factors influencing the accounting students career choice as public accountants. J. Account. Finance Audit. Stud. (JAFAS) **8**, 1–29 (2022). https://doi.org/10.32602/jafas.2022.016
12. Durgut, M., Pehlivan, A.: Analysis of factors that affect the job choice of accounting students. Sos. Bilim. Derg. 16 (2019)
13. Lidiyawati, Sahara, A.: Factors affecting the interest of accounting students. Palarch's J. Archaeol. 12 (2020)
14. Khalid, F.M., Abdul Rauf, F.H., Hanapiah, M.A., Joon, M.I., Razak, N.A., Vijayandran, S.: Factors influencing accounting students' career choice in Malaysia. Glob. Bus. Manag. Res. Int. J. 10 (2021)
15. Hanlin, W.: Factors affecting accounting students' career choice. Wenzhou-Kean University, p. 24 (2020)

16. Supriyadi, S.G., Jatmika, D., Asnawi, A.: Factors affecting career selection of accounting students to become public accountants. Int. J. Econ. Bus. Account. Res. (IJEBAR) 5 (2020)
17. Taha, M.H., Eldalash, S.A., Mohamed, H.Z.: The role of accounting disclosure of sustainable development activities using metaverse in the field of education and training. In: El Khoury, R., Alareeni, B. (eds.) How the Metaverse Will Reshape Business and Sustainability. Contributions to Environmental Sciences & Innovative Business Technology. Springer, Singapore (2023). https://doi.org/10.1007/978-981-99-5126-0_7
18. Khalid, F.M., Abdul Rauf, F.H.: Factors influencing accounting undergraduates' career path: evidence from Malaysia. In: Othman, N.S., Jaaffar, A.H.B., Harun, N.H.B., Buniamin, S.B., Mohamad, N.E.A.B., Ali, I.B.M., Razali, N.H.B.A., Hashim, S.L.B.M. (eds.) Driving Sustainability Through Business-Technology Synergy, vol. 100. European Proceedings of Social and Behavioural Sciences, pp. 246–258. European Publisher (2020). https://doi.org/10.15405/epsbs.2020.12.05.26

Factors Affecting Employees' Performance in the Hospitality Industry in Kingdom of Bahrain

Qais A. ALmaamari and Maya Anwar Majdalawi

Abstract The aim of this paper is to examine the effect of several factors on employee performance in the hospitality industry in Bahrain. The factors include incentives, leadership, and work environment. This study focuses on the literature review which will help examine the factors (incentives, leadership, and work environment) in order to give insight into the influence of these factors and employee performance in the hospitality sector in Bahrain. The finding/hypothesis of this study shows that incentives, leadership, and work environment all have a positive influence on employee performance. This research will aid HR managers in developing innovative methods to help workers perform better and get a deeper knowledge of the factors that influence employee performance.

Keywords Leadership · Work environment · Incentives · Employee performance · Hospitality industry · Bahrain

1 Introduction

Employee performance stands as a vital aspect of every organization's success. Increased employee engagement leads to enhanced productivity, efficiency, and overall corporate success. Well-performing employees contribute to cost savings, heightened productivity, and a competitive edge. Moreover, their involvement and job satisfaction positively impact the work environment and staff morale [4, 13, 35].

An individual is considered a top performer if they efficiently and effectively fulfill their duties while meeting the organization's objectives within specified time frames [32]. For optimal employee performance, every employee must actively contribute;

Paper type: literature review.

Q. A. ALmaamari (✉) · M. A. Majdalawi
Administrative Science Department, College of Administrative and Financial Science, Gulf University, Sanad 26489, Kingdom of Bahrain
e-mail: dr.qais.almaamari@gulfuniversity.edu.bh

B. Awwad (ed.), *The AI Revolution: Driving Business Innovation and Research*, Studies in Systems, Decision and Control 524, https://doi.org/10.1007/978-3-031-54379-1_21

if just one person performs exceptionally well, it may not guarantee overall success for the business.

High-performing individuals often demonstrate innovative problem-solving skills, fostering organizational innovation. Their performance fosters collaboration and teamwork, contributing to a robust workplace culture. Organizations that prioritize employee performance offer opportunities for professional growth, attracting and retaining top talent. By investing in employee development and supporting their performance, organizations can achieve their goals, drive innovation, and maintain a strong market position.

Challenges in employee performance have been identified in developed countries due to factors such as lack of managerial support and leadership issues (Problems of Performance Appraisal in Developed Country, 2023). Developing nations, like South Africa, aim to enhance employee performance by addressing workplace challenges such as insufficient management and leadership, hindering progress.

Despite their expertise in business and economics, GCC countries, including Saudi Arabia, encounter challenges in enhancing employee productivity and performance [19]. Improving employee performance in Bahrain's hotel industry can significantly impact the sector. Focusing on enhancing employee performance can elevate all aspects of the visitor experience, leading to increased client satisfaction and positive ratings. Well-trained, motivated employees offering exceptional service provide hotels with a competitive edge, attracting more visitors and building a strong reputation.

Bahrain, despite its small size, has emerged as a regional tourist hub with around 190 hotels and resorts, and more scheduled to open by 2020 [16]. Major hotels in Bahrain have empowered non-management staff to make decisions in line with organizational goals without requiring manager authorization, emphasizing the importance of factors that boost employee performance [33].

2 Significance of Study

The study of the factors that affect employee performance is necessary in order to know how to increase employee performance in organizations. It is important to know the type of influence of factors such as incentives, work environment, and leadership on employee performance to acknowledge the problem of employee performance in the hospitality sector of Bahrain. Therefore, this study will help HR managers to get a practical implication of the effect of these factors on employee performance. It is important to search the literature review about employee performance and further study the factors that could affect this problem. This study will also help the business field, specifically HR and hospitality to improve their strategies in order to improve employee performance.

3 Literature Review

3.1 The Effect of Leadership on Employee Performance

Leadership encompasses the practice of persuading individuals to willingly achieve collective goals [27]. The impact of leadership on employee performance is considerable. Effective leadership can inspire and motivate staff, resulting in increased performance and productivity. Leaders, by setting clear goals, providing direction, and fostering a positive work environment, can enhance employee engagement and commitment [17]. Communication from leaders ensures that employees understand their roles, aligning their efforts with company objectives.

Leaders who engage in staff development, providing assistance and resources, aid employees in improving their skills and capabilities, ultimately leading to better performance. According to situational leadership theory, effective leadership addresses individual and situational differences, crucial in determining employee performance [12, 18].

Various leadership styles can have different impacts on employee performance, such as transformational, laissez-faire, autocratic, and democratic leadership styles [22]. Research by Alfalayyeh and Alghamdi concluded that leadership has a positive impact on employee performance.

Transformational leadership involves collaboration to recognize the need for change and inspire a vision for such transformation [22]. These leaders motivate workers to reach their potential by encouraging innovation and progress. They facilitate shared awareness, self-management, and motivation among employees [31]. Studies, including one by Tamimi and Sopiah [38], found that transformational leadership positively impacts employee performance. Laissez-faire leadership grants subordinates autonomy with minimal guidance from leaders [7, 22]. While this style empowers team members, it may lead to role ambiguity and reduced group involvement [10]. Autocratic leadership involves centralized decision-making, where leaders dictate assignments and decisions without consulting their team [22]. While effective in quick decision-making, it may stifle creativity and be perceived as controlling.

Democratic leadership emphasizes inclusivity and participation in decision-making [21]. While fostering open communication and inclusivity, it might not be suitable in time-sensitive situations. Study done by Puyri and Pasaribu, confirm that leadership significantly affects employee performance, emphasizing the influence of leadership styles on performance.

H1: leadership has a positive effect on employee performance.

3.2 The Effect of Work Environment on Employee Performance

Employee satisfaction significantly hinges on a welcoming work environment. It can broadly be defined as a workplace meeting diverse expectations, ensuring employees feel content and eager to work daily, without contemplating quitting. As it influences employees' moods, emotions, and actions, a pleasant work environment is critical [11, 36]. The work environment includes job security, safety, performance recognition, motivating benefits, and fostering excellent communication among colleagues and managers. Research by Gu et al. [14] confirms the positive impact of the work environment on employee performance.

A favorable work environment also encourages work-life balance, offers growth opportunities, and recognizes employee achievements. Prioritizing employee well-being contributes to better performance, reduced stress, and increased engagement and productivity. These elements culminate in improved job quality and overall performance, as evidenced in prior research linking the work environment to enhanced productivity [11].

Conversely, a negative work environment characterized by poor communication, lack of support, and high stress adversely affects employee performance. It leads to decreased motivation, higher turnover rates, increased absenteeism, and reduced productivity [37]. Changes in social dynamics, information technology, and flexible work approaches have evolved the office work environment over recent decades, showing that a better workplace environment results in higher productivity.

Heavy workloads, unattainable deadlines, and ongoing pressure can negatively impact employee well-being, hindering concentration, creativity, and problem-solving abilities. This leads to decreased performance and increased error risk (The Effect of Working Environment on Employees Performance—Research Key, 2023).

Research focusing on Bahrain's hospitality sector highlights programs like information sharing, training, and participative decision-making in hotels, factors that can enhance the work environment and consequently improve employee performance [26]. Studies by Shammout concluded that all dimensions of the work environment positively impact employee performance, underlining its significance in organizational success. Fuad and Sankar [26] emphasized the lack of employee empowerment in Bahrain's hospitality industry, signaling a need for an improved work environment to mitigate declining employee performance.

H2: work environment has a positive effect on employee performance.

3.3 The Effect of Incentives on Employee Performance

Incentives wield a significant impact on employee performance within firms. When appropriately developed and executed, incentives can motivate employees, enhance productivity, and drive positive behavior. Financial rewards, recognition, or prospects for career advancement are influential incentives that inspire higher performance levels. Rewards serve as a potent motivator; employees tend to excel when recognized and rewarded based on their performance [28]. Establishing a successful reward system provides employees with purpose and clear objectives, fostering excitement and dedication. Conversely, a lack of an effective reward system can negatively impact organizations, leading to decreased productivity and demotivation among employees [8, 29].

Moreover, rewards significantly impact staff engagement. Acknowledging and rewarding employees' efforts and achievements lead to higher work satisfaction and increased loyalty. Recognizing employees' contributions creates a sense of appreciation within the firm, enhancing engagement and commitment to their roles [29]. Various incentives can be employed within firms to enhance employee performance, such as monetary, non-monetary, performance-based, career development, and recognition incentives [34].

Monetary incentives, including bonuses, commissions, profit-sharing plans, and salary increases, are typical examples. These incentives provide immediate financial benefits, significantly impacting employees' motivation and performance. Non-monetary incentives, however, focus on non-financial benefits to elevate employee happiness and engagement. Examples include recognition programs, employee appreciation events, flexible work arrangements, and achievement certificates [30].

Career development incentives, like promotions, training programs, and mentorship opportunities, aim to create pathways for growth and advancement. Recognition incentives acknowledge employees' efforts and successes, nurturing a positive work environment and boosting morale. Research by Sittenthaler and Mohnen [34] suggests that non-monetary incentives have a more positive impact on employee performance. Scholars like Noorazem et al. [30] affirm that incentives directly affect employee performance.

H3: incentives have a positive effect on the employee performance.

3.4 *Conceptual Framework*

Figure 1 illustrates the relationship between the 3 independent variables, leadership, work environment, and incentives, and the dependent variable which is the problem that this research revolves around. The dependent variable (employee performance) is directly affected by incentives, work environment, and leadership. It presents a positive relationship because when leadership, work environment, and incentives increase, employee performance also increases. On the other hand, when the 3 factors decrease, the employee performance gets negatively affected and also decreases. This shows that these factors are necessary and should be taken into consideration in any organization in order to increase the employee's performance. Although these factors are important to consider and have a significant impact on employee performance,

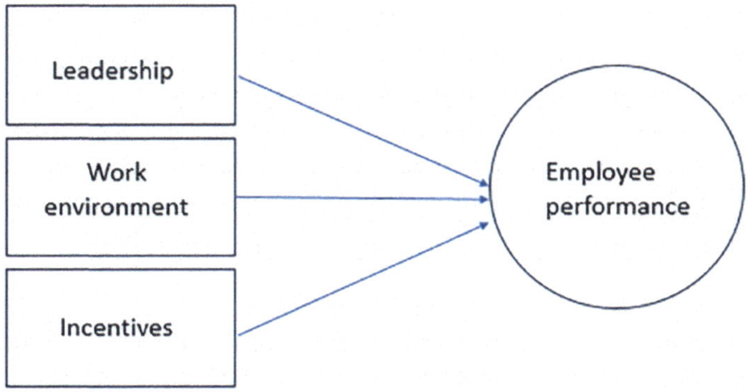

Fig. 1 The research model

they do not fully address the issue of declining employee satisfaction because there are other variables that may be influencing employee performance; however, increasing these variables would significantly improve the overall performance of employees.

4 Limitations and Recommendations

The limitation of this study is that it does not entirely address the factors that affect employee performance, there are many other factors that also affect employee performance so these three factors in this study can't fully solve employees' low performance. Moreover, another limitation of this study is time constraints, there is always room for improvement of the study due to the time limit which makes the study more difficult to execute, Furthermore, another limitation is that the study could have more sources to obtain the information and data from other than secondary sources.

The recommendations of this study are to further research other factors that affect employee performance such as motivation, HR, training, etc. Moreover, using a variety of sources could be helpful for future recommendations such as primary sources. Another recommendation is to make sure there is good time management in order to conduct good research.

5 Conclusion

This study aimed to investigate the impact of various factors on employee performance, including incentives, work environment, and leadership. The findings indicate that all three factors—incentives, work environments, and leadership—positively influence employee performance. Increased incentives result in enhanced employee satisfaction and motivation, consequently elevating performance. Effective leadership within the organization leads to improved employee performance by providing direction and purpose. Moreover, a conducive work environment contributes to increased employee performance as it fosters comfort and enhances productivity. These findings support the validity of all three hypotheses, demonstrating the direct impact of these factors on employee performance. Considering their essential nature, these factors warrant regular consideration within any organization. When an organization effectively integrates these factors, it can significantly elevate employee performance. However, it's crucial to acknowledge that there might be other factors negatively affecting employee performance. Further studies are recommended to thoroughly explore and address any such negative influences on employee performance.

References

1. Ahmed, F., Almaamari, Q.: Understanding the relationship between training, motivation and working environment and employee performance in the banking sector in Bahrain. Int. J. Intellect. Human Resour. Manag. (IJIHRM) **1**(01), 64–68 (2020)
2. Ahmed, Z., Shakoor, Z., Khan, M.A., Ullah, W.: The role of financial risk management in predicting financial performance: a case study of commercial banks in Pakistan. J. Asian Finan. Econ. Bus. **8**(5), 639–648 (2021)
3. Aldoseri, F.I., Almaamari, Q.A.: Factors influencing employee performance at the banking sector in Kingdom of Bahrain: literature review. Int. J. Emerg. Technol. **11**(5), 304–309 (2020)
4. Al-Harethi, A.A.M., Al-Maamari, Q.A.: The impact of strategic planning on improving institutional performance at limkokwing university of creative technology in Malaysia. Int. J. Psychol. Cogn. Sci. **4**(3), 112–129 (2018)
5. Almaamari, Q.A., Ali, B.M., Almeer, S.: Factors influencing organizational performance at petroleum products distribution company in Yemen. Specialusis Ugdymas **1**(43), 2071–2083 (2022)
6. Almaamari, Q.A.: Factors influencing employee turnover in banking sector. In: From Industry 4.0 to Industry 5.0: Mapping the Transitions, pp. 987–997. Springer Nature Switzerland, Cham (2023)
7. Alsaad, L.H., Almaamari, Q.A.: A semi literature review of factors influencing employee's performance at banking service. Int. J. Manag. (IJM) **11**(8) (2020)
8. Almaamari, Q.A., Ali, B.M., Almeer, S.: Factors influencing organizational performance at petroleum products distribution company in Yemen. Specialusis Ugdymas **1**(43), 2071–2083 (2022)
9. Alzoraiki, M., Ahmad, A.R., Ateeq, A.A., Naji, G.M.A., Almaamari, Q., Beshr, B.A.H.: Impact of teachers' commitment to the relationship between transformational leadership and sustainable teaching performance. Sustainability **15**(5), 4620 (2023)
10. Alheet, A.F., Adwan, A.A., Areiqat, A.Y., Zamil, Ahmad, M.A., Saleh, M.A.: The effect of leadership styles on employees' innovative work behavior. Manag. Sci. Lett. **11**(1), 239–246 (2021). https://doi.org/10.5267/j.msl.2020.8.010
11. Badrianto, Y., Ekhsan, M.L.: Effect of work environment and job satisfaction on employee performance in pt. Nesinak industries. J. Bus. Manag. Acc. **2**(1) (2020)
12. Ganapathy, D., Deepak, S.: A study on reskilling and networking on LinkedIn on employee recruitment success and career advancement. In: Aloysius Edward, J., Jaheer Mukthar, K.P., Asis, E.R., Sivasubramanian, K. (eds.) Current Trends in Economics, Business and

Sustainability. ICEBS 2023. Contributions to Environmental Sciences & Innovative Business Technology. Springer, Singapore (2023). https://doi.org/10.1007/978-981-99-3366-2_29
13. George, R., Krishnakumar, S.: Significance of employee engagement and succession planning on organizational development. In: Aloysius Edward, J., Jaheer Mukthar, K.P., Asis, E.R., Sivasubramanian, K. (eds.) Current Trends in Economics, Business and Sustainability. ICEBS 2023. Contributions to Environmental Sciences & Innovative Business Technology. Springer, Singapore (2023). https://doi.org/10.1007/978-981-99-3366-2_35
14. Gu, Z., Chupradit, S., Ku, K.Y., Nassani, A.A., Haffar, M.: Impact of employees' workplace environment on employees' performance: a multi-mediation model. Front. Public Health [online] **10** (2022). https://doi.org/10.3389/fpubh.2022.890400
15. Hamad Aldoseri, A.R.: The impact of ethical leadership on employee performance in the hotel industry in Bahrain: the mediating effect of employee voice (2020)
16. Hamad, R.: The impact of ethical leadership on employee performance in the hotel industry in Bahrain: the mediating effect of employee voice (2020). [online] Scholarworks@UAEU. Available at: https://scholarworks.uaeu.ac.ae/business_dissertations/18/. Accessed 20 Oct. 2023
17. Pancasila, I., Haryono, S., Sulistyo, B.A.: Effects of work motivation and leadership toward work satisfaction and employee performance: evidence from Indonesia. J. Asian Finan. Econ. Bus. **7**(6), 387–397 (2020). [online]. https://doi.org/10.13106/jafeb.2020.vol7.no6.387
18. Pancasila, I., Haryono, S., Sulistyo, B.A.: Effects of work motivation and leadership toward work satisfaction and employee performance: evidence from Indonesia. J. Asian Finan. Econ. Bus. **7**(6), 387–397 (2020). https://doi.org/10.13106/jafeb.2020.vol7.no6.387
19. Abdelmegeed, N.A., Aldoghan, M.: Developing employee productivity and performance through work engagement and organizational factors in an educational society. Societies **13**(3), 65–65 (2023). https://doi.org/10.3390/soc13030065
20. Alheet, A., Adwan, A., Areiqat, A., Zamil, A., Saleh, M.: The effect of leadership styles on employees' innovative work behavior. Manag. Sci. Lett. **11**(1), 239–246 (2021)
21. Ammar: Impact of autocratic and democratic leadership styles on employees' performance and motivation. ResearchGate; unknown. dan (2022). Garuda—Garba Rujukan Digital (2023). [online] Kemdikbud.go.id. Available at: https://garuda.kemdikbud.go.id/documents/detail/2877756. Accessed 20 Oct. 2023
22. Dastane, D.O.: Impact of leadership styles on employee performance: a moderating role of gender. Aust. J. Bus. Manag. Res. **5**(12), 27–52 (2020)
23. Dinastipub.org.: View of the effect of leadership, job satisfaction, organizational citizenship behaviour and competence on employee performance (case study at XYZ Ministry in Indonesia) (2023). [online]. Available at: https://www.dinastipub.org/DIJDBM/article/view/766/484. Accessed 20 Oct. 2023
24. Ejournalsico.com.: View of effect of leadership behavior and work climate on employee performance (2023). [online]. Available at: https://ejournalsico.com/index.php/sico/article/view/15/14. Accessed 20 Oct. 2023
25. Ejournalsico.com.: View of effect of leadership behavior and work climate on employee performance (2023). [online]. Available at: https://ejournalsico.com/index.php/sico/article/view/15/14. Accessed 20 Oct. 2023
26. Fuad, S., Sankar, J.P.: The relationship between employee empowerment and perceived employee job performance among the hospitality. ResearchGate; unknown (2020, September 10). https://www.researchgate.net/publication/344846607_The_Relationship_between_Employee_Empowerment_and_Perceived_Employee_Job_Performance_among_the_Hospitality_Sector_in_the_Kingdom_of_BahrainThe_Case_of_Three_Star_Hotels_in_Bahrain
27. Jahroni, J., Lestari, U.P., Mardikaningsih, R.: Effect of leadership behavior and work climate on employee performance (2020)
28. Liu, W., Liu, Y.: The impact of incentives on job performance, business cycle, and population health in emerging economies. Front. Public Health **9** (2022). https://doi.org/10.3389/fpubh.2021.778101

29. Ngwa, W.T., Adeleke, B.S., Agbaeze, E.K., Ghasi, N.C., Imhanrenialena, B.O.: Effect of reward system on employee performance among selected manufacturing firms in the Litoral region of Cameroon. Acad. Strat. Manag. J. **18**(3), 1–16 (2019)
30. Noorazem, N.A., Md Sabri, S., Mat Nazir, E.N.: The effects of reward system on employee performance. Jurnal Intelek **16**(1), 40–51 (2021). researchgate. https://doi.org/10.24191/ji.v16i1.362
31. Pawar, B.S.: Employee Performance and Well-Being: Leadership, Justice, Support, and Workplace Spirituality. Taylor & Francis. Problems of Performance Appraisal in Developed country (2023). QS Study. https://qsstudy.com/problems-performance-appraisal-developed-country/
32. Rivaldo, Y., Nabella, S.D.: Employee performance: education, training, experience and work discipline. Calitatea **24**(193), 182–188 (2023)
33. Salman, S.F.A., Sankar, J.P.: The relationship between employee empowerment and perceived employee job performance among the hospitality sector in the kingdom of Bahrain: the case of three star hotels in Bahrain. iKSP J. Innov. Writings **1**(2) (2021)
34. Sittenthaler, H.M., Mohnen, A.: Cash, non-cash, or mix? Gender matters! The impact of monetary, non-monetary, and mixed incentives on performance. J. Bus. Econ. **90**(8), 1253–1284 (2020). https://doi.org/10.1007/s11573-020-00992-0
35. Sopiah, S., Kurniawan, D.T., Nora, E., Narmaditya, B.S.: Does talent management affect employee performance? The moderating role of work engagement. J. Asian Finan. Econ. Bus. **7**(7), 335–341 (2020)
36. Sugiarti, E.: The influence of training, work environment and career development on work motivation that has an impact on employee performance at PT. Suryamas Elsindo Primatama in West Jakarta. Int. J. Artif. Intell. Res. **6**(1.2) (2023)
37. Taheri, R.H., Miah, M.S., Kamaruzzaman, M.: Impact of working environment on job satisfaction. Eur. J. Bus. Manag. Res. **5**(6) (2020)
38. Tamimi, M., Sopiah, S.: The influence of leadership style on employee performance: a systematic literature review: a systematic (2022). [online]. ResearchGate. Available at: https://www.researchgate.net/publication/366334093_The_Influence_of_Leadership_Style_on_Employee_Performance_A_Systematic_Literature_Review_A_Systematic_Literature_Review. Accessed 22 Oct. 2023

Literature Review of Factors Influencing Financial Performance of Commercial Banks in Bahrain

Qais A. ALmaamari and Fatema Khaled Mohamed Isa

Abstract The purpose of this paper is to investigate how the Kingdom of Bahrain's commercial banks' financial performance is influenced by three variables: financial risk, intellectual capital, and internal control. This study highlights the key elements affecting the financial performance of commercial banks in the literature review. The finding of this study is Financial risk have negative impact on financial performance, while intellectual capital and internal control have a more significant and positive impact on financial performance. The results of this study may give top managers and decision-makers a theoretical knowledge and understanding of the factors that could impact financial performance in the banking industry. The banking industry will also benefit from this study in terms of other aspects.

Keywords Financial risk · Intellectual capital · Internal control · Financial performance · Commercial banks · Kingdom of Bahrain

1 Introduction

The analysis of financial performance involves consolidating operational and financial data to comprehend a company's activities. This evaluation has significant implications for economic growth as it exposes banks' strengths and weaknesses [10]. Factors like financial risk, intellectual capital, and internal control influence how well banks operate. Managing these risks effectively is crucial for profitability and

Q. A. ALmaamari (✉)
Administrative Science Department, College of Administrative and Financial Science, Gulf University, Sanad 26489, Kingdom of Bahrain
e-mail: dr.qais.almaamari@gulfuniversity.edu.bh

F. K. M. Isa
Accounting and Financial Science Department, College of Administrative and Financial Science, Gulf University, Sanad 26489, Kingdom of Bahrain
e-mail: 210101115506@gulfuniversity.edu.bh

B. Awwad (ed.), *The AI Revolution: Driving Business Innovation and Research*, Studies in Systems, Decision and Control 524, https://doi.org/10.1007/978-3-031-54379-1_22

meeting regulatory obligations [4]. Evaluating financial aspects aids in resource allocation, effective fund utilization, and risk reduction, ensuring increased productivity and profitability [8, 10, 15].

European nations, particularly Nordic financial institutions, are seen as pioneers in sustainable development. Environmental, social, and governance (ESG) practices have gained significance, impacting managerial decisions and offering long-term gains like improved productivity and corporate reputation [25].

Internet finance's impact on Chinese commercial banks' financial performance is the focus of Dong et al.'s [14] research. Internet finance encompasses technology integration in financial services, a rapidly evolving sector that has witnessed substantial growth in China, influencing online banking accessibility and effective advertising methods.

Factors associated with auditing and management significantly affect Yemeni commercial banks' financial performance. Metrics like return on assets and return on shareholders' equity are crucial, with board size affecting these metrics differently. Hazaea et al. [17] emphasize the importance of internal audit (IA) efficiency in enhancing financial performance.

The environmental, social, and governance (ESG) performance of UAE banks has gained attention, impacting financial performance due to changing corporate strategies and concerns about environmental impact. Zaman and Ellili [26] highlight the role of strong corporate governance in maintaining reputation and long-term growth.

Oudat and Ali [22] stress the growing importance of risk management in today's uncertain economy, particularly in handling financial risks such as stock market fluctuations. They underscore the significance of risk assessment in making investment decisions and reducing uncertainty for investors.

2 Significance of the Study

The theoretical significance of the study is to expand on prior research by examining three key elements that influence financial performance: financial risk, intellectual capital, and internal control. Since financial performance directly affects a bank's success or failure, this paper will examine each component in depth in order to develop practical recommendations that can be used to strengthen the areas where financial performance is weak and maintain the areas where it is strong.

3 Literature Review

3.1 *The Impact of Financial Risk on Financial Performance*

Managing financial risk significantly affects the financial performance of commercial banks. Credit risk, stemming from loans and other sources, poses bankruptcy threats to banks. Non-performing loans and provisions are critical variables impacting bank performance, revealing how credit loss can detrimentally affect them [2]. The interest rate is a key factor influencing bank operations. Studies on Pakistan's major banks found an inverse correlation between interest rates and financial success. Mismanaging interest rate risks can lead to declining bank performance [2].

Apart from credit and interest rate risks, liquidity risk also affects banks' vulnerability. Long-term loans funded by short-term deposits expose banks to liquidity shortages [1]. Studies on Bahraini and Malaysian banks highlighted a strong inverse relationship between deposit volatility and liquidity risk, but a positive association between bank capitalization and liquidity risk [2].

Similar findings on Pakistani banks emphasized the negative impact of liquidity risk on performance, showing an adverse effect of higher non-performing loan ratios. While the net stable funding ratio affects liquidity positively, it can adversely impact performance [2, 9].

Operating cost ratios affect financial performance in various ways. Higher operational costs may indicate higher revenue, suggesting a correlation between financial performance and cost-to-income. However, higher operating expense ratios might not always correlate with high retained earnings, potentially affecting overall financial performance [16].

H1: Financial risk have negative impact on financial performance.

3.2 *The Impact of Intellectual Capital on Financial Performance*

The term intellectual capital (IC) has drawn the greatest attention from academics working in a variety of professions as well as professionals working in the business and economic sectors. This is as a result of the crucial part that IC may play in improving financial performance, developing competitive advantages, and ensuring the sustainability of businesses. IC is viewed as a strategic asset that should be carefully managed in order for the business to reap the most benefits [13].

A company's capacity to design, implement, and evaluate its intangible assets in order to boost value and subsequently enhance performance is referred to as intellectual capital. The majority of academics concur that an all-encompassing concept, intellectual capital, is employed to define intangible assets. The exact method of measuring intellectual capital has not yet been established. A researcher discovered the Value Added Intellectual Capital (VAICTM) technique as a way to assess intellectual capital. Value Added Capital Employed (VACA), Value Added Human Capital (VAHU), and Structural Capital Value Added (STVA) are three of the model's primary building blocks [21, 7].

It was stated that managers or investors struggle to comprehend how a firm's resources, when some of them are intangible or IC, might generate value in the future because accounting methodologies failed to take this reality into account. There is proof that IC has a favorable effect on the profits and revenues of 30 Taiwanese enterprises. A similar outcome was noted when researchers looked at how IC affected the Jordanian telecommunications industry's financial performance (FFP). Further research into the relationship between IC and long-term corporate performance revealed that firms with higher IC are more profitable [13, 18].

Several earlier research indicated a significant and favorable association between the main elements of intellectual capital. The primary VAICTM components had an impact on financial performance, according to Ozkan et al. [23], who looked at Turkish banks. Ismail and Karem [19], who looked at the impact of intellectual capital on financial performance in Bahrain between 2005 and 2007 demonstrated significant results, is another study that supports this claim. Additionally, studies by Buallay [11] and Al-Musali and Ismail [3] demonstrate the large positive impact that intellectual capital has on financial success [21].

H2: Intellectual capital have positive impact on financial performance.

3.3 The Impact of Internal Control on Financial Performance

By 1992, the COSO-developed internal control-integrated framework had become widely used globally. The framework emphasizes the variability of internal control based on context, industry, and company type. However, lack of clear definitions in laws or regulations concerning internal control might lead to disagreements. It is generally understood as a process influenced by the board of directors, management, and other staff [5, 24].

Internal control serves as a defense system against threats to an organization's assets. Its objectives include implementing security measures to protect assets and ensuring the accuracy of information for internal or external users. The overall aim is to enhance operational efficiency and encourage compliance with management policies within the entity [12, 20].

Banks typically establish internal control systems to monitor and mitigate risks, strengthening their risk management protocols. Each bank customizes its internal control mechanisms to ensure effective risk management. A well-functioning internal

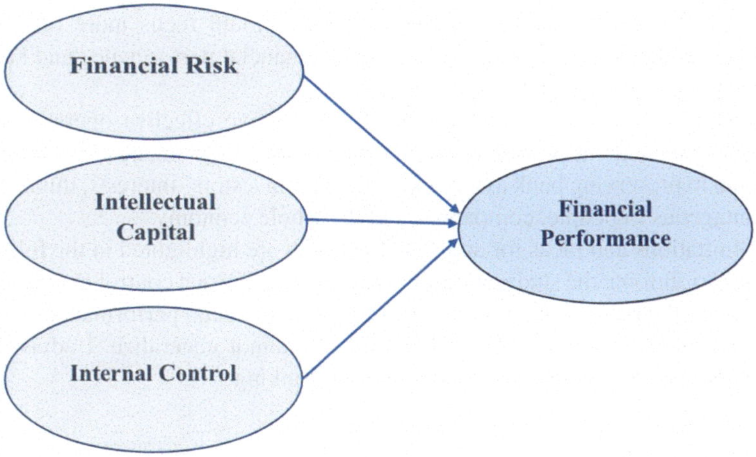

Fig. 1 Research model of this study

control system aims to ensure a bank operates effectively, maintains reliable management data and financial reporting, and complies with relevant laws and regulations. A bank without competent internal controls may face vulnerabilities. The modern concept of internal controls revolves around assisting firms in managing risks and supporting effective governance procedures, encompassing risk identification and mitigation [24].

Internal controls can be preventive or detective. Preventive controls anticipate issues, propose solutions, make adjustments, and prevent errors or omissions. On the other hand, detective controls detect issues, determine their roots, and rectify flawed controls upon identification, minimizing threats and addressing issues promptly [24].

H3: Internal control have positive impact on financial performance.

4 The Conceptual Framework

Figure 1 explains the link between the four variables in Commercial Banks in Bahrain, where financial performance acts as the DV and financial risk, intellectual capital, and internal control act as the IVs which influence the DV.

5 Limitations and Recommendations

The development of the financial performance of commercial banks should take into account aspects such as financial risk, intellectual capital, and internal control.

According to the research, commercial banks should focus more on internal controls since they have a significant impact on financial performance and because they are efficient management tools for banks.

According to the study's findings, banks must have effective operational and credit risk management procedures in place to protect their financial performance. In addition to preserving bank assets and defending investors' interests, this will also be advantageous to people, companies, and the whole economy.

The limitations and ideas for additional research are highlighted in the following sections. To reinforce the study, additional research on internal control may be undertaken. Due to the paucity of studies on its impact on financial performance. We only chose three independent variables, therefore we cannot generalize. I advise other researchers to choose loan defaults and internet banking.

6 Conclusion

The aim of the study is to investigate how financial risk, intellectual capital, and internal control affect the financial performance of Bahraini commercial banks. Credit risk, interest rate risk, and liquidity risk are used to quantify financial risk management, whereas return on equity, return on assets, and return on investment are used to gauge financial success. The study came to the conclusion that financial risk management is a significant element that lowers the financial performance of commercial banks. The performance of the financial industry is likely to suffer from key elements of financial risk management such as credit risk, interest rate risk, and liquidity risk. All of these elements have a negative effect on the banking industry's performance. On the other hand, intellectual capital and internal control have a positive effect on the banking industry's performance.

References

1. Ahmed, F., Almaamari, Q.: Understanding the relationship between training, motivation and working environment and employee performance in the banking sector in Bahrain. Int. J. Intellect. Human Resour. Manage. (IJIHRM) **1**(01), 64–68 (2020)
2. Ahmed, Z., Shakoor, Z., Khan, M.A., Ullah, W.: The role of financial risk management in predicting financial performance: a case study of commercial banks in Pakistan. J. Asian Finan. Econ. Bus. **8**(5), 639–648 (2021)
3. Al-Musali, M. A. K., Ismail, K. N. I. K.: Intellectual capital and its effect on financial performance of banks: evidence from Saudi Arabia. Procedia-Social Behav. Sci. **164**, 201–207 (2014)
4. Aldoseri, F.I., Almaamari, Q.A.: Factors influencing employee performance at the banking sector in Kingdom of Bahrain: literature review. Int. J. Emerg. Technol. **11**(5), 304–309 (2020)
5. Alkadash, T.M., AlZgool, M., Ateeq, A.A., Alzoraiki, M., Alkadash, R.M., Nadam, C., AlMaamari, Q., Milhem, M., Dawwas, M.: The impact of employee satisfaction, emotional intelligence and organizational commitment on marketing service quality in medical equipment companies,

Bahrain. In: Artificial Intelligence and Transforming Digital Marketing, pp. 903–915. Springer Nature Switzerland, Cham (2023)
6. Almaamari, Q.A., Ali, B.M., Almeer, S.: Factors influencing organizational performance at petroleum products distribution company in Yemen. Specialusis Ugdymas **1**(43), 2071–2083 (2022)
7. Almaamari, Q.A.: Factors influencing employees' productivity in Bahraini Alhelli Company—literature review. In: From Industry 4.0 to Industry 5.0: Mapping the Transitions, pp. 383–387 (2023)
8. Almaamari, Q.A.: Factors influencing employee turnover in banking sector. In: From Industry 4.0 to Industry 5.0: Mapping the Transitions, pp. 987–997. Springer Nature Switzerland, Cham (2023)
9. Alsaad, L.H., Almaamari, Q.A.: A semi literature review of factors influencing employee's performance at banking service. Int. J. Manage. (IJM) **11**(8) (2020)
10. Banu, A.R., Santhiyavalli, G.A.: TOPSIS approach to evaluate the financial performance of scheduled commercial banks in India. Int. J. Econ. Res. **21**(1), 24–33 (2019)
11. Buallay, A.: Is sustainability reporting (ESG) associated with performance? Evidence from the European banking sector. Manage. Environ. Qual. Int. J. **30**(1), 98–115 (2019)
12. Dawwas, M.I.: The relationship between HR practices, personality, and perceived organizational support. Baltic J. Law Polit. **15**(1), 305–330 (2022)
13. Desoky, A.: The impact of intellectual capital on firm's financial performance: empirical evidence from Bahrain. Desoky, AM and Mousa, GA. Invest. Manage. Finan. Innov. **17**(4), 189–201 (2020)
14. Dong, J., Yin, L., Liu, X., Hu, M., Li, X., Liu, L.: Impact of internet finance on the performance of commercial banks in China. Int. Rev. Financ. Anal. **72**, 101579 (2020)
15. Eshwari, D., Baby, M.K.: The effect of capital structure on the profitability of FMCG companies. In: Aloysius Edward, J., Jaheer Mukthar, K.P., Asis, E.R., Sivasubramanian, K. (eds.) Current Trends in Economics, Business and Sustainability. ICEBS 2023. Contributions to Environmental Sciences and Innovative Business Technology. Springer, Singapore (2023). https://doi.org/10.1007/978-981-99-3366-2_37
16. Gadzo, S.G., Kportorgbi, H.K., Gatsi, J.G.: Credit risk and operational risk on financial performance of universal banks in Ghana: a partial least squared structural equation model (PLS SEM) approach. Cogent Econ. Finan. **7**(1), 1589406 (2019)
17. Hazaea, S.A., Tabash, M.I., Khatib, S.F., Zhu, J., Al-Kuhali, A.A.: The impact of internal audit quality on financial performance of Yemeni commercial banks: an empirical investigation. J. Asian Finan. Econ. Bus. **7**(11), 867–875 (2020)
18. Hussin, R.M., Putit, L., Subramaniam, G.: Social media as an open innovation: deciphering its relationship with firm performance, compatibility, and security concern. In: Rafiki, A., Dana, L.P., Nasution, M.D.T.P. (eds.) Open Innovation in Small Business. Contributions to Environmental Sciences & Innovative Business Technology. Springer, Singapore (2023). https://doi.org/10.1007/978-981-99-5142-0_6
19. Ku Ismail, K. N. I., Abdul Karem, M.: Intellectual capital and the financial performance of banks in Bahrain. J. Bus. Manage. Account. **1**(1), 65–79 (2011)
20. Mardi, M., Perdana, P.N., Suparno, S., Munandar, I.A.: Effect of accounting information systems, teamwork, and internal control on financial reporting timeliness. J. Asian Finan. Econ. Bus. **7**(12), 809–818 (2020)
21. Muhammad, R., Mangawing, M.A., Salsabilla, S.: The influence of intellectual capital and corporate governance on financial performance of Islamic banks. Jurnal Ekonomi & Keuangan Islam 77–91 (2021)
22. Oudat, M.S., Ali, B.J.: The underlying effect of risk management on banks' financial performance: an analytical study on commercial and investment banking in Bahrain. Ilkogretim Online **20**(5) (2021)
23. Ozkan, N., Cakan, S., Kayacan, M.: Intellectual capital and financial performance: a study of the Turkish Banking Sector. Borsa Istanbul Rev. **17**(3), 190–198 (2017)

24. Pakurár, M., Haddad, H., Nagy, J., Popp, J., Oláh, J.: The impact of supply chain integration and internal control on financial performance in the Jordanian banking sector. Sustainability **11**(5), 1248 (2019)
25. Rahi, A.F., Akter, R., Johansson, J.: Do sustainability practices influence financial performance? Evidence from the Nordic financial industry. Acc. Res. J. **35**(2), 292–314 (2021)
26. Zaman, S., Ellili, N.O.D.: The effect of ESG disclosure on the financial performance of UAE banks. Available at SSRN 4043024 (2022)

Exploring the Relationship Between ESG Practices and Corporate Financial Performance

Ahmad Yahia Mustafa Alastal and Ali Ateeq

Abstract This research provides a critical analysis of the existing literature on Environmental, Social, and Governance (ESG), with a particular focus on the significant importance of environmental policies within various organizational settings. The effectiveness of applying environmental solutions inside organizational structures is highlighted by the use of the Levers of Control framework in this study. In conclusion, this study determines that the Levers of Control framework significantly enhances the implementation of environmental strategies. It suggests that more empirical investigation is needed to explore the complex interplay between this framework and environmental strategies. The aforementioned investigation has the potential to provide significant findings about the intricacies of organizational control systems and their impact on the implementation of sustainable practices.

Keywords ESG · Environmental strategies · Levers of control · Organizational sustainability · Literature review · Governance · Social responsibility · Organizational control mechanisms · Sustainability practices · Empirical research

1 Introduction

The integration of Environmental, Social, and Governance (ESG) aspects has emerged as a prominent topic in current discussions inside organizations [1]. This reflects a growing recognition of the complex relationships between business actions and wider social and environmental issues [2]. This study undertakes a comprehensive examination of the current body of literature on environmental, social, and governance (ESG) factors, with a particular focus on the significant influence of environmental policies in various organizational contexts [3].

A. Y. M. Alastal · A. Ateeq (✉)
Administrative Science Department, College of Administrative and Financial Science, Gulf University, Sanad 26489, Bahrain
e-mail: dr.ali.ateeq@gulfuniversity.edu.bh

A. Y. M. Alastal
e-mail: dr.ahmed.alastal@gulfuniversity.edu.bh

B. Awwad (ed.), *The AI Revolution: Driving Business Innovation and Research*, Studies in Systems, Decision and Control 524, https://doi.org/10.1007/978-3-031-54379-1_23

The significance of adopting ecologically friendly practices by corporations has increased in recent years. The existing body of research emphasizes the need of implementing successful environmental policies, as they are crucial not only for meeting legal requirements but also for improving business reputation and ensuring long-term sustainability [4]. The gradual integration of environmental issues into organizational structures presents a challenge that requires a deliberate and systematic approach. This prompts the use of strategic frameworks, such as the Levers of Control, to effectively address this complexity.

This study employs the Levers of Control framework to examine the effectiveness of implementing environmental strategies within organizational contexts. Developed by Gunawan et al. [5], the framework provides a comprehensive model for organizations to manage and align their strategies, emphasizing the role of interactive and diagnostic controls in strategy execution. By applying this framework to the domain of environmental strategies, the research aims to shed light on how organizational control mechanisms can be leveraged to enhance the adoption and execution of sustainability initiatives.

In December 2022, Bahrain Boursa introduced an ESG reporting guide to aid listed organizations in integrating ESG factors into annual reports, aligning with Bahrain's economic vision and UN SDGs. Simultaneously, the World Federation of Exchanges (WFE) issued guidance in 2015, promoting optimal use of existing ESG reports, and later in June 2018, refined its approach by categorizing ESG into 30 metrics—10 each for environmental, social, and governance aspects. This comprehensive framework supports organizations in effectively assessing and communicating their ESG performance.

The WFE metrics is shown in Table 1.

2 Literature Review

2.1 Environmental Performance Financial Outcomes

The environmental performance of businesses plays a crucial role in resolving several challenges, such as the reduction of greenhouse gas emissions and the efficient use of energy resources [6]. The concept being referred to is the quantification of an organization's influence on natural systems, which includes the assessment of its effects on land, air, water, and ecosystems [7]. The benefits include increased organizational performance, decreased production costs, fortified stakeholder connections, enhanced reputation, and competitive advantages [8].

The prevailing body of research demonstrates a positive correlation between environmental performance and financial results, signifying enhanced reputation and financial prosperity. However, a minority of studies propose an inverse association [9]. This alternative perspective argues that the implementation of environmental programs has the potential to increase management expenses, which might have a

Table 1 ESG metrics based on WFE

ESG categories	Metrics
Environmental (E) 10 metrics	GHG emissions
	Emissions intensity
	Energy usage
	Energy intensity
	Energy mix
	Water usage
	Environmental operations
	Environmental oversight
	Environmental oversight
	Climate risk mitigation
Social (S)	CEO pay ratio
	Gender Pay ratio
	Employee turnover
	Gender diversity
	Temporary worker ratio
	Non-discrimination
	Injury rate
	Global health and safety
	Child and forced labor
	Human rights
Governance (G)	Board diversity
	Board independence
	Incentivized pay
	Collective bargaining
	Supplier code of conduct
	Ethics and anti-corruption
	Data privacy
	Sustainability reporting
	Disclosure practices
	External assurance

negative influence on financial performance. The existing body of research highlights the complex relationship between environmental stewardship and financial performance within corporate settings [10].

The Relationship between Social Performance and Financial Outcomes

The focal point of organizational social performance is on the fulfillment of social duties, which encompasses several dimensions such as the CEO Pay Ratio, Gender

Pay Ratio, and Human Rights [11]. There is a substantial body of evidence that indicates a prevailing positive correlation between corporate social responsibility (CSR) and financial success. The research conducted by Aftab et al. [12] consistently establishes a positive correlation between corporate social responsibility (CSR) efforts and financial performance. This suggests that firms that appropriately handle social elements likely to have improved financial success. Nevertheless, it is essential to do deeper investigation into the particular processes and intricacies of this connection. In general, the current corpus of research emphasizes the relevance of social responsibility in exerting a beneficial impact on financial performance. This aligns with the prevailing pattern of firms acknowledging the importance of societal factors in attaining sustainable and resilient financial results.

Corporate Governance and Financial Performance

The relationship between corporate governance and financial performance has been a topic of significant interest and research in the field of business and management. The operation of a company is greatly influenced by several factors of corporate governance, such as board structure, ownership design, and executive remuneration [13]. The existing body of research highlights a favorable link between Environmental, Social, and Governance (ESG) practices and Corporate Financial Performance (CFP). This relationship has been emphasized in studies such as the one conducted by Chen et al. [14]. Organizations that adopt environmental, social, and governance (ESG) principles have been seen to exhibit enhanced financial performance. This correlation underscores the increasing acknowledgement that the integration of environmental and social factors into company practices is conducive to achieving financial prosperity. Recognizing these interconnections is crucial for firms aiming to adopt ethical and sustainable practices, as the implementation of a robust governance structure that incorporates environmental, social, and governance (ESG) principles may contribute to improved financial performance and bolster long-term resilience.

ESG investment, which places emphasis on environmental, social, and governance considerations, is oriented towards the objective of achieving sustained value appreciation over the long term. According to Flak and Rose [15] the adoption of a pragmatic approach involves the alignment of technology advancements with improvements in governance, profitability, and human wellbeing. ESG reporting has significant importance for companies as it enables them to effectively tackle existing issues, capitalize on forthcoming prospects, and assume a prominent position in promoting sustainable development for the purpose of social inclusion. The enhancement of productivity may be achieved by expanding reporting standards beyond financial aspects to include environmental, social, and governance elements. According to Tunio et al. [16] there is a strong emphasis on the establishment of a connection between the number and quality of environmental disclosures via ESG reporting [17].

Environmental, Social, and Governance in Bahrain (ESG)

According to Bourse [18] ESG reporting can help companies in Bahrain align with the country's Bahrain Economic Vision 2030. This vision aims to reshape the outlook of government, society, and the economy around sustainability, fairness, and competitiveness. ESG reporting allows companies to contribute to national priorities and fulfill economic growth goals set out in the 2030 vision. By adopting ESG practices, companies can work towards the vision's principles as Bahrain strives for long-term prosperity. The Bahrain economic vision 2030 highlighted that by 2030, the private sector will need to independently drive economic growth in Bahrain. The Bahrain Economic Vision 2030 sees prosperity built on a solid foundation of sustainable government finances that invest in the future through education and training, especially in applied sciences. As technology and new global competitors shorten product life cycles, entrepreneurship and innovation will be key to sustaining a vibrant private sector. However, growth cannot compromise the environment or Bahrainis' long-term wellbeing. Protecting Bahrain's environment and cultural heritage will remain top priorities. Furthermore, the main goal for the sustainability 2030 vision is to attain economic prosperity by empowering the private sector, developing human capital, and encouraging entrepreneurship and innovation, while also protecting the environment and preserving cultural heritage [19].

Bahrain's Vision 2030 defines "Sustainability" with these key points:

- By 2030, the private sector should independently drive economic growth in Bahrain.
- Government finances will follow the principle of sustainability, maintaining a stable and forward-looking system.
- Bahrain will invest its resources in the future by improving human capital through education and training, especially in applied sciences.
- Entrepreneurship and innovation will help ensure a vibrant, sustainable private sector.
- Economic growth must not compromise the environment or long-term wellbeing of Bahrainis. Protecting the environment and cultural heritage will be top priorities.

The Kingdom of Bahrain has established important regulations that address Environmental, Social, and Governance matters. Some examples of these key regulations are supreme council for environment, the Bahrain corporate governance code.

The Supreme Council for the Environment in Bahrain is the official body responsible for setting the national environmental strategy and overseeing its implementation across government ministries, agencies, and institutions. The Council aims to spread environmental awareness among citizens and residents, promote green and sustainable values, encourage eco-friendly technologies, and increase community involvement in environmental protection. Its goals are to define environmental policies, monitor their execution, boost environmental awareness, advance sustainability, and foster public participation.

The Bahrain Corporate Governance Code contains nine core principles of corporate governance aligned with international best practices. The Code provides recommendations for implementing the principles and supporting overall good governance. It operates on a "comply or explain" basis, meaning companies should either comply with the recommendations or provide an explanation if they do not comply. The nine principles which aim to guide companies in adhering to top standards of corporate governance. These principles are the company shall be headed by an effective, collegial and informed board, the directors and officers shall have full loyalty to the company, the board shall have rigorous controls for financial audit, internal control and compliance with law, the company shall have rigorous procedures for appointment, training and evaluation of the board, the company shall remunerate directors and officers fairly and responsibly, the board shall establish a clear and efficient management structure, the board shall communicate with shareholders and encourage their participation, company shall disclose its corporate governance, and companies which refer to themselves as "Islamic" must follow the principles of Islamic Shari'a.

The Ministry of Sustainable Development

In June 2022, His Majesty King Hamad bin Isa Al Khalifa issued Royal Decree No. 20 to establish the Ministry of Sustainable Development. This ministry aim to serve as an integrating institution for the government's sustainability policies. It's also advocate for the Kingdom's efforts, accomplishments, and future plans to further the sustainability agenda across all areas.

The Voluntary National Review Report on Sustainable Development Goals (SDGs) 2030 showed that the Bahrain has moved a great step to achieve the economic prosperity by empowering the private sector, developing human capital, and encouraging entrepreneurship and innovation, while also protecting the environment and preserving cultural heritage [20].

Furthermore, the Government Plan (2023–2026) titled "From Recovery to Sustainable Growth" which launched in late 2022 by His Royal Highness Prince Salman bin Hamad Al Khalifa, the Crown Prince and Prime Minister of Bahrain aims to build on the previous Government Action Plan and the economic recovery plan that focused on stabilizing the economy after COVID-19 impacts. The Key priorities outlined in the new 4-year plan include:

1. Fiscal sustainability—Ensuring disciplined government spending and increasing revenue sources.
2. Private sector growth—Creating an attractive business environment and supporting SMEs.
3. Investment in people—Upgrading education and skills to equip Bahrainis for future jobs.
4. Digital transformation—Improving e-services and infrastructure for digital economy growth.
5. Sustainable development—Transitioning to a greener, low-carbon economy.

3 Conclusion

In conclusion, this study provides a comprehensive analysis of the complex correlation between ESG (Environmental, Social, and Governance) practices and business financial performance, with a particular focus on the significant influence of environmental regulations. The use of the Levers of Control framework serves to emphasize the effectiveness of incorporating environmental tactics within organizational structures. The research paper proposes the need for more empirical research to better understand the intricate relationship between this framework and environmental initiatives. The importance of ESG reporting is emphasized, in line with national strategic goals such as Bahrain's Economic Vision 2030. This study provides valuable insights into the convergence of sustainability, governance, and financial results, hence reinforcing the need for firms to adopt responsible practices in order to achieve long-term success.

References

1. de Souza Barbosa, A., et al.: Integration of Environmental, Social, and Governance (ESG) criteria: their impacts on corporate sustainability performance. Humanit. Soc. Sci. Commun. **10**(1), 1–18 (2023)
2. Jámbor, A., Zanócz, A.: The diversity of environmental, social, and governance aspects in sustainability: a systematic literature review. Sustainability **15**(18), 13958 (2023)
3. Alawadi, A., et al.: Diversified boards and the achievement of environmental, social, and governance goals. Bus. Ethics Environ. Responsib. (2023)
4. Mukhtar, B., et al.: Integrating ESG disclosure into the relationship between CSR and green organizational culture toward green innovation. Soc. Responsib. J. (2023)
5. Gunawan, A., et al.: The effect of levers of control, firm's capability, and transformational leadership on competitive advantage: a study on manufacturing firms in Indonesia. J. Infrastruct. Policy Dev. **7**(3) (2023)
6. Benedek, A., Rokicki, T., Szeberényi, A.: Bibliometric evaluation of energy efficiency in agriculture. Energies **16**(16), 5942 (2023)
7. DeAsiain, M.L., Bonaño, J., Borrallo-Jiménez, M.: Urban Socio-Ecosystem Renewal: An Ecosystem Services Assessment Approach. Springer (2023)
8. Zhu, Y., et al.: Understanding corporate green competitive advantage through green technology adoption and green dynamic capabilities: does green product innovation matter? Systems **11**(9), 461 (2023)
9. Rajabalizadeh, J.: CEO overconfidence and financial reporting complexity: evidence from textual analysis. Manag. Decis. **61**(13), 356–385 (2023)
10. Habib, A.M.: Do business strategies and environmental, social, and governance (ESG) performance mitigate the likelihood of financial distress? A multiple mediation model. Heliyon **9**(7) (2023)
11. Simo, C., Tchuigoua, H.T., Nzongang, J.: Does corporate social responsibility pay? Evidence from social ratings in microfinance institutions. Technol. Forecast. Soc. Chang. **187**, 122180 (2023)
12. Aftab, J., et al.: Does corporate social responsibility drive financial performance? Exploring the significance of green innovation, green dynamic capabilities, and perceived environmental volatility. Corp. Soc. Responsib. Environ. Manage. (2023)

13. Fei, W., et al.: The impact of environmental, social, and governance, board diversity and firm size on the sustainable development goals of registered firm in China. Econ. Res.-Ekonomska istraživanja **36**(1), 668–686 (2023)
14. Chen, S., Song, Y., Gao, P.: Environmental, social, and governance (ESG) performance and financial outcomes: analyzing the impact of ESG on financial performance. J. Environ. Manage. **345**, 118829 (2023)
15. Flak, L.S., Rose, J.: Stakeholder governance: adapting stakeholder theory to e-government. Commun. Assoc. Inf. Syst. **16**(1), 31 (2005)
16. Tunio, R.A., et al.: The relationship between corporate social responsibility disclosures and financial performance: a mediating role of employee productivity. Environ. Sci. Pollut. Res. **28**, 10661–10677 (2021)
17. Javed, M., Rashid, M.A., Hussain, G.: When does it pay to be good—a contingency perspective on corporate social and financial performance: would it work? J. Clean. Prod. **133**, 1062–1073 (2016)
18. Bourse, B.: Bahrain Bourse ESG Reporting Guide (2020)
19. Government, B.: From Regional Pioneer to Global Contender: The Economic Vision 2030 for Bahrain (2008)
20. Bahrain, M.o.S.D.: Bahrain voluntary national review 2023, p. 158 (2023)

Internal Control Role in Enhancing Financial Performance of Higher Education

Houda Bessachi, Fouad Announ, Abdesslam Menacer, and Mujeeb Saif Mohsen Al-Absy

Abstract The objective of the study is to know the role of internal control in improving the performance of institutions of higher education by knowing the nature of the relationship between internal control and financial performance in higher education institutions. The Statistical Package for Social Sciences (SPSS) was used in this study to analyze the data to meet the study's goals. The study concluded that financial performance contributes effectively to achieving the basic objectives of the economic institution. This serves the requirements of both managers and shareholders by providing information about the financial function Which will give a clear picture of the financial situation, which helps in identifying the strengths and weaknesses and helps to rationalize financial decisions.

Keywords Financial performance · Internal control · And university institutions

H. Bessachi
Accounting and Financial Sciences Department, Faculty of Economics, Management and Commercial Sciences, University Center of Tipaza, Tipaza, Algeria
e-mail: bessachi.houda@cu-tipaza.dz

F. Announ · A. Menacer
Accounting and Financial Sciences Department, Faculty of Economics, Management and Commercial Sciences, University of Blida 2, El Affroun, Algeria
e-mail: f.announ@univ-blida2.dz

M. S. M. Al-Absy (✉)
Accounting and Financial Science Department, College of Administrative and Financial Science, Gulf University, Sanad 26489, Kingdom of Bahrain
e-mail: dr.mujeeb.saif@gulfuniversity.edu.bh

B. Awwad (ed.), *The AI Revolution: Driving Business Innovation and Research*, Studies in Systems, Decision and Control 524, https://doi.org/10.1007/978-3-031-54379-1_24

1 Introduction

Worldwide financial scandals have had a huge impact on the world economies where many severe accounting scandals are witnessed globally [1–4]. As a result, authorities and governments work to improve and update their corporate governance (CG) codes [5] where it is play a vital role in risk management and advancing best practices in business and monitoring businesses' operations and possess the ability to impact business decision-making [6, 7].

A result of the increase and widening of the scope of activities and the program practiced by economic units of all kinds and forms led to an increase and manifestation of the desire to obtain an independent internal assessment of the effectiveness of management within these units. Further, the users of financial reporting need more data to assess a company's operations and forecast its future performance [8, 9] where management must deliver to stakeholders a reliable and accurate financial report that aids them in making informed and significant decisions [10].

Hence, internal control as an internal mechanism of corporate governance plays an important role in monitoring and controlling the firm's activities and performance. In the corporate world, the idea of "firm performance" has a lot of significance [11]. The institution can evaluate its performance, especially the financial performance, as the latter is considered one of the most important methods that are used in the science of management, as it expresses the ability of the institution to achieve its long-term goals, and the extent of its ability to exploit its resources towards achieving the desired goal, as it is the mirror that reflects the position of the institution from different sides.

The effectiveness of internal controls is positively impacted by the internal control structure, which in turn encourages moral organizational conduct [12]. The internal control improves corporate performance [13]. Similarly, Internal control weaknesses, according to the author, typically have a negative impact on a company's success, and that the impact varies depending on whether important internal control elements are present [14]. There is a positive association between organizational success and internal audit control, risk management, and monitoring processes [15]. Study conducted by Tetteh et al. [16] demonstrates a substantial effect of internal control systems measure by control activity, control environment, information, and communication on the success of the company's success, as measured by total profitability.

This study's primary goal is to understand how financial control contributes to improved financial performance. It seeks to give the most significant main approaches to the idea of financial control and performance improvement in a thorough manner. Additionally, it tries to explain the significance of financial control and financial performance as well as their interrelationship.

2 Literature Review

Internal controls are policies, procedures, and systems put in place by a business to guarantee operational effectiveness, risk management, and legal compliance [17]. Internal control is a procedure carried out by several parties in an organization such as the board of directors, management, and internal auditor and the objective of internal control is to give reasonable certainty that goals pertaining to operations, reporting, and all activities are met and in line with regulations [18]. Additionally, it is described as a group of policies and procedures put in place by management with the intention of assisting in the accomplishment of the organization's primary goals and delivering reassuring assurance about the management of risks associated with activities. Internal control is materialized through organizational choices and role distribution and by the systems of verification conducted by managers according to the particularities of their respective activities [19]. To ensure the accuracy of financial reporting as well as the compliance with regulatory requirements, internal controls are one mechanism of the accounting and auditing which used in the finance department of a corporation [20].

The internal control system is categorized into accounting and administrative controls. Accounting control connotes the plan of the organization, logic, and process which are executed with a view of securing assets and promoting the dependability of financial records. Administrative control consists of the plan of an organization, logic, and process which are executed with a view of securing assets and assuring dependability of operational efficiency and adherence to management policies and directives. This is also referred to as operational controls. When control is weak, operational efficiency can be subverted notwithstanding the nature of the organization. If there is no control in place, the financial management of an organization will end in lawlessness, which can be mounted to a free for all between the management and employees [21].

Internal control aims to achieve many purposes, the most important of which are [22]: (i) Business operations must be carried out effectively. Internal controls must be in place to ensure that processes go off without a hitch. This lessens the risk of inefficiencies and obstacles to the company's capacity to generate value; (ii) Asset protection which refers to the steps that should be done to ensure that resources are efficiently used for the appropriate goals and are not at risk of theft or abuse; (iii) Avoiding and identifying fraud and other illegal activity, even though small enterprises, with straightforward organizational structures may be subject to these infractions, as organizations grow larger and more complicated, the types of fraud increase in variety, therefore controls must be well equipped to counter organization activities; (iv) Completeness and accuracy of financial records: It will be impossible for a company to present trustworthy financial statements if its financial records are questionable. Hence, accounting systems should be efficiently designed to effectively record transactions to ensure that the financial records accurately represent the nature of the business that was transacted; (v) In order for companies to comply with their legal reporting obligations, financial statements should be prepared promptly and accurately. Financial statements should help shareholders to make a wise decision. Internal controls not only applied for accounting procedure but also applied for management practices, which are important for successful strategy planning, decision-making, and performance monitoring of firms.

A range of metrics, e.g., capital adequacy ratio, profitability, solvency, liquidity, and leverage, are used to determine the financial success of a company during a given period. Further, financial performance performs as a determinant of a company's capabilities in managing the company resource as well as control and evaluate its activities [23]. Additionally, it serves as a general measure of a business's long-term financial stability [24]. Part of financial performance reflected in financial statements is the cash flows, profit-loss and capital changes. The financial plan of the organization is decided upon by corporate management using this data. Financial data like as cash flows and retained earnings are included in the financial statements, which also comprise a balance sheet with a profit-and-loss calculation [23].

The factors affecting the financial performance of the economic enterprise are classified into internal and external factors related to the external economic environment [25]. In terms of Internal factors, the most important internal factors affecting financial performance are organizational structure, organizational climate, and the size of the institution. First, the organizational structure affects financial performance by dividing the tasks and responsibilities related to the financial function and then defining the activities and allocating the necessary resources for them, as well as the impact of the nature of the organizational structure on financial decision-making and its suitability for the established financial objectives, and the extent to which it corrects the nature of existing deviations.

Second, the organizational climate which reflects the extent of clarity of the organization in the organization, and employees' awareness of the relationship of the organization's goals, operations, and activities to financial performance. Good financial activity and thus financial performance. It means those modern methods and skills that serve the desired goals, such as customized production technology, continuous improvement technology etc. Therefore, the economic enterprise must pay great attention to the technology used, which must be consistent with its main objectives, through adaptation and absorption. for its developments with the aim of harmonizing technology and financial performance, which puts it before the inevitability of developing the latter in a way that suits the technology used.

Third, the size of the institution. The size of the organization and the way it is classified may have a negative impact on financial performance because a large institution's management becomes more complicated and intertwined, and a positive impact because a large institution necessitates many financial analysts, which contributes to financial performance. This situation is the most plausible in improving its financial performance.

Regarding the external factors, several external factors affect financial performance, the most important of which are market, competition, and economic conditions. In terms of market structure, economic commodity markets can take different shapes, depending on the market framework and the strategies that an organization employs to achieve its primary purpose, which is profit maximization. The market has an impact on financial success due to the rules of supply and demand. Demand will positively affect financial performance, but in the opposite case, we will notice a decline in financial performance.

Secondly, the competition where it is a double-edged sword concerning the financial performance of the economic enterprise, so we see it as a catalyst for enhancing it when the enterprise faces the repercussions of competition and tries hard to improve its image and financial position through its financial performance to keep pace with these repercussions. On the other hand, the enterprise is not eligible for these repercussions. It cannot face the competition, its financial situation deteriorates, and consequently, the financial performance worsens.

Lastly, the economic conditions. The general economic conditions may affect financial performance, whether in a negative way or vice versa. For example, we find it in economic crises, or cases of inflation that negatively affect financial performance, either in the case of high aggregate demand or state support for products that may positively affect financial performance.

3 Methodology

The process of defining the research methodology depends on the main objective that the study seeks to reach by answering the questions posed that reflect the problem of the study and testing the validity of the proposed hypotheses. Its parties and ideas related to it highlight the various aspects of the relationship between it and the financial performance and the analytical approach was used in the applied side. The following are the study hypotheses:

H_1: There are no statistically significant differences for the internal control on the financial performance.

H_2: There is no statistically significant effect between internal control and financial performance at the significant level of 0.05.

The study relies mainly on a field survey of some universities. Authors took a sample of (50) individuals and (40) samples were retrieved, i.e. (80%). 5 samples were excluded due to the failure to complete the conditions and answers in them. (30) sample, equivalent to (60%) of the total distributed sample, according to Table 1.

Table 1 Number of distributed questionnaires

Questionnaire	Number	Percentage
Distributed questionnaires	50	100
Retrieved questionnaires	40	80
Excluded surveys	05	10
Valid questionnaires for analysis	30	60

Source Prepared by researchers based on the results of SPSS 19

To evaluate the role of internal control in improving financial performance—a case study of some universities, a model was relied on based on a study of the dimensions of internal control and the dimension of financial performance, to know the relationship between these two components as well as how internal control contributes to improving financial performance, which represents the main focus of the study.

The preparation of the questionnaire relied on the use of the five-point Likert scale, which is among the most widely used methods in measuring trends, as well as each component of measuring trends that contains many variables associated with it.

4 Results

4.1 Demographic Statistics

Table 2 shows that the respondents between the ages of 40 and 49 represent the largest percentage, at 40%. The number of males is greater than the number of females. Most of the workers who gave their opinion on the subject are holders of higher degrees. Further, the major studied by most of the targeted is accounting. Most of the workers who gave their opinion on the subject work within finance and accounting. Targeted individuals with experience in this field.

4.2 Diagnostic Tests

To ensure the validity of the questionnaire and the appropriateness of its paragraphs, we presented the questionnaire to a group of arbitrators with experience and specialization, who are members of the faculty in the Faculty of Commercial Sciences, Economic Sciences, and Management Sciences at universities of Yahya Fares and Souk Ahras, where they issued their judgment on The degree of suitability of the questions, the extent of the validity and accuracy of the verbal and scientific formulation of the questionnaire's phrases, the extent to which the questionnaire covers the problem of the study and the achievement of its objectives.

Regarding the stability of the resolution variables, authors used the stability coefficient Cronbach Alpha to measure the stability of the resolution, so that the value of Cronbach Alpha is acceptable if its value exceeds 0.60. The results indicated that the stability value was high. Table 3 shows that the alpha coefficient was high by 0.977, and the stability value was also high. This indicates the stability of the resolution variables.

In terms of distribution normality test, the Smirnov-Kolmogorov test (S-K) was used to test whether the data followed a normal distribution. Table 4 shows that the probability value (Sig) for all areas of the study is greater than the level of function

Table 2 Demographic statistics

Demographic	Options	Repetition	The ratio
Gender	Male	19	63.3
	Female	11	36.7
Age	Less than 30 years old	7	23.3
	From 30 to 39 years	7	23.3
	From 40 to 49 years	12	40
	Older than 49 years old	4	13.3
Qualification	Vocational institute	02	06.7
	Bachelor	11	36.7
	Master	2	6.7
	Magister	5	16.7
	Ph.D.	10	33.3
Scientific specialization	Accounting	16	53.3
	Finance	7	23.3
	Economy	3	10
	Business management	4	13.3
Job title	Director of the finance and accounting department	11	36.7
	Vice chancellor of the university	4	13.3
	Director of the internal oversight department	5	16.7
	Dean of the college	3	10
	User interest manager	5	16.7
	User interest manager	2	6.7
Years of experience	Less than 8 years old	13	43.3
	From 9 to 16 years old	8	26.7
	From 17 to 24 years old	7	23.3
	From 25 to 32 years old	2	6.7

Table 3 Cronbach Alpha

Coefficient	Alpha coefficient	Constancy
Questions	0.977	0.988

0.05 (0.344 > 0.05), and thus the distribution of data for these areas follows a normal distribution, whereby parametric tests used to answer the hypotheses of the study.

Table 4 Normal distribution test

Questionnaire themes	Test value	Sig
Internal control	1.209	0.107
Financial performance	0.944	0.334
The questionnaire as a whole	0.937	0.344

4.3 Results

Authors analyzed the results of the questions that represent internal control in some universities. The study used the five-point Likert scale as shown in Table 5.

Table 6 shows that all the statements related to the internal control dimension were at a higher arithmetic mean than the peaceful arithmetic mean (3), and thus they express the individuals' approval of these statements, but there is a large discrepancy in the answers. This is what the standard deviation shows.

Based on the foregoing, we conclude that the level of internal control among university employees was high according to the study scale, as the average response of the respondents for the dimension as a whole was (3.4708), and this indicates that the respondents have the desire to use internal control for the benefit of the university, but they do not know how to apply it on the ground. Reality is better than practice.

Regarding financial performance, we will analyze part of the results of the questions that represent financial performance.

Table 7 shows that all the expressions (from 01 to 08) were with an arithmetic mean greater than the peaceful arithmetic mean (3), and thus they express the individuals' agreement on these expressions, but there is a great discrepancy in the answers. This is what the standard deviation shows. Based on the foregoing, we conclude that the level of financial performance was high according to the study scale, as the average response of the respondents for the dimension was (3.7750). This indicates that the respondents have positive tendencies characterized by approval towards the desire to be active members in improving financial performance.

After the data has been collected and analyzed using appropriate statistical methods and in light of the hypotheses that this study aimed to test, we will discuss, through this part, the testing of hypotheses related to the field study and verifying their validity or not by reviewing the results of an analysis for each hypothesis.

Table 5 Likert five scale

Degree	Strongly disagree	Disagree	Neutral	Agree	Strongly agree
Level	1	2	**3**	4	5
Arithmetic mean	1–1.79	1.80–2.59	2.60–3.39	3.40–4.19	4.20–5

Source Abdel-Fattah Ezz Hussein, Introduction to Descriptive and Inferential Statistics, Algorithm for Scientific Publishing and Distribution, Jeddah—Saudi Arabia, 2007, p. 237

Respondents have five alternatives to select from on a 5-point Likert scale. There are two extremes, two middle positions, and one neutral viewpoint among the alternatives.

Table 6 Degree of approval for statements after internal control

Internal control	Arithmetic mean	Standard deviation	Sample direction
The reports of the internal auditor are considered one of the most important tools that senior management relies on	4.13	0.973	Agree
One of the tasks of the internal auditor is to examine the extent of compliance with internal financial and administrative policies and orders	3.90	0.759	Agree
The independence of the internal auditor enables him to achieve the objectives of the control process with a better level of efficiency in the financial field	3.53	1.074	Agree
Obtaining information indirectly from other interests, and the impartiality of the internal auditor strengthens the credibility of internal control	3.10	0.960	Neutral
One of the tasks of the internal auditor is to verify the extent to which the planned objectives are achieved and the effectiveness of the results	3.23	1.073	Neutral
The Internal Control Department can submit reports to the senior management in which it identifies deviations and strengths and weaknesses	3.50	1.009	Agree
The internal controller, through the authority granted to him, can interfere at any time with the work of other departments, which contributes to discovering manipulations promptly	3.33	1.155	Neutral
The information provided by the Internal Control Department can be relied upon by other departments for its certainty	3.03	1.351	Neutral
Internal Control (the dimension as a whole)	3.4708	0.90775	Agree

In terms of first hypothesis, Table 8 shows that the value of (t) is equal to − 5.30, and the associated probability is 0.000, which is less than the level of significance (0.05), and from it, we accept the null hypothesis H0, which says, "there are no statistically significant differences for the internal control of financial performance" and the possible error of type β.

Regarding the second hypothesis, authors used simple regression analysis to test the second main research hypothesis. It was found through the analysis, which is presented in Table 9, that there is a significant relationship (at the level 5%) between internal control and financial performance. It reached (t = − 1.122) with a level of significance (0.193), and the value of the correlation coefficient (R) indicates that the strength of the relationship between the two previous variables is (0.953), as the financial control variable explained (90.5%) of the variation in financial performance depending on the value of the determination coefficient (R2), and thus we reject the

Table 7 Scores for financial performance statements

Financial performance	Arithmetic mean	Standard deviation	Sample direction
Comparing the actual performance with the planned one contributes to discovering deviations and financial manipulations	4.40	0.894	Not agree
Any defect in the accounting information system is a source of many problems that negatively affect the financial performance of higher education institutions	3.83	1.206	Agree
Identifying imbalances and correcting them promptly contributes to avoiding many problems of financial corruption	3.73	1.363	Agree
Accuracy and correctness of accounting information help the management to carry out proper planning and take the most appropriate decision	3.90	0.923	Agree
Good management and optimal utilization of available resources increase the level of financial performance of higher education institutions	3.57	1.073	Agree
Ensuring that the process of spending money is going in the appropriate direction and is in line with administrative decisions, reduces the waste of money	3.57	0.971	Agree
Working to reduce waste and waste in the use of financial resources increases the profitability and efficiency of higher education institutions	3.60	1.163	Agree
The internal control system is a tool for controlling financial performance and ensuring the achievement of the objectives of higher education institutions	3.60	1.476	Agree
Financial performance (the dimension as a whole)	3.7750	1.01889	Agree

Table 8 Results of testing the first main hypothesis

	Arithmetic mean	Standard deviation	Standard error	T	df	Sig
Internal control and financial performance	− 0.30417	0.31432	0.05739	− 5.30	29	0.000

null hypothesis. The alternative hypothesis H1 was accepted, which states that "there is a statistically significant effect between internal control and financial performance at the level of significance (0.05)".

Table 9 Testing the main hypothesis

		Regression	Error	Total
Sum of squares		21.715	02.181	23.896
Degrees of freedom		01	28	29
Mean of squares		21.715	0.078	
Total morale	F value			
	Sig			
Partial significance (Regression coefficients)		Constant	Financial performance	
	B	0.265	0.849	
	T	1.333	16.696	
	Sig	0.193	0.000	
Interpretive ability	R	0.953		
	R2	0.905		

We have the equation: $y = \beta + \alpha x$ i.e.: $y = 0.265 + 0.849x$
Y: financial performance X: internal control

5 Conclusion

Through the study that we carried out, it became clear to us the importance of the role that internal control plays in improving financial performance, and the study that we conducted showed that relationship. An effective control system is characterized by credibility, as it is a cornerstone of the institution. The internal control system helps ensure the accuracy of the recorded accounting data to determine the degree of reliance on them before making any decisions or drawing any plans.

The large size of the institution is considered an obstacle to improving the financial performance because it is more complex and intertwined, but at the same time it contributes to improving the financial performance of the institution, using several financial analysts to get rid of this obstacle and exploit it in its favor. The study showed that the level of internal control and financial performance was acceptable according to the study scale. Further, there is a statistically significant effect between internal control and financial performance. The value of the correlation coefficient (R) indicates that the strength of the relationship between the two previous variables is (0.953), as the financial control variable explained (90.5%) of the variation in financial performance depending on the value of the coefficient of determination (R2).

The study recommends adopting an internal electronic control program to ensure that individuals adhere to the control procedures established in the institution, and thus increase and improve their financial performance. Further, establishing a modern and advanced control system that affects all parts of the institution and its structures, internally and externally. Hence, there is a need to develop a program to evaluate the performance of the institution and the results achieved and this evaluation should be either monthly, tripartite, or annual, to discover errors and deviations and correct them. Lastly, policy makers could benefit from the experiences of developed and successful countries in the financial aspect in general and oversight in institutions in particular.

References

1. Al-Absy, M.S.M.: Impactful women directors and earnings management. Cogent Bus. Manag. **9**(1), 2148873 (2022)
2. Al-Absy, M.S.M., Ismail, K.N.I.K., Chandren, S.: The association between real activities and accruals earnings management in Malaysian listed companies. Contaduría y administración **66**(3) (2021)
3. Al-Absy, M.S.M., Ismail, K.N.I.K., Chandren, S.: Board chairmen's involvement in audit committees and earnings management practices, pp. 1–26 (2017). Available at SSRN: https://ssrn.com/abstract=3063845
4. Al-Absy, M.S.M.: Board chairman characteristics and real earnings management. Sustainability **14**(22), 15025 (2022)
5. Al-Absy, M., Hasan, M.: Impact of the board of directors'characteristics on firm performance: a case of Bahraini listed firms (2023)
6. Chandren, S., Al-Absy, M.S.M., Qasem, A.: Internal and external governance impact analysis and contribution on operating performance during goods and services tax implementation. A Malaysian evidence. Int. J. Supp. Chain Manag. (IJSCM) **9**(1), 908–920 (2020)
7. Mahdi, S.M.A., Al-Absy, M.S.M., Alastal, A.Y.: Corporate social responsibility and firm's performance: evidence from Bahrain. In: International Conference on Business and Technology. Springer (2023)
8. Al-Absy, M.S.M., Ismail, K.N.I.K., Al-Dubai, S.A.: Accountants' perceptions on the adoption of international financial reporting standards in Yemen. Int. J. Econ. Financ. Issues **6**(4), 1911–1917 (2016)
9. Al-Absy, M.S.M., Ismail, K.N.I.K.: Accountants' perception on the factors affecting the adoption of international financial reporting standards in Yemen. Int. J. Finan. Res. **10**(4), 128–142 (2019)
10. Al-Absy, M.S.M.: Impactful female directors and earnings management: the moderating effect of ownership concentration. Adm. Sci. **13**(5), 129 (2023)
11. Al-Absy, M.S.M., AlMahari, N.H.: The interaction effect of nomination committee's effectiveness on board of directors' characteristics and firm performance. Adm. Sci. **13**(5), 135 (2023)
12. Nguyen, N.P., Hoai, T.T.: Internal controls driven by mindfulness toward enhanced ethical behaviors: empirical evidence from Vietnam. Heliyon **9**(7) (2023)
13. Zhou, H., Chen, H., Cheng, Z.: Internal control, corporate life cycle, and firm performance. In: The Political Economy of Chinese Finance, pp. 189–209. Emerald Group Publishing Limited (2016)
14. Lai, S.-C., Li, H., Lin, H., Wu, F.: The influence of internal control weaknesses on firm performance. J. Acc. Finan. **17**(6), 82–95 (2017)
15. Abiodun, E.A.: Internal control procedures and firm's performance. Int. J. Sci. Technol. Res. **9**(2), 6407–6415 (2020)
16. Tetteh, L.A., Kwarteng, A., Aveh, F.K., Dadzie, S.A., Asante-Darko, D.: The impact of internal control systems on corporate performance among listed firms in Ghana: the moderating role of information technology. J. Afr. Bus. **23**(1), 104–125 (2022)
17. Zhang, H., Dong, S.: Digital transformation and firms' total factor productivity: the role of internal control quality. Financ. Res. Lett. **57**, 104231 (2023)

18. International Finance Corporation, W.B.O., Internal Control Handbook, in Control Environment (2021)
19. Tol Sheet, Practice of internal control, Definition, in Responsibilities and implementation at HI, Internal Audit Toolkit-RMIA (2018)
20. Kenton, W., Julius, M., Kvilhaug, S.: Internal Controls Definition, Types, and Importance (2022)
21. Adelana, O.O.: Internal control system and financial performance of manufacturing firms in Nigeria. IIARD Int. J. Econ. Bus. Manag **6**(3) (2020).
22. Association of Chartered Certified Accountants (ACCA): Internal controls (2023). Retrieved on 10.102023 from https://www.accaglobal.com/gb/en/student/exam-support-resources/fundamentals-exams-study-resources/f1/technical-articles/internal-controls.html#:~:text=Internal%20control%20should%20have%20the,of%20value%20in%20the%20organisation
23. Fatihudi, D., Mochkla, M.: How measuring financial performance. Int. J. Civil Eng. Technol. (IJCIET) **9**(6) (2018)
24. Kenton, W., Murry, C., Courage, A.: Financial Performance: Definition, How it Works, and Example (2023). https://www.investopedia.com/terms/f/financialperformance.asp
25. Nouili, N.: Using management accounting tools to improve the financial performance of the economic institution—a case study of the Great Mills Corporation of the South, Biskra. PhD thesis, third phase. major in accounting, University of Mohamed Kheidar, Biskra (2020)

Maximizing Market Impact: An In-Depth Analysis of the Market Penetration Strategy and Its Effective Tools for Sales Growth and Brand Expansion in the E-commerce Markets of Oman and Bahrain

Boumedyen Shannaq, Imran Saleem, and Mohanaad Shakir

Abstract This study aims to analyze the prevailing consumer behavior in the E-commerce sectors of Oman and Bahrain, with a primary focus on developing sustainable strategies to enhance business technology, specifically within the ecommerce domain. Utilizing exploratory data analysis, data was predominantly obtained through intelligent agents from Terproject.com and official statistics from Statista. The analysis covers various aspects, including Revenue in the eCommerce Market, sales channels, annual growth rate (CAGR 2023–2028), and penetration rate. The findings indicate a substantial revenue increase in Bahrain, expected to rise from 988.10 million to 1458.00 million, and in Oman, from 1695.00 million to 2543.00 million between 2023 and 2028. The projected annual growth rate (CAGR 2023–2028) is estimated to be 8.09% for Bahrain and 8.45% for Oman. User penetration rates suggest the dominance of beverage and fashion products in Bahrain, whereas food, beverage, and fashion are expected to lead in Oman. The resulting recommendations offer the potential for devising an effective and sustainable Market Penetration Strategy.

Keywords Improving business technology · Market penetration strategies · Sustainable initiatives

B. Shannaq (✉) · I. Saleem · M. Shakir
University of Buraimi, CoB, Buraimi, Oman
e-mail: boumedyen@uob.edu.om

I. Saleem
e-mail: imran.saleem@uob.edu.om

M. Shakir
e-mail: mohanaad.t@uob.edu.om

B. Awwad (ed.), *The AI Revolution: Driving Business Innovation and Research*, Studies in Systems, Decision and Control 524, https://doi.org/10.1007/978-3-031-54379-1_25

1 Introduction

Market penetration encompasses strategies to sell existing products or services in a well-established market, aiming to outperform competitors and expand market share. It signifies the percentage of a company's sales in relation to the overall market revenue. Tactics may involve enticing existing or new customers, implementing competitive pricing, and adopting innovative approaches to boost profits. Crucial for business growth, market penetration aligns with the Ansoff Matrix, as developed by Meldrum and McDonald [15]. It involves a company's endeavors to augment its market share within an existing market, outshine competitors, and enhance overall revenue. Market share denotes the portion of total market sales or revenue held by a specific company. For example, in an industry with an annual sale of 50,000 units, a company selling 5000 units claims a 10% market share. Strategies for market penetration encompass encouraging existing customers to make additional purchases, attracting new customers, and luring clients away from competitors. Implementation methods include competitive pricing, intensified marketing efforts, and incentive systems such as gifts and loyalty discounts. The assessment of market penetration involves evaluating the sales volume of an existing product relative to the entire target market. Figure 1 illustrates the business growth strategy based on the Ansoff matrix.

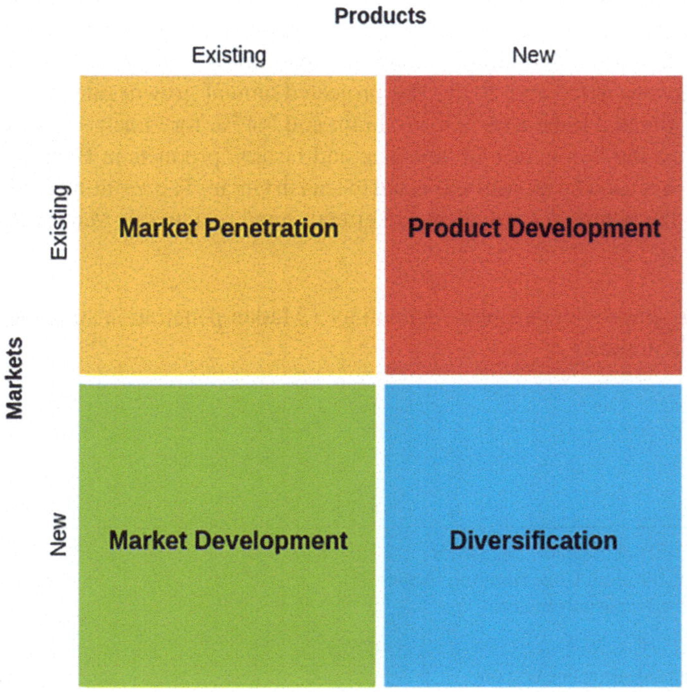

Fig. 1 The Ansoff matrix [15]

Visually, the Ansoff matrix is depicted as a graphic representation featuring two axes: horizontally, classifying products as "old" and "new"; vertically, dividing markets into "old" and "new." It is crucial to understand the context of "old" and "new," referring to existing or potential products and markets for the company. This matrix aids companies in assessing their current position and strategizing for growth stages. The Market Penetration Strategy, recognized as the simplest and most common approach, aims to amplify sales and the percentage of a product or service within the total market. Tools utilized include expanding market share, increasing purchase frequency, and creating new opportunities for consumer product utilization. This strategy involves minimal risks as it operates within a familiar market and deals with familiar products. Calculating the Market Penetration Rate entails evaluating a brand's share in the overall market size, proving beneficial for both new and established brands.

Market Penetration Rate

Assessing the total market size and determining a brand's share of the market pie is highly valuable for both emerging and well-established brands. Market penetration can be computed through a straightforward equation:

$$\text{Market Penetration Rate} = (\text{Number of Customers}/\text{Target Market Size}) \times 100$$

Consider a scenario where a nation boasts a populace of 500 million, among whom 100 million possess Samsung. With a simple calculation (100/500 × 100), the Samsung penetration is 20%, implying that 80% of the population, or 400 million individuals, remain untapped. Noteworthy market penetration for consumer goods is estimated between 2 and 6%, while business products ideally range from 10 to 40%. Some brands assess penetration quarterly, while others do so post advertising and marketing campaigns. Optimal market penetration yields immediate financial benefits. In 2020, Amazon held a 49% share of the US e-commerce market, surpassing the combined share of its top three competitors and constituting 5% of all US retail spending. In the same year, the iPhone claimed a market share between 15 and 20%, with 77.3 million units sold, culminating in a 19.2% share. Brands achieving high market penetration acquire a robust reputation, enabling them to set prices and dictate market trends. By 2017, Apple's global smartphone market share exceeded 50%, growing to 66% of industry profits and 32% of total mobile phone revenue by 2020. In the soft drinks sector, Coca-Cola has retained a dominant 41% share in the US since 2004, with a global range of 30–40%. Market penetration strategies empower brands to capture larger market shares, leaving competitors with fewer opportunities and diminished revenue.

In this work we discover the market penetration strategies, give examples from two gulf regions Bahrain and Oman, and give recommendations on the best penetration strategies.

2 Literature Review

Yang et al. [28] this empirical study investigates the nexus between digital platform service capabilities, cross-border e-commerce firms' digital transformation capabilities, and their performance. Findings reveal that platform service capabilities influence enterprise performance by mediating the digital transformation capability. The growth of cross-border e-commerce platforms is pivotal for the digital trade ecosystem. The study emphasizes considering digital platform service abilities and cross-border e-commerce firms' digital transformation capabilities for sustaining China's digital trade growth. The research underscores the positive impact of both digital platform service and digital transformation capabilities on cross-border e-commerce enterprise performance, with digital transformation acting as a mediator. Presently, enterprise digitization is often synonymous with enterprise informatization, characterized by digitizing enterprise infrastructure [14]. However, the application of digital technology is recognized as the digital transformation of enterprises [11]. While enterprise digitization primarily involves digitizing infrastructure, digital transformation extends beyond, encompassing the integration of digital applications in areas like R&D, production, finance, and talent [12]. Looking ahead, China's high-level opening-up involves shaping a new world trade order, fostering an open world economy, and nurturing a shared future for humanity. The nation's international trade growth propels reform, development, and innovation, establishing a new global relationship framework based on solidarity, cooperation, and mutual benefits [8, 13].

Alhaimer [3] this research explores the impact of diverse risk factors on online shopping behavior in Kuwait during the COVID-19 pandemic. Utilizing online questionnaires distributed across social media, 385 responses were gathered and analyzed using AMOS 21 for structural equation modeling. Results reveal that risk susceptibility, severity, and formal penalties positively influence online buying attitudes. Product, financial, and non-delivery risks exhibit no significant impact. Convenience risk negatively affects attitudes. Additionally, a novel finding indicates that formal penalties for lockdown violations directly and positively shape consumer behavior in favor of online shopping during the pandemic. The study emphasizes the shift in factors influencing user attitudes and behavior during emergencies compared to non-emergency times [1]. This study investigates the influence of online shopping on green product purchasing behavior in Oman, utilizing primary data gathered from a survey of 50 online shoppers in Muscat. Findings highlight the need for improved green product awareness and diverse offerings to encourage environmentally conscious consumer choices [4]. This study examines the impact of online customer satisfaction, focusing on product quality, application safety, delivery guarantee, and offers in online shopping among Omani youth. Using purposive sampling, 120 participants were surveyed, and data were analyzed using SPSS. Results indicate that perceptions of product quality and service guarantee significantly influence online customer satisfaction, emphasizing the importance of the Quality Safety Assurance (QSA) model. The study provides valuable insights for online marketers seeking to enhance customer satisfaction in Oman [7]. This study, within the broader context of examining global trends in consumer protection within e-commerce literature, focuses on analyzing the legal framework and current regulations for e-commerce and consumer protection in Oman. Utilizing a normative legal research approach, the findings reveal a disparity in consumer protection measures for offline and e-commerce transactions in Oman, highlighting the need for comprehensive legislation to address global concerns and enhance consumer confidence in online environments. The study is pioneering in its exploration of legal provisions, regulations, and consumer confidence in Oman, offering valuable insights for policymakers [6]. Existing literature often evaluates Quality of Service (QoS) independently, limiting a comprehensive understanding of service quality management. Analyzing enterprise operations aids decision-making for service improvement. This study compares QoS in the telecommunications sector between private and public companies in Oman, focusing on customer satisfaction. The SERVQUAL model with six dimensions measures QoS. Statistical tools, including ANOVA and t-test, analyze 62 responses. Results reveal similarities in QoS between private and public companies, with a significant difference in service reliability [18]. Despite government efforts, Small and Medium Enterprises (SMEs) face declining survival rates globally. This study utilizes qualitative content analysis, data mining, and SPSS 25 to extract factors affecting SME success and failure. The research generates a conceptual framework based on System Theory, identifying relationships among external, internal, and moral/ethical factors crucial for survival strategies. The resulting framework offers innovative insights for SMEs and potential businesses [9]. This study assesses the awareness and usage of business-to-consumer (B2C) e-commerce in Bahrain through a literature review and a questionnaire distributed in universities. It highlights the low participation in e-commerce among Bahraini residents due to concerns about internet regulations and privacy issues. The research focuses on consumer perspectives, specifically non-technical limitations, offering valuable insights for businesses aiming to establish a successful B2C environment in Bahrain [2]. The study devised a predictive model for post-COVID business demand, aiming to identify resilient industries. It utilized a Web Scraper for competitor data extraction and machine learning to create a classification model. Results highlight key future business sectors like medical services and software automation [21]. This study devised two schemas for cloud-based distributed computing, aiming to enhance dynamic load balancing in complex virtual computing. It analyzed and discussed various approaches, architectures, and applications for user access and application execution in a cloud-based distributed computing environment. The goal is to improve efficiency, ensure secure user access, and develop principles for running demanding applications. The designed methodology enhances overall performance and allows dynamic balancing of computing nodes through migration processes [10]. E-commerce, known as online shopping, enables consumers to purchase goods and services directly from sellers over the internet. Customers use a web browser to submit orders, browse products, and compare prices. Online stores provide product details, images, and payment options, allowing for convenient and efficient transactions, benefiting both buyers and sellers [5]. The surge in online shopping, driven by technology, is pronounced in high-internet-penetration countries like Bahrain, boasting a 99% rate. Anticipated e-commerce revenue in 2022 is USD 1.434 million. This study assesses customer usage and satisfaction with AI features in Amazon and Alibaba, analyzing demographic influences on usage and satisfaction [20]. International companies prioritize product name accuracy for profit maximization. Customers' misspelling during registration poses challenges in product searches. This study introduces the LPINS method, language-independent and superior to existing methods. It outperforms SOUNDX and n-gram methods, accommodating diverse languages and characters, offering flexibility in adjusting score values for precise results. While existing literature extensively explores E-Commerce and its implementation, our research uniquely contributes by proposing a methodology to examine contemporary consumer behavior in the E-commerce markets of Oman and Bahrain. This work is crucial for enriching literature with real cases from the Gulf regions. To gain further insights into the intricacies of the topic addressed in this study, consider referring to the following relevant studies: Al Shamsi and Shannaq [21], Shannaq et al. [22, 24, 25], Segumpan and McAlaney [26], More [16], Vasista and Zamil [27], Kumar and Arthi [19], Alnaseri et al. [17]. Reviewing numerous previous studies is vital to enriching and deepening our understanding of the current topic. These prior studies provide a foundation and framework for the present research, offering valuable insights and results that enhance understanding of the studied issue.

3 Methodology

Figure 2 illustrates the primary stages of the proposed methodology presented in this study.

3.1 Methodology Overview

1. Identifying the Problem and Defining Goals

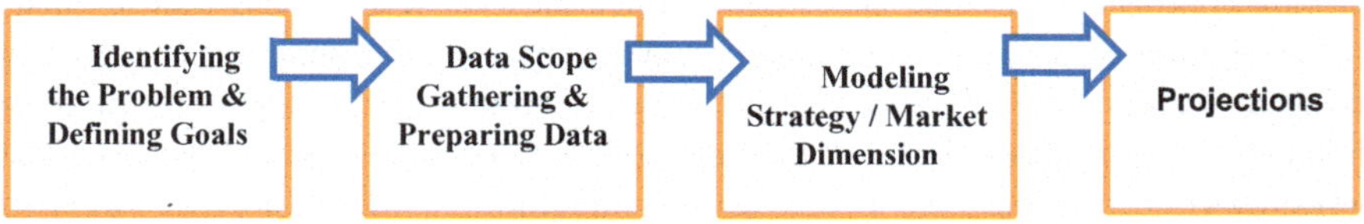

Fig. 2 Proposed methodology

The first stage, it is necessary to clearly define the problem and agree on the objectives of the study. Problem definition is one of the most critical stages of the entire research process, since a well-defined problem is a half-solved problem.

2. Data Scope

Focus on B2C transactions for tangible goods through digital channels to end-users.

Covers desktop and mobile transactions, excluding 'eService's', digital media, B2B digital goods, and 'e-commerce'. Financial data refers to annual total revenue, excluding shipping costs.

3. Modeling Strategy/Market Dimension

Utilizes a combined top-down and bottom-up approach for market size determination.

Relies on financial disclosures, industry reports, primary research, purchasing trends, and performance metrics.

Incorporates key indices and country-specific data for individual market size estimates.

4. Projections

Employs diverse forecasting methods based on market behavior.

Utilizes S-curve function and exponential trend smoothing for digital products and services.

Primary drivers include GDP per capita, consumer spending, internet penetration, and population.

3.2 *Evaluation of Present Consumer Behavior in the E-Commerce Market Bahrain Case: (Bahrain—e-commerce, 2022)*

Bahrain is experiencing a rapid surge in e-commerce, with an anticipated regional market value exceeding $28 billion by 2022. High computer literacy rates and widespread internet usage, especially among the youth, contribute to elevated online shopping rates. Social media platforms like WhatsApp, YouTube, Facebook, and Instagram are popular for advertising retail services. The government's strategic plans, such as the National E-Government Strategy 2020, align with the shift towards a digital economy, evidenced by recent regulations like the Electronic Communications and Transactions Law and the Personal Data Protection Law. The Information and eGovernment Authority (IGA) plays a pivotal role in enhancing government services, focusing on increased participation, partnerships, e-literacy, information protection, efficient government, quality service, e-government channels, and innovation. The COVID-19 pandemic prompted the establishment of virtual marketplaces and legislation requiring online businesses to obtain commercial registration. Bahrain encourages e-commerce growth with initiatives like Start Up Bahrain and

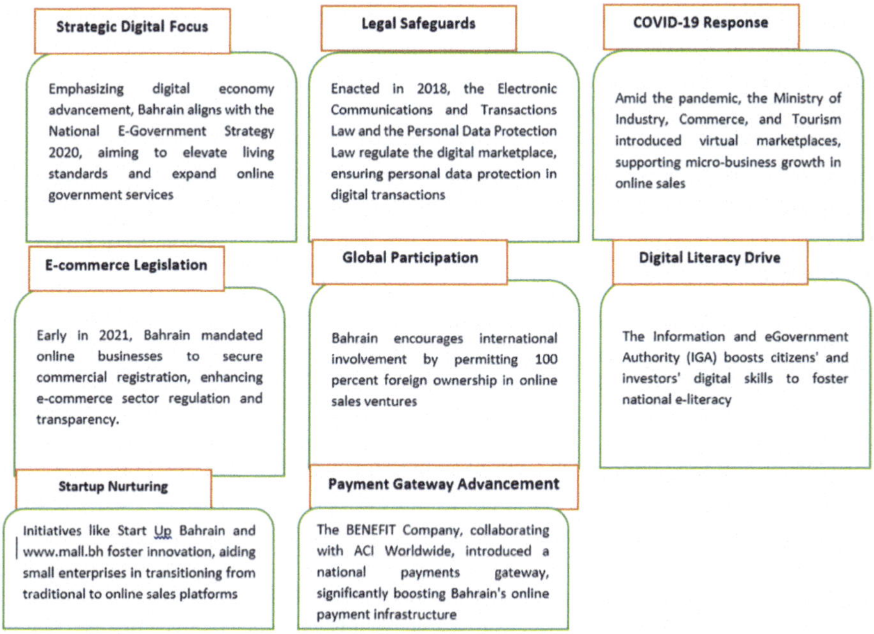

Fig. 3 A comprehensive strategy for online shopping advancements in Bahrain [29]

Mall of Bahrain [31]. Despite these efforts, challenges persist, including low local business adoption of online retail, a preference for cash on delivery, and limited consumer acceptance compared to global benchmarks. However, international platforms like Souq.com, Amazon, and eBay remain popular among Bahraini consumers. The benefit company, serving as the national gateway, plays a crucial role in facilitating electronic financial transactions for example in Bahrain actively bolsters its eCommerce sector and online shopping with a multifaceted approach as illustrated in Fig. 3:

Despite challenges like low local business adoption and a preference for cash on delivery, ongoing initiatives strive to propel Bahrain's e-commerce sector forward.

Oman Case: Evaluation of Present Consumer Behavior in the e-commerce Market.

3.3 Oman's E-Commerce Landscape: A Comprehensive Overview

The E-Commerce Market in Oman is projected to reach USD 1.09 billion by 2028, with an estimated value of USD 0.58 billion in 2023, experiencing a growth rate of 13.54% during 2023–2028. Despite the nation's economic contraction of 3.5% in 2020, attributed to lower oil prices and the global impact of COVID-19, Oman, with a population exceeding 5.1 million and a 75% internet penetration rate, presents a burgeoning e-commerce market. Although the market is in its early stages, opportunities abound for expansion. While Oman exhibits high smartphone usage, only 8% engage in mobile shopping. Grocery and music purchases dominate the current online shopping landscape, with over a quarter of Oman's population favoring foreign websites. E-commerce constitutes a mere 1% of total retail sales, but the government, through the Information Technology Authority (ITA), actively promotes digital society and e-government services. Notably, local businesses, particularly Omani SMEs, increasingly leverage social media for product promotion and sales. Challenges persist, notably the evolving consumer mindset, as customers express a preference for traditional in-store purchases.

3.4 E-Commerce Trends Shaping Oman's Market

The Sultanate's increasing internet penetration, reaching over 75%, and near-universal smartphone use underscore the potential for market growth. The country's robust IT infrastructure development, surging mobile broadband and internet subscribers, and the expanding adoption of 4G and 5G technologies contribute to this upward trajectory. However, the ongoing COVID-19 pandemic has impacted consumer spending on telecom services and devices, posing a potential barrier to market expansion. Despite challenges, major market players are innovating to tap into this burgeoning market, with the Oman Telecommunication company planning a substantial increase in 5G sites by 2024. Source: Mordor Intelligence [30].

As of January 2022, there is a lack of specific details regarding recent endeavors in Oman aimed at enhancing the e-commerce sector and online shopping. However, governments traditionally employ diverse strategies to uplift these domains as illustrated in Fig. 4.

For the most recent and detailed information regarding Oman's endeavors to enhance its eCommerce sector and online shopping landscape, it is recommended to consult up-to-date sources and official announcements.

4 Finding and Results

Figures 5, 6, 7, 8 and 9 demonstrate the revenue in US$ from 2023 to 2028, the revenue is in millions.

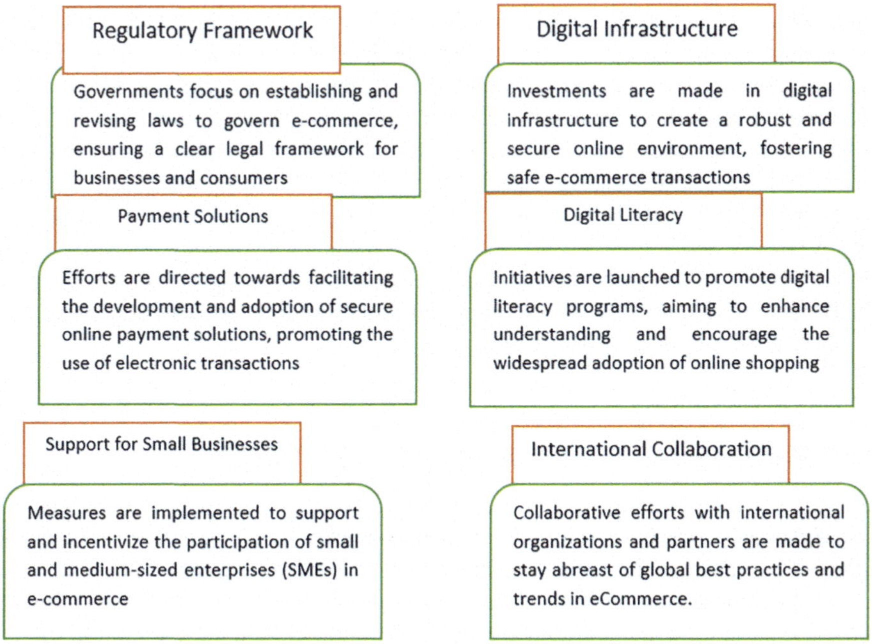

Fig. 4 Oman focuses on boosting e-commerce and online shopping strategies [30]

Fig. 5 Revenue 23–28 Oman-Bahrain

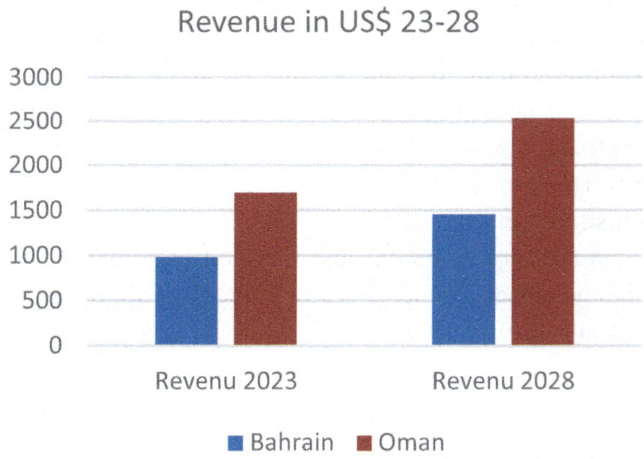

Fig. 6 Annual rate 23–28 Oman-Bahrain

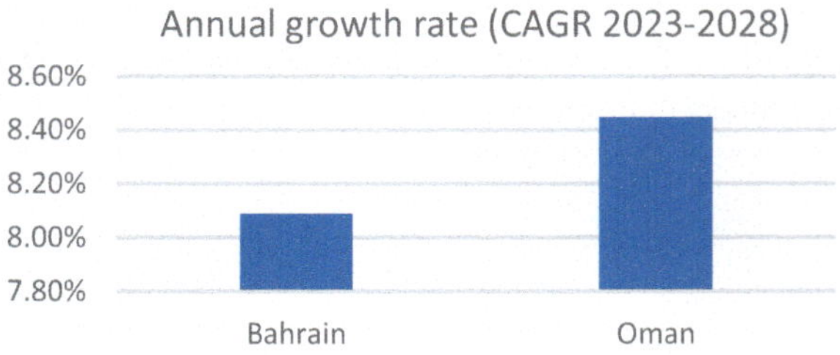

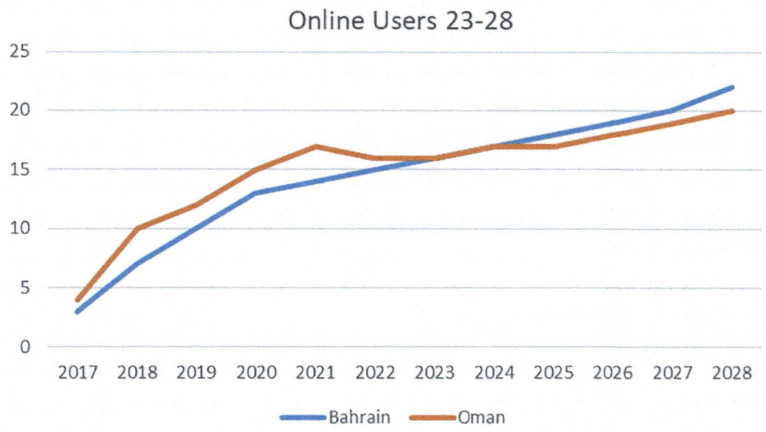

Fig. 7 Online users 23–28 Oman-Bahrain

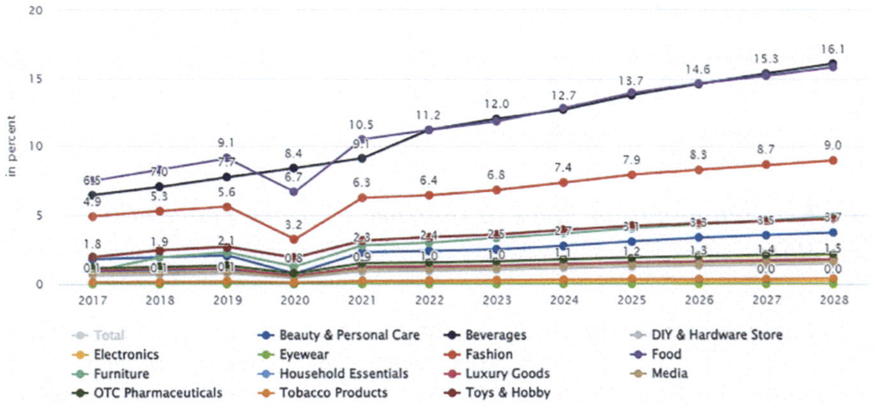

Fig. 8 Penetration rate worldwide [32]

5 Recommendations

In the new economic conditions, companies must constantly improve the quality of service in order to retain customers. To do this, it is necessary to optimize logistics processes—increase speed and flexibility, reduce logistics costs and increase customer satisfaction.

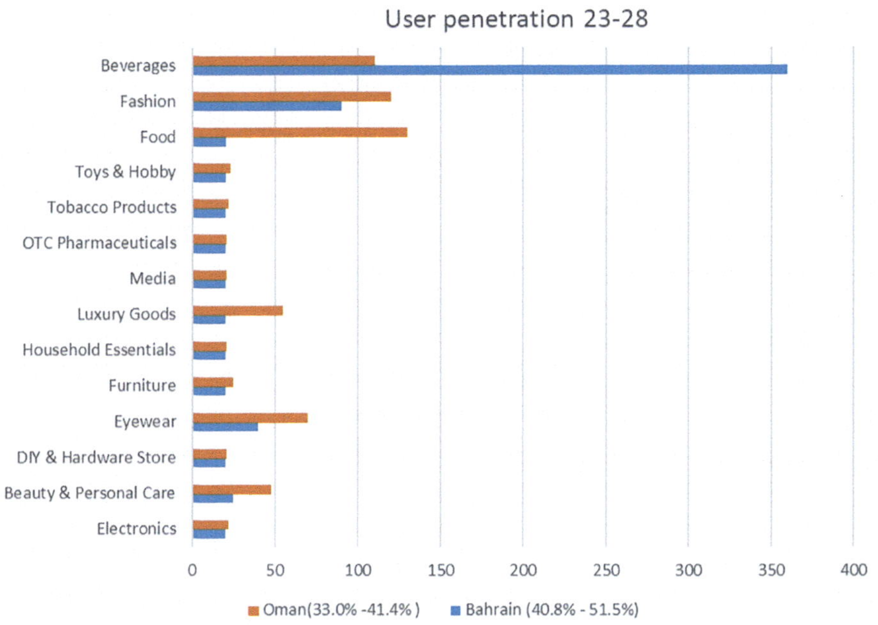

Fig. 9 User penetration rate Oman-Bahrain

5.1 Recommendation Steps to Improve Country Penetration Rate

1. **Enhance Service Quality**: In the current economic landscape, companies should prioritize continuous improvement in service quality to retain customers.
2. **Optimize Logistics Processes**: Focus on optimizing logistics processes to increase speed and flexibility, reduce costs, and enhance customer satisfaction. Key tools for optimization include Warehouse Management Systems (WMS), Route and Vehicle Management (TMS), Order Management Systems (OMS), and Supply Chain Management (SCM) systems.
3. **Utilize Advanced Software Solutions**: Implement sophisticated software solutions, such as microservices-based systems, for nationwide parcel terminal networks. An exemplary case is SHIPDAY.Com, which successfully caters to various user groups, including online stores, logistics companies, banks, tenants, and individuals, serving a vast clientele of over 50 million users.
4. **Leverage Customer Data Platforms (CDPs)**: Employ Customer Data Platforms like Mindbox in customer relationship management (CRM) to consolidate and manage customer data from diverse sources. This enables businesses to gain a deeper understanding of customers and personalize communications across channels, ultimately enhancing user experience, engagement, and the likelihood of purchases.

5. **Implement Conversion Rate Optimization (CRO)**: Address the intensifying competition in e-commerce by implementing Conversion Rate Optimization strategies. Analyze data, test website elements, and optimize the user interface to boost conversions, revenue, and the overall effectiveness of marketing campaigns.
6. **Adopt Progressive Web Applications (PWAs)**: Embrace the trend of Progressive Web Applications (PWAs) for mobile optimization in e-commerce. PWAs offer advantages such as fast page loading and uninterrupted access, providing cost savings compared to developing separate mobile applications.
7. **Explore Niche Marketplaces and D2C Trends**: Stay abreast of changes in the e-commerce landscape, including the emergence of niche marketplaces and the Direct-to-Consumer (D2C) trend. While universal marketplaces continue to dominate, specialized "vertical" marketplaces are gaining traction, and manufacturers are increasingly shifting towards D2C models. Consider exploring and adapting to these evolving trends for sustained market penetration.

6 Conclusion

In this research, we examine Oman's efforts to enhance E-commerce and online shopping, along with a comprehensive strategy for advancing online shopping in Bahrain. Additionally, the study explores market penetration strategies, showcasing examples from Bahrain and Oman, and evaluates their performance in E-commerce. Based on the analysis, this paper provides recommendations for effective penetration strategies. The trends and case studies presented offer valuable insights into the evolving E-commerce landscape. As businesses face increasing challenges, adaptability becomes crucial for success, and the decision to embrace or resist these challenges remains a strategic choice for each business.

References

1. Al Sabahi, L., Al Kharusi, M., Al Hinai, A.F.: Online shopping in Oman—its influence and opportunities. Int. J. Res. Innov. Soc. Sci. (IJRISS) **VI**(VI), 2454–6186 (2022)
2. Al Shamsi, I., Shannaq, B.: Predictive management model—the leading business model with the coronavirus (COVID-19) epidemic. Int. J. Adv. Sci. Technol. **29**(3), 14975–14983 (2020)
3. Alhaimer, R.: Fluctuating attitudes and behaviors of customers toward online shopping in times of emergency: the case of Kuwait during the COVID-19 pandemic. J. Internet Commer. **21**(1), 26–50 (2022). https://doi.org/10.1080/15332861.2021.1882758
4. Al-Jahwari, N.S., Rahman Khan, M.F., Al Kalbani, G.K., Al Khansouri, S.S.: Factors influencing customer satisfaction of online shopping in Oman—youth perspective. Humanit. Soc. Sci. Rev. **6**(2), 64–73 (2018). https://doi.org/10.18510/hssr.2018.628
5. Almohsen, E., Henari, T., Al Rawi, A., Salman, H., Fardan, M.: Evaluating customer usage and satisfaction in Bahrain towards AI-enabled e-commerce websites. In: 2022 International Conference on Innovation and Intelligence for Informatics, Computing, and Technologies (3ICT), pp. 117–123 (2022). https://doi.org/10.1109/3ICT56508.2022.9990846

6. Al-Shamsi, I.R., Shannaq, B., Devarajanayaka, K.M.: A comparative analysis of the service quality in public and private company telecommunication services. In: Arai, K. (ed.) Advances in Information and Communication, vol. 651, pp. 167–186. Springer Nature Switzerland (2023). https://doi.org/10.1007/978-3-031-28076-4_15
7. Belwal, R., Al Shibli, R., Belwal, S.: Consumer protection and electronic commerce in the Sultanate of Oman. J. Inf. Commun. Ethics Soc. **19**(1), 38–60 (2021). https://doi.org/10.1108/JICES-09-2019-0110
8. Chen, S.-H., Xiao, H., Huang, W., He, W.: Cooperation of cross-border e-commerce: a reputation and trust perspective. J. Glob. Inf. Technol. Manag. **25**(1), 7–25 (2022). https://doi.org/10.1080/1097198X.2021.2022396
9. Fryad Henari, T., Mahboob, R.: E-commerce in Bahrain: the non-technical limitations. Educ. Bus. Soc. Contemp. Middle East. Issues **1**(3), 213–220 (2008). https://doi.org/10.1108/17537980810909832
10. Gopalan, G.: Exploring the influential factors of online consumer shopping habits and intention in GCC. Int. J. Innov. Sci. Res. Technol. **5**(4) (2020). https://www.ijisrt.com/assets/upload/files/IJISRT20APR480.pdf
11. Hunady, J., Pisár, P., Vugec, D.S., Bach, M.P.: Digital transformation in European union: north is leading, and south is lagging behind. Int. J. Inform. Syst. Project Manag. **10**(4), 39–56 (2022). https://doi.org/10.12821/ijispm100403
12. Khin, S., Ho, T.C.: Digital technology, digital capability and organizational performance: a mediating role of digital innovation. Int. J. Innov. Sci. **11**(2), 177–195 (2019). https://doi.org/10.1108/IJIS-08-2018-0083
13. Khmeleva, G.A., Kurnikova, M.V., Nedelka, E., Tóth, B.I.: Determinants of sustainable cross-border cooperation: a structural model for the Hungarian context using the PLS-SEM methodology. Sustainability **14**(2), 893 (2022). https://doi.org/10.3390/su14020893
14. Ma, S., Liang, Q.: Industry competition, life cycle and export performance of China's cross-border e-commerce enterprises. Int. J. Technol. Manag. **87**(2/3/4), 171 (2021). https://doi.org/10.1504/IJTM.2021.120926
15. Meldrum, M., McDonald, M.: The Ansoff matrix. In: Meldrum, M., McDonald, M. (eds.) Key Marketing Concepts, pp. 121–126. Macmillan Education UK (1995). https://doi.org/10.1007/978-1-349-13877-7_24
16. More, A.B.: Implementing digital age experience marketing to make customer relations more sustainable. In: Nayyar, A., Naved, M., Rameshwar, R. (eds.) New Horizons for Industry 4.0 in Modern Business. Contributions to Environmental Sciences & Innovative Business Technology. Springer, Cham (2023). https://doi.org/10.1007/978-3-031-20443-2_5
17. Alnaseri, M., Örs, M., Sheker, M., Shakir, M., Muttar, A.K.: Factors affecting online shopping intention through verified webpages: a case study from the Gulf Region. In: Recent Advances in Intelligent Systems and Smart Applications, pp. 75–95 (2021)
18. Sadriwala, K., Shannaq, B., Khan, L.: Clustering technology for analysing small and medium enterprises to develop strategies for sustainable development. Int. J. Innov. Creativity Change **14**(9) (2020). https://www.ijicc.net/images/Vol_14/Iss_9/14917_Sadriwala_2020_E_R.pdf
19. Prasanna Kumar, U., Arthi, R.: A study on utility and feasibility of digital marketing tools with lead acquisition, lead nurturing and client engagement. In: Aloysius Edward, J., Jaheer Mukthar, K.P., Asis, E.R., Sivasubramanian, K. (eds.) Current Trends in Economics, Business and Sustainability. ICEBS 2023. Contributions to Environmental Sciences & Innovative Business Technology. Springer, Singapore (2023). https://doi.org/10.1007/978-981-99-3366-2_24
20. Shannaq, B., Alexandrov, V.: Using product similarity for adding business. Glob. J. Comput. Sci. Technol. **10**(12) (2010). https://www.sial.iias.spb.su/files/386-386-1-PB.pdf
21. Al Shamsi, I.R., Shannaq, B.: University of Buraimi in Oman: pandemic lessons and initiatives for education reform from instructors' perspectives. In: Challenges and reforms in gulf higher education: confronting the COVID-19 pandemic and assessing future implications, pp. 137–156. Taylor & Francis Group. Abingdon, Oxon, Routledge (2023). eBook ISBN:9781003457299. https://www.taylorfrancis.com/chapters/edit/10.4324/9781003457299-13/university-buraimi-oman-ibrahim-rashid-al-shamsi-boumedyen-shannaq

22. Shannaq, B., Devarajanayaka, K.M., Shakir M., Abbas A.D.: Generating an integrated SWOT strategy from the SERVQUAL survey results-the need for a comparative assessment of telecommunication companies in Oman. In: The Second International Conference on Emerging Technology Trends in Internet of Things and Computing, pp 020001. Ramadi, Iraq (2023). https://doi.org/10.1063/5.0188360
23. Shannaq, B., Al Shamsi, I.: Visual search technology for Omani student mindset management using e-learning library problem-oriented search. Int. J. Control Autom. **13**(2), 1413–1427 (2020)
24. Shannaq, B., Al Shamsi, I., Majeed, S.N.A.: Management information system for predicting quantity martials. TEM J. **8**(4), 1143–1149 (2019)
25. Shannaq B., Talab, M.A., Shakir, M., Sheker M.T., and Farhan, A.M.: Machine learning model for managing the insider attacks in big data. In: The Second International Conference on Emerging Technology Trends in Internet of Things and Computing. Ramadi, Iraq, pp 020013 (2023). https://di.org/10.1063/5.0188358
26. Segumpan, R.G., McAlaney, J.: Introduction. In: Segumpan, R.G., McAlaney, J. (eds.) Challenges and Reforms in Gulf Higher Education, 1st ed., pp. 1–4. Routledge (2023). https://doi.org/10.4324/9781003457299-1
27. Vasista, T.G., Zamil, A.M.A.: Role of metaverse in the fourth industrial revolution for providing customer experiences. In: El Khoury, R., Alareeni, B. (eds.) How the Metaverse Will Reshape Business and Sustainability. Contributions to Environmental Sciences & Innovative Business Technology. Springer, Singapore (2023). https://doi.org/10.1007/978-981-99-5126-0_14
28. Yang, Y., Chen, N., Chen, H.: The digital platform, enterprise digital transformation, and enterprise performance of cross-border e-commerce—from the perspective of digital transformation and data elements. J. Theor. Appl. Electron. Commer. Res. **18**(2), 777–794 (2023). https://doi.org/10.3390/jtaer18020040

Links

29. Bahrain—e-commerce (2022). Retrieved from https://www.trade.gov/country-commercial-guides/bahrain-ecommerce#:~:text=In%20early%202021%2C%20the%20Bahraini,online%20sales%20activities%20in%20Bahrain
30. Mordor Intelligence (2023). https://www.mordorintelligence.com/industry-reports/e-commerce-market-in-oman
31. Mall (2023). https://www.mall.bh/
32. Stistica (2023). https://www.statista.com/outlook/emo/ecommerce/worldwide#revenue

Influence of Leverage on the Relationship Between Institutional Investors and Performance: Productivity and Financial Performance of Saudi Listed Firms

Reem Alshagri

Abstract In this paper, we aim to evaluate how leverage would reshape the effect of institutional investors on mitigating agency cost and enhancing productivity and financial performance of Saudi manufacturing listed firms over the period 2010–2020. Additionally, we aim to construct an optimal financial frontier that captures the accounting and market sides of performance using the stochastic output distance function (ODF) approach. We used the stochastic frontier approach (SFA) to estimate the production frontier. Our findings suggest institutional investors positively impact firms' productivity and financial performance. However, this positive effect weakens as the level of leverage increases. Our results indicate that institutional investors and leverage act as substitutive monitoring devices to mitigate agency costs and enhance firms' performance. From a policy perspective, to reduce agency costs and improve firm performance, policymakers may find it optimal to recommend using a lower level of debt and encourage institutional investors as a monitoring device.

Keywords Technical efficiency (TE) · Total factor productivity (TFP) · Financial performance · Institutional investors · Leverage · Agency costs · Stochastic frontier approach (SFA) · Output distance function approach (ODF)

1 Introduction

Numerous studies have attempted to explain the link between institutional investors and firm performance. This research is crucial because institutional investors are one of the primary tools for mitigating agency costs [1]. According to agency theory, agency costs arise when ownership is dispersed, which implies that managers have the ultimate authority to run the company and prioritize their interests over those

R. Alshagri (✉)
Economics Department, King Saud University, Riyadh, Saudi Arabia
e-mail: Rshegri@ksu.edu.sa

B. Awwad (ed.), *The AI Revolution: Driving Business Innovation and Research*, Studies in Systems, Decision and Control 524, https://doi.org/10.1007/978-3-031-54379-1_26

of shareholders, which may affect the firm's value performance negatively. Concentrated ownership can alleviate this problem, as large shareholders (such as institutional investors) are more inclined to monitor management decisions, ensuring that the company is engaging in activities that are likely to boost share value. Thus, the presence of large shareholders should reduce managerial opportunism, leading to less conflict of interest between managers and shareholders, lower agency costs, and increased firm performance [2].

Theoretically, three main hypotheses describe institutional investors' effect on firms' performance: (i) the efficient monitoring hypothesis or 'active monitoring view, which suggests that institutional investors exert a positive influence on firm performance. (ii) The strategic alignment hypothesis indicates a negative relationship between institutional investors and the firm's performance. (iii) The efficiency abatement hypothesis considers institutional investors passive investors; thus, their effect might be insignificant.

The lack of consensus regarding the effect of institutional investors on firm performance in the theoretical literature has followed with mixed results empirically. However, the mixed results of empirical investigations could be attributed to the tendency of prior studies to ignore the effects of certain key factors that affect the relationship between institutional investors and firm performance. For instance, Jensen and Warner [3] stated that the precise impact of institutional ownership on performance remains unclear, indicating the need to investigate the interrelationships among institutional ownership, firm characteristics and firm performance in more depth. For instance, McConnell and Servaes [4] found that the effect of institutional investors on performance differed according to firms' growth opportunities and called for further exploration of the effects of firm characteristics on the association between institutional investors and performance. In addition, Liang et al. [5] found solid evidence of the effect of the life cycle of firms on the relationship between institutional investors and firm performance. Furthermore, according to Fung and Tsai [6], the failure to account for interactions between institutional investors and other corporate governance measures could explain the mixed findings in earlier empirical studies. Therefore, this paper contributes to the existing literature by assessing how leverage would reshape the effect of institutional investors on mitigating agency costs and enhancing performance.

The decision to incorporate the effect of leverage stems from the fact that both institutional investors and leverage are one of the main tools used to mitigate agency costs and improve performance. According to Jensen and Meckling [1], leverage can serve to decrease agency costs and conflicts between managers and shareholders because it can be used as a tool to protect cash flow from waste by managers since it will place more pressure on managers to generate cash flows to pay off debt and improve firm performance [7]. Moreover, there is an argument in the literature that leverage, and institutional investors can act as substitute tools for mitigating agency costs, thereby affecting firm performance. For instance, studies such as Bathala et al. [8], Michaely and Vincent [9], and Chung and Wang [10] found that the relationship between institutional investors and leverage is negative. Thus, the present paper examines how the level of leverage affects the relationship between institutional investors and firm performance.

Prior studies have used different performance measures to assess the relationship between institutional investors and firm performance. This paper focuses on two aspects of performance: financial performance and productivity. Financial performance is considered a multifaceted concept, and the literature has developed various observed measures and indicators to explain it, including the return on assets (ROA), return on equity (ROE) and Tobin's Q. Generally speaking, these indicators are divided into two main proxies, namely accounting-based measures and market-based measures.[1] The main advantages of accounting-based measures are their availability for listed firms and the fact that they can provide insights into the economic rates of return. However, scholars have criticised such measures for being built-in short-term bias and allowing for manipulation by managers. In light of these criticisms, market-based measures have become increasingly popular with researchers because they are safe from managerial manipulation.

However, according to Seth [11], market-based measures can capture firm performance more accurately because they measure the present value of future income. In contrast, accounting-based measures reflect past performance [12, 13]. However, Pervan et al. [14] argued that it could prove problematic to use market-based measures in emerging markets due to variations in the volatility of stocks. Furthermore, Chen et al. [15] stated that Tobin's Q, which is a market-based measure, is influenced by unstable factors related to the psychology of investors and market projections. When estimating the financial performance of firms, previous studies (e.g. Cornett et al. [16], Martins [17], Eshwari and Baby [18], Thirumagal et al. [19]) have commonly used one observable indicator as a proxy for the financial performance of firms. Yet, the preceding discussion suggests that using a single indicator may result in only a partial understanding of the relationship between institutional investors and the financial performance of firms, as a single proxy is unlikely to capture the true financial performance [12, 20]. Thus, since financial performance is a complex concept that lacks a clear and unique measure, this paper contributes to the literature by constructing a single financial performance that considers accounting and market-based aspects. To achieve this, the study employs the ODF approach, which is a commonly used tool in the field of productivity economics. To the best of our knowledge, this is the first study to utilize the ODF approach to estimate financial performance. Moreover, ODF approach is advantageous because it allows for the estimation of a financial frontier that represents the optimal financial performance that firms can attain. Any deviation from this optimal performance is considered an indicator of agency costs or financial inefficiency. Thus, Using ODF approach helps to measure agency costs and evaluate how institutional investors can help reduce them and enhancing firm performance.

---

[1] Accounting-based indicators measure a company's profitability in the short term, and they reflect the company's performance in previous years. The ROA, ROE and profit margin are all examples of accounting-based measurements. By contrast, the market-based measures take a long-term view and focus on predicting future performance based on current and past performance. Tobin's Q and the market-to-book value ratio (MBVR) are examples of market-based measures.

While most prior empirical studies focused on the influence of institutional investors on the financial performance of firms, a few studies sought to assess the effect of institutional investors on total factor productivity [TFP] (technical efficiency [TE]) as a measure of performance[2] (e.g. Lichtenberg and Pushner [21], Chowdhury and Geringer [22] and You et al. [23]). Scholars have long debated the best indicator to use as a proxy for performance. For instance, Hanousek et al. [13] and Chiang and Lin [24] have argued that financial measures represent inaccurate proxies for performance because they are sensitive to the choice of accounting methods and can be easily manipulated, especially in countries with weak legal systems. In light of this, Hanousek et al. [13] and Chiang and Lin [24] suggested productivity measures TFP or TE to be more reliable tools for assessing firm performance since they measure the maximum production capacity of firms given the inputs, which makes them more complex and less beneficial to manipulate. This debate suggests that no single proxy can accurately measure performance. Furthermore, agency costs manifest not only in the loss of firm value but also in a decline in productivity [25]. Thus, the present paper bridges a gap in the literature by assessing institutional investors' impact in mitigating agency costs not only on financial performance but also on productivity.

Most prior studies focused on examining the relationship between institutional investors and firm performance in developed markets and, to some extent, emerging markets such as China, India and many transition economies. Unlike developed economies, emerging markets are less efficient, with low levels of transparency and weak legal systems [26]. This paper uses unbalanced panel data from Saudi-listed manufacturing firms spanning 2010–2020. In the Saudi context, only the study of Amin and Hamdan [27] and the study of Hamdan [28] have sought to analyse the relationship between institutional investors and Saudi-listed firms' performance.[3] This paper extends the findings of such studies by using a longer sample period and assessing how the level of leverage affects the relationship between institutional investors and latent financial performance and productivity.

Findings of this paper suggest that institutional investors play a significant role in decreasing agency costs and, therefore, increasing production performance and improving the financial performance of firms. However, it should be recognised that

[2] Productivity can be measured using TFP or TE. The TFP represents the difference between the total output and the total input. In another context, the TFP is defined as a technological change or progress. Moreover, several studies have stated that the TFP has two components: the change in technical efficiency and technological progress (Mahadevan 2003). The TE captures the gap between the optimal output and the actual output, where a company can produce more output without increasing the inputs due to the improved efficiency of current inputs, resulting from learning by doing or experience, advances in technology and improvements in industrial organisation (Coelli et al. 2005).

[3] Although both studies assessed the impact of institutional investors on the financial performance of Saudi-listed firms as measured by the ROA (an accounting-based measure), they produced mixed results. Hamdan [28] found that institutional investors had an insignificant impact on the ROA, whereas Amin and Hamdan [27] found that institutional investors had a significant positive impact on the ROA. This inconsistency could be attributed to the fact that Hamdan [28] assessed the link between institutional investors and the ROA for listed firms using cross-sectional data for 2016, while Amin and Hamdan [27] analysed listed firms from 2013 to 2014.

the positive impact of institutional investors on firms' performance weakens as the level of leverage increases. Thus, the findings suggest that institutional investors and leverage can substitute for each other to mitigate agency costs and affect firm performance.

The remainder of this paper is organised as follows. Section 2 provides a brief overview of the relevant literature. Section 3 describes the data and econometric methodologies. Sections 4 and 5 present the results and robustness check. Section 6 presents the conclusion.

2 Literature Review

2.1 The Effect of Institutional Investors Performance

In their seminal work, Berle and Means (1932) found that the separation between ownership and control of firms negatively impacts firm performance since this separation would make monitoring and controlling management activities more difficult for owners, reducing managers' incentive to maximise the firm's value.

The divergence of interests between shareholders and managers has created the agency theory, which has emerged with establishing the stock market and expanding public ownership in firms. This diffusion of corporate shares led to the separation between the control of corporate entities and their shareholders, which forms the core of the agency problem. Jensen and Meckling [1] provided the initial theoretical background to the problem, claiming that owners delegate full power to managers to manage firms and, further, that the agency problem emerges when there is a conflict between the interests of owners and managers. While owners aim to maximise their wealth, managers may not fulfil that as they seek to maximise their welfare by focusing on increasing the firm's size through increasing sales, which increases managers' power, compensation and the likelihood of promotion [7]. Moreover, managers may be motivated to invest in projects that yield high short-term profits rather than projects that accumulate more wealth in the long run [29]. One key reason behind agency conflicts is that managers and owners have different attitudes toward risks. Furthermore, this gap is further widened by the information asymmetry between them, where managers are more exposed to information than owners [30].

This divergence of interests highlights the importance of shareholders when monitoring managers' activities. However, such monitoring is expensive and gives rise to agency costs. Given the significant value of their shares, one decisive way to monitor management operations involves institutional investors, who can monitor management activities more effectively and diminish the risks facing firms [31]. Hirschman [32] stressed the importance of the role of institutional investors in ensuring that managers are working toward maximising shareholders' benefits. According to Hirschman [32], institutional investors can complain about management strategies by relating complaints directly to managers or selling their shares. Yet, the latter option is rarely convenient, given the large size of shares they own. Furthermore, Mallin and Melis [33] stated that, as institutional investors are risking their capital by investing in a particular corporate entity, they must exert a direct impact on managers' actions to ensure that they are pursuing the welfare of both the corporate entity and its owners [30].

Different theories exist regarding the impact of institutional investors on firm performance: the efficient monitoring hypothesis, the strategic alignment hypothesis and the efficiency abatement hypothesis. According to the efficient monitoring hypothesis, institutional shareholders positively affect firms' performance. On the one hand, by using their professional knowledge, expertise and advanced managerial skills, institutional shareholders can monitor management practices more efficiently and at a lower cost when compared with atomistic shareholders, mainly due to economies of scale concerning research and monitoring activities. Furthermore, they can use their voting rights to influence managers' decisions to increase the share value and provide the firm with the funds required to expand. In addition, the existence of institutional investors could reduce information asymmetries[4] [34–40].

However, as posited by the strategic alignment hypothesis, institutional shareholders can negatively impact firm performance through cooperating with managers in the use of firm resources to pursue mutual benefits and maximise their profits at the expense of minor or small shareholders. Thus, the influence of institutional investors on firm performance is negative [35, 39, 40]. By contrast, the efficiency abatement hypothesis claims that institutional investors are passive investors if they prioritize short-term trading profits over monitoring managers. Thus, their impact on firm performance is deemed insignificant [38]. Yet, Maug [41] argued that institutional shareholders can be passive or active depending on the size of their shares. If their shares are relatively small, they can quickly liquidate them if the firm's performance is poor. However, if their shares are relatively large, they will have a greater incentive to monitor managers [16].

Several studies have examined the effect of institutional investors on financial performance, mainly in developed markets, with mixed results. Only a few studies have been conducted in emerging markets, with similarly mixed results. For instance, several studies found that institutional investors have a positive impact on financial performance, thereby supporting the efficient monitoring hypothesis (e.g. Li et al. [42], Cornett et al. [16]; Gürbüz et al. [43]; Lin and Fu [40]). In contrast to such studies, Wei and Varela [44], Charfeddine and Elmarzougui [45], and Liang et al. [5] found the impact of institutional investors on financial performance to be negative, which supports the strategic alignment hypothesis. In addition, Agrawal and Knoeber [46] and Martins [17] found strong evidence supporting the efficiency abatement hypothesis. Moreover, Lin [47] found that institutional investors' effect on firms' observed financial performance is insignificant, thereby supporting the efficiency abatement hypothesis. However, when the shares held by institutional investors reach a certain limit, their effect becomes significant and positive. Yet, Chaudhuri et al.

[4] Foster and Viswanathan [36] and Back et al. [37] show that information about firms' performance is disclosed quickly when there are a greater number of institutional investors.

[20] suggested a U-shaped relationship between institutional investors and financial performance.

That inconsistency in findings may be due to the impact of certain factors not considered in previous studies. McConnell and Servaes [4] found that the positive impact of institutional ownership on firm performance is valid for low-growth firms, while it is insignificant for high-growth firms. Their findings support Jensen [7] and Stulz [48] conclusions, which state that the conflict of interests between managers and shareholders is greatest in low-growth firms. In addition, Liang et al. [5] found that institutional investors have a negative effect on firm performance. However, when incorporating the effect of firms' age, they found that the impact of institutional investors is only significant for mature firms. Furthermore, Fung and Tsai [6] found that institutional investors' monitoring role becomes increasingly crucial when internal governance is inadequate.

Limited research has been conducted to explore the relationship between technical efficiency or productivity and firm ownership structure, particularly institutional investors. The results of the studies have been inconsistent. For instance, You et al. [23] discovered that institutional investors had a negative impact on the technical efficiency of pharmaceutical companies, while Lichtenberg and Pushner [21] found that institutional investors had a positive impact on the total factor productivity. On the other hand, Chiang and Lin [24] and Chowdhury and Geringer [22] found that the impact of institutional investors on total factor productivity was insignificant.

2.2 Leverage, Institutional Investors and Performance

Jensen and Meckling [1] suggested that high leverage reduces agency costs and inefficiency [49]. Using leverage can help protect cash flows from being wasted by managers and put more pressure on management to generate cash flows to pay off debt, thereby improving firm performance [7]. Grossman and Hart [50] argued that debt financing increases managers' awareness of the importance of maximizing firm profits to avoid bankruptcy and reputational damage. However, debt financing can also have a negative impact since it may increase the cost of bankruptcy. Additionally, since managers are obligated to pay off debts, their financial resources are limited, so they may be forced to refrain from investing in profitable projects [4].

Several researchers have studied the relationship between leverage and firm performance, but the results are inconclusive. McConnell and Servaes [35] found that leverage positively affects firms' performance. However, in their later study, McConnell and Servaes [4] found that the positive effect of leverage is related to low-growth firms, whereas high-growth firms are negatively associated with leverage. On the other hand, Margaritis and Psillaki [49] found a significant curvilinear relationship between technical efficiency and leverage. Meanwhile, Isik and Soykan [51] and Lin and Fu [40] all concluded that leverage has a negative impact on firm performance.

As the preceding discussion has shown, a considerable body of literature indicates that leverage and institutional investors could be an important mechanism for controlling managerial behaviour, mitigating agency problems and increasing firm performance. While most prior empirical works focused on either the effect of the ownership structure on performance or the effect of leverage on performance, little attention has been paid to examining the interrelationships among ownership structure, leverage and firm performance.

Studies by Chaganti and Damanpour [52], Bathala et al. [8] and Grier and Zychowicz [53] represent the first attempts to evaluate the impact of institutional investors on leverage. Assuming that institutional investors play an active role in mitigating the agency problem and since leverage is another mechanism that is used to reduce agency costs and, hence, improve firm performance, it is predicted that an increase in institutional investors might offset the need for leverage to mitigate the agency problem, which would indicate that these tools can substitute for one another in mitigating agency costs. In this regard, several empirical studies have found the relationship between institutional investors and leverage is negative [9, 54]. Thus, firms with greater monitoring by institutional investors may find it optimal to utilize lower debt levels.

Yet, an increase in institutional investors might improve shareholders' influence and rights, enabling institutional investors to force managers to increase the leverage level to decrease their discretion. As a result, management discretion would be limited, and the agency costs of free cash flow would decrease. Hence, in this scenario, institutional investors and leverage may complement each other when mitigating agency costs [10, 53, 55].

3 Methodology and Data

3.1 Performance Measured by Production Efficiency (Stochastic Frontier Approach SFA)

The main aim of this paper is to investigate how institutional investors affect the performance of firms. This section measures performance as technical efficiency, which is estimated based on the production function. Technical efficiency means that firms can produce the maximum output given their resources and production technology. The SFA, which was initially developed by Aigner et al. [56] and Meeusen and van Den Broeck [57], is used to estimate the production frontier, with any deviation from it being due to technical inefficiency. In other words, technical inefficiency means a firm produces less than its optimal level. First, consider the following production function:

$$y_{it} = \beta_0 + \sum_{i=1}^{k} \beta_i x_{it} + \sum_{j=1}^{N-1} \gamma_j D_j + \sum_{t=1}^{T-1} \theta_t D_t + v_{it} - u_{it} \tag{1}$$

where y_{it} denotes the output of i-th industry at year t, X_{it} is the input of i-th industry at year t (labour, capital and intermediate inputs), D_j is an industry-specific dummy variable, and D_t is the years dummy. The year and industry dummies are included in the model to control for any unobserved heterogeneity at the year and industry levels. The error term in Eq. (1) has two parts: v_{it}, which is a random variable distributed as $N\left(0, \sigma_v^2\right)$, and u_{it}, which represents technical inefficiency and is distributed as a half-normal distribution with a pre-truncated mean zero and variance equal to σ_u^2. We model σ_u^2 as a function of the number of institutional investors, leverage, size, age and tangible assets. The presence of these variables in σ_u^2 represents the determination of technical inefficiency [58].

3.2 *Optimal Financial Frontier (ODF Approach)*

The main objective of firms and their shareholders is to maximise firm value. This paper uses the ODF approach, which operates under the assumption that inputs are non-allocable. In other words, an increase in one input increases all outputs [59]. Utilising the ODF approach allows for establishing a frontier (or benchmark) representing the maximum (optimal) financial performance that firms could achieve. In contrast, any deviation from that frontier represents a measurement of agency costs.

In the present model, both the ROA, which is an accounting-based measure, and Tobin's Q, which is a market-based measure, are used to estimate the optimal financial performance as follows:

$$\frac{Roa_{it}}{Tobin's\,Q_{it}} = \beta_0 + \beta_1 \frac{Size_{it}}{Tobin's\,Q_{it}} + \beta_2 \frac{Age_{it}}{Tobin's\,Q_{it}} + \beta_3 \frac{TangibleAssets_{it}}{Tobin's\,Q_{it}} + \sum_{j=1}^{N-1} \gamma_j D_j + \sum_{i=1}^{k} \theta_t D_t + v_{it} - u_{it} \tag{2}$$

where D_j denotes an industry-specific dummy and D_t represents a year dummy. The error term in Eq. (2) features two parts: v_{it}, which is a random variable distributed as $N\left(0, \sigma_v^2\right)$, and u_{it}, which represents technical inefficiency (agency costs) and is distributed as a half-normal distribution with a pre-truncated mean zero and variance equal to σ_u^2. We model σ_u^2 as a function of the number of institutional investors, leverage and revenue. The presence of these variables in σ_u^2 represents the determination of technical inefficiency or agency costs [20, 60].

Table 1 Definitions of key variables

Dependent variables	Description
Output (Revenue)	Log of real total revenue
ROA	Return on sales (net income/sales) multiplied by asset utilization (sales/assets)
Tobin's Q	(Market value of equity + book value of assets − book value of equity)/book value of assets
Independent variables	
Capital	Log of real fixed assets and long-term investments
Intermediate Inputs	Log of real Inventories
Labour	Log of real wages
Size	Log of real assets
Age	Log of the number of years since the establishment
Leverage	Total debts to total assets
Institutional investors	Log of the number of institutional investors
Tangible assets	Log of ratio of fixed assets to total assets

3.3 Data

The data used in this paper is an unbalanced panel of Saudi manufacturing firms listed on the stock market from 2010 to 2020. The data used are annual and obtained from the Bloomberg database.[5,6] The consumer price index has been used to convert the nominal variables into real values. All variables are defined in Table 1.

4 Empirical Findings

The results from Model 1 in Tables 2 and 3 indicate that the coefficient of the number of institutional investors is negative but significant, suggesting that as the number of institutional investors increases, there is a decrease in production and financial technical inefficiency, which in turn leads to an increase in technical efficiency. These findings align with the efficient monitoring hypothesis and are consistent with the results obtained by Lichtenberg and Pushner [21], Cornett et al. [16] and Lin and Fu [40]. Therefore, an increase in institutional investors plays an active role in mitigating agency costs and improving firm performance, decreasing agency costs and resulting in better production and financial performance. However, the results also indicate that

[5] The Saudi capital market follows the Global Industry Classification Standard (GICS), which features four levels. The industries used here located at level two according to the GICS classification.

[6] The analysis starts in 2010 because data concerning institutional investors are only available from 2010 onwards.

while leverage does not significantly impact production performance, it positively (negatively) impacts financial technical inefficiency (technical efficiency).

Moreover, the findings from Model 2 in Tables 2 and 3 show that the coefficient of the interaction between institutional investors and leverage is positive and statistically significant, indicating that the positive impact of institutional investors on technical efficiency weakens as the leverage level increases. That implies that leverage and institutional investors act as substitute tools in mitigating agency costs and affecting performance.

Based on Table 4, the average production technical efficiency of firms in the sample is approximately 78%, indicating that they are only 22% away from the production frontier. The table also reveals that production efficiencies across industries vary between 74 and 88%. Retail, consumer durables, and apparel industries are the most efficient, while the food and staples retailing industry is the least efficient. Further,

Table 2 SFA production function

	Model 1	Model 2
Variables	Coef.	Coef.
Capital	– 0.084[0.293]	– 0.166[0.287]
Intermediate inputs	0.722***[0.111]	0.742***[0.101]
Wage	0.318**[0.158]	0.320**[0.130]
0.5*Capital^2	0.182*[0.097]	0.230***[0.089]
0.5* Intermediate inputs^2	– 0.026[0.028]	– 0.031[0.024]
0.5*Wage^2	– 0.131*[0.073]	– 0.134*[0.070]
Wage*Capital	– 0.115**[0.049]	– 0.116**[0.047]
Wage*Intermediate inputs	0.136***[0.045]	0.145***[0.047]
Intermediate inputs*Capital	– 0.042[0.039]	– 0.061*[0.035]
Constant	1.773***[0.536]	1.776***[0.513]
Log of σ_u^2 [Technical inefficiency variance determinations]		
Number of institutional investors	– 0.899**[0.354]	– 2.404***[0.705]
Leverage	0.029[1.981]	– 4.493**[2.080]
Number of institutional investors*Leverage	–	4.894***[1.257]
Size	0.219[0.381]	0.306[0.256]
Age	– 0.714**[0.281]	– 0.923***[0.188]
Tangible assets	161.751**[72.115]	182.353***[65.501]
Constant	1.565[2.056]	2.946**[1.257]
Log of σ_v^2 [Error term variance]		
Constant	– 1.656***[0.191]	– 1.662***[0.143]
Obs.	555	555
Log pseudolikelihood	– 410.6899	– 396.910

***, ** and * donate significance at the 1%, 5% and 10% respectively. Robust standards errors are in brackets

Table 3 Stochastic output distance function results ODF (financial performance)

	Model 1	Model 2
Variables	Coef.	Coef.
Size/Tobin's Q	− 0.008***[0.002]	− 0.007***[0.001]
Age/Tobin's Q	− 0.001[0.003]	− 0.002[0.003]
Tangible assets/Tobin's Q	− 0.005[0.004]	− 0.004[0.004]
Constant	0.143***[0.017]	0.135***[0.016]
Log of σ_u^2 [Technical inefficiency variance determinations]		
Number of institutional investors	− 0.601***[0.195]	− 1.488***[0.217]
Leverage	6.032***[0.542]	3.872***[0.657]
Number of institutional investors*Leverage	–	2.987***[0.498]
Revenue	− 0.883***[0.108]	− 0.860***[0.093]
Constant	− 5.43***[0.274]	− 5.036***[0.270]
Log of σ_v^2 [Error term variance]		
Constant	− 7.930***[0.197]	− 7.954***[0.195]
Obs.	613	613
Log pseudolikelihood	1249.102	1263.845

***, ** and * donate significance at the 1%, 5% and 10% respectively. Robust standards errors are in brackets

the average financial technical efficiency of the firms, estimated through the ODF, is approximately 97%, indicating that the firms in the sample are only 3% away from the financial frontier. Results in Table 4 reveal that the financial efficiencies of industries vary between 97.8 and 94.9%, with retail being the most efficient and food and staples retailing being the least efficient.

Table 4 Mean of technical efficiency (TE) by industry

Industry	Mean of production TE (SFA)	Mean of financial TE (ODF)
Materials	0.757	0.969
Retailing	0.883	0.978
Food and staples retailing	0.743	0.949
Consumer durables and apparel	0.821	0.951
Food and beverage	0.782	0.959
Capital goods	0.800	0.954
Average TE score	0.780	0.965

5 Robustness Check

To assess the robustness of our empirical findings, we lagged independent variables by one-period to alleviate potential endogeneity concerns that might arise from reserve causality between dependent and independent variables. Results from Tables 5 and 6 align with our findings in Tables 2 and 3.

Table 5 SFA production function (controlling for endogeneity)

	Model 1	Model 2
Variables	Coef.	Coef.
Capital	– 0.043[0.296]	– 0.160[0.292]
Intermediate inputs	0.714***[0.110]	0.747***[0.099]
Wage	0.296*[0.171]	0.310**[0.140]
0.5*Capital^2	0.166*[0.097]	0.231**[0.091]
0.5*Intermediate inputs^2	– 0.021[0.029]	– 0.03[0.025]
0.5*Wage^2	– 0.147*[0.087]	– 0.144*[0.082]
Wage*Capital	– 0.115**[0.054]	– 0.121**[0.050]
Wage*Intermediate inputs	0.141***[0.047]	0.154***[0.050]
Intermediate inputs*Capital	– 0.033[0.043]	– 0.061[0.040]
Constant	1.835***[0.573]	1.878***[0.542]
Log of σ_u^2 [Technical inefficiency variance determinations]		
Number of institutional investors	– 1.061***[0.377]	– 2.696***[0.781]
Leverage	– 0.909[1.721]	– 5.236***[1.721]
Number of institutional investors*Leverage	–	5.649***[1.495]
Size	0.211[0.375]	0.261[0.270]
Age	– 0.579**[0.293]	– 0.717***[0.175]
Tangible assets	148.846***[41.087]	161.32***[40.299]
Constant	1.541[2.087]	2.631**[1.268]
Log of σ_v^2 [Error term variance]		
Constant	– 1.723***[0.208]	– 1.715***[0.162]
Obs.	487	– 342.732
Log pseudolikelihood	– 355.690	– 342.732

***, ** and * donate significance at the 1%, 5% and 10% respectively. Robust standards errors are in brackets

Table 6 Stochastic ODF results (controlling for endogeneity)

	Model 1	Model 2
Variables	Coef.	Coef.
Size/Tobin's Q	− 0.008***[0.002]	− 0.006***[0.002]
Age/Tobin's Q	− 0.002[0.003]	− 0.003[0.003]
Tangible assets/Tobin's Q	− 0.007**[0.004]	− 0.004[0.004]
Constant	0.143***[0.015]	0.135***[0.013]
Log of σ_u^2 [Technical inefficiency variance determinations]		
Number of institutional investors	− 0.483**[0.219]	− 1.454***[0.312]
Leverage	4.79***[0.719]	2.908***[0.830]
Number of institutional investors*Leverage	–	3.231***[0.828]
Revenue	− 0.831***[0.158]	− 0.805***[0.126]
Constant	− 5.038***[0.362]	− 4.728***[0.362]
Log of σ_v^2 [Error term variance]		
Constant	− 8.089***[0.246]	− 8.099**[0.203]
Obs.	564	564
Log pseudolikelihood	1123.758	1136.963

***, ** and * donate significance at the 1%, 5% and 10% respectively. Robust standards errors are in brackets

6 Conclusion

Institutional investors play a significant role in the capital markets. The increasing number of institutional investors accessing the stock market has attracted the interest of academics and policymakers. However, there is a need to investigate the relationship between institutional ownership, firm characteristics, and firm performance. Previous studies have shown that the impact of institutional investors on firm performance varies based on a firm's growth opportunities and lifecycle. The financial literature highlights that institutional investors and capital structure are essential for monitoring and controlling opportunistic managerial behaviour, leading to improved firm performance. This paper aimed to analyse the relationship between institutional investors and firm performance while considering the impact of leverage on that relationship.

The empirical results presented in this paper highlight several important issues for policymakers. According to the findings, an increase in institutional investors leads to increased production and financial performance. However, while leverage has an insignificant impact on production performance, it has a negative impact on financial performance. Furthermore, an increase in leverage hinders the positive impact of institutional investors on production performance and financial performance. Thus, to mitigate agency costs and improve firm performance, policymakers may find it optimal to recommend using a lower level of debt and encourage institutional investors as a monitoring device.

References

1. Jensen, M.C., Meckling, W.H.: Theory of the firm: managerial behavior, agency costs and ownership structure. J. Financ. Econ. **3**(4), 305–360 (1976)
2. Shleifer, A., Vishny, R.W.: Large shareholders and corporate control. J. Polit. Econ. **94**(3, Part 1), 461–488 (1986)
3. Jensen, M.C., Warner, J.B.: The distribution of power among corporate managers, shareholders, and directors. J. Financ. Econ. **20**, 3–24 (1988)
4. McConnell, J.J., Servaes, H.: Equity ownership and the two faces of debt. J. Financ. Econ. **39**(1), 131–157 (1995)
5. Liang, C.-J., Lin, Y.-L., Huang, T.-T.: Does endogenously determined ownership matter on performance? Dynamic evidence from the emerging Taiwan market. Emerg. Mark. Financ. Trade **47**(6), 120–133 (2011)
6. Fung, S., Tsai, S.C.: Institutional ownership and corporate investment performance. Can. J. Adm. Sci./Revue Canadienne des Sciences de l'Administration **29**(4), 348–365 (2012)
7. Jensen, M.: Agency costs of free cash flow, corporate finance, and takeovers. Am. Econ. Rev. **76**(2), 323–329 (1986)
8. Bathala, C.T., Moon, K.P., Rao, R.P.: Managerial ownership, debt policy, and the impact of institutional holdings: an agency perspective. Finan. Manag., 38–50 (1994)
9. Michaely, R., Vincent, C.: Do institutional investors influence capital structure decisions. Johnson School Research Paper Series (2012)
10. Chung, C.Y., Wang, K.: Do institutional investors monitor management? Evidence from the relationship between institutional ownership and capital structure. North Am. J. Econ. Finan. **30**, 203–233 (2014)
11. Seth, A.: Value creation in acquisitions: a re-examination of performance issues. Strateg. Manag. J. **11**(2), 99–115 (1990)
12. Rowe, W.G., Morrow, J.L., Jr.: A note on the dimensionality of the firm financial performance construct using accounting, market, and subjective measures. Can. J. Adm. Sci./Revue Canadienne des Sciences de l'Administration **16**(1), 58–71 (1999)
13. Hanousek, J., Kočenda, E., Shamshur, A.: Corporate efficiency in Europe. J. Corp. Finan. **32**, 24–40 (2015)
14. Pervan, M., Pervan, I., Todoric, M.: Firm ownership and performance: evidence for Croatian listed firms. World Acad. Sci. Eng. Technol. **2012**(61), 964–970 (2012)
15. Chen, J., Blenman, L., Chen, D.-H.: Does institutional ownership create values? The New Zealand case. Q. J. Finan. Acc. 109–124 (2008)
16. Cornett, M.M., et al.: The impact of institutional ownership on corporate operating performance. J. Bank. Finan. **31**(6), 1771–1794 (2007)
17. Vasconcelos, L.N.C., Martins, O.S.: Corporate governance through stock trading in Brazil: empirical evidence with institutional investors. In: XIX UPS International Conference in Accounting (2019)
18. Eshwari, D., Baby, M.: The effect of capital structure on the profitability of FMCG companies. In: International Conference on Economics, Business and Sustainability. Springer (2023)
19. Thirumagal, P., Tirkey, A., Suresh, S.: Moderating effect of board size and board independence between women on board and ESG. In: International Conference on Economics, Business and Sustainability. Springer (2023)
20. Chaudhuri, K., Kumbhakar, S.C., Sundaram, L.: Estimation of firm performance from a MIMIC model. Eur. J. Oper. Res. **255**(1), 298–307 (2016)
21. Lichtenberg, F.R., Pushner, G.M.: Ownership structure and corporate performance in Japan. Jpn. World Econ. **6**(3), 239–261 (1994)
22. Chowdhury, S.D., Geringer, M.J.: Institutional ownership, strategic choices and corporate efficiency: evidence from Japan. J. Manage. Stud. **38**(2), 271–292 (2001)
23. You, T., Chen, X., Holder, M.E.: Efficiency and its determinants in pharmaceutical industries: ownership, R&D and scale economy. Appl. Econ. **42**(17), 2217–2241 (2010)

24. Chiang, M.H., Lin, J.H.: The relationship between corporate governance and firm productivity: evidence from Taiwan's manufacturing firms. Corp. Govern. Int. Rev. **15**(5), 768–779 (2007)
25. Depken, C.A., Nguyen, G.X., Sarkar, S.K.: Agency costs, executive compensation, bonding and monitoring: a stochastic frontier approach. In: Annual Meeting of American Economic Association (AEA), Boston, MA (2006)
26. Khanna, T., Palepu, K.G.: Winning in Emerging Markets: A Road Map for Strategy and Execution. Harvard Business Press (2010)
27. Amin, A.A., Hamdan, A.M.: Evaluating the effect of ownership structure on firm performance: evidence from Saudi Arabian listed companies. J. Econ. Cooperation Dev. **39**(3), 65–92 (2018)
28. Hamdan, A.: Board interlocking and firm performance: the role of foreign ownership in Saudi Arabia. Int. J. Manag. Finan. (2018)
29. Watson, D., Head, A.: Corporate Finance: Principles and Practice, 3rd ed. Pearson Education, Essex (2010)
30. Mallin, C.A.: Corporate Governance. Oxford University Press (2019)
31. Roberts, G., Yuan, L.E.: Does institutional ownership affect the cost of bank borrowing? J. Econ. Bus. **62**(6), 604–626 (2010)
32. Hirschman, A.O.: Exit, Voice, and Loyalty: Responses to Decline in Firms, Organizations, and States, vol. 25. Harvard University Press (1970)
33. Mallin, C., Melis, A.: Shareholder rights, shareholder voting, and corporate performance. J. Manag. Govern. **16**(2), 171–176 (2012)
34. Pound, J.: Proxy contests and the efficiency of shareholder oversight. J. Financ. Econ. **20**, 237–265 (1988)
35. McConnell, J.J., Servaes, H.: Additional evidence on equity ownership and corporate value. J. Financ. Econ. **27**(2), 595–612 (1990)
36. Foster, F.D., Viswanathan, S.: Strategic trading when agents forecast the forecasts of others. J. Financ. **51**(4), 1437–1478 (1996)
37. Back, K., Cao, C.H., Willard, G.A.: Imperfect competition among informed traders. J. Financ. **55**(5), 2117–2155 (2000)
38. Elyasiani, E., Jia, J.: Distribution of institutional ownership and corporate firm performance. J. Finan. Quant. Anal. **34**(3), 606–620 (2010)
39. Nashier, T., Gupta, A.: The effect of institutional ownership on firm performance. IUP J. Corp. Govern. **15**(3), 36 (2016)
40. Lin, Y.R., Fu, X.M.: Does institutional ownership influence firm performance? Evidence from China. Int. Rev. Econ. Finan. **49**, 17–57 (2017)
41. Maug, E.: Large shareholders as monitors: is there a trade-off between liquidity and control? J. Financ. **53**(1), 65–98 (1998)
42. Li, J., et al.: The effects of institutional ownership on corporate governance and performance: an empirical assessment in Hong Kong. Manag. Int. Rev. **46**(3), 259–276 (2006)
43. Gürbüz, A.O., Aybars, A., Kutlu, Ö.: Corporate governance and financial performance with a perspective on institutional ownership: empirical evidence from Turkey. J. Appl. Manag. Acc. Res. **8**(2) (2010)
44. Wei, Z., Varela, O.: State equity ownership and firm market performance: evidence from China's newly privatized firms. Glob. Financ. J. **14**(1), 65–82 (2003)
45. Charfeddine, L., Elmarzougui, A.: Institutional ownership and firm performance: evidence from France. J. Behav. Finan. **7**(4) (2010)
46. Agrawal, A., Knoeber, C.R.: Firm performance and mechanisms to control agency problems between managers and shareholders. J. Finan. Quant. Anal. **31**(3), 377–397 (1996)
47. Lin, F.-L.: A panel threshold model of institutional ownership and firm value in Taiwan. Int. Res. J. Financ. Econ. **42**(1), 54–62 (2010)
48. Stulz, R.: Managerial discretion and optimal financing policies. J. Financ. Econ. **26**(1), 3–27 (1990)
49. Margaritis, D., Psillaki, M.: Capital structure, equity ownership and firm performance. J. Bank. Finan. **34**(3), 621–632 (2010)

50. Grossman, S.J., Hart, O.D.: Corporate financial structure and managerial incentives. In: The Economics of Information and Uncertainty, pp. 107–140. University of Chicago Press (1982)
51. Isik, O., Soykan, M.E.: Large shareholders and firm performance: evidence from Turkey. Eur. Sci. J. **9**(25) (2013)
52. Chaganti, R., Damanpour, F.: Institutional ownership, capital structure, and firm performance. Strateg. Manag. J. **12**(7), 479–491 (1991)
53. Grier, P., Zychowicz, E.J.: Institutional investors, corporate discipline, and the role of debt. J. Econ. Bus. **46**(1), 1–11 (1994)
54. Crutchley, C.E., et al.: Agency problems and the simultaneity of financial decision making: the role of institutional ownership. Int. Rev. Financ. Anal. **8**(2), 177–197 (1999)
55. La Porta, R., et al.: Agency problems and dividend policies around the world. J. Financ. **55**(1), 1–33 (2000)
56. Aigner, D., Lovell, C.K., Schmidt, P.: Formulation and estimation of stochastic frontier production function models. J. Econometrics **6**(1), 21–37 (1977)
57. Meeusen, W., van Den Broeck, J.: Efficiency estimation from Cobb-Douglas production functions with composed error. Int. Econ. Rev. 435–444 (1977)
58. Kumbhakar, S.C., Wang, H., Horncastle, A.P.: A Practitioner's Guide to Stochastic Frontier Analysis Using Stata. Cambridge University Press (2015)
59. Kumbhakar, S.C.: Efficiency measurement with multiple outputs and multiple inputs. J. Prod. Anal. **7**(2–3), 225–255 (1996)
60. Habib, M.A., Ljungqvist, A.: Firm value and managerial incentives: a stochastic frontier approach. J. Bus. **78**(6), 2053–2094 (2005)
61. Berle, A.A.. Jr., Means, G.C.: The modern corporation and private property. Macmillan, New York (1932)
62. Mahadevan, R.: To measure or not to measure total factor productivity growth? Oxford Dev. Stud. **31**, 365–378 (2003). https://doi.org/10.1080/1360081032000111742
63. Coelli, T.J., et al.: An introduction to efficiency and productivity analysis. Springer Science & Business Media (2005)

Integrating Occupational Health and Safety with Human Resource Management: A Comprehensive Approach

Marwan Melhem, Ali Ateeq, Mohammed Alzoraiki, and Baligh Beshr

Abstract The present research aims to examine the incorporation of Occupational Health and Safety (OHS) within the framework of Human Resource Management (HRM) and its subsequent influence on the overall welfare of organizations. A quantitative methodology was used to gather data by means of a survey that evaluated the present condition of integration, commitment from leadership, efficacy of training, efforts related to health and wellbeing, compliance with regulations, and management of risks. The results indicate that there are different degrees of integration seen within firms, highlighting the need of implementing customized methods. The presence of a strong commitment to leadership significantly enhances the overall safety culture. The inclusion of training initiatives within Human Resource Management (HRM) programs has been shown to have a substantial impact on the enhancement of employee knowledge. The convergence of health and wellness efforts with occupational health and safety (OHS) has a beneficial effect on overall well-being. One potential recommendation is to establish standardized protocols for regulatory compliance, ensuring consistent adherence to regulatory requirements across different entities. Another recommendation is to promote the development of coordinated risk management strategies, facilitating the effective identification, assessment, and mitigation of risks via collaborative efforts among relevant stakeholders. These observations serve to improve worker safety and bolster organizational resilience.

M. Melhem · A. Ateeq (✉) · M. Alzoraiki · B. Beshr
Administrative Science Department, College of Administrative and Financial Science, Gulf University, Sanad 26489, Kingdom of Bahrain
e-mail: dr.ali.ateeq@gulfuniversity.edu.bh

M. Melhem
e-mail: dr.marwan.milhem@gulfuniversity.edu.bh

M. Alzoraiki
e-mail: dr.mohammed.saied@gulfuniversity.edu.bh

B. Beshr
e-mail: baligh.beshr@gulfuniversity.edu.bh

B. Awwad (ed.), *The AI Revolution: Driving Business Innovation and Research*, Studies in Systems, Decision and Control 524, https://doi.org/10.1007/978-3-031-54379-1_27

Keywords Occupational health and safety · Human resource management · Safety culture · Training effectiveness · Health and wellness initiatives · Regulatory compliance · Organizational well-being

1 Introduction

The integration of Occupational Health and Safety (OHS) with Human Resource Management (HRM) has become a crucial strategic need in the dynamic field of organizational management [1]. The integration described signifies a fundamental change in the manner in which firms address employee well-being, safety, and overall operational efficiency [2, 3]. In the pursuit of sustainability and resilience, organizations recognize the need of integrating occupational health and safety (OHS) practices with human resource management (HRM) strategies. This integration plays a crucial role in cultivating a workplace atmosphere that places a high value on safety, fosters a good organizational culture, and safeguards the well-being and contentment of employees [4]. The justification for this integration arises from the acknowledgement that the welfare of workers is linked with the achievement of organizational prosperity [5, 6]. Multiple studies have emphasized the interdependent association between a robust safety culture and improved organizational performance [7, 8]. The establishment of a culture, achieved through the collaborative efforts of Occupational Health and Safety (OHS) and Human Resource Management (HRM), serves to not only reduce workplace risks but also enhance productivity, boost employee morale, and strengthen the overall resilience of the business [9].

The integration of training and development activities within the field of Human Resource Management (HRM) plays a crucial role in ensuring that personnel possess the essential knowledge and abilities to effectively execute safety regulations [10]. In addition, the integration of health and wellness programs, facilitated by human resource management (HRM), with occupational health and safety (OHS) efforts brings about a comprehensive perspective on the well-being of employees [11]. The aforementioned strategy encompasses not just physical health but also encompasses mental well-being, hence leading to increased work satisfaction and engagement [12, 13]. The integration of occupational health and safety (OHS) and human resource management (HRM) is essential when dealing with the intricate regulatory framework [14]. Through the coordination of activities related to regulatory compliance, firms are able to not only meet their legal requirements but also optimize their operational efficiency by streamlining operations [15].

The objective of this research is to explore the intricacies of this integration, analyzing the fundamental elements and processes that contribute to its effectiveness. This study aims to enhance our comprehension of how firms may effectively integrate occupational health and safety (OHS) with human resource management (HRM) to achieve long-term success and promote the well-being of employees. This will be achieved by combining insights from both quantitative and qualitative research methodologies.

2 Research Objectives

The research objectives of this study are as follows:

1. To Assess current OHS and HRM integration within organizations.
2. To Explore how integration influences a positive safety culture and leadership's role.
3. To Investigate the effectiveness of OHS training integrated into HRM programs

3 Literature Review

The combination of Occupational Health and Safety (OHS) and Human Resource Management (HRM) has emerged as a central area of interest for both scholars and professionals in the field of modern organizational management. This literature review presents a comprehensive analysis of the integration of Occupational Health and Safety (OHS) and Human Resource Management (HRM), emphasizing the several aspects that contribute to the achievement of organizational performance and the well-being of employees.

Many research investigations have shown that a strong safety culture has an important role in maintaining workplace health and minimizing accidents [16–18]. The incorporation of OHS principles into HRM procedures has been acknowledged as a significant aspect in the establishment and maintenance of such a culture [19]. According to Al Halbusi et al. [20] the importance of leadership commitment and employee engagement in fostering an environment where safety is more than just a set of standards but an established component of business principles cannot be overstated.

Education as well as growth are increasingly important components of the integration process. According to the study, including OHS training into broader staff development programs enhances not just safety awareness but also skill acquisition and job satisfaction [21–23]. This is in line with the widely held belief that investing in employee well-being increases organizational success.

Researchers emphasize the need of a coordinated strategy between OHS and HRM in order to negotiate the complicated terrain of legislation and standards [24, 25]. This partnership assures not only legal compliance, but also operational efficiency by seamlessly incorporating safety standards into HRM policy.

Finally, the research supports the idea that integrating OHS and HRM is more than a theoretical concept, but a strategic need for modern enterprises. Businesses may lay the groundwork for long-term success, employee well-being, and resilience in an ever-changing working environment by understanding and implementing these findings.

4 Method

The quantitative approach used in this conceptual study is used to build a framework for OHS-HRM integration [26–28]. A organized survey with a wide range of participants is guided by operationalized variables [29–32]. Statistical analysis evaluates data, providing empirical backing for theoretical claims.

5 Discussion

Occupational Health and Safety (OHS) and Human Resource Management (HRM) integration is a complicated process that is critical for corporate resilience and employee well-being. The quantitative results from our study shed light on essential facets of this integration.

To begin with, the evaluation of the present condition of integration demonstrates diverse degrees seen across different businesses. The presence of diverse elements emphasizes the need of using customized approaches to effectively increase complete integration. The dedication of leaders plays a crucial role in shaping the development of a safety culture. Organizations that demonstrate a robust dedication to leadership exhibit elevated levels of safety awareness, which is consistent with the results reported in the existing research [33, 34].

The efficacy of training, which serves as a fundamental aspect of the integration between Occupational Health and Safety (OHS) and Human Resource Management (HRM), is clearly shown in our findings [35, 36]. The inclusion of occupational health and safety (OHS) training into human resource management (HRM) programs has a substantial impact on the enhancement of workers' knowledge and skill acquisition [37]. This statement is consistent with the viewpoint expressed by Organization [38] who highlight the significance of providing adequate training in the field of occupational health and safety education.

The integration of health and wellness programs, under the guidance of human resource management (HRM), in conjunction with occupational health and safety (OHS) activities, has a significant effect [36]. According to Organization [38], those who took part in the study had heightened levels of well-being and satisfaction. This finding supports the notion that there is a strong correlation between the health of employees and the overall performance of a business.

The need for attention is indicated in the domain of regulatory compliance alignment. While several firms demonstrate proficiency in aligning occupational health and safety (OHS) standards with human resource management (HRM) procedures, others have difficulties, underscoring the need of implementing standardized ways.

The results of our study emphasize the need of implementing comprehensive risk management and incident response procedures. Organizations that use a synchronized strategy demonstrate enhanced readiness and adaptability, in accordance with the suggestions put out by Thomas and Suresh [39].

In summary, this discourse underscores the crucial significance of leadership, efficient training, and the convergence of health and wellness initiatives in achieving successful integration of occupational health and safety with human resource management. The optimization of integration for sustained performance necessitates firms to address difficulties related to regulatory compliance and establish integrated methods for risk management.

6 Recommendation

Several significant suggestions arise to better the integration of Occupational Health and Safety (OHS) with Human Resource Management (HRM). To begin with, it is crucial to emphasize the significance of enhancing leadership commitment, as it plays a central role in establishing a favourable safety culture. The standardization of regulatory compliance methods is of utmost importance as it guarantees consistent adherence throughout a wide range of enterprises. It is advisable to optimize occupational health and safety (OHS) training within the frameworks of human resource management (HRM), with a particular focus on fostering ongoing skill development. There is a need for increased emphasis on the integration of health and wellness activities, under the guidance of Human Resource Management (HRM), in order to enhance the overall well-being and happiness of employees. Organizations are advised to promote the development of integrated risk management and incident response methods, so enhancing their ability to withstand and recover from unexpected occurrences.

7 Conclusion

In conclusion, the incorporation of Occupational Health and Safety (OHS) within the framework of Human Resource Management (HRM) necessitates a comprehensive and all-encompassing perspective. The dedication of leaders plays a pivotal role in creating a good safety culture within an organization, whilst the implementation of successful training programs and health initiatives significantly contribute to the overall well-being of employees. The standardization of regulatory compliance is a crucial aspect, as it promotes consistency and adherence to established guidelines. Additionally, establishing a coordinated approach to risk management plays a significant role in enhancing the resilience of organizations. The aforementioned guidelines emphasize the significant impact that the combination of occupational health and safety (OHS) and human resource management (HRM) may have on the development of workplaces that are not only safer and healthier, but also more resilient.

References

1. Kandemir, H., Yildirim, D., Tezcan, M.: The effect of human resource management practices on open innovation: a research on blue collar employees in IT business. Uluslararası Yönetim Akademisi Dergisi **6**(2), 345–356 (2023)
2. Ramdhan, R.M., et al.: Internal corporate social responsibility as a microfoundation of employee well-being and job performance. Sustainability **14**(15), 9065 (2022)
3. Al-refaei, A.A.-A., et al.: An integrated mediating and moderating model to improve service quality through job involvement, job satisfaction, and organizational commitment. Sustainability **15**(10), 7978 (2023)
4. Sorensen, G., et al.: Measuring best practices for workplace safety, health and wellbeing: the workplace integrated safety and health assessment. J. Occup. Environ. Med. **60**(5), 430 (2018)
5. Haricharan, S.J.: Leadership, management and organisational implications for public service employee well-being and performance. SA J. Hum. Resour. Manag. **21**, 2080 (2023)
6. Akenttä, P., Virtaharju, J.: A metatheoretical framework for organizational wellbeing research: toward conceptual pluralism in the wellbeing debate. Organization **30**(3), 551–572 (2023)
7. Dai, L., et al.: SEM-based study on the impact of safety culture on unsafe behaviors in Chinese nuclear power plants. Nucl. Eng. Technol. (2023)
8. Ofori, E.K., et al.: Exploring new antecedent metrics for safety performance in Ghana's oil and gas industry using partial least squares structural equation modelling (PLS-SEM). Resour. Policy **81**, 103368 (2023)
9. Yesilyaprak, T., Korkmaz, F.D.: The relationship between surgical intensive care unit nurses' patient safety culture and adverse events. Nurs. Crit. Care **28**(1), 63–71 (2023)
10. Pandey, A., Balusamy, B., Chilamkurti, N.: Disruptive Artificial Intelligence and Sustainable Human Resource Management: Impacts and Innovations—The Future of HR. CRC Press (2023)
11. Roehl, M.T.: Putting implementation into context: exploring the influence of physical, social, and task contexts on the implementation of health promotion programs. Ger. J. Human Resour. Manage. **37**(2), 79–106 (2023)
12. Chirico, F., Giorgi, G., Magnavita, N.: Addressing all the psychosocial risk factors in the workplace requires a comprehensive and interdisciplinary strategy and specific tools. J. Health Soc. Sci. **8**(3), 170–174 (2023)
13. Ateeq, A.: The mediating role of engagement in the relationship between. Management **21**(3), 435–447 (2023)
14. Fan, D., et al.: Using the past to map out the future of occupational health and safety research: where do we go from here? Int. J. Human Resour. Manage. **31**(1), 90–127 (2020)
15. Hastig, G.M., Sodhi, M.S.: Blockchain for supply chain traceability: business requirements and critical success factors. Prod. Oper. Manag. **29**(4), 935–954 (2020)
16. Fernández-Muñiz, B., Montes-Peón, J.M., Vázquez-Ordás, C.J.: Safety culture: analysis of the causal relationships between its key dimensions. J. Saf. Res. **38**(6), 627–641 (2007)
17. Deepak, M.D., Mahesh, G.: A framework for enhancing construction safety through knowledge-based safety culture indicators. Int. J. Constr. Manage. **23**(12), 2039–2047 (2023)
18. Hagos, E., et al.: Safety culture among transport companies in Ethiopia: are they ready for emerging fleet technologies? Sustainability **15**(4), 3232 (2023)
19. Lestari, Y.A.P.: The application of maintenance functions in the human resources management of Islamic financial institutions. Invest J. Sharia Econ. Law **3**(1), 70–83 (2023)
20. Al Halbusi, H., et al.: Individual and psychological factors influencing hotel employee's work engagement: the contingent role of self-efficacy. Cogent Bus. Manage. **10**(3), 2254914 (2023)
21. Ateeq, A.: Knowledge management's effect on learning quality in Bahrain's private universities (2023)
22. Alzoraiki, M., et al.: Impact of teachers' commitment to the relationship between transformational leadership and sustainable teaching performance. Sustainability **15**(5), 4620 (2023)

23. Ali, A.A.A., et al.: Moderating effect of strategic planning on the relationship between career path planning and job performance. Sustainability **15**(11), 8490 (2023)
24. Wilkinson, A.: Human Resource Management: A Very Short Introduction. Oxford University Press (2022)
25. Nankervis, A., et al.: Human Resource Management 11e. Cengage AU (2022)
26. Al Balushi, F.S.M., Ali, A.A.A., Alzoraiki, M.: The impact of forecasting strategy and training and development strategy on preparing leadership competencies administrative support in the Sultanate of Oman. Int. J. Intellect. Human Resour. Manage. (IJIHRM) **3**(01), 34–41 (2022)
27. Ateeq, A., Ebrahim, S., Al-Ghatam, R.: The impact of stress and its influencing factors among dentists during the COVID-19 pandemic in Kingdom of Bahrain. F1000Research **11**, 930 (2022)
28. Barakat, B., et al.: Assessing the impact of green training on sustainable business advantage: exploring the mediating role of green supply chain practices. Sustainability **15**(19), 14144 (2023)
29. Habtoor, N.H., Ali, E.S.: Human side of total quality management: a critical review. Int. J. Intellect. Human Resour. Manage. (IJIHRM) **3**(02), 27–32 (2022)
30. Ali, A.A.A., Habtoor, N.: Islamic work ethics and its impact on enhancing organizational commitment in Bahraini private educational universities. Int. J. Intellect. Human Resour. Manage. (IJIHRM) **3**(02), 33–42 (2022)
31. Alkadash, T.M., et al.: The effects of leadership style on employee sustainable behaviour: a theoretical perspective. In: Artificial Intelligence and Transforming Digital Marketing, pp. 205–213. Springer (2023)
32. Ali, A.A.A., Yusoff, S.H., Kalsom, A.: The relationship between equal opportunities and employee performance: a special reference to with spesific refence to BDF hospital in the Kingdom of Bahrain. IJASOS-Int. E-J. Adv. Soc. Sci. **5**(14), 1200–1206 (2019)
33. Arboleda, A., et al.: Management practices as antecedents of safety culture within the trucking industry: similarities and differences by hierarchical level. J. Saf. Res. **34**(2), 189–197 (2003)
34. Farokhzadian, J., Nayeri, N.D., Borhani, F.: The long way ahead to achieve an effective patient safety culture: challenges perceived by nurses. BMC Health Serv. Res. **18**(1), 1–13 (2018)
35. Benkarim, A., Imbeau, D.: Investigating the implementation of Toyota's human resources management practices in the aerospace industry. Merits **2**(3), 126–145 (2022)
36. Wissemann, A.K., et al.: Strategic guidance and technological solutions for human resources management to sustain an aging workforce: review of international standards, research, and use cases. JMIR Hum. Factors **9**(3), e27250 (2022)
37. Yanmaz, K., Yentür, D.: Importance of Human Resources in Occupational Health and Safety
38. Organization, W.H.: Caring for those who care: guide for the development and implementation of occupational health and safety programmes for health workers (2022)
39. Thomas, A., Suresh, M.: Readiness for agility adaptability and alignment in healthcare organizations. IISE Trans. Healthcare Syst. Eng. **13**(2), 161–174 (2023)

AI-Powered Websites in e-commerce, Healthcare, Education

Mohammad Allaymoun and Samer Shorman

Abstract Websites that use artificial intelligence (AI) have transformed several sectors, including e-commerce, healthcare, and education. These websites make use of artificial intelligence (AI) technologies to improve user interactions, streamline workflows, and offer tailored services. Machine learning algorithms are used by AI-enabled e-commerce platforms to analyze massive volumes of customer data and present individualized product recommendations. AI-driven websites have revolutionized patient care and medical research in the healthcare industry. These websites use AI algorithms to examine patient data, make disease diagnoses, and make therapy recommendations. AI-driven websites have transformed learning processes in the realm of education. Platforms for personalized learning use AI to evaluate student performance, modify the content, and offer specialized recommendations. Overall, AI-powered websites in e-commerce, healthcare, and education hold immense potential to revolutionize industries, improve services, and enhance user experiences. As AI technologies continue to advance, these websites will play an increasingly vital role in shaping the future of these sectors.

Keywords Artificial Intelligence (AI) · E-commerce · Healthcare · Education · AI-powered websites · SDG11

1 Introduction

AI-powered websites are being utilized across various industries to enhance user experiences, automate tasks, and provide personalized services. Here are some industries that are currently leveraging AI-powered websites:

M. Allaymoun (✉)
Administrative Science Department, College of Administrative and Financial Science, Gulf University, Sanad 26489, Kingdom of Bahrain
e-mail: alkarak1@yahoo.com

S. Shorman
Department of Computer Science, Applied Science University, Eker, Kingdom of Bahrain
e-mail: samer.shorman@asu.edu.bh

B. Awwad (ed.), *The AI Revolution: Driving Business Innovation and Research*, Studies in Systems, Decision and Control 524, https://doi.org/10.1007/978-3-031-54379-1_28

AI-powered websites have made significant advancements in various industries, transforming processes, and improving user experiences. Here are examples of AI-powered websites in different sectors, along with relevant references: Healthcare provides telemedicine platforms, medical diagnosis assistance, personalized treatment recommendations, and patient monitoring systems [1]. Finance and Banking: such as robo-advisory services, fraud detection systems, algorithmic trading platforms, and personalized financial recommendations [2]. Retail and e-commerce offer recommendation engines, virtual shopping assistants, personalized marketing campaigns, and dynamic pricing algorithms [3]. Manufacturing and Supply Chain: optimize production processes, enable predictive maintenance, enhance inventory management, and streamline supply chain operations [4]. Education and e-learning: AI-powered websites offer personalized learning platforms, intelligent tutoring systems, automated grading, and adaptive educational content [5]. Travel and Hospitality: AI-powered websites provide personalized travel recommendations, chatbot-based customer support, dynamic pricing, and intelligent booking systems [6]. Automotive: AI-powered websites provide virtual showrooms, vehicle customization tools, and virtual test drives to enhance the car buying experience [7]. Real Estate: offers property search platforms, virtual property tours, and automated property valuation systems [8]. Agriculture provides crop monitoring and analysis, precision farming recommendations, and automated irrigation systems [9]. Energy and Utilities: enable energy consumption analysis, predictive maintenance of infrastructure, and optimization of energy distribution networks [10]. Media and Entertainment: AI-powered websites offer personalized content recommendations, video and audio content analysis, and automated content tagging [11]. Legal Services: AI-powered websites provide legal research assistance, contract analysis, and predictive analytics for case outcomes [12].

The application of AI in websites is expanding rapidly, and its potential is being harnessed in numerous sectors to drive innovation and improve efficiency. The use of AI on websites, however, comes with several difficulties. It's imperative to handle issues like privacy worries, data security, and ethical problems to make sure that tasks are completed efficiently. To retain user confidence and reduce biases, it is also crucial to ensure openness, fairness, and accountability in AI systems [13, 14].

This paper is structured into seven sections, commencing with the introduction, and concluding with the conclusion. Section 2 provides a review of the DOMAINS OF AI-POWERED WEBSITES IN E-COMMERCE, whereas Sect. 3 delves into an explanation of AI-POWERED WEBSITES IN E-COMMERCE. Moving forward, Sects. 4 and 5 comprehensively examine AI-POWERED WEBSITES IN HEALTHCARE. Lastly, Sect. 6 scrutinizes AI-POWERED WEBSITES IN EDUCATION.

2 Domains of AI-Powered Websites in the e-commerce

AI-powered websites have significantly impacted the e-commerce industry, revolutionizing the way businesses operate and enhancing the online shopping experience. These websites can improve user engagement, boost conversion rates, and spur revenue growth by comprehending consumer preferences and behaviors. Furthermore, chatbots and virtual assistants powered by AI play a key role in customer service by offering immediate assistance and responding to inquiries, thereby enhancing customer happiness and lowering operational expenses.

Recommendation Engines: AI-powered recommendation engines analyze user behavior, purchase history, and preferences to provide personalized product recommendations, improving customer satisfaction and boosting sales. Visual Search and Image Recognition: AI-powered websites utilize computer vision techniques to enable visual search, allowing users to find products by uploading images or using real-time camera captures [15].

Virtual Try-On: AI-powered websites offer virtual try-on capabilities, allowing customers to virtually try on products such as clothing, eyewear, or makeup, enhancing the online shopping experience [16]. Chatbots and virtual assistants: AI-powered chatbots and virtual assistants provide personalized customer support, answer product-related queries, assist in the purchasing process, and handle customer inquiries 24/7.

Dynamic Pricing: AI-powered websites use algorithms to dynamically adjust product prices based on factors like demand, competitor pricing, and customer behavior, optimizing revenue and increasing competitiveness. Fraud Detection and Prevention: AI-powered websites employ machine learning algorithms to detect and prevent fraudulent activities, such as payment fraud, account takeovers, and fake reviews [17].

3 AI-Powered Websites in the e-commerce

Here are some additional examples of AI-powered websites in the e-commerce industry as shown in Fig. 1.

Amazon: Amazon utilizes AI algorithms to provide personalized product recommendations based on user browsing and purchase history. Its AI-powered virtual assistant, Alexa, enables voice-based shopping and provides personalized assistance to users.

eBay: eBay utilizes AI algorithms to provide personalized product recommendations based on user browsing and purchase history. The website also employs AI for image recognition and categorization to enhance product search and filtering. ASOS: ASOS, an online fashion retailer, uses AI-powered recommendation systems to suggest clothing items to users based on their browsing behavior and purchase history. The website also offers a visual search feature that allows users to find similar

Fig. 1 AI-powered websites in the e-commerce

products based on uploaded images. Stitch Fix: Stitch Fix is an online personal styling service that leverages AI algorithms to curate personalized clothing recommendations for users. The website collects information about users' style preferences, sizes, and budgets to provide tailored fashion choices. Farfetch: Farfetch is a luxury fashion platform that employs AI to enhance the user experience. The website uses AI algorithms to provide personalized product recommendations based on user preferences, browsing history, and fashion trends.

Alibaba: Alibaba, a global e-commerce giant, uses AI extensively on its platforms. The website employs AI-powered chatbots for customer support, recommendation systems for personalized product suggestions, and image recognition for visual search capabilities. Walmart: Walmart utilizes AI-powered websites for various purposes, including inventory management, demand forecasting, and personalized product recommendations. The website employs algorithms that analyze customer data to provide tailored recommendations and optimize the supply chain.

Best Buy: Best Buy, a consumer electronics retailer, uses AI-powered websites to enhance the customer experience. The website employs recommendation systems to suggest products based on user preferences and browsing history. AI is also utilized for dynamic pricing and inventory management. Sephora: Sephora's website utilizes AI-powered technologies to offer personalized beauty recommendations. The website uses algorithms to analyze user preferences, skin type, and previous purchases to suggest relevant products and provide virtual try-on experiences.

Walmart Labs: Walmart Labs, the technology division of Walmart, focuses on AI-powered e-commerce solutions. Their website showcases various AI-driven initiatives, such as image recognition for product search, chatbots for customer service, and dynamic pricing algorithms. Nike: Nike's website employs AI-powered personalization to enhance the customer experience. The website utilizes data analysis and machine learning algorithms to recommend personalized products, offer customized design options, and provide tailored fitness advice.

These examples highlight how AI-powered websites in the e-commerce industry leverage algorithms and user data to enhance the customer experience, provide personalized recommendations, optimize inventory management, and streamline various aspects of online retail.

4 Domains of AI-Powered Websites in the Healthcare

AI-powered websites have made significant advancements in the healthcare industry, transforming processes, improving patient care, and enabling new medical discoveries. Here are examples of AI-powered websites in healthcare, along with relevant references:

AI-powered websites offer early diagnosis of medical disorders, improve diagnostic accuracy, and maximize treatment outcomes by utilizing predictive analytics and pattern recognition. These websites also provide telemedicine services, enabling remote consultations, monitoring, and the provision of individualized healthcare. Figure 2 shows the domains of AI-powered websites in healthcare.

i. **Medical Image Analysis**

   a. utilize deep learning algorithms to analyze medical images, such as X-rays, MRIs, and CT scans, for accurate diagnosis, detection of abnormalities, and treatment planning.

ii. **Clinical Decision Support**

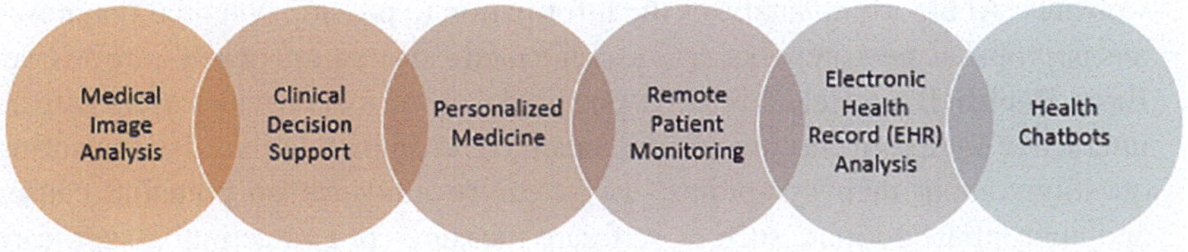

Fig. 2 Domains of AI-powered websites in healthcare

   a. provide clinical decision support systems that leverage machine learning to assist healthcare professionals in making accurate diagnoses, suggesting treatment options, and predicting patient outcomes [18].

iii. **Personalized Medicine**

   a. analyze patient data, including genetic information, medical history, and lifestyle factors, to provide personalized treatment recommendations, drug dosage optimization, and targeted therapies [19].

iv. **Remote Patient Monitoring**

   a. enable remote monitoring of patient vital signs, activity levels, and health parameters through wearable devices, facilitating early detection of health issues and timely interventions [20].

v. **Electronic Health Record (EHR) Analysis**

   a. leverage natural language processing and machine learning techniques to extract valuable insights from EHRs, supporting clinical decision-making, identifying trends, and improving patient outcomes.

vi. **Health Chatbots**

   a. employ chatbots to provide automated triage, answer healthcare-related questions, and offer basic medical advice, enhancing access to healthcare information and support [21].

5 AI-Powered Websites in the Healthcare

a. **WebMD**: WebMD is a popular healthcare website that utilizes AI algorithms to provide users with personalized health information and recommendations. The website offers symptom checkers, medical content, drug information, and health-related news tailored to individual users' needs.
b. **Ada Health**: Ada Health is an AI-powered healthcare platform that offers an online symptom assessment tool. Users can input their symptoms, and the website's AI algorithm analyzes the information to provide potential diagnoses and appropriate next steps, such as seeking medical attention or self-care advice.
c. **Buoy Health**: Buoy Health is an AI-powered website that provides users with an interactive symptom checker and personalized recommendations. By answering questions about their symptoms, users receive guidance on potential causes and appropriate actions, such as self-care measures or contacting a healthcare professional.
d. **Your.MD**: Your.MD is an AI-driven website that offers a symptom checker and a personal health assistant. Users can input their symptoms, and the platform's AI algorithm provides potential diagnoses, information on conditions, and guidance on seeking appropriate medical care.

e. **Zocdoc**: Zocdoc is a healthcare appointment booking website that incorporates AI to help users find available doctors, dentists, and specialists in their area. The platform considers factors like location, insurance coverage, and availability to provide users with suitable options for booking appointments.
f. **PathAI**: PathAI is an AI-powered website that assists pathologists in analyzing and diagnosing medical images, particularly in the field of pathology. The platform uses machine learning algorithms to aid in the detection and characterization of diseases, such as cancer, from digitized pathology slides.
g. **23andMe**: 23andMe is a genetic testing and analysis website that leverages AI algorithms to provide users with detailed genetic reports and insights about their ancestry, health predispositions, carrier status for certain conditions, and traits.
h. **IBM Watson Health**: IBM Watson Health offers a range of AI-powered solutions for the healthcare industry. Their website provides access to various tools, including AI-powered clinical decision support systems, medical image analysis, population health management, and drug discovery.
i. **DeepMind Health**: DeepMind Health, a subsidiary of Google's DeepMind, focuses on leveraging AI technologies for healthcare applications. Their website showcases projects involving AI-driven solutions, such as image analysis for detecting eye diseases and predictive models for patient deterioration.
j. **Tempus**: Tempus is a healthcare technology company that utilizes AI-powered websites for precision medicine. Their platform analyzes clinical and molecular data to generate personalized treatment recommendations for cancer patients.
k. **Butterfly Network**: Butterfly Network develops handheld, AI-powered ultrasound devices and offers a corresponding website platform. The website provides tools for image interpretation, analysis, and sharing, enabling healthcare professionals to leverage AI for diagnostic support.
l. **Covera Health**: Covera Health is an AI-powered radiology platform that aims to improve the accuracy and quality of medical imaging. Their website provides radiologists with AI tools for image analysis, assisting in the detection and diagnosis of abnormalities.
m. **Aidoc**: Aidoc offers an AI-powered radiology platform that integrates with existing picture archiving and communication systems (PACS). Their website provides radiologists with AI algorithms that can automatically detect and prioritize critical findings in medical images.
n. **Buoy Labs**: Buoy Labs provides an AI-powered website and virtual assistant for symptom checking and triage. Users can input their symptoms, and the platform's AI algorithm helps guide them to appropriate healthcare resources or advice.
o. **Medtronic**: Medtronic, a medical technology company, utilizes AI-powered websites for various purposes. Their website showcases AI-driven solutions for remote patient monitoring, predictive analytics, and personalized care management.

p. **PathAI**: PathAI offers an AI-powered pathology platform that assists pathologists in diagnosing diseases from digitized pathology slides. Their website provides tools for image analysis, helping pathologists improve accuracy and efficiency in their diagnoses.
q. **Zebra Medical Vision**: Zebra Medical Vision uses AI algorithms to analyze medical imaging data. Their website offers tools for radiologists and healthcare providers to assist in the detection and diagnosis of various conditions, such as lung cancer and cardiovascular diseases.
r. **Google Health**: Google Health utilizes AI and machine learning in various healthcare initiatives. Their website showcases projects involving AI-powered image analysis, predictive modeling, and data analytics to improve disease detection, treatment outcomes, and population health.

6 The Dimensions of AI-Powered Websites in the Education

AI-powered websites have also made significant contributions to the education sector, transforming the way students learn, providing personalized experiences, and assisting educators in various aspects of teaching. Here are examples of AI-powered websites in education, along with relevant references. Figure 3 shows the dimensions of AI-powered websites in the education:

As a result, these websites improve student engagement and academic results by enabling personalized learning paths, adaptive exams, and intelligent tutoring systems. Additionally, automatic grading, plagiarism detection, and administrative activities are made possible by AI-powered websites, expediting educational procedures, and lightening the load on teachers.

Adaptive Learning Platforms: AI-powered websites offer adaptive learning platforms that use machine learning algorithms to personalize learning experiences,

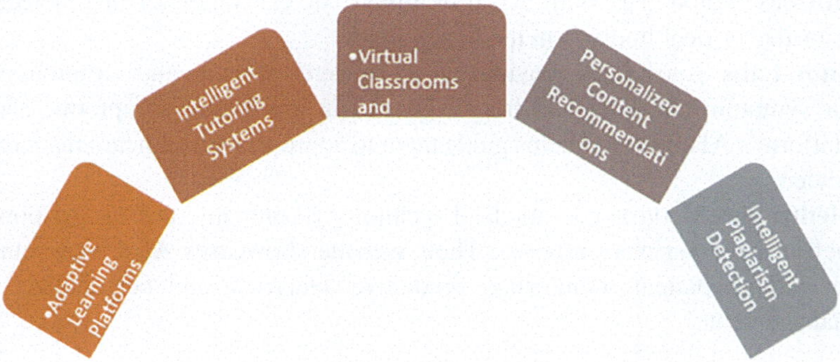

Fig. 3 The dimensions of AI-powered websites in the education

adapt content delivery, and provide targeted feedback to students [22, 26]. Intelligent Tutoring Systems: AI-powered websites provide intelligent tutoring systems that use natural language processing and machine learning to deliver personalized instruction, provide feedback, and track student progress. AI-powered websites employ natural language processing and machine learning techniques to automatically grade assignments, essays, and quizzes, offering timely feedback to students.

Virtual Classrooms and Simulations: AI-powered websites provide virtual classrooms and simulations that facilitate interactive learning experiences, allowing students to engage in virtual experiments, simulations, and collaborative activities [23]. Personalized Content Recommendations: AI-powered websites utilize machine learning algorithms to recommend relevant educational resources, such as articles, videos, and online courses, based on individual student preferences and learning goals [24]. Intelligent Plagiarism Detection: AI-powered websites employ machine learning algorithms to detect plagiarism in student assignments by comparing text similarity and identifying potential instances of copying or paraphrasing [25, 27].

Examples of AI-powered websites in the education

(a) **Knewton Alta**: Knewton Alta is an adaptive learning platform for higher education. The website uses AI algorithms to personalize instruction, provide targeted practice activities, and offer real-time feedback to help students master course content. https://www.knewton.com/.
(b) **DreamBox Learning**: DreamBox Learning is an adaptive math learning platform for K-8 students. The website utilizes AI algorithms to provide personalized math lessons based on students' individual needs and learning progress. https://www.dreambox.com/
(c) **Cognii**: Cognii is an AI-powered virtual tutor that offers interactive learning experiences and assessments. The website utilizes natural language processing and machine learning to provide personalized feedback, formative assessment, and automated grading. https://www.cognii.com/.
(d) **Duolingo**: Duolingo is a language learning platform that incorporates AI to personalize language instruction. The website uses AI algorithms to adapt lessons based on learners' performance, provide targeted feedback, and optimize the learning experience. https://www.duolingo.com/.
(e) **Edmodo**: Edmodo is a social learning platform that incorporates AI to enhance collaboration and personalized learning. The website uses AI algorithms to recommend educational resources, facilitate communication, and provide analytics to educators.
(f) **Smart Sparrow**: Smart Sparrow is an adaptive learning platform that enables educators to create interactive, AI-powered online courses. The website utilizes AI algorithms to deliver personalized instructional content, adaptive assessments, and real-time feedback.
(g) **McGraw-Hill ALEKS**: ALEKS (Assessment and Learning in Knowledge Spaces) is an adaptive learning platform offered by McGraw-Hill Education. The website utilizes AI algorithms to assess students' knowledge, provide personalized instruction, and track their learning progress.

(h) **Cerego**: Cerego is an adaptive learning platform that uses AI to enhance memory retention and knowledge acquisition. The website employs algorithms to personalize study plans, optimize learning schedules, and provide spaced repetition for efficient learning.
(i) **Gradescope**: Gradescope is an AI-powered grading platform that streamlines the grading process for educators. The website employs AI algorithms to assist in grading subjective assessments, provide consistency, and save time.

7 Conclusion

In conclusion, websites that use AI have significantly advanced many different businesses. AI-driven websites have changed product discovery and recommendation systems in the e-commerce industry. They may offer tailored product recommendations by analyzing enormous volumes of user data, which will boost consumer engagement, conversion rates, and revenue growth. Beyond e-commerce, the effects of AI-powered websites may be seen in fields like healthcare and education. These websites help with disease diagnosis, treatment suggestions, and telemedicine services in the healthcare industry, enhancing patient care and supporting medical research. Websites with artificial intelligence (AI) powering them make it possible for individualized learning experiences, adaptive testing, and intelligent tutoring systems, improving student engagement and academic results. However, the adoption of AI on websites also raises challenges and considerations. Privacy, data security, and ethical concerns are paramount, requiring robust measures to protect user information and ensure transparency. Overall, AI-powered websites have transformed industries by leveraging AI technologies to enhance user experiences, optimize processes, and deliver personalized services.

References

1. Obermeyer, Z., et al.: Predicting the future—big data, machine learning, and clinical medicine. New England J. Med. (2016)
2. Arner, D.W., et al.: Fintech and Regtech in a nutshell, and the future in a sandbox. J. Finan. Regul. (2020)
3. Verhoef, P.C., et al.: Creating value with big data analytics: making smart marketing decisions. Int. J. Res. Mark. (2015)
4. Tao, F., et al.: Industrial artificial intelligence: challenges and opportunities. Engineering (2020)
5. Anderson, J.R., et al.: Cognitive tutor: applying cognitive science to education. Cogn. Instr. (1995)
6. Xiang, Z., et al.: Artificial intelligence and its future in tourism. J. Travel Res. (2017)
7. Chen, H., et al.: The impact of artificial intelligence on the automotive industry. Technol. Forecast. Soc. Change (2020)
8. Chen, Y., et al.: Artificial intelligence in real estate: applications and opportunities. J. Real Estate Lit. (2019)
9. Kamilaris, A., et al.: Deep learning in agriculture: a survey. Comput. Electron. Agric. (2017)

10. Taylor, J.E., et al.: Artificial intelligence and machine learning for energy efficiency. Ann. Rev. Environ. Resour. (2019)
11. Porter, D., et al.: The impact of artificial intelligence on media and journalism. Dig. Journalism (2018)
12. Ashley, K.D., et al.: Artificial intelligence and legal analytics: new tools for law practice in the digital age. Law Libr. J. (2018)
13. Allaymoun, M.H., Hamid, O.A.H.: Business intelligence model to analyze social network advertising. In: 2021 International Conference on Information Technology (ICIT), pp. 326–330. IEEE (2021, July)
14. Dawwas, M., Allaymoun, M., Alzgool, M.: Enhancing green recruitment through implementing artificial intelligence: Zoho recruitment system. In: Alareeni, B.A.M., Elgedawy, I. (eds.) Artificial Intelligence (AI) and Finance. Studies in Systems, Decision and Control, vol. 488. Springer, Cham (2023). https://doi.org/10.1007/978-3-031-39158-3_1
15. Kang, L., et al.: TasselNet: counting tassels in the wild via local counts regression network. In: Proceedings of the IEEE/CVF Conference on Computer Vision and Pattern Recognition (2020)
16. Han, H., et al.: DeepFashion2: a versatile benchmark for detection, pose estimation, segmentation, and re-identification of clothing images. In: Proceedings of the IEEE/CVF Conference on Computer Vision and Pattern Recognition (2020)
17. Bhattacharya, S., et al.: DeepFraudGuard: a deep learning framework for automated fraud detection in e-commerce (2020). arXiv preprint arXiv:2005.02614
18. Chen, J.H., et al.: A machine learning approach to predicting clinical deterioration in hospitalized patients. J. Clin. Med. (2019)
19. Gligorijevic, J., et al.: AI in personalized medicine: lessons learned from collaborative filtering. Mol. Pharm. (2018)
20. Ding, X., et al.: Deep learning for healthcare monitoring systems: a review. IEEE Access (2020)
21. Denecke, K., et al.: Conversational agents for health and wellbeing: a state-of-the-art review. J. Med. Internet Res. (2020)
22. Corbett, A.T., et al.: A large-scale study of the impact of assessment and practice on student learning. Proc. Natl. Acad. Sci. (2015)
23. Sengupta, P., et al.: Integrating computational thinking with K-12 science education using agent-based computation: a theoretical framework. Educ. Inform. Technol. (2013)
24. Chen, G., et al.: Personalized learning recommendation model for K-12 students. IEEE Access (2019)
25. Li, J., et al.: A comprehensive survey of text similarity evaluation using machine learning techniques. ACM Comput. Surv. (2020)
26. Al-Shoqran, M., Shorman, S.: Applications of artificial intelligence in COVID-19 pandemic. In: European, Asian, Middle Eastern, North African Conference on Management and Information Systems, pp. 313–321. Springer International Publishing, Cham (2021, March)
27. Shorman, S.: Internet of Things Application to development of smart classroom system. Int. Rev. Comput. Softw. (IRECOS) **14**(1), 22–26 (2019)

Assessing the Influence of Work-Related Pressure on Employee Performance: A Case Study of Solidarity Insurance Company in the Kingdom of Bahrain

Ali Ateeq, Abdulla Yusuf Abdulrazaq, Mohammed Alzoraiki, Marwan Milhem, and Ahmad Al Astal

Abstract The current research investigates the effects of work-related stress on workers at Solidarity Insurance Company in Bahrain, including an examination of both advantageous and disadvantageous consequences. This research utilizes predetermined goals to examine the complex nature of work pressure, which includes psychological experiences, adaptive reactions, dynamic conditions, and physiological pressures. The research takes into account the larger worldwide context of organizational issues, highlighting the need of a comprehensive knowledge of the dynamic links between work pressure and performance. The primary objective of this study is to provide significant knowledge on the impact of work-related stresses on employee performance. This research endeavour seeks to establish a comprehensive and structured approach for conducting investigations in this domain. The primary objective is to provide pragmatic recommendations for establishing work environments that are both healthier and more productive, so helping not just Solidarity Insurance Company but also enterprises grappling with the complexities of the contemporary global workplace.

Keywords Work-related stress · Job pressure · Psychological experiences · Adaptive responses · Dynamic situations · Physiological stresses · Employee performance

---

A. Ateeq (✉) · A. Y. Abdulrazaq · M. Alzoraiki · M. Milhem · A. Al Astal
Administrative Science Department, College of Administrative and Financial Science, Gulf University, Sanad 26489, Kingdom of Bahrain
e-mail: dr.ali.ateeq@gulfuniversity.edu.bh

A. Y. Abdulrazaq
e-mail: 180104114617@gulfuniversity.edu.bh

M. Milhem
e-mail: dr.marwan.milhem@gulfuniversity.edu.bh

A. Al Astal
e-mail: dr.ahmed.alastal@gulfuniversity.edu.bh

B. Awwad (ed.), *The AI Revolution: Driving Business Innovation and Research*, Studies in Systems, Decision and Control 524, https://doi.org/10.1007/978-3-031-54379-1_29

1 Introduction

In the contemporary global context, organizations are confronted with a multitude of issues, among which work-related pressures have emerged as a salient problem. The aforementioned pressures have received considerable scrutiny owing to their substantial influence on workers, including both social and health aspects [1]. Employees from various backgrounds face a range of pressures originating from different sources [2]. It is important to acknowledge that these demands extend beyond the scope of individual individuals and instead cover the whole of the workforce inside a business [3]. Moreover, the repercussions of these circumstances transcend beyond the confines of the professional environment, into the personal lives of workers and detrimentally impacting their interpersonal connections with loved ones and acquaintances [4]. As a result, this has a negative impact on their performance inside the company, limiting their capacity to achieve organizational goals [5]. The attainment of such a milestone is closely connected to the caliber of the human capital and their adeptness in carrying out designated duties.

The topic of work-related stress is often discussed by workers at their organizational meetings. Individuals often face the daunting job of navigating through a wide range of pressures coming from many sources, which are linked to the responsibilities done by both individuals and groups within the organizational context [6].

The personal lives of employees can generate a multitude of stressors [7]. The pressures discussed in this context are the result of a complex interaction of several influences and causes [8]. Therefore, it is crucial to carefully examine this research and consider the possible negative consequences it may have on both people and organizations. These consequences have the ability to hinder the achievement of desired objectives. It is worth noting that the phenomenon of work-related pressures has become a subject of legal scrutiny, especially in Western cultures. The lack of clear communication from companies on the underlying causes of increased stress may result in the termination of an employee or the imposition of financial restitution.

The purpose of this research was to examine the impact of work-related stress on employee performance in the specific setting of Solidarity Insurance Company in the Kingdom of Bahrain. This has particular relevance in an industry marked by fierce rivalry within the labour market. The organization need to adopt proactive actions to address and alleviate factors that may hinder employee performance due to external pressures inside the enterprise.

2 Research Questions

Based on the problem discussed, the main research question for the current study has been formulated as follows:

- Is there an impact of work-related pressures on the performance of employees at Solidarity Insurance Company in the Kingdom of Bahrain?

From the main research question, the following sub-questions emanate:

2.1 Is there a positive impact of work-related pressures on the performance of employees at Solidarity Insurance Company in the Kingdom of Bahrain?
2.2 Is there a negative impact of work-related pressures on the performance of employees at Solidarity Insurance Company in the Kingdom of Bahrain?

Research Objectives

1. **Main Objective**: To assess the impact of work-related pressures on the performance of employees at Solidarity Insurance Company in the Kingdom of Bahrain.
   Specific Objectives:
2. To determine the presence and extent of any positive impact of work-related pressures on the performance of employees at Solidarity Insurance Company in the Kingdom of Bahrain.
3. To investigate the presence and extent of any negative impact of work-related pressures on the performance of employees at Solidarity Insurance Company in the Kingdom of Bahrain.

3 Literature Review

3.1 Employee Performance

The concept of employee performance has been a subject of considerable scholarly attention, and while multiple definitions have been proposed, a consensus remains elusive. Harishchandra [9] offer a fundamental interpretation, defining performance as the execution and achievement of work, emphasizing the outcomes resulting from specific employee efforts. Alternatively, others view performance as the level of achievement with regards to both the quality and quantity of work undertaken [10–12]. This perspective centers on the employee's ability to meet assigned tasks efficiently.

Moreover, Hassan's perspective on performance places emphasis on task completion, aligning the concept with the degree of task fulfillment in an employee's job description. In this view, performance reflects the employee's ability to meet job requirements and involves a dynamic interplay between effort and outcomes, with outcomes being the basis for performance assessment [13–15].

In summary, the definition of employee performance varies, encompassing the idea of achieving tasks, the quantity and quality of work, as well as outcomes. These definitions underline the significance of understanding an employee's role, their capabilities, and the effort they put in, and emphasize the results achieved as the core determinant of performance. However, despite these varied definitions, a shared understanding of employee performance remains a critical aspect in organizational and management literature.

3.1.1 Determinants of Employee Performance

The task of recognizing and understanding the many elements that affect employee performance, together with the complex interconnections between these factors, is a significant difficulty owing to their wide range and the complexity involved in measuring their effect on [16, 17]. Numerous academic researchers have posited that employee performance is influenced by a wide range of factors, encompassing.

1. **Motivation**: Motivation refers to the intrinsic drive that workers possess, influencing their commitment and effort towards their task [18]. The phenomenon may be often classified into two distinct categories: those who lack motivation often display below-average performance, while those who are driven to complete their duties tend to show elevated levels of success [19].
2. **Competence** refers to the talents and capacities possessed by employees, which have a substantial impact on their ability to perform duties efficiently [20]. According to Kossyva and Theriou [21] workers who possess the necessary skills and abilities are more inclined to perform exceptionally in their designated positions and actively contribute to the success of their firm.

The aforementioned criteria are interrelated and demonstrate complex dynamics that jointly impact the performance of employees [22]. A comprehensive understanding of the significance of motivation and competence is essential for assessing and enhancing employee performance within the framework of a company [23].

3.2 Work Pressure in the Workplace

Work pressure is a multidimensional construct that encompasses the psychological and physical stress experienced by employees in response to various demands and challenges in the workplace [24]. This literature review delves into the concept of work pressure, providing an overview of different definitions and its potential impact on employees.

Psychological Experience of Work Pressure

Work pressure is often described as a psychological experience, resulting from a myriad of factors within the work environment or from individual characteristics [25]. These factors can lead to emotional responses such as anxiety, tension, or frustration, which may manifest physically, affecting an individual's health and well-being [26]. Such experiences can be detrimental to employees, as they may lead to burnout, reduced job satisfaction, and decreased performance.

Adaptive Responses to External Situations

Another perspective defines work pressure as adaptive responses to external situations [27]. In this context, individuals react to the demands and challenges posed by their work environment, leading to deviations in behavior, psychological well-being, and physical health [28]. While these responses may serve as coping mechanisms, excessive or chronic work pressure can have adverse effects on employees and their job performance.

Dynamic Circumstances and Potential Losses

Work pressure is also understood as dynamic circumstances that involve potential gains and losses [29]. It arises from the competition between work needs and external job demands [30]. Employees face the challenge of achieving their objectives within the constraints of their resources, with the risk of losing what they aspire to achieve [31]. The implications of this dynamic pressure on employee performance are significant and warrant further investigation.

Physiological and Psychological Stressors

Work pressure can be viewed as physiological and psychological stressors that make it challenging for individuals to adapt to their work situations [32]. These stressors often stem from the nature of their job, the importance of their role within the organization, and external factors such as organizational structures and job responsibilities [33]. The impact of these stressors on employees' overall well-being and performance necessitates an in-depth understanding.

In summary, work pressure is a complex and multifaceted phenomenon, encompassing various dimensions and factors. The psychological, physical, and behavioral consequences of work pressure emphasize its importance in the field of organizational psychology and management. Further research is needed to explore the moderating and mediating variables that influence the relationship between work pressure and employee performance. Understanding work pressure is crucial for organizations seeking to create a healthier and more productive work environment for their employees.

The study's hypothesized model is depicted in Fig. 1.

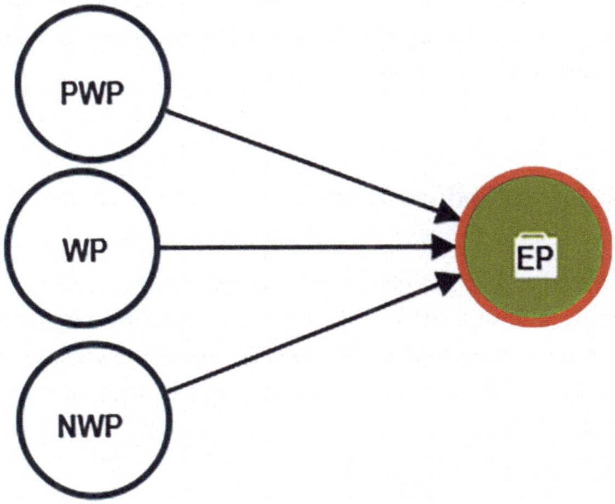

Fig. 1 Conceptual model. *Note* Negative work pressures = NWP, positive work pressures = PWR, employee performance = EP, work pressure = WP

Method: In our methodology for this critical review paper, we sourced literature to Assessing the Influence of Work-Related Pressure on Employee Performance: A Case Study of Solidarity Insurance Company in the Kingdom of Bahrain [14, 16, 34, 35]. Through a systematic analysis of the collected literature, we pinpointed salient trends in areas like recruitment, engagement, and HRIS utilization [11, 36–38]. This analytical process informed the evidence-based recommendations we subsequently proposed for enhancing HRM in a digital context [2].

4 Discussion and Conclusion

This research focuses at how stress at work affects workers at one company in particular, Solidarity Insurance Company in Bahrain. The study uses established objectives to investigate both positive and negative outcomes.

The literature study digs into the varied nature of job pressure, that extend psychological experiences, adaptive responses, dynamic situations, and physiological stresses. The analysis highlights the need for a detailed knowledge of these dynamic relationships between job strain and performance.

Understanding the broader international backdrop of organizational difficulties, whereby work-related stressors have substantial influence on workers' social and health dimensions, the study's particular emphasis on Solidarity Insurance Company assumes critical importance, considering the fiercely competitive personality of the business. The study inquiries seek to elucidate the effects of work-related pressures on employee performance, offering a methodical framework for systematic investigation.

The main goal of this study is to provide valuable insights to both Solidarity Insurance Company and other businesses that are facing the challenges of the modern global workplace. The research seeks to illuminate the intricate connection between job-related stressors and employee performance, with the goal of offering practical suggestions for establishing work environments that are both healthier and more conducive to productivity.

5 Conclusion

The current research study conducted at Solidarity Insurance Company in Bahrain investigates the effects of work-related stress, including both advantageous and detrimental consequences. The present literature review underscores the need of developing a sophisticated comprehension of the intricate interplay between stress and performance. The research seeks to provide significant insights for the creation of healthy work environments, with a specific emphasis on the competitive business landscape and larger international organizational difficulties. In essence, this document functions as a strategic framework for effectively navigating the intricate dynamics of contemporary work environments.

References

1. Anasori, E., De Vita, G., Küçükergin, K.G.: Workplace bullying, psychological distress, job performance and employee creativity: the moderating effect of psychological resilience. Serv. Ind. J. **43**(5–6), 336–357 (2023)
2. Ateeq, A., Ebrahim, S., Al-Ghatam, R.: The impact of stress and its influencing factors among dentists during the COVID-19 pandemic in Kingdom of Bahrain. F1000Research **11**, 930 (2022)
3. Ruehle, R.C.: The moral permissibility of digital nudging in the workplace: reconciling justification and legitimation. Bus. Ethics Q. 1–30 (2023)
4. Rahman, M., Aluko, O.I.S., Abu, Z.: Work related stress management and the performance of workers in public health facilities in Kwara State, Nigeria. Kwara State University, Nigeria (2023)
5. Madi Odeh, R.B., et al.: The transformational leadership role in achieving organizational resilience through adaptive cultures: the case of Dubai service sector. Int. J. Product. Perform. Manag. **72**(2), 440–468 (2023)
6. Manroop, L., Petrovski, D.: Exploring layers of context-related work-from-home demands during COVID-19. Pers. Rev. **52**(6), 1708–1727 (2023)
7. Abbas, G., et al.: The moderation model of employee optimism: the impact of empower leadership on the employee organizational stress. J. ISOSS **9**(2), 343–350 (2023)
8. Zhu, S., et al.: The influencing factors and mechanisms for urban flood resilience in China: from the perspective of social-economic-natural complex ecosystem. Ecol. Ind. **147**, 109959 (2023)
9. Harishchandra, N.: Impact of Workplace Environmental Factors on Employee Performance
10. Korhonen, T., et al.: How performance measurement can support achieving success in project-based operations. Int. J. Project Manage. **41**(1), 102429 (2023)
11. Alkadash, T.M., et al.: The effects of leadership style on employee sustainable behaviour: a theoretical perspective. In: Artificial Intelligence and Transforming Digital Marketing, pp. 205–213. Springer (2023)
12. Ateeq, A.: The mediating role of engagement in the relationship between. Management **21**(3), 435–447 (2023)
13. Utomo, H.J.N., et al.: Investigating the role of innovative work behavior, organizational trust, perceived organizational support: an empirical study on SMEs performance. J. Law Sustain. Dev. **11**(2), e417–e417 (2023)
14. Al-refaei, A.A.-A., et al.: An integrated mediating and moderating model to improve service quality through job involvement, job satisfaction, and organizational commitment. Sustainability **15**(10), 7978 (2023)
15. Ateeq, A., et al.: Impact of employee loyalty on job performance: mediating role of job satisfaction on the example of Zain Company, Bahrain. Management **21**(2), 470–481 (2023)
16. Alzoraiki, M., et al.: Impact of teachers' commitment to the relationship between transformational leadership and sustainable teaching performance. Sustainability **15**(5), 4620 (2023)
17. Ali, A.A.A., Yusoff, S.H., Kalsom, A.: The relationship between equal opportunities and employee performance: a special reference to with spesific refence to BDF hospital in the Kingdom of Bahrain. IJASOS-Int. E-J. Adv. Soc. Sci. **5**(14), 1200–1206 (2019)
18. Hartnell, C.A., et al.: Manufacturing motivation in the Mundane: servant leadership's influence on employees' intrinsic motivation and performance. J. Bus. Ethics 1–20 (2023)

19. Ali, A.A.A., Yusoffii, S.H., Aliiii K.: Fundamental of Equal Opportunities in Employee's Performance: A Critical Review
20. Khang, A., Jadhav, B., Birajdar, S.: Industry revolution 4.0: workforce competency models and designs. In: Designing Workforce Management Systems for Industry 4.0, pp. 11–34. CRC Press (2023)
21. Kossyva, D., et al.: Retaining talent in knowledge-intensive services: enhancing employee engagement through human resource, knowledge and change management. J. Knowl. Manage. (2023)
22. Boccoli, G., Gastaldi, L., Corso, M.: The evolution of employee engagement: towards a social and contextual construct for balancing individual performance and wellbeing dynamically. Int. J. Manag. Rev. **25**(1), 75–98 (2023)
23. Balen, J., et al.: Essential determinants for assessing the strategic agility framework in small and medium-sized enterprises (SMEs). Eur. J. Sci. Appl. Sci. **2023**(10), 2124–2129 (2023)
24. Savarimalai, R., et al.: Stress and coping among police personnel in South India. Ind. Psychiatry J. (2023)
25. El Ayadi, A.M., et al.: The social, economic, emotional, and physical experiences of caregivers for women with female genital fistula in Uganda: a qualitative study. Glob. Public Health **18**(1), 2242458 (2023)
26. Adekeye, O., et al.: Exploring the well-being of people affected by skin NTDs in Kaduna and Kwara States, Nigeria: a photovoice and scoping review study. Int. Health **15**(Supplement_1), i100–i109 (2023)
27. Reynolds, R.B.: A crisis-adaptive approach to resilience-building in pre-service teaching and librarianship education: learning about and learning to be. In: Emerging Adulthood in the COVID-19 Pandemic and Other Crises: Individual and Relational Resources, pp. 315–334. Springer (2023)
28. Delmo, C.A., Dequito, E.R.: A study on work-life balance, technostress creators and psychological well-being of librarians. EPRA Int. J. Multidisc. Res. (IJMR) **9**(6), 151–163 (2023)
29. Kearney, C., Lichtenstein, B.: Generative emergence: exploring the dynamics of innovation and change in high-potential start-up ventures. Br. J. Manag. **34**(2), 898–914 (2023)
30. Bunjak, A., et al.: Job demands and burnout: the multilevel boundary conditions of collective trust and competitive pressure. Human Relat. **76**(5), 657–688 (2023)
31. Madhavan, V., et al.: Addressing agency problem in employee training: the role of goal congruence. Sustainability **15**(4), 3745 (2023)
32. Javaid, Z.K., Mahmood, K., Ali, A.A.: Mediating role of mindfulness between quality of life and workplace stress among working women: quality of life and workplace stress among working women. J. Workplace Behav. **4**(1), 68–80 (2023)
33. Dodanwala, T.C., Santoso, D.S., Yukongdi, V.: Examining work role stressors, job satisfaction, job stress, and turnover intention of Sri Lanka's construction industry. Int. J. Constr. Manag. **23**(15), 2583–2592 (2023)
34. Ali, A.A.A., et al.: Moderating effect of strategic planning on the relationship between career path planning and job performance. Sustainability **15**(11), 8490 (2023)
35. Ali, A.A.A., Habtoor, N.: Islamic work ethics and its impact on enhancing organizational commitment in Bahraini private educational universities. Int. J. Intellect. Human Resour. Manage. (IJIHRM) **3**(02), 33–42 (2022)
36. Ali, A.A.A., Alzoraiki, M., Al Aghbas, N.S.M.: Working remotely, social isolation and access to support and communication during Covid-19 its influence on job performance. Int. J. Intellect. Human Resour. Manage. (IJIHRM) **3**(01), 23–33 (2022)
37. Ali, A.A.A., et al.: Association between Islamic work ethics and employee commitment among medical staff in the public health sector in Bahrain. Int. J. Intellect. Human Resour. Manage. (IJIHRM) (2023)
38. Ateeq, A.: Knowledge Management's Effect on Learning Quality in Bahrain's Private Universities (2023)

Do Corporate Renewable Energy Initiatives Influence Sustainable Development?

Neethu Suraj, N. Abhishek, Swarn G. Kanchan, Delecta Jenifer Rajendran, Swapna Raghupathi, and Sween Kuriakose

Abstract The corporate efforts in eradicating environmental negativities are a significant aspect that is playing a vital role in sustainable development. Renewable energy initiatives by corporates are also a trending effort through which corporates can promote the process of generation and consumption of clean energy and that helps to conserve non-renewable natural resources and contributes to the same for future generations. The present paper intends to analyze the perception of administrators, managers, and academicians on renewable energy initiatives by corporates and also to study the impact of renewable energy initiatives on sustainable development with the mediating effect of environmental, social, and economic aspects. Therefore, the study followed quantitative and qualitative approaches to achieve the objectives. The study employed descriptive statistics, structured equation modelling, and mediation analysis techniques to draw conclusions based on objectives. The results of the one-sample t-test revealed that the respondents positively opined that renewable energy initiatives by corporations influence sustainable development. Further, the results of simple linear regression analysis confirmed that such initiatives have an influence on sustainable development. The outcome of this study helps industries to know the importance of such initiatives and provides inputs to design their business models based on such initiatives.

Keywords Renewable energy · Corporate initiatives · Clean energy · Clean environment · Sustainable development · UNSDG-7

N. Suraj · N. Abhishek (✉) · S. G. Kanchan · S. Raghupathi · S. Kuriakose
Institute of Management and Commerce, Srinivas University, Mangalore, India
e-mail: abhishekalmighty93@gmail.com

D. J. Rajendran
Department of Management Studies, Saveetha Engineering College, Chennai, India

B. Awwad (ed.), *The AI Revolution: Driving Business Innovation and Research*, Studies in Systems, Decision and Control 524, https://doi.org/10.1007/978-3-031-54379-1_30

1 Introduction

Due to rapid industrialization, natural resource exploitation is increasing daily [1]. Natural disasters like earthquakes, landslides, tsunamis, and hurricanes are also frequently occurring [2]. Further, man-made disasters such as anthropogenic acts such as various pollution disturb the environment and create uncertainty for future generations [3]. As a corporate citizen, it is the duty and responsibility of the company to undertake activities and initiatives that certainly minimize such negative externality on the environment and future generations [4]. Therefore, the promoting environmental-friendly business activities by inculcating various initiatives, for instance, Renewable Energy Initiatives (REI) such as the production, consumption, and distribution of renewable sources of energy, use of electric vehicles, production and distribution of RE products for instance solar panels, solar roofs and electric vehicles are necessary [5]. There are many rules and regulations governing to protection of the environment that have emerged in recent times which have taken an initiation through ties with corporate bodies, NGOs, and Government initiatives that slowly eradicate environmental pollution, which in turn will protect the resources for future generations [6, 7].

Today corporates are running under a triple-bottom-line line business model which includes profit, people, and planet. By undertaking REI corporates can lower the overall business costs which in turn enhances profitability [8]. On the other hand, they also lower toxic emissions/carbon footprint which in turn promotes a pollution-free environment that leads to the good health of people [9, 10].

REIs by corporates ensure energy security and reduce the dependency on external energy suppliers. This aspect particularly matters a lot during energy supply disruptions.

In other words, REI by corporations has a socio-economic and environmental impact that positively influences sustainable development [11, 12].

In this paper, an attempt has been made to evaluate the impact of REI by corporations on sustainable development through the mediation effect of environmental, social, and economic aspects.

The succeeding part of the paper is organized as a Literature review, Conceptual Model, Methods, results, discussions, and conclusion.

2 Literature Review

2.1 Green Business Practices

Pang et al. [13] observed that many corporates including financial institutions started to promote e-business operations to minimize the negativities on the environment that occurred through carbon footprints. This aspect is also noted by Salman and

Ismael [14]. In addition to this, [15] highlighted that a firm's green practices are positively correlated with improved financial performance.

Babon-Ayeng et al. [16] noted that the prior experience in green projects plays a vital role in implementing green business practices. In support of this, [17, 18] highlighted that undertaking green business practices supports following circular economy principles such as repair, reduce, reuse, and recycling in a business model that creates value for the business itself in the long run. Further, many researchers in their study for instance, [19, 20], highlighted that the inculcation of a model that promotes the re-use of industry waste is a trend among many green business practices. Further, [21–23] as part of green business initiatives inducing artificial knowledge and digitization of supply chain management enhance the accountability, transparency, and sustainable performance of the organization at the core. On the other hand, [24] highlighted that the involvement of corporates along with governments in minimizing carbon emissions is essential to promote a sustainable socio-economic and natural environment.

2.2 Renewable Energy Initiatives

A study by Jia et al. [25] recommended that the corporates effort to promote renewable energy initiatives is essential specifically in underdeveloped and developing countries.

The environmental, economic, and social environmental development in an industrial world directly depends on the level of involvement of government and corporates in promoting renewable energy production and consumption [26–28]. The national policy on energy consumption certainly influences the nature of energy production and consumption in the particular environment [29].

Obrecht et al. [30] argued that as renewable energy market is very complex and focuses on the environmental and economic aspect, and that negatively impact SDGs as they least bother on social aspects. On the contrary, [31] climate risks that emerged out of industrial practices disturb socio-economic development, but the inducement of technology-based renewable energy initiatives helps in alleviating their effects. In support of this [32, 33] recommended that the corporates needed to be involved in undertaking photovoltaic solar power plants and use of electric vehicles for promoting sustainable performance.

Interestingly, [34, 35] highlighted that the corporates that are involved in renewable energy projects have more profitability than conventional firms.

Zhou [36] in a study argued the prominence of involvement in renewable energy generation and consumption by companies specifically, transportation organizations like railways to promote socio-economic sustainable development. Authors [15, 37] noted that by installing renewable energy projects corporate carbon emission footprint due to coal thermal power generation can be minimized.

From the literature review, it is observed that the majority of studies have focused on exploring the significance of green energy practices, CSR practices, and renewable energy initiatives by corporates but no studies have attempted to evaluate the role renewable energy initiatives on sustainable development through the mediating effect of environment, social and economic aspects. Therefore, the present study intends to address the following research questions:

RQ1—What is the perception of various stakeholders on renewable energy initiatives by corporates?
RQ2—What is the impact of renewable energy initiatives by corporates on sustainable development?

3 Methodology

3.1 Research Design

The present study employed both quantitative and qualitative research as recommended by Authors [38–40]. A qualitative approach is followed for the development conceptual model, and research instrument through a focus group study. A quantitative approach is followed for the collection of data through a structured questionnaire and analysis and interpretation of the same.

3.2 Instrument Development

Since there is no standard measurement tool in the published literature to evaluate the impact of REIs on sustainable development, the researcher developed a questionnaire through a systematic and scientific procedure and also tested the same as per a well-accepted reliability and validity procedure [41].

3.3 Item Generations

With the help of previous literature and focus group studies the researcher constructed items to measure pre-determined constructs such as REIs and SD from the Managers, Administrators, and Academicians' perceptions.

The opinions of experts from the focus group study were recorded and later content analysis was performed to generate specific items for constructs. In the first stage, the researcher identified 16 items for REIs and 13 items for SD.

In the second stage after considering the opinion of experts 12 items for the REI construct and 8 items each for SD. Table 1 shows item generation and sources.

Table 1 Item generation and sources

Concept	Dimensions	Items	Criteria/description	Source
Impact REIs of corporates on sustainable development	Need for REIs by corporates	NREI1	REI is needed to have a cleaner and healthier environment	[42]
		NREI2	The REI aligns with the company's CSR goals	[43]
		NREI3	It enhances the company's reputation	[44]
		NREI4	It is required to reduce and maintain the cost	[45]
		NREI5	REIs ensure energy security	[46]
		NREI6	It improves innovation	[47]
		NREI7	It leads to a competitive advantage	[48]
		NREI8	Embracing REIs can help companies comply with the regulations and avoid potential penalties	[49]
		NREI9	It is needed to have long-term stability	[50]
		NREI10	REIs are required to improve employee morale and attract and retain top talent	[51]
		NREI11	By investing in RE measures, companies can build resilience against climate-related risks	[52]
		NREI12	REIs provide opportunities for companies to collaborate with governments, NGOs, and other organizations to work towards a shared goal of sustainable development	[53]
	The overall impact of REIs by corporates on sustainable development	IREI1	It creates jobs and can lead to shifts in the labor market and attract a diverse workforce	[54]
		IREI2	RE projects influence population distribution by leading to increased urbanization in areas with suitable resources for renewable energy generation	[55, 56]
		IREI3	Investment in renewable energy leads to growth in per capita income as people move to these areas for better economic prospects	[57]
		IREI4	The transition to renewable energy can lead to shifts in population due to changes in job opportunities	[58]
		IREI5	Cleaner environments can attract and retain residents, leading to demographic changes over time	[59]

(continued)

Table 1 (continued)

Concept	Dimensions	Items	Criteria/description	Source
		IREI6	RE projects provide new income streams for rural communities resulting in rural revitalization	[60, 61]
		IREI7	It leads to increased investment in education and research, attracting students and professionals and potentially affecting demographics	[60, 62]
		IREI8	It helps to have good international relations	[63]

Source Authors compiled

3.4 Response Format

The questionnaire is divided into 6 parts, the first part concerns to demographic profile of the respondents through the closed-ended questions. From second part to the sixth part is related to REIs by corporates and their influence on SD and questions were asked on the Likert 0.5 scale where 5 (Strongly Agree) and 1 (Strongly Disagree).

3.5 Sources of Data and Their Collection Procedures

The present study relied both on primary and secondary sources of data. Many authors [64–66] Recommended the perception survey method in stakeholder studies therefore, the present research gathered primary data through an online survey between June and September 2023. A total of 271 responses were received at a response rate of 271 (45.17%). Out of the collected responses, 253 are completed responses and all are considered for further analysis. The participant's profile is indicated in Table 2.

Table 2 depicts the demographic profile of the target respondents, which shows the gender distribution, age groups, positions, and years of experience of the surveyed respondents.

3.6 Construct Validity

Within the context of a research paper, the construct validity test constitutes a pivotal methodology aimed at evaluating the extent to which a measurement instrument effectively aligns with the underlying theoretical construct it is designed to assess [67].

The reliability of the variables was assessed using Cronbach's Alpha and Composite Reliability. Initially, the factor loading for all the factors was calculated

Table 2 Profile of participants

Profile	Classification	No. of respondents	Percentage
Gender	Male	193	76.28
	Female	60	23.72
	Total	**253**	**100.00**
Age	Below 25 years	117	46.25
	25–35	72	28.46
	35–45	32	12.65
	45–55	23	9.09
	55 years and above	9	3.56
	Total	**253**	**100.00**
Position	Production manager	33	13.04
	Service manager	18	7.11
	Other administrative Position	88	34.78
	Academicians	114	45.06
	Total	**253**	**100.00**
Experience	Less than 10 years	174	68.77
	10–20	50	19.76
	20–30	22	8.70
	30 years and above	7	2.77
	Total	**253**	**100.00**
Experience/knowledge of REIs	Solar energy	94	37.15
	Wind energy initiatives	58	22.92
	Solar and wind power	5	1.98
	Hydro	68	26.88
	Biopower	14	5.53
	Nuclear	6	2.37
	Municipal solid waste energy initiative	8	3.17
	Total	**253**	**100.00**

Source Survey data

and items having factor loading less than 0.600 were removed. The results of reliability and validity for the rest of the items are depicted in Table 3. Alpha and CR values for all the variables were greater than the recommended value of 0.700 by Ilias et al. [68]. The Average Variance Extracted (AVE) values and CR values were higher or close to 0.500 and 0.700, respectively, which confirms convergent validity [68]. Multi-collinearity was also assessed with Variance Inflation Factor (VIF) for each indicator the value of which was less than 5 therefore, there is no problem with discriminant validity [68].

Table 3 Reliability and validity

Dimensions	Items	λ	Alpha	CR	AVE	VIF
Need for REIs by corporates	NREI1	0.698	0.853	0.913	0.585	2.439
	NREI2	0.819				
	NREI3	0.762				
	NREI4	0.709				
	NREI5	0.758				
	NREI6	0.780				
	NREI7	0.757				
	NREI8	0.675				
	NREI9	0.650				
	NREI10	0.612				
	NREI11	0.660				
	NREI12	0.750				
The overall impact of REIs by corporates on sustainable development	IREI1	0.758	812	788	0.622	1.632
	IREI2	0.728				
	IREI3	0.652				
	IREI4	0.757				
	IREI5	0.651				
	IREI6	0.738				
	IREI7	0.756				
	IREI8	0.792				

Source Authors compiled

3.7 Discriminant Validity

Discriminant validity is a critical aspect of construct validity, ensuring that a measurement instrument can effectively differentiate between different constructs [69, 70]. Table 4 shows that Discriminant values which are bold and italic are greater than the correlation between the constructs, discriminant validity is not an issue.

Table 4 Discriminant validity using the criterion by Fornell & Larcker and the Heterotrait-Monotrait Method (HTMT)

Variables	NREI	IREI
NREI	**0.816**	
IREI	0.647	**0.847**

Source Authors compiled

3.8 Tools

The researchers employed descriptive statistics, one sample t-test and simple linear regression analysis for evaluating the perception of various stakeholders on REIs by corporates and their influence on sustainable development.

3.9 Results of Descriptive Statistics and One Sample t-test

Table 5 shows the results of one sample t-statistics on the opinion of managers and academicians on REIs by corporations. The respondents strongly agreed that (M = 4.00, SD = 1.266) REIs by corporates are needed to have a cleaner and healthier environment. The respondents agreed that by undertaking REIs corporates can match with their CSR goals (M = 3.81, SD = 1.149). From the opinion of managers and academicians, it is noted that the involvement in promoting RE production and consumption by corporates favours gaining a reputation in the market (M = 3.82, SD = 1.161). The respondents agreed that renewable energy initiatives are needed to maintain and reduce the overall cost of energy consumption (M = 3.79, SD = 1.171). Respondents agreed that these initiatives by corporates help ensure energy security in society (M = 3.78, SD = 1.170). The majority of respondents also agreed that by inducing REIs corporates can promote innovation in the energy consumption process (M = 3.92, SD = 1.110). From the opinions of the respondents, it is found that such initiatives help companies gain a competitive advantage in the market (M = 3.87, SD = 1.133). The respondents also agreed that by adopting such initiatives companies may comply with regulatory requirements and minimize penalties (M = 3.78, SD = 1.019). From the perception of various respondents, it was found that REIs promote the long-term sustainability of the organizations (M = 3.85, SD = 1.115). The respondents also agreed that REIs are required to improve employee morale and attract and retain top talent (M = 3.70, SD = 1.118). From the opinion of respondents, it is observed that such initiatives by corporates can build resilience against climate-related risks (M = 3.85, SD = 1.106). Lastly, respondents also agreed that REIs provide opportunities for companies to collaborate with governments, NGOs, and other organizations to work towards a shared goal of sustainable development (M = 3.83, SD = 1.209).

Table 6 the perception of respondents confirms that RE creates jobs and can lead to shifts in the labour market and it will also attract a diverse workforce (M = 3.93, SD = 1.040).

The majority of the respondents state that RE projects will influence population distribution by leading to increased urbanization in areas with suitable resources for RE generation (M = 3.93, SD = 1.040). The majority of the respondents believe that investing in renewable RE will lead to growth in per capita income as people move to these areas for better economic prospects (M = 3.93, SD = 1.040). The majority of the respondents state that the transition to RE can lead to shifts in population due

Table 5 Need for REIs by corporates (NREI)

Statements	N	Df	Mean	Std. deviation	t-value	*P*-value
NREI1	253	252	4.00	1.266	12.562	0.001
NREI2	253	252	3.81	1.149	11.214	0.001
NREI3	253	252	3.82	1.161	11.210	0.001
NREI4	253	252	3.79	1.171	10.795	0.001
NREI5	253	252	3.78	1.170	10.638	0.001
NREI6	253	252	3.92	1.110	13.198	0.001
NREI7	253	252	3.87	1.133	12.153	0.001
NREI8	253	252	3.78	1.019	12.156	0.001
NREI9	253	252	3.87	1.115	12.346	0.001
NREI10	253	252	3.70	1.118	10.013	0.001
NREI11	253	252	3.85	1.106	12.221	0.001
NREI12	253	252	3.83	1.209	10.868	0.001

Source Authors compiled

Table 6 The overall impact of REIs by corporates on sustainable development

Statements	N	Df	Mean	Std. deviation	t-value	*P*-value
IREI1	253	252	3.65	1.171	8.803	0.001
IREI2	253	252	3.68	1.146	9.494	0.001
IREI3	253	252	3.67	1.140	9.371	0.001
IREI4	253	252	3.70	1.082	10.347	0.001
IREI5	253	252	3.74	1.204	9.713	0.001
IREI6	253	252	3.81	1.178	10.890	0.001
IREI7	253	252	3.72	1.159	9.925	0.001
IREI8	253	252	3.82	1.052	12.428	0.001

Source Authors compiled

to which changes in job opportunities are possible (M = 3.93, SD = 1.040). The respondents opined that cleaner environments can attract and retain residents which will in turn lead to demographic changes over time (M = 3.93, SD = 1.040). The majority of the respondents agreed that RE projects provide new income streams for rural communities resulting in Rural Revitalization (M = 3.93, SD = 1.040). The majority of the respondents find that RE will lead to increased investment in education and research thereby it will attract students and professionals by potentially affecting demographics (M = 3.93, SD = 1.040). The majority of the respondents state that RE helps to have good International Relations (M = 3.93, SD = 1.040).

Table 7 Simple linear regression results

Model		Unstandardized coefficients		Standardized coefficients	t	Sig.
		B	Std. error	Beta		
1	(Constant)	0.476	0.137		3.471	0.001
	REI	0.847	0.035	0.838	24.346	0.001

a. Dependent variable: SD
Source Authors compiled

3.10 Simple Linear Regression Results

Table 7 to analyze the influence of REIs on SD simple linear regression analysis has been performed (see Table 7). The results revealed that REI by corporates positively influences SD ($\beta = 0.838$, $t = 24.346$, $p = 0.001$) which is statistically significant at 1%. These results are supported by Authors [71–73].

4 Conclusion

The present research aimed to analyze the perception of various stakeholders on REI by corporates and their impact on sustainable development. The results of descriptive statistics and one sample t-test confirmed that the stakeholders positively opined that REIs by corporations influence sustainable development through environmental, social, and economic aspects. Further, the results of simple linear regression analysis revealed that REI influence on SD. These findings suggest that in general REI undertaken by corporates significantly plays a vital role in the betterment of the environment, society, economy, and sustainable development at large. By considering this the corporates which are not yet undertaken REI can move forward with this evidence. These results and observations add value to the rapidly expanding field of corporate initiatives on socio-economic, environmental, and sustainable development. A major limitation of the study is that it focuses only on the stakeholder's perception and does not study the company's REI by companies directly. Therefore, a further study could assess the long-term effects of REI by companies by analyzing the information revealed on REI by companies in their reports. This could also help in validating the results of the present study. These findings assist policymakers in framing a guideline, and framework on REIs to be undertaken by companies further it also supports the government in offering subsidies, incentives, and tax benefits for the companies that are undertaking REIs. Further, by the outcome of this study corporate can get input to undertake clean energy projects and technology. This in turn contributes to achieving UNSDG-7 i.e., affordable clean energy set by UNO.

References

1. Usman, M., Balsalobre-Lorente, D.: Environmental concern in the era of industrialization: can financial development, renewable energy and natural resources alleviate some load? Energ. Policy **162**, 112780 (2022)
2. Barbier, E.B.: Scarcity and Frontiers: How Economies have Developed Through Natural Resource Exploitation. Cambridge University Press (2010)
3. Naidu, R., Biswas, B., Willett, I.R., Cribb, J., Singh, B.K., Nathanail, C.P., Coulon, F., Semple, K.T., Jones, K.C., Barclay, A., Aitken, R.J.: Chemical pollution: a growing peril and potential catastrophic risk to humanity. Environ. Int. **156**, 106616 (2021)
4. Reinhardt, F.L., Stavins, R.N., Vietor, R.H.: Corporate social responsibility through an economic lens (2008)
5. Majid, A., Yasir, M., Yasir, M., Javed, A.: Nexus of institutional pressures, environmentally friendly business strategies, and environmental performance. Corp. Soc. Responsib. Environ. Manag.Responsib. Environ. Manag. **27**(2), 706–716 (2020)
6. Doh, J.P., Guay, T.R.: Corporate social responsibility, public policy, and NGO activism in Europe and the United States: an institutional-stakeholder perspective. J. Manage. Stud. **43**(1), 47–73 (2006)
7. Székely, F., Knirsch, M.: Responsible leadership and corporate social responsibility: metrics for sustainable performance. Eur. Manag. J. **23**(6), 628–647 (2005)
8. Deason, W.: Comparison of 100% renewable energy system scenarios with a focus on flexibility and cost. Renew. Sustain. Energ. Rev. **82**, 3168–3178 (2018)
9. Masondo, Z.N.: Environmental pollution and climate change: an ethical interrogation of the payment of carbon tax as a means to reduce greenhouse gas emission in South Africa. Doctoral dissertation (2020)
10. Khan, H., Weili, L., Khan, I.: Environmental innovation, trade openness and quality institutions: an integrated investigation about environmental sustainability. Environ. Dev. Sustain. 1–31 (2022)
11. Michalek, M.: Sustainable development in the retail sector: REI's stewardship efforts (2008)
12. Pineda, S., Bock, A.: Renewable-based generation expansion under a green certificate market. Renew. Energ. **91**, 53–63 (2016)
13. Pang, D., Li, K., Wang, G., Ajaz, T.: The asymmetric effect of green investment, natural resources, and growth on financial inclusion in China. Resour. Policy. Policy **78**, 102885 (2022)
14. Salman, D., Ismael, D.: The effect of digital financial inclusion on the green economy: the case of Egypt. J. Econ. Dev. (2023) (ahead-of-print).
15. Lin, Z., Liao, X., Jia, H.: Could green finance facilitate low-carbon transformation of power generation? Some evidence from China. Int. J. Clim. Change Strat. Manag. **15**(2), 141–158 (2023)
16. Babon-Ayeng, P., Oduro-Ofori, E., Owusu-Manu, D.G., Edwards, D.J., Kissi, E., Kukah, A.S.K.: Socio-political factors underlying the adoption of green bond financing of infrastructure projects: the case of Ghana. J. Capital Mark. Stud. **6**(3), 304–319 (2022)
17. Mishra, S., Jain, S., Malhotra, G.: The anatomy of circular economy transition in the fashion industry. Soc. Responsib. J. **17**(4), 524–542 (2021)
18. Arena, M., Azzone, G., Piantoni, G.: Uncovering value creation in innovation ecosystems: paths towards shared value. Eur. J. Innov. Manag.Innov. Manag. **25**(6), 432–451 (2022)
19. Pujiati, A., Nurbaeti, T., Damayanti, N.: What are the factors that determine differing levels of environmental quality? Evidence from Java and other islands in Indonesia. Manag. Environ. Qual. Int. J. (2022) (ahead-of-print)
20. Stocco, L.C., Cezarino, L.O., Liboni, L.B., Venkatesh, V.G.: Wholesaler echelon and Industry 4.0 in circular supply chains—a systematic review. Mod. Supp. Chain Res. Appl. **4**(2), 141–158 (2022)
21. Di Vaio, A., Latif, B., Gunarathne, N., Gupta, M., D'Adamo, I.: Digitalization and artificial knowledge for accountability in SCM: a systematic literature review. J. Enterprise Inform. Manag. (2023) (ahead-of-print)

22. Han, Y., Tan, S., Zhu, C., Liu, Y.: Research on the emission reduction effects of carbon trading mechanism on power industry: plant-level evidence from China. Int. J. Clim. Change Strat. Manag. (2022) (ahead-of-print)
23. Leal Filho, W., Sima, M., Sharifi, A., Luetz, J.M., Salvia, A.L., Mifsud, M., Olooto, F.M., Djekic, I., Anholon, R., Rampasso, I., Kwabena Donkor, F.: Handling climate change education at universities: an overview. Environ. Sci. Eur. **33**, 1–19 (2021)
24. Herth, A., Blok, K.: Quantifying universities' direct and indirect carbon emissions—the case of Delft University of Technology. Int. J. Sustain. High. Educ. **24**(9), 21–52 (2022)
25. Jia, J., Rong, Y., Chen, C., Xie, D., Yang, Y.: Contribution of renewable energy consumption to CO_2 emissions mitigation: a comparative analysis from the income levels' perspective in the belt and road initiative (BRI) region. Int. J. Clim. Change Strat. Manag. **13**(3), 266–285 (2021)
26. Slacik, J., Greiling, D.: Coverage of G4-indicators in GRI-sustainability reports by electric utilities. J. Public Budg. Account. Financ. Manag.Budg. Account. Financ. Manag. **32**(3), 359–378 (2020)
27. Ghosh, S.: COVID-19, clean energy stock market, interest rate, oil prices, volatility index, geopolitical risk nexus: evidence from quantile regression. J. Econ. Dev. **24**(4), 329–344 (2022)
28. Lahrech, A., Abu-Hijleh, B., Aldabbas, H.: The impact of global renewable energy demand on economic growth—evidence from GCC countries. Arab Gulf J. Sci. Res. (2023)
29. Huang, X., Khan, Y.A., Arshed, N., Salem, S., Shabeer, M.G., Hanif, U.: Increasing social resilience against climate change risks: a case of extreme climate affected countries. Int. J. Clim. Change Strat. Manag. **15**(3), 412–431 (2023)
30. Obrecht, M., Singh, R., Zorman, T.: Conceptualizing a new circular economy feature—storing renewable electricity in batteries beyond EV end-of-life: the case of Slovenia. Int. J. Product. Perform. Manag. **71**(3), 896–911 (2021)
31. Wei, J., Liang, G., Alex, J., Zhang, T., Ma, C.: Research progress of energy utilization of agricultural waste in China: bibliometric analysis by citespace. Sustainability **12**(3), 812 (2020)
32. Valtteri, K., Gonzalez-Perez, M.A.: Renewable energy in international business (2020)
33. Ndzibah, E., Andrea Pinilla-De La Cruz, G., Shamsuzzoha, A.: End of life analysis of solar photovoltaic panel: roadmap for developing economies. Int. J. Energ. Sect. Manag. **16**(1), 112–128 (2022)
34. Westerman, W., De Ridder, A., Achtereekte, M.: Firm performance and diversification in the energy sector. Manag. Financ.Financ. **46**(11), 1373–1390 (2020)
35. Caceres, A.L., Jaramillo, P., Matthews, H.S., Samaras, C., Nijssen, B.: Hydropower under climate uncertainty: characterizing the usable capacity of Brazilian, Colombian and Peruvian power plants under climate scenarios. Energ. Sustain. Dev. **61**, 217–229 (2021)
36. Zhou, X.: Path selection and mechanism innovation of improving railway energy efficiency—from foreign experience and Chinese practice. Railway Sci. **1**(2), 324–344 (2022)
37. Chang, K.H., Lin, G.: Optimal design of hybrid renewable energy systems using simulation optimization. Simul. Model. Pract. Theor. **52**, 40–51 (2015)
38. Patton, M.Q.: Enhancing the quality and credibility of qualitative analysis. Health Serv. Res. **34**(5 Pt 2), 1189 (1999)
39. Cooper, D.R., Schindler, P.S., Sun, J.: Business Research Methods, vol. 9, pp. 1–744. Mcgraw-Hill, New York (2006)
40. Tashakkori, A., Creswell, J.W.: The new era of mixed methods. J. Mixed Methods Res. **1**(1), 3–7 (2007)
41. Roberts, P., Priest, H.: Reliability and validity in research. Nurs. Stand.. Stand. **20**(44), 41–46 (2006)
42. Bekun, F.V., Alola, A.A., Gyamfi, B.A., Ampomah, A.B.: The environmental aspects of conventional and clean energy policy in sub-Saharan Africa: is the N-shaped hypothesis valid? Environ. Sci. Pollut. Res.Pollut. Res. **28**, 66695–66708 (2021)
43. Rangan, K., Chase, L., Karim, S.: The truth about CSR. Harv. Bus. Rev. **93**(1/2), 40–49 (2015)
44. Issa, A., Hanaysha, J.R.: Powering profits: how renewable energy boosts financial performance in European non-financial companies. Int. J. Acc. Inform. Manag. (2023)
45. Stram, B.N.: Key challenges to expanding renewable energy. Energ. Policy **96**, 728–734 (2016)

46. Valentine, S.V.: Emerging symbiosis: renewable energy and energy security. Renew. Sustain. Energ. Rev. **15**(9), 4572–4578 (2011)
47. He, A., Xue, Q., Zhao, R., Wang, D.: Renewable energy technological innovation, market forces, and carbon emission efficiency. Sci. Total. Environ. **796**, 148908 (2021)
48. Amran, M.E., Muhtazaruddin, M.N.: Renewable energy optimization review: variables towards competitive advantage in green building development. In: Progress in Energy and Environment, pp. 1–15 (2019)
49. Punitha, S., Rasdi, R.M.: Corporate social responsibility: adoption of green marketing by the hotel industry. Asian Soc. Sci. **9**(17), 79 (2013)
50. Adebayo, T.S.: Renewable energy consumption and environmental sustainability in Canada: does political stability make a difference? Environ. Sci. Pollut. Res.Pollut. Res. **29**(40), 61307–61322 (2022)
51. Davidescu, A.A., Apostu, S.A., Paul, A., Casuneanu, I.: Work flexibility, job satisfaction, and job performance among Romanian employees—implications for sustainable human resource management. Sustainability **12**(15), 6086 (2020)
52. In, S.Y., Weyant, J.P., Manav, B.: Pricing climate-related risks of energy investments. Renew. Sustain. Energ. Rev. **154**, 111881 (2022)
53. Palit, D.: Solar energy programs for rural electrification: experiences and lessons from South Asia. Energ. Sustain. Dev. **17**(3), 270–279 (2013)
54. Madanizadeh, S.A., Pilvar, H.: The impact of trade openness on labour force participation rate. Appl. Econ. **51**(24), 2654–2668 (2019)
55. Grodzicki, T., Jankiewicz, M.: The impact of renewable energy and urbanization on CO_2 emissions in Europe—spatio-temporal approach. Environ. Dev. **44**, 100755 (2022)
56. Ritu, R.K., Kaur, A.: Towards environmental sustainability: nexus of ecological footprint, human capital, economic growth and energy consumption in India. Manag. Environ. Qual. Int. J. (2023)
57. Overland, I., Reischl, G.: A place in the Sun? IRENA's position in the global energy governance landscape. Int. Environ. Agreements Polit. Law Econ. **18**(3), 335–350 (2018)
58. Hsu, D., Andrews, C.J., Han, A.T., Loh, C.G., Osland, A.C., Zegras, C.P.: Planning the built environment and land use towards deep decarbonization of the United States. J. Plann. Literature **38**(3), 426–441 (2023)
59. Baloch, Q.B., Shah, S.N., Iqbal, N., et al.: Impact of tourism development upon environmental sustainability: a suggested framework for sustainable ecotourism. Environ. Sci. Pollut. Res. **30**, 5917–5930 (2023). https://doi.org/10.1007/s11356-022-22496-w. J. Policy Anal. Manag. **21**(1), 45–62 (18 pages) (Winter, 2002). Published By: Wiley
60. Owusu, P.A., Asumadu-Sarkodie, Dubey, S. (Reviewing Editor): A review of renewable energy sources, sustainability issues and climate change mitigation. Cogent Eng. **3**, 1 (2016). https://doi.org/10.1080/23311916.2016.1167990
61. Clausen, L.T., Rudolph, D.: Renewable energy for sustainable rural development: synergies and mismatches. Energ. Policy **138**, 111289 (2020). ISSN 0301-4215
62. Wang, Z., Le Hoa Pham, T., Wang, B., et al.: The simultaneous impact of education and financial development on renewable energy consumption: an investigation of Next-11 countries. Environ. Sci. Pollut. Res. **29**, 85492–85509 (2022). https://doi.org/10.1007/s11356-022-21330-7
63. Lei, X., Alharthi, M., Ahmad, I., Aziz, B., ul Abdin, Z.: Importance of international relations for the promotion of renewable energy, preservation of natural resources and environment: empirics from SEA nations. Renew. Energ. **196**, 1250–1257 (2022). ISSN 0960-1481
64. Hider, P., Pymm, B.: Empirical research methods reported in high-profile LIS journal literature. Libr. Inf. Sci. Res.. Inf. Sci. Res. **30**(2), 108–114 (2008)
65. Connaway, L.S., Connaway, L.S., Powell, R.R., Powell, R.R.: Basic Research Methods for Librarians. ABC-CLIO (2017)
66. Ilias, A., Ghani, E.K., Baidi, N., Abdul, R.: XBRL adoption: an examination on the Malaysian business reporting system (MBRS). Humanities **8**(2), 202–214 (2020)
67. Strauss, M.E., Smith, G.T.: Construct validity: advances in theory and methodology. Annu. Rev. Clin. Psychol. **5**, 1–25 (2009)

68. Sürücü, Maslakci: Validity and reliability in quantitative research. Bus. Manag. Stud. Int. J. **8**(3), 2694–2726 (2020)
69. Henseler, J., Ringle, C.M., Sarstedt, M.: A new criterion for assessing discriminant validity in variance-based structural equation modeling (2015)
70. Fornell, C., Larcker, D.F.: Evaluating structural equation models with unobservable variables and measurement error (1981)
71. Taha, M.H., Eldalash, S.A., Mohamed, H.Z.: The role of accounting disclosure of sustainable development activities using metaverse in the field of education and training. In: El Khoury, R., Alareeni, B. (eds.) How the Metaverse Will Reshape Business and Sustainability. Contributions to Environmental Sciences & Innovative Business Technology. Springer, Singapore (2023). https://doi.org/10.1007/978-981-99-5126-0_7
72. Shaju, A.: Opportunities and risks of the "metaverse" for environmental sustainability. In: El Khoury, R., Alareeni, B. (eds.) How the Metaverse Will Reshape Business and Sustainability. Contributions to Environmental Sciences & Innovative Business Technology. Springer, Singapore (2023). https://doi.org/10.1007/978-981-99-5126-0_10
73. Karthick, A.V., Gopalsamy, S.: Role of IoT in business sustainability. In: Aloysius Edward, J., Jaheer Mukthar, K.P., Asis, E.R., Sivasubramanian, K. (eds.) Current Trends in Economics, Business and Sustainability. ICEBS 2023. Contributions to Environmental Sciences & Innovative Business Technology. Springer, Singapore (2023). https://doi.org/10.1007/978-981-99-3366-2_2

ESG Metrics, Company Reputation, and Financial Performance: An Examination of Intel Corporation

Sanaa Malaka, Amal Alsuwaidi, Juma Alyazia, Inkara Nazarbekova, Ragni Kishore, Elgilani Elshareif, and Ikhlaas Gurrib

Abstract This study investigates the relationships between financial performance, ESG strategies and reputation at Intel Corporation, to gain insights into its market dynamics and competitive strategies. The research methodology involves an analysis of variance, ordinary least square regressions, and reliability tests. Qualitative and quantitative data were examined to assess market innovation initiatives and response to industry challenges. A survey was carried out to examine the connection between a company's reputation and its financial performance. Findings support a strong correlation between company reputation and financial performance. Despite challenges, including manufacturing delays and increasing competition, Intel's adaptability and commitment to innovation have allowed it to remain a key player in the semiconductor industry.

Keywords Intel Corporation · ESG · Financial performance · Reputation · Sustainability Semiconductor Industry

1 Introduction

In 1968 Gordon Moore and Robert Noyce formed the worldwide technology behemoth Intel Corporation. To date, the company had a significant impact on how computing and technologies have evolved. Intel achieved ground-breaking strides from its early emphasis on electronic memory technologies, including the creation of an initial MOS silicon gate in 1969. However, its most important contribution

S. Malaka (✉) · A. Alsuwaidi · J. Alyazia · I. Nazarbekova · R. Kishore · E. Elshareif · I. Gurrib
Canadian University in Dubai, Dubai, United Arab Emirates
e-mail: 20230002864@students.cud.ac.ae

E. Elshareif
e-mail: elgilani@cud.ac.ae

I. Gurrib
e-mail: ikhlaas@cud.ac.ae

B. Awwad (ed.), *The AI Revolution: Driving Business Innovation and Research*, Studies in Systems, Decision and Control 524, https://doi.org/10.1007/978-3-031-54379-1_31

occurred in 1971 with the release of the 4004 microprocessors, a technological advance that enabled smaller and more potent computers, revolutionizing computing.

With the introduction of the 8086 microprocessors in 1978, Intel cemented its leadership in the personal computers (PC) market and provided the groundwork for the widely used x86 CPU architecture. Intel sustained its sector dominance through a mix of technological advancement and intense marketing, despite the challenge from rivals like Advanced Micro Devices (AMD). In addition to microprocessors, Intel's growth also included chipsets, Graphics Processing Units (GPU), networking hardware, and memory technology. The firm's expansion was greatly impacted by strategic purchases such as Altera and Mobileye. Recent difficulties, including increased competition and production snags, compelled Intel to investigate alternative technologies. To be at the forefront of innovation, the corporation made significant investments in AI, 5G, quantum computing, and edge computing. Additionally, Intel showed a dedication to environmental sustainability and social responsibility by working toward sustainable operations and encouraging multicultural and inclusive workplaces. Intel maintains a significant influence on shaping the future of computers and associated technologies due to its vast worldwide network of offices, manufacturing plants, and research and development centres. Intel's influence within the tech industry remains substantial as it adapts to evolving sectors and responds to changing market dynamics.

Intel is listed under the Nasdaq stock exchange (Nasdaq: INTC). With a 52-week price range of $24–$40, it currently has a market capitalization value of $164 billion, positioning it among the world's leading semiconductor chip manufacturers. From the fiscal years 2007–2016, Intel consistently held the 45th position in the 2020 Fortune 500 list of the largest American companies by total revenue for nearly a decade. Key competitors include Acer, Lenovo, HP, and Dell which all provide computer processors. Along with producing embedded processors, flash memory, graphics, chips, network interface controllers, and motherboard chipsets, Intel also produces additional computing and communication-related products.

Intel, originally named Integrated Electronics, played a substantial role in the development of Silicon Valley as a major center for advanced technology. The computer chip (microchip) was primarily invented by a Noyce. Intel initially focused on producing SRAM and DRAM memory chips, which formed the bulk of its business until 1981. It wasn't until the personal computer (PC) gained widespread popularity that Intel shifted its main line of business, despite having developed the world's first competitive microprocessor chip in 1971. The Strengths, Weaknesses, Opportunities and Threats (SWOT) of Intel are summarized in Table 6 in the appendix section. Among other factors, Intel gives significant importance to technologies such as artificial intelligence and metaverse. This is backed by studies like Joseph [12] and Navid and Gupta [13] who supported that such technologies help to reshape business and sustainability.

As per [7], Intel has a low ESG risk rating from Morningstar Sustainalytics, and one of the highest scores in governance quality from Institutional Shareholder Services (ISS). Thirumagal et al. [14] highlighted that woman on board contributes to governance factor, with however no impact on environmental and social factors. This is commendable to major competitors such as Broadcom incorporated, Advanced Micro Devices incorporated and Nvidia corporation. Finally, but not least, major indicators in the Sustainable Finance Disclosure Regulation (SFDR) such as renewable energy consumed, human right policy, and female to male ratios point to Intel positioning itself as an ESG leader in its respective semiconductor industry.

In the last decade, there has been an increased attention and investments in greener energies, particularly in the U.S., EMDE, and China [8]. Although numerous financial strategies demonstrated evidence of success in traditional markets including cryptocurrencies, currencies markets, fixed income, and equity markets (for e.g., [9–11]), the area of ESG and financial performance has attracted even more attention due to climate change events such as Conference of Parties of the United Nations Framework Convention on Climate Change (UNFCCC) COP28 which will be hosted in UAE in December 2023. Studies like [4], which provide supporting views on different energy sources, enhance the importance of analysing ESG related factors determining financial performance on different financial instruments.

The research objective is to study the complex factors such as ESG Framework and Corporate Reputation and how they affect financial performance of the company on the example of Intel Corporation. The implications of the study shed further light in understanding the interaction between ESG metrics, company reputation, and financial performance which is becoming increasingly vital in today's business climate and long-term viability and success. Specifically, findings provide ground evidence as to (i) why companies with strong ESG metrics often attract more socially responsible investors resulting in better access to capital, and (ii) whether companies in specific industries like semi-conductors, or sectors like technology are more concerned with ESG scores and/or company reputation to drive financial performance.

The rest of this paper is structured as follows: An overview of the ESG framework related to financial performance, ESG activities and reputation, including current practices at Intel is provided in the literature review section. The methodology and data section then provides a breakdown of the research design, and quantitative/qualitative data used, followed by the analysis section. We rest our case with some concluding remarks.

2 Literature Review

ESG framework—ESG, which stands for Environmental, Social, and Governance, is a framework that evaluates a company's performance at these critical areas, reflecting the commitment to responsible and sustainable business practices. The concept of ESG appeared at the early 2000s when it gained prominence to assess non-financial risks and opportunities in the corporate world. It emerged in response to growing concerns about environmental sustainability, social responsibility, and ethical governance, driven by factors such as climate change awareness, evolving societal expectations, and corporate scandals. While ESG was not invented by a single individual or entity, it gained traction through various initiatives, guidelines, and reporting frameworks developed by organizations such as the Global Reporting Initiative (GRI) and the United Nations Principles for Responsible Investment (PRI). ESG has become increasingly important for the corporate world because it not only aligns businesses with ethical and sustainable values, but also mitigates risks, enhances reputation, attracts investors, and fosters long-term profitability. Recently, ESG considerations have evolved into a fundamental aspect of corporate strategy, influencing decision-making, stakeholder engagement, and ultimately, a company's competitiveness and resilience in a rapidly changing global landscape.

Implementation of ESG practices at Intel Corp—Intel was one of the first corporations that recognized the importance of implementing ESG practices. In 2020 the company also created a special ESG framework—"RISE", which stands for "Responsible", "Inclusive", "Sustainable", "Enabled". By analyzing the various initiatives mentioned at Intel's Annual Reports, the balance between environmental, social and governance initiatives is defined. Energy efficiency, water stewardship and reduction, waste reduction and supply chain sustainability are the main environmental objectives. Regarding social goals, Intel focuses on diversity, inclusion, employee well-being, community engagement. And to reach governance objectives the company enhances board diversity, ethical business practices, shareholder engagement, data privacy and security. In 2022 Intel also released Green Bonds with a fixed interest rate of 4.15% annually. Out of total $1.25 billion issued, $0.425 billion were already allocated by the end of 2022. Green bonds are released to finance sustainable and environmentally friendly projects and initiatives. As per Intel's Annual Green Bond Report 2023, there were defined six areas to fund such as Circular Economy, Green Buildings, Waste Management, Pollution Prevention and Control, renewable Energy and Water Stewardship.

Intel ESG Score—ESG scores measure a company's approach to environmental, social, and governance issues, offering valuable insights for socially responsible investors. There are many agencies that provide information regarding ESG scores, however, they are not standardized and scores of the same company may vary across different agencies. Therefore, to get a comprehensive understanding of Intel's ESG performance were used Dow Jones Sustainability World Index and Sustainalytics Rating Agency. According to the Dow Jones Sustainability World Index, in 2022, Intel's score was 68, while the industry's mean was only 22. This shows that Intel is doing a lot to make a positive impact in these important areas. Even compared to the companies in the industry, with a score around 90, Intel is doing exceptionally well. However, according to Sustainalytics, Intel's ESG performance has been classified as "Low Risk", getting 18.9 points. That score is close to "Medium" which starts from 20 points. In the context of the Semiconductor industry, Intel is standing at the 57th position out of 334 companies evaluated.

Financial performance and implementation of ESG—The relationship between a company's financial performance and the ESG score has become a subject of extensive research in recent years. However, the findings in this area remain controversial and can vary significantly depending on factors such as industry and geographic location. While some studies suggest a positive correlation between strong ESG practices and enhanced financial performance, others indicate a more complex and context-dependent relationship. For example, [2] that ESG-focused firms with high green innovation intensity tend to improve their financial performance, potentially due to cost reduction and product differentiation through green innovation. The study reveals a positive correlation between ESG practices and financial performance in England and Germany. Amadio et al. assessed ESG strategies separately and discovered that social and governance practices have a greater impact on financial performance than environmental practices.

However, in emerging markets the situation is opposite. As per research of Duque-Grisales et al. [6], there is negative correlation between ESG scores and financial performance in Latin American companies. The negative nature of this correlation suggests that companies in Latin America with superior ESG scores tend to exhibit lower profitability. The costs associated with implementing ESG initiatives are high, and that is affecting a cash flow of companies and budgeting, and there is lack of institutional support in those regions. This difference can be explained by the different characteristics of developed and developing countries when it comes to their financial markets. Daugaard et al. [5] stated that emerging markets feature micro-market structures that significantly differ from those of developed markets. These differences encompass various facets, including the percentage of retail investors, the extent of legal protection of investors, and the government's stance regarding ESG practices. Amel-Zadeh et al. [1] proposed that institutional investors are the main drivers of ESG practices.

Correlation between financial performance and company reputation—There is a positive correlation between financial performance and company reputation. As businesses increasingly recognize the dependence of financial success on their public image and reputation, they are investing more in building and maintaining a positive brand image. A strong reputation can translate into improved financial performance through increased customer loyalty, higher sales, a competitive advantage in attracting top talent, and greater access to capital. Therefore, nowadays a positive brand image is becoming a strategic asset. Positive Financial Performance and a good reputation are interconnected terms and that has been proven by various research papers, for example, [3] concluded that good financial performance improved company reputation.

3 Methodology and Data

This study employs a mixed-method approach, combining qualitative and quantitative approaches to examine the relationship among the following factors: a firm's ESG practices and its reputation in relation to financial performance in Dubai's public sphere. Quantitative data is collected through surveys that feature both closed-ended questions on topics like satisfaction, preferences, and demographics, as well as open-ended questions for detailed opinions and suggestions. Surveys are chosen due to their efficiency in collecting data from a large sample size. The link to the survey is available upon request. For the purpose of this study, Firm's ESG practices (F'sESG) and Firm's Reputation (F's R) are independent variables and the firm's Financial Performance (F'sFP) is the dependent variable. The two hypotheses tested are whether (i) there is a positive correlation between a Firm's ESG Activities and its Financial Performance, and (ii) there is a positive correlation exists between a Firm's Reputation and its Financial Performance. For the qualitative model, online surveys with respondents from Dubai were carried out. The sample size consists of 50 respondents, chosen through convenience sampling, which selects participants based on their accessibility and convenience. The research design explores causal relationships between variables and uses statistical tests like *t*-tests, ANOVA, two-factor replication tests, and simple linear regression to uncover significant differences or relationships among the variables. Primary data is collected through online surveys via Google Forms, containing 15 questions are closed-ended, Likert-scale were implemented ranging from 1 to 5 to gauge opinions. Additionally, one open-ended question captures qualitative insights. A total of 50 questionnaires were distributed randomly to Dubai residence. Secondary data such as financial performance indicators are sourced from Marketline and Factset.

4 Analysis

To evaluate whether there is a no table difference between two groups of people concerning their workplace's approach to social responsibility and its connection to firms' financial performance (referred to as "F'sFP DV"), *t*-tests have been applied. This analysis considers variations in responses. It assists in determining if endorsing social responsibility at work significantly impacts one's belief in its importance for a firm's financial performance (F'sFP DV). To deem a variable as significant, its *p*-value must be below 0.05, which equates to a 95% level of significance. In a *t*-test, adhering to the rule means the (*t*-statistics) needs to be less than the *t*-critical value to support the null hypothesis. As reported in Table 1, the *p*-value, which gauges the importance of the distinction between whether the workplace stresses social responsibility and ethical conduct concerning firms' financial performance (F'sFP DV), exceeds 0.002. Given that 0.002 is below the threshold of 0.05, this suggests that there is a significant connection between workplace efforts related to ethical practices and their impact on an individual's belief in the significance of social responsibility in achieving positive outcomes.

Table 1 *t*-test of two-samples (assuming unequal variances)

	Yes = 1	*No* = 2
Mean	4.026	2.727
Variance	1.120	1.058
Observations	39	11
Hypothesized mean difference	0	
Df	16	
t Stat	3.673	
P(T < = t) one-tail	0.001	
t Critical one-tail	1.746	
P(T < = t) two-tail	0.002	
t Critical two-tail	2.119	

Source Authors

4.1 Analysis of Variance

The purpose of the ANOVA is to examine variations among two or more groups concerning the Dependent Variable F'sFP. In this context, the ANOVA test has been applied to analyse how responses to the survey question about individuals' media and news exposure levels relate to the Dependent Variable F'sFP. The null hypothesis is that responses of all levels of exposure to media and news are equal in toward F'sFP, and the alternate responses of all levels of exposure to media and news are not equal in toward F'sFP. As per Table 2, F-value surpasses the critical F-value (17.48 > 2.81), suggesting the acceptance of the alternate hypothesis. The *p*-value falls below 0.05 (0.00 < 0.05), indicating that the difference among groups with varying levels of media and news exposure towards the dependent variable, Firm's Financial Performance is significant.

Table 2 Analysis of variance

Groups	Count	Sum	Average	Variance		
Low	13.00	34.20	2.63	1.27		
Moderate	8.00	24.00	3.00	0.80		
High	22.00	95.00	4.32	0.50		
Very high	7.00	33.80	4.83	0.03		
ANOVA						
Source of variation	SS	df	MS	F	P-value	F crit
Between groups	36.03	3.00	12.01	17.48	0.00	2.81
Within groups	31.59	46.00	0.69			
Total	67.62	49.00				

Source Authors

Table 3 ESG and financial performance relationship

ANOVA								
	Df	*SS*	*MS*	*F*	*Significance F*			
Regression	1.00	67.58	67.58	84,432.78	0.00			
Residual	48.00	0.04	0.00					
Total	49.00	67.62						
	Coefficients	*Standard error*	*t stat*	*P-value*	*Lower 95%*	*Upper 95%*	*Lower 95.0%*	*Upper 95.0%*
Intercept	– 0.02	0.01	– 1.24	0.22	– 0.04	0.01	– 0.04	0.01
Mean F's ESG	1.00	0.00	290.57	0.00	1.00	1.01	1.00	1.01

Source Authors

4.2 Relationships Between ESG/Reputation and Financial Performance

4.2.1 Firm's ESG Towards Firm's Financial Performance

The primary objective of conducting a simple linear regression analysis is to assess both the importance and impact of the independent variable, which in this case is "Firm's ESG," on the dependent variable, namely "Firm's Financial Performance". For the influence of the independent variable (Firm's ESG) on the dependent variable (Firm's Financial Performance) to be deemed significant, the *p*-value must be less than 0.05. In line with Table 3, with a *p*-value of 0, this indicates the significance of the effect of the independent variable (FESG) on the dependent variable (F'sFP).

4.2.2 Firm's Reputation Towards Firm's Financial Performance

The purpose of conducting a simple linear regression (SLR) analysis is to assess the importance and impact of the independent variable (Firm's Reputation) on the dependent variable (Firm's Financial Performance). To establish the significance of the influence of the second independent variable (Firm's Reputation) on the dependent variable (Firm's Financial Performance), the p-value must be below 0.05 in accordance with the testing guideline. As reported in Table 4, with a *p*-value of 0.0, which is less than 0.05, it indicates the significance of the impact of the mediating variable (FR) on the dependent variable (F'sFP).

Table 4 Firm's reputation and financial performance relationship

ANOVA								
	df	*SS*	*MS*	*F*	*Significance F*			
Regression	1.00	60.05	60.05	380.79	0.00			
Residual	48.00	7.57	0.16					
Total	49.00	67.62						
	Coefficients	*Standard error*	*t stat*	*P-value*	*Lower 95%*	*Upper 95%*	*Lower 95.0%*	*Upper 95.0%*
Intercept	− 0.01	0.20	− 0.04	0.96	− 0.41	0.39	− 0.41	0.39
Mean F's R (IV-2)	0.95	0.05	19.51	0.00	0.85	1.05	0.85	1.05

Source Authors

4.3 Reliability Test

A reliability test ensures each item is linked to its variable. Reliability, measured by Cronbach Alpha, should exceed 0.6. We use a two-factor ANOVA without replication for data analysis. Cronbach Alpha is calculated using the formula: 1 − (mean square errors/mean square rows) to measure each variable's reliability. As per Table 5, Firm's ESG (independent variable 1): The Cronbach Alpha stands at a robust 0.96, well above the 0.6 threshold. This firmly establishes the reliability of the independent variable (Firm's ESG) in the study, reflecting strong internal consistency among its related items or questions. Firm's Reputation (independent variable 2): The Cronbach Alpha score of 0.97 comfortably surpasses the 0.6 threshold. Consequently, the independent variable "Firm's Reputation" is confidently considered reliable and suiTable for analysis. Firm's Financial Performance (dependent variable): The Cronbach Alpha value of 0.964 exceeds the 0.6 threshold, affirming the reliability of the variable "F'sFP" (Firm's Financial Performance) for use in the study.

5 Conclusive Remarks

Intel has played a pioneering role in technology, introducing innovations like microprocessors and chipsets. The literature review emphasizes Intel's dedication to ESG through initiatives such as the "RISE" framework and the issuance of Green Bonds. While research on the financial impact of ESG practices varies, Intel's strong ESG score highlights its commitment to sustainable growth. Furthermore, there is a positive correlation between financial performance and corporate reputation, aligning with Intel's efforts to cultivate a strong brand image. In summary, the research illuminates Intel's historical journey, its ESG endeavors, and the interplay between ESG,

Table 5 Reliability tests (ANOVA: two-factor without replication)

Firm's ESG						
ANOVA						
Source of variation	*SS*	*Df*	*MS*	*F*	*P-value*	*F crit*
Rows	282.25	49.00	5.76	24.75	0.00	1.44
Columns	2.54	3.00	0.85	3.63	0.01	2.67
Error	34.21	147.00	0.23			
Total	319.00	199.00				
Firm's reputation						
ANOVA						
Source of variation	*SS*	*df*	*MS*	*F*	*P-value*	*F crit*
Rows	274.81	49.00	5.61	34.39	0.00	1.44
Columns	0.77	3.00	0.26	1.58	0.20	2.67
Error	23.98	147.00	0.16			
Total	299.56	199.00				
Firm's financial performance						
ANOVA						
Source of variation	*SS*	*df*	*MS*	*F*	*P-value*	*F crit*
Rows	284.50	49.00	5.81	27.91	0.00	1.44
Columns	2.42	3.00	0.81	3.88	0.01	2.67
Error	30.58	147.00	0.21			
Total	317.50	199.00				

Source Authors

Table 6 SWOT analysis of intel corporation

Strengths	Opportunities
• Significant investment in R&D	• Data centre growth
• Global brand recognition	• Investments in AI startups
• Diverse product portfolios	• Introduction of 5G networks
• Enhanced manufacturing capabilities	• Acquisitions and partnerships
• Economies of scale	• Sustainability of the environment
Weaknesses	Threats
• Manufacturing delays	• Supply chain disruptions
• Aggressive competition	• Technological shifts
• Declining PC market	• Regulatory challenges
• Dependency on few key customers	• Cybersecurity risks
• Security vulnerabilities	
• Limited scope of business	

Source Marketline database

company reputation, and financial performance. Statistical analysis showed strong connections between workplace social responsibility, media exposure, and ESG practices with financial performance perceptions. The firm's ESG significantly affects firm's financial performance, highlighting ESG's importance in financial outcomes. Similarly, the firm's reputation positively influences firm's financial performance, emphasizing the role of reputation in financial success. All variables (ESG, Reputation, and Financial Performance) demonstrated strong internal consistency in reliability tests.

The research assumes a causal relationship between ESG, Corporate Reputation and financial performance, however that causation is very complex, and other unmeasured variables could be influencing the observed correlations. Finally, it's essential to acknowledge that it is difficult to measure ESG score due to its complexity. The agencies have different metrics and benchmarks to assess ESG scores, therefore the data might be not reliable. Drawing from the research findings and Intel's dedication to ESG practices, corporate reputation, and financial performance, the following recommendations are proposed:

Tailor Strategies to Geographic Context: Intel should adopt a strategy of tailoring its approach to the specific geographic context in which it operates. Recognizing that ESG practices can vary significantly across regions due to diverse regulatory frameworks, cultural perspectives, and societal expectations, Intel should prioritize localization in its sustainability efforts. This approach entails aligning its ESG practices with the unique needs and priorities of each market it serves. By doing so, Intel can navigate the intricacies of differing regional landscapes, staying compliant with local regulations and customizing its sustainability initiatives to resonate with the values and preferences of local stakeholders. This tailored approach not only ensures compliance but also fosters stronger engagement with local communities, enhancing the company's reputation and impact on a global scale.

Perpetuate Research Efforts: It is important to continue investing in research and development, outperform the competitors in terms of performance, energy efficiency, and special features. These innovations should be effectively communicated to consumers through targeted marketing campaigns that emphasize the value proposition. In parallel, it is essential to study the interplay between ESG practices and financial performance across diverse markets to obtain valuable insights for refining strategies.

Strategic Alliances: The company should form more strategic partnerships to jointly develop and promote products or technologies. This can help Intel to gain a competitive advantage and reduce costs. Additionally, partnerships can help to gain access to new markets, resources, and expertise.

Expand Community Engagement: Extend and enrich community engagement initiatives. Demonstrating dedication to the communities in which Intel operates can further enhance the company's reputation and cultivate goodwill.

Appendix

References

1. Amel-Zadeh, A., Serafeim, G.: Why and how investors use ESG information: evidence from a global survey. Financ. Anal. J. **74**(3), 87–103 (2018). https://doi.org/10.2469/faj.v74.n3.2
2. Chouaibi, S., Chouaibi, J., Rossi, M.: ESG and corporate financial performance: the mediating role of green innovation: UK common law versus Germany civil law. Euro. Med J. Bus. **17**(1), 46–71 (2022). https://doi.org/10.1108/EMJB-09-2020-0101
3. Cocis, A.-D., Batrancea, L., Tulai, H.: The link between corporate reputation and financial performance and equilibrium within the airline industry. Mathematics **9**(17), 2150 (2021). https://doi.org/10.3390/math9172150
4. Contu, D., Elshareif, E.E., Gurrib, I.: Assessing views towards energy sources with social media data: the case of nuclear energy in the UAE. Processes **9**, 1945 (2021). https://doi.org/10.3390/pr9111945
5. Daugaard, D.: Emerging new themes in environmental, social and governance investing: a systematic literature review. Account. Financ. **60**(2), 1501–1530 (2020). https://doi.org/10.1111/acfi.12479
6. Duque-Grisales, E., Aguilera-Caracuel, J.: Environmental, social and governance (ESG) scores and financial performance of multinationals: Moderating effects of geographic international diversification and financial slack. J. Bus. Ethics 1–20 (2019). https://doi.org/10.1007/s10551-019-04177-w
7. Factset.: ESG Peer Comparison. Factset database (2023). [Retrieved: 8th October 2023]
8. Gurrib, I., Kamalov, F., Starkova, O., Makki, A., Mirchandani, A., Gupta, N.: Performance of equity investments in sustainable environmental markets. Sustainability **15**(9), 7453 (2023). https://doi.org/10.3390/su15097453
9. Gurrib, I.: Momentum in low carbon and fossil fuel free equity investing. Int. J. Energy Econ. Policy **13**(5), 461–47 (2023). https://doi.org/10.32479/ijeep.14342
10. Gurrib, I., Kamalov, F., Elshareif, E.: Can the leading US energy stock prices be predicted using the Ichimoku cloud? Int. J. Energy Econ. Policy **11**(1), 41–51 (2021). https://doi.org/10.32479/ijeep.10260
11. Gurrib, I., Nourani, M., Bhaskaran, R.K.: Energy crypto currencies and leading U.S. energy stock prices: are Fibonacci retracements profiTable? Financ. Innov. **8**, 8 (2022). https://doi.org/10.1186/s40854-021-00311-8
12. Joseph, A.: An exploratory study on metaverse and SDGs. In: El Khoury, R., Alareeni, B. (eds.) How the Metaverse Will Reshape Business and Sustainability. Contributions to Environmental Sciences & Innovative Business Technology. Springer, Singapore (2023). https://doi.org/10.1007/978-981-99-5126-0_9
13. Naved, M., Gupta, A.K.: Metaverse as a tool for the achievement of SGDs: challenges, opportunities, and applications. In: El Khoury, R., Alareeni, B. (eds.) How the Metaverse Will Reshape Business and Sustainability. Contributions to Environmental Sciences & Innovative Business Technology. Springer, Singapore (2023). https://doi.org/10.1007/978-981-99-5126-0_11
14. Thirumagal, P.G., Tirkey, A., Suresh, S.: Moderating effect of board size and board independence between women on board and ESG. In: Aloysius Edward, J., Jaheer Mukthar, K.P., Asis, E.R., Sivasubramanian, K. (eds.) Current Trends in Economics, Business and Sustainability. ICEBS 2023. Contributions to Environmental Sciences & Innovative Business Technology. Springer, Singapore (2023). https://doi.org/10.1007/978-981-99-3366-2_3

Investors' Intent to Invest in Stock Market: An Exploratory Post-COVID Study for Oman

Dharmendra Singh, Bashir Fida, S. Anand, and Tamanna Dalwai

Abstract The COVID-19 pandemic has drastically changed the landscape of investment. This study investigates the investors' intent to invest in the Oman stock market in general and the COVID-19 pandemic. The study used semi-structured interviews to gather opinions from investors and experts. A set of 20 questions was formulated, including an open-ended question regarding investment intent and behaviour. The interview questions cover the respondents' demographics, factors affecting investment decisions, investment challenges, COVID-19 lessons learned, and digital and technology asset investment intentions. The interviews were recorded using MS Team, and transcripts were created and analysed by NVivo 14. The results revealed that investors have a positive outlook on the Oman stock market and plan to invest in technology assets in the future. The study found that ESG considerations, market regulations, and economic conditions significantly influence investor behaviour. COVID-19 has altered investment strategies, with investors now prioritizing healthcare and technology stocks, adjusting assets, and learning to tolerate risk.

Keywords Investment · Intent · Qualitative study · Stock market · Oman

D. Singh · B. Fida (✉)
Modern College of Business and Science, Muscat, Oman
e-mail: bashir@mcbs.edu.om

S. Anand
College of Banking and Financial Studies, Muscat, Oman

T. Dalwai
Muscat College, Muscat, Oman

B. Awwad (ed.), *The AI Revolution: Driving Business Innovation and Research*, Studies in Systems, Decision and Control 524, https://doi.org/10.1007/978-3-031-54379-1_32

1 Introduction

Investment plays a crucial role in the development of a country's economy. Understanding investment intention is critical in today's volatile financial environment as it sheds light on how individuals approach asset management and financial decision-making. With the financial world becoming more complex and diverse, investment intention plays an increasingly crucial role in financial planning and investment decisions. Investment intention refers to an individual's willingness and readiness to allocate financial resources toward various investment routes to achieve specific financial results [2, 14]. This complex phenomenon is influenced by financial literacy, risk perception, economic conditions, and personal circumstances [10]. Therefore, investment intention is crucial in personal finance and financial decision-making [6]. Comprehending these intents holds significant importance for individual investors striving to enhance their portfolios and financial institutions, policymakers, and researchers endeavouring to grasp and predict market trends.

Over time, there has been an increase in personal investments in financial markets, particularly in the stock market. The trend could be due to increased financial market awareness, asset availability, social trends, and a desire for return. The stock market is a lucrative place for investors due to its profitability and diversification benefits (Celerier et al. 2016).

The examination of investor intentions is a complex undertaking, involving a wide range of elements that jointly influence investment choices. Traditional economic theories commonly depict investors as rational actors who make decisions primarily based on projected returns and risk evaluations. However, the actual dynamics of investor behavior are considerably more complex and multifaceted [9]. A complex interplay of emotions, perceptions, biases, societal influences, and broader macroeconomic pressures shapes the complicated mosaic of investor intentions. Investor intentions are primarily focused on the study of human behavior in the financial domain, where cognitive biases, emotional reactions, and social interactions influence investment decisions. Behavioral finance, an emerging discipline within the realm of economics, acknowledges the significant impact of psychological elements on financial decision-making, questioning the conventional assumptions of rationality inherent in classical economic frameworks [11].

The motivations underlying investor intentions exhibit a wide range of diversity, reflecting the heterogeneous nature of market participants. Various factors influence an investor's intentions, including financial goals, risk tolerance, time perspectives, and income levels. Investors frequently want to attain a range of objectives that extend beyond purely financial benefits. These objectives may include funding retirement, safeguarding wealth, promoting humanitarian endeavors, or aligning investments with personal beliefs and ethical considerations. Furthermore, it is imperative to consider the social and psychological aspects when making investment decisions. Social influences, such as peer pressure and herd mentality, frequently influence investor moods, resulting in market fluctuations that deviate from conventional economic forecasts. Behavioral biases, including herd behavior, overconfidence, loss aversion, and anchoring, substantially influence investor intentions, resulting in irrational market behaviors and the mispricing of assets [8].

Apart from personal-level considerations, broader macroeconomic pressures significantly impact investment intentions. Various factors, such as global economic trends, interest rates, geopolitical events, and government regulations, have a significant impact on financial markets, influencing the way investors perceive and assess risk and potential returns. The intricate nature of investor intentions requires the adoption of a qualitative methodology to comprehensively examine the various factors, including motivations, perceptions, and considerations, that influence investment choices [7]. Qualitative research methodologies, such as in-depth interviews and thematic analysis, provide a comprehensive approach to examining the complexities of investor behavior. Qualitative research reveals subjective nuances that quantitative analyses sometimes fail to capture by directly engaging with investors, recording their narratives, and investigating the underlying rationales behind their decisions.

This qualitative study explores the complex landscape of investor intentions, examining the various elements that influence investment choices. This study sheds light on the intricate motivations, perspectives, and considerations that drive investors in their quest for financial goals. It utilizes comprehensive interviews with various investors and employs a meticulous thematic analysis. This research endeavors to enhance comprehension of investor behavior in financial markets by elucidating the intricacies involved.

The complex ramifications of the COVID-19 pandemic on investor intentions highlight the necessity of thoroughly comprehending how worldwide crises modify investment environments and impact decision-making processes within financial markets. The research questions are constructed from the essence of the Planned Behaviour Theory proposed by Ajzen and Fishbein [1]. The constructs of PBT are translated into well-structured questions and below research questions are formulated.

Research questions

The study aims to answer these research questions.

1. What factors influence investors 'intent to invest in the Oman stock market, and identify hurdles to do so?
2. Investigate the various factors that affect investment decisions and the process of creating a portfolio.
3. How did investors respond to the COVID-19 pandemic, and what lessons were learned?
4. What other types of investments are investors considering for the future, besides investing in the stock market?

2 Literature Review

The COVID-19 pandemic caused significant disruptions in global financial markets, substantially influencing investment intentions [13]. The advent of the global pandemic brought about unprecedented uncertainty, leading investors to reevaluate their perceptions of risk, investment strategies, and long-term financial objectives. According to existing research, it has been found that the occurrence of the pandemic has led to an increased inclination toward risk aversion among investors [5]. The occurrence of heightened volatility led to extensive market changes and triggered a phenomenon known as flight to safety, as investors sought shelter in historically stable assets such as government bonds and gold. The economic consequences of lockdowns, supply chain disruptions, and business closures influenced the revaluation of company performance and profitability expectations. This, in turn, impacted investor mood and the decisions made about capital allocation [4].

Moreover, the COVID-19 pandemic has expedited pre-existing patterns in digital transformation and innovation, leading investors to gravitate towards sectors such as technology and healthcare that possess resilience during unpredictable circumstances [3]. The COVID-19 pandemic has led to significant societal changes, including a shift towards remote labor and increased reliance on technology. As a result, there has been a notable increase in investments in digital infrastructure and technology-driven enterprises. This trend reflects the shifting tastes of investors in response to the pandemic-induced circumstances. During these societal changes, the COVID-19 epidemic has brought the significance of sustainable investing to the forefront, garnering heightened focus on environmental, social, and governance (ESG) considerations [12]. Investors have started integrating environmental, social, and governance (ESG) factors into their decision-making processes, acknowledging the importance of resilient and socially responsible investments in effectively managing unexpected crises.

3 Methodology

The present study employs a rigorous methodology to investigate the research question. The primary objective of this qualitative study is to investigate the complex aspects of investor intentions by conducting in-depth interviews with a broad group of individuals. The qualitative methodology provides a comprehensive comprehension of the intricate elements that contribute to investment choices, enabling the examination of subjective interpretations, incentives, and contextual impacts. A snowball sampling technique has been utilized to ensure a diverse representation of investor profiles. This ensures individuals with varied levels of investment experience, a wide array of financial objectives, and varying degrees of risk tolerance. The selection process has ensured participation from diverse demographic backgrounds, including

Table 1 Details of respondents

S. No.	Respondents	Number of respondents
1	Individual investors with experience more than 10 years	5
2	Individual investors with experience of less than 5 years	2
3	Individual investors with experience of less than 1 year	1
3	Representatives from the institutional investment sector	2
4	Representative from academia	2
Total		12

individuals of various ages, genders, and socioeconomic status. This approach aims to gather a comprehensive and diverse dataset.

Semi-structured interviews have been used to collect the responses from the investors in Oman. The online interviews have been performed in a one-on-one format, facilitating a more comprehensive examination of individual viewpoints, incentives, and cognitive processes. The interview guide is shared in the annexure, which captures financial objectives, perceptions of risk, sociocultural influences, ethical deliberations, and reactions to macroeconomic determinants. The average duration of each interview was approximately 25–35 min, which allowed participants sufficient opportunity to express their opinions and share their experiences pertaining to investing decisions.

Thematic analysis has been employed for data analysis using Nvivo software, which enables the identification of repeating patterns, themes, and insights throughout the interviews. The transcripts were edited and cleaned to correct the spelling and delete words like 'OK" and "hmm." Scripts were skimmed, and the coding was created based on the respondents' answers and thoughts. The interviews were conducted using a snowball sampling method. Representative individual investors, institutional investors, and experts were selected to conduct the interviews. Respondents sought formal consent, and the questionnaire was sent to them in advance enough to read and prepare their answers. The details of the respondents are given in Table 1.

4 Results Analysis

This section presents research findings from audio recordings transcriptions and analysis using MS Teams. The demographic characteristics of respondents are provided in Table 2. A total of 12 interviews were conducted, and nine interview scripts were coded and analyzed based on four research questions after the screening. The respondents belong to eight nationalities.

Table 2 Demographic characteristics of respondents

S. No.	Respondents	Age	Profession	Gender	Nationality	Qualification
1	Respondent 1	43	Portfolio manager	Male	Omani	ACCA
2	Respondent 2	35	Finance manager	Male	Pakistani	ACCA and CMA
3	Respondent 3	37	Banker	Female	Omani	Postgraduate
4	Respondent 4	48	Professor	Male	American	Doctorate
5	Respondent 5	46	Investor	Male	Norway	Doctorate
6	Respondent 6	41	Investor	Male	Indian	Postgraduate
7	Respondent 7	51	Investor and Professor	Male	Denmark	Doctorate
8	Respondent 8	44	Investor	Male	Canadian	Doctorate
9	Respondent 9	45	Academician	Male	Indian	Postgraduate
10	Respondent 10	55	Investor	Female	Omani	Graduate
11	Respondent 11	39	Banker	Female	England	Postgraduate
12	Respondent 12	44	Investor	Male	Indian	Graduate
Total					12	

Research question: Motivating Factors to invest and hurdles for the stock market

The word cloud derived from interview transcripts for motivating factors to invest in the Oman stock market is below. According to the word cloud, significant factors that affect the intent are geopolitical conditions, stable economic situation, law and order, the advantage of a small economy, the opportunity to earn a premium return, the strong presence of the public sector, the oil sector, and strong currency. The respondents were asked to mention three to five main hurdles for investment. They identified the low trade volume, less information, lack of awareness, number of IPOs, and limited investment platforms as the main obstacles for stock investment in Oman (Figs. 1 and 2).

According to respondent 1, an Omani portfolio manager, investment in stock market depend on economic conditions, government policies, geopolitical events and global trends of investment. He described these factors as below.

> Let me put four points at least. So first are the economic conditions, second would be the government policies, third would be the geopolitical events and the 4th one I would give to global trends.

Respondent two, a finance manager of a private healthcare service provider with an Islamic background, highlighted the below factors:

> Comparatively, the stock market is a halal investment option per the Sharia guidelines, so it is a good option in particular; the Oman stock market is a sound currency exchange value.

Fig. 1 Word cloud for RQ1

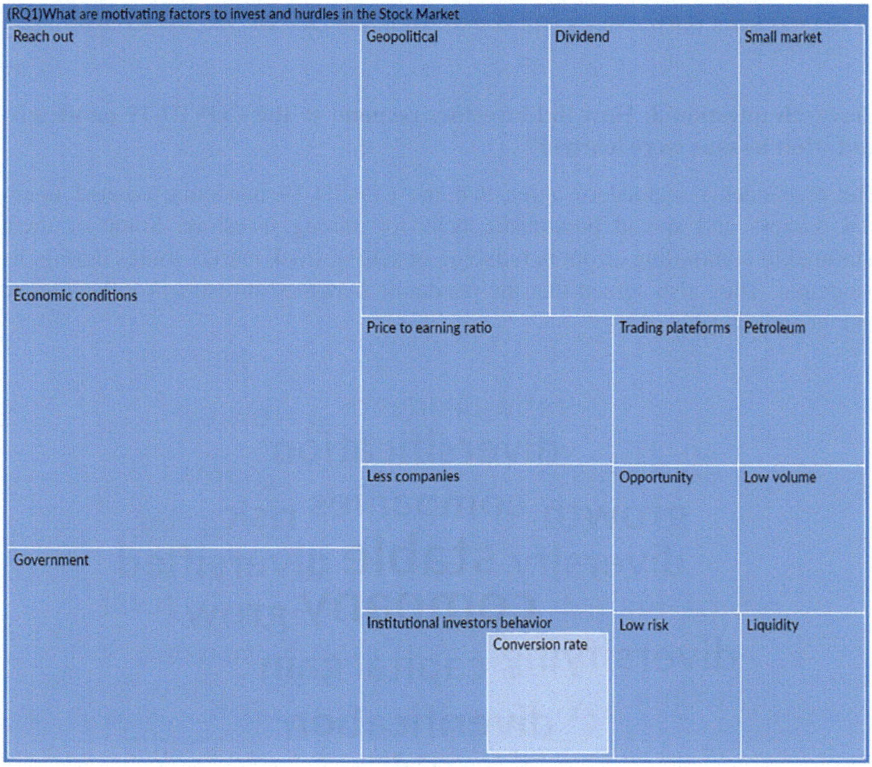

Fig. 2 Hierarchy chart for RQ1

Economic conditions, dividends, government policies, and geopolitics are the top factors that influence investors' investment intent in the Oman stock market. A smaller number of public companies, awareness, and limited trade volume remain among the highest quoted hurdles for investment.

Research question: Factors that affect investment decisions and portfolio creation

Figures 3 and 4 report the word cloud and hierarchy chart for the second research question, respectively. Investors create portfolios and make investment decisions based on various factors. Some of the most quoted factors are dividend, return, diversification, stable company, growing stock, capital gain, liquidity, personal and financial need, advice of financial brokers and advisors, and influence of friends and family members. In addition to that, respondent 7, a professor at a public sector university and an individual investor, highlighted the importance of mobile applications that enabled him to invest efficiently.

> I think the biggest reason for me investing in the stock market was that it is very easy nowadays, you know, with the help of these smart apps, you can very quickly make the investment.

Research question 3: How did investors respond to the COVID-19 pandemic, and what lessons were learned?

The respondents seemed to agree that the COVID-19 pandemic created financial distress and spread pessimistic behavior among investors. Some of them admitted to committing errors in reducing or selling stock market shares during the pandemic. They also agreed that the pandemic benefited investment in healthcare and technological assets.

Fig. 3 Word cloud for RQ2

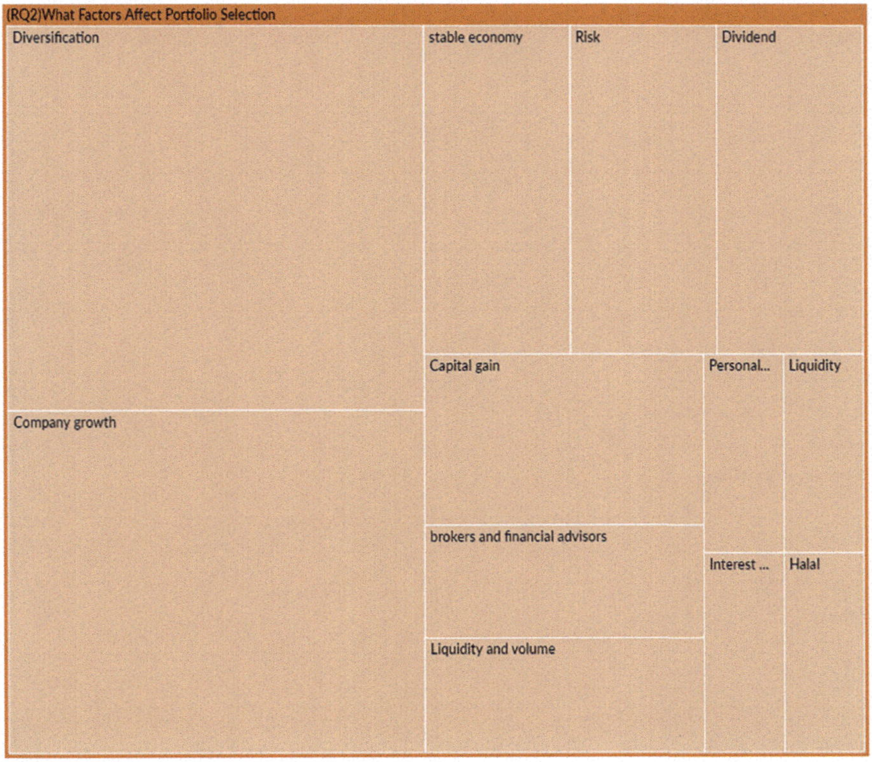

Fig. 4 Hierarchy chart for RQ2

Respondent 9: A banker and investor stated as "Well, to tell the truth, I made a mistake by exiting from the portfolio very early. I bought. I knew that these were cheap stocks".

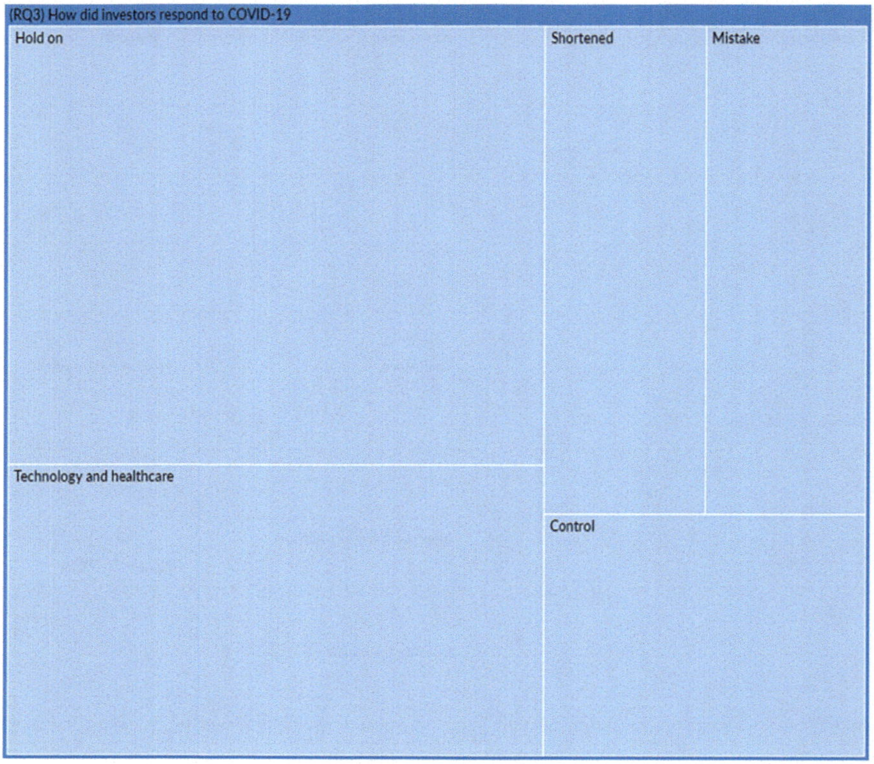

Lessons learned.

> One is that people have learned or may have. I mean, rather, they would have moved toward investing through a digital platform. Now, digital transformation has been accepted by many organizations.

He kept on saying that.

> The lesson that we have seen is especially the things that are beyond your control, managing risk to some extent is possible beyond that point. You don't have to stress too much to yourself

Research question no. 4: What other types of investments are investors considering for the future besides investing in the stock market?

The survey respondents were asked about their willingness to invest in technological assets and other forms of investments, including cryptocurrencies and crowdfunding. Furthermore, their opinion was sought regarding future investment in stocks of companies that take ESG considerations into account. Eight out of nine advocated for ESG consideration while making future investments. Figure 5 shows the word cloud for potential investments in the future. Finally, Fig. 6 reports the word cloud for all research questions.

The respondent 6, highlighted the importance of ESG as".

Fig. 5 Word cloud for future investment

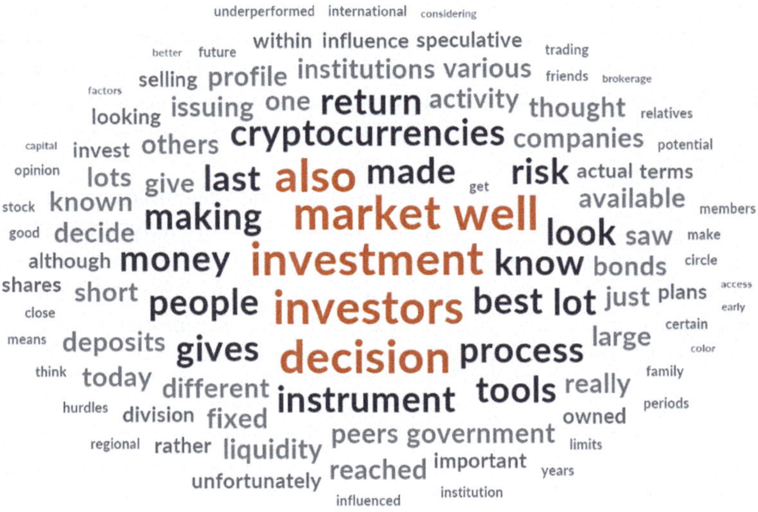

Fig. 6 Word cloud for all research questions

> I think ESG would be much more important. People are realizing that we need to be more environmentally friendly. So, this green investing is becoming very popular, and this is how it should be because even if we make money and you know the companies are polluting the environment we all will die. I think we should go for green investing

The respondent 5, an investor and academician mentioned in these words.

> Personally, I like to invest in the companies which are having DCS ESG considerations and of course now these information are also available in two these digital apps where they provide information about how green is a company if they are following ESG considerations or not.

5 Conclusion

The study explores investors' intent to invest in the Oman stock market, focusing on the impact of the COVID-19 pandemic. A well-structured set of questions was constructed. Twelve in-depth interviews were conducted. Codes were constructed from the transcripts and analyzed using NViVo14 software. The word cloud and hierarchy charts were created for each research question. The results showed that investors mentioned economic conditions, stability, strong currency, and government policies as significant factors to investment in the Oman stock market. The limited public stock, low trade volume, and lack of awareness are among the hurdles to investing in the Oman stock market. Company growth, dividends, and capital gains are factors to consider while creating an investment portfolio. COVID-19 impacted investment in how investors withdrew or reduced their investments during the pandemic. However, they mentioned that investors should not worry about such incidents and hold their investments. The investors also opined to invest in technology assets, and ESG-supported stocks in the future.

This study has implications for future investors and policymakers. The government policies may continue to support an investment-friendly environment by providing information and paving ways for the expatriate populations to invest in future IPOS. Moreover, future investment is inclined towards technological and crowdfunding avenues; policymakers may capitalize on the sentiments and formulate policies to ease the investment paths for potential investors.

References

1. Ajzen, I.: From intentions to actions: a theory of planned behaviour. In: Kuhl, J., Beckman, J., (eds.) Action-Control: From Cognition to Behavior, pp. 11–39. Heidelberg, Springer (1985)
2. Akhtar, F., Das, N.: Predictors of investment intention in Indian stock markets: extending the theory of planned behaviour. Int. J. Bank Market. **37**(1), 97–119 (2019). https://doi.org/10.1108/IJBM-08-2017-0167
3. Katsiampa, P., Yarovaya, L., Zięba, D.: High-frequency connectedness between Bitcoin and other top-traded crypto assets during the COVID-19 crisis. J. Int. Finan. Markets. Inst. MoneyFinan. Markets. Inst. Money **79**, 101578 (2022). https://doi.org/10.1016/j.intfin.2022.101578
4. Kiruba, A., Vasantha, V.: Determinants in investment behaviour during the COVID-19 pandemic. The Indonesian Capital Market Rev. **13**(2) (2021). https://doi.org/10.21002/icmr.v13i2.13351
5. Kumari, J.S., Senani, K.G.P., Ajward, R.: Predicting investors' intention to invest in the stock market during COVID-19: can we use an extended theory of planned behavior? J. Asia Bus. Stud. **17**(4), 681–700 (2022). https://doi.org/10.1108/JABS-12-2021-0517
6. Oppong C, Salifu Atchulo A, Akwaa-Sekyi EK, Grant DD, Kpegba SA (2023) Financial literacy, investment and personal financial management nexus: empirical evidence on private sector employees. Cogent Bus. Manage. **10**, 2. https://doi.org/10.1080/23311975.2023.2229106

7. Palamida, E., Papagiannidis, S., Xanthopoulou, D.: Linking young individuals' capital to investment intentions: comparing two cultural backgrounds. Eur. Manag. J.Manag. J. **36**(3), 392–407 (2018). https://doi.org/10.1016/j.emj.2017.06.004
8. Rathee, P., Aggarwal, S.: Understanding impact investment intention using the extended theory of planned behaviour. Global Bus. Rev. (2022). https://doi.org/10.1177/09721509221115001
9. Saliya, C.A.: Driving forces of individual investors in stock market participation. Saliya, CA (2021b). Driving Forces of Individual Investors in Stock Market Participation, Review of Economics and Finance **19**, 73–79 (2021)
10. Sivaramakrishnan, S., Srivastava, M., Rastogi, A.: Attitudinal factors, financial literacy, and stock market participation. Int. J. Bank Market. **35** (2017). https://doi.org/10.1108/IJBM-01-2016-0012
11. Thanki, H., Shah, S., Rathod, H.S., Oza, A.D., Burduhos-Nergis, D.D.: I am ready to invest in socially responsible investments (SRI) options only if the returns are not compromised: Individual investors' intentions toward SRI. Sustainability **14**(18), 11377 (2022)
12. Vicente-Ortega Martínez, C.: ESG Investments and their Evolution during the COVID-19 Pandemic (2021). https://repositorio.comillas.edu/xmlui/handle/11531/51851
13. Xu, X., Xie, Y., Xiong, F., Li, Y.: The impact of COVID-19 on investors' investment intention of sustainability-related investment: evidence from China. Sustainability, **14**(9), Article 9 (2022). https://doi.org/10.3390/su14095325
14. Yadav, R., Pathak, GS.: Determinants of consumers' green purchase behavior in a developing nation: applying and extending the theory of planned behavior. Ecolog. Econ. **134**(C), 114–122 (2017)

A Literature Review on Creative Accounting in Governments from Various Countries

Dedi Rusdi and Edy Suprianto

Abstract The aim of this research is to collect evidence of creative accounting in local governments from various countries. This research is interesting to carry out because this research is the first to examine earnings management carried out by governments. Second, this research can add to scientific studies, especially financial accounting, through empirical evidence about earnings management carried out in governments around the world. This research uses an interpretive approach by systematically reviewing several articles discussing earnings management in the public sector. It is concluded that term numbers manipulation in financial statements in the government sector has different terms. Some governments use creative accounting, surplus-deficit management, and earnings management. Theories that are widely used are public choice theory and agency theory. Meanwhile, the factors that influence the government to carry out creative accounting are political, regional government characteristics, and financial performance.

Keywords Creative accounting · Literature review · Local government

1 Background

Several researchers, such as [1, 2], stated that accrual accounting is considered to increase the transparency and accountability of government financial statements. However, several studies show evidence that the government also carries out earnings management, for example, [3–5] about the implementation of accrual accounting in government entities in Indonesia since 2010. However, several regional governments have obtained Unqualified Opinions from the Financial Audit Board of Indonesia. The Financial Audit Board also found several problems related to system weaknesses, non-compliance, inefficiency, extravagance, ineffectiveness, and budget markups which impacted state losses.

D. Rusdi · E. Suprianto (✉)
Universitas Islam Sultan Agung, Semarang, Indonesia
e-mail: edysuprianto@unissula.ac.id

B. Awwad (ed.), *The AI Revolution: Driving Business Innovation and Research*, Studies in Systems, Decision and Control 524, https://doi.org/10.1007/978-3-031-54379-1_33

The implementation of accrual accounting in the government sector has opened up opportunities for government accountants to carry out creative accounting. Some evidence of accrual accounting in the government sector includes budget inflation or markups. Budget inflation is a mode of corruption in the government. Milesi-Ferretti [5] found evidence that European government accountants use accrual accounting to hide budget fiscal policy deviations. Buti et al. [6] stated that local governments cover government debt by using accrual accounting. Alt et al. [7] and De Castro et al. [8] also stated that there had been creative accounting practices in local governments. The evidence illustrates that the government uses accounting methods to manipulate financial statements like the approach used by [9].

This research is interesting to review for several reasons; (1) there are still some arguments regarding whether earnings management or the term of creative accounting in the government sector is still relevant. Some parties who oppose this term argue that the government is not profit-oriented, so it is impossible to manipulate profits. However, those who support this opinion state that there are loopholes or attempts by the government to manipulate the numbers in financial statements to cover up the government's negative motivations. (2) There is still little research that discusses the behavior of government accountants in manipulating numbers in financial statements. There has been no previous research in Indonesia that proves this statement.

2 Research Methods

This research method used an interpretive approach by systematically reviewing the literature on several articles discussing earnings management in the public sector, earnings management when changing company managers, and the theory and practice of manipulation in government accounting. The selected articles were published in accredited journals indexed by Sinta 2 and international journals indexed by Scopus (Table 1). The research stages that will be carried out this year are as follows: first, looking for sources for literature review or library study materials from various reference sources. Second, evaluate the sources' contents to ensure the data obtained is valid and reliable. Third, create a summary of the contents of these sources in the form of an Excel table to map research data. Fourth, conduct a literature review on earnings management in governments from various countries by looking at its weaknesses and strengths. Fifth, publish the results of this research in an international journal indexed by Scopus.

Table 1 Articles sample

No	Journal name	Group
1	ABACUS	1
2	Accounting and Business Research	1
3	Advances in Accounting	1
4	American Review of Public Administration	1
5	Applied Economics Letters	2
6	Brazilian Administration Review	1
7	Clémenceau and Soguel Swiss Journal of Economics and Statistics	1
8	Defence and Peace Economics	1
9	Finance Research Letters	1
10	Financial Accountability and Management	4
11	German Economic Review	1
12	Int. J. Accounting, Auditing and Performance Evaluation	1
13	International Journal of Professional Business Review	1
14	International Political Science Review	1
15	Journal of Accounting and Economics	1
16	Journal of Accounting and Public Policy	2
17	Journal of Applied Accounting Research	1
18	Journal of Banking and Finance	1
19	Journal of Comparative Economics	1
20	Journal of Governmental and Nonprofit Accounting	1
21	Journal of Management and Governance	2
22	Journal of Public Budgeting, Accounting and Financial Management	3
23	Public Management Review	1
24	Public Money and Management	5
25	Review of Accounting Studies	1
26	Review of Economics	1
27	The British Accounting Review	1
28	The International Journal of Accounting	1

3 Accrual Accounting and Creative Accounting

The implementation of accrual accounting in government is considered to increase the transparency and accountability of government Braam financial reports [1, 10]. However, several previous studies have shown evidence of the negative side of implementing accrual accounting. Milesi-Ferretti [3, 11, 12] concluded that local governments tend to carry out manipulation practices accounting numbers for a specific purpose.

The term manipulation of numbers in financial reports in the private sector is known as earnings management. However, governments that are characterized by not seeking profit have different terms. Hagen and Wolff [13] and Clémenceau and Soguel used the term creative accounting, [14] used the term surplus-deficit management, while Schettini and Terra used the term financial statement manipulation. However, some researchers still use earnings management, for example, Ibrahim et al. [15], and Marco Bisogno and Donatello [16].

Bisogno and Donatello [16] explained that creative accounting in the public sector has different terms. Hagen et al. [13] and Clémenceau and Soguel used the term creative accounting, [14] used the term surplus-deficit management, while Schettini and Terra used the term financial statement manipulation. However, some researchers still use earnings management, for example, Ibrahim et al. [15] and Bisogno and Donatello [16]. According to Noronha et al. creative accounting is a continuum of deliberate intervention in the external financial reporting process from something appropriate to a violation of GAAP rules, with the intention of misleading several stakeholders regarding the company's economic conditions and performance. Sevin et al. defined creative accounting as an attempt by company management to influence reported income in the short term. Meanwhile, Gakhar stated that earnings management is a term in the field of accounting practice that may follow the applicable accounting standard rules but deviate from the objectives of these rules. It is different from Ronen and Yaari, who define creative accounting into three classifications, namely white, gray, and black, depending on the goals to be achieved by the manager. It can be concluded that creative accounting is a deliberate action taken by company managers to change the values in the financial statements for certain purposes.

According to Clémenceau and Soguel, creative accounting is an opportunistic action carried out by preparers of financial reports in the public sector with specific objectives. Although this creative accounting practice is considered less critical than in the private sector, it is possible that local governments experiencing significant budget surpluses may be interpreted as a sign of excessive tax collection, while local governments experiencing budget deficits may be considered an indication of poor financial management which can invite criticism from political opponents [3, 12]. Conditions like this allow local governments to carry out creative accounting if the requirements need financial statements not to have a budget surplus or deficit.

Based on the literature review, research topics on creative accounting in the public sector were developed based on methods or approaches used in private sector organizations. For example, [3] used the method of Jones, and Salma et al. [15] used the method of Dechow et al. Meanwhile, [17] used Dechow and Dichev method. They tried to develop existing methods by adjusting the characteristics of the public sector organizations that became the research sample.

4 Factors Causing Creative Accounting in Government

Several factors influence local governments to carry out creative accounting behavior. One of them is the political factor. Previous researchers, for example, [3], found that creative accounting practices in regional governments replaced regional heads. These results are supported by [18, 19]. Donatella [12] also stated that political competition is the cause of local governments carrying out creative accounting. Cohen et al. [3] stated that playing accrual accounting numbers is an option.

Political factors can also influence financial performance [20–22]. A regional government led by a second-term regional head will have better financial performance. On the other hand, a regional government led by a first-term regional head will have poor financial performance. This is because the incumbent regional head has wider experience and relationships than the new regional head. Cohen [22] proved that state finances are influenced by political competition in a country. These results are supported by Boubakri et al. [20] and Boubakri et al. [21].

Local government characteristics also influence creative accounting in public sector organizations [23, 24]. Regional government characteristics include regional size, welfare, functional differentiation, regional age, regional head educational background, regional leverage, and intergovernmental revenue. Suprianto [24] states that the characteristics of regional governments in Indonesia influence creative accounting.

Previous studies stated that financial performance triggers the government to carry out creative accounting. Research conducted by Zakariya stated that budget inflation or markup is a mode of corruption in government. Milesi-Ferretti (2004) also finds evidence that European government accountants used accrual accounting to hide fiscal policy deviations in the budget. Buti et al. [6] stated that local governments cover government debt by using accrual accounting. Alt et al. [7] and De Castro et al. [8] also stated that local governments have adopted creative accounting practices. Different results were shown by [15], who were unable to prove the effect of financial targets on public sector organizations (Hospitals) in Italy.

5 Underlying Theory

Research that adopts the perspective of agency theory in government agency conflicts occurs between society. Based on agency theory, the main characteristic of an agency relationship lies in the contract of delegation of authority and responsibility from the principal to the agent. Contracts can originate from custom, common interests to achieve common goals, and formal legal ties. In terms of formal rules, government entities are run by referring to a set of rules that specify each participant's duties, authority, and responsibilities. Although the workings and mechanisms of relations between participants in government organizations are different from those in the corporate sector, these formal ties indicate the existence of contracts within government organizations in Indonesia. This justifies that there is an agency relationship in government organizations in Indonesia.

Boyne attempts to explain how a municipality works to offer an understanding of the philosophy and motives behind its existence. He considered municipalities as industries consisting of "buyers" and "sellers." He also pointed out the importance of two aspects of public choice theory. The first concerns the "axiom of self-interest", in which politicians would pursue their own interests rather than the public; the second concerns "competitive pressures," which motivate politicians to act in the public interest rather than their own interests. Therefore, it is easy to understand that there are conflicting interests between politicians and stakeholders.

According to Zimmerman, this conflict of interest aligns with asymmetric information, a creating fertile ground for earnings management practices because both parties will try to satisfy their interests. Politicians will pursue re-election and voters will monitor how politicians act to assess whether social welfare goals are being met. Likewise, Guarini found that accounting information is a key factor in achieving politicians' interests and motives.

According to [25], the basic assumption is that politicians engage in opportunistic behavior to satisfy their own interests when pressure from stakeholders is intense. Excessive pressure may be on the election. Garcia-Sánchez et al. argued that elections motivate politicians to behave opportunistically to satisfy their own interests and get re-elected. They also found that proximity to elections can affect local governments' financial position. Finally, a Ferreira et al. [4] found a relationship between the stage of the pre-election period and earnings management when political competition is intense. As stated previously, accruals are used to measure managers' discretion in accounting, as they are susceptible to opportunistic behavior.

Meanwhile, according to public choice theory, Cohen et al. [3] stated that playing accrual accounting numbers is a choice. Local governments having significant budget surpluses may be interpreted by citizens and constituents as a sign of excessive taxation compared to the quality and quantity of services provided. High surpluses can also confuse the central government about the level of subsidies provided. However, on the other hand, a regional government budget deficit can indicate poor financial management, leading to criticism from political opponents, dissatisfaction with citizens and ultimately sanctions from the central government and supervisory authorities.

6 Conclusion

This research is the first research to explain creative accounting practices in Indonesia Based on the results of this research, it is concluded that term numbers manipulation in financial statements in the government sector has different terms. Some governments use creative accounting, surplus-deficit management, and earnings management.

Theories that are widely used are public choice theory and agency theory. Meanwhile, the factors that influence the government to carry out creative accounting are political, regional government characteristics, and financial performance. The results of this research can be used as a reference and reading material for further research. The limitation of this research is that there is still very little research that discusses earnings management in public sector organizations.

References

1. Harun, H.: Obstacles to Indonesian public sector accounting reforms. Bull. Indones. Econ. Stud.Indones. Econ. Stud. **43**(3), 365–375 (2007)
2. Harun, H., Van-Peursem, K., Eggleton, I.R.C.: Indonesian public sector accounting reforms: dialogic aspirations a step too far? Acc. Auditing Account. J. **28**(5), 706–738 (2015)
3. Cohen, S., Bisogno, M., Malkogianni, I.: Earnings management in local governments: the role of political factors. J. Appl. Acc. Res. **20**(3), 331–348 (2019)
4. Ferreira, A., Carvalho, J., Pinho, F.: Earnings management around zero: a motivation to local politician signaling competence. Public Manag. Rev.Manag. Rev. **15**(5), 657–686 (2013)
5. Milesi-Ferretti, G.: Good, bad or ugly? on the effects of fiscal rules with creative accounting. J. Public Econ. **88**, 377–394 (2003)
6. Buti, M., Martins, J.N., Turrini, A.: From deficits to debt and back: political incentives under numerical fiscal rules. CESifo Econ. Stud. **53**(1), 115–152 (2007)
7. Alt, J., Lassen, D.D., Wehner, J.: Moral hazard in an economic union: politics, economics and fiscal gimmickry in Europe. In: Political Science and Political Economy. Working Paper 05/ 2012. Department of Government, London School of Economics (2012)
8. De Castro, F., Pérez, J.J., Rodríguez-Vives, M.: Fiscal data revisions in Europe. J. Money, Credit, Bank **45**(6), 1187–1209 (2013)
9. Ananda, R.F., Rahmadhani, S.N., Rahman, F., Dewi, S.: A review on innovation audits. In: Rafiki, A., Dana, LP., Nasution, M.D.T.P. (eds) Open Innovation in Small Business. Contributions to Environmental Sciences & Innovative Business Technology. Springer, Singapore (2023). https://doi.org/10.1007/978-981-99-5142-0_12
10. Taha, M.H., Eldalash, S.A., Mohamed, H.Z.: The role of accounting disclosure of sustainable development activities using metaverse in the field of education and training. In: El Khoury, R., Alareeni, B. (eds) How the Metaverse Will Reshape Business and Sustainability. Contributions to Environmental Sciences & Innovative Business Technology. Springer, Singapore (2023). https://doi.org/10.1007/978-981-99-5126-0_7
11. Beck, A.W.: Opportunistic financial reporting around municipal bond issues. Rev. Acc. Stud. **23**(3), 785–826 (2018)
12. Donatella, P.: Is political competition a driver of financial performance adjustments? an examination of Swedish municipalities. Public Money Manage. **40**(2), 122–130 (2020). https://doi.org/10.1080/09540962.2019.1667684
13. von Hagen, J., Wolff, G.B.: What do deficits tell us about debt? empirical evidence on creative accounting with fiscal rules in the EU. J. Bank. Financ. **30**(12), 3259–3279 (2006)
14. Stalebrink, O.J.: An investigation of discretionary accruals and surplus-deficit management: evidence from Swedish municipalities. Financ. Acc. Manage. **23**(4), 441–458 (2007)
15. Ibrahim, S., Noikokyris, E., Fabiano, G., Favato, G.: Manipulation of profits in Italian publicly-funded healthcare trusts. Public Money Manage. **39**(6), 428–435 (2019). https://doi.org/10.1080/09540962.2019.1578539
16. Bisogno, M., Donatella, P.: Earnings management in public-sector organizations: a structured literature review. J. Public Budg. Account. Financ. Manag.Budg. Account. Financ. Manag. **34**(6), 1–25 (2021)

17. Anagnostopoulou, S.C., Stavropoulou, C.: Earnings management in public healthcare organizations: the case of the English NHS hospitals. Public Money Manage. **43**(2), 95–104 (2021). https://doi.org/10.1080/09540962.2020.1866854
18. Capalbo, F., Lupi, C., Smarra, M., Sorrentino, M.: Elections and earnings management: evidence from municipally-owned entities. J. Manage. Govern. **25**(3), 707–730 (2021). https://doi.org/10.1007/s10997-020-09523-z
19. Stalebrink, O.J., Donatella, P.: Public pension governance and opportunistic accounting choice: a politico-economic approach. Am. Rev. Public Admin. **51**(3), 227–245 (2021)
20. Boubakri, N., Cosset, J.C., Saffar, W.: The impact of political connections on firms' operating performance and financing decisions. J. Financ. Res. **35**(3), 397–423 (2012)
21. Boubakri, N., Guedhami, O., Mishra, D., Saffar, W.: Political connections and the cost of equity capital. J. Corp. Finan.Finan. **18**(3), 541–559 (2012). https://doi.org/10.1016/j.jcorpfin.2012.02.005
22. Cohen, S.: Cash versus accrual accounting measures in Greek municipalities: proxies or not for decision-making? Int. J. Account. Audit. Perform. Eval.Eval. **8**(3), 203–222 (2012)
23. Clemenceau, M., Soguel, N.: Does personal background influence a finance minister to cook the books? an investigation of creative accounting in swiss cantons. Appl. Econ. **49**(10), 941–953 (2017)
24. Suprianto, E.: Creative accounting in local governments to obtain funds for covid-19 management. Int. J. Prof. Bus. Rev. **8**(3), 1–17 (2023)
25. Pilcher, R.: Local governmental management of discretionary and specific accruals. Int. J. Account. Audit. Perform. Eval.Eval. **7**(1–2), 32–60 (2011)

Merger, Capital Adequacy, and Earning Management in Conventional or Sharia Banks

Edy Suprianto

Abstract The research purpose is to test whether banks in Indonesia choose mergers or perform earnings management to meet the minimum capital adequacy requirements. In addition, this study also distinguishes how the earnings management behavior in both conventional banks and Sharia banks. The research population was all Banks in Indonesia. The sample in this study was all banks (conventional and Sharia banks) listed on the Indonesian stock exchange from 2019 to 2022. We use purposive sampling method to collect samples. The data analysis technique used in the study was the STATA. The results showed a positive effect between earnings management and the capital adequacy ratio in conventional and Sharia banks. Sharia banks were less likely to perform earnings management than conventional banks. These results also indicated that banks prefer mergers over earnings management to meet the minimum capital adequacy requirements.

Keywords Merger · Bank · Capital adequacy · Earning management

1 Introduction

Research on earnings management in the banking industry has been conducted by many previous researchers [11, 14, 18, 21, 22]. Banks have several motivations for earnings management, including meeting minimum capital requirements [7]. Collins et al. [7] found that banks that do not comply with regulatory limits on capital must do earnings management to get around these regulations. Other studies on the same motivation have been conducted by [2, 3, 9]. Research on the motivational reasons for banks to conduct earnings management to meet the minimum capital adequacy limit in Indonesia will be very interesting to study due to the new requirements that have been issued by the Financial Services Authority regulation (POJK) Number 12/POJK.03/2020 concerning Consolidation of Commercial Banks.

E. Suprianto (✉)
Universitas Islam Sultan Agung, Semarang, Indonesia
e-mail: edysuprianto@unissula.ac.id

B. Awwad (ed.), *The AI Revolution: Driving Business Innovation and Research*, Studies in Systems, Decision and Control 524, https://doi.org/10.1007/978-3-031-54379-1_34

The Indonesian Financial Services Authority (OJK) regulation states that the minimum core capital for commercial banks is 1 trillion rupiahs in 2020, 2 trillion rupiahs in 2021, and 3 trillion rupiahs in 2022. Chairman of the OJK Board of Commissioners Mahendra Siregar states that OJK will consolidate if the banks do not meet the company's core capital. This case will potentially create mergers or acquisitions. Fulfilling core capital requirements aims to establish a more solid Indonesian banking architecture. This condition will be interesting to examine in order to show whether banks that do not meet the requirements choose to merge with other banks or perform earnings management. As explained by Healy and Wahlen and Ahmed et al. [2], banks will try to manipulate the capital adequacy ratio in accordance with the requirements by increasing or decreasing reported profits to mislead stakeholders.

Research in Indonesia that examined capital adequacy in the banking sector has been conducted by previous researchers [23] dan [16] stating that capital adequacy affects earnings management. Pambekti [17] also found evidence that the banking industry performs earnings management, but Sharia banks have lower earnings management than conventional banks. Research on earnings management practices in the Indonesian banking industry is still fascinating for several reasons; First, Indonesian Financial Services Authority (OJK) regulations regarding capital adequacy in the last three years will motivate banks to comply with these regulations. Second, this study will compare the earnings management practices of Sharia and conventional banks. Third, previous research has never linked capital adequacy with earnings management and banking mergers. This is an interesting study because seeing the phenomenon in Indonesia in the last four years, many banks have merged with other banks and other companies. Therefore, this study will examine whether banks in Indonesia choose mergers or perform earnings management to increase their capital adequacy.

2 Theory and Hypothesis

In Indonesia, banks are divided into rural, central, and commercial banks based on their function. Rural Banks (BPR) conduct business activities conventionally or based on Sharia principles. Their activities do not provide services in payment traffic. Meanwhile, a country's central bank is generally an institution responsible for monetary policy in the country's territory. In Indonesia, the function of the central bank is held by Bank Indonesia (BI). Moreover, commercial banks conduct business activities conventionally or based on Sharia principles, which provide services in payment traffic. The duties of Commercial Banks consist of; collecting funds from the public in the form of deposits, channeling funds to the public in the form of loans, issuing money through credit payments and investments, offering financial services such as credit cards, traveler's checks, ATMs, interbank money transfers, and other services, providing facilities for international trade, and providing storage of valuable goods.

Mbat [15] also explains that the difference between commercial banks and other banks is that deposit receipts can be withdrawn by check. Commercial banks' function as a monetary function (creation of deposits and means of payment) makes them unique among financial intermediaries [5, 8, 10]. They can perform this function by providing an absolute promise to convert demand deposits into legal currency upon request. Other functions of commercial banks that align with those of other financial intermediaries are (a) credit functions (credit and investment activities) and (b) non-credit services (e.g., advisory services). Commercial banks perform the first function (monetary function) in a monopolistic manner but share other functions competitively with other financial intermediaries. Banks that do not meet the minimum Core Capital are subjected to administrative sanctions. First, they must adjust the form and business activities of the Bank to become Rural Banks (BPR)/Sharia Rural Banks (BPRS) or submit applications for revocation of business licenses as requested by the Bank in accordance with the provisions of laws and regulations concerning Commercial Banks or Sharia Banks.

Several authors studying earnings management in banking sectors have found empirical support fora the claim that banks use loan loss allowances for earnings management [2]. Suripto [21] examined the characteristics of companies as determinants of conventional and Sharia bank earnings management in several ASEAN (Association of Southeast Asian Nations) countries from 2014 to 2020. The results showed that Capital Adequacy Ratio, Profit Before Tax and Interest, Non-Performing Loans and Changed, Firm Size, CPI (Consumer Price Index), and Exchange rates determine earnings management between Sharia and conventional banks. Proença et al. [18] found evidence of earnings management for banking efficiency purposes. Fitri and Siswantoro [13] used 28 Sharia banks in Asia listed on the stock exchange from 2013 to 2017, proving that only audit committee size has a significant negative effect on earnings management. Meanwhile, an independent audit committee has a negative effect but is insignificant. The results of this study are supported by the research of [9].

The motivation for performing earnings management is mostly associated with utilizing loan loss provisions (LLP) to increase or maintain capital adequacy in accordance with minimum capital regulatory requirements. Several researchers such as Ahmed et al. [2], Anandarajan et al. [3], Leventis et al. (2011), Belal et al. [5] and Elnahass et al. [9] found that evidence of earnings management performed by both Sharia and conventional banks utilized loan loss provisions (LLP) with the motivation to avoid minimum capital adequacy requirements. PSAK No. 50 (IAI, 2013) and PSAK No. 55 (IAI, 2013) state that banks may provide non-performing loans.

Abdel and Archer [1]; Safieddine tated that Sharia banks tend to have opportunistic management behavior due to limited contractual arrangements between banks and depositors. Depositors are not directly involved in the financial and business decisions made by the bank, so the risk is very high. Abdullah et al. and Gusrianti and Putri Hendra Sari found that Sharia board supervision of Sharia banks in Indonesia and Malaysia was less effective. These results align with research conducted by Nurfarahin Hari and Ahmad F. S. Hassan and Yusuf Karbhari. However, earnings management by Sharia banks is predicted to be smaller than conventional banks. Based on the explanation above, this study hypothesizes that:

H1: Sharia banks and conventional banks have the same behavior in earning management to meet the minimum capital adequacy requirements.
H2: Earning management in Sharia banks is smaller than earning management in conventional banks.

Erickson and Wang and Suprianto [19] showed evidence that the acquiring company performs accrual earnings management to increase profits before merging. Similar results were also proven by [6] Njah and Jarboui. Zhang found that companies perform accrual and real earnings management before and during mergers. Baik et al. [4] also found evidence that the acquiring company performs earnings management to cover the additional costs of the deal due to the acquisition process. The acquired company gains advantages in terms of capital.

However, Heron and Lie found no evidence that companies perform earnings management before acquisitions. Pungaliya and Vijh also found no evidence of earnings management during the merger. This is because the costs of such manipulation can exceed the benefits. The undercapitalized banks have two options to increase profits: performing earnings management and mergers. Companies that have merged are predicted to do less earnings management, and conversely, if companies choose earnings management to increase capital, they tend not to merge. Based on the explanation above, this study hypothesizes that:

H3: Sharia banks and conventional banks that merge will have a negative effect on earnings management.

3 Methodology

The population in this study were all banks listed on the Indonesian stock exchange. The sample was selected using the purposive sampling method with the following criteria: first, banking companies must report financial statements from 2019 to 2022. Then, all the data needed for this research could be obtained. The data in this study were the annual financial statements of banking companies which can be obtained through the company's website and the IDX website (www.idx.co.id). The dependent variable used in this study was earning management as measured using the residual value (DLLP) of Eq. 1:

$$LLP_{it} = \propto_0 + \propto_1 NPL_{it-1} + \propto_2 \Delta NPL_{it} + \propto_3 \Delta TL_{it} + € \quad (1)$$

where LLP_{it} is the total loan loss provision for bank i in year t deflated by beginning loans, NPL_{it-1} is the beginning balance of non-performing loans for bank i in year t, ΔNPL_{it} is the change in the value of non-performing loans for bank i in year t deflated by beginning loans ΔTL_{it} is the change in the value of total loans for bank i in year t deflated by beginning loans.

The research model used to test the hypothesis is as follows:

$$DLLP_{it} = \beta_0 + \beta_1 TIER1_{it-1} + \beta_2 MERGER_{it} + \beta_3 EBTLLP_{it} + \beta_4 LEV_{it} + \beta_5 Cov_{it} + \beta_6 GDP_{it} + \beta_7 IB_{it} + \beta_8 Size_{it} + \text{€} \quad (2)$$

DLLP is a bank earnings management variable. TIER 1t − 1 ratio is the ratio of the total bank Tier 1 capital to risk weighted assets. The MERGER variable is a bank variable that merges with other banks. This variable was measured using a dummy variable where 1 is for banks that are merging and 0 for banks that are not merging.

This study used several control variables including EBTLLP (Profit before tax and LLP). It is the bank's ability to generate profit before contractual obligations and LLP. LEV is the leverage variable (total debt to total equity). This variable is to see to what extent the savings bank's potential capital is affected by risk reduction (Kiema and Jokivuolle 2014). The GDP variable is the annual growth rate of national income. The Cov variable is the year Covid occurred in Indonesia, measured by one-year Covid-19 occurred and 0 for the year before Covid-19. IB is a dummy variable fora the type of bank (take a value of 1 for Sharia banks and 0 for conventional banks).

4 Results

Table 1 shows the number of samples in this study with 180 observations (company-year). The number of conventional banks from 2019 to 2022 has not changed, i.e., 42 banks, while the number of sharia banks from 2019 to 2022 has also not changed, i.e., 4 banks. In 2021, there will be a bank merger between PT. BRI Syariah Tbk with Bank BNI Syariah and Bank Mandiri Syariah to become PT. Indonesian Sharia Bank Tbk. However, it does not change the number of banks used in the sample of this study. Nevertheless, the number of observations decreased by 2 observations for PT. BRI Syariah 2021. In 2022, no data was available due to a merger and in 2019 and 2020, 2 observations for the financial statements of PT. Bank Syariah Indonesia Tbk.

Table 1 Research sample

No	Information	2019	2020	2021	2022	Total
1	Conventional bank	42	42	42	42	168
2	Sharia bank	4	4	4	4	16
3	Incomplete data	–	–	(2)	(2)	(4)
	Total sample	46	46	44	44	180

Based on Table 2, it can be seen that the average DLLP value is 2.65. This positive value indicates that it is possible for the bank to perform earnings management by increasing the LLP value so that the bank's profit becomes higher. It is also supported by the correlation test results between the DLLP (earnings management) variable and the EBTLLP variable (earnings before taxes and interest) which shows a value of 0.1098. It means there is a positive correlation between earnings management and company profits. The banking size value in this sample is an average of 83.2 trillion rupiah. Tier value is Tier-1 banking with an average of 3.29 meaning that banks in Indonesia have a good level of core capital adequacy due to the standard set by Bank Indonesia as a regulator is 15%. The Merger variable value is 0.133, meaning that only 13% of the observations merged in the year of observation. The average Covid value is 0.75, meaning that only in 2019, this study included when Covid-19 occurred in Indonesia. While 2020–2022 is the year of the Covid-19 pandemic. The average profit value of the banking samples in this study was 231 billion Rupiah, while the average banking leverage rate was 69.37%. Table 3 reports the correlation between variables.

The hypothesis test results in Table 4 show that the coefficient value of the tier variable is 0.0007 with a significance level of 0.096. These results prove that there is a positive effect between earnings management and the capital adequacy ratio in banking, both conventional and Sharia banks. Thus, the first hypothesis is accepted at a significance level of 10%. Banks in Indonesia (both Sharia and conventional) have been proven to perform earning management to meet the minimum capital adequacy requirements. The results of this study support previous research conducted by [1], Safieddine and Abdullah et al. and Elnahass et al. [9] which proves that Sharia banks and conventional banks also perform profit management for certain purposes. Sharia banks and conventional banks aim to increase the amount of their capital so that earnings management is still carried out.

The coefficient value of the variable ib shows – 0.069 with a significance level of 0.031. These results indicate that Sharia banks are less likely to perform earning management than conventional banks. Thus, the second hypothesis is accepted at the

Table 2 Descriptive statistics

Variable	Obs	Mean	Std. dev	Min	Max
DLLP	180	2.65e + 07	2.48e + 10	– 7.57e + 10	1.79e + 11
Size	180	8.32e + 11	3.22e + 12	2,173,162	2.32e + 13
Tier	180	32.29358	26.27108	0.84	188.31
Ebtllp	180	2.34e + 10	9.94e + 10	– 4.24e + 10	6.69e + 11
Lev	180	69.37797	43.1862	0.01	435.096
Gpd	180	37.36507	0.0928368	37.27536	37.51371
Covid	180	0.75	0.4342205	0	1
Merger	180	0.1333333	0.3408828	0	1
Ib	180	0.1555556	0.3634445	0	1

Table 3 Correlation between variables

	DLLP	Tier	Ib	Merger	Size	Ebtllp	Lev	Gdp	Covid
DLLP	1.0000								
Tier	0.0779	1.0000							
Ib	− 0.2130	0.0530	1.0000						
Merger	− 0.1797	− 0.0569	0.2588	1.0000					
Size	− 0.1977	0.0243	0.0348	− 0.1246	1.0000				
Ebtllp	− 0.1501	0.0019	0.0102	− 0.1575	0.9413	1.0000			
Lev	− 0.0309	− 0.1237	− 0.1912	− 0.0205	0.2648	0.2728	1.0000		
Gdp	− 0.0932	0.2400	− 0.0000	− 0.0568	0.3576	0.3546	0.2239	1.0000	
Covid	− 0.1962	0.1750	0.0000	− 0.1509	0.2784	0.2790	0.1240	0.4000	1.0000

Source	SS	df	MS
Model	3.4654e + 23	3	1.1551e + 23
Residual	1.1017e + 23	176	6.2598e + 20
Total	4.5671e + 23	179	2.25515e + 21

Llp	Coef.	Std. Err.	t	P > | t |	[95% conf. intercal]	
Npll	1.557995	0.0914547	17.04	0.000	1.377506	1.738484
ANPL	1.178978	0.1137345	10.37	0.000	0.9545191	1.403437
ALOAN	0.0095611	0.0022987	4.16	0.000	0.0050246	0.0140976
_cons	5.66e + 09	1.97e + 09	2.87	0.005	1.77e + 09	9.55e + 09

5% significance level. These results are supported by previous research conducted by Elnahass et al. [9] and Suprianto et al. [20]. Abdelsalam et al. also found that Sharia banks did not perform earning management. Even though Sharia banks do profit management, the percentage is smaller than conventional banks. It is because Sharia banks have Sharia supervisory boards and apply Sharia principles.

Table 4 shows that the coefficient value is 0.096 at a significance level 0.022. Therefore, it can be concluded that Sharia banks and conventional banks that merge will have a negative effect on earning management. Banks prefer mergers over earning management to meet the minimum capital adequacy requirements. Thus, merged banks are predicted to be less likely to perform earnings management. Conversely, if banks do not merge, they will choose earnings management to increase capital as an alternative. These results support the research by Heron and Lie which found no evidence that companies perform earnings management before acquisitions. Pungaliya and Vijh also found no evidence of earnings management during the merger. It is because the costs of such manipulation can exceed the obtained benefits. Therefore, earnings management and mergers are two options for companies to increase the amount of their capital. There is a trade-off between these two things depending on the costs and benefits obtained by the company.

Table 4 Regression test results

Random-effectc GLS regression				Number of Obs	=	180
Group Variable	:	Pers		Number of groups	=	46
R-sq:				Obs per group	:	
Within	=	0.0936		Min	=	1
Between	=	0.3623		Avg	=	3.9
Overall	=	0.1564		Max	=	4
				Wald chi2 (8)	=	31.71
Corr(u_i, X)	=	0 (assumed)		Prob > chi2	=	0.0001
DLLP	Coef	Std. err	Z	P > | t |	[95% conf. intercal]	
Tier	0.0007237	0.0004351	1.66	0.096	– 0.0001291	0.0015765
Ib	– 0.0691188	0.0320763	– 2.15	0.031	– 0.1319873	– 0.0062503
Merger	– 0.0960175	0.0420387	– 2.28	0.022	– 0.1784119	– 0.0136231
Size	– 0.0170087	0.0084885	– 2.00	0.045	– 0.0336458	– 0.0003716
Ebtllp	0.0099462	0.0076537	1.30	0.194	– 0.0050548	0.0249472
Lev	– 0.000438	0.0002765	0.16	0.874	– 0.0004981	0.0005857
Gdp	– 0.0005807	0.1391057	– 0.00	0.997	– 0.2732229	0.2720615
Covid	– 0.073353	0.0282276	– 2.60	0.009	– 0.128678	– 0.018028
_conz	0.3696697	5.169668	0.07	0.943	– 9.762693	10.50203
Sigma_u	0					
Sigma_e	0.14998293					
rho	0	(fraction of variance due to u_i)				

5 Conclusion

The results conclude that Sharia and conventional banks have similar behavior in conducting earning management to meet the minimum capital adequacy requirements. However, earnings management carried out by Sharia banks is smaller than conventional banks. Furthermore, this study also concludes that both Sharia and conventional banks that did mergers have a negative effect on earnings management. This research has several limitations; First, this research does not distinguish the company's age between old and new banks. This might create interesting results if further research focuses on bank age and earnings management. Second, this research did not focus on corporate governance which might affect these results. Future research should focus on governance variables associated with banks' capital adequacy and earnings management.

References

1. Abdel Karim, R., Archer, S.: Islamic finance: Innovation and growth. Euromoney Publications, London (2002)
2. Ahmed, A.S., Takeda, C., Thomas, S.: Bank loan loss provisions: a reexamination of capital management, earnings management and signaling effects. J. Account. Econ. **28**(1), 1–25 (1999)
3. Anandarajan, A., Hasan, I., McCarthy, C.: Use of loan loss provisions for capital, earnings management and signaling by Australian banks. Account. Financ. **47**, 357–379 (2007)
4. Baik, B., Cho, K., Choi, W., Kang, J.-K.: The role of institutional environments in cross-border mergers. a perspective from bidder's earnings management behavior. Manage. Int. Rev. **55**(5), 615–646 (2015). https://doi.org/10.1007/s11575-015-0249-4
5. Belal, A.R., Abdelsalam, O., Nizamee, S.S.: Ethical reporting in Islamic bank Bangladesh limited (1983–2010). J. Bus. Ethics **129**, 769–784 (2015)
6. Botsari, A., Meeks, G.: Do acquirers manage earnings prior to a share for share bid? J. Bus. Financ. Acc. **35**(5–6), 633–670 (2008). https://doi.org/10.1111/j.1468-5957.2008.02091.x
7. Collins, J.H., Shackelford, D.A., Wahlen, J.M.: Bank differences in the coordination of regulatory capital, earnings, and taxes. J. Account. Res. **33**(2), 263 (1995). https://doi.org/10.2307/2491488
8. D'Souza, A.F.N., Sai Sundar, H.: Overcoming the challenges of establishing Islamic banks in india: a case study of Al-Arafah Islamic investment bank. In: Aloysius Edward, J., Jaheer Mukthar, K.P., Asis, E.R., Sivasubramanian, K. (eds.) Current Trends in Economics, Business and Sustainability. ICEBS 2023. Contributions to Environmental Sciences & Innovative Business Technology. Springer, Singapore (2023). https://doi.org/10.1007/978-981-99-3366-2_23
9. Elnahass, M., Izzeldin, M., Steele, G.: Capital and earnings management: evidence from alternative banking business models. Int. J. Account. **53**(1), 20–32 (2018). https://doi.org/10.1016/j.intacc.2018.02.002
10. Eshwari, D., Baby, M.K: The effect of capital structure on the profitability of FMCG companies. In: Aloysius Edward, J., Jaheer Mukthar, K.P., Asis, E.R., Sivasubramanian, K. (eds.) Current Trends in Economics, Business and Sustainability. ICEBS 2023. Contributions to Environmental Sciences & Innovative Business Technology (2023). Springer, Singapore. https://doi.org/10.1007/978-981-99-3366-2_37
11. Fan, Y., Huang, Y., Jiang, Y., Liu, F.H.: Watch out for bailout: TARP and bank earnings management. J. Financ. Stabil. **51**(March 2009), 100785 (2020). https://doi.org/10.1016/j.jfs.2020.100785
12. Fan, Y., Jiang, Y., Zhang, X., Zhou, Y.: Women on boards and bank earnings management: from zero to hero. J. Bank. Financ. **107**, 105607 (2019). https://doi.org/10.1016/j.jbankfin.2019.105607
13. Fitri, V., Siswantoro, D.: Can corporate governance mechanisms reduce earnings-management practices in Islamic banks? J. Islamic Account. Bus. Res. **13**(1), 16–31 (2021). https://doi.org/10.1108/JIABR-04-2019-0081
14. Kim, Y.S., Kim, Y., Yi, H.C.: Vice or virtue? the impact of earnings management on bank loan agreements. Int. Rev. Econ. Financ. **73**(December 2020), 303–324 (2021). https://doi.org/10.1016/j.iref.2020.12.028
15. Mbat, D.O.: Commercial banks and rural banking policy in a developing economy: policy implications. Serv. Develop. **6**(2), 367–380 (1982)
16. Mulyaningtyas, Handayani, E.F.: Pengaruh profitabilitas, Kecukupan Modal Dan Likuiditas Terhadap Manajemen Laba Pada E Ntitas Perbankan Yang Terdaftar Di Bursa Efek Indonesia. Akuntansi Dan Teknologi Informasi **15**(1), 22–33 (2022). https://doi.org/10.24123/jati.v15i1.4731
17. Pambekti, G.T.: Analisis perbandingan manajemen laba dengan metode discretionary accrual pada bank syariah dan bank konvensional. J. Ekonomi Keuangan Islam **3**(2), 81–89 (2017). https://doi.org/10.20885/jeki.vol3.iss2.art4

18. Proença, C., Augusto, M., Murteira, J.: The effect of earnings management on bank efficiency: evidence from ECB-supervised banks. Financ. Res. Lett. **51**(August 2022) (2023). https://doi.org/10.1016/j.frl.2022.103450
19. Suprianto, E.: Creative accounting in local governments to obtain funds for Covid-19 management. Int. J. Prof. Bus. Rev. **8**(3), 1–17 (2023). https://doi.org/10.26668/businessreview/2023.v8i3.726
20. Suprianto, E., Rusdi, D., Salim, A.: The models of improving the quality of government financial reporting. In: Lecture Notes on Data Engineering and Communications Technologies, vol. 161. Springer International Publishing (2023). https://doi.org/10.1007/978-3-031-26281-4_5
21. Suripto.: Earnings management determinants: comparison between islamic and conventional banks across the ASEAN region. Asia Pacific Manage. Rev. 1–9 (2022). https://doi.org/10.1016/j.apmrv.2022.01.005
22. Wu, Y.C., Wei Kiong Ting, I., Lu, W.M., Nourani, M., Kweh, Q.L.: The impact of earnings management on the performance of ASEAN banks. Econ. Model. **53**, 156–165 (2016). https://doi.org/10.1016/j.econmod.2015.11.023
23. Yogi, L.M.D.P., Damayanthi, I.G.A.E.: Pengaruh Arus Kas Bebas, capital adequacy ratio dan good corporate governance pada manajemen laba. E-J. Akuntansi Univ. Udayana **15**(2), 1056–1085 (2016). https://ojs.unud.ac.id/index.php/Akuntansi/article/view/19869

The Existence of Creditor on the Internet Financial Reporting Timeliness

Dina Alafi Hidayatin and Habiburrochman Habiburrochman

Abstract The present study examined the influence of company characteristics on the timing of internet-based financial reporting. The dataset utilized in this inquiry consisted of 850 manufacturing firms that were listed on the Indonesian Stock Exchange (IDX) between the years 2013 and 2020. The study utilized logistic regression analysis to determine the impact of leverage on the timeliness of online financial reporting, and it was shown that leverage has a substantial influence on this aspect. In contrast, the variables of company size, profitability, and liquidity did not exhibit a statistically significant link with reporting timeliness. This observation underscores the intricate nature of the variables in the digital era. Furthermore, it was shown that the existence of creditors serves as a driving force for companies to accelerate their adoption of internet-based financial reporting. This highlights the significance of creditor supervision in bolstering corporate responsibility and promoting financial openness in the era of digitalization.

Keywords Internet financial reporting · Company attributes · Leverage · Creditors · Corporate accountability · Financial transparency

1 Introduction

Internet financial reporting helps organizations share financial data. In the current paradigm change, firms can submit financial reports to stakeholders without direct involvement [1]. Accessing financial information about their invested capital online is also an option for stakeholders. It also entices investors to invest in the company. This technique has gained popularity in Indonesia's business community since the

D. A. Hidayatin
Department of Accounting, Sekolah Tinggi Ilmu Ekonomi Cendekia Bojonegoro, East Java 62115, Indonesia

H. Habiburrochman (✉)
Department of Accounting, Faculty of Economics and Business, Airlangga University, East Java, Indonesia
e-mail: habib@feb.unair.ac.id

B. Awwad (ed.), *The AI Revolution: Driving Business Innovation and Research*, Studies in Systems, Decision and Control 524, https://doi.org/10.1007/978-3-031-54379-1_35

Financial Services Authority's instruction (OJK, number 29/POJK.04/2016) on the Annual Report of Issuers or Public Companies. The 'new normal' makes internet financial reporting an attractive topic for investigation.

Internet financial reporting increases transparency and accessibility, making financial information more accessible to more people. This benefits investors and the community. This helps governments, regulators, and advocacy groups analyze financial data, improving corporate responsibility. By adopting internet financial reporting, organizations may foster transparency and confidence. This can attract socially concerned investors and boost their environmental responsibility reputation [2]. In conclusion, online financial reporting improves efficiency, encourages environmentally sustainable practices, and reinforces a company's transparency and accountability [3–5].

1.1 The Financial Report Quality

Financial report quality can be assessed using different factors. Financial statements must be impartial. Additionally, timely financial report delivery affects report quality. The sooner stakeholders receive financial accounts, the more current the decision-making information. In contrast, delayed release may render financial statement data obsolete. By protecting present and future investors, timely financial statement delivery helps preserve the capital market process [6–8].

Many multinational studies have explored internet financial reporting. Inconsistencies persist in earlier investigations. A Ghana Stock Exchange study found a positive correlation between internet financial reporting index and profitability and leverage [8]. No explanation was given by firm size, liquidity, or auditor size. Company size and scattered ownership affected online financial reporting in New Zealand [9]. Liquidity and profitability did not affect internet financial reporting. Additionally reviewed the company's online financial communication. The study examined Nigerian Stock Exchange companies. The internet financial reporting index was strongly associated with company size and industry. Profitability, auditor type, and firm age did not alter the internet financial reporting index.

Examined how corporate governance and company factors affect Egyptian stock exchange-listed companies' internet financial reporting. Disclosure indexes for businesses' three internet financial reporting components were created using an unweighted checklist [10]. Internet financial reporting total, content, and presentation were substantially connected with firm size, ownership diffusion, business model, profitability, auditor type, institutional ownership, and board size. The level of internet financial statement disclosure on London Stock Exchange-listed companies was positively affected by company size but not industrial sector [11].

In London, [12] examined if the intensive corporate environment could affect website financial transparency and its factors. This study found that companies having a big environmental impact prefer to report it online. It also showed that corporate size affects website financial disclosure. Leverage and ownership concentration hurt financial disclosure on the company's website, while profitability and auditor type didn't. Abdullah et al. [13] analyzed internet financial reporting on 273 Indonesian Sharia Stock Index companies by firm size, age, public ownership, and audit quality. Firm size, age, and public ownership positively and considerably affected internet financial reporting, while audit quality did not. 300 Shanghai and Shenzhen Stock Exchange-listed Chinese companies were investigated for voluntary financial statement disclosure based on company characteristics [14]. Auditors, ownership structures, international listings, and government rules boosted Internet-based voluntary disclosure.

This study differs from others. This research analyzes how corporate variables affect internet financial reporting timeliness. Indonesia Stock Exchange manufacturing companies from 2013 to 2020 were examined. This study will add to the literature on using internet financial reporting to increase organizational efficiency and effectiveness by expanding business networks.

2 Literature Review

2.1 Agency Theory

Agency theory emerged from principal-agent information ownership disparities [15, 16]. Two sides perceive disproportionate information differently. Principal and agent have firm interests yet disagree. Agency has more information, but principle only gets it when agent discloses it. By concealing information from the principal, the agent may commit fraud. Fraudulence is usually done to retain a good reputation for managing principal's funds, which might alter agent incentives.

2.2 Signaling Theory

Due to distinct interests from the management and company owner, signaling theory varies from information asymmetry [17]. The agent's communication to the principal is key to this notion. Companies with good quality and reputation who intentionally give good information to the public signal positive or bad, according to the notion. The company hopes investors will understand and interpret the signal to predict future events and make investment decisions.

2.3 Company Characteristics

Companies' internal situations are described by their qualities. This analysis uses company size, profitability, liquidity, and leverage [18–20].

a. Size of company

The quantity of assets owned usually indicates company size. Companies with many assets might gain investor and creditor trust. Thus, such companies can raise cash to grow.

b. Profitability

Company profitability is used to evaluate its ability to make money. Companies with high profitability manage their assets and liabilities well to produce profits.

c. Leverage

Leverage measures the company's debt-financed assets. High leverage makes a corporation dependent on creditors for operational funding. Low leverage allows the corporation to finance its operations autonomously. Lever-age also forces managers to risk losing control if they don't pay their debts.

d. Liquidity

Companies' financial health is typically predicted by liquidity ratios. Liquidity is used to evaluate a company's financial situation and bankruptcy risk. The company's financial position and bankruptcy risk improve with a higher liquidity ratio.

2.4 Internet Financial Reporting

Internet financial reporting (IFR) is a mechanism for presenting financial reports through internet media or displaying financial reports on the company's website page. IFR has been used since the 1990s and has provided many conveniences for business people [12]. This convenience is obtained because IFR has many advantages, such as delivering non-stop availability of financial information, presenting information in various languages, building cooperation among stakeholders, providing faster and low-cost information distribution, and so on [21]. In addition, [22] said that the internet can be used as a resource with low distribution costs, but can provide many innovations in delivering financial information.

3 The Framework of Thinking and Hypothesis Development

3.1 The Impact of Company Size on the Timeliness of Internet Financial Reporting

Large asset-based companies have many investors. Companies must use technology to manage and provide information [23]. Technology speeds up information preparation and dissemination to decision-makers. Ze et al. [24] said big corporations may employ advanced information technology to address stakeholder information delivery needs. Small companies cannot submit internet financial reporting as quickly as large companies.

Many similar research have yielded varied findings. Example: [6] found that firm size strongly affected internet financial reporting ratings for Croatian Stock Exchange companies but not for Slovak Stock Exchange companies. The internet financial reporting index was not significantly affected by company size in [7]'s research. Contrary to prior studies, [25, 26] found that company size significantly affected internet financial reporting disclosure. Furthermore, [12] found a strong correlation between company size and internet financial reporting disclosure. This study hypothesizes the following from the above findings:

H1: company's size affects the timeliness of internet financial reporting.

3.2 The Impact of Profitability on the Timeliness of Internet Financial Reporting

A high profitability ratio reflects corporate health. Thus, sharing good news with investors immediately might be a positive sign for management. Conversely, low profitability will push management to delay financial statement submission to investors. According to [9], the signaling theory states that profitable organizations have an incentive to rapidly disclose information to signal investors to compensate management. According to agency theory, profitable organizations have an incentive to rapidly share knowledge to improve management compensation [27].

Many similar investigations yielded different results. Profitability is unrelated to voluntary internet financial reporting [9]. In contrast, [8] discovered a strong correlation between profitability and internet financial disclosure. Ismail [26] also linked profitability to online financial information distribution. According to [7], profitability greatly affected the internet financial reporting index. These findings lead this study to hypothesize:

H2: company's profitability affects the timeliness of internet financial reporting.

3.3 The Impact of Leverage on the Timeliness of Internet Financial Reporting

High-leverage companies have higher agency costs [15]. Internet financial reporting can reduce agency costs. The sooner internet financial reporting is submitted, the more stockholders can learn about the company's condition. Ismael and Roberts [11] suggested that leverage lowers agency cost because firm creditors care more about principal and interest. To prevent insolvency, creditors monitor managers more.

Management actually cares about leverage circumstances differently. Management hides high leverage since it shows the company's inability to satisfy short-term obligations. Thus, management performance is poor. According to [12], internet financial reporting reduces corporate leverage. Thus, high leverage scares investors/creditors and vice versa. Leverage also influenced voluntary financial statement disclosure online [9]. Agyei-Mensah [7] also discovered that leverage significantly affects internet financial reporting. This study proposes the following idea based on previous research:

H3: company's leverage affects the timeliness of internet financial reporting.

3.4 The Impact of Liquidity on the Timeliness of Internet Financial Reporting

The liquidity ratio measures financial performance. A well-performing corporation has high liquidity. It shows that the corporation can meet its short-term obligations. A greater liquidity ratio indicates a better ability to pay short-term debt [20]. This usually allows organizations to attract investors and creditors. Thus, significant liquidity will encourage corporations to release their financial statements immediately. Signaling theory suggests that high-liquidity organizations are more likely to communicate information rapidly [28]. High liquidity ratios allow organizations to supply stakeholders with more immediate information, according to agency theory [29]. Previous research were inconclusive. According to [29], liquidity did not affect the company's internet financial reporting disclosure. In contrast, [30] revealed that liquidity negatively affected the company's annual report disclosure index. Similarly, [9] observed that liquidity affected voluntary inter-net financial reporting disclosure. This study hypothesizes the following based on past studies:

H4: company's liquidity affects the timeliness of internet financial reporting.

4 Research Methodology

This study uses secondary data from Indonesia Stock Exchange-listed manufacturing companies' 2013–2020 annual reports. This study uses purposive sampling to select a sample from this population. In this study, corporate characteristics are independent factors while internet financial reporting timeliness is dependent. Size, profitability, leverage, and liquidity represent corporate traits. This study analyses data using a quantitative research approach with multivariate logistic regression. Formula for data analysis:

$$\ln \frac{Y}{1-Y} = a + bX_1 + cX_2 + dX_3 + eX_4 + \varepsilon \quad (1)$$

where:

Y = dependent variable (timeliness of internet financial reporting).
X1 = 1st independent variable (company size).
X2 = 2nd independent variable (profitability).
X3 = 3rd independent variable (leverage).
X4 = 4th independent variable (liquidity).
ε = error.

The research steps are as follows:

1. Determining manufacturing companies listed on the Indonesia Stock Exchange in 2013–2020 to find out the total population
2. Determining the research sample using purposive sampling method, with the following criteria:

   a. Listed on the IDX in 2013–2020,
   b. Publishing financial reports for 2013–2020 on its website,
   c. Using the rupiah currency unit in its financial statements,
   d. The publication date of the annual financial statements is observable on the website
   e. Not being late in submitting internet financial reporting

The initial data of the total population included 1.258 companies. However, after screening the data using the predetermined criteria, this study only uses a sample of 850 companies.

3. Tabulating and measuring each variable with predetermined measurement instruments:

   a. Independent variable: company characteristics
      i. Company size → Ln total assets
      ii. Profitability → return on equity = net profit after tax divided by total stock capital
      iii. Leverage → debt to assets ratio = total liabilities divided by total assets
      iv. Liquidity → current ratio = current assets divided by current liabilities

b. Dependent variable: timeliness of IFR

An IFR timeliness dummy variable is used. Companies who provide financial statements online within 90 days were rated 1. Company financial statements submitted online for more than 90 days were given a value of 0. Even though OJK decree 29/POJK.04/2016 requires enterprises to publish their financial statements within four months of closing, most lenders require early submission.

c. Performing data analysis and hypothesis testing using multiple logistic regression

5 Results and Discussion

5.1 Results

Based on data screening using predetermined criteria, resulting in a sample of 850 data companies, with the following details in Table 1.

The audited financial statements are obtained by downloading from each company's website or IDX website, while the IFR publication date was obtained by requesting data to the Indonesia Stock Exchange (IDX) via email. As a result, the estimated IFR publication date on the IDX website was the same as the IFR publication date on the company's website.

Table 1 Research sample

	2013	2014	2015	2016	2017	2018	2019	2020	Σ
Population	146	148	149	150	160	168	168	169	1258
No website/website can't be accessed	3	1	5	3	3	0	0	0	15
There are no financial reports on the website	10	15	5	6	5	0	0	0	39
Do not use rupiah	7	9	21	21	22	30	28	25	163
Unknown IFR publication date	2	10	6	2	12	8	8	9	57
Not on time (late) in submitting IFR	8	7	2	8	2	2	54	51	134
Sample	116	108	110	110	116	128	78	84	850

Source www.idx.co.id (processed), 2021

Table 2 Logistics regression results

	Iteration history	Hosmer and Lemeshow test		Nagelkerke R square		Variables in the equation		
	-2 Likelihood logs	Chi square	Sig	Cox and Snell R square	Nagelkerke R square	B	Sig	Exp (B)
Step 0	990.544	–	–	–	–	–	–	–
Step 1	978.147	18.925	0.015	0.014	0.021	–	–	–
X1						0.022	0.337	1.022
X2						– 0.078	0.297	0.925
X3						– 0.322	0.008	0.724
X4						– 0.015	0.241	0.985
Constant	–	–	–	–	–	0.623	0.340	1.865

Source SPSS output (processed), 2021

5.2 *Multiple Logistics Regression Analysis*

5.2.1 Testing the Model Fit

Testing the model fit is conducted by comparing the initial – 2 log likelihood results with the final – 2 log likelihood. For example, the initial – 2 Log Likelihood value is 990,544, while the final – 2 Log Likelihood value is 978,147. On this value, the difference is 12,397 (See Table 2). Thus, it indicates that the hypothesized model will fit the data [31]. In other words, the addition of X1, X2, X3, and X4 as independent variables into the model is able to influence the model fit.

5.2.2 Testing the Regression Model Fit

Hosmer and Lemeshow's Goodness of Fit Test and Chi-Square values assessed regression model fit. Chi-Square reveals 18.925 with 0.015 significance. The anticipated and observed classifications are not significantly different if the significance value is over 0.05. Therefore, the regression model was fit for the next test [31].

5.2.3 Regression Equation

Based on the results of the calculation of the statistical tool, this study wrote the regression equation as follows:

$$\ln \frac{Y}{1 - Y} = 0.623 + 0.022X_1 - 0.078X_2 - 0.322X_3 - 0.015X_4 \qquad (2)$$

If the variables company size (X1), profitability (X2), leverage (X3), and liquidity (X4) are zero, then the variable timeliness of internet financial reporting (Y) equals 0.623. If the other independent variables are constant, the variable of timeliness of internet financial reporting (Y) will increase or decrease by about 0.022 for each increase or decrease in the company size variable (X1) by 1. The coefficient value of the profitability variable (X2) is − 0.078, indicating that the variable of internet financial reporting timeliness (Y) will increase or decrease by about 0.078 for each increase or decrease in X2 by 1. The coefficient value of the leverage variable (X3) is − 0.322, meaning that if the other independent variables remain constant, the timeliness of internet financial reporting (Y) will increase or decrease by about 0.322. For each increase or decrease in the value of the liquidity variable (X3) by 1, assuming the other independent variables are constant, the timeliness of internet financial reporting (Y) will increase or decrease by about 0.015.

5.2.4 Discussion

Statistics demonstrate that only leverage (X3) has a significance value below 0.05. Therefore, only hypothesis 3 may be accepted out of four. The significance values for company size (X1), profitability (X2), and liquidity (X4) are more than 0.05, hence those three hypotheses should be rejected.

Company size (X1) does not affect online financial reporting timeliness (Y). This matches [6, 7]. A corporation with many assets can distribute financial statements cheaply. However, since financial statements are distributed online, it is cost-effective. However, organizations with more total assets, especially those with a variety of assets, take longer to prepare their financial statements due to the accounting treatment required. Submitted financial disclosures to current and potential investors may take longer. The enormous number of investors for some companies may motivate them to deliver financial reports immediately. However, huge corporations often have many investors, therefore this criterion is overlooked.

Hypothesis 2 is rejected because profitability variable (X2) does not affect internet financial reporting timeliness (Y). Similar results were found [9]. Company profitability is its ability to make money. High profits don't prompt corporations to publish fast internet financial reporting. The corporation believes investors care more about stability than profitability. Investors want companies that create steady profits and assure long-term business stability.

Hypothesis testing demonstrates leverage (X3) significantly influences internet financial reporting timeliness (Y). Thus, the study [7] that found leverage affects internet financial reporting supports this result. Creditors might hasten up financial report release. Leverage measures the company's debt-financed assets. High leverage makes a corporation dependent on creditors for operational funding. However, modest debt allows the corporation to finance its operations independently. Signalling theory suggests that corporations publish their financial figures soon to maintain creditors' faith and keep lending money. This study shows that creditors can force listed corporations to release financial data immediately. Agency theory also suggests that highly indebted corporations should disclose more and publish their financial reports to lessen information asymmetry with lenders.

Internet financial reporting timeliness is unaffected by liquidity (X4). This supports [29]'s observation that liquidity does not affect internet financial disclosure. Company liquidity is the ability to manage current assets to meet short-term obligations. Signalling theory says a corporation with great liquidity can publish financial statements instantly. Statistical tests in this study showed otherwise. Companies with high or low liquidity want to submit internet financial reporting immediately.

6 Conclusion, Limitations, and Recommendations

Based on logistic regression hypothesis tests, this study found that only leverage affects internet financial reporting timeliness, while company size, profitability, and liquidity do not. Leverage's large effect suggests creditors can influence listed corporations to provide financial disclosures immediately.

However, this study has drawbacks. First, this study only measures the effect of company size, profitability, debt, and liquidity on internet financial reporting timeliness using four independent variables. Other factors can be used to proxy firm characteristics in future study. Second, this analysis only samples manufacturing companies with five years of reporting. By including more time and firms from various sectors/industries, future research can contribute.

References

1. Setiawan, A.B., Karman, Nugroho, A.C., Dunan, A., Mudjiyanto, B.: Digital strategies and policy approach for small medium micro business development in Indonesia BT—open innovation in small business: creating values for sustainability. In: Rafiki, A., Dana, L.-P., Nasution, M.D.T.P. (eds.), pp. 71–86. Singapore, Springer Nature Singapore (2023)
2. Ananda, R.F., Rahmadhani, S.N., Rahman, F., Dewi, S.: A review on innovation audits bt—open innovation in small business: creating values for sustainability. In: Rafiki, A., Dana, L.-P., Nasution, M.D.T.P., (eds.), pp. 185–194. Singapore, Springer Nature Singapore (2023)
3. Shenkoya, T.: Can digital transformation improve transparency and accountability of public governance in Nigeria? Transform. Gov. People, Process Policy **17**(1), 54–71 (2023). https://doi.org/10.1108/TG-08-2022-0115
4. Bharti, P., Park, B.: Internet International Journal of Internet, Broadcasting and Communication. Int. J. Internet, Broadcast. Commun. **15**(5), 227–239 (2023). https://doi.org/10.7236/IJIBC.2023.15.2.227
5. Amalia, M.M., Napu, F., Kraugusteeliana, K.: Improving financial transparency and accountability of bumdes through a savings and loan information system. J. Ekonomi **12**(02), 1664–1670 (2023) [Online]. Available: http://ejournal.seaninstitute.or.id/index.php/Ekonomi/article/view/2000%0Ahttp://ejournal.seaninstitute.or.id/index.php/Ekonomi/article/download/2000/1585
6. Pervan, I.: Voluntary financial reporting on the internet analysis of the practice of stock-market listed croatian and slovene joint stock companies. Financ. Theory Pract. **30**(1), 1–27 (2006)

7. Agyei-Mensah, B.K.: Corporate financial reporting: firm characteristics and the use of internet as a medium of communication by listed firms in Ghana **6**(6) (2012)
8. Umoren, A.O., Asogwa, I.E.: Internet financial reporting and company characteristics : a case of quoted companies in Nigeria **4**(12), 72–81 (2013)
9. Laswad, F., Fisher, R., Oyelere, P.: Determinants of voluntary Internet financial reporting by local government authorities. J. Account. Public Policy **24**(2), 101–121 (2005). https://doi.org/10.1016/j.jaccpubpol.2004.12.006
10. Taha, M.H., Eldalash, S.A.E., Mohamed, H.Z.: The role of accounting disclosure of sustainable development activities using metaverse in the field of education and training BT—How the metaverse will reshape business and sustainability. In: El Khoury, R., Alareeni, B., (eds.), pp. 63–74. Singapore: Springer Nature Singapore (2023)
11. Ismael, H.R., Roberts, C.: Factors affecting the voluntary use of internal audit: evidence from the UK. Manag. Audit. J. **33**(3), 288–317 (2018). https://doi.org/10.1108/MAJ-08-2016-1425
12. Dâmaso, G., Lourenço, I.C.: Internet financial reporting: environmental impact companies and other determinants. Int. Conf. Enterp. Syst. Accent. Logist. **2011**(July), 331–359 (2011)
13. Abdullah, M.D.F., Ardiansah, M.N., Hamidah, N.: The effect of company size, company age, public ownership and audit quality on internet financial reporting. Sriwij. Int. J. Dyn. Econ. Bus. **1**(2), 153 (2017). https://doi.org/10.29259/sijdeb.v1i2.153-166
14. Franchi, C.: Internal control reporting by Chinese listed companies: an empirical analysis. dspace.unive.it (2019)
15. Jensen, M.C., Meckling, W.H.: Theory of the firm: managerial behavior, agency costs and ownership structure. Corp. Gov. Values, Ethics Leadersh. **3**, 77–132 (2019). https://doi.org/10.2139/ssrn.94043
16. Fama, E.F., Jensen, M.C.: Agency problems and residual claims. SSRN Electron. J. **26**(6), 1983 (2005). https://doi.org/10.2139/ssrn.94032
17. Ross, S.A.: Determination of financial structure: the incentive-signalling approach. Bell J. Econ. **8**(1), 23–40 (1977). https://doi.org/10.2307/3003485
18. Surachyati, E., Abubakar, E., Daulay, M.: Analysis of factors that affect the timeliness of submission of the financial statements on transportation companies in Indonesia stock exchange. Int. J. Res. Rev. **6**(1), 190–201 (2019). [Online]. Available: http://www.gkpublication.in/IJRR_Vol.2_Issue6_June2015/IJRR0066.pdf
19. Mahadwartha, P.: Predictability power of dividend policy and leverage policy to managerial ownership in indonesia: an agency theory perspective. J. Ekon. dan Bisnis Indones. **18**(3), 288–297 (2004)
20. Amr, A.M.: Analyzing the effect of firm liquidity on the quality of financial reporting: an empirical study of firms listed in the Egyptian stock exchange. Int. J. Soc. Sci. Econ. Res. **1**(10), 1604–1628 (2016)
21. Mitchell Williams, S., Pei, C.A.H.W.: Corporate social disclosures by listed companies on their web sites: an international comparison. Int. J. Account. **34**(3), 389–419 (1999). https://doi.org/10.1016/s0020-7063(99)00016-3
22. Hedlin, P.: The internet as a vehicle for investor relations: the swedish case. Int. J. Phytoremediation **21**(1), 373–381 (1999). https://doi.org/10.1080/096381899336104
23. Ashbaugh, H., Johnstone, K.M., Warfield, T.D.: Corporate reporting on the internet. Account. Horizons **13**(3), 241–257 (1999). https://doi.org/10.2308/acch.1999.13.3.241
24. Ze, Z.X., Dyson, J.R., Powell, P.L.: The impact of information technology on corporate financial reporting: a contingency perspective. Br. Account. Rev. **28**(3), 203–227 (1996). https://doi.org/10.1006/bare.1996.0015
25. Craven, B.M., Marston, C.L.: Financial reporting on the internet by leading UK companies. Int. J. Phytoremediation **21**(1), 321–333 (1999). https://doi.org/10.1080/096381899336069
26. Ismail, D.T.H.: An empirical investigation of factors influencing voluntary disclosure of financial information on the internet in the GCC countries. SSRN Electron. J. (7) (2005). https://doi.org/10.2139/ssrn.420700
27. Abd-Elsalam, O.H.: The introduction and application of international accounting standards to accounting disclosure regulations of a capital market in a developing country: the case of egypt dedication Thesis Submitt. degree Dr. Philos., pp. 1–336 (1999)

28. Naser, K., Al Kandari, A., Al-Mutairi, A., Nuseibeh, R.: Can substitution and signaling theories explain the relationship between external audit fees and the effectiveness of internal corporate governance? Glob. J. Manag. Bus. Res. **13**(5), 41–49 (2013). [Online]. Available: https://pdfs.semanticscholar.org/8eb5/74e2fd6040675c08177e022fc096ac6f133c.pdf
29. Aly, D., Simon, J., Hussainey, K.: Determinants of corporate internet reporting: evidence from Egypt. Manag. Audit. J. **25**(2), 182–202 (2010). https://doi.org/10.1108/02686901011008972
30. Olusegun Wallace, R.S., Naser, K., Mora, A.: The relationship between the comprehensiveness of corporate annual reports and firm characteristics in Spain. Account. Bus. Res. **25**(97), 41–53 (1994). https://doi.org/10.1080/00014788.1994.9729927
31. Ghozali, I.: *Aplikasi Analisis Multivariate Dengan Program SPSS*, IV. Badan Penerbit Universitas Diponegoro, Semarang (2006)

The Impact of BASEL III Liquidity Requirements on Bank Performance: Lesson from Indonesia

Mohamad I. Modjo and Syntha R. A. Giannina

Abstract This research examines the impact of Basel III liquidity requirements on the performance of publicly traded commercial banks in Indonesia throughout 2016–2022. It further scrutinizes the differential effects of these requirements on small and large banks. Using a sample set of 39 banks, the study reveals that enforcing the Liquidity Coverage Ratio (LCR) requirement imposes regulatory constraints, thereby diminishing banks' profitability within the Indonesian market. Interestingly, the study uncovers that larger banks are less affected by implementing Basel III than their smaller counterparts. These findings offer crucial insights that could aid banks and regulatory bodies in tackling the challenges of liquidity regulations. This study underscores the importance of considering bank size when implementing regulatory changes, as the effects can vary significantly.

Keywords Basel III · Liquidity coverage ratio · Profitability

1 Introduction

Prudential authority requires banks to maintain a set of liquidity requirements. These liquidity requirements are put in place to ensure they have enough cash and liquid assets to meet their financial obligations and handle unexpected events. By maintaining these requirements, banks can safeguard the financial system's stability and protect depositors' funds [1, 3].

The Basel Committee on Banking Supervision, a global forum for banking regulators, initiated a program of regulatory reforms known as Basel III, which explicitly provides a liquidity framework comprising liquidity constraints: Liquidity Coverage Ratio (LCR) and Net Stable Funding Ratio (NSFR). Basel III LCR and NSFR aim to enhance the banking system's resilience to financial turmoil and reduce maturity mismatches.

M. I. Modjo (✉) · S. R. A. Giannina
Finance International Program, Binus University International, Jl. Hang Lekir I No. 6, Senayan Jakarta 10270, Indonesia
e-mail: mohamad.modjo@binus.edu

B. Awwad (ed.), *The AI Revolution: Driving Business Innovation and Research*, Studies in Systems, Decision and Control 524, https://doi.org/10.1007/978-3-031-54379-1_36

This study examines the impact of Basel III liquidity requirements on Bank performance in Indonesia. Moreover, it will explore the gradation of the effects of BASEL III on the performance of small and large banks. This study contributes by enhancing our insight into the impact of bank prudential regulation on its performance, particularly the impact of bank liquidity requirement regulation and its ramifications on bank profitability. Therefore, it will bridge the existing literature gap and provide further evidence from Indonesia's case. Further, this finding contributes to the current theoretical literature about the impact of Basel III on the performance of commercial banks in Indonesia.

2 Literature Review

2.1 Bank Risk and Prudential Regulation

The banking business is risky due to its asymmetric information and agency problems inherent in its operations. Therefore, the risk is natural to the operations of banks. Berger and Bouwman [16] point out that banks are more vulnerable to risk, illiquidity, risk, and insolvency. In general, four distinct types of risk are considered the main risks for bank operations. These risks are liquidity, credit, capital, and operational risks. Liquidity risk refers to a bank's failure to fulfill its debts, which might jeopardize its financial situation and survival [8]. It has reputational consequences by eroding the depositors' trust and leading to a decline in the bank's overall financial performance [23].

Furthermore, Credit risk can be defined as the possibility that a corporate entity cannot repay its loans. Credit risk for banks is obtainable by external credit ratings, which cannot provide direct and intermediate credit information on a bank [25]. Market risk is the risk caused by the market conditions in which the company operates. The financial statements of banks reflect the diversity of their income sources, which indicates that trading activities constitute a growing percentage of the revenue for many banks. The fluctuation of non-traditional income could significantly amplify market risk and the risks inherent in traditional lending activities. Lastly, operational risk encompasses the potential uncertainties stemming from a financial institution's daily operating procedures, activities, and systems. It is commonly regarded as a residual risk, as any risks faced by banks that do not fall under market or credit risk categories are classified under operational risk [2].

All these risks have the potential to adversely affect a company's financial performance, which highlights the critical importance of risk mitigation. Failure to mitigate risks can lead to decreased profits, while successfully mitigating all risks can lead to increased profits.

2.2 BASEL III Requirements and Bank Performance

There has been a growing apprehension regarding the relationship between liquidity requirements and risk (profitability), mainly how the liquidity requirement affects the positive and negative impacts on bank performance. Several scholarly investigations have explored the association between bank performance and Basel III regulations, with varying outcomes. One of the studies conducted by [28] revealed that applying Basel III standards improved commercial banks' operational effectiveness and performance in Bangladesh and other emerging markets. In line with the previous study, [26] noted that commercial banks in Tanzania that complied with strict capital regulations exhibited greater operational efficiency than their peers, contributing to a steady economic expansion. Bibi and Mazhar [17] also suggest a positive impact of Basel III on banks' profitability.

Nevertheless, a study by [11] found that the profitability of 90 banks in the United Kingdom was negatively affected by adopting the Liquidity requirement, which required banks to hold low-yielding liquid assets, resulting in a shift in their investment portfolios. Similarly, [15] stated that the introduction of the Basel III framework harmed the short-term lending activities of large European banks, leading to a reduction in their growth rate.

In Indonesia, studies regarding the impact of implementing Basel III liquidity requirements on the performance of commercial banks are still limited. Nonetheless, several studies did exist. For example, [30] surveyed the impact of implementing Basel III liquidity requirements on the liquidity and stability of commercial banks in Indonesia and found a favorable outcome. In line with this, [18] analyzed the impact of BASEL III on the Bank's Capital in Indonesia and reported significant effects. However, these studies did not specifically investigate the potential impact on bank profitability. Another study conducted by [20] focuses on the impact of implementing the Basel 3 liquidity coverage ratio using the DSGE model. However, this study focuses on a specific period from 2005 to 2017, during which the Basel III framework was not fully implemented in Indonesia. Some empirical research is also outdated regarding data used and modeling approaches [7, 27, 31]. This study adopts a dynamic panel data approach to address these gaps. We also employ the latest available data on liquidity coverage ratio (LCR) pre- and post-complete implementation of Basel III.

2.3 Conceptual Framework and Hypotheses

As discussed above, the relationship between bank liquidity regulation in BASEL III and their performance still requires further investigation. There are still some contrasting views on the impact of the rule. On the one hand, Nedorezova and

Maraval [29] observed that implementing Basel III affects banks' performance negatively. Similarly, [6] found that Basel III had adversely and considerably affected the performance of India's public-sector banks.

On the other hand, [28] showed that Basel III standards improved commercial banks' operational efficiency and profitability in Bangladesh and other emerging countries. Furthermore, [26] noted that Tanzanian commercial banks that complied with strict capital and liquidity regulations exhibited greater operational efficiency than their peers. Thus, it is clear that the issue of BASEL III liquidity requirement impact on performance still needs to be explored further [10]. In addition, to the author's knowledge, no study has investigated this particular issue in the Indonesian context. However, most studies show the unfavorable outcome of implementing BASEL III liquidity requirements on performance.

While it is important for banks to maintain liquidity to meet their short-term obligations and mitigate potential liquidity risks, these assets typically have lower returns than long-term assets such as loans or investments. For these purposes, the present study formulates the following hypothesis:

H1: The implementation of BASEL III liquidity requirements has a significant negative effect on banks' profitability performance.

This study will also look at the gradation impact of the BASEL III liquidity requirement based on banks' size and adopt the following hypothesis:

H2: The implementation of BASEL III has a less pronounced effect on the profitability of larger banks compared to smaller banks.

3 Methods

3.1 Data

Data for this study is obtained from the annual financial statements of commercial banks in Indonesia listed in Bursa Efek Indonesia (IDX) from 2016 to 2022. The study also utilized consolidated financial information from FactSet and Bloomberg data sources. The period from 2016 to 2022 was chosen for this research study to capture the pre-and post-implementation phases of Basel III.

3.2 Empirical Model

Following [28, 33], the following dynamic panel model is used for the analysis.

$$\begin{aligned} \mathrm{ROA}_{\mathrm{it}} = \alpha &+ \beta_1 \mathrm{ROA}_{\mathrm{it}-1} + \beta_2 \mathrm{LCR}_{\mathrm{it}} + \beta_3 \mathrm{LCR}_{\mathrm{it}-1} \\ &+ \beta_4 \mathrm{LCR}^2_{\mathrm{it}} + \beta_5 \mathrm{Size}_{\mathrm{it}} + \beta_6 \mathrm{Cap}_{\mathrm{it}} + \beta 7 \mathrm{LCRAsset}_{\mathrm{it}} + \mathrm{u}_{\mathrm{it}} \end{aligned} \tag{1}$$

$$\begin{aligned} ROA_{it} &= \alpha + \beta_1 ROA_{it-1} + \beta_2 LCR\_P_{it} + \beta_3 LCR\_P_{it-1} + \beta_4 LCR\_P^2_{it} \\ &\quad + \beta_5 Size_{it} + \beta_6 Cap_{it+} + \beta 7 LCR\_PAsset_{it} + u_{it} \end{aligned} \quad (2)$$

3.3 *Variables Measurements*

This study uses return on assets (ROA) as banks' performance measurement. In addition, this study uses banks' log of total assets to measure their sizes and the ratio of capital to assets as the measure of banks' capital. Moreover, the Lcr_p ratio is determined by dividing the total loans by the total deposits, providing insight into how effectively banks can meet their short-term obligations. Lastly, the LcrAsset variable is an independent variable for hypothesis II, which represents the combined effect of liquidity and asset size on the bank's profitability performance [5]. Table 1 presents an overview of the research variables with detailed information on each examined variable.

Table 1 List of variables, type, and description

Variable name	Type	Description
ROA	DV	The ratio is calculated by dividing the earnings before interest and tax (EBIT) by the average total assets. This ratio explains how efficiently a company utilizes its assets to generate earnings
LCR	IV	The liquidity coverage ratio (LCR) is a measure that assesses a bank's ability to meet its obligations over thirty days during crises. It is calculated following BASEL guidelines by dividing the High-Quality Liquid Assets (HQLAs) by the expected net cash inflows
LCR^2	IV	The squared term of the liquidity coverage ratio (LCR) is obtained by multiplying LCR by itself (LCR * LCR)
LCR_P	IV	The ratio is determined by dividing the total loans by the total deposits, providing insight into how effectively banks can meet their short-term obligations. A higher ratio suggests that banks may have lower liquidity levels, which can lead to increased risk and potentially lower profitability
SIZE	CV	The size of a bank is determined by the bank's total assets
CAP	CV	It represents the net worth or the proportion of assets owned by the bank. This metric provides insights into the bank's financial strength
LCRAsset	IV	This independent variable for hypothesis II represents the combined effect of liquidity and asset size on the bank's profitability performance. By multiplying the liquidity ratio with the asset variable, it provides an indication of the proportion of liquid assets in relation to the bank's total assets, as well to cater for possible nonlinear relationship between size and performance

4 Result

4.1 Descriptive Statistics

The descriptive statistics of the variables employed in the study are examined in this result section. Return on assets (*ROA*) is the dependent variable to measure management's capacity to generate profits from the firm's assets. The statistics are summarised in Table 2.

The table conveys that the mean value of ROA is 0.807, suggesting that banks' management has generated an average favorable return for shareholders. The dataset's minimum and maximum ROA values are − 9.720 and 3.280, respectively, demonstrating a broad range of profitability across the banks. Positive skewness in the variables liquidity coverage ratio (*LCR*) and LCR^2 suggests a right-skewed distribution, indicating that these variables exhibit a longer right tail than the left one. The positive skewness of the *LCR* variable, which indicates the liquidity coverage ratio, demonstrates that a substantial number of banks in the dataset have a lower LCR ratio. This finding implies that most banks have lower liquidity buffers or are less equipped to meet short-term liquidity needs than those with higher liquidity coverage ratios. Conversely, negative skewness in the variables ROA, ROE, *CAPRatio,* and *SIZE*, implies a left-skewed distribution, indicating a longer left tail relative to the right tail.

Table 2 Descriptive analytics

	Mean	Std .dev	Min	Max	Skewness	Kurtosis
ROA	0.807	1.996	− 9.72	3.28	− 2.79	12.751
ROE	5.599	13.409	− 61.13	20.42	− 2.507	9.845
Incapital	23.872	1.629	20.772	26,868	− 253	2.222
Inasset	25.608	1.583	22.037	28.32	− 471	2.706
Capratio	221	0.105	0.016	0.594	0.043	3.952
Leratio	2.078	1.133	88	9.415	3.59	20.053
Leratio2	5.593	9.833	774	88.642	5.997	43.183
Inasset2	658.27	79.954	485.645	802,047	− 333	2.578
Ler_	921	217	217	1.631	565	5.179
Cr_p2	894	435	0.047	2.659	1.681	6.57
LerAsset	52.842	26.271	21.783	212.073	2.872	14.851
Ler pAsset	23.566	5.642	5.791	42.273	366	4.557

4.2 Dynamic Panel Data Result

Our models have successfully passed the Arellano and Bond tests for second-order autocorrelation in residuals and the Hansen test for instrument validity. The statistical significance of the *p-values*, which exceed the 5% significance level, confirms that the reported estimates are not affected by autocorrelation or instrument over-identification issues that could potentially lead to a bias or error in the conclusion. Thus, the study can confidently assert that the obtained estimates are reliable and free from such concerns.

Table 3 reveals the results of Model 1. The empirical findings indicate an inverse relationship between bank liquidity and the profitability of commercial banks in Indonesia, as evidenced by the coefficient estimate of – 7.056 for the variable lcratio in Model 1. The large magnitude of the coefficient for this variable suggests that Basel III liquidity constraint substantially impacts the *ROA* compared to the other variables in the regression model.

This result indicates that as the *LCR* levels increase, there is a corresponding decrease in the *ROA* of banks. Given this, the bank's performance will suffer when higher liquidity levels are required. The association can be attributed to the opportunity cost factor. Holding a greater proportion of liquid assets, such as cash or highly liquid securities, implies that banks have forgone investment opportunities in assets with the potential for higher returns. Typically, liquid assets offer lower returns than hazardous or longer-term investments. Therefore, banks that maintain a significant proportion of their liquid assets forego the opportunity to earn higher returns by investing in other asset classes.

Table 3 Dynamic panel data result model 1

ROA	Coef	St. Err	t-value	p-value	[95% Conf]	Interval	Sig
L1	0.299	0.2	1.50	0.134	– 0.092	0.69	
Lcratio	– 7.056	4.135	– 1.71	0.088	– 15.161	1.048	*
L1	– 0.335	0.179	– 1.87	0.062	– 0.686	0.017	*
Lcratio2	0.117	0.04	2.89	0.004	0.037	0.196	***
Capratio	– 2.729	3.422	– 0.80	0.425	– 9.437	3.979	
Lnasset	– 0.157	0.418	– 0.38	0.707	– 0.976	0.662	
LcrAsset	0.247	0.159	1.55	0.121	– 0.065	0.56	
Mean dependent var		0.851	SD dependent var		1.856		
Number of Obs		115	Chi-square		–		
***$p < 0.01$, **$p < 0.05$, *$p < 0.1$							
Arellano-Bond test for AR (1) in first differences: z = -1.73 Pr > z = 0.083							
Arellano-Bond test for AR (1) in first differences: z = 0.54 Pr > z = 0.592							
Hansen test of overid, restrictions: chi2(19) = 19.70 Prob > chi2 = 0.413							
(Robust, but weakened by many instruments)							

Furthermore, this negative relationship could be due to the trade-off between risk and return, as banks prioritize stability and safety over higher profitability. Liquid assets are considered less risky and provide portfolios with stability and liquidity. However, the trade-off for this lower-risk exposure is the possibility of lower returns, as riskier assets typically offer greater yield opportunities.

The regression coefficient of − 0.335 for the lagged one variable of *LCR* (*L1*) is statistically significant, suggesting that when the lagged liquidity coverage ratio increases, there is a corresponding decrease in the return on assets (ROA). The significant negative relationship implies that a shift in the lagged liquidity coverage ratio may impact future ROA. Modifying the liquidity coverage levels in previous periods can potentially affect the ability of banks to generate returns on their assets in subsequent periods. A higher proportion of liquid assets in banks' portfolios in previous periods is associated with yielding lower returns in subsequent periods.

On the other hand, the quadratic term of LCR coefficients has a positive relationship with the ROA with a value of 0.102. The quadratic term is derived by squaring the liquidity coverage ratio values, which facilitates examining possible nonlinear relationships between the liquidity coverage ratio and the dependent variable. The significant positive coefficient of LCR^2 indicates that the liquidity coverage ratio and the dependent variable have a curvilinear relationship. It further suggests that as the liquidity coverage ratio deviates from a certain optimal level, its effect on the dependent variable becomes increasingly pronounced. The positive coefficient indicates that an increase in the liquidity coverage ratio above a specified threshold positively affects the dependent variable. This finding aligns with the outcome of prior research conducted in both emerging market economies [32] and advanced economies [19], the empirical findings demonstrate a nonlinear association between the square of bank liquidity and bank profitability.

Banks may deliberately increase their liquidity coverage ratio as a precaution against possible market volatility or financial disturbances. This strategy entails expanding the allocation of liquid assets, such as cash or highly liquid securities, to ensure that banks can strengthen their resilience and maintain stability. This business decision could negatively impact the current return on assets (ROA) due to the lower returns associated with holding a larger proportion of liquid assets to meet liquidity requirements. However, this prospective strategy could position the bank for enhanced financial outcomes in the future.

Table 4 reveals the outcome of Model 2, which includes the dummy year variable. The purpose of including the year variable in the second table is to capture the potential time effects or time-specific factors that may influence the dependent variable (ROA) in the dynamic panel-data estimation model. The analysis can account for any year-specific systematic variations in the dependent variable by including the year variable as a predictor. However, the significance of other variables in a model can be affected by including specific variables. Furthermore, the years 2016 and 2021 have dropped due to collinearity.

Based on the findings presented in Table 4, it can be observed that the coefficient for *L1* is positive, indicating a potential lagged effect on *ROA*. The evidence suggests that the previous *ROA* positively influences the current ROA. The lcratio has

Table 4 Dynamic panel data result model 2

ROA	Coef	St. err	t-value	p-value	[95% Conf]	Interval	Sig
L1	0.363	− 0.178	2.04	0.041	0.15	0.711	**
Lcratio	− 7.864	4.499	− 1.75	0.08	− 16.682	0.953	*
L1	− 0.356	0.13	− 2.73	0.006	− 0.612	− 0.1	***
Lcratio2	0.098	0.036	2.74	0.006	0.028	0.168	***
Capratio	− 3.342	2.657	− 1.26	0.209	− 8.55	1.866	
Lnasset	− 0.139	0.446	− 0.31	0.755	− 1.014	0.736	
LcrAsset	0.295	0.177	1.67	0.095	− 0.051	0.642	*
2017	0.497	0.28	1.77	0.076	− 0.053	1.046	*
2018	0.43	0.346	1.24	0.214	− 0.248	1.107	
2019	0.458	0.343	1.33	0.182	− 0.215	1.131	
2020	− 0.452	0.239	− 1.89	0.059	− 0.921	0.017	*
2022	0.346	0.252	1.37	0.171	− 0.149	0.84	
Mean dependent var		0.851	SD dependent var		1.856		
Number of Obs		115	Chi-square		–		

****p < 0.01, **p < 0.05, *p < 0.1*

Arellano-Bond test for AR (1) in first differences: z = − 1.61 Pr > z = 0.107

Arellano-Bond test for AR (1) in first differences: z = 0.85 Pr > z = 0.393

Hansen test of overid, restrictions: chi2(14) = 17.10 Prob > chi2 = 0.251

(Robust, but weakened by many instruments)

a significant coefficient value, showing that banks' liquidity coverage ratio substantially negatively impacts Return on Assets. The significant negative coefficients of LCR in Model 2 indicate that variables are strongly correlated. This finding aligns with results from [24, 21], who argued that liquidity regulations would decrease the banks' performance.

Furthermore, the *SIZE* variable as a proxy for the bank size shows an insignificant correlation with ROA. However, the relationship became significantly positive when the variable interreacted with the *LCR*. The *LCRAsset* variable represents the interaction between a bank's LCR and the total assets of banks. By multiplying the *LCR* and asset variables, the authors attempted to determine the synergistic effect of these two factors on the return on assets. Specifically, the estimated coefficient of 0.295 indicates a positive relationship between the variables lcrAsset and roa, implying that a greater interaction between liquidity coverage ratio and bank size is associated with higher asset returns.

This finding suggests that the impact of implementing BASEL III on the profitability of larger banks is relatively less significant compared to smaller banks. In other words, larger banks tend to experience a smaller decline or have a better ability to mitigate the negative effects on their profitability when compared to smaller banks.

In contrast, smaller banks with lower liquid assets may face more challenges in maintaining adequate liquidity and managing their profitability effectively. This results in line with [35, 22], who suggest that larger banks may face greater exposure to the requirements of LCR than smaller banks as they have greater access to external sources of funding. Due to their scale, reputation, and established relationships, larger banks may have an advantage when raising capital and obtaining funding to satisfy BASEL III's liquidity requirements. This funding access may enable them to optimize their liquidity, maintain lower financing costs, and potentially increase their profitability in the future.

All in all, the models show that we could accept H1 and H2, indicating a significant negative impact on the performance of banks in Indonesia following the implementation of the BASEL 3 liquidity requirement. Additionally, it was found that larger banks exhibit a less pronounced effect on profitability compared to smaller banks. Thus, BASEL 3's liquidity requirement only slightly affected their profits. The following section will discuss the study's conclusion and provide recommendations based on the findings.

5 Chapter V: Conclusion

5.1 Conclusion

The onset of the financial market crisis in 2007/2008 resulted in the banking system's solvency deterioration. In response, the Basel Committee on Banking Supervision (BCBS) introduced Basel III, a set of more rigorous rules and regulations to address banks' liquidity positions. One primary concern regarding Basel III liquidity requirements is their potential unfavorable effects on banks' profitability. While previous literature has examined the influence of liquidity regulations on bank performance, there is a lack of studies exploring this relationship specifically within Indonesia.

In light of the research findings, the study presents evidence that implementing liquidity regulations would initially harm banks' performance in Indonesia. Maintaining higher levels of liquid assets is considered favorable for the overall banking industry as it reduces the risk of individual banks facing insolvency during periods of liquidity stress. However, banks often find liquid assets less appealing due to their tendency to generate lower returns than more illiquid assets, particularly during normal economic conditions. Therefore, although the liquidity requirements mitigate the liquidity risk faced by banks, it is accompanied by a trade-off in profitability.

However, as banks adapt to regulatory changes and implement suitable strategies, the correlation between liquidity requirements and profitability becomes positive. A successful adaptation strategy may improve banks' liquidity positions while generating higher returns on their assets, thereby reversing the previously observed negative correlation.

This evidence underlines the importance of effective liquidity management in banking institutions. Therefore, it is recommended that bank executives in Indonesia develop a liquidity optimization model to maximize the use of their liquid assets. Additionally, national regulators should consider the trade-off between safety and the opportunity costs of holding low-yield securities when implementing liquidity regulations within their jurisdictions. This finding highlights the need for a nuanced approach to liquidity regulation to balance ensuring financial stability and promoting profitability in the banking sector.

References

1. Abbas, F., Younas, Z.I.: How do bank capital and capital buffer affect risk: empirical evidence from large US commercial banks. J. Central Banking Theory Prac. **10**(2), 109–131 (2021). https://doi.org/10.2478/jcbtp-2021-0016
2. Abdullah, M., Shahimi, S., Ismail, A.: Operational risk in Islamic banks: examination of issues. Qual. Res. Financ. Markets (2011)
3. Adu-Gyamfi, M.: The Bankruptcy of Lehman brothers: causes, effects and lessons learnt. J. Insur. Financ. Manage. **1** (2016)
4. Ahamed, F.: Determinants of liquidity risk in the commercial banks in Bangladesh. Euro. J. Bus. Manage. Res. **6**(1), 164–169 (2021). https://doi.org/10.24018/ejbmr.2021.6.1.729
5. Ali, C., Sofoklis, V.: Determinants of bank profitability: evidence from the greek banking sector. Econ. Ann. **54**(182), 93–118 (2009). https://doi.org/10.2298/EKA0982093A
6. Ali, M.A., Pervez, A., Bansal, R., Khan, M.A.: Analyzing performance of banks in india: a robust regression analysis approach. Disc. Dyn. Nat. Soc. (2022). https://doi.org/10.1155/2022/8103510
7. Andre, O., Taqwa, S.: Pengaruh Profitabilitas, Likuiditas, dan Leverage Dalam Memprediksi Financial Distress (Studi Empiris Pada Perusahaan Aneka Industri yang Terdaftar di BEI Tahun 2006–2010) (2013)
8. Antoniades, A.: Liquidity risk and the credit crunch of 2007–2008: evidence from micro-level data on mortgage loan applications (2014)
9. Asongu, S.A., Odhiambo, N.M.: Size, efficiency, market power, and economies of scale in the African banking sector. Financ. Innov. **5**(1) (2019). https://doi.org/10.1186/s40854-019-0120-x
10. Bai, J., Krishnamurthy, A., Weymuller, C.-H., Berger, A., Bouwman, C., Brunnermeier, M., Beom Choi, D., Copeland, A., Fleming, M., Martin, A., Schaeck, K.: Measuring liquidity mismatch in the banking sector (2014)
11. Banerjee, R.N., Mio, H.: The impact of liquidity regulation on banks. J. Financ. Intermed. **35**, 30–44 (2018). https://doi.org/10.1016/j.jfi.2017.05.008
12. Bank for International Settlements.: Evaluation of the impact and efficacy of the Basel III reforms (2022). https://www.bis.org/bcbs/publ/d544.pdf
13. Bank of Canada.: Financial system review—2022 (2022). https://www.bankofcanada.ca/2022/06/financial-system-review-2022/
14. Basel Committee on Banking Supervision (BCBS).:. Basel III: International Framework for Liquidity Risk Measurements, Standards and Monitoring (2010)
15. Ben Naceur, S., Marton, K., Roulet, C.: Basel III and bank-lending: evidence from the United States and Europe. J. Financ. Stab. **39**, 1–27 (2018). https://doi.org/10.1016/j.jfs.2018.08.002
16. Berger, A.N., Bouwman, C.H.S.: How does capital affect bank performance during financial crisesα. J. Financ. Econ. **109**(1), 146–176 (2013). https://doi.org/10.1016/j.jfineco.2013.02.008

17. Bibi, S., Mazhar, F.: Determinants of banks profitability & liquidity and the role of basel III in islamic & conventional banking sector of Pakistan: a case study of NBP. Econ. Financ. Lett. **6**(1), 40–56 (2019). https://doi.org/10.18488/journal.29.2019.61.40.56
18. Bonnario, M.: Dampak implementasi Basel III terhadap permodalan pada bank di indonesia (studi kasus pada PT Bank Negara Indonesia tbk). Jurnal Manajemen Bisnis **13** (2018)
19. Bordeleau, É., Graham, C.: The impact of liquidity on bank profitability (2010)
20. Chawwa, T.: Impact of reserve requirement and liquidity coverage ratio: a DSGE model for Indonesia. Econ. Anal. Policy **71**, 321–341 (2021). https://doi.org/10.1016/j.eap.2021.05.002
21. Giordana, G.A., Schumacher, I.: An empirical study on the impact of basel III standards on banks' default risk: the case of luxembourg. J. Risk Financ. Manage. 8–29 (2017)
22. Gržeta, I., Žiković, S., Tomas Žiković, I.: Size matters: analyzing bank profitability and efficiency under the Basel III framework. Financ. Innov. **9**(1) (2023). https://doi.org/10.1186/s40854-022-00412-y
23. Hakimi, A., Zaghdoudi, K.: Liquidity risk and bank performance: an empirical test for Tunisian Banks. Bus. Econ. Res. **7**(1), 46 (2017). https://doi.org/10.5296/ber.v7i1.10524
24. King, M.R.: The Basel III net stable funding ratio and bank net interest margins. J. Banking Financ. **37** (2013)
25. Liao, H.H., Chen, T.K., Lu, C.W.: Bank credit risk and structural credit models: agency and information asymmetry perspectives. J. Bank. Financ. **33**(8), 1520–1530 (2009). https://doi.org/10.1016/j.jbankfin.2009.02.016
26. Lotto, J.: The empirical analysis of the impact of bank capital regulations on operating efficiency. Int. J. Financ. Stud. **6**(2) (2018). https://doi.org/10.3390/ijfs6020034
27. Masdupi, E., Defri.:. Pengaruh capital adequacy ratio (CAR), Likuiditas dan Efisiensi operasional terhadap Profitabilitas Perusahaan Perbankan yang Terdaftar di BEI. J. Manajemen **1** (2012)
28. Mashamba, T.: The effects of Basel III liquidity regulations on banks' profitability. J. Govern. Regul. **7**(2), 34–48 (2018). https://doi.org/10.22495/jgr_v7_i2_p4
29. Nedorezova, E., Maraval, C.: (n.d.) The basel III liquidity requirements and banks' stock returns: a quantitative study of the impact of the Basel III liquidity requirements on the banks' stock returns
30. Pratama, P.: Pengaruh Nim, Npl, Roa,Ldr, Dan Bopo Terhadap Capital Adequacy Ratio (CAR) (2018)
31. Ruwanti, G.: Pengaruh Rasio Likuiditas Terhadap Kinerja Keuangan Perbankan Pada Bank-Bank Swasta Go-Public Di Bursa Efek Indonesia **12**(2) (2011)
32. Shahchera, M.: The impact of liquidity assets on Iranian bank profitability. In: International Conference on Management, Behavioral Sciences and Economics Issues (2012)
33. Sidhu, A.V., Rastogi, S., Gupte, R., Bhimavarapu, V.M.: Impact of liquidity coverage ratio on performance of select Indian banks. J. Risk Financ. Manage. **15**(5) (2022). https://doi.org/10.3390/jrfm15050226
34. SMBC.: The LCR disclosure (2016). https://www.smbc.co.jp/asia/thailand/TH_LCR_Disclosure.pdf
35. Veeramoothoo, S., Hammoudeh, S.: Impact of basel III liquidity regulations on U.S. Bank performance in different conditional profitability spectrums. North Am. J. Econ. Financ. **63** (2022). https://doi.org/10.1016/j.najef.2022.101826

Moving Towards Industry 5.0: Opportunities and Challenges in Bahrain Higher Education

Rumpa Roy and Ali Ateeq

Abstract This study aimed to investigate the challenges and opportunities of industry 5.0 in Bahrain higher education sector. The research questions are related to students' motivation and engagement in learning, IT infrastructure and resources affecting implementation of digital technologies, social and emotional wellbeing of students, and synergy between industry 5.0 and society 5.0. The research is qualitative in nature. Focus group interview was conducted with eight faculty members and content analysis was done to analyze the interview transcripts. Qualitative analysis reveals that application of industry 5.0 in education enhances students' engagement and motivation. With collaborative and experiential learning, social belongingness and emotional wellbeing of the students are improved. The human centric approach promotes acquisition of new skills and competencies aligned with the future jobs and focuses on linking activities with sustainable development goals. Challenges and opportunities of industry 5.0 in education have been identified. Based on the findings, a framework has been proposed for the universities to deal with the era 5.0. The framework shows how the universities should respond to the rapid changes brought by disruptive technologies with human centric approach. The universities need to implement digital transformation strategy, focus on building industry partnership, international collaboration, training, robust IT infrastructure and resources, and promote the culture of innovation and entrepreneurship.

Keywords Sustainable development goals · Learning experience · Social and emotional well-being · Society 5.0 · Learning environments

R. Roy · A. Ateeq (✉)
Administrative Science Department, College of Administrative and Financial Science, Gulf University, Sanad 26489, Kingdom of Bahrain
e-mail: dr.ali.ateeq@gulfuniversity.edu.bh

B. Awwad (ed.), *The AI Revolution: Driving Business Innovation and Research*, Studies in Systems, Decision and Control 524, https://doi.org/10.1007/978-3-031-54379-1_37

1 Introduction

Industry 5.0 refers to the era of robots along with the integration of technology. It is a transition to revolutionize the industrial framework ensuring human well-being and sustainability. It is also a pursuit to use innovative solutions for generating employment opportunities and economic growth which contributes to Sustainable Development Goals (SDGs). In education, industry 5.0 focuses on how human technology adopts human emotions in order to increase the efficiency and effectiveness of teaching and learning in a dynamic environment. Education 5.0 focuses on disruptive technologies in order to offer personalized learning experience that supports the social and emotional well-being of the students. This results in creative and innovative solutions that improve quality of life in the society [1]. Moving towards industry 5.0 envisions the transition to society 5.0 which will lead the future world. Industry 5.0 requires highly skilled professionals to drive towards society 5.0. This enhances learning capabilities and life skills of the students. This further allows them to mitigate risk, create sustainable solutions, and embrace digital transformation [2]. Future educators and practitioners will be supported with smart collaborative Robots to undertake tasks with human touch. It is noteworthy that industry 5.0 has ushered in an array of educational applications.

Within the context of Bahrain educational sector, it relates to using smart classrooms and learning environments ranging for interactive projector, smart board, virtual reality to offer immersive learning experience. Hands on learning skills of the students are enhanced through the use of virtual reality solutions in several disciplines. Notable application of industry 5.0 relates to using artificial intelligence and machine learning in providing personalized learning experience to students with diverse needs, capability and access to technology. This in turn contributes to massification of higher education which has become new paradigm amidst Covid-19 pandemic. There is a gap in the existing literature about the challenges and opportunities in embracing industry 5.0 within Bahrain higher education context. This research provides an overview of Industry 5.0 in education and investigates the challenges and opportunities of its application in Bahrain higher education sector. The research questions are:

- How does industry 5.0 facilitate students' motivation and engagement in learning?
- How does the IT infrastructure and resources affect implementation of industry 5.0 technologies?
- How does the implementation of industry 5.0 contribute to social and emotional well-being of students?
- How does industry 5.0 bring together society 5.0?

Focus group interview has been conducted to address the above-mentioned research questions. Subsequently, a framework has been proposed to lead the higher education providers towards industry 5.0.

2 Literature Review

The development of the era 5.0 in industry, society and education is underpinned by the fact that innovation and technological revolution do not replace the role of humans. It is important to integrate technology and emotional intelligence of human beings to increase efficiency in production process. Thus the concept of industry 5.0 thrives towards creativity, human touch and complementary relation between machine and human [3]. In Industry 5.0, there will be a primary emphasis on customized mass production with humans assuming the role of guiding robots [1]. History of industrial revolution in the context of societal revolution shows the origin of human beings to Hunter-gatherer Society (Society 1.0). Society 2.0 was characterized by agrarian society followed by Industrial Society (Society 3.0). This is the starting point of industrial revolution which was known as Industry 1.0 (mechanization). Industry 2.0 led to the use of electricity followed by Industry 3.0 (IT system) which resulted in emergence of Society 4.0 (Information Society). Industry 4.0 brought cyber-physical systems and other technologies. Latest transformation is known as Industry 5.0 which blends automation and human touch. This further gave birth to super-smart society (Society 5.0) [2].

Industry 5.0 ushers in rapidly changing digital technologies and artificial intelligence-based solutions in various sectors. Robots act as collaborator with human brain leading to value added production process. Further, this version has considered the aspect of environment protection and other aspects of sustainable future [4]. Human machine interaction has been evolved within social, technical, legal, environmental, organizational and ethical perspectives. This has been dominated by human needs and market requirements thus shifting from mass production to mass customization [5]. Human–robot co-working has its own consequences related to legal, regulatory, psychological, social, and ethical perspectives [6]. Industrial revolution has brought transformative development in all sectors of the economy. With the popularity in concepts like sustainability, human-centricity, carbon footprint, industries are struggling to adapt to the expectations of smart and innovative society. Industry 5.0 seeks to bridge the gap between manufacturing and social needs [7]. With the advent of fifth industrial revolution, organizations with innovative practices and responsive to market needs are more prone to success in terms of providing customer focused goods and services. Industry 5.0 emphasizes a paradigm shift towards enhancing the skills and abilities of people relying on personalization and customization of products [8].

In education industry 5.0 reflects the synergy between instructor, students and technology to improve the efficiency in learning process. The digital technologies have high impact on how the students learn, unlearn and relearn along with teaching pedagogies [9]. Society 5.0 envisions that universities, companies and others who are contributing to innovation ecosystem collaborate significantly and bring together techno-centric and human-centric approaches. Within the context of digital and green transformation, the universities are required to show their commitment for sustainable development goals through social innovation, partnering with multiple stakeholders

[10]. The interrelation between industrial revolution and curriculum innovation is evident in the field of vocational education and training through knowledge transfer, skills and attitude linked to production process [11].

Industry 5.0 (version of fifth industrial revolution) was coined by European Commission which emphasizes research and innovation; sustainable development; resilience, and human centricity. This approach integrates social and environmental imperatives with technological revolution. This era focuses on systemic approach, social objectives, emotion, ethics, and values. The research conducted within the context of Engineering students of Lublin University of Technology; Poland shows the nature of competencies in the field of sustainable development [12]. The research conducted by Mitchell and Guile have introduced the concept of fusion skills to cope up with the challenges of industry 5.0 especially in engineering curriculum. This is because by 2030, there are some threats on employability and jobs such as, automation, creation of new technology jobs, new jobs in health care and energy sector. It is obvious that digital skills are not enough to meet the requirement of future jobs. Future jobs arising out of human–machine collaboration have an impact on the economic research and reporting mechanism at work. These type of hybrid activities need fusion skills which results in better outcome in any business process [13]. Industry 5.0 signifies a change in emphasis from prioritizing economic value to prioritizing societal value and focusing on wellbeing rather than welfare. This aims to promote economic prosperity beyond the job creation and growth. Digitalization ensures inclusivity in diverse domains of society such as education, culture, sport and finance [14]. People are expected to develop innovative solutions to newly arising societal issues and make use of technology, economic resources, and mental capabilities in the pursuit of SDGs. Academic resilience of social science students in one of the Indonesian universities in facing industry 5.0 refer to the following parameters: "competence, self-confidence, character, commitment, interest, and self-control" [15]. The study conducted by Diestro concludes that learning readiness of students is impacted by value creation and productivity in the context of education 5.0 in a state university at Philippines [16]. Higher education institutions are required to formulate strategies to deal with the challenges of era 5.0 such as lifelong learning and multidisciplinary program; practical courses, sustainability driven curriculum and delivery, embedding skills and competencies for the future [17].

3 Methodology

The research is qualitative in nature. A focused group interview was conducted with eight faculty members contributing to different academic programs at Gulf University. Structured questions were prepared and participants in the interview attended the face to face session to share their insights. Content analysis was performed to interpret the interview transcripts within identified categories, reflecting the perceptions of the interviews on the industry 5.0 and how it is interrelated to education 5.0 and society 5.0. Based on the findings of the focus group interview, a framework has been proposed to provide direction to the universities to deal successfully with the era 5.0. The questions covered the following themes: students' motivation and engagement in learning through industry 5.0; Role of ICT infrastructure and resources to facilitate this revolution; social and emotional well-being of students; integration of Industry 5.0 and Society 5.0; opportunities and challenges to embrace this change.

4 Results

The researcher conducted a focus group interview and the interview transcript was analysed to interpret the results within the context of era 5.0, using deductive qualitative analysis [18]. A summary of the interview transcript is presented below:

- Industry 5.0 facilitates students' motivation and engagement in learning through the integration between technology and teaching pedagogy. "By integrating advanced technologies such as artificial intelligence, virtual reality, and augmented reality into educational settings, Industry 5.0 enables interactive and immersive learning experiences". Perception of the student has been changed with the advent of education 5.0 which emphasizes humanized aspect of learning. "With the optimistic view for futuristic education, each era/version is preparing the users for the next version". The students are exposed to reflect creativity and enhance their practical learning skills through simulation, projects, problem solving, and scenario based tasks. Students are able to understand relevance of the topic in a collaborative environment through virtual reality, augmented reality, gamification and so on. These innovative pedagogies not only make learning enjoyable but prepare work-ready graduates.
- "A robust and reliable IT infrastructure is essential for supporting the connectivity, data exchange, and integration of various technologies". Widespread use of educational technologies has been supporting the institutions to expand the reach in social media through You tube, Instagram, TikTok. "New applications specially in the field of design and engineering have given competitive advantage to the smart users, staying updated with the applications has become a challenge". The universities invest in establishing labs, procuring cutting-edge hardware/devices, software, and digital tools subscribing to software to respond to the digital transformation. "Data storage, back up and server to secure data facilitate the automation in teaching, learning, research and campus operation". "Transition to education 6.0 will follow similar disruptive technologies and IT infrastructure but with more human centric approach". Technology has become the context for learning rather than a tool.
- Industry 5.0 encourages students for collaborative learning, teamwork, sharing ideas, networking with peers. This explores opportunities for social interaction and belongingness beyond the classroom contact hours. Students are able to develop interpersonal skills and competencies through active engagement via synchronous and asynchronous sessions. "The acceleration in the version has

become imperative in order to social and emotional wellbeing of the students and faculty. The gap between user interface and user experience gets reduced with the human centric approach". Industry 5.0 and education 5.0 result in democratizing higher education with flexibility, inclusivity and accessibility. "Students' participation in group project, assignment, role play, brainstorming, simulation and the like not only provides opportunities for socialization but reduces stress, builds self-confidence and motivates for learning". "Universities around the world have started exploring metaverse through self-developed platform in various directions such as, removing space constraints and conducting virtual classes anywhere and everywhere; providing experiential learning and augmented learning, experiencing 3D perspective views of human body parts, creating own content and avatar, offering healthcare and bio engineering courses".

- "Industry 5.0 brings together society 5.0 through sustainable, inclusive and human centric societies". It promotes equality, diversity, cross border cultural exchange, empathy for colleagues and so on. Regular updates on these educational technologies underpinned by industry 5.0 support the higher education providers to cater to the needs of the market both for current and future jobs. The trend to map all the activities of industry and universities with sustainable development goals has been evolved since industry 5.0 ignored human interaction and emotion. Aligning digital technologies with societal needs brings in societal progress in an era of digitization. The collaboration between industry and academia plays a pivotal role to bring this paradigm of the new era namely, industry 5.0, education 5.0 and society 5.0.
- Within the context of Bahrain higher education sector, industry 5.0 ushers in significant opportunities in terms of preparedness of stakeholders, IT infrastructure, resources and facilities, partnership between industry and academia towards integrating technology into learning. Industry 5.0 explores opportunities for personalized learning experience which allows fast and slow learners to engage in the learning process as per the needs and preferences. This era of disruptive technologies coupled with human emotions and actions facilitate learning in terms of gained knowledge, skills and competencies. "It opens up opportunities for research collaboration, innovation, internationalization of higher education through collaborative activities between international students, staff and students exchange program. This results in contributing to economic growth and improving quality of life".
- The challenges faced by higher education are related to speed of acceleration in transiting from one era to another; readiness of the users; coping up with the changes in application and its implementation; training for faculty and students etc. "Once entered into the era there is no opportunity to retrospect and go back to the previous era even though people need time to adapt to the changes brought by the previous era". Another challenge faced by the university is to allocate fund for bringing new technologies and training. "Because of the fast change in the application, it becomes difficult to predict the future skills in the job market". There is another issue related to ethical consideration of the new technologies.

- As a way forward, the universities must align their strategies with digital transformation, SDGs. There is a change in the culture towards showing readiness for digitalization; invest in providing cutting edge hardware and software; high speed internet

Based on the above transcript, content analysis was done to depict the interpretation of the narratives obtained from the focus group session; the summary results are presented in Table 1. The content analysis identified the categories, perception of the participants, and the interpretation therein.

Content analysis of the interview transcripts reveal that interviewees are very much aware about the era 5.0 and how industry 5.0 is portrayed to promote sustainable development, having resilience, environmental sustainability and human centricity as its core characteristics. The participants shared their opinion about the challenges and opportunities in higher education sector in Bahrain to respond to the industry 5.0. Regarding the way ahead for the universities, participants came out with specific measures like building trust between the stakeholders, national and international collaboration, network with industry, digitization and so on.

5 Discussion

Qualitative analysis of the interview results categorically illustrates that within the disruptive context, the organizations are vulnerable to uncertainty, unpredictability, and complexity. As per their opinion, industry 5.0 provides opportunities for interactive and immersive learning experience, collaborative learning and personalized learning experiences such that it increases student engagement in the digital platform. This motivates the students to participate in the learning and personalized learning allows him/her to learn at own speed. Industry 5.0 coupled with society 5.0 increases social belongingness and emotional wellbeing of the students. The human centric approach promotes acquisition of new skills and competencies to be at par with the future jobs and focus on aligning activities with sustainable development goals. This era explores opportunities for internationalization, partnership with industry and digital transformation in teaching, research and campus operation. These comments are also confirmed by the existing research which covers issues like internationalization and industry partnership [19, 20]. Based on the introspection of the participants on the future imperatives of the universities to embrace digitization and social innovation, the researcher proposed a framework applicable for any higher education provider in the kingdom. The framework for the universities in the era 5.0 has been presented in Fig. 1.

Figure 1 illustrates how the universities should respond to the rapid changes brought by disruptive technologies with human centric approach. At the strategic level, the university management should have the vision for digital transformation, internationalization, collaboration with industry, promoting the culture of innovation and entrepreneurship. At the operational level, learner centric education, experiential

Table 1 Content analysis of interview transcripts

Category	Identified issues	Interpretation
Students' engagement and motivation in learning	Engagement through integration between technology and teaching pedagogy; acquisition of skills and competencies; reflect creativity; hands on learning experience; innovative pedagogies; collaborative learning environment; motivated to learn through customized learning opportunity	• Interactive and immersive learning experiences • Collaborative learning • Personalized learning • Career opportunities
Social interaction and belongingness	Bonding between classmates/ colleagues, interaction between student and instructor via digital platform; leadership and teambuilding; communication with stakeholders	• Teamworking and empathy • Interpersonal relation • Multidisciplinary team • Social connectedness
IT infrastructure and resources	Robust and reliable IT infrastructure; cutting-edge hardware, software, and digital tools; automation in teaching, learning, research and campus operation	• Reliable network • Cloud based platforms • Data security and privacy • Digital fluency
Social and emotional well-being	Developing interpersonal skills and competencies through active engagement; participation in group work/ assignment/project; peer interaction; reduces stress, builds self-confidence; democratizing higher education with flexibility, inclusivity and accessibility; takes care of mental well-being of students	• Human centric approach • Learning for enjoyment and pleasure • Experiential learning • Opportunities for Metaverse education
Link between industry 5.0 and society 5.0	Promoting sustainable, inclusive and human centric societies; catering to the needs of the market both for current and future jobs; mapping all the activities with SDGs, synergy between digital transformation and human emotions	• Partnership between industry and academia • Societal progress • Contributing to SDGs through its endeavors

(continued)

Table 1 (continued)

Category	Identified issues	Interpretation
Opportunities to adapt to industry 5.0 in Bahrain higher education	Building industry-academia network; joint research project with international universities; providing international exposure to staff and students; raising the quality of life; contributing to economic diversification	• University's strategy towards digital transformation • Fostering digital well-being • Empowering students with new set of skills and competencies • Promoting innovation and entrepreneurship • International collaboration •Staff and student exchange
Challenges in moving towards industry 5.0	Allocating funds for bringing new technologies and training; fast change in the application; difficulty to predict the future skills in the job market; adaptability to cope up with the digital transformation trends; ethical consideration	• Resistance to change • Lack of training in using digital tools and technologies • Redesigning the curriculum and teaching pedagogies
Way ahead for the universities	There is a change in the culture towards showing readiness for digitalization; invest in providing cutting edge hardware and software, high speed internet	• Build the trust between new generation and the disruptive technologies • Articulate digital transformation strategy • Promote national and international collaboration • Build network between industry and university

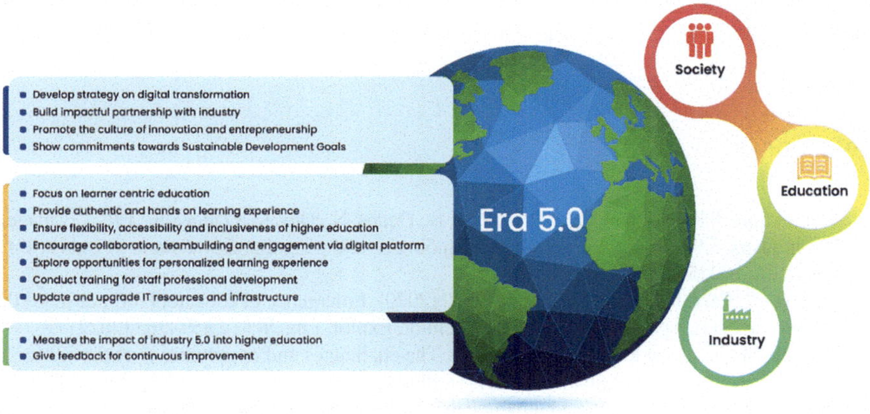

Fig. 1 Proposed framework for universities in era 5.0

learning, flexibility, accessibility and inclusion, robust IT resources and infrastructure, training, innovative teaching pedagogies are pivotal to adapt to the new era. Finally, measuring outcome and feedback for improvement ensure closing the loop. This framework is equally applicable to other countries around the world through transformation in vocational education, strengthening the need for STEAM education and interdisciplinary programs, upskilling and reskilling through robotics and automation.

6 Conclusion

Industry 4.0 relied on automation at the expense of human, social and environmental dimensions. Industry 5.0 brings these dimensions back into consideration focusing on sustainable development, human centricity, and resilience. Education 5.0 speaks about academic revolution but prioritizes humans, not technology. It does not focus only on providing better connectivity, infrastructure or electronic gadgets but to prepare competitive global graduates with intellectual, social and emotional skills who meet the expectations of the stakeholders. Society 5.0 aims to improve people's lives, encourage social responsibility, and promote sustainability by aligning technological revolution with societal needs. Within this context, the research investigates the challenges and opportunities of industry 5.0 in Bahrain higher education sector. Qualitative analysis of the focused group interview responses depict that organizations embrace technological revolution. Industry 5.0 ensures better engagement and motivation of students, improves social and emotional wellbeing contributing to sustainable development goals. The study proposed a framework to be adapted by universities to respond to the new era of 5.0 focusing on digital transformation strategy, internationalization, partnership with industry, collaborative learning, innovation and entrepreneurship, training, robust IT resources and infrastructure.

References

1. Maddikunta, P.K.R., Pham, Q.V., Prabadevi, B., Deepa, N., Dev, K., Gadekallu, T.R., Liyanage, M.: Industry 5.0: a survey on enabling technologies and potential applications. J. Indust. Inf. Integr. **26**, 100257 (2022)
2. Saxena, A., Pant, D., Saxena, A., & Patel, C. (2020). Emergence of educators for industry 5.0: an indological perspective. Int. J. Innov. Technol. Explor. Eng. **9**(12), 359–363 (2022)
3. Tavares, M.C., Azevedo, G., Marques, R.P.: The challenges and opportunities of era 5.0 for a more humanistic and sustainable society—a literature review. Societies, **12**(6), 149 (2022)
4. Nahavandi, S.: Industry 5.0—a human-centric solution. Sustainability **11**(16), 4371 (2019)
5. Pizoń, J., Gola, A.: Human–machine relationship—perspective and future roadmap for industry 5.0 solutions. Machines **11**(2), 203 (2023)
6. Demir, K.A., Döven, G., Sezen, B.: Industry 5.0 and human-robot co-working. Proc. Comput. Sci. **158**, 688–695 (2019)

7. Leng, J., Sha, W., Wang, B., Zheng, P., Zhuang, C., Liu, Q., Wang, L.: Industry 5.0: prospect and retrospect. J. Manuf. Syst. **65**, 279–295 (2022)
8. Paschek, D., Mocan, A., Draghici, A.: (2019, May) Industry 5.0—the expected impact of next industrial revolution. In: Thriving on future education, industry, business, and society, proceedings of the makelearn and TIIM international conference, pp 15–17. Piran, Slovenia
9. Al-Emran, M., & Al-Sharafi, M.A.: Revolutionizing education with industry 5.0: challenges and future research agendas. Int. J. Inf. Technol. Lang. Stud. **6**(3)
10. Carayannis, E.G., Morawska-Jancelewicz, J.: The futures of Europe: society 5.0 and industry 5.0 as driving forces of future universities. J. Knowl. Econ. 1–27 (2022)
11. Marinič, P., Pecina, P.: Industry 5.0 in vocational education. In: INTED2023 Proceedings, pp. 3805–3811. IATED (2023)
12. Mazur, B., Walczyna, A.: Sustainable development competences of engineering students in light of the industry 5.0 concept. Sustainability **14**(12), 7233 (2022)
13. Mitchell, J., Guile, D.: Fusion skills and industry 5.0: conceptions and challenges. Insights Into Global Eng. Educ. After Birth Indust. **5**(0), 53 (2022)
14. Dovleac, L., Chițu, I.B., Nichifor, E., Brătucu, G.: Shaping the inclusivity in the new society by enhancing the digitainability of sustainable development goals with education. Sustainability **15**(4), 3782 (2023)
15. Rachmawati, I., Multisari, W., Triyono, T., Simon, I.M., da Costa, A.: Prevalence of academic resilience of social science students in facing the industry 5.0 era. Int. J. Eval. Res. Educ. **10**(2), 676–683 (2021)
16. Diestro, D.: Learning readiness in education 5.0 as influenced by value creation and academic productivity. EPRA Int. J. Res. Develop. (IJRD) **7**(8), 179–184 (2022)
17. Broo, D.G., Kaynak, O., Sait, S.M.: Rethinking engineering education at the age of industry 5.0. J. Indust. Inf. Integr. **25**, 100311 (2022)
18. El Marsafawy, H., Roy, R., Ali, F.: Measuring learning outcomes: bridging accreditation requirements and LMS functionalities. Quality Assur. Educ. (ahead-of-print) (2022)
19. Bevanda, V., Brenko, D., Franković, P.: University-industry collaboration: a case study of enhancing the quality of information technology education. In: INTED2014 Proceedings, pp 252–261. IATED
20. Alhalwaki, H., Hamdan, A.M.M.: Factors affecting the implementation of internationalisation strategies in higher education institutions: evidence from Bahrain. Int. J. Manage. Educ. **13**(1), 1–27 (2019)

Factors Affecting the Trading Performance of Cryptocurrency Investors in Bahrain

Tasnim Khalid, Mujeeb Saif Mohsen Al-Absy, and Mohammad Omar Farooq

Abstract Cryptocurrencies have begun to gain popularity and are now a viable alternative asset class for investors. The purpose of this study is to investigate the effect of behavioral factors (e.g., herding and overconfidence), demographic factors (e.g., education), and investment experience on trade performance. This is quantitative research where a questionnaire was distributed to 270 cryptocurrency investors in Bahrain. The results show that behavioral factors such as herding, and overconfidence significantly affect the trading performance of cryptocurrency investors. However, demographic indicators such as educational level, investor age, and investment experience have no impact on cryptocurrency investors' trading performance. Based on the knowledge of the researchers, this is the first study of this type in Bahrain. Hence, it will significantly contribute to the body of knowledge about cryptocurrency investment. Further, the study helps policymakers reevaluate the performance of cryptocurrency investment and strengthen the current policies and approach in this regard. The study advances knowledge of the characteristics and practices of individual Bitcoin investors that influence investment choices.

Keywords Cryptocurrency · Trading · Investment · Performance · Behavioral factors · Demographic factors

1 Introduction

The use of technology in modern life has grown essential [1, 2]. Further, developments in financial technology have led to the creation of a new kind of asset known as cryptocurrency. Currently, it is expected that there will be over 106 millions cryptocurrency buyers and sellers worldwide, making it one of the marketplaces with

T. Khalid · M. S. M. Al-Absy (✉)
Accounting and Financial Science Department, College of Administrative and Financial Science, Gulf University, Sanad 26489, Kingdom of Bahrain
e-mail: dr.mujeeb.saif@gulfuniversity.edu.bh

M. O. Farooq
Department of Economics, United International University, Dhaka 1212, Bangladesh

B. Awwad (ed.), *The AI Revolution: Driving Business Innovation and Research*, Studies in Systems, Decision and Control 524, https://doi.org/10.1007/978-3-031-54379-1_38

the fastest growth [3]. Cryptocurrency is based on Blockchain, an electronic ledger system. Blockchains are distributed ledger systems that connect to one another at each node through a peer-to-peer network. Each network component meticulously validates every transaction.

Over 500 virtual currencies are currently in competition, including well-known ones like Bitcoin, Litecoin, Ripple, and Ethereum. The most valuable cryptocurrency by market capitalization is now Bitcoin [4]. Bitcoin was established in 2009 by Satoshi Nakamoto. According to legend, Bitcoin was the first cryptocurrency to put a strong emphasis on decentralization, which implies that the user controls these currencies independently of any rules set by other parties regulating the transaction [5].

The three main factors that affect the price of Bitcoin and other cryptocurrencies are the supply and demand of cryptocurrencies, macroeconomic and financial developments, and investor interest according to study conducted by Ciaian, Rajcaniova [6]. Further, investors find it challenging to choose the best situation in trading and investing to optimize returns because of widely erratic price movements. When investing in cryptocurrencies, the three main concerns are operations, security, and currency rate risk.

Trading cryptocurrency also broadly resembles trading on the present stock market [7]. Investors of all experience levels are drawn to cryptocurrencies. Both large institutional investors and small individual investors are impacted by market fluctuations. Additionally, digital trading systems that are incredibly automatic handle the buy and sell orders. Values can drastically alter overnight as a result of shifts in sentiment, celebrity, or a single note receiving a lot of attention on websites like Reddit [8].

The high volatility of the cryptocurrency market is a crucial problem. A single coin's price could rise by above 100% within hours and then quickly decline again. Even the most common cryptocurrency, Bitcoin, can experience dramatic price swings. For instance, in just two days in February 2021, it dropped 20% of its value [9]. Markets also go through more serious boom-and-bust cycles than the conventional stock market does. For instance, throughout well-known "bull runs" (as in 2017), the price of Bitcoin would occasionally fall between 30 and 40%.

On the other hand, cryptocurrency price changes are frequently even more unpredictable. Opportunities for trading are generated, but the risk is extraordinarily high. An investment of $100 made at the proper time during the so-called "crypto bubble" of 2017 witnessed a return of over 100,000%, increasing it to over $1 million in less than a year. One such popular coin is the privacy coin Verge. During the bust cycle of 2017–2018, this coin and others dropped by over 80%, costing many altcoin investors who did not sell at the peak a large portion of their capital gains.

In order to turn the Kingdom into a regional fintech hub, the Central Bank of Bahrain (CBB) has put in place a number of key regulatory measures aimed at recruiting new bitcoin and blockchain businesses. Since Bahrain is considered the first Middle Eastern country to establish a central bank-licensed cryptocurrency exchange, it is anticipated that Bahrain will continue to be at the forefront of regional fintech growth as cryptocurrencies develop. Around the world, there are more than 4300 sites for trading cryptocurrencies. Only two trading platforms, the Rain platform (where seven digital currencies are traded) and the CoinMENA platform (where nine digital currencies are traded), were granted licenses by the Central Bank of Bahrain. With more than 50,000 digital currencies in use today, Bahrainis are increasingly requesting cryptocurrency platforms and traders to invest in this industry, according to their reports.

Gaining profit from an investment is its main goal. But investing in cryptocurrencies remains highly speculative and therefore carries significant risk with the potential for major losses. According to anecdotal evidence, many investors are losing money in Bahrain. Poor trading performance, which has produced losses rather than gains, is therefore the primary issue in this study. Analyzing the variables affecting investor trading performance in Bahrain is essential. The success or failure of cryptocurrency trading remains a topic under investigation in Bahrain, despite the fact that numerous studies focused on other areas or countries have been carried out to identify the factors that cause investors to profit or lose money while investing in cryptocurrencies.

Herding occurs when investors choose to follow observable market movements or other people's judgments rather than their own convictions and knowledge. Although it may not always result in optimal market results, such behavior may be perceived as being sensible from an individual perspective for a variety of reasons [10]. According to Humra [11], herding behavior occurs when a group of investors base their judgments on information shared by the group and ignore other information. Therefore, significant market price differences result when the group majority takes the incorrect course of action [12].

Additionally, according to Baker and Nofsinger [13], overconfidence is a type of high self-confidence that one can use their information efficiently since they have great analytical ability. This inexperience and incapacity to assess the available information merely amount to an illusion of believing [14]. A person's conduct when making financial decisions is thought to be influenced by demographic characteristics, and changes in these elements may have an impact on how a person acts.

Thus, the study aims to examine the influence of the behavioral factors (e.g., herding and overconfidence), demographic factors (e.g., education), and investment experience on cryptocurrency traders in the Kingdom of Bahrain. This study is extremely important for investors and traders since everyone who trades or invests in cryptocurrencies wants to make money from their investments. To prevent losses, cryptocurrency investors must be aware of the key elements that will affect their investments before making any. This research will help them identify and understand these factors. It can also be extensively used by businesses that provide financial services and investments to understand the key elements influencing trading in cryptocurrency, which could assist them in advising their clients on how to maximize gains and minimize losses.

Investment choices are more crucial than ever today. This study was conducted to learn more about behavioral and demographic factors that affect investment choices and outcomes in Bahrain's developing cryptocurrency industry. The study's main goal is to comprehend how investors behave while choosing investments and how these factors impact their success. Furthermore, research in this area would considerably advance the body of knowledge because it hasn't been done in Bahrain.

This study makes a new contribution combining a number of independent factors with a single dependent variable. The researchers and academicians will profit from this research because it will give them more information and value. Since there isn't such a robust study that examined the relationship between all five of those factors and cryptocurrency investments in the Kingdom of Bahrain, researchers and academics can identify trends using realistic information from the current study concerning how herding, overconfidence, age, level of education, and investment experience influence the investments of investors in cryptocurrency.

The following section will review the literature in terms of the trading performance of cryptocurrency. Further, Sect. 3 will introduce the method used in the study. In Sect. 4, the authors will summarize the descriptive Statistics and explain the results of the study. Lastly, the conclusion of the study will be placed in Sect. 5.

2 Literature Review

The term "trader" describes a person who executes swift trades on a specific instrument, such as stocks, foreign exchange, or virtual money. The golden rule of trading is to buy when prices are low and sell when they are high, or if trading short, sell high and then buy low. The trader will make capital gains if the price rises (or falls respectively). Trading is a career that necessitates in-depth investigation to reach the objective and obtain the greatest profit. Prihatini and Widakdo [15] encouraged traders to comprehend the proper trading kind and application before beginning to trade. The performance of digital currency among traders and investors has not yet been the subject of considerable research [16].

Due to decentralization and computerization, blockchain can significantly improve the efficiency of stock trades [17]. Prihatini and Widakdo [15] described trading performance as a technique for determining a trader's return and risk tolerance—or lack thereof. Trading success changes over time and under different conditions as a result of several behavioral biases [18]. Several elements that determine whether an investment in cryptocurrencies is successful or unsuccessful have an impact on trading performance. The herding effect has an impact on the choice of investment, according to Wamae [19]'s study of the behavioral factors influencing investment decisions in the stock market. This study included variables risk aversion, conducting prospecting, anchoring, and herding. Additionally, risk aversion affects how investment decisions are made, as does prospecting. Furthermore, anchoring has an impact on financial choices.

Regarding herding, a few research have looked at how it affects the trading success of bitcoin investors. The Greater Jakarta region's choice to make investments in cryptocurrencies is not significantly influenced by herd behavior, according to research by Nurbarani and Soepriyanto [20]. As a result of their investigation, Kengatharan and Kengatharan [21] found that the herding factor and its dimensions had little influence on the selection of an investing strategy. However, studies by Silwal and Bajracharya [22], Wamae [19], and Le Luong and Thi Thu Ha [23] revealed that herding behaviors had a beneficial and significant effect on investment success. Herding behavior, as demonstrated by the fact that investors typically rely on collective knowledge from other traders rather than individual information, has a substantial impact on investment decisions, claim Kartini and Nahda [12]. Herding, according to Calderón [24], has the greatest beneficial effect on investment success. Similar findings were made by Cao, Nguyen [25], who discovered that herding influences investment decision-making favorably.

According to research by Nurbarani and Soepriyanto [20] on the element of overconfidence and its impact on trading success, overconfidence significantly influences people's decisions to invest in cryptocurrencies in the Greater Jakarta region. Also coming to the same conclusion were Le Luong and Thi Thu Ha [23], Kartini and Nahda [12], Aziz and Khan [26] and [27]. Additionally, the research by Silwal and Bajracharya [22] shows that overconfidence had the strongest positive and substantial influence on the performance of investments. Although the study by Kengatharan and Kengatharan [21] shows that overconfidence has a detrimental significant impact on investing success.

Prior research has been done on the impact of demographic characteristics, such as education, gender, age, and work experience, on performance. According to Anggita and Kawedar [28], generation had no bearing on performance, but education level and length of employment did. Guterresa, Armanu [29] found evidence demonstrating the impact of education on team member performance. Many research, including those by Das and Jain [30], Chavali and Mohanraj [31], and Tanusdjaja [32], have attempted to answer this question with varied degrees of success. Studies by Khanam [33], Artina and Cholid [34], Christanti and Mahastanti [35], and Putri and Isbanah [36], among others, have found a variety of results regarding the relationship between education level and investment decisions.

According to the hypothesis of herd behavior, investors make judgments based on the advice of others rather than their own research. According to the herd behavior idea, an investor will decide based on all the information at hand. People who are interested in it are influenced by this notion of thinking and behave like everyone else. Investors follow what they see other investors doing rather than relying just on their own judgment [22]. Unquestionably, as was observed in the literature review, the demographic and behavioral factors affect the investing performance of cryptocurrency investors. This study examines the levels of impact that demographic and behavioral characteristics have on the decisions that cryptocurrency investors make and the outcomes of their investments in the Kingdom of Bahrain. The investigation came up with the following hypothesis:

H_1: There is a positive relationship between age and trading performance of cryptocurrency investor in Bahrain.

H_2: There is a positive relationship between educational qualification and trading performance of cryptocurrency investor in Bahrain.

H_3: There is a positive relationship between investment experience and trading performance of cryptocurrency investor in Bahrain.
H_4: There is a positive relationship between herding and trading performance of cryptocurrency investor in Bahrain.
H_5: There is a positive relationship between overconfidence and trading performance of cryptocurrency investor in Bahrain.

3 Methodology

This study is quantitative research where it distributed surveys for collecting evidence of the research and determining the effect of behavioral factors (e.g., herding and overconfidence), demographic factors (e.g., education), and investment experience on trading performance of cryptocurrency investors. The population of the study are cryptocurrency investors in Bahrain. Based on the statistics of Start.io audiences, the active population of cryptocurrency investors in Bahrain are 75,136 [3]. Based on the Raosoft's website, the study targeted 270 cryptocurrency traders in Bahrain.

In terms of measurements, the sampling technique used for this research is non-probability sampling. Three items, 5-point agreement Likert Scale, used to measure trading performance following Silwal and Bajracharya [22] and Le Luong and Thi Thu Ha [23]. In term of herding behavior, the study used seven items, 5-point agreement Likert Scale, following Silwal and Bajracharya [22] and Le Luong and Thi Thu Ha [23]. Concerning overconfidence behavior, the study used five items, 5-point agreement Likert Scale, to measure the overconfidence following Silwal and Bajracharya [22] and Le Luong and Thi Thu Ha [23]. The Statistical Package for Social Sciences (SPSS) was used to examine the answers. On social media channels, the survey was sent to 270 cryptocurrency investors in Bahrain. Of them, 118 investors responded, yielding a respectable response rate of 43.7%.

4 Results

4.1 Descriptive Statistics

Table 1 reveals that, with a ratio of 60.2, traders and investors are overwhelmingly male. 48 investors, or 40.7% of the sample, are cryptocurrency investors who are between the ages of 25 and 35. Typically, these investors are newcomers. Because the sample appropriately reflects the fact that a sizable portion of cryptocurrency investors are between the ages of 25 and 35, this study could accurately represent the performance of these individuals' investments. The following table demonstrates that more than half of the study sample had a bachelor's degree (60.2%), while investors with only a high school diploma or an equal educational background had the lowest number (5.1%). The data also reveals that a sizable portion of the sample investors have been active in cryptocurrency trading for three to five years; of the respondents, 42.4% had been doing so for under three years, 46.6% had been active for three to five years, and 5.9% had been active for six to ten years.

Table 2 shows the descriptive statistics for all items related to herding behavior (HERD), overconfidence behavior (OVR) and cryptocurrency traders (TRD). In terms of herding behavior, results of the tables above show that the percentage of Average Agreed for all items is above 76.2%. Overall, the majority of those surveyed strongly agreed with the herding behavior-related questions, as "strongly agree" received the highest percentage of responses across all questions. The lowest response rate was "strongly disagree" across the board. This showed that a large majority of investors firmly support the herding variable. Regarding overconfidence behavior, the results show that the percentage of average agreed for all items is above 71.2%. In general, the majority of respondents agreed or strongly agreed with the statements made in the questions about overconfident behaviors. In contrast, "strongly disagree" or "agree" responses had the lowest percentage across all questions. This demonstrated that the majority of investors are in strong agreement.

Questions about the prior success of bitcoin investors in the Kingdom of Bahrain are presented in Table 2. The results show that on average more than half of cryptocurrency investors strongly agree or agree on all questions. The level of average is higher the 50% for all items of the cryptocurrency investors performance.

Table 1 Demographic statistics

Variable	Categorize	Frequency	Percent
Gender	Male	71	60.2%
	Female	29	39.8%
Age	< 25	34	28.8
	25–35	48	40.7
	36–45	22	18.6
	46–55	9	7.6
	> 55	5	4.2
Educational qualification	High school or equivalent	6	5.1
	Diploma or equivalent	15	12.7
	Bachelor's degree	71	60.2
	Master's degree	18	15.3
	Doctor of philosophy degree (PhD)	8	6.8
Investment experience	< 3	50	42.4
	3–5 years	55	46.6
	6–10 years	7	5.9
	> 10 years	6	5.1

Table 2 Descriptive statistics

No	Item	Strongly agree %	Agree %	Neutral %	Disagree %	Strongly disagree %	Average agreed %
HERD1	Your investment decisions are influenced by the actions of other investors generally	59.3	31.4	7.6	1.7	0	90.7
HER2	Your investment decisions are influenced by the choice of digital currencies made by other investors	50.8	35.6	13.6	0	0	86.4
HER3	Your investment decisions are influenced by the buying and selling decisions of other investors	50.8	33.1	14.4	1.7	0	83.9
HER4	You usually react rapidly to changes in other investors' decisions and follow their market reactions	50.8	25.4	22.9	0.8	0	76.2
HER5	If your peers are investing in the same cryptocurrency, your intention to do so will increase	55.9	28.8	11.9	3.4	0	84.7
HER6	If your friends prefer to invest in bitcoin, you are inclined to do so as well	47.5	35.6	13.6	3.4	0	83.1

(continued)

Table 2 (continued)

No	Item	Strongly agree %	Agree %	Neutral %	Disagree %	Strongly disagree %	Average agreed %
OVR1	You feel that your knowledge and talents in the bitcoin market will enable you to outperform the market	48.3	30.5	17.8	2.5	0	78.8
OVR2	Normally, you can predict whether the performance will be good or bad	42.4	39	13.6	4.2	0.8	81.4
OVR3	You believe yourself to be an experienced investor	43.2	28	15.3	11	2.5	71.2
OVR4	Compared to those of your coworkers or acquaintances, you feel more confident in your own investing opinions	45.8	28.8	19.5	5.1	0.8	74.6
OVR5	Based on the current pricing, you can predict future changes in cryptocurrency prices	46.6	29.7	11	11.9	0.8	76.3
TRD1	The return on your most recent investment is in line with your expectations	38.1	16.9	24.6	18.6	1.7	55

(continued)

Table 2 (continued)

No	Item	Strongly agree %	Agree %	Neutral %	Disagree %	Strongly disagree %	Average agreed %
TRD2	Recently, your rate of return has been on par with or higher than the market's average return rate	34.7	16.1	26.3	19.5	3.4	50.8
TRD3	You're satisfied with the investments you made in the previous year	35.6	16.9	20.3	20.3	6.8	52.5

4.2 *Regression Results*

The study starts with examining the reliability of the questionnaire which measured using a reliability test. The reliability test for this study used the Cronbach Alpha coefficient, and it concluded that questions were reliable if they met the requirement of > 0.6. In cases where the Cronbach Alpha coefficient is below 0.6, the question is regarded as unreliable.

Table 3 shows that results of Cronbach's alpha, which is 0.881, 0.896 and 0.934 for the variable of herding, overconfidence, and trading performance, respectively. Hence it is higher than 0.6 indicates that the scale has outstanding dependability.

Concerning the correlations between variables, the study conducted the test of correlations and found that there is no high correlation between variables. Table 4 shows that most of the correlations between variables are less than 30% and few are within 30–70%. However, there is no correlation between higher than 70%. Hence, results indicate that there is not a correlation issue in the data of the study.

Regarding the model of the study, it seems to be fit and significant. Table 5 shows that R square is 56% and the model is significant at 1% level.

The regression results in presented in Table 6, which show that overconfidence has a positive significant impact on trading performance. Therefore, it is agreed that the overconfidence factor has a favorable impact on Bahraini cryptocurrency investors' trading success. The study found that overconfidence bias had a big impact

Table 3 Cronbach alpha

Variable	Item	Cronbach's alpha
Herding	6	0.881
Overconfidence	5	0.896
Trading performance	3	0.934

Table 4 Correlation test

		HER	OVR	TRD	Educational qualification	Age	Investment experience
HER	Pearson correlation	1	0.455**	0.446**	− 0.068	0.049	− 0.084
	Sig. (2-tailed)		0.000	0.000	0.464	0.596	0.367
	N	118	118	118	118	118	118
OVR	Pearson correlation	0.455**	1	0.688**	− 0.071	− 0.171	− 0.313**
	Sig. (2-tailed)	0.000		0.000	0.446	0.063	0.001
	N	118	118	118	118	118	118
TRD	Pearson correlation	0.446**	0.688**	1	− 0.064	− 0.326**	− 0.381**
	Sig. (2-tailed)	0.000	0.000		0.489	0.000	0.000
	N	118	118	118	118	118	118
Educational qualification	Pearson correlation	− 0.068	− 0.071	− 0.064	1	− 0.087	− 0.036
	Sig. (2-tailed)	0.464	0.446	0.489		0.347	0.696
	N	118	118	118	118	118	118
Age	Pearson correlation	0.049	− 0.171	− 0.326**	− 0.087	1	0.523**
	Sig. (2-tailed)	0.596	0.063	0.000	0.347		0.000
	N	118	118	118	118	118	118
Investment experience	Pearson correlation	− 0.084	− 0.313**	− 0.381**	− 0.036	0.523**	1
	Sig. (2-tailed)	0.367	0.001	0.000	0.696	0.000	
	N	118	118	118	118	118	118

**. Correlation is significant at the 0.01 level (2-tailed)

Table 5 Model summary

Model	R	R square	Adjusted R square	Std. error of the estimate	Change statistics				
					R square change	F change	df1	df2	Sig. F change
1	0.747a	0.558	0.539	0.80680	0.558	28.305	5	112	0.000

on trading performance. There is a good chance that cryptocurrency holders will be more passionate and driven to improve their investing skills.

Herding has a considerable beneficial impact on trading success, as evidenced by the P-value of the herding variable, which is below 0.05 and equal to 0.005. Therefore, it is believed that the herding element has a favorable impact on the trading results of bitcoin investors in Bahrain. As a result, because they value other people's investing decisions over their own, investors react impulsively to changes in other people's opinions. They give more weight on what the market decides and pay less attention to the company's future.

Concerning qualifications, the result shows that education qualifications do not significantly impact trading performance. The educational qualification variable's P-value is 0.595, which is more than 0.05. Thus, the claim that a cryptocurrency investor's trading success is positively influenced by their educational background in Bahrain is disproved.

In terms of age, the result shows that age variable has insignificant impact on trading performance as the P-value of age variable is 0.090 which is more than 0.05. Further, it has negative insignificant impact. As a result, it is disproved that age has a positive significant impact on the trading success of cryptocurrency investors in Bahrain.

Regarding investment, the result shows that there is an insignificant negative relationship between investment experience and trading performance as the P-value of investment experience variable is 0.214 which is more than 0.05. Hence, the hypothesis that there is a positive significant relationship between investment experience and the trading performance of cryptocurrency investor in Bahrain is rejected.

Table 6 Regression results

Model		Unstandardized coefficients		Standardized coefficients	t	Sig
		B	Std. error	Beta		
1	(Constant)	1.168	0.409		2.857	0.005
	OVR	0.755	0.106	0.528	7.090	0.000
	HER	0.391	0.136	0.205	2.873	0.005
	Educational qualification	– 0.082	0.153	– 0.034	– 0.533	0.595
	Age	– 0.221	0.083	– 0.199	– 2.670	0.090
	Investment experience	– 0.144	0.115	– 0.095	– 1.249	0.214

Dependent variable: TRD

5 Conclusion

In recent times, cryptocurrencies have gained popularity and established themselves as an alternative asset class. However, not all cryptocurrency investors will see a profit from their investment and a high rate of return because a savvy investor should be knowledgeable about the market and all the circumstances and dangers associated with trading cryptocurrencies.

The purpose of this research is to investigate the effect of behavioral factors (e.g., herding and overconfidence), demographic factors (e.g., education), and investment experience on trade performance. The results show that behavioral traits such as herding, and overconfidence significantly affect the trading performance of cryptocurrency investors. However, demographic criteria including an investor's age, educational level, and financial experience have little to no impact on how well they do in trading cryptocurrencies.

The paper provides thorough policy recommendations that will assist financial institutions, financial professionals, and politicians in making lucrative Bitcoin investment decisions. The results of the study have a number of implications for investors: (i) Investors are educated about the elements necessary to increase their income; (ii) institutions that are dealing with cryptocurrencies will be able to take these elements into account when advising investors in order to appeal to their interests for the companies to stay relevant and competitive; and (iii) The decision-makers will offer guidelines against any kind of price manipulation intended to deceive naïve investors and ensure that trading on cryptocurrency assets is conducted in an orderly manner.

The current study's main limitation is that it only looked at the effects of three demographic and two behavioral factors on the trading performance of cryptocurrency traders. Hence, more demographic details such as marital status, work history, and informal education might be included in future investigations. Future research may also include a bigger number of samples and a more varied sample of participants. It is also advised that future research explore the behaviors impacting the performance of cryptocurrency traders in many different markets using behavioral characteristics. If behavioral factors can be utilized in a variety of markets and investor types, they can be effectively determined using the study's findings.

References

1. Al-Absy, M.S.M.: Effects of COVID-19 pandemic on accounting students' capability to use technology. Int. J. Learn. Teach. Educ. Res. **22**(1), 247–267 (2023)
2. Al-Absy, M.S.M., Liandu, N.D.: The impact of corona pandemic on the students widespread use of technology: a case study of Gulf University. Chart. **1200**, 1400
3. Crypto.: Cryptocurrency buyers Retrative on 25 May 2023 (2023)
4. Yermack, D.: Is Bitcoin a real currency? An economic appraisal. In: Handbook of digital currency, pp. 31–43. Elsevier (2015)

5. Koppelmans, V., Bloomberg, J.J., Mulavara, A.P., Seidler, R.D.: Brain structural plasticity with spaceflight. npj Microgravity **2**(1), 2 (2016)
6. Ciaian, P., Rajcaniova, M., Kancs, D.A.: The digital agenda of virtual currencies: can BitCoin become a global currency? Inf. Syst. e-Bus. Manage. **14**, 883–919 (2016)
7. Kim, H.J., Hong, J.S., Hwang, H.C., Kim, S.M., Han, D.H.: Comparison of psychological status and investment style between bitcoin investors and share investors. Front. Psychol. **11**, 502295 (2020)
8. Kraaijeveld, O., De Smedt, J.: The predictive power of public Twitter sentiment for forecasting cryptocurrency prices. J. Int. Finan. Markets. Inst. Money **65**, 101188 (2020)
9. Dahham, A.Z.D., Ibrahim, A.A.: Effects of volatility and trend indicator for improving price prediction of cryptocurrency. In: IOP conference series: materials science and engineering. IOP Publishing (2020)
10. Hwang, S., Salmon, M.: Market stress and herding. J. Empir. Financ. **11**(4), 585–616 (2004)
11. Humra, Y.: Behavioral finance: an introduction to the principles governing investor behavior in stock markets. Int. J. Financ. Manage. **5**(2), 23–30 (2014)
12. Kartini, K., Nahda, K.: Behavioral biases on investment decision: a case study in Indonesia. J. Asian Financ. Econ. Bus. **8**(3), 1231–1240 (2021)
13. Baker, H.K., Nofsinger, J.R.: Psychological biases of investors. Financ. Serv. Rev. **11**(2), 97–116 (2002)
14. Pompian, M.M., Wood, A.S.: Behavioral finance and wealth management: how to build optimal portfolios for private clients. John Wiley and Sons, New Jersey Inc (2006). Retrieved from http://library
15. Prihatini, D., Widakdo, D.S.W.P.J.: Demographic factors, personality traits, and the performance of cryptocurrency traders. Jurnal Minds: Manajemen Ide dan Inspirasi **9**(1), 53–64 (2022)
16. Garay, U., Pulga, F.: The performance of retail investors, trading intensity and time in the market: evidence from an emerging stock market. Heliyon **7**(12) (2021)
17. Srivastava, S.: Prospective application of blockchain in mutual fund industry, in new horizons for industry 4.0 in modern business, pp. 121–148. Springer (2023)
18. Hossain, M.F., Nasrin, S.: Factors affecting selection of equity shares: the case of retail investors in Bangladesh. Eur. J. Bus. Manage. **4**(20), 110–124 (2012)
19. Wamae, J.N.: Behavioural factors influencing investment decision in stock market: a survey of investment banks in Kenya. Int. J. Soc. Sci. Entrepreneurship **1**(6), 68–83 (2013)
20. Nurbarani, B.S., Soepriyanto, G.: Determinants of investment decision in cryptocurrency: evidence from Indonesian investors. Univ. J. Acc. Financ. **10**(1), 254–266 (2022)
21. Kengatharan, L., Kengatharan, N.: The influence of behavioral factors in making investment decisions and performance: study on investors of Colombo stock exchange, Sri Lanka. Asian J. Financ. Acc. **6**(1), 1 (2014)
22. Silwal, P.P., Bajracharya, S.: Behavioral factors influencing stock investment decision of individuals. Int. Res. J. Manage. Sci. **6**(1), 53–73 (2021)
23. Le Luong, P., Thi Thu Ha, D.: Behavioral factors influencing individual investors decision-making and performance: a survey at the Ho Chi Minh stock exchange (2011)
24. Calderón, O.P.: Herding behavior in cryptocurrency markets. arXiv preprint arXiv:1806.11348 (2018)
25. Cao, M.M., Nguyen, N.-T., Tran, T.T.: Behavioral factors on individual investors' decision making and investment performance: a survey from the vietnam stock market. J. Asian Financ., Econ. Bus. **8**(3), 845–853 (2021)
26. Aziz, B., Khan, A.: Behavioral factors influencing individual investor's investment decision and performance, evidence from Pakistan stock exchange. Int. J. Res. Financ. Market. **6**(7), 74–86 (2016)
27. Mahalakshmi, T., Anuradha, N.: Factors affecting Investment decision making and investment performance among individual investors in India. Int. J. Pure Appl. Math. **118**(18), 1667–1776 (2018)

28. Anggita, A., Kawedar, W.: Pengaruh Knowledge Sharing dan Faktor Demografi terhadap Kinerja Pegawai. Diponegoro J. Acc. **6**(3), 381–392 (2017)
29. Guterresa, L., Armanu, A., Rofiaty, R.: The role of work motivation as a mediator on the influence of education-training and leadership style on employee performance. Manage. Sci. Lett. **10**(7), 1497–1504 (2020)
30. Das, S., Jain, R.: A study on the influence of demographical variables on the factors of investment-a perspective on the Guwahati region. Int. J. Res. Hum. Arts Lit. **2**(6), 97–102 (2014)
31. Chavali, K., Mohanraj, M.P.: Impact of demographic variables and risk tolerance on investment decisions–an empirical analysis. Int. J. Econ. Financ. Issues **6**(1), 169–175 (2016)
32. Tanusdjaja, H.: Keputusan Investasi Investor Individu Berdasarkan Kompetensi, Overconfidence, Dan Pendidikan. Jurnal Muara Ilmu Ekonomi Dan Bisnis **2**(1), 234–244 (2018)
33. Khanam, Z.: The impact of demographic factors on the decisions of investors during dividend declaration: a study on dhaka stock exchange, bangladesh. IOSR J. Bus. Manage. (IOSR-JBM). **19**(8), 1–7 (2017)
34. Artina, N., Cholid, I.: Pengaruh Tingkat Literasi Keuangan dan Faktor Demografi terhadap Pengambilan Keputusan Investasi (Studi Kasus Pegawai Kantor Badan Kepegawaian Daerah Sumatera Selatan). Jurnal Keuangan dan Bisnis **16**(1), 84–99 (2018)
35. Christanti, N., Mahastanti, L.A.: Factors considered by investors in investing. J. Theory Appl. Manage. **4**(3), 37–51 (2011)
36. Putri, R.A., Isbanah, Y.: Faktor-Faktor Yang Mempengarui Keputusan Investasi Pada Investor Saham Di Surabaya. Jurnal Ilmu Manajemen **8**(1), 197–209 (2020)

People's Perceptions of the Benefits, Risks, and Acceptance of Services Delivered by Drones in the United Arab Emirates

Nasser A. Saif Almuraqab, Sami Miniaoui, and Sajjad Jasimuddin

Abstract Recently, commercial unmanned aerial vehicles, also known as drones, have received substantial interest in public and civil environments because of their potential to deliver services and products. These are relatively new applications of drone technology. This study was conducted to understand people's perspectives on the benefits, challenges, and attitudes towards the use of this technology for delivering these services in the UAE. We conducted a descriptive survey with 238 respondents who were students and employees at the University of Dubai and collected data using an online questionnaire. The present study reveals people's perceptions of the benefits and risks of drone services. Additionally, this study gauged attitudes and concerns related to the use of drones. These results will enable scholars and decision-makers to conduct further investigations on the adoption of services delivered by drones.

Keywords Drones · Services · Technology acceptance · User perception · Smart city

1 Introduction

Digitalization is the mainstream nowadays [18]. Drones are becoming increasingly affordable, and their increasing capabilities have triggered the widespread application of this technology. Moreover, researchers have conducted case studies on the various applications of drones in several industries. Ayamga et al. [2] discussed the

N. A. S. Almuraqab (✉)
Dubai Business School, University of Dubai, Dubai, UAE
e-mail: nasser@ud.ac.ae

S. Miniaoui
College of Engineering and IT, University of Dubai, Dubai, UAE
e-mail: sminiaoui@ud.ac.ae

S. Jasimuddin
Kedge Business School, Paris, France
e-mail: sajjad.jasimuddin@kedgebs.com

B. Awwad (ed.), *The AI Revolution: Driving Business Innovation and Research*, Studies in Systems, Decision and Control 524, https://doi.org/10.1007/978-3-031-54379-1_39

application of drones in the agriculture, military, and healthcare industries. Additionally, the authors have attempted to identify the weaknesses and threats, such as the lack of training, skills to manipulate drones, safety, and security related to drone use. They also identified the opportunities and strengths related to drone use, such as cost-effectiveness and improved operations. Healthcare applications of drones were explored by [12] where the authors studied the use of drones for delivering healthcare and related services in North America. The study concluded that drones can be a cost-effective solution for delivering healthcare services to patients in remote areas. [23] reported several applications of drones in warehouses, including inventory management, inspection, and surveillance. The authors demonstrated that smart factories cannot be achieved unless drones are used. In addition, drones can perform repetitive and dangerous tasks that workers cannot. In another study, [4] reported several disruptive implications of using drones in marketing, as a new sustainable method [17]. The authors focused on the impact of drone use on business models, consumers, societies, and public policies [4]. Although they did not provide any solutions to the issues mentioned in their study, they provided directions for future research. The authors proposed that the perceived risks (e.g., safety) and benefits (e.g., efficiency) of drone use should be investigated. In the next section, we investigate the benefits and risks of drone use reported in the literature as well as the adoption of this technology by users and their perceptions of drone use. The results of this study can pave the way toward understanding users' perceptions in the UAE regarding this technology.

2 Literature Review

2.1 Benefits of Using Drones

Numerous studies have addressed the benefits of using drones in different scenarios and applications. According to [3], drones have several benefits, such as continuous functioning, low maintenance and labor costs, and high accuracy; additionally, drone use can improve the safety of factories. Berninzon et al. [5] identified that drones in the healthcare sector can be used to deliver emergency medical supplies and vaccines, as well as transport blood to remote locations such as mountainous areas. The main benefits of this technology can be summarized as follows: (1) reaching remote areas that are not accessible using regular modes of transport, (2) supplying medicines agilely and quickly, and (3) reducing the wastage of blood and medicines that have fast-approaching expiry dates.

According to [7], the majority of employees in construction companies believe that drones improve documentation on construction sites, whereas clients perceive the use of drones during construction to be positive. Additionally, as a result of drone use, almost 50% of the interviewed employees reported improved communication and job coordination.

2.2 Risks of Using Drones

Maghazei and Netland [15] highlighted the challenges of using drones in factories. The authors identified five categories of these challenges, which included technological, operational, organizational, legislative/societal, and mental [22]. Additionally, [7] identified the barriers to using drones in construction, which mostly involved obtaining drone licenses, uncertainties regarding drone regulations, and the lack of drone flying skills. [9] explored the adoption of drones in Malaysia. The author found that people perceived the risks associated with drone use to be moderate to high. Moreover, users perceive drones as having promising values and mostly think drones are moderately beneficial. Acording to the authors, the United States government provides adequate support for the use of this technology, whereas countries such as Canada do not provide sufficient support. In another study, [8] explored the economic concerns related to drone use and established a conceptual framework for the acceptance of this technology. The findings show that civil drone acceptance is impacted by social and economic risks. The social risks of drone use are associated with security, safety, and privacy, whereas the economic risks of drone use are associated with job substitution and skill degradation.

Since the main objective was to introduce drone delivery services for commercial purposes, an investigation of customer acceptance was conducted. [19] used the theory of planned behaviour to explore consumers' acceptance of service-delivery drones. The research results indicated that if a user perceives drones as a means of privacy erosion and expresses safety concerns, then it will negatively impact their attitude towards using drones. Conversely, if users perceive the functional performance of drones in delivery tasks, then they will have a favourable attitude toward using them. Therefore, the authors suggested that drones in delivery services can be accepted and adopted by cultivating a positive attitude towards commercial drones.

In contrast, [16] proposed an extended technology adoption model to explain the adoption model of drones by German farmers. The relevance of drones in agriculture, perceived usefulness, perceived ease of use, and attitude of confidence in using drones were the factors in this model. This study indicated that communicating and demonstrating the benefits of using drones to every farmer according to their needs and challenges is the only way to change their perception and increase their intention to use this technology.

The following sections will explain the study conducted regarding the use of drones for public services in the UAE. The research methodologies will be described and include the sample size, research instruments, and data collection methods. Subsequently, the paper will present the data analysis and research findings. The paper will conclude with the discussion and conclusion sections that will summarize the contributions of this study and provide directions for future research.

3 Methodology

This study used a descriptive survey design and a quantitative research approach. This research was conducted at the University of Dubai, which has an estimated population of 1200 (employees and students), representing over 41 nationalities. The online questionnaire was distributed via email and WhatsApp to all participants using snowball sampling as these participants were encouraged to share the questionnaire with their connections (friends, relatives). Using online tools such as emails and social media applications was an effective means of reaching participants.

3.1 Sample Size and Sampling Technique

We used simple random sampling to allocate the research sample, as well as snowball sampling to obtain more responses. The total sample size was 238, which was an adequate sample size for the statistical analysis [6]. In addition, simple random sampling was used to ensure that all participants had an equal chance of being selected for this study [10].

3.2 Research Instruments

A structured questionnaire was prepared to collect the data. The questionnaire comprised of three sections where the first section introduced the research topic and its objectives, the second section contained simple demographic questions related to age, gender, city, qualification, and education, and the final section included fourteen items drawn from relevant studies. The questions were close-ended, which, according to [10], are quick to compile, straightforward to code, and do not discriminate based on the respondents' ability to articulate responses. The Likert scale was used to record responses, which, according to [1], is one of the most widely used techniques for analyzing descriptive survey results. The survey contained the following options: strongly agree (SA), agree (A), neutral (N), disagree (D), and strongly disagree (SD).

3.3 Data Collection

Microsoft Forms was used to design the questionnaire, which was subsequently shared using online tools (e.g., email and WhatsApp). Using a social media messenger facilitated and expedited the snowball technique as it was used to not only provide every student and staff member with an equal opportunity of being selected for the study but also enhance the response rate.

4 Data Analysis

The study had 238 respondents have completed the survey. Table 1 shows the demographic information of the respondents including age, gender, level of education, experience with drones, and nationality. Regarding gender, 127 (53%) respondents were male, and 111 (47%) were female. As for nationality, 53% of the respondents were UAE nationals and 47% were expatriates. Furthermore, in terms of age, 82 respondents (34%) were 35–44 years old, 63 (26%) respondents were 25–34 years old, and 60 (25%) respondents were between the ages of 18 and 24. The majority of respondents held a bachelor's level of education (51%), and 39% of respondents held a postgraduate qualification such as a master's or PhD. The majority of the respondents (78%) had no experience with drones.

Table 2, Respondents' answers, lists the responses to the questionnaire on the services delivered by drones. The table includes information about the benefits and risks of drone use, as well as the respondents' attitudes towards using these services.

To facilitate the reading and understanding of the results, we merged the 'agree' and 'strongly agree' responses and categorized these as 'agree.' Similarly, 'disagree' and 'strongly disagree' were merged and categorized as 'disagree.' The respondents were asked which sector they preferred to obtain their services using drones, and the answers were interesting. Of the respondents, 58% were willing to accept drone services from the private and public sectors, whereas 12% of the respondents were willing to accept these services solely from the public sector and only 2% of the respondents were willing to accept drone services solely from the private sector. Moreover, 28% of the respondents were not willing to accept these services from either sector.

Table 1 Demographic information of the respondents

Demographic variables	Statistics
Age	18–24: 60 (25%) 25–34: 63 (26%) 35–44: 82 (34%) 45–54: 27 (11%) 55–64: 5 (2%)
Gender	Male: 127 (53%) Female: 111 (47%)
Education level	High school: 11 (5%) Diploma: 13 (13%) Bachelor's: 122 (51%) Postgraduate: 92 (39%)
General experience with drones	None: 186 (78%) Less than 1 year: 21 (9%) 1–3 years: 12 (5%) More than 3 years: 19 (8%)
Are you UAE national?	Yes: 126 (53%) No: 112 (47%)

Table 2 Respondents' answers

	Survey statement	SA F (%)	A F (%)	N F (%)	D F (%)	SD F (%)
Benefits	Drones are useful for civil operations or tasks	103 (43%)	104 (44%)	20 (8%)	3 (1%)	8 (3%)
	End-users will enjoy using drones	77 (32%)	117 (49%)	32 (13%)	4 (2%)	7 (3%)
	Business operations will be enhanced by using drones	73 (31%)	103 (43%)	45 (19%)	12 (5%)	5 (2%)
	Drone technology will support economic movement	52 (22%)	108 (45%)	61 (26%)	14 (6%)	3 (1%)
	In general, I think drones are useful	79 (33%)	113 (47%)	40 (17%)	2 (1%)	4 (2%)
Risks	I am worried that drones may affect my privacy	63 (26%)	101 (42%)	50 (21%)	19 (8%)	5 (2%)
	I am worried that drones may be harmful	23 (10%)	54 (22%)	87 (37%)	65 (27%)	9 (4%)
	I am worried about the reliability of drones	10 (4%)	64 (27%)	103 (43%)	55 (23%)	6 (3%)
	I am worried that drones may be used to spy on my family	38 (16%)	101 (42%)	59 (25%)	31 (13%)	9 (4%)
Attitudes	I think drones should be used in the UAE	45 (19%)	118 (50%)	59 (25%)	13 (5%)	3 (1%)
	I feel positive about the use of drones in the UAE	37 (16%)	127 (53%)	59 (25%)	11 (5%)	4 (2%)
	I believe using drones is positive	39 (16%)	124 (52%)	64 (27%)	8 (3%)	3 (1%)
	I think the use of drones is invaluable	40 (17%)	124 (52%)	62 (26%)	9 (4%)	3 (1%)
Accepting product/service delivery using drones?		• Government sector only: 29 (12%) • Private sector only: 5 (2%) • Both government and private sectors: 138 (58%) • No, not at all: 67 (28%)				

4.1 Benefits of Using Drones

In fact, 87% of the respondents agreed that drones may be useful in civil operations such as searching, monitoring, security, and other police tasks; 8% were neutral on this topic; and 4% disagreed with the use of drones for these operations. In addition, 81% responded that using drones brought fun and entertainment and 74% of the respondents believed that drones could facilitate business operations and provide cost savings. Furthermore, 67% of the respondents agreed that the use of drones will increase the economic movement in the country as this technology will provide new and unexpected opportunities for businesses. However, 26% of the respondents were neutral regarding the statement on economic movement. Nevertheless, 80% of the respondents agreed that drones were useful in general.

4.2 Risks of Using Drones

The responses regarding the risks of drone use revealed that 68% of the respondents were worried about the privacy issues pertaining to the use of this technology for public and private services; however, 21% were neutral about these concerns. In addition, 32% agreed that drones may harm them physically when used inappropriately or programmed incorrectly. However, 37% were neutral about this issue, and 31% disagreed with these concerns. Hence, 31% agreed that drones are not reliable in general, whereas 43% were natural, and 26% disagreed. In addition, approximately 58% of respondents were concerned that drones may be used to spy on them or their loved ones, whereas 25% were neutral on this matter and 17% disagreed with this concern.

4.3 Attitude Towards Drone Use

Most of the respondents had a positive attitude toward availing public and private sector services delivered by drones. For instance, 69% agreed that drones should be used in the UAE, 25% were neutral on this matter, and 6% disagreed with drone use in the UAE. In addition, approximately 79% expressed a positive attitude towards using drones in the UAE, and 68% agreed with the statement that using drones is positive. Finally, 69% believed that the use of drones is valuable, 26% were neutral on this matter, and only 5% disagreed with this.

5 Discussion

Customers nowadays are technologically savvy, and their shifting preferences for products vary regionally [18]. In many countries, small-sized drones have been adopted for personal use, primarily for capturing photos or videos. However, because drones are advancing and becoming larger, faster, and more capable than they were before, it has opened doors for unexpected new opportunities. This has attracted the attention of both the public and private sectors. In particular, during the digital era or the era of smart cities, governments are attempting to provide citizens with new experiences [14], and policymakers have started to adopt, embrace, and react to these technologies by proposing new ways to deliver services using drones. For instance, in Dubai, the drone transportation program was launched in 2017 [13, 24].

In addition to transportation and delivery, other services, such as healthcare [12] and railway infrastructure monitoring [11], have been implemented using drones.

Drones have the potential to disrupt businesses by presenting alternatives to existing procedures or channel partners. For example, last-mile drone delivery is gaining traction, bringing the idea of customers seeing drones at their doorsteps; therefore, drones are becoming customer-facing agents and touchpoints. This is especially true for customers who reside in remote places, as drones do not require traditional infrastructure. Consequently, drones enable marketers to contact and engage with hitherto inaccessible client categories. Additionally, drones are being integrated into warehouses and retail environments to save operating costs and provide benefits to customers [11]. The main role of commercial drones in the retail sector is related to delivery services [19]. For example, Walmart filed a patent for employing drones to transport goods between departments in shops.

The delivery of products and services is one of the most direct ways in which drones alter company structures. Uber, for example, focuses on developing people-carrying drones, and passenger drones are currently in operation at EHang in China. These new services may have a significant impact on mobility, cost, transit, and delivery [13]. Similarly, countries such as the UAE, particularly Dubai, have started using taxi drones. Sheikh Hamdan bin Mohammed has articulated that the Dubai Program for Facilitating Drone Transportation is set to establish a state-of-the-art infrastructure. This infrastructure will facilitate innovators and relevant organizations in conducting trials of unmanned aerial vehicle prototypes or drones in specified zones, as well as in formulating legislation that enhances the efficiency of drone deployment. The primary objective of the program is to unlock novel economic prospects and fortify Dubai's position as a trailblazer in advanced technological research and development across a wide spectrum of forward-thinking industries.

The majority of the respondents agreed that drones are beneficial and that they will boost the economy in general. Additionally, they agreed that drones will positively impact business operations. Although [19] presented the benefits of drone use, the respondents expressed concerns regarding their privacy when using services delivered by drones. Furthermore, the respondents believed that the unreliable operations of drones may harm them physically or be misused to spy on them and their loved ones. The approval of drone use rightly raises consumer concerns regarding privacy and security because this technology is thought to be used for surveillance and cyber spying by collecting large amounts of data via facial recognition cameras, digital imprints, thermal images, mobile device snooping, and other means [19, 21].

Drone makers, security businesses, governments, and regulatory authorities must collaborate to restrict flying permits to limit aerial risks and protect customer privacy before consumers can embrace drones as a necessary instrument for the delivery of products and services [21]. Furthermore, as increasing numbers of drones are being tested and authorized for commercial use, the possibility of mishaps and mid-air crashes is no longer speculative. As a result, the media is increasingly reporting incidents related to drones all over the world [20]. These reports have caused politicians and residents, especially those who perceive automation to be overwhelming and threatening, to raise further concerns.

Despite the related risks and concerns raised by the respondents, the responses recorded in Table 2 reveal that services delivered by drones are likely to be adopted by UAE nationals and expats. As explained previously, many respondents agree that drones are valuable and that they should be used in the UAE, and the respondents have positive feelings regarding services delivered by drones in the UAE. The benefits of drone use were especially understood during the COVID-19 pandemic since precautionary activities and the lockdown forced economic operations to rely more on smart services, online ordering, and home delivery. Nowadays, the number of delivery service providers, such as Talabat, Deliveroo, Jeebly, and Zomato, etc., is increasing, and if these services use drones instead of regular vehicles, such as cars and motorbikes, there will be less traffic, fewer accidents, and an improved environment because of fewer carbon emissions.

6 Conclusion

This study revealed the customers' perspective, considering the benefits, and risks of implementing public services using drones. The literature revealed several social and economic risks related to drone use; however, people's perceptions of the use of drones were not investigated. Instead, the literature revealed some industries such as building, healthcare, and agriculture. The study also discussed the adoption of public drone services, such as taxis and delivery drones. The present study reveals people's perceptions of the benefits and risks of drone services. Additionally, this paper gauged attitudes and concerns related to the use of drones. These results will enable scholars and decision-makers to conduct further investigations into the adoption of services delivered by drones. Recent studies have broadly discussed the existing economic and social aspects of drone use. Based on literature, we developed a questionnaire to evaluate the attitudes of the residents of the UAE towards this technology.

Although many respondents agreed that the adoption of drones would be beneficial for them and the overall economy, they expressed some concerns and apprehensions regarding privacy and physical harm. In general, people in the UAE feel positive about drones and want the government to launch public services that use drones, especially taxi and delivery services. This is a positive step toward drone use, and we hope that policymakers, scholars, and practitioners will utilize this study when considering the adoption of drones in smart cities. Additionally, drones have a significant influence on customers and businesses, and the impact of this technology is expected to increase in the future.

This study has certain limitations. The research sample was gathered from one university and did not represent the entire country. Therefore, the responses obtained may not be an accurate representation of UAE residents' perspectives and preferences regarding services delivered by drones. Future research must examine and discover more insights on this topic. Furthermore, as drone technology advances, more research is required to reveal the wide range of theoretical and practical consequences of drone use for taxis and product deliveries. The effects of drone use on business models, customers, and society must be investigated in more detail in the future.

References

1. Ary, D., Jacobs, L.C., Razavieh, A.: Introduction to research in education, 6th edn. Wadsworth Group, Belmont, California (2002)
2. Ayamga, M., Akaba, S., Nyaaba, A.A.: Multifaceted applicability of drones: a review. Technol. Forecast Soc. Change **167**(120677) (2021)
3. Bechtsis, D., Tsolakis, N.: Trends in industrial informatics and supply chain management. Int. J. New Technol. Res. **4**, 91–93 (2018)
4. Beninger, S., Robson, K.: The disruptive potential of drones. Mark. Lett. **31**(4), 315–319 (2020)
5. Berninzon, A.L., Vongasemjit, O., Russell, T., Sheffi, Y.: Potential benefits of drones for vaccine last-mile delivery in Nepal. Master degree dissertation, MIT (2021)
6. Bullen, P.B., Brack T.: [Internet]. How to choose a sample size (for the statistically challenged); c2013. Available from: www. tools4dev.org/resources/how-to-choosea-sample-size.
7. Burgett, J.: An industry focused course to address barriers and benefits of drones in construction. EPiC Ser. Built Environ. **2**, 210–218 (2021)
8. Chamata, J.: Factors delaying the adoption of civil drones: a primitive framework. Int. Technol. Manage. Rev. **6**(4), 125–132 (2017)
9. Chamata, J.: A proposal for the adoption of unmanned aerial technology in Malaysia. In: 4th Borneo Research Education Conference (BREC), Aug 17–18. Sarawak, Malaysia (2016)
10. Cohen, L., Manion, L., Morrison, K.: Research methods in education (Kiranakis, S., Mavraki, M., Mitsopoulou, C., Bithara, P., Filopoulou, M. Trans.). Athens, Greece: Metechmio. (Original work published 2000) (2008)
11. Flammini, F., Pragliola C., Smarra, G.: Railway infrastructure monitoring by drones. In: International Conference on Electrical Systems for Aircraft, Railway, Ship Propulsion and Road Vehicles & International Transportation Electrification Conference (ESARS-ITEC), pp. 1–6 (2016). https://doi.org/10.1109/ESARS-ITEC.2016.7841398
12. Hiebert, B., Nouvet, E., Jeyabalan, V., Donelle, L.: The application of drones in healthcare and health-related services in North America: a scoping review. Drones **4**(3), 30 (2020)
13. Hussein, H., Aamir M.: Hamdan bin Mohammed Lauches Dubai programme to enable drone transportation; c2021 (2021). Available from: http://wam.ae/en/details/1395302994758
14. Kuttikaden, H.S., Daniel, J.C.T.: A study on user experience of Amazon pay. In: International Conference on Economics, Business and Sustainability, pp. 321–327. Singapore: Springer Nature Singapore (2023). https://doi.org/10.1007/978-981-99-3366-2_38
15. Maghazei, O., Netland, T.: Drones in manufacturing: Exploring opportunities for research and practice. J. Manuf. Technol. Manag. **31**(6), 1237–1259 (2019)
16. Michels, M., von Hobe, C.F., von Ahlefeld, P.J.W., Musshoff, O.: The adoption of drones in German agriculture: a structural equation model. Precis. Agric. **22**(6), 1728–1748 (2021)
17. More, A.B.: Implementing digital age experience marketing to make customer relations more sustainable. In: New Horizons for Industry 4.0 in Modern Business, pp. 99–119. Cham: Springer International Publishing (2023). https://doi.org/10.1007/978-3-031-20443-2_5
18. Raju, V.: A case study on the buying behaviour of online customers in Bangalore. In: International Conference on Economics, Business and Sustainability, pp. 54–65. Singapore: Springer Nature Singapore (2023). https://doi.org/10.1007/978-981-99-3366-2_7
19. Ramadan, Z.B., Farah, M.F., Mrad, M.: An adapted TPB approach to consumers' acceptance of service-delivery drones. Technol. Anal. Strat. Manage. **29**(7), 817–828 (2017)
20. Ravich, T.M.: Commercial drones and the phantom menace. J. Int'l. Media. Ent. L. **5**, 175 (2013)
21. Ravich, T.M.: Courts in the drone age. N Ky L Rev. **42**, 161 (2015)

22. Scott, J., Scott, C.: Drone delivery models for healthcare. In: Proceedings of the 50th Hawaii International Conference on System Sciences, Village, HI, USA, 4–7 January 2017, pp. 3297–3304 (2017). Available online: http://hdl.handle.net/10125/41557
23. Tubis, A.A, Ryczyński, J., Żurek, A.: Risk assessment for the use of drones in warehouse operations in the first phase of introducing the service to the market. Sensors **21**(20), 6713 (2021)
24. Wakefield, J.: Dubai tests drone taxi service, c2017 (2017). Available from: https://www.bbc.com/news/technology-41399406

The Effect of Stress at Workplace on Employee Turnover on Hyundai Company

Ali Ateeq and Sara Abdulaziz Albureshedaid

Abstract In today's fast-paced corporate world, stress, turnover, and engagement shape contemporary businesses. Academic academics and industry people must understand these factors' complex interactions. Stress is a primary cause of work unhappiness, low productivity, and absenteeism. Employee turnover may also hurt corporate effectiveness by limiting talent retention. Employee engagement is increasingly linked to employee well-being and corporate success. Although stress, turnover, and engagement have been studied individually, there is a rising interest in their interrelationship. Employee involvement may mediate the association between workplace stress and turnover, according to one research. Stress from rising corporate expectations has led to burnout, job dissatisfaction, and more turnover. Understanding how workplace stress affects employee turnover is crucial to establishing effective stress management measures to reduce turnover. Employee involvement boosts productivity, work happiness, and organizational commitment. Engagement's significance in stress-turnover moderation is unclear. Investigating this intermediate role may help improve employee engagement and minimize stress-related attrition.

Keywords Workplace stress · Hyundai company · Employee turnover · Employee engagement · Organizational performance · Stress management

1 Introduction

The dynamics of modern firms are significantly influenced by workplace stress, employee turnover, and employee engagement. The ability of the workforce to effectively adjust to a rapidly changing environment plays a crucial role in the achievement of organizational success. The comprehension of the interconnectedness between these factors is a matter of great importance for both academic researchers and professionals in the field. According to [28] workplace stress has been identified

A. Ateeq (✉) · S. A. Albureshedaid
College of Administrative and Financial Sciences, Gulf University, Building 1964, Road 4363, Sanad 743, Kingdom of Bahrain
e-mail: dr.ali.ateeq@gulfuniversity.edu.bh

B. Awwad (ed.), *The AI Revolution: Driving Business Innovation and Research*, Studies in Systems, Decision and Control 524, https://doi.org/10.1007/978-3-031-54379-1_40

as a significant factor contributing to reduced levels of job satisfaction, diminished productivity, and heightened rates of absenteeism. Moreover, it is important to note that a significant increase in staff turnover may have negative consequences on an organization's overall performance, as it might hinder the organization's capacity to retain a proficient and knowledgeable workforce [3, 25, 29]. In recent years, there has been a growing acknowledgment of the significance of employee involvement in promoting employee welfare and improving organizational results [17, 34].

Nevertheless, whereas much research has been conducted on the separate impacts of stress, turnover, and engagement, there is an increasing scholarly focus on comprehending the intricate interconnections among these variables [18, 22]. The current focus is on the possible mediating effect of employee engagement in the relationship between workplace stress and employee turnover.

The prevalence of elevated levels of stress in the workplace is a significant worry. The contemporary demands of the corporate environment might result in employee burnout and fatigue, hence leading to work discontent and the possibility of turnover. Scholars have extensively acknowledged the detrimental effects of stress on both work performance and the overall well-being of employees [15]. It is vital to comprehend the inherent correlation between workplace stress and employee turnover, as such understanding may provide insights into strategies that businesses can apply to decrease turnover rates via the effective management and mitigation of stressors [32].

Furthermore, the significance of employee engagement in achieving organizational performance is being increasingly recognized [7, 9, 34]. Employees that are actively involved in their work are more inclined to show higher levels of productivity, job satisfaction, and organizational commitment. According to [24] there is a significant correlation between employee engagement and work performance, job satisfaction, and retention. Nevertheless, there is still a lack of clarity about the specific role of participation in mediating the association between stress and turnover. The examination of this intermediary function might contribute to a more holistic comprehension of the underlying mechanisms, perhaps yielding valuable insights into strategies that businesses can adopt to enhance employee engagement as a means of mitigating the impact of stress on employee turnover.

Employee turnover: According to [38] turnover is the process by which employees leave the company and need to be replaced. Turnover is the cessation of people joining an organization along with the supply of financial benefits by the organization in question [20]. There are three categories of employee turnover: (1) involuntary and (2) voluntary; (2) functional and (3) dysfunctional; and (3) uncontrollable and (4) manageable. (1) Environmental factors, such as kinship obligations to the environment and employment opportunities, (2) Individual factors, such as job satisfaction, commitment to the organization, job-seeking behavior, intention to stay, general training/competency enhancement, willingness to work hard, and negative or positive feelings toward work, (3) Structural factors, such as autonomy, fairness, and opportunity, are those that affect the likelihood of turnover.

Stress: Stress, which forces someone to deviate from normal function due to changes in their psychological and physiological conditions [36]. Emotional pattern and physiological reactions that happen in response to demands from within or outside the organization. According to [33] work stress is a feeling of pressure that employees endure because of their jobs. Work stress results from a combination of many stressors, including pressure-filled work environments. Employees may experience stress at work from a variety of sources, including: (1) environmental factors, such as economic, political, and technological uncertainty, (2) organizational factors, such as task demands, role demands, and interpersonal demands; and (3) individual factors [35]. In the meantime, the physical factors that contribute to work stress are: (1) noise, weariness, temperature, and humidity; (2) work load; (3) the nature of the job, which includes novel situations and personal risks; and (4) freedom (Lin, Khin, Lee, and Tee).

2 Literature Review

The issue of workplace stress has emerged as a prominent global problem for enterprises, exerting adverse effects on both the physical and mental health of workers, as well as exerting an influence on employee attrition rates. The primary objective of this literature review is to investigate the correlation between workplace stress and employee turnover in the specific organizational setting of Hyundai, a well-established automotive firm. This research will also provide substantiated arguments and include pertinent citations to support the analysis.

2.1 The Impact of Workplace Stress on Employee Turnover

Numerous research have unequivocally shown a causal relationship between workplace stress and heightened rates of employee departure. The term of "stress" was first established by [14] describe the physiological and psychological reactions that occur in response to excessive expectations. Subsequent study conducted by [27] revealed a significant association between high levels of stress and work unhappiness, leading individuals to express a desire to disengage from the organization. Within the realm of Hyundai, the aforementioned correlation may be substantiated by the considerable production demands and rigorous work schedules that are prevalent within the automobile sector [23].

2.2 The Relationship Between Organizational Commitment and Stress

The idea of organizational commitment was established by [30] including emotional, normative, and continuous commitment. Research findings indicate that there exists an inverse relationship between elevated levels of workplace stress and affective commitment, wherein individuals experiencing high levels of stress are less like to develop emotional attachment towards the business. The absence of dedication may be seen as a rationale for the higher rates of staff turnover at Hyundai, since it leads to discontent and disengagement caused by stress [4, 12, 31].

2.3 The Impact of Coping Mechanisms on Employee Turnover Rates

According to the research conducted by [21] it has been shown that people use a range of coping strategies in order to effectively manage and address stress [21]. When workers at Hyundai experience elevated levels of stress and have difficulties in properly managing it, they are more inclined to contemplate departure from the firm as a means of escaping the stressful milieu. The use of the stress-coping-turnover connection may be used to provide justification for the condition seen at Hyundai.

2.4 The Impact of Stress on Health Outcomes

Extensive research has shown a strong association between stress and health issues, particularly chronic stress, which has been found to have detrimental impacts on health [1, 5]. In the automobile manufacturing business, characterized by extended work hours and physically strenuous tasks, the presence of stress may give rise to many health-related concerns among the workforce. Consequently, individuals may choose to resign from their positions at Hyundai in order to safeguard their personal welfare.

2.5 The Relationship Between Leadership Styles and Stress Management

The function of leadership styles in the management of workplace stress is of utmost importance. [19] has shown that the implementation of transformational leadership, which is distinguished by its provision of emotional support and encouragement, may effectively alleviate the adverse consequences of stress. The stress levels of workers at Hyundai may be considerably influenced by the leadership style of supervisors and managers, therefore highlighting the need of implementing good stress management measures [13, 31].

According to the Job Demands-Resources (JD-R) model proposed by [26] stress may arise from job demands, such as workload and time pressure, while job resources, such as social support and autonomy, can alleviate this stress. The working environment at Hyundai may potentially be characterized by elevated job expectations, which, in the absence of adequate resources and support, might give rise to heightened levels of stress and, therefore, a higher rate of employee turnover.

2.6 Job Demands-Resources Model

In summary, the literature analysis offers a thorough comprehension of the correlation between workplace stress and employee turnover specifically within the context of Hyundai. Numerous studies have repeatedly shown that workplace stress serves as a substantial predictor of employee attrition. The rationales are derived from theoretical frameworks such as organizational commitment, coping mechanisms, health outcomes, leadership styles, and the JD-R model. These frameworks emphasize the necessity of implementing stress management strategies and support systems within the organization in order to alleviate the negative impact of stress on employee turnover.

3 Methodology

The present research utilizes a critical review technique to evaluate the influence of workplace stress on employee turnover within the context of Hyundai [6, 8, 11, 37]. The present study aims to conduct a comprehensive analysis of existing scholarly literature, research studies, and pertinent data sources in order to critically assess the correlation between stress and employee turnover within the organizational context [2, 10]. The primary objective of this study is to conduct a thorough evaluation of the existing body of information pertaining to this subject matter, with the intention of informing further research endeavors and prospective treatments [16].

4 Conclusion

Hyundai's employee turnover and workplace stress are linked in complicated and nuanced ways, as shown by a study of the relevant literature and studies. Stress has been linked to higher employee turnover, according to the data. Employees' physical and mental health, as well as their level of dedication to the company, are also crucial. In addition, stress management and social support are essential for reducing the severity of these consequences. These results highlight the need for Hyundai and other companies in a similar position to implement stress-reduction methods, enhance leadership practices, and provide their staff with the tools they need to successfully deal with stress. Reducing stress in the workplace is important for a number of reasons, including the health and happiness of workers as well as their contributions to the bottom line. Additional empirical research is required to give insights more relevant to Hyundai's setting.

References

1. Adamopoulos, I.P., Syrou, N.F.: Associations and correlations of job stress, job satisfaction and burn out in public health sector. Euro. J. Environ. Pub. Health **6**(2), em0113 (2022)
2. Al-refaei, A.A.-A., Ali, H.B.M., Ateeq, A.A., Alzoraiki, M.: An integrated mediating and moderating model to improve service quality through job involvement, job satisfaction, and organizational commitment. Sustainability **15**(10), 7978 (2023)
3. Al Mamun, C.A., Hasan, M.N.: Factors affecting employee turnover and sound retention strategies in business organization: a conceptual view. Prob. Perspect. Manage. **15**(1), 63–71 (2017)
4. Ali, A.A.A., Alzoraiki, M., Al Aghbas, N.S.M.: Working remotely, social isolation and access to support and communication during Covid-19 its influence on job performance. Int. J. Intell. Hum. Res. Manage. (IJIHRM) **3**(01), 23–33 (2022)
5. Ali, A.A.A., Alzoraiki, M., Al-shaibah, M., Almaamari, Q.: Enhancing contextual performance through Islamic work ethics with mediating role of normative commitment. Math. Stat. Eng. Appl. **71**(4), 8668–8683 (2022)
6. Ali, A.A.A., Habtoor, N.: Islamic work ethics and its impact on enhancing organizational commitment in Bahraini private educational universities. Int. J. Intell. Hum. Reso. Manage. (IJIHRM) **3**(02), 33–42 (2022)
7. Ali, A.A.A., Alzoraiki, M., AL-Tahitah, A.N., Almaamari, Q., Al-shaibah, M.: Association between Islamic work ethics and employee commitment among medical staff in the public health sector in Bahrain. Int. J. Intell. Hum. Res. Manage. (IJIHRM) (2023)
8. Ali, A.A.A., Yusoffii, S.H., Aliiii, K.: Fundamental of equal opportunities in employee's performance: a critical review
9. Alkadash, T.M., Nagi, M., Ateeq, A.A., Alzoraiki, M., Alkadash, R.M., Nadam, C., Dawwas, M.: The effects of leadership style on employee sustainable behaviour: a theoretical perspective. In: Artificial Intelligence and Transforming Digital Marketing, pp. 205–213. Springer (2023)
10. Alzoraiki, M., Ahmad, A.R., Ateeq, A.A., Naji, G.M.A., Almaamari, Q., Beshr, B.A.H.: Impact of teachers' commitment to the relationship between transformational leadership and sustainable teaching performance. Sustainability **15**(5), 4620 (2023)
11. Ateeq, A.: The mediating role of engagement in the relationship between. Management **21**(3), 435–447 (2023)
12. Ateeq, A., Alzoraiki, M., Milhem, M., Al-Absy, M.: Impact of employee loyalty on job performance: mediating role of job satisfaction on the example of zain company. Bahrain. Manage. **21**(2), 470–481 (2023)
13. Ateeq, A.: Knowledge management's effect on learning quality in Bahrain's private universities (2023a)
14. Ateeq, A., Ebrahim, S., Al-Ghatam, R.: The impact of stress and its influencing factors among dentists during the COVID-19 pandemic in Kingdom of Bahrain. F1000Research **11**, 930 (2022)

15. Aubouin-Bonnaventure, J., Chevalier, S., Lahiani, F.-J., Fouquereau, E.: Well-being and performance at work: a new approach favourable to the optimal functioning of workers through virtuous organisational practices. Int. J. Organiz. Anal. (2023)
16. Barakat, B., Milhem, M., Naji, G.M.A., Alzoraiki, M., Muda, H.B., Ateeq, A., Abro, Z.: Assessing the impact of green training on sustainable business advantage: exploring the mediating role of green supply chain practices. Sustainability **15**(19), 14144 (2023)
17. Björk-Fant, J.M.: Mental well-being at work: Support and protective factors of work engagement and work-life balance in contemporary working life (2023)
18. Bright, L.: Does stress type matter? clarifying the relationships between public service motivation, work-related stress, and employee attitudes (2023)
19. Chen, J., Ghardallou, W., Comite, U., Ahmad, N., Ryu, H.B., Ariza-Montes, A., Han, H.: Managing hospital employees' burnout through transformational leadership: the role of resilience, role clarity, and intrinsic motivation. Int. J. Environ. Res. Public Health **19**(17), 10941 (2022)
20. Domurath, A., Taggar, S., Patzelt, H.: A contingency model of employees' turnover intent in young ventures. Small Bus. Econ. **60**(3), 901–927 (2023)
21. Guszkowska, M., Dąbrowska-Zimakowska, A.: Coping with stress during the second wave of the COVID-19 pandemic by Polish university students: Strategies, structure, and relation to psychological well-being. Psychol. Res. Behav. Manage. 339–352 (2022)
22. Hafeez, S., Memon, M.A., Mirza, M.Z., Raziq, M.M., Sarwar, N., Ting, H.: The dual impact of job variety on employee happiness and stress: the mediating role of employee engagement and burnout. J. Manage. Develop (2023)
23. Hazeen Fathima, M., Umarani, C.: A study on the impact of role stress on engineer intention to leave in Indian construction firms. Sci. Rep. **12**(1), 17576 (2022)
24. Jamaludin, N.L., Kamal, S.A.: The relationship between remote work and job satisfaction: the mediating role of perceived autonomy. Inf. Manage. Bus. Rev. **15**(3 (SI)), 10–22 (2023)
25. Lazar, I., Osoian, C., Ratiu, P.I.: The role of work-life balance practices in order to improve organizational performance (2010)
26. Liao, H., Liang, R., He, H., Huang, Y., Liu, M.: Work stress, burnout, occupational commitment, and social support among Chinese pediatric nurses: a moderated mediation model. J. Pediatr. Nurs.Pediatr. Nurs. **67**, e16–e23 (2022)
27. Mat Rifin, H., Danaee, M.: Association between burnout, job dissatisfaction and intention to leave among medical researchers in a research organisation in Malaysia during the COVID-19 pandemic. Int. J. Environ. Res. Public Health **19**(16), 10017 (2022)
28. Mishra, A.: Withdrawal behaviours and absenteeism at work post covid-19. Organiz. Hum. Capital Develop. **1**(2), 62–75 (2022)
29. Muthuswamy, V.V.: Economic impact of HRM practices on organizational economic performance: does employee retention mediates? Cuadernos de Economía **46**(130), 31–41 (2023)
30. Noor, M.T., Rudini, A., Christiana, A., Prasetyo, D., Susanto, H.: Job embeddedness and organizational commitment in controversial industry: mediating eole of collectivism. Quality-Access Success **24**(196) (2023)
31. Park, Y., Kim, J., Lee, H.: The influences of supportive leadership and family social support on female managers' organizational effectiveness: the mediating effect of positive spillover between work and family. Behav. Sci. Sci. **13**(8), 639 (2023)
32. Sajjad, R., Ata, G., Rizwan, A.: Employee turnover intentions in ostracized work environment and organizational cynicism. J. Develop. Soc. Sci. **4**(3), 740–755 (2023)
33. Saleem, M.S., Isha, A.S.N.B., Benson, C., Awan, M.I., Naji, G.M.A., Yusop, Y.B.: Analyzing the impact of psychological capital and work pressure on employee job engagement and safety behavior. Front. Public Health **10**, 1086843 (2022)
34. Salem, N.H., Ishaq, M.I., Yaqoob, S., Raza, A., Zia, H.: Employee engagement, innovative work behaviour, and employee wellbeing: Do workplace spirituality and individual spirituality matter? Bus. Ethics, Environ. Responsib. **32**(2), 657–669 (2023)
35. Selander, K., Nikunlaakso, R., Laitinen, J.: Association between work ability and work stressors: cross-sectional survey of elderly services and health and social care service employees. Arch. Public Health **80**(1), 1–9 (2022)

36. Srivastava, M., Singh, G., Kharwar, P.S., Jaiswal, S.: Occupational stress among armed forces and police personnel: a review. Int. J. Community Med. Public Health **10**(3), 1 (2023)
37. Thakral, P., Srivastava, P.R., Dash, S.S., Jasimuddin, S.M., Zhang, Z.: Trends in the thematic landscape of HR analytics research: a structural topic modeling approach. Manage. Decis. (2023)
38. Yue, C.A., Thelen, P.D., Walden, J.: How empathetic leadership communication mitigates employees' turnover intention during COVID-19-related organizational change. Manag. Decis. Decis. **61**(5), 1413–1433 (2023)

Leveraging Human Resource Information Systems and Artificial Intelligence in Predicting Employee Satisfaction

Mohammad H. Allaymoun, Tamer Alkadash, Samer Shorman, and Moufaq Yousef

Abstract This paper examines the use of human resource information systems (HRIS) and artificial intelligence (AI) in predicting employee satisfaction. The literature review identifies variables that can be used to measure employee satisfaction and their corresponding measurement methods, including absences, complaints, delays in working hours, completing tasks, and managers' feedback. The Mixed methodology approach referred for this, a quantitative approach, and the study uses a qualitative approach to gather data on HR information system integration and AI-based employee satisfaction prediction in HR decision-making processes. The study adheres to ethical guidelines and principles, and the data collection process involves both primary and secondary data sources. The results of the study indicate that the J48 decision tree algorithm achieved a 72.92% accuracy rate in predicting employee satisfaction based on several attributes. The J48 pruned tree model had 12 nodes and 7 leaves, with absences being the root node. The study concludes that further research is needed to validate the model's performance on larger datasets and with additional attributes.

Keywords Human resource information systems · Employee satisfaction · Artificial intelligence · Machine learning · Prediction

---

M. H. Allaymoun (✉) · T. Alkadash
College of Administrative and Financial Sciences, Gulf University, Manamah, Kingdom of Bahrain
e-mail: alkarak1@yahoo.com

T. Alkadash
e-mail: dr.tamer.alkadash@gulfuniversity.edu.bh

S. Shorman
Computer Science Department, College of Art and Science, Applied Science University, Al Ekir, Kingdom of Bahrain
e-mail: Samer.shorman@asu.edu.bh

M. Yousef
Ministry of Health, Amman, Hashemite Kingdom of Jordan

B. Awwad (ed.), *The AI Revolution: Driving Business Innovation and Research*, Studies in Systems, Decision and Control 524, https://doi.org/10.1007/978-3-031-54379-1_41

1 Introduction

Employee satisfaction is a critical aspect of any organization's success, as it directly impacts employee retention, productivity, and overall performance [1]. Traditional methods of measuring employee satisfaction through surveys and feedback mechanisms can be time-consuming, subjective, and may not always yield accurate results [1–3]. With the advancement of technology, human resource information systems (HRIS) and artificial intelligence (AI) can be leveraged to predict employee satisfaction with greater accuracy and efficiency. Moreover, the field of business has been significantly impacted by the continuous rise in the amount of data, with the emergence of new tools and technologies that greatly aid in maintaining competitiveness [1]. However, this increase in the volume and variety of data presents a challenge. To address this, technical tools capable of analyzing big data have been developed, yielding impressive results that aid departments in assessing and improving company operations [4].

The topic of HRIS (Human Resource Information Systems) and data science has become increasingly relevant in recent years due to the growing importance of data in human resource management [2]. HRIS refers to the technology and systems used to manage employee data, including employee records, payroll, benefits, performance management, and more [3]. Data science, on the other hand, involves the use of statistical methods, algorithms, and machine learning techniques to extract insights and knowledge from large data sets [5].

In the context of HR, data science can be used to analyze employee data to identify patterns, trends, and relationships that can inform HR decision-making [6]. For example, data science can be used to analyze employee turnover rates, performance metrics, and engagement surveys to identify areas for improvement and develop effective retention strategies. It can also be used to identify key predictors of employee success, such as specific skills or personality traits, which can be used to inform hiring and training decisions [7].

The integration of HRIS and data science has the potential to revolutionize human resource management by enabling organizations to make more data-driven decisions [8]. By analyzing large amounts of employee data, organizations can gain valuable insights into their workforce, which can help them identify areas for improvement, develop effective HR policies and practices, and ultimately improve employee satisfaction and performance. Additionally, the use of HRIS and data science can help organizations automate HR processes, reduce errors, and improve efficiency, saving time and resources [9].

Employee satisfaction is a critical factor that can significantly influence the performance and success of an organization. When employees are satisfied with their work, they tend to be more motivated, engaged, and committed to their jobs. This can lead to increased productivity, better quality of work, and higher levels of customer satisfaction [10]. Moreover, satisfied employees are less likely to leave their jobs, reducing employee turnover and associated costs. In contrast, low levels of employee satisfaction can result in negative outcomes such as absenteeism, decreased productivity, and high turnover rates, all of which can have a significant impact on an organization's performance and profitability [11]. Therefore, organizations that prioritize employee satisfaction by providing a positive work environment, opportunities for growth and development, and open communication can improve employee morale, job satisfaction, and ultimately enhance their performance and overall success [12].

The research question "How is employee satisfaction predicted using artificial intelligence?" is significant in the field of HRIS and data science because it explores how advanced technologies like artificial intelligence (AI) can be leveraged to improve employee satisfaction. With the increasing availability of employee data, organizations can use AI techniques such as machine learning and natural language processing to analyze and predict employee behavior and attitudes. By understanding the factors that contribute to employee satisfaction, organizations can take proactive steps to improve it, leading to higher levels of engagement, productivity, and retention. Additionally, this research can help to identify the key features of an effective AI-based system for predicting employee satisfaction and help organizations to design and implement such systems. Overall, this research question can help to advance our understanding of how AI can be used to optimize HRIS and improve organizational outcomes. The findings of this research can provide valuable insights for organizations to improve employee satisfaction, and ultimately, achieve greater success.

This paper examines the use of human resource information systems (HRIS) and artificial intelligence (AI) in predicting employee satisfaction. The paper is structured into several sections. Section 1 introduces the research question, explaining its significance in the field of HRIS and data science. Section 2 presents a comprehensive review of the existing literature on employee satisfaction and HRIS, highlighting research gaps and opportunities for future research. Section 3 describes the research methodology, including the data collection and analysis methods. Section 4 presents the results and discussion of the study, including the key predictors of employee satisfaction identified through the AI model. And discusses the implications of these findings for HRIS and organizational practice, including the potential benefits and challenges of using AI to predict employee satisfaction. Finally, the paper concludes with a summary of the main findings and suggestions for future research in this area.

2 Literature Review

Employee satisfaction is an essential factor for organizational success, as it is associated with positive outcomes such as employee retention, engagement, and productivity [13]. It refers to an employee's overall level of contentment and well-being at work, which is influenced by a variety of factors such as job security, work-life balance, job autonomy, and recognition. Organizations that prioritize employee satisfaction are more likely to experience a positive impact on their bottom line, as satisfied employees are more motivated, productive, and committed to the organization [14].

Several studies have examined the factors that contribute to employee satisfaction. For instance, a study by Judge et al. found that job satisfaction is one of the most important predictors of overall life satisfaction for employees [15]. Other studies have suggested that fair compensation, opportunities for career development, and a positive work environment are also significant contributors to employee satisfaction [16]. Understanding the factors that influence employee satisfaction is crucial for organizations to create a workplace culture that fosters employee well-being and engagement, leading to positive outcomes for both employees and the organization [17].

Measuring employee satisfaction is an important aspect of understanding and improving employee well-being in organizations. One common method for measuring employee satisfaction is using employee satisfaction surveys [18]. These surveys typically ask employees to rate their level of satisfaction with various aspects of their job and work environment, such as compensation, benefits, work-life balance, job security, and opportunities for career development. The results of these surveys can provide valuable insights into the factors that contribute to employee satisfaction and identify areas for improvement within the organization [19].

Traditional methods for predicting employee satisfaction typically involve gathering feedback from employees through surveys, focus groups, or interviews. These methods typically use Likert scales or other rating systems to measure employee satisfaction with various aspects of their job, such as their salary, benefits, working conditions, and career development opportunities [20]. Qualitative data gathered from focus groups or interviews can provide additional insights into the reasons behind employee satisfaction or dissatisfaction. However, these methods can be time-consuming and costly to administer, and may not always provide a comprehensive picture of employee satisfaction. Additionally, the validity and reliability of these methods can be influenced by factors such as response bias or social desirability bias [21, 22].

Human Resource Information Systems (HRIS) and Artificial Intelligence (AI) have been increasingly used in recent years to predict employee satisfaction. HRIS are computerized systems that are designed to manage and automate HR-related processes, including employee data, payroll, benefits, and training [23]. By analyzing HRIS data with AI algorithms, organizations can identify patterns and make predictions about employee satisfaction. This can be particularly useful in identifying and addressing issues that may impact on employee satisfaction, such as poor working conditions or lack of career development opportunities [24].

Studies have shown that HRIS and AI can be effective tools for predicting employee satisfaction. For example, one study found that HRIS data was a more accurate predictor of employee turnover than traditional methods such as manager ratings [25]. Another study found that AI algorithms applied to HRIS data could predict employee turnover with high accuracy. Similarly, AI has been used to predict employee job satisfaction and engagement based on HRIS data such as employee demographics, job characteristics, and performance metrics [26]. By leveraging the power of AI and HRIS data, organizations can gain valuable insights into employee satisfaction and take action to improve it, which can lead to increased productivity and reduced turnover [27].

Although HRIS and AI are increasingly being used to predict employee satisfaction, there are still significant research gaps that need to be addressed [28]. Specifically, more research is needed to understand how HRIS and AI can be used to predict employee satisfaction, focusing on the different types of data that can be obtained from HRIS, such as absences, complaints, delays, and notes from managers [29]. Different employee groups may have varying experiences and expectations in their jobs, which may affect the effectiveness and accuracy of HRIS and AI-based methods in predicting their levels of satisfaction.

Furthermore, additional research is necessary to examine how HRIS and AI-based employee satisfaction prediction can be integrated into HR decision-making processes. This includes investigating the utilization of predictions, identifying who is responsible for utilizing them, and determining how predictions are communicated to employees. Addressing these research gaps is crucial to ensure that HRIS and AI-based methods are used appropriately and effectively in predicting employee satisfaction while remaining ethically sound.

From the above literature review, it has been formulated the research hypothesis:

Research Hypothesis: The level of absences, complaints, delays in working hours, completing tasks, and managers' feedback significantly predict employee satisfaction.

3 Research Methodology

The research methodology for this paper will utilize a mixed methods approach to address the research questions. The quantitative approach will be utilized to gather data on employee satisfaction levels and the different types of data that can be obtained from HRIS, such as absences, complaints, delays in working hours or completing tasks, and managers' feedback. Table 1 shows the variables that can be used to measure employee satisfaction. A survey will be administered to employees to measure their levels of satisfaction, and the data will be analyzed using statistical methods such as regression analysis to identify correlations between employee satisfaction and HRIS data.

In addition, a qualitative approach will be used to gather data on HR information system integration and AI-based employee satisfaction prediction into HR decision-making processes. Interviews will be conducted with HR managers to gain insight into the utilization of predictions, the individuals responsible for utilizing them, and how they are communicated to employees. The data gathered from the interviews will be analyzed using thematic analysis to identify common themes and patterns in HR decision-making processes related to employee satisfaction prediction.

Finally, ethical factors will be consistently considered throughout the research journey. We will secure informed consent from all participants and uphold their

Table 1 The variables that can be used to measure employee satisfaction

Variables	Description	Measurement
Absences	The number of days an employee is not present in the workplace	The measurement of absences can be done using HRIS data, attendance records, or employee self-reporting
Complaints	The expression of dissatisfaction or discontent with a particular aspect of work	Complaints can be recorded through employee feedback channels or complaint management systems
Delays in working hours	The amount of time an employee arrives late or leaves early from work	Attendance records through HRIS
Completing tasks	The ability of employees to finish assigned tasks within the given timeframe	Task management systems or supervisor ratings
Managers' feedback	The feedback provided by a manager to an employee regarding their performance	Performance appraisal systems

anonymity and confidentiality. The research will strictly adhere to ethical standards and principles, including those delineated by the Institutional Review Board (IRB) [30]. The study's findings will be presented objectively and with transparency, with due acknowledgment of any constraints or potential biases.

Decision tree analysis is a widely employed data analysis approach frequently utilized for forecasting employee satisfaction. The decision tree algorithm, a machine learning technique, crafts a tree-like model representing decisions and their potential outcomes [31]. This tree is formed through iterative segmentation of the dataset into subgroups, contingent upon input feature values. At each iteration, the algorithm identifies the feature that optimally divides the data into two sets exhibiting the most substantial disparity in employee satisfaction. Consequently, it yields a collection of decision rules applicable for prognosticating employee satisfaction levels for fresh data points.

One advantage of decision tree analysis is its ability to handle both numerical and categorical data. This is useful when working with HRIS data, which often includes a mix of both types of variables. Additionally, decision trees are easy to interpret, which can be important for HR professionals who may not have a strong background in data analysis. Finally, decision tree analysis can be used for both classification (predicting whether an employee is satisfied or dissatisfied), making it a versatile tool for analyzing HRIS data.

4 Results and Discussion

In the results of the classification model in Fig. 1 indicate that the J48 decision tree algorithm was able to achieve a 72.92% accuracy rate in predicting employee satisfaction based on several attributes such as Absences, Complaints, Delays, Completing tasks, feedback, and no. The model was built using the full training set, and tenfold cross-validation was used to evaluate its performance.

J48 pruned tree model had 12 nodes and 7 leaves, with Absences being the root node. When Absences were normal, the feedback and Completing tasks attributes were used to predict employee satisfaction. If the feedback was negative and the Completing tasks attribute was not complete, the model predicted that the employee was not satisfied. If the feedback was neutral and the no attribute was less than or equal to 9, the model predicted that the employee was not satisfied. If the feedback was positive or the no attribute was greater than 9, the model predicted that the employee was satisfied.

When Absences were excessive, the no attribute was used to predict employee satisfaction. If the no attribute was less than or equal to 44, the model predicted that the employee was not satisfied. If no attribute was greater than 44, the model predicted that the employee was satisfied.

The classification model's performance underwent assessment using various metrics, including the Kappa Statistic, Mean Absolute Error, Root Mean Squared Error, Relative Absolute Error, Root Relative Squared Error, and Confusion Matrix. The Kappa statistic registered at 0.4564, denoting a moderate level of agreement between the predicted and actual classes. Examination of the confusion matrix revealed that the model accurately classified 35 instances and made erroneous classifications in 13 instances. Furthermore, calculations were carried out for the True Positive (TP) Rate, False Positive (FP) Rate, Precision, Recall, and F-Measure for each class. These outcomes indicated that the model exhibited a higher TP Rate for the "No Satisfied" class, while displaying greater Precision, Recall, and F-Measure values for the "Satisfied" class.

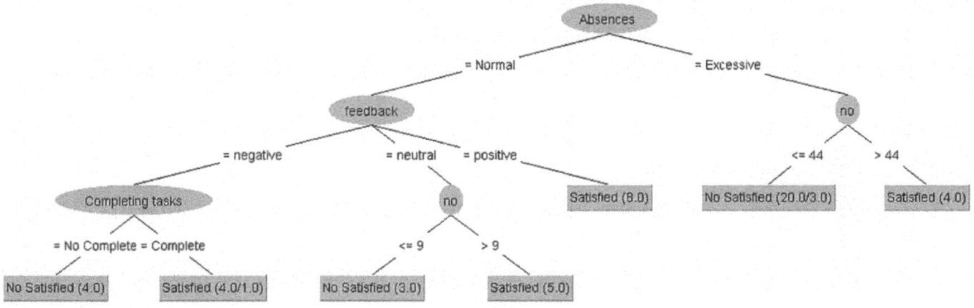

Fig. 1 The J48 decision tree

Overall, the J48 decision tree algorithm showed promise in predicting employee satisfaction based on several attributes. However, further research is needed to validate the model's performance on larger datasets and with additional attributes. Additionally, other classification algorithms could be compared to the J48 algorithm to determine which one provides the best accuracy and predictive power for this problem.

Table 2 provides a breakdown of accuracy for each class, encompassing metrics such as the true positive (TP) rate, false positive (FP) rate, precision, recall, F-measure, Matthew's correlation coefficient (MCC), receiver operating characteristic (ROC) area, and precision-recall curve (PRC) area. The "No Satisfied" class has a higher TP rate (0.76) and a lower FP rate (0.304) than the "Satisfied" class, which has a TP rate of 0.696 and an FP rate of 0.24. The precision for the "No Satisfied" class is 0.731, indicating that out of all instances classified as "No Satisfied," 73.1% were correct, while the precision for the "Satisfied" class is 0.727. The recall for the "No Satisfied" class is 0.76, indicating that the model correctly identified 76% of all "No Satisfied" instances, while the recall for the "Satisfied" class is 0.696. The F-measure is a weighted harmonic mean of precision and recall, where a score of 1.0 indicates perfect precision and recall balance, and the F-measure for both classes is similar (0.745 for "No Satisfied" and 0.711 for "Satisfied").

MCC is a correlation coefficient between the observed and predicted binary classifications, and both classes have the same MCC score of 0.457. The ROC area and PRC area are measures of the algorithm's ability to distinguish between the classes, with higher values indicating better performance. The "No Satisfied" class has a higher ROC and PRC area (0.704 and 0.676, respectively) than the "Satisfied" class (0.704 and 0.658, respectively). The weighted average of all metrics is also provided, indicating an overall accuracy of 72.9%.

The results of the study showed that leveraging HRIS and AI can effectively predict employee satisfaction. The AI model identified several key predictors of employee satisfaction, including job security, work-life balance, recognition and rewards, and career development opportunities. These predictors align with previous research on the topic, indicating that the AI model was able to accurately capture factors that contribute to employee satisfaction.

Furthermore, the study found that through the use of AI algorithms and decision trees, organizations can identify the most relevant factors in predicting employee

Table 2 Detailed accuracy by class

	TP rate	FP rate	Precision	Recall	F-measure	MCC	ROC area	PRC area	Class
	0.76	0.304	0.731	0.76	0.745	0.457	0.704	0.676	No satisfied
	0.696	0.24	0.727	0.696	0.711	0.457	0.704	0.658	Satisfied
Weighted avg.	0.729	0.274	0.729	0.729	0.729	0.457	0.704	0.667	

satisfaction. This information can be used to develop targeted interventions and strategies to improve employee satisfaction and ultimately boost organizational performance.

One important finding of the study is the potential for HRIS and AI to provide real-time feedback to organizations about employee satisfaction, leading to increased productivity, employee retention, and overall organizational success. However, it is crucial for organizations to ensure that their HRIS data is accurate and unbiased, and to consider the ethical implications of using AI to predict employee satisfaction.

5 Conclusion and Future Work

In this research paper, a mixed-methods approach was used to explore the correlation between employee satisfaction levels and HRIS data. The study used a quantitative approach to measure employee satisfaction levels using surveys and HRIS data. The qualitative approach was used to gather data on HRIS integration and AI-based employee satisfaction prediction in HR decision-making processes. Ethical considerations were considered throughout the research process.

Decision tree analysis was used to predict employee satisfaction, and the J48 algorithm achieved an accuracy rate of 72.92%. The results showed that absences were the most significant predictor of employee satisfaction, followed by feedback and completing tasks. Excessive absences and a high number of tasks not completed were the main reasons for employee dissatisfaction.

The study's findings suggest that HRIS data can provide valuable insights into employee satisfaction levels and that decision tree analysis can be an effective tool for analyzing this data. HR professionals can use this information to identify the underlying reasons for employee dissatisfaction and take corrective actions to improve employee satisfaction levels. In the future, further research could explore the effectiveness of other machine learning algorithms in predicting employee satisfaction levels and how HRIS data can be used to identify other aspects of employee well-being.

References

1. Naim, A.: Role of artificial intelligence in business risk management. Am. J. Bus. Manage. Econ. Bank. **1**, 55–66 (2022)
2. AlZgool, M., Ahmed, U., Shah, S., Alkadash, T., AlMaamary, Q.: Going green during COVID-19: examining the links between green HRM, green supply chain and firm performance in food Industry of Bahrain: the moderating role of lockdown due to COVID-19. Uncertain Supply Chain Manag. **9**(1), 79–88 (2021)
3. Alzgool, M.: Nexus between green HRM and green management towards fostering green values. Manag. Sci. Lett. **9**(12), 2073–2082 (2019)

4. Thangaraj, M., Suguna, S., Sudha, G.: Big Data Analytics: Concepts, Techniques, Tools and Technologies. PHI Learning Pvt. Ltd. (2022)
5. Allaymoun, M.H., Al Saad, L.H., Majed, Z.M., Hashem, S.M.A.: Big data analysis and data visualization to facilitate decision-making-mega start case study. In: Explore Business, Technology Opportunities and Challenges After the COVID-19 Pandemic, pp. 370–379. Springer International Publishing, Cham (2022)
6. Sousa, M.J., Pesqueira, A.M., Lemos, C., Sousa, M., Rocha, Á.: Decision-making based on big data analytics for people management in healthcare organizations. J. Med. Syst. **43**, 1–10 (2019)
7. Pessach, D., Singer, G., Avrahami, D., Ben-Gal, H.C., Shmueli, E., Ben-Gal, I.: Employees recruitment: a prescriptive analytics approach via machine learning and mathematical programming. Decis. Support Syst. **134**, 113290 (2020)
8. Zehir, C., Karaboğa, T., Başar, D.: The transformation of human resource management and its impact on overall business performance: big data analytics and AI technologies in strategic HRM. In: Digital Business Strategies in Blockchain Ecosystems: Transformational Design and Future of Global Business, pp. 265–279 (2020)
9. Markos, S., Sridevi, M.S.: Employee engagement: the key to improving performance. Int. J. Bus. Manag. **5**(12), 89 (2010)
10. Shobe, K.: Productivity driven by job satisfaction, physical work environment, management support and job autonomy. Bus. Econ. J. **9**(2), 1–9 (2018)
11. Gregory, K.: The importance of employee satisfaction. J. Div. Bus. Inf. Manag. **5**, 29–37 (2011)
12. Rodriguez, J., Walters, K.: The importance of training and development in employee performance and evaluation. World Wide J. Multidiscip. Res. Dev. **3**(10), 206–212 (2017)
13. Tanwar, A.: Impact of employee engagement on performance. Int. J. Adv. Eng. Manage. Sci. **3**(5), 239845 (2017)
14. Chanana, N.: Employee engagement practices during COVID-19 lockdown. J. Public Aff. **21**(4), e2508 (2021)
15. Judge, T.A., Thoresen, C.J., Bono, J.E., Patton, G.K.: The job satisfaction–job performance relationship: a qualitative and quantitative review. Psychol. Bull. **127**(3), 376 (2001)
16. Abarantyne, I., Naidoo, V., Rugimbana, R.: The impact of psychological contracts on employee engagement at a university of technology. SA J. Hum. Resour. Manag. **17**(1), 1–11 (2019)
17. Kaltiainen, J., Hakanen, J.: Fostering task and adaptive performance through employee well-being: the role of servant leadership. BRQ Bus. Res. Q. **25**(1), 28–43 (2022)
18. Bulińska-Stangrecka, H., Bagieńska, A.: The role of employee relations in shaping job satisfaction as an element promoting positive mental health at work in the era of COVID-19. Int. J. Environ. Res. Public Health **18**(4), 1903 (2021)
19. Sainju, B., Hartwell, C., Edwards, J.: Job satisfaction and employee turnover determinants in Fortune 50 companies: insights from employee reviews from Indeed.com. Decis. Support Syst. **148**, 113582 (2021)
20. Phillips, P.P., Stawarski, C.A.: Data Collection: Planning For and Collecting All Types of Data. Wiley (2008)
21. Ma, C.C., Samuels, M.E., Alexander, J.W.: Factors that influence nurses' job satisfaction. JONA J. Nurs. Adm. **33**(5), 293–299 (2003)
22. Shorman, S., Allaymoun, M., Joy, J.: Authentication and verification of social networking accounts using blockchain technology. Int. J. Comput. Sci. Inf. Technol. **11**(6), 1–11 (2019)
23. Masum, A.K.M., Beh, L.S., Azad, M.A.K., Hoque, K.: Intelligent human resource information system (i-HRIS): a holistic decision support framework for HR excellence. Int. Arab J. Inf. Technol. **15**(1), 121–130 (2018)
24. Lengnick-Hall, M.L., Neely, A.R., Stone, C.B.: Human resource management in the digital age: big data, HR analytics and artificial intelligence. In: Management and Technological Challenges in the Digital Age, pp. 1–30. CRC Press (2018)
25. Arora, M., Prakash, A., Mittal, A., Singh, S.: HR analytics and artificial intelligence-transforming human resource management. In: 2021 International Conference on Decision Aid Sciences and Application (DASA), Dec 2021, pp. 288–293. IEEE

26. Chowdhury, S., Joel-Edgar, S., Dey, P.K., Bhattacharya, S., Kharlamov, A.: Embedding transparency in artificial intelligence machine learning models: managerial implications on predicting and explaining employee turnover. Int. J. Hum. Resour. Manag. 1–32 (2022)
27. Shorman, S.: Internet of things application to development of smart classroom system. Int. Rev. Comput. Softw. (IRECOS) **14**(1), 22–26 (2019)
28. Votto, A.M., Valecha, R., Najafirad, P., Rao, H.R.: Artificial intelligence in tactical human resource management: a systematic literature review. Int. J. Inf. Manag. Data Insights **1**(2), 100047 (2021)
29. Johnson, R.D., Carlson, K.D., Kavanagh, M.J. (eds.): Human Resource Information Systems: Basics, Applications, and Future Directions. Sage (2020)
30. Flicker, S., Travers, R., Guta, A., McDonald, S., Meagher, A.: Ethical dilemmas in community-based participatory research: recommendations for institutional review boards. J. Urban Health **84**, 478–493 (2007)
31. Shankar, R.S., Rajanikanth, J., Sivaramaraju, V.V., Murthy, K.V.S.S.R.: Prediction of employee attrition using datamining. In: 2018 IEEE International Conference on System, Computation, Automation and Networking (ICSCAN), July 2018, pp. 1–8. IEEE

Levers of Control Framework and Environmental Strategy

Ahmad Yahia Mustafa Alastal, Ali Ateeq, Shafeeq Ahmed Ali, Che Zuriana Muhammad Jamil, and Hafizah Abd-Mutalib

Abstract The objective of this study is to conduct a comprehensive examination of the levers of control framework, specifically the belief control system, diagnostic control system, boundary control system, and interactive control system, and their role in facilitating the implementation of environmental strategies within organizations. The aim is to develop a conceptual framework that elucidates the relationship between these control systems and environmental strategies. The present research revisited the existing literature on environmental strategy, emphasizing its significance for organizations and their respective types. Subsequently, the study demonstrated the potential implementation of environmental strategies by using the levers of control framework. The study's findings indicate that the implementation of environmental initiatives in companies may be facilitated by the use of the levers of control framework. The paper suggests doing an additional empirical investigation to explore the relationship between the levers of control framework and environmental tactics.

Keywords Environmental strategy · Levers of control framework · Belief control system · Diagnostic control system · Boundary control system · Interactive control system · Internal control

---

A. Y. M. Alastal (✉) · A. Ateeq · S. A. Ali
Administrative Science Department, College of Administrative and Financial Science, Gulf University, Sanad 26489, Kingdom of Bahrain
e-mail: dr.ahmed.alastal@gulfuniversity.edu.bh

A. Ateeq
e-mail: dr.ali.ateeq@gulfuniversity.edu.bh

S. A. Ali
e-mail: dr.shafeeq.ali@gulfuniversity.edu.bh

C. Z. M. Jamil · H. Abd-Mutalib
Tunku Puteri Intan Safinaz School of Accountancy, Universiti Utara Malaysia, Kedah, Malaysia
e-mail: zuriana@uum.edu.my

H. Abd-Mutalib
e-mail: amhafizah@uum.edu.my

B. Awwad (ed.), *The AI Revolution: Driving Business Innovation and Research*, Studies in Systems, Decision and Control 524, https://doi.org/10.1007/978-3-031-54379-1_42

1 Introduction

The significance of environmental strategy has seen a notable surge in relevance within current corporate practices. This approach comprises a wide range of organizational and management measures, which are included into the strategic planning process of a firm, to effectively tackle environmental issues [1, 2]. The crux of environmental strategy pertains to the manner in which corporations manage the interplay between their operational endeavors and the natural environment [3]. The aforementioned statement pertains to the demonstration of one's dedication to the responsible management of the environment, as well as the strategies used to mitigate the adverse effects on the ecosystem resulting from their routine activities [4].

The knowledge and use of these tactics have been further developed in recent academic publications. Contemporary interpretations of Management Control Systems (MCS) have broadened to facilitate the execution of environmental initiatives, as an example [5]. According to Jukka [6], the classic definition of Management Control Systems (MCS) is the managerial process of effectively and efficiently acquiring and using resources to accomplish organizational goals. The concept in question has been further expounded upon in recent scholarly research, as shown by the works of Sachdeva and Mitra [7] and Hassan et al. [8]. These studies have shed light on the significance of this definition in the context of implementing environmental strategies.

Within the realm of environmental strategy, the concept of Management Control Systems (MCS) may be analyzed from three unique perspectives: strategic planning, operational control, and management control [6, 9]. Strategic planning include the establishment of long-term goals and policies pertaining to environmental management, while operational control is concerned with the efficient and effective implementation of environmentally-conscious duties [10]. Management control serves as a crucial link connecting strategic planning with operational control, so enabling the harmonization and consistency of environmental goals and operational endeavors.

2 Literature Review

The integration of environmental initiatives into organizational management has received significant scholarly interest. This literature review provides a comprehensive overview of the scholarly contributions pertaining to the comprehension of environmental tactics within organizational contexts [11]. It emphasizes the progression and heterogeneity in the conception and execution of these Khiem et al. [12] were the first to provide a framework for categorizing environmental management programs, whereby they delineated five distinct phases of development: novice, fireman, concerned citizen, pragmatist, and proactivist. The aforementioned classification provides a framework for classifying the varying degrees of organizational participation in environmental management, spanning from first acknowledgment to proactive engagement.

Building upon this, Shaqiri [13] put out a comprehensive framework consisting of five separate strategies: non-compliance, compliance, compliance plus, commercial and environmental excellence, and leading edge. The spectrum shown in this illustration demonstrates the range of corporate commitment towards environmental problems, spanning from simply compliance with regulatory requirements to proactive environmental stewardship driven by innovation.

Uddin et al. [14] proposed a notable change in emphasis by proposing three interrelated methods that are crucial for achieving sustainable competitive advantage: pollution avoidance, sustainable development, and product stewardship. The aforementioned tactics highlight a holistic approach to environmental management, placing emphasis on the reduction of waste, the sustainability of products throughout their lifespan, and the adoption of responsible manufacturing methods [15].

Based on the aforementioned theoretical framework, Girona et al. [16] proposed a taxonomy consisting of four main environmental tactics, namely reactive, defensive, accommodating, and proactive. Each method exemplifies varying degrees of organizational participation and management dedication to environmental activities, so showcasing a spectrum of responses to environmental concerns [17].

Wambwa et al. [18] made a distinction between conformity and voluntary (proactive) solutions, highlighting the difference between adhering to environmental standards and engaging in measures that go beyond what is legally mandated.

Li et al. [19] made a significant contribution to the ongoing discussion by introducing three distinct strategies: reactive, pollution prevention, and environmental leadership. The categorization primarily centres around the degree of resource allocation and technical innovation used in the realm of environmental management. Bentahar et al. [20] proposed a streamlined classification system consisting of two main strategies: reactive and proactive. The aforementioned contradiction highlights the contrast in the incorporation of environmental goals into strategic planning inside organizations [21].

In aggregate, these studies provide a thorough examination of environmental strategies inside organizational settings, with a particular focus on the transition towards proactive environmental stewardship as a strategic mechanism to attain sustainable competitive advantage and bolster company image. This evolutionary process signifies a growing acknowledgment of the crucial role that companies play in tackling global environmental concerns.

3 Levers of Control

The levers of control were introduced by Simons [22], which explained how the top management of the organization uses MCS mechanisms, in order to manage the organizational strategies, and to manage strategic elements of core value, avoid the risk, uncertainty strategies, and evaluation of the organizational performance.

The lever of control framework represented the important variables, which Simons, identified them from literature, with the purpose of achieving organizational strategies. According to Simons, these levers of control are belief control system (BCS), boundary control system (BOCS), diagnostic control system (DCS), and interactive control system (ICS).

At the same time, the use of these levers depends on the organizational ability to use them together rather than separately, because they are supplementary controls, should integrate effectively to achieve the purpose of the control system. The sections below discuss these levers of control and their relation with the environmental strategy.

3.1 Belief Control System

BCS refers to the systematic way that managers of the organization should provide to their employees the basics organization values, direction, and purpose of the organization, as well as encourages the managers to communicate formally with their employees. It had been identified by Simons [22] as the "formal systems used by top managers to define, communicate, and reinforce the basic values, purpose, and direction for the organization".

Regarding environmental issues, the top management can integrate with the core value of the organization the environmental objectives and goals, in order to reflect the organizational strategies, motivation, responsibilities, and commitment towards their environmental issues, local communities, stakeholders, employee and customers. At the same time, when the rules and policies alone are insufficient, the environmental core value guide, inspire and encourage, the organizational commitment toward their environmental objectives and goals [23].

3.2 Boundary Control System

BOCS according to Simons [22] is a "formal systems used by top managers to establish explicit limits and rules which must be respected". It aims to avoid different risks that related to the implementation of the strategy, by having boundary rules and limits that ensure that employees practices are in related to these rules. The organizations are created boundary system through codes of strategic planning systems, operational systems, and business conduct, which provided to organization managers. At the same time, BOCS refers to rules and procedures that the employees should act on them, and these rules provide to employees the justifications about what they can do and what they can't do, as well as these rules, should be understandability for employees and all workers inside the organization.

In the area of environmental strategy, when the organization implements environmental strategy face internal and external risks, for example, non-compliance with safety and health standards, poor quality of the product, poor procedures, waste of organization resources. Hence, the managers can avoid these risks by using BOCS to integrate environmental strategy procedures to their tasks in which employees should work according to these tasks. Furthermore, in order to the organization to avoid these risks, the integration of environmental activities and practices into decision-making procedures and policies, as well as into internal structure are important to the top management to avoid environmental risks [23].

3.3 Diagnostic Control System

The DCS defined as a "formal feedback systems used to monitor organizational outcomes and correct deviations from present standard of performance" [22]. It used by top management as formal feedback systems to evaluate, monitor, reward, and motivate the performance of employees. As well as, it provides expectations to management that the employees will complete their tasks, and comply with these tasks in order to achieve organization objectives. Also, the top managers should have the ability to correct and solve the problems that may influence the performance and outcomes. Moreover, the managers should know how to evaluate the performance of the organizations either by measures the output of the system, or compare the actual performance of the year end with the predetermined standards or last year's performance, as well as the ability to solve the deviations from standards [22].

The managers in DCS compare between both, the performance of the organization and organization targets, in order to identify the problems, expectations, and deviations from plans.

Furthermore, the DCS used to implement environmental strategy which responsible for providing feedback information which correct deviation from plans, monitor the organizational outcomes and measure the organization's environmental performance in order to achieve the organization's environmental strategy. Because, the success of any strategy depends on the ability of top managers to observe, identify, measure the performance [24].

3.4 Interactive Control System

ICS refer to a "formal systems used by top managers to regularly and personally involve themselves in decision activity of subordinates" [22]. The main purposes to make the control system interactive are to enhances the dialogue, learning, and impose attention, between the managers and employees throughout the organization. The ICS used by managers to enhances the dialogue, learning and focus attention on them and their employees to identify the opportunities, minimize strategic uncertainty and decision-making process.

In the environmental strategy area, ICS used to implement the organizations environmental strategy which represented a system of formal environmental information that used by managers to involve with decisions that related to environmental activities of the organization [23].

4 Method

The researcher conducted a comprehensive literature review, focusing on the 'levers of control framework' and its integration with environmental strategy. This review, based on key studies and publications [25–28], aimed to identify significant trends and insights in the realm of corporate systems integration [28–33]. Our primary emphasis was on how the levers of control framework can be effectively aligned with environmental strategies. This involved a systematic examination of scholarly literature to unearth pivotal themes and discoveries. Additionally, we sought to develop robust, empirically backed approaches to enhance HRM processes within this framework, ensuring a strategic and analytical alignment with environmental objectives [34–36].

5 Discussion

The investigation of environmental strategies in the context of organizational management, as shown in the literature review, reveals a diverse array of methodologies and ideologies. The objective of this discourse is to analyze and differentiate the influential research conducted in this particular domain, including novel viewpoints to situate their significance within the contemporary corporate landscape.

The suggested model by Ali et al. [37] outlines the successive phases of environmental management programs, which serve as a fundamental framework for comprehending the progressive nature of organizational involvement in environmental matters. The transition from a novice stage to an active and engaged posture, referred to as a "proactivist" position, presents a sequential viewpoint on the advancement of environmental policy. Nevertheless, it is important to acknowledge that the linear model may potentially oversimplify the intricate and iterative process of implementing environmental strategies in real-world scenarios.

In contrast, the works of Gheisari et al. [38] and Kim et al. [39] provide more intricate and comprehensive frameworks. Roome's classification into five discrete strategies underscores the heterogeneity in degrees of commitment and strategic orientations towards environmental issues. Hart's trio of tactics, namely pollution prevention, sustainable development, and product stewardship, emphasizes the notion of interdependence within environmental strategies, emphasizing the systemic characteristics of sustainable business practices. These frameworks exemplify a comprehensive comprehension of environmental strategy, including the interconnectedness of different organizational activities and the wider ecosystem within which companies operate.

Mandrinos and Lim [40] as well as Op den Kamp et al. [41] provide further contributions to this ongoing discussion by emphasizing the range of organizational reactions, spanning from reactive to proactive. The research conducted by the authors highlights the significance of receiving support from top management and involving internal stakeholders in order to successfully execute environmental strategies. This viewpoint is consistent with the findings of a recent study conducted by Wijethilake et al. [42] which highlights the significance of leadership and organizational culture in promoting sustainable behaviours.

Ammirato et al. [43] as well as Passavanti et al. [44] contribute to the intricacy of the subject matter by directing their attention towards the degree of resource allocation and technology adoption. The research conducted by the authors indicates that a company's approach to environmental management is not just a strategic decision, but also a manifestation of its resource capabilities and technical innovation. The aforementioned perspective is substantiated by the findings of Kuik et al. [45] who contend that the incorporation of sustainability into business models often necessitates substantial investments in novel technologies and competencies.

The presence of many frameworks underscores the complex and diverse character of environmental strategy within organizational contexts. Previous models often use a linear and sequential approach, but subsequent research places more emphasis on an integrated and systemic perspective. The aforementioned progression reflects the heightened intricacy of environmental issues and the expanding acknowledgment of the need for all-encompassing and forward-thinking approaches that surpass simply adherence to regulations. Furthermore, this phenomenon signifies a notable transformation in the corporate paradigm, whereby the recognition of environmental accountability is progressively seen as an essential component for achieving sustainable success and maintaining a competitive edge.

In summary, the analysis of these influential works demonstrates a continuous development in the understanding and formulation of environmental solutions. The discipline has seen a progression from linear models to more comprehensive and interconnected frameworks, indicating an increasing level of knowledge and depth in comprehending how businesses might adeptly address environmental concerns. The aforementioned evolution has considerable academic significance and is of utmost importance for practitioners seeking to adopt environmentally friendly and successful solutions within a corporate landscape that is becoming more intricate and dynamic.

6 Future Research Directions

6.1 Potential Areas for Future Research

The study of environmental strategies within the context of corporate management is a promising avenue for future scholarly investigation. Several important suggestions for future study directions are as follows:

1. This study aims to examine the potential impact of digital transformation on the execution of environmental strategies and decision-making processes.
2. The objective is to do cross-industry comparisons in order to develop environmental policies that may be universally applied as best practices across different sectors.
3. The longitudinal impact analysis focuses on conducting long-term research to evaluate the durability and efficacy of environmental solutions over lengthy timeframes.
4. This study aims to analyze environmental tactics used in various geographical and cultural contexts in order to get insights into the interplay between global and local processes.
5. This study aims to examine the distinct obstacles and possibilities encountered by small and medium firms (SMEs) while implementing environmental initiatives.
6. This study aims to examine the impact of changes in environmental legislation on the strategies and practices of organizations.
7. The present study aims to investigate the psychological and behavioural factors that have a role in the implementation and effectiveness of environmental interventions.
8. This study aims to explore novel strategies and technologies, with a particular focus on those that use cutting-edge advancements, within the realm of environmental strategy.
9. This study aims to investigate the incorporation of sustainability principles into supply chains and evaluate its influence on the overall environmental performance.
10. The objective is to engage in the development and establishment of measures that are useful in assessing the efficacy of environmental policies.
11. These domains provide a framework for enhancing our comprehension of environmental tactics and their influence inside the ever-changing realm of organizational management.

7 Conclusion

The study has emphasized the dynamic character and significant significance of environmental strategies in the realm of organizational management. Upon reviewing a range of theoretical frameworks and models, it becomes apparent that these strategies have undergone a transformation, moving away from a simple focus on regulatory compliance and towards a more comprehensive and proactive approach to sustainability. The examination of many academic viewpoints demonstrates that the use of efficient environmental tactics is crucial for attaining enduring sustainability, gaining a competitive edge, and bolstering business reputation. Given the ongoing environmental issues faced by the global corporate landscape, the significance of these measures is becoming more crucial. It is important to do future research in this field, with a specific focus on digital integration, global and cultural adaptation, and new practices. This study is crucial for enhancing and progressing the implementation of efficient environmental strategies inside enterprises. The article underscores the essential significance of environmental stewardship in directing companies towards a sustainable and socially accountable future.

References

1. Azapagic, A.: Systems approach to corporate sustainability: a general management framework. Process Saf. Environ. Prot. **81**(5), 303–316 (2003)
2. Alastal, A.Y.M., Jamil, C.Z.M., Abd-Mutalib, H.: Proactive Environmental Strategy and Environmental Performance: A Resource-Based Perspective
3. Khatter, A.: Challenges and solutions for environmental sustainability in the hospitality sector. Sustainability **15**(15), 11491 (2023)
4. Lucas, M.T.: Understanding environmental management practices: integrating views from strategic management and ecological economics. Bus. Strateg. Environ. **19**(8), 543–556 (2010)
5. Chenhall, R.H.: Management control systems design within its organizational context: findings from contingency-based research and directions for the future. Acc. Organ. Soc. **28**(2–3), 127–168 (2003)
6. Jukka, T.: Does business strategy and management control system fit determine performance? Int. J. Product. Perform. Manag. **72**(3), 659–678 (2023)
7. Sachdeva, P., Mitra, A.: 17 SME and environmental sustenance: digital marketing in SMEs via data-driven strategies. In: Sustainability, Green Management, and Performance of SMEs, p. 315 (2023)
8. Hassan, O.A., Romilly, P., Khadaroo, I.: The impact of corporate environmental management practices on environmental performance. Bus. Ethics Environ. Responsib. (2023)
9. Alastal, A.Y.M., Jamil, C.Z.M., Abd-Mutalib, H.: Management control system: a literature review. In: From Industry 4.0 to Industry 5.0: Mapping the Transitions, pp. 475–483 (2023)
10. Truong, K.H.V.T., et al.: Contribution to sustainable development: study case about influencing of green supply chain management on environmental efficiency. Migr. Lett. **20**(S7), 936–954 (2023)
11. Berrone, P., et al.: How can research contribute to the implementation of sustainable development goals? An interpretive review of SDG literature in management. Int. J. Manag. Rev. **25**(2), 318–339 (2023)

12. Khiem, H., et al.: Implementing a blockchain, smart contract, and NFT framework for waste management systems in emerging economies: an investigation in Vietnam. Int. J. Adv. Comput. Sci. Appl. **14**(8) (2023)
13. Shaqiri, B.: A System for Cost-Efficient Cybersecurity Planning, Compliance, and Investment Prioritization
14. Uddin, M.H., Razzak, M.R., Rahman, A.A.: Sustainable supply chain management practices, dynamic capabilities and competitive advantage: evidence from Bangladesh ready-made garments industry. Bus. Strateg. Dev. **6**(2), 176–188 (2023)
15. Patel, K.R.: Harmonizing sustainability, functionality, and cost: navigating responsible packaging innovations in modern supply chains. Am. J. Econ. Manage. Bus. (AJEMB) **2**(8), 287–300 (2023)
16. Girona, S.N., et al.: Environmental movements shaping the landscape in Genoa and Girona: from reactive to regenerative local mobilizations? Prog. Plann. 100777 (2023)
17. Baba, S., Brunet, M.: Project legitimacy: towards a theoretical framework. Eur. Manag. Rev. (2023)
18. Wambwa, D., Mundike, J., Chirambo, B.: Environmental protection fund audits in Zambia: a strategy for ensuring regulatory compliance in Zambia's mining sector. Extr. Ind. Soc. **16**, 101353 (2023)
19. Li, G., et al.: Piezoelectric polarization coupled with photoinduced catalytic oxidation technology for environmental pollution control: recent advances and future prospects. Sci. Total Environ. 167284 (2023)
20. Bentahar, O., Benzidia, S., Bourlakis, M.: A green supply chain taxonomy in healthcare: critical factors for a proactive approach. Int. J. Logist. Manag. **34**(1), 60–83 (2023)
21. Hina, M., et al.: Circular economy business models as pillars of sustainability: where are we now, and where are we heading? Bus. Strateg. Environ. (2023)
22. Simons, R.: Levers of Control: How Managers Use Innovative Control Systems to Drive Strategic Renewal. Harvard Business Press (1994)
23. Alastal, A.Y.M.: Mediating Effect of an Eco-Control System on the Relationship Between a Proactive Environmental Strategy and Environmental Performance in Jordan. Universiti Utara Malaysia (2021)
24. Wijethilake, C.: Proactive sustainability strategy and corporate sustainability performance: the mediating effect of sustainability control systems. J. Environ. Manage. **196**, 569–582 (2017)
25. Ali, A.A.A., et al.: Moderating effect of strategic planning on the relationship between career path planning and job performance. Sustainability **15**(11), 8490 (2023)
26. Al-Refaei, A.A.-A., et al.: An integrated mediating and moderating model to improve service quality through job involvement, job satisfaction, and organizational commitment. Sustainability **15**(10), 7978 (2023)
27. Alzoraiki, M., et al.: Impact of teachers' commitment to the relationship between transformational leadership and sustainable teaching performance. Sustainability **15**(5), 4620 (2023)
28. Ali, A.A.A.: The Relationship Between Islamic Work Ethics and Employee Performance Among Medical Staff in the Public Health Sector in Bahrain: Employee Commitment as Mediating Variable. Universiti Sains Islam Malaysia (2022)
29. Ali, A.A.A., Alzoraiki, M., Al Aghbas, N.S.M.: Working remotely, social isolation and access to support and communication during COVID-19 its influence on job performance. Int. J. Intellect. Hum. Resour. Manag. (IJIHRM) **3**(01), 23–33 (2022)
30. Ali, A.A.A., et al.: Association between Islamic work ethics and employee commitment among medical staff in the public health sector in Bahrain. Int. J. Intellect. Hum. Resour. Manag. (IJIHRM) (2023)
31. Ateeq, A.: Knowledge Management's Effect on Learning Quality in Bahrain's Private Universities (2023)
32. Alkadash, T.M., et al.: The effects of leadership style on employee sustainable behaviour: a theoretical perspective. In: Artificial Intelligence and Transforming Digital Marketing, pp. 205–213. Springer (2023)

33. Barakat, B., et al.: Assessing the impact of green training on sustainable business advantage: exploring the mediating role of green supply chain practices. Sustainability **15**(19), 14144 (2023)
34. Ateeq, A.: The mediating role of engagement in the relationship between equal opportunities and employee performance. Probl. Perspect. Manag. **21**(3), 435–447 (2023)
35. Al Balushi, F.S.M., Ali, A.A.A., Alzoraiki, M.: The impact of forecasting strategy and training and development strategy on preparing leadership competencies administrative support in the Sultanate of Oman. Int. J. Intellect. Hum. Resour. Manag. (IJIHRM) **3**(01), 34–41 (2022)
36. Rahardjo, K.P., Kurniawati, K.: Caroll pyramid implementation of environmental responsibility programs of PT Pertamina EP Subang Field, Tambun Field, and Jatibarang Field in organic agriculture. J. Soc. Res. **2**(9), 3331–3340 (2023)
37. Ali, K., Kausar, N., Amir, M.: Impact of pollution prevention strategies on environment sustainability: role of environmental management accounting and environmental proactivity. Environ. Sci. Pollut. Res. **30**(38), 88891–88904 (2023)
38. Gheisari, M., et al.: Deep learning: applications, architectures, models, tools, and frameworks: a comprehensive survey. CAAI Trans. Intell. Technol. (2023)
39. Kim, K.-T., et al.: hIPPYlib-MUQ: a Bayesian inference software framework for integration of data with complex predictive models under uncertainty. ACM Trans. Math. Softw. (2023)
40. Mandrinos, S., Lim, W.M.: De-internationalization: an organizational institutionalism perspective. Glob. Bus. Organ. Excell. **42**(3), 58–73 (2023)
41. Op den Kamp, E.M., et al.: Creating a creative state of mind: promoting creativity through proactive vitality management and mindfulness. Appl. Psychol. **72**(2), 743–768 (2023)
42. Wijethilake, C., Upadhaya, B., Lama, T.: The role of organisational culture in organisational change towards sustainability: evidence from the garment manufacturing industry. Prod. Plan. Control **34**(3), 275–294 (2023)
43. Ammirato, S., et al.: Still our most important asset: a systematic review on human resource management in the midst of the fourth industrial revolution. J. Innov. Knowl. **8**(3), 100403 (2023)
44. Passavanti, C., et al.: The evolution of student entrepreneurship: state of the art and emerging research direction. Int. J. Manag. Educ. **21**(2), 100820 (2023)
45. Kuik, S., et al.: A systematic literature review on the transition to circular business models for small and medium-sized enterprises (SMEs). Sustainability **15**(12), 9352 (2023)

How Does Earnings Manipulation Influence the Firm Performance? The Moderating Effect of the Board Gender Diversity

Ahmed Elmashtawy, Ali Ateeq, Mohamed Salaheldeen, Marwan Milhem, Mohammed Alzoraiki, Zuhair Al Ani, and Rumpa Roy

Abstract The study's purpose is to examine the influence of real earnings management (REM) on firm performance (FP). Further, it investigates the moderating impact of board gender diversity (BDIV), which is represented by the percentage of females on the board in this relationship. The study employed a quantitative approach in examining 87 non-financial firms in Egypt over a period spanning from 2014 to 2021. The findings concluded a positive influence of REM on FP. Additionally, this relationship was moderated by BDIV, where female existence on the board weakened the positive influence of REM on FP. While male existence on the board strengthened the positive impact of REM on FP. The study has several implications for the theoretical explanation of the relationship between REM, FP, and BDIV in the Egyptian context. The study has practical implications for regulators, boards, firms, and stakeholders.

A. Elmashtawy
Faculty of Business, Economics and Social Development, Universiti Malaysia Terengganu (UMT), Kuala Nerus, Malaysia

Faculty of Commerce, Menoufia University, Menoufia, Egypt

A. Ateeq (✉) · M. Milhem · M. Alzoraiki · R. Roy
Faculty of Administrative and Financial Sciences, Gulf University, Sanad 26489, Kingdom of Bahrain
e-mail: dr.ali.ateeq@gulfuniversity.edu.bh

M. Milhem
e-mail: dr.marwan.milhem@gulfuniversity.edu.bh

R. Roy
e-mail: dqa@gulfuniversity.edu.bh

M. Salaheldeen
Faculty of Commerce, Menoufia University, Shebin El-Kom, Egypt

Z. Al Ani
College of Engineering, Gulf University, Sanad 26489, Kingdom of Bahrain
e-mail: coo@gulfuniversity.edu.bh

B. Awwad (ed.), *The AI Revolution: Driving Business Innovation and Research*, Studies in Systems, Decision and Control 524, https://doi.org/10.1007/978-3-031-54379-1_43

Keywords Board gender diversity · Real earnings management · Firm performance · Egyptian Exchange

1 Introduction

The management's goal is to develop and improve firm performance (FP) while also serving the interests of shareholders, which is accomplished through managing the firm's resources in the most efficient manner [1, 2]. As a result, FP reflects the performance of its management in numerous aspects, such as earnings manipulation (EM), debt repayment, selecting the proper financing structure that results in lower capital costs, and confronting the dangers to which the firm is exposed [3, 4].

Many studies have discussed the negative effects resulting from the practice of EM [5–8], and among these effects are, for example, the weakening in the FP and the deception of the market in general about the real earnings of the firm, and thus the exposure of investors and other related parties to poor awareness of the size of future risks to which the firm may be exposed [9, 10]. Real Earnings Management (REM) expresses the practices and decisions that firms follow to manage their earnings through actual activities, with the aim of delivering a picture to stakeholders that the firm has achieved the objectives of the financial report through the real activities of the firm [3, 11].

Corporate governance [1] represents the framework that provides protection for the parties involved with the firm, especially shareholders and management [12]. Investors and other stakeholders believe in the strength of CG mechanisms to reduce such negative effects as well as mitigate any conflicts of interest between shareholders and management [1]. The Board of Directors is considered a vital mechanism for CG, as it is the shareholders' agent in managing the firm. The board is responsible for managing, monitoring, and improving performance, in addition to implementing governance rules. The more effective the board of directors, the lower the agency cost, and the better the FP [13]. The board of directors' monitoring duty is an important CG mechanism. The gender mix of this monitoring function can affect FP. The participation of women on boards of directors leads to greater financial performance in firms [14].

Based on extrapolating the accounting literature that focused on studying the relationship between board gender diversity (BDIV), EM, and FP [1, 15–17], the study notes that most firms that have women on their boards have no more EM practices when issuing their financial statements [18, 19]. This has a favorable impact on restricting EM, lowering agency issues, and improving the FP [20, 21].

The research gap is that, while literature reviews have made significant progress in the study and analysis of REM practices and their influence on FP in the foreign environment [1, 2, 4, 6, 9], there are still some research gaps in this field that can be studied in the Egyptian context, specifically the moderating effect of BDIV on the influence of REM on FP. As a result, the current study aims to investigate the extent to which REM affects FP. Furthermore, in order to better understand the variables that influence this relationship, the study investigated the moderating effect provided by BDIV for Egyptian non-financial firms. The results demonstrated that REM had a significant positive effect on FP. Furthermore, the study discovered substantial evidence that BDIV has a significant effect on the link between REM and FP.

The current study adds to the body of knowledge about the factors that influence the link between REM and FP. The study focuses on the moderating influence of BDIV on this relationship and gives new evidence of the consequences of BDIV in diminishing manager opportunistic behaviors. This study makes at least two distinct contributions. The study makes numerous theoretical additions to the literature on REM, FP, and BDIV in the context of developing countries. The second contribution is to provide practical implications and effective tools for CG in diverse economic sectors to regulars, boards, firms, and stakeholders.

2 Literature Review

The study was founded on agency theory and resource dependence theory. There are two types of relationships: (1) REM and FP; and (2) the moderating effect of BDIV on the relation between REM and FP.

2.1 Studies Related to REM and FP

Some studies [11, 22, 23] have found that there is a negative impact of REM practices on the financial indicators related to the FP in a way that reflects the decline of the FP and the weak financial position of the firms. There is also a statistically significant impact of REM on the risks of declining operating cash flows in financial reports, as it is one of the most important financial risks facing firms and affects the FP [10]. Implicitly, the negative impact of REM practices on the FP appears in the light of the studies [7, 17, 20], which showed the extent of the impact of CG mechanisms on reducing REM.

The studies [1, 6, 22] also found a negative correlation between REM practices and return on share, and the studies indicated that limiting REM enhances the firm's ability to borrow and limits opportunistic behavior to manage the earnings of these firms, which positively affects the improvement of the FP. Wenfang and Ayisi [2] examined the relationship between REM and FP in the emerging market. The study concluded that REM has a positive but not significant effect on agency costs and that agency costs have a positive and significant effect on the FP. While REM has a negative and insignificant effect on the FP. While the study [21] aimed to determine the relationship between REM and the FP. The study concluded that REM through sales or optional expenses does not have an influence on the FP. Therefore, the following hypothesis has been formulated:

H1: REM has a negative influence on FP.

2.2 Studies Related to REM, FP, and BDIV

The study of [3] is considered the pioneer study in the matter of measuring REM, as it provided a quantitative model for measuring management manipulation by practicing the real activities of the firm. The study found evidence that managers manipulate earnings through REM to avoid reporting annual losses by reducing prices through discounts to temporarily increase sales. Tulcanaza-Prieto and Lee [24] provided a framework relying on Roychowdhury's [3] model to measure the REM. The study indicated the possibility of actual testing of the proposed framework in future studies. Several previous studies findings [16, 18, 19] indicate that BDIV is favorably connected to restriction REM.

Governance regulators are interested in the diversity of the board members. It might thus be argued that gender diversity improves the board's monitoring and supervision functions [25]. A large body of research demonstrates that women are more conservative and risk-averse than men [26]. According to [15], the inclusion of women on boards of directors has an important role in fostering FP.

The relationship between REM and the FP is affected by CG [1, 12]. Previous literature agreed that most firms with BDIV have lower REM practices [16, 18, 19], which contributes to the enhanced reliability of financial statements, better predictability of risks of financial hardship and bankruptcy, and a better level of creditworthiness. Hence, improving FP. Some studies, however, show the opposite results [26–28], which found no evidence that BDIV has an impact on FP and REM.

The study proposes that gender-diverse boards are likely to improve FP based on proven gender disparities in ethics, risk-taking behavior, reputational concerns, diligence level, and monitoring intensity [15, 25]. The empirical evidence is conflicting. On the one hand, female executives and directors are renowned for having a positive impact on FP [14, 29]. According to [13], boards with female directors require more monitoring in the form of additional audit efforts from the external auditor. Conversely, [30] contends that a firm may select a woman for its board of directors just for "window dressing" (to demonstrate diversity) rather than increasing performance. Based on the preceding discussion, the following hypothesis has been proposed:

H2: The BDIV moderates the negative relationship between REM and FP.

3 Methodology

3.1 Data and Sample

The study focuses on the listed non-financial firms on the Egyptian Exchange during the period from 2014 to 2021. The final sample contained 87 firms listed on the Egyptian Exchange, distributed to 13 sectors, with 696 yearly observations. Moreover, the study relied on secondary data to collect sample data. The study uses ordinary least squares panel data regression models with fixed effects.

3.2 Variables Definition

The independent variable is REM, the dependent variable is FP, and the moderator variable is BDIV. Moreover, the control variables are firm size (LS), board size (BS), leverage [3], and sales [9, 15, 22, 31]. FP is measured based on price to earnings per share (PER). The value of PER is measured based on the stock price divided by earnings per share [10, 20, 24]. The BDIV is measured as the number of women on the board divided by the total number of board members [15, 18, 32]. Regarding REM, the majority of previous studies relied on the model presented by. To measure REM practices, with some studies making modifications to the model based on the availability of data in the applied study environment, with the most important of these studies [5, 24]. REM can be determined on this basis by estimating the values of abnormal operating activities in each of the sales activities, production, and optional expenses as follows:

$$\mathrm{CFO_{it}/A_{ti-1}} = \alpha_{\mathrm{it}} + \mathrm{B_1(1/A_{it-1}) + B_2(S_{it}/A_{it-1}) + B_3(\Delta S_{it}/A_{it-1}) + E_{it}}$$

$$\begin{aligned}\mathrm{PRO_{it}/A_{it-1}} = \alpha_{\mathrm{it}} &+ \mathrm{B_1(1/A_{it-1}) + B_2(S_{it}/A_{it-1}) + B_3(\Delta S_{it}/A_{it-1})} \\ &+ \mathrm{B_3(\Delta S_{it-1}/A_{it-1}) + E_{it}}\end{aligned}$$

$$\mathrm{DISEXP_{it}/A_{it-1}} = \alpha_{\mathrm{it}} + \mathrm{B_1(1/A_{it-1}) + B_2(S_{it}/A_{it-1}) + E_{it}}$$

Concerning the control variables, firm size (LS) is measured by the natural logarithm of total assets at the end of the year [9, 33]. Board size (BS) is measured by the total number of directors on a board [13, 32]. While the leverage [3] is measured by the total liabilities to total assets ratio [31, 34]. Meanwhile, sales are measured by the sales growth rate for years $t - 1$ to t [21, 35].

3.3 Regression Models Specifications

Based on the above, the following statistical models were formulated to determine whether REM influences FP and whether the BDIV moderates the relation between REM and FF.

Model 1: The direct effect model assesses how REM can influence FP in the Egyptian non-financial sector. This model answers hypothesis 1.

$$PER_{it} = \alpha + .\beta_1 REM + \beta_2 LS_{it} + \beta_3 BS_{it} + \beta_4 LEV_{it} + \beta_5 SALES_{it} + \varepsilon_{it}$$

Model 2: The moderator effect model is used to investigate the moderator influence of BDIV on the association between REM and FP in the Egyptian non-financial sector. This model answers hypothesis 2.

$$PER_{it} = \alpha + .\beta_1 REM_{it} + \beta_2 BDIV_{it} + \beta_3 REM * BDIV_{it} + \beta_4 LS_{it} + \beta_5 BS_{it} + \beta_6 LEV_{it} + \beta_7 SALES_{it} + \varepsilon_{it}$$

4 Results and Discussion

4.1 Descriptive and Correlation Analysis

The descriptive statistics investigate whether variables follow the normal distribution using the Kolmogorov–Smirnov and Shapiro–Wilk tests, where the significance values were greater than 0.05, indicating that the variables follow the normal distribution. The correlation coefficient results reveal that there are no multicollinearity problems in the study, as the highest correlation value is 0.65. To detect the extent of the multicollinearity problems, the study used collinearity diagnostics to identify the values of the variance inflation factor [4] and tolerance. Where the VIF values for each variable were less than 1.30 and tolerance values were greater than 0.05, which refers to the fact that the analysis does not suffer multicollinearity problems within variables and the ability of study models to explain the effect on FP [5–7]. To avoid any kind of autocorrelation problem, the study ran an autocorrelation test using Durbin–Watson (D-W). Where the values of D-W are within the specified range of test 1.5:2.5. This showed that variables do not suffer from autocorrelation problems between them.

4.2 Direct Effect Regression

The findings in Table 1 present the direct and moderating effects of the regression models. All models were indicated in Table 1, and they were all controlled by firm size, board size, financial leverage, and sales. In Model 1, the study tested the extent of the direct causal impact of REM on the FP measure, especially PER. The results in Model 1 concluded the existence of a significant positive impact of REM on PER (B = 0.267; Sig. = 0.000 < 0.05). This result revealed that firms with a high level of REM have a higher FP, and these firms can improve their performance through REM. Furthermore, REM can be used as a protective measure, which reduces a firm's

Table 1 Direct and moderating effect regression result

Variables	Model 1	Model 2
	Direct effect regression PER	Moderating effect regression PER
C	− 0.014	− 0.002
	0.000	0.000
REM	0.267	− 0.372
	0.000	0.049
BDIV		− 0.081
		0.020
BDIV * REM		− 0.025
		0.014
LS	0.071	0.012
	0.031	0.501
BS	0.023	0.384
	0.022	0.046
LEV	− 0.024	0.024
	0.047	0.028
SALES	0.047	0.049
	0.000	0.000
R-squared	31.12	41.04
Adjusted R-squared	30.07	37.09
F-statistic	29.74	29.17
Prob (F-statistic)	0.00	0.00
Durbin–Watson stat	1.67	1.89

Source Table by authors

riskiness and enhances its performance. This result was consistent with the results of [1, 2, 8, 9, 22].

4.3 Moderating Effect Regression

The Model 2 findings in Table 1 present the moderating role of BDIV. The results expose that the interaction between BDIV and REM moderates the relationship between REM and FP. The interaction term is statistically negative and significant. In Model 2, the findings reported that the moderating impact was negatively significant on PER ($B = -0.025$; Sig. $= 0.014 < 0.05$). This result indicates that the existence of females on the board leads to a reduction in opportunistic management behavior to manipulate earnings through real activities in order to increase the FP. This indicates the vital role of the BDIV. These results were consistent with those revealed by [10–14, 28, 29].

Overall, it was concluded from the results of assessing the accuracy of Models 1 and 2 that the values of R^2 reached 31.12 and 41.04, respectively; these are consistent with the values of the adjusted R^2 that reached 30.07 and 37.09, respectively, which indicates the accuracy of the models and independence of the factors affecting the FP. Also, the results revealed that the regression models were highly significant, as the significance values were 0.00.

5 Conclusion

The study investigated the influence of REM on FP and the moderating impact of BDIV on the relationship between REM and FP. The study relied on a sample of Egyptian non-financial firms from 2014 to 2021. The findings revealed that REM had a significant positive influence on FP. Additionally, the study found strong evidence that BDIV significantly impacts the relation between REM and FP, where the women's existence on the board weakens the positive impact of REM on FP. The results also revealed that the BDIV plays an important and significant role in monitoring and preventing management from engaging in REM, which positively affects improving FP.

The current study contributes to the literature on the determinants of the relation between REM and FP. Specifically, the study focuses on the moderating influence of BDIV on this relationship. This study presents at least two distinct contributions. Theoretically, the study makes several contributions to the literature on REM, FP, and BDIV in the Egyptian context. The second contribution is to provide regulars, boards, firms, and stakeholders with practical implications and effective mechanisms for CG in various economic sectors. So, the study suggests a new moderator related to BDIV that impacts the relationship between REM and FP.

There are some limitations to the study. First, the analysis was conducted over an eight-year period and applied to one country; therefore, the conclusions cannot be generalized all the variables affecting the results cannot be controlled. Thus, for a larger interpretation, the analysis can be conducted over a longer period of time and applied to and compared across multiple countries. Secondly, the FP can be measured in several ways. PER was used as a measure of FP, and therefore, this measure may not include all aspects of FP. Additionally, the main limitation of the current study is the restriction of the sample to Egyptian non-financial firms. Future research may investigate the causes of a positive relationship between REM and FP. In addition, future research may analyze the influence of BDIV on the relation between REM and FP using other measures of FP. Future research may compare the expertise of male members to that of female members and analyze the effects of both. Additionally, evaluating the impact of the accrual EM on FP. Future research may conduct more studies in the financial sectors, such as banks and non-bank financial services, in order to determine the extent of the difference between those sectors and the moderating impact of BDIV on the relationship between REM and FP. Finally, future research may attempt to expand the study's results further by focusing on diverse markets and revealing generalizable conclusions.

References

1. Boachie, C., Mensah, E.: The effect of earnings management on firm performance: the moderating role of corporate governance quality. Int. Rev. Financ. Anal. **83**, 102270 (2022). https://doi.org/10.1016/j.irfa.2022.102270
2. Wenfang, Z., Ayisi, A.L.: Earnings management and corporate performance: an empirical evidence from Ghana. Res. J. Financ. Acc. **11** (2020)
3. Roychowdhury, S.: Earnings management through real activities manipulation. J. Account. Econ. **42**(3), 335–370 (2006)
4. Kouaib, A., Lacombe, I.: CEO overconfidence and subsequent firm performance an indirect effect via earnings manipulations. J. High Technol. Manag. Res. **34**(1), 100452 (2023)
5. Al-Shattarat, B.: Real Earnings Management Activities, Meeting Earnings Benchmarks and Future Performance: UK Evidence. University of Plymouth (2017)
6. Aqabna, S.M., Aga, M., Jabari, H.N.: Firm performance, corporate social responsibility and the impact of earnings management during COVID-19: evidence from MENA region. Sustainability **15**(2), 1485 (2023)
7. Khuong, N.V., Ha, N.T.T., Thu, P.A.: The relationship between real earnings management and firm performance: the case of energy firms in Vietnam. Int. J. Energy Econ. Policy **9**(2), 307–314 (2019)
8. Xu, R.Z., Taylor, G.K., Dugan, M.T.: Review of real earnings management literature. J. Account. Lit. **26**, 195 (2007)
9. Ding, R., Li, J., Wu, Z.: Government affiliation, real earnings management, and firm performance: the case of privately held firms. J. Bus. Res. **83**, 138–150 (2018)
10. Dakhlallh, M.M., Rashid, N., Abdullah, W.A.W., Qawqzeh, H.K., Dakhlallh, A.M.: Accrual-based earnings management, real earnings management and firm performance: evidence from public shareholders listed firms on Jordanian's stock market. J. Adv. Res. Dyn. Control Syst. **12**(1) (2020)
11. Lim, H.J., Mali, D.: An analysis of the positive effect of real earnings management on financial performance. Asian Rev. Account. **31**(2), 284–316 (2023)
12. Singh, S., Tabassum, N., Darwish, T.K., Batsakis, G.: Corporate governance and Tobin's Q as a measure of organizational performance. Br. J. Manag. **29**(1), 171–190 (2018)
13. Li, H., Chen, P.: Board gender diversity and firm performance: the moderating role of firm size. Bus. Ethics A Eur. Rev. **27**(4), 294–308 (2018)
14. Brahma, S., Nwafor, C., Boateng, A.: Board gender diversity and firm performance: the UK evidence. Int. J. Financ. Econ. **26**(4), 5704–5719 (2021)
15. Gharbi, S., Othmani, H.: Threshold effects of board gender diversity on firm performance: panel smooth transition regression model. Corp. Gov. Int. J. Bus. Soc. **23**(1), 243–261 (2023)
16. Zalata, A.M., Ntim, C.G., Alsohagy, M.H., Malagila, J.: Gender diversity and earnings management: the case of female directors with financial background. Rev. Quant. Financ. Account. **58**(1), 101–136 (2022)
17. Kahloul, I., Grira, J., Hlel, K.: The trilogy of economic policy uncertainty, earnings management and firm performance: empirical evidence from France. J. Econ. Financ. **47**(1), 184–206 (2023)
18. Orazalin, N.: Board gender diversity, corporate governance, and earnings management: evidence from an emerging market. Gend. Manag. Int. J. **35**(1), 37–60 (2020)
19. Li, X., Than, E.T., Ahmed, R., Ishaque, M., Huynh, T.L.D.: Gender diversity of boards and executives on real earnings management in the bull or bear period: empirical evidence from China. Int. J. Financ. Econ. **28**(3), 2753–2771 (2023)

20. Osma, B.G., Gomez-Conde, J., Lopez-Valeiras, E.: Management control systems and real earnings management: effects on firm performance. Manag. Account. Res. **55**, 100781 (2022)
21. Hessian, M.I.: The impact of managerial ability on the relation between real earnings management and future firm's performance: applied study. Int. J. Bus. Ethics Gov. 57–85 (2019)
22. Rahman, J.M., Xiong, N.: Real earnings management through sales manipulation and firm performance: evidence from China. Account. Bus. Public Interest (2021)
23. Kumar, M., Vij, M., Goswami, R.: Effect of real earnings management on firm performance: evidence from India. Vision **27**(3), 386–396 (2021)
24. Tulcanaza-Prieto, A.B., Lee, Y.: Real earnings management, firm value, and corporate governance: evidence from the Korean market. Int. J. Financ. Stud. **10**(1) (2022). https://doi.org/10.3390/ijfs10010019
25. Zhang, L.: An institutional approach to gender diversity and firm performance. Organ. Sci. **31**(2), 439–457 (2020)
26. Duppati, G., Rao, N.V., Matlani, N., Scrimgeour, F., Patnaik, D.: Gender diversity and firm performance: evidence from India and Singapore. Appl. Econ. **52**(14), 1553–1565 (2020)
27. Sun, J., Liu, G., Lan, G.: Does female directorship on independent audit committees constrain earnings management? J. Bus. Ethics **99**(3), 369–382 (2011). https://doi.org/10.1007/s10551-010-0657-0
28. Masmoudi Mardessi, S., Makni Fourati, Y.: The impact of audit committee on real earnings management: evidence from Netherlands. Corp. Gov. Sustain. Rev. **4**(1), 33–46 (2020). https://doi.org/10.22495/cgsrv4i1p3
29. Lee, K.W., Thong, T.Y.: Board gender diversity, firm performance and corporate financial distress risk: international evidence from tourism industry. Equal. Divers. Incl. Int. J. **42**(4), 530–550 (2023)
30. Simionescu, L.N., Gherghina, Ş.C., Tawil, H., Sheikha, Z.: Does board gender diversity affect firm performance? Empirical evidence from Standard & Poor's 500 Information Technology Sector. Financ. Innov. **7**(1), 1–45 (2021)
31. Vakilifard, H., Mortazavi, M.S.: The impact of financial leverage on accrual-based and real earnings management. Int. J. Acad. Res. Account. Financ. Manag. Sci. **6**(2), 53–60 (2016)
32. Elmashtawy, A., Che Haat, M.H., Ismail, S., Almaqtari, F.A.: The moderating effect of joint audit on the association between board effectiveness and audit quality: empirical evidence from a developing country. Bottom Line. Ahead-of-print (2023)
33. Elmashtawy, A., Salaheldeen, M.: Big data techniques and internal control: evidence from Egypt. In: Proceedings of the 2nd International Conference on Emerging Technologies and Intelligent Systems: ICETIS 2022. Lecture Notes in Networks and systems, vol. 584, pp. 14–23 (2023). https://doi.org/10.1007/978-3-031-25274-7_2
34. Elmashtawy, A., Che Haat, M.H., Ismail, S., Almaqtari, F.A.: Audit committee effectiveness and audit quality: the moderating effect of joint audit. Arab Gulf J. Sci. Res. Ahead-of-print (2023). https://doi.org/10.1108/AGJSR-09-2022-0202
35. Elmashtawy, A., Salaheldeen, M.: Big data and business analytics: evidence from Egypt. In: International Conference on Information Systems and Intelligent Applications: ICISIA 2022. Lecture Notes in Networks and Systems, vol. 550, pp. 503–512 (2022). https://doi.org/10.1007/978-3-031-16865-9_40

Advances in AI and Their Effects on Finance and Economic Analysis

Naseem Hassan Abu Jamie, Tanseem N. Abu-Jamie, and Mujeeb Said Mohsen Al-Absy

Abstract This review of a paper explores how artificial intelligence (AI) is revolutionizing the fields of economics and finance. Traditional methods in these fields are undergoing a seismic shift as a result of the quick advancements in AI technologies, which will improve efficacy, accuracy, and inclusivity. This review offers crucial insights into the implications of these advancements for the financial and economic landscape in addition to providing an overview of the main points discussed in the paper. The article also emphasizes the various ways that AI is being used in economic and financial analysis, from the rise of robo-advisors revolutionizing wealth management to the use of sentiment analysis for more precise consumer behavior forecasting. The paper investigates how AI is enabling more accurate economic forecasting, improving market trend understanding, and driving data-driven decision-making. Financial institutions are better positioned to address unmet needs, like expanding credit availability and enhancing risk assessment methodologies, by utilizing AI algorithms. This review provides invaluable insights into the ongoing transformation of these important industries through a thorough analysis of the opportunities and difficulties presented by the integration of AI in finance and economic analysis. It highlights the necessity of a well-rounded strategy that makes use of AI's potential while maintaining moral considerations and human expertise. This review emphasizes the crucial role of ongoing research and careful implementation in maximizing the potential benefits of AI for the financial and economic landscape as the boundaries of innovation continue to be pushed.

Keywords Finance · Economics · Artificial intelligence

N. H. Abu Jamie
Department of Economics, Al-Azhar University Gaza, Gaza, Palestine
e-mail: naseem.jamie@alazhar.edu.ps

T. N. Abu-Jamie (✉)
Department of Artificial Intelligence, Palestine University, Gaza, Palestine
e-mail: tanseemnas1997@gmail.com

M. S. M. Al-Absy
Accounting and Financial Science Department, College of Administrative and Financial Science, Gulf University, Sanad 26489, Kingdom of Bahrain
e-mail: dr.mujeeb.saif@gulfuniversity.edu.bh

B. Awwad (ed.), *The AI Revolution: Driving Business Innovation and Research*, Studies in Systems, Decision and Control 524, https://doi.org/10.1007/978-3-031-54379-1_44

1 Introduction

The incorporation of artificial intelligence (AI) has emerged as a ground-breaking force in the dynamic world of contemporary business and economics with the potential to transform established practices and methodologies [41]. This paper review explores the profound effects of these developments on decision-making, forecasting, and general operational paradigms [12]. It examines the remarkable convergence of AI and the fields of finance and economic analysis [18]. The financial and economic sectors are on the verge of a transformation marked by previously unheard-of levels of efficiency, accuracy, and inclusivity due to the rapid advancement of AI technologies [30]. The incorporation of artificial intelligence (AI) has emerged as a ground-breaking force in the dynamic world of contemporary business and economics with the potential to transform established practices and methodologies [33]. This paper review explores the profound effects of these developments on decision-making, forecasting, and general operational paradigms. It examines the remarkable convergence of AI and the fields of finance and economic analysis. The financial and economic sectors are on the verge of a transformation marked by previously unheard-of levels of efficiency, accuracy, and inclusivity due to the rapid advancement of AI technologies. Long-standing problems in economic analysis and finance can now be solved thanks to the rapid advancement of AI technologies. In this review, we explore the various ways that these domains are utilizing AI's transformative potential. This paper aims to shed light on the crucial interaction between technology and decision-making by exploring how AI-driven tools are changing financial landscapes and enhancing economic perspectives. The implementation of AI into the finance industry has given rise to a variety of innovations that rethink established procedures. The rise of robo-advisors [9], which represent the fusion of AI and finance, is one notable example [13]. These digital advisors upend the conventional advisory model dominated by human experts by using AI algorithms to provide customized investment recommendations. The development of robo-advisors opens up access to financial advice to all people, breaking down socioeconomic barriers that previously only allowed the wealthy to use these services [27]. This democratization, which has its roots in AI's data-driven capabilities, represents a move toward financial inclusivity and gives people the power to make wise investment choices regardless of their upbringing or level of financial literacy.

1.1 Artificial Intelligence (AI) Improvement in Finance

The implementation of AI into the finance industry has given rise to a variety of innovations that rethink established procedures [34]. The rise of robo-advisors, which represent the fusion of AI and finance, is one notable example. These digital advisors upend the conventional advisory model dominated by human experts by using

AI algorithms to provide customized investment recommendations [36]. The development of robo-advisors opens up access to financial advice to all people, breaking down socioeconomic barriers that previously only allowed the wealthy to use these services [37]. This democratization, which has its roots in AI's data-driven capabilities, represents a move toward financial inclusivity and gives people the power to make wise investment choices regardless of their upbringing or level of financial literacy [8]. According to simulations by the McKinsey Global Institute, if artificial intelligence is adopted and assimilated at the global average rate, by 2030 there could be an additional $13 trillion in global economic activity, or about 16% more than cumulative GDP from today. This represents additional GDP growth of 12% annually.

Additionally, the incorporation of AI has accelerated the development of algorithmic trading, in which computers execute trades at rates and volumes that are incomprehensible to human traders [33]. Automated and effective trading strategies are made possible by AI-powered algorithms that analyze market data and trends in real-time. With this automation, market liquidity is improved, price discovery mechanisms are strengthened, and it is less likely that traders will make emotional trading decisions based on personal biases. The proliferation of AI-driven trading systems has raised issues with algorithmic stability, the potential for market manipulation, and the requirement for regulatory oversight [31].

1.2 The Influence of AI on Economic Analysis

AI has the same revolutionary potential in the field of economic analysis [14]. The analytical prowess of AI has proven to be a match for traditional economic models, which are frequently characterized by assumptions that may not accurately reflect the complexity of real-world scenarios. Machine learning algorithms and AI-enabled data analytics give economists the tools they need to process and synthesize enormous datasets, enabling more precise and dynamic economic forecasts [29]. Policymakers can now make data-driven decisions and anticipate new trends, which has broad repercussions for them. AI-driven productivity growth is influenced by a number of variables, including labor automation, innovation, and new competition. The rate of AI adoption is a micro factor; the size of the impact is influenced by macro factors like a country's labor market structure or the degree of global connectivity. Seven potential impact channels were looked at in a simulation by the McKinsey Global Institute. The first three concern how the adoption of AI will affect the variety and amount of production factors that directly affect business productivity. The remaining four are side effects of AI adoption related to the general economic environment and the shift to AI.[1] We acknowledge that these seven channels are merely a starting point based on our current knowledge and ongoing trends, rather than being definitive or necessarily exhaustive (Fig. 1).

[1] Notes from the AI frontier: Modeling the impact of AI on the world economy.

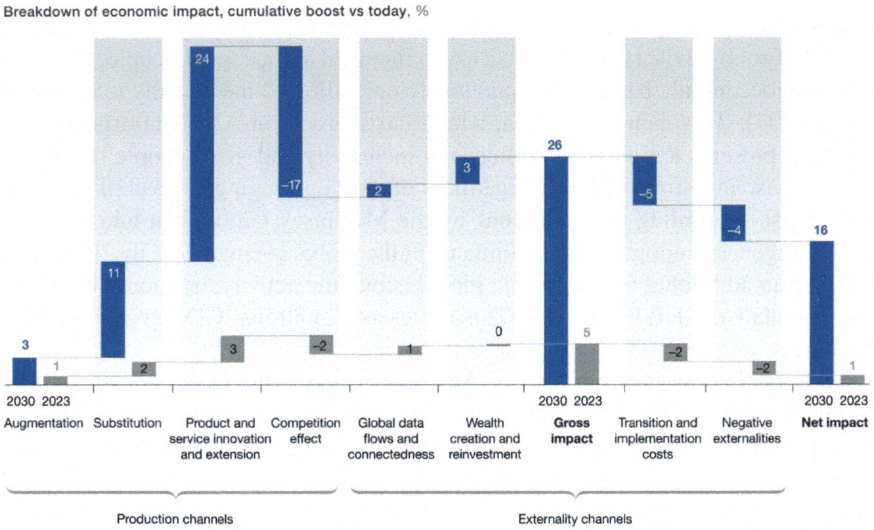

Fig. 1 Artificial intelligence net economic impact for seven channels

1.3 Horizons of Ethical and Regulatory Navigation

Ethics and regulatory frameworks are crucial in determining how quickly AI is incorporated into financial and economic analysis as well as how these transformations will proceed [32]. Important discussions about the responsible use of AI technologies have been sparked by worries about data privacy, algorithmic bias, transparency, and accountability [3]. A balance between innovation and ethical stewardship is crucial as financial institutions and economic analysts adopt AI-driven tools. With the help of AI's data-driven insights, predictive abilities, and computational prowess, the changing landscape of finance and economics is about to undergo a seismic shift. However, despite this technological renaissance, the ethical implications and legal requirements highlight the need for a productive partnership between human ingenuity and artificial intelligence [2]. The following sections of this paper review will analyze important applications of AI in finance and economic analysis while keeping in mind these complex factors. This review aims to offer a thorough perspective on how AI is reshaping, redefining, and reinvigorating the worlds of finance and economic analysis, from the nuances of algorithmic trading to the ethical implications of AI-driven decision-making.

2 Literature Review

Artificial intelligence (AI) has become a disruptive force that is changing the way that economic analysis and finance are practiced. This review examines the significant influence of AI in these fields, highlighting significant developments and their effects. The cited references offer in-depth perceptions of many facets of AI adoption in economic and financial analysis. A paradigm shift in the fields of economic analysis and finance is being sparked by artificial intelligence (AI). A global survey by the Cambridge Centre for Alternative Finance (CCAF) and the World Economic Forum highlights AI's transformative influence, emphasizing how it alters business models, competitive environments, and workforce dynamics [35]. Machine learning algorithms improve market efficiency and trading accuracy in the context of algorithmic trading, fostering improved liquidity and the detection of trading opportunities [1]. By lowering costs and extending access to wealth management services, robo-advisors run on AI algorithms that democratize financial advice [39]. According to Ghasemaghaei [16], predictive analytics strengthens data-driven investment strategies by examining previous data and market trends. According to Kijewski et al. [24], NLP-driven sentiment analysis offers insights into market sentiment and encourages proactive decision-making. According to Jha et al. [23], predictive analytics strengthens data-driven investment strategies by examining previous data and market trends. According to Almeida et al. [3], NLP-driven sentiment analysis offers insights into market sentiment and encourages proactive decision-making. Additionally, Gruetzemacher and Paradice [17] research on AI-driven economic forecasting improves predictions through the analysis of enormous datasets. Despite ethical limitations, AI also revolutionizes risk assessment and credit scoring [26], and it lowers fraud losses by spotting irregularities in user behavior and transactional patterns [4]. Regulatory oversight addresses algorithmic stability and flash crash prevention, while portfolio management benefits from AI's efficiency and risk control capabilities [10]. At the least, a fair deployment is required due to the financial industry's wider societal and economic impact [21]. This extensive literature review presents a foundational framework for further investigation and research in this dynamic area while outlining the transformative potential of AI in finance and economic analysis.

3 Enhancing Forecasting and Decision-Making with AI in Finance and Economic Analysis

Enhancing Forecasting and Artificial intelligence (AI) has the power to completely alter how businesses run and impact the world economy. A McKinsey Global Institute report claims that AI has the potential to significantly boost global economic activity. According to the report, AI may cause profound changes in how people live and work, which could result in previously unheard-of economic growth.[2] But the adoption of AI might also widen disparities between nations, businesses, and workers. AI is already changing business operations and opening up new growth prospects. It can increase key performance metrics like revenue, productivity, business growth, digital transformation, and efficiency thanks to its capacity for processing enormous amounts of data.[3] According to a study by the consulting firm Accenture that looked at 12 developed economies, artificial intelligence (AI) could double annual global economic growth rates by 2035 [38].

3.1 Algorithmic Trading: Revolutionizing Market Dynamics

One of the most prominent applications of AI in the financial sector is algorithmic trading. By combining AI algorithms, real-time market data analysis, and automated execution, it has made trading activities possible to reach previously unattainable levels of efficiency and precision.[4] In order to spot patterns, execute trades, and quickly adapt to changing market conditions, algorithmic trading systems quickly sift through enormous amounts of data. These AI-driven systems minimize trading latency while also reducing emotional bias, which frequently affects human trading decisions.[5] However, due to the rapid growth of algorithmic trading, worries have been raised about market stability, potential flash crashes, and regulatory oversight.[6]

3.2 Robo-Advisors

Robo-advisors are a key development in the financial sector because they seamlessly combine wealth management services and artificial intelligence (AI) [7]. These digital financial advisors make personalized, automated investment recommendations to clients by utilizing cutting-edge AI algorithms. Robo-advisors build diversified portfolios and optimize asset allocations while requiring little human involvement by studying a client's financial objectives, risk tolerance, and investment horizon. A distinguishing feature of robo-advisors is their ability to democratize financial advice by removing barriers associated with high advisory fees and a select clientele and making professional financial advice available to a wider range of investors. A new era of inclusivity in financial planning is being ushered in by this fusion of technology and finance, which improves accessibility while also reducing the biases present in human decision-making [39]. Several algorithms are used by

[2] Notes from the AI frontier: Modeling the impact of AI on the world economy.

[3] The impact of AI on business.

[4] Active Trading in ETFs: The Role of High-Frequency Algorithmic Trading.

[5] 4 Big Risks of Algorithmic High-Frequency Trading.

[6] Flash crashes explained.

Table 1 Algorithms used in portfolio management

Algorithm	Application	Benefits	How it works
Modern portfolio theory	Optimize asset allocations within a portfolio	Diversified portfolios, risk-return optimization	Analyzes historical data, correlations, and risk profiles to create efficient portfolios. Investors choose allocations based on risk tolerance
Mean-variance optimization	Find portfolios with the highest return for risk	Balances risk and return for investor preferences	Calculates expected returns and variances of asset combinations, optimizing for the highest Sharpe ratio
Risk parity	Allocate based on risk rather than weights	Reduces concentration risk, equalizes risk contributions	Calculates and allocates funds to achieve equal risk contribution from all assets
Monte Carlo simulations	Model future portfolio performance	Provides a range of possible outcomes, aids in risk assessment	Generates thousands of scenarios based on historical data to assess the probability of achieving specific investment goals
Tactical asset allocation	Adjust allocations based on market conditions	Adapts to changing market trends, reduces exposure to losses	Analyzes market indicators and economic data to make real-time adjustments to asset allocation
Tax-loss harvesting	Minimize capital gains taxes	Enhances after-tax returns by offsetting gains with losses	Monitors the portfolio for investments with losses and sells them to offset gains and reduce tax liability

robo-advisors to generate automated and data-driven investment recommendations. Some of the important algorithms frequently used in robo-advisory services are listed in Table 1.

This table has been compiled to provide a clear overview of each algorithm's application, the benefits, and how it works it offers in the context of robo-advisory services.

3.3 *Analytics for Investment Strategies*

The landscape of investment strategies is also changing as a result of AI's predictive analytics capabilities. Artificial intelligence (AI)-powered algorithms can produce predictive models that help investors make wise decisions by examining historical data, economic indicators, and market trends [42]. These models provide information on potential dangers, chances, and the best points of entry or exit. Consequently, less reliance is placed on intuition and guesswork, and investment decisions become more data-driven. However, the relevance and quality of the data are what determine how accurate the predictive models are, raising concerns about model generalization and the effects of unforeseen events.

3.4 Understanding Market Sentiment Through Sentiment Analysis

Sentiment analysis is a tool that interprets societal sentiment and emotions from textual data sources like social media platforms, news articles, and financial reports. It was made possible by the union of AI and natural language processing [42]. Sentiment analysis offers economists and analysts timely insights into market sentiment and consumer behavior in the context of financial markets. Sentiment analysis helps predict market trends, consumer spending patterns, and shifts in investor sentiment by measuring public sentiment [15]. To ensure accurate sentiment predictions, it is difficult to understand the context, sarcasm, and nuances of language.

3.5 Economic Prediction and Risk Evaluation

Economic forecasting has advanced thanks to AI, and economists have been able to create models that predict economic indicators more precisely [6]. AI algorithms help produce more accurate economic forecasts by analyzing large datasets and spotting intricate patterns. These predictive models can be used by policymakers to help them make decisions that are in line with new economic trends [31]. AI also significantly contributes to risk assessment by improving the assessment of creditworthiness and the detection of potential financial vulnerabilities [5]. By incorporating AI into financial and economic analysis, one can make better decisions by increasing efficiency and gaining a deeper understanding of market dynamics [43]. To ensure ethical and sustainable implementation, it is crucial to address ethical issues, regulatory compliance, and the need for human expertise in addition to these technological advancements. Table 2 provides a thorough overview of the profound influence of artificial intelligence (AI) on economic research and finance. It briefly describes a variety of AI algorithms and their various financial applications. AI has emerged as a key player in contemporary finance, with applications ranging from algorithmic trading, which increases market efficiency, to robo-advisors, which increase financial accessibility, and predictive analytics, which improve investment strategies. Regulatory oversight, algorithmic stability, and ethical issues are just a few of the challenges that this transformative journey faces. However, AI is still reshaping the financial landscape by offering real-time insights, enhancing decision-making, and promptly anticipating changes in the economy, underscoring its crucial role in determining the direction of financial services and economic analysis.

3.6 Advancing Financial and Economic Insights

The landscape of forecasting and decision-making is undergoing a transformative evolution as AI applications in the fields of finance and economic analysis continue to proliferate [14]. Predictive analytics, sentiment analysis, algorithmic trading, and improved economic forecasting are just a few examples of how AI is changing conventional approaches. But as these applications become more widespread, it becomes critical to have strong ethical standards, openness, and legal frameworks [40].

The following sections of this paper review will delve further into the difficulties, possibilities, and potential outcomes of integrating AI into financial and economic analysis, helping to provide a thorough understanding of this changing environment.

4 Enhancing Financial Institutions: Unmet Needs and AI Solutions

The incorporation of artificial intelligence (AI) stands as a beacon of innovation in the constantly changing landscape of financial institutions, addressing unmet needs with previously unheard-of efficiency [22]. Traditional financial practices frequently struggled with issues like restricted credit availability for underserved populations, outdated risk assessment techniques, and the onerous burden of regulatory compliance [20]. However, AI solutions have become powerful allies that are launching these organizations into a new era. Financial institutions are currently increasing credit availability through alternative data assessments, improving risk assessment methodologies with real-time analytics, streamlining regulatory compliance with automation, personalizing customer experiences through AI-driven chatbots, strengthening fraud detection and prevention, and optimizing portfolio management by leveraging the power of AI [28]. In addition to filling unmet needs, these solutions pave the way for financial institutions that are more inclusive, effective, and resilient, placing them at the forefront of the rapidly evolving financial landscape. This Table 3 offers a succinct summary of the key issues that artificial intelligence (AI) is tackling in the financial sector. It highlights important financial institution needs that have not been met and the corresponding AI solutions. By increasing credit availability, strengthening fraud detection and prevention efforts, streamlining regulatory compliance,

Table 2 A brief summary of AI impact in finance and economic analysis

Algorithm	Application	Benefits	Challenges
Algorithmic trading	Real-time market data analysis using AI-driven algorithms identifies trading opportunities and executes trades with astounding speed and accuracy	Improved market liquidity, decreased emotional biases, increased trading efficiency, and improved price discovery mechanisms	Regulatory oversight, algorithmic stability, and potential flash crashes are all issues
Robo-advisors	Robo-advisors use AI algorithms to give clients automated and tailored investment advice	Expert financial advice is more widely available, there are lower fees than with traditional advisors, and wealth management is becoming more democratic	Regulatory issues, the accuracy of risk assessments, and the need to balance AI and human counsel
Predictive analytics	Assisting in investment decisions, AI-powered predictive models analyze historical data, economic indicators, and market trends	Improved returns on investments, risk reduction, and data-driven investment strategies	Dependencies on the accuracy of the data, the generalizability of the model, and the effects of unforeseen circumstances
Sentiment analysis	Textual data sources like social media and the news are used to analyze public sentiment using natural language processing and AI techniques	Insights into market sentiment, predictions of consumer behavior, and proactive decision-making in real time	Reliable sentiment predictions require the accurate interpretation of context, sarcasm, and linguistic nuance
Economic forecasting	By analyzing enormous datasets and spotting intricate patterns, AI improves economic forecasting to produce more precise and dynamic predictions	Proactive responses to new economic developments and a better understanding of economic trends	Interpreting the model, accounting for unforeseen events, and relevance and quality of the data
Risk assessment and credit scoring	Compared to conventional methods, AI algorithms more accurately assess creditworthiness and risk factors, assisting in lending decisions	Increased credit availability, improved risk assessment, and expanded credit access for previously underserved populations	Ethical issues, data privacy, and algorithmic transparency in risk assessment

(continued)

Table 2 (continued)

Algorithm	Application	Benefits	Challenges
Fraud detection	AI algorithms examine user behavior and transaction data to quickly spot fraudulent activity	Fraud losses are decreased, financial institutions are more secure, and client confidence is increased	Achieving a balance between accuracy and reducing false positives, adaptability to changing fraud schemes, and data privacy
Portfolio management	Asset allocations are dynamically changed by AI-driven portfolio management systems in response to risk and market conditions	Enhanced portfolio efficiency, risk control, and asset allocation planning	Decisions made by AI that can be interpreted, client trust, and regulatory compliance

personalizing customer experiences, and improving portfolio management effectiveness, these AI-driven solutions are revolutionizing the industry. Financial institutions are reaping the rewards in the form of greater financial inclusion, improved risk management, cost savings, increased customer engagement, increased security, and improved portfolio performance. AI is playing a crucial part in reshaping the financial landscape to satisfy changing customer and institutional demands.

5 Navigating Opportunities and Challenges: Insights from AI Integration

The adoption of artificial intelligence (AI) in the world of financial institutions creates a dual landscape of ground-breaking opportunities and difficult challenges [19]. Financial institutions are navigating a dynamic environment where innovation holds the promise of greater efficiency, improved customer experiences, and improved risk management as they eagerly adopt AI technologies to improve their operations [34]. This journey does have some challenges, though. Institutions must carefully navigate through issues with data privacy, legal compliance, algorithmic biases, and the requirement for ongoing skill development [25]. The success and sustainability of financial institutions in the AI-driven era will be largely dependent on their ability to effectively manage these challenges while also utilizing the vast potential of AI. We delve into both the remarkable opportunities that AI integration presents and the intricate challenges that demand careful consideration and strategic solutions Maximizing AI Benefits: The Role of Ongoing Research and Implementation exploration of both opportunities and challenges related to AI integration in financial institutions.

Table 3 Enhancing financial institutions: unmet needs and AI solutions

Enhancing financial institutions	Unmet need	AI solution	Benefits
Expanding credit availability	Traditional credit assessment models may exclude certain individuals or businesses due to limited credit histories or unconventional financial profiles	AI algorithms analyze alternative data sources (e.g., social media activity, online behavior) to assess creditworthiness, expanding access to credit for underserved populations	Enhanced financial inclusion, broader customer base, and reduced lending risk through improved assessment
Enhancing risk assessment methodologies	Traditional risk assessment methods may not adequately capture evolving risks, leading to vulnerabilities	Machine learning models continuously analyze large datasets to identify emerging risks, enabling proactive risk management	Improved risk assessment, early warning systems for potential crises, and more resilient financial institutions
Streamlining regulatory compliance	Regulatory compliance can be cumbersome and costly for financial institutions, diverting resources from core functions	AI-powered compliance tools automate regulatory checks, monitor transactions for suspicious activities, and ensure adherence to evolving regulations	Cost savings, reduced compliance errors, and improved efficiency in meeting regulatory requirements
Personalizing customer experiences	Providing personalized financial advice to a diverse customer base is challenging for traditional institutions	AI-driven chatbots and virtual assistants offer personalized financial guidance, responding to customer queries in real-time	Enhanced customer engagement, improved user experiences, and more effective financial advice delivery
Fraud detection and prevention	Traditional fraud detection methods are often reactive and may not keep pace with evolving fraud tactics	AI algorithms analyze transaction patterns and user behaviors in real-time, detecting fraudulent activities promptly	Reduced financial losses due to fraud, enhanced security, and improved customer trust
Portfolio management efficiency	Managing complex investment portfolios efficiently and cost-effectively can be challenging	AI-driven portfolio management systems dynamically adjust asset allocations based on market conditions and investor goals	Improved portfolio performance, risk management, and cost-efficiency

6 Opportunities

- Enhanced Efficiency: AI can automate routine tasks, streamline processes, and lessen the amount of manual labor required, which increases efficiency and lowers costs [11].
- Improved Customer Experiences: Chatbots and virtual assistants powered by AI provide personalized interactions that increase customer satisfaction and engagement.
- Advanced Risk Management: AI can analyze enormous datasets continuously in real-time, improving risk assessment, early warning systems, and the resilience of financial institutions.
- Data-Driven Decision-Making: AI enables organizations to make wise choices and optimize their tactics by providing insightful data analysis results.
- Financial Inclusion: AI-based credit assessment models can increase underserved populations' access to credit, promoting financial inclusion.

7 Challenges

- Data Privacy: Being compliant with laws and safeguarding customer information, handling sensitive financial data calls for dependable data privacy measures.
- Regulatory Compliance: Financial institutions are responsible for ensuring that AI systems abide by the complex and ever-evolving regulatory frameworks.
- Algorithmic Bias: AI algorithms may carry over prejudices from prior data, which could result in unfair results or discriminatory practices that need to be carefully monitored and mitigated.
- Ethical Concerns: The ethical application of AI in finance, such as responsible lending and treating customers fairly, is a subject of growing concern.
- Skills Gap: To ensure that employees can effectively use AI technologies, financial institutions need to invest in workforce upskilling.
- Cybersecurity Risks: As AI is used more frequently, institutions are exposed to potential cybersecurity risks that necessitate strong security measures.

A comprehensive approach is necessary to navigate these opportunities and difficulties. Strong governance, ethical AI principles, cybersecurity investments, and ongoing regulatory compliance efforts are required of financial institutions. Leveraging AI's full potential while ensuring the integrity and reliability of financial operations requires striking a balance between innovation and responsible and ethical use of the technology.

8 Results

The application of artificial intelligence (AI) to the study of economics and finance has produced outstanding outcomes and revolutionary changes. AI technologies have revolutionized traditional methods, enhancing efficiency, accuracy, and inclusivity in a variety of contexts.

8.1 AI in Finance

In the world of finance, AI-driven tools like robo-advisors have decentralized access to financial advice, dismantling socioeconomic barriers and empowering investors from all socioeconomic backgrounds. Due to algorithmic trading's reduction of emotional biases and enhancement of price discovery mechanisms, market liquidity and efficiency have significantly increased. Predictive analytics has also given investors access to data-driven strategies, enhancing returns and lowering risks. Asset allocations have been dynamically optimized by AI-driven portfolio management systems, increasing overall portfolio efficiency.

8.2 AI in Economic Analysis

Economic analysis has been significantly impacted by AI. AI-driven economic forecasting has given decision-makers more accurate forecasts, enabling them to be proactive in responding to economic developments. Sentiment analysis has given economists insights into consumer and market sentiment, facilitating quick decisions. Additionally, AI algorithms have improved credit scoring and risk analysis, improving the accuracy and accessibility of lending decisions.

9 Conclusion

A new era of opportunities and difficulties has begun with the incorporation of artificial intelligence into economics and finance. In financial institutions, AI has increased productivity, individualized customer experiences, and advanced risk management. However, it also raises issues like the need for ongoing workforce upskilling, algorithmic bias, regulatory compliance, and data privacy.

It is crucial to strike a balance between utilizing AI's potential and abiding by ethical and legal requirements as these industries continue to develop. Financial institutions must have strong governance, abide by moral AI principles, implement effective cybersecurity measures, and maintain ongoing regulatory compliance in order to fully benefit from AI while maintaining the public's confidence in their business practices.

In summary, decision-making, forecasting, and operational paradigms have changed as a result of the incorporation of AI into economics and finance. Even though there are difficulties, the potential for increased effectiveness, inclusivity, and data-driven judgment make AI an important tool for the future of these sectors.

References

1. Ahmed, A.A.A., Agarwal, S., Kurniawan, I.G.A., Anantadjaya, S.P., Krishnan, C.: Business boosting through sentiment analysis using artificial intelligence approach. Int. J. Syst. Assur. Eng. Manag. **13**(Suppl 1), 699–709 (2022)
2. Allioui, H., Mourdi, Y., et al.: Unleashing the potential of AI: investigating cutting-edge technologies that are transforming businesses. Int. J. Comput. Eng. Data Sci. (IJCEDS) **3**(2), 1–12 (2023)
3. Almeida, D., Shmarko, K., Lomas, E.: The ethics of facial recognition technologies, surveillance, and accountability in an age of artificial intelligence: a comparative analysis of US, EU, and UK regulatory frameworks. AI Ethics **2**(3), 377–387 (2022)
4. Aziz, L.A.-R., Andriansyah, Y.: The role artificial intelligence in modern banking: an exploration of AI-driven approaches for enhanced fraud prevention, risk management, and regulatory compliance. Rev. Contemp. Bus. Anal. **6**(1), 110–132 (2023)
5. Aziz, S., Dowling, M.M.: AI and machine learning for risk management (2018). Published as: Aziz, S., Dowling, M.: Machine learning and AI for risk management. In: Lynn, T., Mooney, G., Rosati, P., Cummins, M. (eds.) Disrupting Finance: FinTech and Strategy in the 21st Century, pp. 33–50. Palgrave (2019)
6. Aziz, S., Dowling, M.: Machine learning and AI for risk management. In: Disrupting Finance: FinTech and Strategy in the 21st Century, pp. 33–50 (2019)
7. Belanche, D., Casaló, L.V., Flavián, C.: Artificial intelligence in FinTech: understanding robo-advisors adoption among customers. Ind. Manag. Data Syst. **119**(7), 1411–1430 (2019)
8. Biallas, M., O'Neill, F.: Artificial Intelligence Innovation in Financial Services (2020)
9. Brenner, L., Meyll, T.: Robo-advisors: a substitute for human financial advice? J. Behav. Exp. Financ. **25**, 100275 (2020)
10. Buckley, R.P., Zetzsche, D.A., Arner, D.W., Tang, B.W.: Regulating artificial intelligence in finance: putting the human in the loop. Syd. Law Rev. **43**(1), 43–81 (2021)
11. Cao, L.: AI in finance: challenges, techniques, and opportunities. ACM Comput. Surv. (CSUR) **55**(3), 1–38 (2022)
12. Clancy, M.: Reflections on the financial and ethical implications of music generated by artificial intelligence. Ph.D. thesis, Trinity College, Dublin (2021)
13. D'Acunto, F., Rossi, A.G.: Robo-Advising. Springer (2021)
14. Dwivedi, Y.K., Hughes, L., Ismagilova, E., Aarts, G., Coombs, C., Crick, T., Duan, Y., Dwivedi, R., Edwards, J., Eirug, A., et al.: Artificial intelligence (AI): multidisciplinary perspectives on emerging challenges, opportunities, and agenda for research, practice and policy. Int. J. Inf. Manage. **57**, 101994 (2021)
15. Ferguson-Cradler, G.: Narrative and computational text analysis in business and economic history. Scand. Econ. Hist. Rev. **71**(2), 103–127 (2023)

16. Ghasemaghaei, M.: Does data analytics use improve firm decision making quality? The role of knowledge sharing and data analytics competency. Decis. Support Syst. **120**, 14–24 (2019)
17. Gruetzemacher, R., Paradice, D.: Deep transfer learning & beyond: transformer language models in information systems research. ACM Comput. Surv. (CSUR) **54**(10s), 1–35 (2022)
18. Guenole, N., Feinzig, S.: The Business Case for AI in HR. With Insights and Tips on Getting Started. IBM Smarter Workforce Institute, IBM Corporation, Armonk (2018)
19. Gwagwa, A., Kachidza, P., Siminyu, K., Smith, M.: Responsible artificial intelligence in Sub-Saharan Africa: landscape and general state of play (2021)
20. Hassan, M.K., Rabbani, M.R., Ali, M.A.M.: Challenges for the Islamic finance and banking in post COVID era and the role of FinTech. J. Econ. Coop. Dev. **41**(3), 93–116 (2020)
21. Hoppmann, J., Peters, M., Schneider, M., Hoffmann, V.H.: The two faces of market support—how deployment policies affect technological exploration and exploitation in the solar photovoltaic industry. Res. Policy **42**(4), 989–1003 (2013)
22. Innes, C.R., Andrieu, J.: Banking on FinTech in emerging markets. This publication may be reused for noncommercial purposes if the source is cited as IFC, a member of the World Bank Group (2022). www.ifc.org/thoughtleadership
23. Jha, J., Vishwakarma, A.K., Chaithra, N., Nithin, A., Sayal, A., Gupta, A., Kumar, R.: Artificial intelligence and applications. In: 2023 1st International Conference on Intelligent Computing and Research Trends (ICRT), pp. 1–4. IEEE (2023)
24. Kijewski, M., Ślepaczuk, R., et al.: Predicting prices of S&P500 index using classical methods and recurrent neural networks. Work. Pap. Fac. Econ. Sci. Univ. Wars (2020)
25. Kumar, J., Gupta, S.S.: Impact of artificial intelligence towards customer relationship in Indian banking industry. Gyan Manag. J. **17**(1), 105–115 (2023)
26. Kumar, A., Sharma, S., Mahdavi, M.: Machine learning (ML) technologies for digital credit scoring in rural finance: a literature review. Risks **9**(11), 192 (2021)
27. Lam, J.W.: Robo-advisors: a portfolio management perspective. Senior thesis, Yale College, p. 20 (2016)
28. Lau, T., Leimer, B.: The era of connectedness: how AI will help deliver the future of banking. J. Digit. Bank. **3**(3), 215–231 (2019)
29. Lee, M., Yun, J.J., Pyka, A., Won, D., Kodama, F., Schiuma, G., Park, H., Jeon, J., Park, K., Jung, K., et al.: How to respond to the fourth industrial revolution, or the second information technology revolution? Dynamic new combinations between technology, market, and society through open innovation. J. Open Innov. Technol. Mark. Complex. **4**(3), 21 (2018)
30. Madhav, A.S., Tyagi, A.K.: The world with future technologies (post-COVID-19): open issues, challenges, and the road ahead. In: Intelligent Interactive Multimedia Systems for e-Healthcare Applications, pp. 411–452 (2022)
31. Milana, C., Ashta, A.: Artificial intelligence techniques in finance and financial markets: a survey of the literature. Strateg. Change **30**(3), 189–209 (2021)
32. Munoko, I., Brown-Liburd, H.L., Vasarhelyi, M.: The ethical implications of using artificial intelligence in auditing. J. Bus. Ethics **167**, 209–234 (2020)
33. Pallathadka, H., Ramirez-Asis, E.H., Loli-Poma, T.P., Kaliyaperumal, K., Ventayen, R.J.M., Naved, M.: Applications of artificial intelligence in business management, e-commerce and finance. Mater. Today Proc. **80**, 2610–2613 (2023)
34. Perifanis, N.-A., Kitsios, F.: Investigating the influence of artificial intelligence on business value in the digital era of strategy: a literature review. Information **14**(2), 85 (2023)
35. Ryll, L., Barton, M.E., Zhang, B.Z., McWaters, R.J., Schizas, E., Hao, R., Bear, K., Preziuso, M., Seger, E., Wardrop, R., et al.: Transforming paradigms: a global AI in financial services survey (2020)
36. Sifat, I.: Artificial intelligence (AI) and future retail investment
37. Susskind, R.E., Susskind, D.: The Future of the Professions: How Technology Will Transform the Work of Human Experts. Oxford University Press, USA (2015)
38. Szczepanski, M.: Economic impacts of artificial intelligence (AI) (2019)
39. Tan, G.K.S.: Robo-advisors and the financialization of lay investors. Geoforum **117**, 46–60 (2020)

40. Weber-Lewerenz, B.: Corporate digital responsibility (CDR) in construction engineering—ethical guidelines for the application of digital transformation and artificial intelligence (AI) in user practice. SN Appl. Sci. **3**, 1–25 (2021)
41. Yaseen, S.G.: Cutting-Edge Business Technologies in the Big Data Era: Proceedings of the 18th SICB "Sustainability and Cutting-Edge Business Technologies" Volume 1, vol. 135. Springer Nature (2023)
42. Zhang, H., Song, M., He, H.: Achieving the success of sustainability development projects through big data analytics and artificial intelligence capability. Sustainability **12**(3), 949 (2020)
43. Žigienė, G., Rybakovas, E., Alzbutas, R.: Artificial intelligence based commercial risk management framework for SMEs. Sustainability **11**(16), 4501 (2019)

Green Entrepreneurship and Sustainability in Developing Countries: Opportunities, Challenges, and Future Research Directions

Mohamed Salaheldeen, Ali Ateeq, Zuhair Al Ani, Mohammed Alzoraiki, Shafeeq Ahmed Ali, and Marwan Milhem

Abstract The development of green entrepreneurship plays a crucial role in addressing environmental challenges and promoting the principles of sustainability. This study provides an overview of current trends and future research directions in the field of green entrepreneurship and sustainability. It offers significant insights for both scholars and practitioners in this area. The key issues that are included by the circular economy are the concepts of resource circulation, the adoption of clean energy solutions, the implementation of sustainable farming methods, and the promotion of eco-innovation efforts. In addition, this study highlights developing areas of inquiry, including regulatory frameworks, the importance of social entrepreneurship, and the impact of technology on the promotion of environmentally friendly business initiatives. By directing their attention towards these specific domains, academics and practitioners have the potential to significantly expedite the transition towards a more environmentally conscious and sustainable future. This overview serves as a guiding framework for those committed to tackling environmental concerns and advocating for a more environmentally friendly and sustainable global context.

Keywords Green entrepreneurship · Sustainability · Environmental challenges · Environmental sustainability · Sustainable development

---

M. Salaheldeen
Faculty of Commerce, Menoufia University, Shebin El-Kom, Egypt

A. Ateeq (✉) · M. Alzoraiki · S. A. Ali · M. Milhem
Faculty of Administrative and Financial Sciences, Gulf University, Sanad 26489, Kingdom of Bahrain
e-mail: dr.ali.ateeq@gulfuniversity.edu.bh

S. A. Ali
e-mail: dr.shafeeq.ali@gulfuniversity.edu.bh

M. Milhem
e-mail: dr.marwan.milhem@gulfuniversity.edu.bh

Z. Al Ani
College of Engineering, Gulf University, Sanad 26489, Kingdom of Bahrain
e-mail: coo@gulfuniversity.edu.bh

B. Awwad (ed.), *The AI Revolution: Driving Business Innovation and Research*, Studies in Systems, Decision and Control 524, https://doi.org/10.1007/978-3-031-54379-1_45

1 Introduction

Green entrepreneurship represents a dynamic and transformative approach to business that places environmental sustainability at its core [1]. Entrepreneurs in this realm seek innovative solutions to pressing environmental challenges while simultaneously driving economic growth. These individuals and ventures adopt environmentally friendly practices, develop sustainable products and services, and often operate with a triple-bottom-line perspective, considering not only profits but also the planet and people [2]. Green entrepreneurs are at the forefront of shaping a more sustainable future, harnessing their creativity, vision, and determination to address critical issues such as climate change, resource depletion, and pollution [3]. Their endeavors not only contribute to the preservation of our planet but also serve as catalysts for positive change, inspiring others to embrace sustainable practices and demonstrating that business success can go hand in hand with environmental responsibility [4].

Sustainability is a fundamental principle that underpins our collective responsibility to safeguard the planet for current and future generations [5]. It encompasses the concept of meeting our present needs without compromising the ability of future generations to meet their own needs [6]. Sustainability transcends environmental considerations and extends to social and economic dimensions as well [7]. It calls for responsible stewardship of natural resources, the promotion of equitable societies, and the pursuit of economic systems that are both prosperous and environmentally friendly. Embracing sustainability necessitates a holistic approach, recognizing that our choices and actions today have far-reaching consequences, and it serves as a guiding beacon in the pursuit of a more balanced and harmonious coexistence with our environment and fellow beings [8].

Developing countries have witnessed a significant increase in green startups and eco-friendly businesses. According to the Global Entrepreneurship Monitor (GEM), in 2020, several developing countries, including India, Brazil, and Kenya, reported substantial growth in the number of startups focused on sustainability and environmental solutions [9]. The importance of research exploring the nexus between green entrepreneurship and sustainability cannot be overstated in our current era defined by environmental crises and the urgent need for sustainable solutions [10]. This research not only sheds light on innovative business practices that are essential for mitigating climate change, conserving natural resources, and reducing environmental degradation but also provides crucial insights into fostering economic growth while maintaining ecological integrity. Understanding how green entrepreneurs navigate challenges and seize opportunities within the sustainability landscape is instrumental in shaping policies, investments, and societal attitudes towards a more sustainable future.

2 Literature Review

2.1 Green Entrepreneurship and Sustainability in Developing Countries

Green entrepreneurship and sustainability are becoming increasingly important in the context of developing countries, where the need for economic growth and environmental protection is paramount [4]. These nations often face unique challenges, including high levels of poverty, resource scarcity, and vulnerability to climate change [3, 11]. Green entrepreneurship, which focuses on environmentally conscious business practices and innovations, can play a pivotal role in addressing these challenges and promoting sustainable development [12].

In developing countries, green entrepreneurship offers a pathway to economic growth that is not only financially rewarding but also environmentally responsible [13]. It encourages the creation of businesses that prioritize resource efficiency, waste reduction, and sustainable practices [14]. These businesses can generate employment opportunities, stimulate local economies, and reduce poverty while simultaneously minimizing their ecological footprint [15]. In this way, green entrepreneurship contributes to a more balanced and sustainable development trajectory.

Sustainability, in the context of green entrepreneurship, is not limited to environmental concerns but extends to social and economic dimensions as well [16]. It entails creating businesses that benefit local communities, conserve natural resources, and foster resilience against climate change impacts [17]. For instance, green entrepreneurs in agriculture may introduce sustainable farming practices that improve crop yields, reduce environmental harm, and enhance food security for local populations [18].

Furthermore, green entrepreneurship in developing countries can help bridge the technology gap by introducing innovative solutions and promoting knowledge transfer [19]. It often leverages emerging technologies and creative approaches to address local challenges, such as access to clean energy, safe drinking water, and affordable healthcare. By doing so, green entrepreneurs not only provide immediate solutions but also contribute to long-term sustainable development [20].

Green entrepreneurship and sustainability are interlinked in the context of developing countries, offering a promising avenue for inclusive and environmentally responsible economic growth [21]. These endeavors not only address pressing challenges but also promote a holistic approach to development that encompasses economic, social, and environmental dimensions [22]. As the world seeks sustainable solutions to global issues, supporting and fostering green entrepreneurship in developing nations is crucial for achieving lasting positive change [23].

2.2 Green Entrepreneurship and Sustainability in Developing Countries: Opportunities, Challenges

Green entrepreneurship and sustainability in developing countries present a complex landscape filled with both opportunities and challenges. Understanding these dynamics is crucial for fostering environmentally responsible economic development in these regions.

Opportunities

1. **Local Resource Utilization**: Developing countries often possess rich biodiversity and natural resources. According to a McKinsey survey in 2020, 66% of respondents reported that their companies were taking steps to improve the sustainability of their supply chains [24]. This demonstrates a growing awareness of sustainability in the business world. Green entrepreneurs can leverage these resources to create sustainable businesses, such as eco-tourism, sustainable agriculture, and herbal medicine production [25].
2. **Job Creation**: Green entrepreneurship can generate employment opportunities, especially in rural areas where traditional job markets may be limited. This can lead to poverty reduction and improved living standards. The International Labour Organization (ILO) reported in 2020 that the transition to a green economy could create 24 million new jobs globally by 2030 [26]. This highlights the employment potential of sustainability-focused entrepreneurship.
3. **Access to Clean Energy**: Many developing countries face energy access challenges. Green entrepreneurs can introduce affordable and sustainable energy solutions, such as solar panels and small-scale wind turbines, providing reliable electricity to off-grid communities. According to the International Energy Agency (IEA), in 2020, renewable energy accounted for nearly 90% of the increase in global electricity capacity, indicating a significant shift towards sustainable energy sources [27].
4. **Innovation and Adaptation**: The necessity for sustainable solutions in resource-constrained environments often drives innovation. Green entrepreneurs have the capability to cultivate and foster innovative, cost-effective, and highly influential technologies and strategies that aim to tackle and ameliorate the diverse set of challenges faced by local communities. These individuals possess the ingenuity and resourcefulness necessary to conceptualize and develop sustainable solutions that not only provide a positive environmental impact, but also offer tangible benefits to the immediate vicinity. By leveraging their unique skill set and creative thinking, green entrepreneurs are able to devise and implement measures that effectively address the specific needs and circumstances of the local context, ultimately contributing to the overall betterment and enhancement of the community.
5. **Global Market Access**: A survey by Nielsen in 2019 found that 81% of global respondents felt strongly that companies should help improve the environment [28]. Consumers are increasingly choosing products and services from companies

that demonstrate a commitment to sustainability. Sustainable and ethical products from developing countries, such as fair-trade coffee and artisanal crafts, can access lucrative global markets, providing economic benefits to local producers.

6. **Circular Economy**: Circular economy and green entrepreneurship are intricately linked concepts that are gaining prominence in the quest for sustainable and environmentally responsible economic models [29]. Circular economy principles challenge the traditional linear "take-make-dispose" approach to production and consumption by emphasizing the reduction of waste and the efficient use of resources. Green entrepreneurship, on the other hand, refers to the pursuit of innovative business ventures that prioritize environmental sustainability and social responsibility. When these two concepts intersect, they create a powerful synergy with several key implications.

Challenges

1. **Access to Capital**: Despite the growth of green entrepreneurship, access to finance remains a significant challenge. A report by the International Finance Corporation (IFC) noted that up to 78% of micro, small, and medium-sized enterprises (MSMEs) in developing countries face challenges in accessing the necessary funding for sustainable initiatives [30]. Securing funding for green startups in developing countries can be challenging due to limited access to venture capital, angel investors, and traditional financing options.
2. **Infrastructure and Technology Gaps**: Inadequate infrastructure and technology limitations present substantial hurdles for green entrepreneurship, particularly in developing regions [31]. Without reliable access to essential infrastructure such as energy, transportation, and communication networks, the establishment and growth of green ventures become exceptionally challenging. Furthermore, technology limitations can hinder the adoption of innovative eco-friendly solutions and hinder a business's ability to optimize resource use or reduce environmental impacts. Overcoming these barriers requires concerted efforts from governments, investors, and the private sector to invest in infrastructure development and technological advancement, ensuring that green entrepreneurs have the necessary tools and support to thrive in their pursuit of sustainability-driven initiatives.
3. **Policy and Regulatory Barriers**: Policy and regulatory barriers pose significant challenges for green entrepreneurship in developing countries [32]. Often characterized by complex bureaucratic processes, unclear environmental regulations, and inconsistent enforcement, these barriers can stifle innovation and investment in sustainable ventures. Additionally, the lack of targeted incentives and financial support for green startups can deter potential entrepreneurs from entering the sector. Addressing these challenges necessitates a concerted effort from governments to create clear and supportive regulatory frameworks that encourage sustainable practices and provide incentives for green initiatives. By streamlining processes, offering financial incentives, and promoting transparency and consistency in regulations, developing countries can unlock the immense potential of green entrepreneurship as a catalyst for economic growth and environmental sustainability.

4. **Awareness and Education**: A lack of awareness and education can significantly hinder the growth and impact of green entrepreneurship [4]. Without adequate knowledge about environmental challenges and the potential benefits of sustainability-focused ventures, aspiring entrepreneurs may not recognize the opportunities for innovation and positive environmental impact. Moreover, potential investors, consumers, and policymakers may not fully understand the value of green products and services. Bridging this awareness gap is crucial to fostering a supportive ecosystem for green entrepreneurship [21]. Investment in environmental education, awareness campaigns, and training programs can empower individuals and communities to engage in sustainable practices, drive demand for eco-friendly products, and encourage policymakers to create favorable conditions for green startups. In this way, education and awareness become essential catalysts for the success and expansion of green entrepreneurship, both in developing and developed regions.
5. **Market Access**: Market access profoundly influences the success and growth of green entrepreneurship. Limited market access can be a formidable barrier, particularly for startups in the sustainable sector, as they often face challenges in reaching a critical mass of consumers. Inadequate distribution channels, high entry barriers, and consumer preferences that may not prioritize green products can hinder market penetration [33]. However, effective market access strategies can amplify the impact of green entrepreneurship. Collaborations with established businesses, e-commerce platforms, and innovative marketing tactics can help green startups expand their reach. Moreover, favorable trade policies and consumer education campaigns can stimulate demand for eco-friendly products and services, ultimately fostering a more conducive environment for green entrepreneurship to flourish.
6. **Social Inequity**: Social inequity is a critical consideration in the realm of green entrepreneurship [34, 35]. While the green sector holds promise for addressing environmental challenges, it must also confront disparities in access and participation. Historically marginalized communities often lack the resources, education, and opportunities to engage fully in green entrepreneurship initiatives, perpetuating social inequalities. To foster true sustainability, it is essential to prioritize inclusive policies and support mechanisms that ensure a more equitable distribution of the benefits of green entrepreneurship. By providing access to training, funding, and mentorship for underrepresented groups and disadvantaged communities, green entrepreneurship can become a powerful tool for not only environmental stewardship but also social empowerment and equity.

Method: In conducting an article review titled "Green Entrepreneurship and Sustainability in Developing Countries: Opportunities, Challenges, and Future Research Directions," the research methodology will encompass a systematic examination of pertinent literature drawn from academic databases such as Scopus and Web of Science [36, 37]. Key information and findings from selected peer-reviewed articles and reports will be extracted and subjected to thematic content analysis to identify recurrent themes, opportunities, and challenges in the context of green entrepreneurship in developing countries [38–41]. The synthesized data will offer a comprehensive overview of the current landscape, and citations will be meticulously included to ensure the credibility of the review process, shaping future research agendas [35, 42, 43].

Future Research Directions

Green entrepreneurship and sustainability are rapidly evolving fields with increasing significance in addressing global environmental challenges. Future research directions in these domains are pivotal to advancing our understanding and fostering innovative solutions.

Firstly, exploring the intersection of green entrepreneurship and emerging technologies is a promising area for future research. With the ongoing development of technologies such as artificial intelligence, blockchain, and the Internet of Things, their applications in promoting sustainability and green business models warrant comprehensive investigation. Understanding how these technologies can be leveraged for eco-friendly innovations, supply chain transparency, and sustainable resource management is crucial. Moreover, research could delve into the potential risks and ethical considerations associated with the use of these technologies in green entrepreneurship.

Secondly, there's a growing need for research on the role of policy frameworks in nurturing green entrepreneurship. Governments worldwide are increasingly recognizing the importance of fostering sustainable business practices. Investigating the impact of policies, regulations, and incentives on the growth of green startups and their long-term sustainability is essential. Researchers can also explore the effectiveness of international agreements and collaborations in promoting global sustainability initiatives, such as the Paris Agreement and the Sustainable Development Goals.

Lastly, social entrepreneurship within the realm of green initiatives presents a fertile area for exploration. Studying the motivations, challenges, and successes of social entrepreneurs dedicated to environmental causes can shed light on effective strategies for driving sustainable change. Research can also delve into the intersectionality of social and environmental issues, emphasizing how green social enterprises can contribute to both ecological and societal improvements, particularly in marginalized communities.

In summary, the future research directions in green entrepreneurship and sustainability encompass an array of exciting opportunities. By delving into the synergy between emerging technologies, policy frameworks, and social entrepreneurship, researchers can provide invaluable insights to guide businesses, policymakers, and society towards a more sustainable and environmentally conscious future.

3 Conclusion

In conclusion, the dynamic intersection of green entrepreneurship and sustainability in developing countries presents a landscape brimming with opportunities and challenges. As we navigate the path towards a more sustainable global future, it is evident that these regions play a pivotal role. Developing countries, often rich in natural resources, can harness their potential to foster green entrepreneurship and sustainable practices. Opportunities abound for innovative solutions addressing pressing environmental issues, job creation, and economic growth while simultaneously mitigating the negative impacts of rapid industrialization. However, the path is laden with challenges, including limited access to capital and resources, inadequate infrastructure, and policy hurdles that necessitate careful navigation.

To unlock the full potential of green entrepreneurship and sustainability in developing countries, concerted efforts are required. Policymakers, international organizations, and stakeholders must collaborate to create an enabling environment, offering incentives, reducing regulatory barriers, and providing access to financing. Moreover, education and capacity-building programs can empower local entrepreneurs and communities to engage in sustainable practices. While the road ahead is complex, it is also one teeming with transformative opportunities, where the fusion of entrepreneurship and sustainability can pave the way for more equitable, prosperous, and environmentally resilient societies in the developing world.

References

1. Makhloufi, L., et al.: Impact of green entrepreneurship orientation on environmental performance: the natural resource-based view and environmental policy perspective. Bus. Strateg. Environ. **31**(1), 425–444 (2022)
2. Habib, M.A., Bao, Y., Ilmudeen, A.: The impact of green entrepreneurial orientation, market orientation and green supply chain management practices on sustainable firm performance. Cogent Bus. Manag. **7**(1), 1743616 (2020)
3. Tien, N.H., et al.: Green entrepreneurship understanding in Vietnam. Int. J. Entrep. **24**(2), 1–14 (2020)
4. Tien, N.H., et al.: Green entrepreneurship: a game changer in Vietnam business landscape. Int. J. Entrep. Small Bus. **48**(4), 408–431 (2023)
5. Battour, M., et al.: Halal tourism: what is next for sustainability? J. Islam. Tour. **1**(inaugural issue), 80–91 (2021)
6. Demastus, J., Landrum, N.E.: Organizational sustainability schemes align with weak sustainability. Bus. Strateg. Environ. (2023)
7. Sancak, I.E.: Change management in sustainability transformation: a model for business organizations. J. Environ. Manage. **330**, 117165 (2023)
8. Salaheldeen, M., et al.: Prospects for achieving the Sustainable Development Goals 2030 through a proposed Halal Entrepreneurship Success Index (HESI). SHS Web Conf. **124**, 08001 (2021)
9. Hill, S., et al.: Global entrepreneurship monitor 2021/2022 global report: opportunity amid disruption. In: Expo 2020 Dubai (2022)
10. Soomro, B.A., et al.: Going green with the green market and green innovation: building the connection between green entrepreneurship and sustainable development. Kybernetes (2023)

11. Al-Refaei, A.A.-A., et al.: An integrated mediating and moderating model to improve service quality through job involvement, job satisfaction, and organizational commitment. Sustainability **15**(10), 7978 (2023)
12. Soomro, B.A., Ghumro, I.A., Shah, N.: Green entrepreneurship inclination among the younger generation: an avenue towards a green economy. Sustain. Dev. **28**(4), 585–594 (2020)
13. Khan, S.A.R., et al.: A green ideology in Asian emerging economies: from environmental policy and sustainable development. Sustain. Dev. **27**(6), 1063–1075 (2019)
14. Khan, S.A.R., et al.: Determinants of economic growth and environmental sustainability in South Asian Association for Regional Cooperation: evidence from panel ARDL. Environ. Sci. Pollut. Res. **27**, 45675–45687 (2020)
15. Gherghina, ȘC., et al.: Small and medium-sized enterprises (SMEs): the engine of economic growth through investments and innovation. Sustainability **12**(1), 347 (2020)
16. Liu, Y., Lee, J.M., Lee, C.: The challenges and opportunities of a global health crisis: the management and business implications of COVID-19 from an Asian perspective. Asian Bus. Manag. **19**, 277–297 (2020)
17. Yi, G.: From green entrepreneurial intentions to green entrepreneurial behaviors: the role of university entrepreneurial support and external institutional support. Int. Entrep. Manag. J. **17**(2), 963–979 (2021)
18. Sher, A., et al.: Green entrepreneurial farming: a dream or reality? J. Clean. Prod. **220**, 1131–1142 (2019)
19. Skordoulis, M., et al.: The mediating role of firm strategy in the relationship between green entrepreneurship, green innovation, and competitive advantage: the case of medium and large-sized firms in Greece. Sustainability **14**(6), 3286 (2022)
20. Salaheldeen, M., Battour, M.: Fostering innovation capability and sustainable innovation in halal industry: the role of halal entrepreneurs' success. J. Islam. Mark. ahead of print (2023)
21. Le Loarne Lemaire, S., et al.: Becoming a green entrepreneur: an advanced entrepreneurial cognition model based on a practiced-based approach. Int. Entrep. Manag. J. 1–28 (2022)
22. Salaheldeen, M., et al.: The perception of success in the halal market: developing a halal entrepreneurship success scale. J. Islam. Mark. **14**(3), 799–825 (2022)
23. Salaheldeen, M.: Opportunities for halal entrepreneurs in the Islamic digital economy: future and trends from a cultural entrepreneurship perspective. In: Ratten, V. (ed.) Cultural Entrepreneurship, pp. 95–107. Springer Nature Singapore, Singapore (2022)
24. Granskog, A., et al.: Survey: Consumer Sentiment on Sustainability in Fashion, p. 17. McKinsey & Company (2020)
25. Battour, M., et al.: Conceptualizing a model for the effect of entrepreneurial digital competencies and innovation capability on the tourism entrepreneurship performance in UAE. In: International Conference on Information Systems and Intelligent Applications: ICISIA 2022. Springer International Publishing, Cham (2023)
26. World Employment and Social Outlook Trends 2020 (2020)
27. World Energy Outlook: International Energy Agency (2020)
28. The Nielsen total audience report. In: Total Audience Report Series—Insights. Nielsen Holdings Plc (NYSE: NLSN) (2019)
29. Mondal, S., Singh, S., Gupta, H.: Assessing enablers of green entrepreneurship in circular economy: an integrated approach. J. Clean. Prod. 135999 (2023)
30. International Finance Corporation: Investing for Impact: The Global Impact Investing Market 2020. World Bank (2021)
31. Ayadi, R., Forouheshfar, Y.: Green Economy in Egypt, Tunisia and Lebanon: Building an Enabling Environment for Green Entrepreneurs. EMEA policy paper, Under INVESTMED (2023)
32. Demirel, P., et al.: Born to be green: new insights into the economics and management of green entrepreneurship. Small Bus. Econ. **52**, 759–771 (2019)
33. Yin, C., et al.: Green entrepreneurship and SME performance: the moderating effect of firm age. Int. Entrep. Manag. J. **18**(1), 255–275 (2022)

34. Drago, C., Gatto, A.: An interval-valued composite indicator for energy efficiency and green entrepreneurship. Bus. Strateg. Environ. **31**(5), 2107–2126 (2022)
35. Ali, A.A.A., Yusoffii, S.H., Aliiii, K.: Fundamental of equal opportunities in employee's performance: a critical review. In: Proceedings of the 5th World Conference on Integration of Knowledge, Bangi, Malaysia (2019)
36. Ali, A.A.A., Alzoraiki, M., Al Aghbas, N.S.M.: Working remotely, social isolation and access to support and communication during COVID-19 its influence on job performance. Int. J. Intellect. Hum. Resour. Manag. (IJIHRM) **3**(01), 23–33 (2022)
37. Al Balushi, F.S.M., Ali, A.A.A., Alzoraiki, M.: The impact of forecasting strategy and training and development strategy on preparing leadership competencies administrative support in the Sultanate of Oman. Int. J. Intellect. Hum. Resour. Manag. (IJIHRM) **3**(01), 34–41 (2022)
38. Alzoraiki, M., et al.: Impact of teachers' commitment to the relationship between transformational leadership and sustainable teaching performance. Sustainability **15**(5), 4620 (2023)
39. Ali, A.A.A., et al.: Moderating effect of strategic planning on the relationship between career path planning and job performance. Sustainability **15**(11), 8490 (2023)
40. Habtoor, N.H., Ali, E.S.: Human side of total quality management: a critical review. Int. J. Intellect. Hum. Resour. Manag. (IJIHRM) **3**(02), 27–32 (2022)
41. Ateeq, A.: The mediating role of engagement in the relationship between equal opportunities and employee performance. Probl. Perspect. Manag. **21**(3), 435–447 (2023)
42. Ateeq, A.: Knowledge Management's Effect on Learning Quality in Bahrain's Private Universities (2023)
43. Ali, A.A.A., Yusoff, S.H., Ali, K.: The relationship between equal opportunities and employee performance: a special reference to with specific refence to BDF hospital in the Kingdom of Bahrain. Int. E-J. Adv. Soc. Sci. **5**(14), 1200–1206 (2019)

Smart Cities Vision: A Prototype for Future Cities

Sandra Job and Ali Ateeq

Abstract Rapid urbanization has caused several issues in cities today. Some of these include overcrowding, pollution, housing problems, an increase in diseases, crime, etc. Smart cities have been proposed as the solution to complex urbanization challenges. It depends heavily on the intelligent use of ICT. A key component of developing a smart city plan is the urban development area prototype. There are different indexes and criteria to evaluate smart cities. Learning from neighborhoods and using literature from other cities in different settings are ways to develop an innovation model. In this research, indicators are compared, and evaluated and a significant set of criteria is produced. Based on relevant information collected from the literature and the new criteria recommended, this research provides insight into developing a prototype for future cities including the indicators and strategies.

Keywords Smart cities · IoT · Urban design · Built environment

1 Transition to Smart Cities

The smart city is often defined as one that connects its business, social, physical, and information technology infrastructure to enhance city intelligence [1]. It is highlighted that people living in traditional cities experience various concerns, including poor traffic management, inadequate energy, water problems, poor lifestyle choices, a traditional school system, poorly equipped hospitals, a lack of work, a lack of safety and security, a lack of awareness of facilities, rules, and regulations, as well as the uncontrollable population expansion. Traditional cities undergo a metamorphosis to become smart cities in order to solve all of these issues [2].

S. Job
College of Engineering, Gulf University, Sanad 26489, Kingdom of Bahrain
e-mail: sandra.job@gulfuniversity.edu.bh

A. Ateeq (✉)
Administrative Science Department, College of Administrative and Financial Science, Gulf University, Sanad 26489, Kingdom of Bahrain
e-mail: dr.ali.ateeq@gulfuniversity.edu.bh

B. Awwad (ed.), *The AI Revolution: Driving Business Innovation and Research*, Studies in Systems, Decision and Control 524, https://doi.org/10.1007/978-3-031-54379-1_46

Eco-friendly building materials and decision-making algorithms are proposed to be utilized in the creation of smart cities. Smart cities require meticulous planning [3–5]. Early on, it's crucial that the city meets its citizens' and the government's needs. Both "functions" and "purposes," which pertain to a city's operations and aesthetic appeal, must be addressed in a defined plan [6]. "Purposes" refers to the advantages a smart city model is supposed to bring about [7]. Therefore, a transition from traditional cities to smart cities is necessary.

2 Smart Cities

Smart cities are defined differently by researchers. It is noted that smart cities are spaces that combine innovative solutions with a wide variety of techniques that improve infrastructure performance in order to achieve Sustainable Urban Development [8]. Moreover, through participation and cooperation between all citizens, an intelligent city can take advantage of the data it produces daily in order to gather new information that will allow its administration to be improved so as to become more sustainable, and competitive and provide better living conditions. In addition, a Smart City considers the needs of its inhabitants and responds to these by making citizens' interactions with public knowledge systems and elements of knowledge more efficient [9–11]. The intelligent city bases its decisions and management on this knowledge, acting immediately or even in advance of potential events [12, 13].

There are many definitions of the term "smart city," but one popular one describes it as the environment that connects physical, social, business, and ICT infrastructure to increase the city's intelligence [14]. In addition, the definition of a smart city also varies depending if it is urban living labs or mesh metropolitan information and communications technologies (ICT) environments as well as based on other ICTs in the city [15].

A smart city is also described as a contemporary, technologically advanced urban area that uses ICT and other technologies to improve residents' quality of life, the effectiveness of their services, and their competitiveness while preserving the supply of resources for both current and future generations [16]. There are many ways to define the term "Smart City," but in general, it refers to the implementation and deployment of ICT infrastructures to support social and urban growth by enhancing the economy, citizen engagement, and administrative effectiveness [17].

From urban simulations and knowledge bases, the term "smart city" first appeared in literature in 1998 and it is still evolving into eco-cities [18]. It is an urban setting that makes use of ICT and other technologies to enhance the effectiveness of regular city operations and the caliber of services offered to the urban population [19, 20]. By incorporating ICT into environmentally friendly and sustainable projects, ICTs improve the quality of life for their citizens. Most of them use the buzzword "smart city" to promote these initiatives. Projects involving "smart mobility and transportation" are included in smart city initiatives, as seen in Shanghai, China, and Tokyo, Japan [21].

Significant characteristics of smart cities discussed are based on the following [22, 23]. The term "smart economy" describes a city's overall capacity for innovation, productivity, and labor market flexibility [24]. Interestingly, smart people are establishing human capital and social interaction between individuals through shared interests in lifelong learning, civic engagement, creativity, and adaptability [25, 26]. This limited definition of "smartness" is closely related to energy and natural resources, transportation and mobility, and living conditions, or, to put it another way, to the green and livable city [27, 28].

In order to ensure a high quality of life, good health, safety, and economic development, the New Urban Agenda places a high priority on proper planning and management in urban areas. It advocates for social participation, the smart city concept, and sustainable urban mobility, among other things. Sustainable transportation has been cited by researchers as one of the factors promoting the growth of smart cities [29, 30].

Through participation in decision-making and transparent governance, smart government promotes citizen participation in governance. Furthermore, Smart Mobility prepares the infrastructure and transportation to support the local ICT infrastructure, accessibility, and innovative, safe, and sustainable transportation systems. The attractiveness of the natural environment, environmental protection, and sustainable resource management all contribute to the preservation of natural resources [31]. By providing cultural amenities, favorable health conditions, high-quality housing, and social cohesion, smart living raises the quality of life [9]. In agreement, the concept of the "smart city," which addresses a variety of issues related to efficient, sustainable city management, is a major force behind sustainable urban development worldwide [32].

It is highlighted that the two defining characteristics of the smart city are cityness and informativeness. Cityness is made up of components that primarily show how a city fits into the global network of other cities, particularly the indicators of flow and place indicators and location indicators [33]. The concept of informativeness focuses on indicators of the information and knowledge-related infrastructures, the knowledge-based sectoral mix and its labor market, the political willingness to build an informational city and to encourage citizen participation and transparency, the maturity of e-government, the stakeholders' information behavior, and, finally, problem areas. Yet another study shows that the composition of a smart city architecture proposed include Smart Factories, Smart Grid, Smart Community, Smart Warehouse, Smart Healthcare, Smart Transportation, and Smart Hospitality [34].

3 Methods

In order to conduct this study, the major characteristics of a smart city is collected from different literature, and the key indicators are identified and developed. Further, under each characteristic, indicators are allocated [35, 36]. The conceptual framework for the study is represented in Fig. 1.

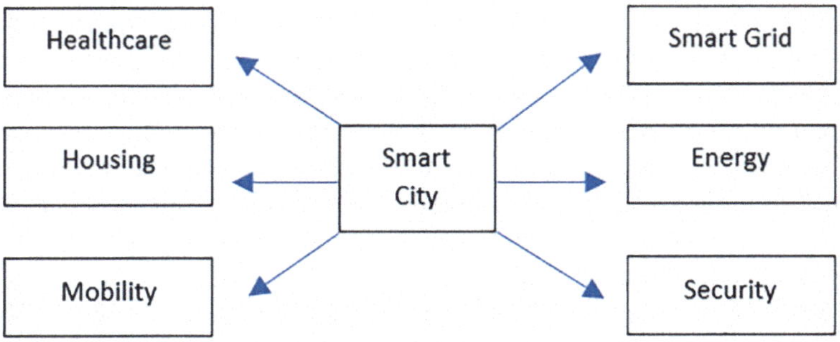

Fig. 1 Identified major characteristics of a smart city

4 Results

4.1 Healthcare

The prototype city features healthcare facilities within each grid. This includes low-rise residential units, mid-rise and high-rise buildings. The aim is to allow better navigation of healthcare through the city from different points using ICTs.

4.2 Housing

With regard to housing, as mentioned previously, 3 levels of classification exists. High-rise, Mid-rise, and Low-rise residential units. It is classified and arranged in a manner that high-rise towers are located in the center of the circular grid which has higher density as density is an urban quality, followed by mid-rise and low-rise houses. The low-rise units are located away from the densely populated area. This is explained further under Smart Grids and Fig. 2.

4.3 Smart Mobility

Some of the features in the smart city prototype include the incorporation of Smart Buses Smart charging stations and digital smart light poles as shown in Fig. 2. The aim of the city is to foster better navigation and movement throughout the grid. This includes the incorporation of different types of streets and sidewalks, proper transportation networking, separate pathways for buses, bicycles, and pedestrians, improved safety for the users using sensors on streetlights and traffic lights detecting pedestrians, charging facilities for vehicles, functional highways, etc.

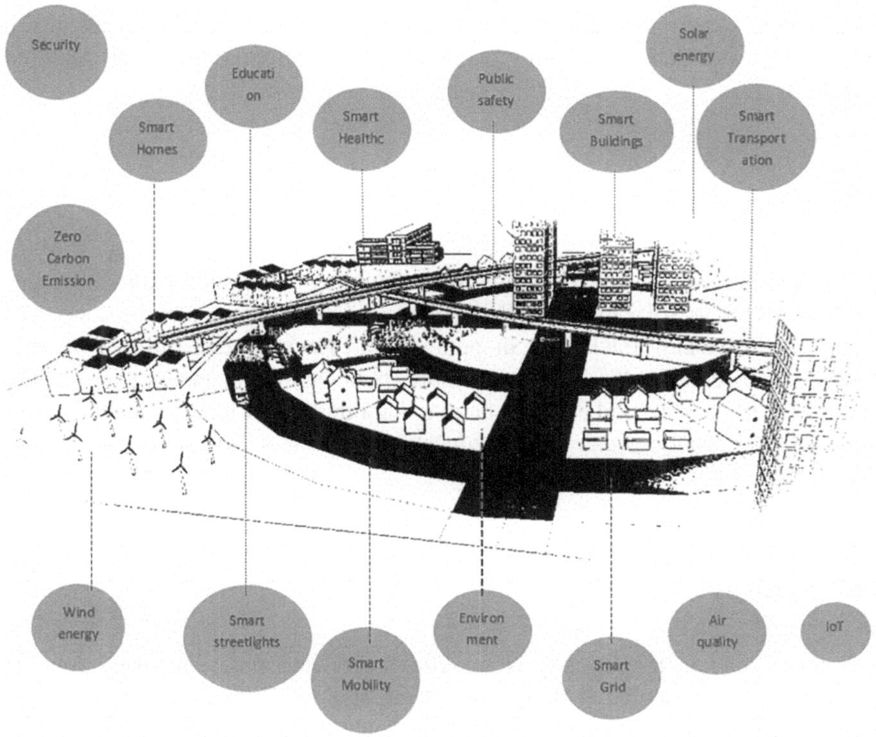

Fig. 2 Smart grid features characteristics and indicators

4.4 Smart Grid

The concept of the grid revolves around the 15-min city. The city is planned in a radial grid to foster maximum efficiency and functionality. Saving time for travel, ensuring safety, fostering socialization, reducing traffic congestion, improving effective surveillance, and enhancing walkability.

Better navigation through the city, featuring separate sections for community living residences and smart homes adjacent to water bodies away from the pollution are other features. The smart grid also ensures that the taller buildings are located towards the Centrepoint by not affecting the air quality, ventilation, and lighting to residences making them different entities that are self-sufficient. The community residences come with smart technologies and have their green spaces as courtyards which can be used for community living and add to sustainability. The taller buildings also are places with sufficient distances between them providing green spaces, in such a way residents can have breathable spaces. It features solar panels, smart energy, smart benches, smart water management, and so on. Away from the city center is the placement of windmills producing wind energy placed on either side of the city, a proper water source, better mobility, hospitals, schools, metro rails, etc. are featured in this city.

4.5 Energy

The urban design of this city features smart buildings integrated with landscapes and features Smart Solar Benches. The city features a combination of the Internet of Things, Smart Health, Intelligent Shopping, Smart Environment, Buildings, traffic management, reduced Air Pollution, Smart street lights and landscaping, Smart Energy, Smart parking, and electric charging facilities. Other features include water and gas leak detection, water quality, education, better mobility and transportation, etc.

4.6 Security

A primary concern for any city is ensuring safety measures effectively. Several planned cities have failed at doing so and the causes of this is adverse, such as theft and crime. This city aims to provide a prototype to tackle the issue of safety using a smart grid system combined with the effective implementation of ICTs.

Figure 2 portrays a vision of a smart city grid applying the characteristics and indicators defined. It represents various indicators of smart cities applied in the 15-min smart grid model.

5 Conclusion

Overcrowding, pollution, crimes, diseases, and accidents are some of the commonly known adverse effects of poor urban design in a city. Developing a smart city plan or a grid is derived as the most effective method to tackle the issues that exist. In order to achieve this, successful implementation of IoTs is necessary. By identifying the constraints and issues that exist around different neighborhoods from relevant literature and by identifying the common indicators of a smart city, a conceptual framework of major characteristics and indicators are produced. The major characteristics identified in this research include Smart Healthcare, Mobility, Housing, Smart Grid, Energy, and Security. A prototype of a radial 15-min city grid featuring these characteristics is represented.

To conclude, a smart city that is sustainable and incorporates the use of ICTs is an effective method to improve the citizen's well-being and improve their quality of life. A smart city can help reduce long-term issues related to maintenance, and improve energy efficiency and functionality. It also contributes to the liveability of a place.

Successful implementation of smart city characteristics and indicators in cities today is necessary to impact the sustainability and liveability of cities of the future.

6 Recommendations

- Future studies require additional classification of characteristics and identifying indicators and how these can be implemented in a real city.
- Implementation of ICTs and constrains are to be analysed to gain deeper understanding on the applicability of a smart grid.
- Future studies can study the effectiveness of a vertical city compared the conceptual model proposed in this paper.

Acknowledgements The author would like to thank the reviewers for their feedback that can help enhance the research paper.

References

1. Komninos, N., et al.: Projects for smart cities: ecosystems, connected intelligence and innovation for the radical transformation of cities. In: Smart Cities and Smart Communities: Empowering Citizens Through Intelligent Technologies, pp. 33–68. Springer (2022)
2. Carvalho, J.P., Bragança, L., Mateus, R.: Automating building sustainability assessment using building information modelling: a case study. J. Build. Eng. **76**, 107228 (2023)
3. Javed, A.R., et al.: Future smart cities: requirements, emerging technologies, applications, challenges, and future aspects. Cities **129**, 103794 (2022)
4. Rai, H.M., et al.: Use of Internet of Things in the context of execution of smart city applications: a review. Discov. Internet Things **3**, 8 (2023)
5. Sutriadi, R.: Soft infrastructure in smart sustainable cities: a literature review. J. Indones. Sustain. Dev. Plan. **4**(2), 195–215 (2023)
6. Halecki, W., et al.: Assessment and planning of green spaces in urban parks: a review. Sustain. Cities Soc. 104280 (2022)
7. Caputo, F., et al.: Rethinking the role of technology for citizens' engagement and sustainable development in smart cities. Sustainability **15**(13), 10400 (2023)
8. Bandari, V.: Beyond technology: a holistic framework for smart urbanization in developing countries. Tensorgate J. Sustain. Technol. Infrastruct. Dev. Countries **5**, 1–13 (2022)
9. Villegas-Ch, W., et al.: Analysis of data from surveys for the identification of the factors that influence the migration of small companies to eCommerce. Future Internet **14**(11), 303 (2022)
10. Ayala-Chauvin, M., Avilés-Castillo, F., Buele, J.: Exploring the landscape of data analysis: a review of its application and impact in Ecuador. Computers **12**(7), 146 (2023)
11. Phanichsiti, S., Thamnita, O., Thipphayasaeng, P.: Internet of Things and big data for a sustainable smart university. J. High. Educ. Theor. Pract. **23**(2), 117–128 (2023)
12. Allam, Z., et al.: The metaverse as a virtual form of smart cities: opportunities and challenges for environmental, economic, and social sustainability in urban futures. Smart Cities **5**(3), 771–801 (2022)

13. Javed, A.R., et al.: A survey of explainable artificial intelligence for smart cities. Electronics **12**(4), 1020 (2023)
14. Jilcha, L.A., et al.: Network intrusion detection in internet of blended environment using ensemble of heterogeneous autoencoders (E-HAE). Comput. Syst. Sci. Eng. **46**(3) (2023)
15. Xiong, G., et al.: Cyber–physical–social systems for smart city. In: Cyber–Physical–Human Systems: Fundamentals and Applications, pp. 511–542 (2023)
16. Almihat, M.G.M., et al.: Energy and sustainable development in smart cities: an overview. Smart Cities **5**(4), 1389–1408 (2022)
17. Gracias, J.S., et al.: Smart cities—a structured literature review. Smart Cities **6**(4), 1719–1743 (2023)
18. Dashkevych, O., Portnov, B.A.: Criteria for smart city identification: a systematic literature review. Sustainability **14**(8), 4448 (2022)
19. Davlyatov, S.: Smart-city ecosystem using block-chain technology. In: 2023 3rd International Conference on Advance Computing and Innovative Technologies in Engineering (ICACITE). IEEE (2023)
20. Alahi, M.E.E., et al.: Integration of IoT-enabled technologies and artificial intelligence (AI) for smart city scenario: recent advancements and future trends. Sensors **23**(11), 5206 (2023)
21. Shafiullah, M., et al.: Review of smart city energy modeling in Southeast Asia. Smart Cities **6**(1), 72–99 (2022)
22. Ramu, S.P., et al.: Federated learning enabled digital twins for smart cities: concepts, recent advances, and future directions. Sustain. Cities Soc. **79**, 103663 (2022)
23. Bellini, P., Nesi, P., Pantaleo, G.: IoT-enabled smart cities: a review of concepts, frameworks and key technologies. Appl. Sci. **12**(3), 1607 (2022)
24. Kazanci, G.: (Re)considering smart city approach in smart economy perspective: evaluation of Konya case. Kent Akad. **15**(1), 277–297 (2022)
25. Jackson, D., Shan, H., Meek, S.: Employer development of professional capabilities among early career workers and implications for the design of work-based learning. Int. J. Manag. Educ. **20**(3), 100692 (2022)
26. Toimbek, D.: Building human capital in Kazakhstan: information-processing skills and socio-economic outcomes of the population. J. Adult Contin. Educ. 14779714221145861 (2022)
27. Fachinelli, A.C., et al.: Urban smartness and city performance: identifying Brazilian smart cities through a novel approach. Sustainability **15**(13), 10323 (2023)
28. Javidroozi, V., et al.: Smart, sustainable, green cities: a state-of-the-art review. Sustainability **15**(6), 5353 (2023)
29. Wawer, M., Grzesiuk, K., Jegorow, D.: Smart mobility in a smart city in the context of generation Z sustainability, use of ICT, and participation. Energies **15**(13), 4651 (2022)
30. Freestone, R., Favaro, P.: The social sustainability of smart cities: a conceptual framework. City Cult. Soc. **29**, 100460 (2022)
31. Bastos, D., et al.: Smart city applications to promote citizen participation in city management and governance: a systematic review. Informatics (2022)
32. Xia, L., Semirumi, D., Rezaei, R.: A thorough examination of smart city applications: exploring challenges and solutions throughout the life cycle with emphasis on safeguarding citizen privacy. Sustain. Cities Soc. **98**, 104771 (2023)
33. Trincado-Munoz, F., et al.: Digital transformation in the world city networks' advanced producer services complex: a technology space analysis. Geoforum 103721 (2023)
34. Naaz, S., Parveen, S.: An in-depth study of 5G-based eco-friendly smart city. In: International Conference on Smart Trends in Computing and Communications. Springer (2023)
35. Ali, A.A.A., Alzoraiki, M., Al Aghbas, N.S.M.: Working remotely, social isolation and access to support and communication during COVID-19 its influence on job performance. Int. J. Intellect. Hum. Resour. Manag. (IJIHRM) **3**(01), 23–33 (2022)
36. Al-Refaei, A.A.-A., et al.: An integrated mediating and moderating model to improve service quality through job involvement, job satisfaction, and organizational commitment. Sustainability **15**(10), 7978 (2023)

Qualitative Investigation of Green Building Rating System Development in Bahrain's Affordable Housing Industry

Husam M. Abdulrahim, Ali Ateeq, Fay A. Al-Khalifa, Najib Al-Aghbas, Sandra Jo, Muskan Nagi, and Ahmed Alastal

Abstract This article presents an investigation into choosing a suitable methodology to explore the development of a Green Building Rating System (GBRS) in Bahrain, with justification for the tools and methodology used based on Bahraini and field circumstances. Qualitative approaches are increasingly used in similar studies to provide a deeper understanding of perceptions, motivations, and perspectives that underpin stakeholder decisions. Based on the exploratory nature of the study and demonstrated effectiveness in the literature, one-on-one interviews were deemed the most effective method for gathering information. The study focuses on stakeholders in the affordable housing industry, utilizing expert sampling to select participants and incorporating a qualitative-methodological approach to investigate the potential and constraints of GBRS deployment in Bahrain's affordable housing sector. This article highlights the effectiveness of using a qualitative approach in presenting the

H. M. Abdulrahim · F. A. Al-Khalifa
College of Science, University of Bahrain, Isa Town, Bahrain

A. Ateeq (✉) · M. Nagi · A. Alastal
Administrative Science Department, College of Administrative and Financial Science, Gulf University, Sanad 26489, Kingdom of Bahrain
e-mail: dr.ali.ateeq@gulfuniversity.edu.bh

M. Nagi
e-mail: dr.muskan.nagi@gulfuniversity.edu.bh

A. Alastal
e-mail: dr.ahmed.alastal@gulfuniversity.edu.bh

N. Al-Aghbas
University of Saba Region, Marib, Yemen
e-mail: n.alaghbas@usr.ac

S. Jo
Architectural and Interior Design Engineering Department, College of Engineering, Gulf University, Sanad 26489, Kingdom of Bahrain
e-mail: sandra.job@gulfuniversity.edu.bh

B. Awwad (ed.), *The AI Revolution: Driving Business Innovation and Research*, Studies in Systems, Decision and Control 524, https://doi.org/10.1007/978-3-031-54379-1_47

new exploratory area of GBRS in Bahrain, comprehensively representing stakeholder perspectives. The rationale and justification for using these tools and methodologies are presented, emphasizing their importance in understanding the motivations and influences behind public and private development.

Keywords Green Building Rating System · Stakeholder · Affordable housing industry · One-on-one interviews · Bahrain

1 Introduction

The development of a Green Building Rating System (GBRS) is increasingly recognized as a key strategy for promoting sustainable built environments. In Bahrain, the implementation of GBRS in the affordable housing market/sector is critical to achieving sustainable development goals. However, choosing a suitable methodology to explore the development of GBRS in Bahrain is a challenging task for researchers. This article aims to investigate the selection of a suitable methodology to explore the development of GBRS in Bahrain, with a particular focus on the effectiveness of qualitative approaches.

This article aims to present a suitable methodology used to explore the implementation of a GBRS in Bahrain, showing the rationale and justification for the use of the tools and methodology employed. The research focuses on stakeholders in the affordable housing industry, identifying their roles and responsibilities within the housing sector, developers and end-users of affordable housing, and the entities that regulate these processes. The purposive sampling technique is used in the selection of participants, particularly expert sampling. A qualitative-methodological approach is adopted to investigate the potential and constraints of GBRS deployment and operation in Bahrain's affordable housing market/sector.

2 Methodology

The most common challenge for researchers is figuring out the best strategy to use when trying to answer the research question [1].

Qualitative approaches have increasingly been used in similar studies, as reflected in the literature, as they tend to provide keen insight into perceptions, motivations, and perspectives that underpin key stakeholder action. This insight is particularly important since field specialists and developers determine the motivations and influences underlying public and private development.

Qualitative methods are characterized by their distinctive interpretation of social sciences, through embracing a perspective of reality that is context sensitive [2, 3]. Alkadash, et al. [4] points out that the purpose of conducting a qualitative approach in a study is not to arrive at definitive conclusions or absolute truth, but rather to serve as a starting point for future research in the field.

There has been a rapid expansion of qualitative research in the social sciences in recent years, relying largely upon techniques that seek to understand, predict and explain human behavior, and the way humans perceive their social world [5, 6]. It is generally accepted that qualitative research consists of the use of methods "that make the world visible" [7]. This includes, at the very least, the use of in-depth interview techniques, the knowledge and skill of researchers to accomplish procedures, the interpretation of results, and the ability to answer questions such as 'why?' and 'how?' [8–10].

Therefore, it can be argued that the objective of qualitative research is to 'understand and explore accounts of social events occurring in the community', whereby open questions and flexible procedures are chosen. These may even be altered during the investigation [11].

The main characteristics of qualitative research include a focus on the dynamic nature of reality, a thorough investigation of different points of view, the distinctions it makes between situations, and how it encourages the researcher to consider these factors in a thoughtful and reflective manner.

Consequently, qualitative research should be conducted with a holistic approach, considering social phenomena as one indivisible unit with which one can identify the "patterns, circumstances, repetitions, manifestations and meanings associated with the research questions" and also contribute to the creation of new theories [12].

Among the built environment research community, qualitative research has experienced a rapid increase in application, including in doctoral research as well as mainstream social science research. Originating from the social sciences, qualitative studies are understandably the preferred method for those conducting research that involves human subjects [13, 14]. The use of qualitative methods in the built environment and building industry field has been identified as a successful method as evidenced by exemplary success stories. This has led to many doctoral students and researchers in the field becoming more inclined to utilize qualitative methods [15, 16].

In recent years, qualitative research methods have been gaining a lot of attention as effective ways to study complex situations and to produce rich findings, while at the same time acknowledging the researchers' role in shaping the research findings [17]. It has been demonstrated in previous studies that qualitative methods are useful when it comes to assessing the efforts and contributions toward the implementation and deployment of new strategies. In addition, qualitative methods have been shown as useful tools when evaluating the most practical implementation supporting the objectives, as well as assessing the opinions of the experts and decision makers engaged in questions that are exploratory in nature.

According to McCrudden and Rapp [18], the qualitative approach is the most appropriate method when undertaking exploratory research as it allows the identification of emerging themes during the research process and subsequent investigations of these themes.

Moreover, as has been pointed out by Amann et al. [19], a qualitative approach can help in understanding the reasons for the conclusions drawn by participants as well as providing a better understanding of their attitudes and opinions. Further, a qualitative approach can enable the discussion to extend into other areas that had not previously been considered, but which may be critical for understanding the issue at hand. It has also been observed by Carter and Henderson (2005) that qualitative research, by its very nature, centers on perceptions, opinions, beliefs, and attitudes.

Friedrich [21] recommend that the qualitative approach be used to address problems that are context-specific, have unknown variables, and that require an exploratory research approach. This is to ensure that the research is reliable and verifiable in view of the concrete nature of the context in which it is conducted and done in a way that is relevant to this environment.

There are several qualitative techniques that can be employed depending on the objectives and complexity of the research. It has been shown that a qualitative approach is among the best approaches that can be used to answer the question of research and to reach an understanding of a particular phenomenon [22–24]. Tracy [15] suggest that the qualitative methodology acts as a sort of umbrella both in terms of the type of data collected and the way that it is analyzed.

One-on-one interviews were deemed to be the most effective method for gathering information in this research, given its exploratory nature and as demonstrated in the literature. The study aims to present a suitable methodology for exploring the implementation developments and representations of GBRS in Bahrain's affordable housing market/sector, adopting a qualitative-methodological approach to investigate the potential and constraints of GBRS deployment and operation. The study focuses on stakeholders in the affordable housing industry, including developers and end-users of affordable housing and entities that regulate these processes, using purposive sampling, particularly expert sampling, in the selection of participants based on their level of expertise in the field.

The rationale for using a qualitative approach is that it provides keen insight into perceptions, motivations, and perspectives that underpin key stakeholder action. Field specialists and developers determine the motivations and influences underlying public and private development, making it essential to understand their views on GBRS and the way in which the concept of GB is characterized in Bahrain.

3 Results

The qualitative approach used in this study provided a deeper understanding of perceptions, motivations, and perspectives that underpin stakeholder decisions. The one-on-one interviews enabled the researchers to explore the potential and constraints of GBRS deployment and operation in Bahrain's affordable housing market/sector. The expert sampling technique ensured that participants had the necessary knowledge and expertise to provide valuable insights into the development of GBRS in Bahrain. The results of the study emphasized the importance of understanding stakeholder perceptions and motivations in the development of GBRS in Bahrain.

The study also reveals that the implementation of GBRS in Bahrain's affordable housing industry faces several challenges, including limited public awareness of GB and green development, in general. Key components of GB processes in Bahrain cannot be understood without knowing the identities and motivations of the stakeholders involved, making it crucial to understand their perspectives and concerns.

4 Discussion

Qualitative approaches are increasingly used in similar studies to provide a deeper understanding of perceptions, motivations, and perspectives that underpin stakeholder decisions. The effectiveness of one-on-one interviews in exploring the potential and constraints of GBRS deployment and operation in Bahrain's affordable housing market/sector was demonstrated in this study. The expert sampling technique ensured that participants had the necessary knowledge and expertise to provide valuable insights into the development of GBRS in Bahrain. The results of the study highlight the importance of understanding stakeholder perceptions and motivations in the development of GBRS in Bahrain, which can inform policy and decision-making processes.

The study's findings highlight the importance of stakeholder engagement in the implementation of GBRS in Bahrain's affordable housing industry. The limited knowledge of GBRS among stakeholders in the affordable housing industry highlights the need for training and awareness-raising initiatives to promote the adoption of these systems. The study's results also show that the implementation of GBRS in Bahrain's affordable housing industry faces several challenges, including limited public awareness of GB and green development, in general.

The study's adoption of a qualitative-methodological approach provides keen insight into stakeholder perspectives and motivations, which is crucial to assessing the potentials and challenges of implementing a GBRS in Bahrain's affordable housing sector. The study serves as a starting point for future research in this field, and the findings can inform policy and decision-making processes related to sustainable development in Bahrain's affordable housing industry.

5 Conclusion

Choosing a suitable methodology to explore the development of GBRS in Bahrain is a challenging task for researchers. This study demonstrates the effectiveness of a qualitative approach, particularly one-on-one interviews, in exploring the potential and constraints of GBRS deployment and operation in Bahrain's affordable housing market/sector.

In order to present the context for the research, identifying the when, where, and who are involved the real estate sector and particularly the affordable housing projects in Bahrain, it is important to realize that the affordable housing market is a pivotal part of the economy and has influenced numerous fields in Bahrain. Accordingly, the sector integrates a broad variety of industries and sectors. This research focuses on stakeholders in the affordable housing industry, taking into account their roles and responsibilities within the housing sector, identifying the developers and end-users of affordable housing, and the entities that regulate these processes.

The expert sampling technique ensured that participants had the necessary knowledge and expertise to provide valuable insights into the development of GBRS in Bahrain. The results of the study highlight the importance of understanding stakeholder perceptions and motivations in the development of GBRS in Bahrain, which can inform policy and decision-making processes.

References

1. Lin, C.-C., Huang, A.Y., Yang, S.J.: A review of ai-driven conversational chatbots implementation methodologies and challenges (1999–2022). Sustainability **15**(5), 4012 (2023)
2. Quintana, R.: Embracing complexity in social science research. Qual. Quant. **57**(1), 15–38 (2023)
3. Ali, A.A.A., Habtoor, N.: Islamic work ethics and its impact on enhancing organizational commitment in bahraini private educational universities. Int. J. Intell. Human Resour. Manage. **3**(02), 33–42 (2022)
4. Alkadash, T.M., et al.: The effects of leadership style on employee sustainable behaviour: a theoretical perspective. In: Artificial Intelligence and Transforming Digital Marketing, pp. 205–213. Springer (2023)
5. Corbetta, P.: Social Research: Theory, Methods and Techniques. Sage (2003)
6. Ateeq, A., Ebrahim, S., Al-Ghatam, R.: The impact of stress and its influencing factors among dentists during the COVID-19 pandemic in Kingdom of Bahrain. F1000Research **11**, 930 (2022)
7. Saputri, N.L., Sofiani, Y., Hadi, I., Widiyarto, S., Suyana, N.: Representation of acehnese king braveness in pocut Muhammad Saga. J. Sci. Res. Educ. Technol. **2**(4), 1531–1542 (2023)
8. Milhem, M., Ayyash, M.M., Ateeq, A.A., Alzoraiki, M.: Examining the relationship between transactional leadership style and leader's emotional intelligence within the palestinian ICT sector. In: Intelligent Systems, Business, and Innovation Research, pp. 753–762. Springer (2024)
9. Ali, A.A.A., Alzoraiki, M., Al Aghbas, N.S.M.: Working remotely, social isolation and access to support and communication during Covid-19 its influence on job performance. Int. J. Intell. Human Resour. Manage. **3**(01), 23–33 (2022)
10. Ali, A.A.A., Yusoffii, S.H., Aliiii, K.: Fundamental of equal opportunities in employee's performance: a critical review. In: Proceedings of the 5th World Conference on Integration of Knowledge, vol. 29. Bangi, Malaysia (2019)
11. Smith, M.J.: Social Science in Question: Towards a Postdisciplinary Framework. Sage (1998)
12. Marshall, S.P.: Schemas in Problem Solving. Cambridge University Press (1995)
13. Orcher, L.: Conducting Research: Social and Behavioral Science Methods. Routledge (2016)

14. Al-Fahim, N.H., et al.: Factors influencing the mobile banking usage: mediating role of perceived usefulness. In: Digital Technology and Changing Roles in Managerial and Financial Accounting: Theoretical Knowledge and Practical Application, vol. 36, pp. 115–128. Emerald Publishing Limited (2024)
15. Tracy, S.J.: Qualitative Research Methods: Collecting Evidence, Crafting Analysis, Communicating Impact. Wiley (2019)
16. Rahmatika, D.F.: Employee loyalty: Peran trust in management dan peers dengan mediasi employee satisfaction. Publik: Jurnal Manajemen Sumber Daya Manusia, Administrasi dan Pelayanan Publik **11**(1), 288–302 (2024)
17. Gisby, A., Ross, C., Francis-Smythe, J., Anderson, K.: The 'rich pictures' method: Its use and value, and the implications for HRD research and practice. Human Resour. Dev. Rev. 15344843221148044 (2023)
18. McCrudden, M.T., Rapp, D.N.: Inquiry worldviews, approaches to research, and mixed methods in educational psychology. In: Handbook of Educational Psychology, pp. 48–70. Routledge (2024)
19. Amann, J., Vayena, E., Ormond, K.E., Frey, D., Madai, V.I., Blasimme, A.: Expectations and attitudes towards medical artificial intelligence: a qualitative study in the field of stroke. PLoS ONE **18**(1), e0279088 (2023)
20. Khoa, B.T., Hung, B.P., Hejsalem-Brahmi, M.: Qualitative research in social sciences: data collection, data analysis and report writing. Int. J. Public Sector Perform. Manage. **12**(1–2), 187–209 (2023)
21. Friedrich, A.: The Impact of Additive Manufacturing on Supply Chains and Business Models: Qualitative Analyses of Supply Chain Design, Governance Structure, and Business Model Change (2023)
22. Qerimi, I., Kastrati, S., Maloku, A., Gabela, O., Maloku, E.: The importance of theory and scientific theories for the scientific study of genocide in the context of the contribution to the development of the science of genocide. Acad. J. Interdisc. Stud. **12**(1), 183 (2023)
23. Milhem, M., Ayyash, M.M., Ateeq, A., Alzaghal, Q., Alzoraiki, M.: Emotional Intelligence in Leadership: Its Role in Fostering Employee Engagement In Palestine's ICT Sector (2024)
24. Barakat, B., et al.: Assessing the impact of green training on sustainable business advantage: exploring the mediating role of green supply chain practices. Sustainability **15**(19), 14144 (2023)

Digitalization and Its Impact on HRM: Leveraging ICT for Efficient Workforce Management

Ali Ateeq, AbdA-Aziz Al-refaei, Mohammed Alzoraiki, Marwan Milhem, and Shafeeq Ahmed Ali

Abstract The article analyzes the influence of digitization and ICT on HRM practices, with a particular emphasis on efficient staff management. Technology adoption has altered recruiting, communication, HR information systems, training, and performance management. Organizations may use digital technologies to expedite operations, increase employee engagement, improve data management, boost e-learning, and incorporate real-time feedback systems. However, issues like data security and worker adaptability must be addressed. To harness the advantages of ICT in HRM, HR professionals must stay current on evolving technologies.

Keywords Digitalization · ICT · HRM practices · Efficient workforce management

---

A. Ateeq (✉) · M. Alzoraiki · M. Milhem · S. A. Ali
Faculty of Administrative and Financial Sciences, Gulf University, Sanad, Kingdom of Bahrain
e-mail: dr.ali.ateeq@gulfuniversity.edu.bh

M. Milhem
e-mail: dr.marwan.milhem@gulfuniversity.edu.bh

S. A. Ali
e-mail: dr.shafeeq.ali@gulfuniversity.edu.bh

A.-A. Al-refaei
Research Management Centre, International Islamic University Malaysia (IIUM), 53100 Jalan Gombak, Malaysia
e-mail: ah_alrefaei@yahoo.com

A. Ateeq · A.-A. Al-refaei · M. Alzoraiki · M. Milhem · S. A. Ali
Administrative Science Department, College of Administrative and Financial Science, Gulf University, Sanad 26489, Kingdom of Bahrain

B. Awwad (ed.), *The AI Revolution: Driving Business Innovation and Research*, Studies in Systems, Decision and Control 524, https://doi.org/10.1007/978-3-031-54379-1_48

1 Introduction

Digitalization and the fast growth of Information and Communication Technology (ICT) have resulted in substantial changes in a variety of fields [1, 2]. Human Resource Management (HRM) is one area where technology improvements have had a significant influence [3]. The use of ICT technologies and digital solutions has transformed HRM practices, allowing firms to expedite operations [4]. Numerous studies have investigated the effects of digitalization and ICT adoption on HRM processes. For instance, Haque and Nishat [5] study on "Digital HRM" contrasts organizations that have adopted digital technologies with those that have not, shedding light on the benefits and challenges of technology adoption in HRM processes. This study illuminates how digitalization has altered recruitment, training, and performance management procedures.

According to Arifin et al. [6] study titled "The Role of Social Media in Recruitment," the digital revolution has had a significant impact on the recruitment landscape. This research investigates the efficacy of social media platforms in attracting and selecting candidates. This study investigates the impact of social media usage on job advertisements, employer branding, and candidate screening, casting light on the potential of social media to improve recruitment outcomes. Kaputa et al. [3] studies on "Digital Learning in the Workplace" indicated that digital learning has emerged as a significant tool for employee development. This study looks on the use and influence of e-learning platforms on employee skill development. It investigates the variables that influence e-learning adoption, the efficacy of e-learning programs, and the long-term link between e-learning use and employee performance. Furthermore, the integration of HR Information Systems (HRIS) has transformed HRM processes, as shown by Mahadik and Imed [7] research on "The Impact of HRIS on HRM Effectiveness." The influence of HRIS implementation on HRM effectiveness in small and big firms is compared in this study. It investigates the effects of HRIS deployment on HR procedures, data management, decision-making, and overall HRM results in various corporate situations. Finally, Thompson's "Digital Performance Management" case study dives into the use of real-time feedback systems in performance management. It looks on how digital performance management technologies affect employee performance, goal alignment, and continuous development. The research illuminates the obstacles and advantages of implementing real-time feedback systems in businesses. These studies provide us with important insights into the influence of digitalization and ICT adoption on HRM practices. Organizations may improve workforce management efficiency and achieve better HRM results by embracing technology in recruiting, learning and development, HRIS adoption, and performance management. Understanding these insights is critical for firms looking to traverse the digital terrain and maximize their human resource management processes in today's technology-driven world.

2 Literature Review

2.1 Impact of Digitalization and ICT on HRM Practices

The integration of digitalization and Information and Communication Technology (ICT) has dramatically altered numerous areas of Human Resource Management (HRM) practices [8, 9]. This section includes a thorough evaluation of significant research that have investigated the influence of digitalization and ICT adoption in HRM.

2.1.1 Recruitment and Talent Acquisition

Digitalization has transformed the recruiting market, allowing firms to attract and choose individuals by using digital platforms and technology more effectively [9, 10]. Holland et al. [11] underlines the advantages of technology use in recruiting procedures in their report on "Digital HRM." The research focuses on how digital solutions, such as human resource information systems and e-recruitment platforms, have expedited candidate sourcing, automated application screening, and enhanced overall recruiting results.

Makhija et al. [12] also investigates the influence of social media in recruiting and talent acquisition. The research demonstrates how businesses use social media channels for job postings, company branding, and applicant involvement. Organizations may tap into a larger talent pool, improve their employer brand image, and allow direct engagement with possible candidates by embracing social media.

2.1.2 Employee Engagement and Communication

Employee involvement and communication inside firms have been altered by digitalization and ICT. Internal communication channels have been transformed by collaboration platforms, instant messaging systems, and intranets [13–15]. Yu et al. [16] shows how technological tools promote seamless communication, information exchange, and collaboration across teams and departments in their report on "The Role of ICT in Enhancing Employee Engagement." This improved communication increases employee engagement and fosters a feeling of belonging inside the firm.

2.1.3 HR Information Systems (HRIS)

HR Information Systems (HRIS) have transformed HRM practices through improving data management and simplifying HR operations[17]. Boro and Sharma [18] compares small and big enterprises to investigate the influence of HRIS on HRM effectiveness. The research demonstrates how HRIS technologies consolidate employee data, automate payroll, and create real-time reporting. HR professionals have access to accurate and up-to-date information by adopting HRIS, resulting in more effective workforce planning, decision-making, and overall HRM results [19].

2.1.4 Training and Development

Through the introduction of e-learning platforms, digitalization has altered employee training and development programs [20]. Authors [21, 22] conducts a long-term research on "Digital Learning in the Workplace," focusing on the effect of e-learning adoption on employee skill development. The study investigates the usefulness of e-learning systems in supporting continuous learning, lowering training expenses, and providing workers with accessible learning opportunities. The research highlights the advantages of e-learning platforms in accommodating different learning styles and guaranteeing consistent delivery of training programs across geographically team members.

2.1.5 Performance Management

Through the adoption of real-time feedback systems, digitalization has transformed performance management procedures [23, 24]. Bankar and Shukla [25] provides a case study on "Digital Performance Management," demonstrating the advantages of real-time feedback techniques. The research investigates how digital performance management technologies allow firms to create objectives, measure progress, and offer workers with ongoing feedback. Organizations that use digital performance management systems encourage a culture of continuous improvement, increase goal alignment, and assist people in their development path [26–28].

Overall, the papers examined illustrate the revolutionary influence of digitalization and ICT adoption on human resource management techniques. The use of digital technologies in recruiting, employee engagement, HRIS, training and development, and performance management has resulted in higher efficiency, better decision-making, and better employee experiences. These results provide useful insights for firms looking to improve their human resource management processes and employ technology to promote organizational success.

In light of this, the following hypotheses are put forward:

Hypothesis 1: Digitalization and Information and Communication Technology (ICT) Positively Influences on HRM Practices

H1a: Digitalization and (ICT) Positively Influences Recruitment and Talent Acquisition.

H1b: Digitalization and (ICT) Positively Influences Employee Engagement and Communication.

H1c: Digitalization and (ICT) Positively Influences HR Information Systems (HRIS).

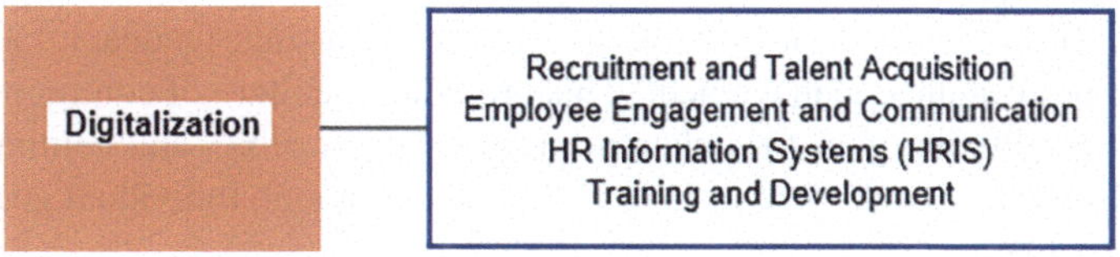

Fig. 1 Conceptual framework

H1d: Digitalization and (ICT) Positively Influences Training and Development.
H1e: Digitalization (ICT) Positively Influences Performance Management.

The study's hypothesized model is depicted in Fig. 1.

Method: In our methodology for this critical review paper, we sourced literature to discern the implications of digitalization and ICT on HRM practices [22, 28, 29]. Through a systematic analysis of the collected literature, we pinpointed salient trends in areas like recruitment, engagement, and HRIS utilization [14, 30–32]. This analytical process informed the evidence-based recommendations we subsequently proposed for enhancing HRM in a digital context [33].

3 Discussion and Conclusion

The discussion section focuses on summarizing and analyzing the key findings from the literature review, highlighting the implications and significance of digitalization and ICT adoption in HRM practices.

The studies reviewed demonstrate the transformative impact of digitalization and ICT adoption in various areas of HRM. The integration of digital tools in recruitment and talent acquisition processes has streamlined candidate sourcing, automated screening processes, and expanded the talent pool. The utilization of social media platforms for job advertisements and employer branding has enhanced candidate engagement and improved recruitment outcomes.

Digitalization and ICT have also transformed employee engagement and organizational communication. Adoption of collaboration platforms, instant messaging systems, and intranets has facilitated interdepartmental connectivity, knowledge sharing, and collaboration. This has resulted in increased employee engagement, enhanced internal communication channels, and a sense of organizational belonging.

HR Information Systems (HRIS) adoption has greatly enhanced HRM effectiveness. HRIS technologies streamline payroll operations and create real-time reporting by centralizing employee data. This allows HR practitioners to make data-driven choices, better workforce planning, and overall HRM results.

In addition, the advent of e-learning platforms has altered training and development programs. These platforms make learning more accessible and adaptable, lower training expenses, and cater to a variety of learning styles. Continuous learning is encouraged, so that workers may obtain new skills and information for their positions.

Digitalization has made it possible to integrate real-time feedback methods in the field of performance management. This promotes a culture of continuous development by facilitating goal formulation, progress monitoring, and continual feedback. Real-time feedback systems help with goal alignment, individual growth, and employee success.

Overall, the debate emphasizes the tremendous advantages and positive effect of HRM digitization and ICT adoption. Organizations that embrace digital transformation and use ICT solutions for recruiting, employee engagement, HRIS deployment, training and development, and performance management may increase efficiency, decision-making, and employee experiences.

Conclusion

Ultimately, digitization and ICT adoption have resulted in dramatic changes in HRM processes. The studies examined provide useful insights on the influence of digitalization and ICT adoption in areas including as recruiting, employee engagement, HRIS deployment, training and development, and performance management.

Digitalization has transformed recruiting procedures, allowing firms to attract and choose individuals more effectively. By enabling seamless connection and collaboration platforms, it has also enhanced employee engagement and communication. HRIS implementation has improved data management and decision-making in HRM, while e-learning platforms have transformed training and development activities. In performance management, real-time feedback techniques have promoted continuous improvement and goal alignment.

Organizations must embrace digital transformation and deliberately use ICT solutions to improve HRM procedures. They may then use the advantages of digitization to improve recruiting results, employee engagement, HR procedures, stimulate continuous learning, and drive performance improvement.

It is critical for enterprises to keep current on developing technology and trends as they traverse the digital world. Organizations that embrace digitalization and successfully use ICT may place themselves at the forefront of HRM practices, driving organizational performance and establishing a competitive edge in the current digital age.

Limitations and Recommendations for Future Research

Limitations

1. Findings may be restricted to certain sectors or organizational environments, lowering generalizability.
2. **Timeline**: Research may not reflect the most recent technology advances in digitalization and ICT usage.
3. **Methodological Variations**: Differences in study methodology make direct comparisons difficult.

Recommendations for Future Research

Longitudinal Research: Conduct longitudinal research to better understand the long-term consequences of digitalization and ICT adoption in human resource management.

Comparative Studies: Evaluate the advantages and difficulties of businesses with and without digitalization and ICT adoption.

Technological Innovations: Examine the influence of developing technology on human resource management techniques.

Employee views on technology adoption in HRM: Gather employee perspectives on technology adoption in HRM to better understand their experiences and attitudes.

Management of Organizational Change: Investigate techniques and best practices for managing organizational change throughout digital transformation.

Addressing these constraints and researching these topics will help us better understand the link between digitalization, ICT adoption, and HRM results.

References

1. Brodny, J., Tutak, M.: Analyzing the level of digitalization among the enterprises of the European Union member states and their impact on economic growth. J. Open Innov. Technol. Mark. Complex. **8**(2), 70 (2022)
2. Snasel, V., Platos, J., El-Qawasmeh, E.: In: Proceedings of International Conference on Digital Information Processing and Communications (ICDIPC 2011), Ostrava, Czech Republic, 7–9 July 2011, vol. 188. Springer Science & Business Media (2011)
3. Kaputa, V., Loučanová, E., Tejerina-Gaite, F.A.: Digital transformation in higher education institutions as a driver of social oriented innovations. In: Social Innovation in Higher Education, pp. 61–85 (2022)
4. Shen, Z., et al.: Digital transition and green growth in Chinese agriculture. Technol. Forecast. Soc. Chang. **181**, 121742 (2022)
5. Haque, M.A., Nishat, S.S.: The impact of HRM digitalization on employee performance in the RMG industry of Bangladesh. Eur. J. Bus. Manage. Res. **7**(4), 192–198 (2022)
6. Arifin, Z., Basundara, A.R., Wiyarni, W.: The Effect of Recruitment and Selection on Employee Productivity with Social Media as Intervening Variable (Case Study on the Clients of Professional Leadership Center (PROLC)
7. Mahadik, R.V., Imed, P.: Human Resource Information Systems: An Effective tool for Human Resource Functions
8. Tapiwa, G.G., et al.: Digitalisation of human resources systems and process necessary for public sector transformation in Zimbabwe. In: Transformational Human Resources Management in Zimbabwe: Solutions for the Public Sector in the 21st Century, pp. 61–74. Springer (2022)
9. Halid, H., Halim, S.N.A., Ravesangar, K.: Human resource management practices in the digital era. In: Technological Challenges: The Human Side of the Digital Age, pp. 109–158. Springer (2022)
10. Amor, A.M.: Reinventing human resources through digitalization. In: Managing Technology Integration for Human Resources in Industry 5.0, pp. 115–130. IGI Global (2023)
11. Holland, P., et al.: The Emerald Handbook of Work, Workplaces and Disruptive Issues in HRM. Emerald Group Publishing (2022)
12. Makhija, M., Pradhan, I., Atri, R.: Social media for talent acquisition: win-win situation for recruiters and jobseekers. J. Inf. Optim. Sci. **43**(6), 1335–1342 (2022)
13. Santoso, N.R., Sulistyaningtyas, I.D., Pratama, B.P.: Transformational leadership during the COVID-19 pandemic: strengthening employee engagement through internal communication. J. Commun. Inq. 01968599221095182 (2022)

14. Ali, A.A.A., Alzoraiki, M., Al Aghbas, N.S.M.: Working remotely, social isolation and access to support and communication during Covid-19 its influence on job performance. Int. J. Intellect. Human Resour. Manage. (IJIHRM) **3**(01), 23–33 (2022)
15. Ali, A.A.A., Yusoffii, S.H., Aliiii, K.: Fundamental of equal opportunities in employee's performance: a critical review. In: Proceedings of the 5th World Conference on Integration of Knowledge, Bangi, Malaysia (2019)
16. Yu, Y., et al.: Circular economy in the construction industry: a review of decision support tools based on Information & Communication Technologies. J. Clean. Prod. 131335 (2022)
17. Pomperada, J.R.: Human resource information system with machine learning integration. Qubahan Acad. J. **2**(2), 5–8 (2022)
18. Boro, M., Sharma, L.: A systematic review with bibliometric analysis on human resource information system (HRIS): implications for sustainable banking
19. Ngulugulu, M.M., Rwela, E.G., Mkwizu, N.Y.: Adoption of human resource information systems on public sectors in Tanzania. East Afr. J. Inform. Technol. **6**(1), 66–76 (2023)
20. Eton, M., Chance, R.: University e-learning methodologies and their financial implications: evidence from Uganda. Asian Assoc. Open Univ. J. (2022) (ahead-of-print)
21. Dooley, S.: The effectiveness of digital learning in the workplace as a form of training in an Irish context. National College of Ireland, Dublin (2022)
22. Ali, A.A.A., et al.: Moderating effect of strategic planning on the relationship between career path planning and job performance. Sustainability **15**(11), 8490 (2023)
23. Agrawal, P., Narain, R.: Analysis of enablers for the digitalization of supply chain using an interpretive structural modelling approach. Int. J. Product. Perform. Manag. **72**(2), 410–439 (2023)
24. Schoop, S.: Digitalization of corporate learning. In: Digital Management in Covid-19 Pandemic and Post-Pandemic Times: Proceedings of the International Scientific-Practical Conference (ISPC 2021). Springer (2023)
25. Bankar, S., Shukla, K.: Performance management and artificial intelligence: a futuristic conceptual framework. In: Contemporary Studies of Risks in Emerging Technology, Part B, pp. 341–360. Emerald Publishing Limited (2023)
26. Flug, J.A., et al.: Kaizen process improvement in radiology: primer for creating a culture of continuous quality improvement. Radiographics **42**(3), 919–928 (2022)
27. Atmaja, D.S., Zaroni, A.N., Yusuf, M.: Actualization of performance management models for the development of human resources quality, economic potential, and financial governance policy in Indonesia ministry of education. Multicult. Educ. **9**(01), 1–15 (2023)
28. Alzoraiki, M., et al.: Impact of teachers' commitment to the relationship between transformational leadership and sustainable teaching performance. Sustainability **15**(5), 4620 (2023)
29. Al-refaei, A.A.-A., et al.: An integrated mediating and moderating model to improve service quality through job involvement, job satisfaction, and organizational commitment. Sustainability **15**(10), 7978 (2023)
30. Ali, A.A.A., et al.: Association between Islamic work ethics and employee commitment among medical staff in the public health sector in Bahrain. Int. J. Intellect. Human Resour. Manage. (IJIHRM) (2023)
31. Ateeq, A.: Knowledge Management's Effect on Learning Quality in Bahrain's Private Universities (2023)
32. Alkadash, T.M., et al.: The effects of leadership style on employee sustainable behaviour: a theoretical perspective. In: Artificial Intelligence and Transforming Digital Marketing, pp. 205–213. Springer (2023)
33. Ateeq, A., Ebrahim, S., Al-Ghatam, R.: The impact of stress and its influencing factors among dentists during the COVID-19 pandemic in Kingdom of Bahrain. F1000Research **11**, 930 (2022)

Exploring the Implementation Efforts and Contributions of Green Building Rating Systems (GBRS) Among Stakeholders in Bahrain's Real Estate Sector: A Qualitative Study

Husam M. Abdulrahim, Ali Ateeq, Fay A. Al-Khalifa, Mohammed Alzoraiki, Marwan Milhem, and Sara Almeer

Abstract The implementation of Green Building Rating Systems (GBRS) is gaining popularity worldwide as a tool for promoting sustainable development in the building sector. Bahrain is interested in implementing GBRS in its real estate sector. This study explores the efforts and contributions made by all stakeholders in Bahrain to deploy GBRS, recognizes the relationship between the strategy and implementation of GBRS, and suggests corrective actions and future needs in the real estate sector. The data were collected through semi-structured, in-depth interviews with stakeholders in Bahrain's real estate sector. The data were analyzed using NVivo software, with a thematic analysis approach. The study's findings reveal that the implementation of GBRS faces several challenges, including limited public awareness and understanding of GBRS, high initial costs, and lack of incentives for developers to adopt GBRS. The study identifies developers, government entities, and end-users as key stakeholders in the implementation of GBRS in Bahrain's real estate sector. The study suggests that more collaboration and coordination among stakeholders are needed to develop clear strategies and policies for the implementation of GBRS. The study concludes that the implementation of GBRS in Bahrain's real estate sector requires more awareness-raising initiatives, incentives for developers, and clear strategies and policies to ensure consistency and effectiveness.

H. M. Abdulrahim · F. A. Al-Khalifa
College of Science, University of Bahrain, Zallaq, Bahrain

A. Ateeq (✉) · M. Alzoraiki · M. Milhem · S. Almeer
Administrative Science Department, College of Administrative and Financial Science, Gulf University, Sanad 26489, Kingdom of Bahrain
e-mail: dr.ali.ateeq@gulfuniversity.edu.bh

M. Milhem
e-mail: dr.marwan.milhem@gulfuniversity.edu.bh

S. Almeer
e-mail: sara.almeer@gulfuniversity.edu.bh

B. Awwad (ed.), *The AI Revolution: Driving Business Innovation and Research*, Studies in Systems, Decision and Control 524, https://doi.org/10.1007/978-3-031-54379-1_49

Keywords Green Building Rating Systems · Sustainable development · Real estate sector · Bahrain

1 Introduction

The main goal of this research is to examine the implementation and consequences of Green Building Rating Systems (GBRS) within Bahrain's real estate industry. These systems play a pivotal role in advancing sustainable development practices within the building sector. This study examines the roles and endeavours of various stakeholders in Bahrain with regards to the implementation of the Green Building Rating System (GBRS). The objective of this study is to comprehensively analyze the interplay between the strategic aspects of GBRS and their implementation in real-world scenarios, while also identifying specific areas that need improvement and potential future advancements within the sector.

To collect data, a series of semi-structured, in-depth interviews were carried out with pertinent stakeholders. The participants' replies were accurately transcribed verbatim and then analyzed using NVivo, a software tool designed for qualitative data analysis. The procedure included a rigorous examination and categorization of the interview transcripts, whereby codes were carefully assigned and then organized into separate groups based on their similarities. Subsequently, a comprehensive analysis was conducted on each discovered topic within its corresponding category.

In order to augment comprehension and facilitate the communication of the results, visual aids such as word clouds and maps were used. The use of visual representations proved to be effective in elucidating the fundamental themes and patterns that were evident in the data. This research offers a complete analysis of the present condition of Government-Brokered Real Estate Sales (GBRS) in Bahrain's real estate industry. It includes an examination of the problems encountered, advancements achieved, and suggestions for future initiatives. This report provides a comprehensive examination of Bahrain's progress towards achieving sustainable development in the construction sector, with a specific focus on the Green Building Rating System (GBRS).

2 Methodology

The study uses a qualitative research design, and the data were collected through semi-structured, in-depth interviews conducted with stakeholders in Bahrain's real estate sector [1–4]. The purposive sampling technique was used to select participants based on their level of expertise and knowledge of GBRS [4–9]. The participants were from private, government, and non-governmental organizations involved in the real estate sector. The interviews were conducted between 2020 and 2021, during which time the Covid-19 pandemic was ongoing, necessitating the use of online meetings with participants.

The data were analyzed using NVivo software. The data coding was done using a thematic analysis approach, where the data were searched for similar codes and categorized into themes [10–12]. The themes were then examined and analyzed to determine patterns and relationships between them.

3 Results

The results of the study reveal that stakeholders in Bahrain's real estate sector are aware of GBRS and its potential benefits. However, the implementation of GBRS faces several challenges, including limited public awareness and understanding of GBRS, high initial costs, and lack of incentives for developers to adopt GBRS. The study identifies developers, government entities, and end-users as key stakeholders in the implementation of GBRS in Bahrain's real estate sector.

The study also reveals that the strategy and implementation of GBRS are closely related. A lack of clear strategies and policies for the implementation of GBRS leads to a lack of consistency in its implementation across the real estate sector. The study suggests that there is a need for more collaboration and coordination among stakeholders in the real estate sector to develop clear strategies and policies for the implementation of GBRS.

4 Discussion

The study's findings highlight the importance of stakeholder engagement and collaboration in the implementation of GBRS in Bahrain's real estate sector. The study suggests that there is a need for more awareness-raising initiatives to promote the adoption of GBRS and incentives for developers to adopt GBRS. The study also suggests that clear strategies and policies for the implementation of GBRS are necessary to ensure consistency and effectiveness in its implementation.

5 Conclusion

This study has explored the drivers and risks associated with the implementation of Green Building Rating Systems (GBRS) in Bahrain's real estate sector. The drivers and risks are significant factors when considering how to close the gaps between strategy and implementation of GBRS in Bahrain. The main drivers behind GBRS in Bahrain, as outlined by stakeholders, include the government's interest in reducing environmental impact while fulfilling the needs for development, global pressure to focus on sustainability, and the potential for increased profits and cost-effectiveness in operation.

The current market trends show that the implementation of GB measures in Bahrain is growing rapidly, with new government policies promoting or mandating energy efficiency, indoor environmental quality, water efficiency, materials, and sustainable site management and planning. However, to accelerate the adoption of GBRS, stakeholders suggest that the government should provide incentives such as tax credit schemes, regulatory mechanisms, favorable banking and insurance products, low-interest rates, and low insurance rates that support sustainable development.

The study has identified that developers, owner-occupiers, and tenants are the three most important parties involved in the implementation of GB. Developers are motivated by the potential profit they can make from selling properties with eco-labeling, while owner-occupiers and tenants are interested in the low operating cost of GB, as well as the increased productivity, improved image, and reduced risks associated with operating the building.

Stakeholders consider cost–benefit and cost implications when making decisions. As long as they believe they will benefit from their investment, they will act in a socially responsible manner, in line with government aspirations. Therefore, presenting the results of multiple studies conducted on the efficacy of GBs will raise awareness and encourage adoption.

In conclusion, the drivers and risks associated with the development of GB need to be mitigated to accelerate its growth in Bahrain's real estate sector. The government should provide incentives to promote sustainable development, and stakeholders should be convinced that GBs will generate higher profits and tend to be more cost-effective to operate. The presentation of the findings from multiple studies on GB effectiveness will raise awareness and encourage adoption, contributing significantly to faster adoption of GBs in Bahrain.

The study concludes that stakeholders in Bahrain's real estate sector are aware of GBRS and its potential benefits. However, the implementation of GBRS faces several challenges, including limited public awareness and understanding of GBRS, high initial costs, and lack of incentives for developers to adopt GBRS. The study suggests that there is a need for more collaboration and coordination among stakeholders in the real estate sector to develop clear strategies and policies for the implementation of GBRS. Further research is needed to establish the effectiveness of GBRS in Bahrain's real estate sector and the potential benefits it can bring to the country's economy and sustainability.

References

1. Adnan, S., Jalala, A.B.U.: Net Zero Energy District NZED: A Strategy Towards Achieving Sustainable Urban Development in Qatar (2016, January)

2. Agency, U.S.E.P., Brownfields, O., Revitalization, L.: Environmental Benefits of Brownfields Redevelopment—A Nationwide Assessment (2020, May)
3. Ahmed, A., Almanea, A., Alsuliman, A.: A case study on Green Buildings and its applicability in Bahrain (2021). Ieeexplore.Ieee.Org. Retrieved from https://ieeexplore.ieee.org/abstract/document/9668090/
4. Ahmed, F.A.: Development of Green Building Rating System for Evaluating Existing Office Buildings in Dubai Based on Al Sa'fat Rating System. Fatima Ahmed Ahmed Diss. (2019, March)
5. Ajmi, A.N., El Montasser, G., Nguyen, D.K.: Carbon emissions—income relationships with structural breaks: the case of the Middle East and North African countries (2014)
6. Akadiri, P.O., Chinyio, E.A., Olomolaiye, P.O.: Design of a sustainable building: a conceptual framework for implementing sustainability in the building sector. Buildings **2**(2), 126–152 (2012). https://doi.org/10.3390/buildings2020126
7. Akdag, H.C., Beldek, T.: Waste management in green building operations using GSCM. Int. J. Supp. Chain Manag. **6**(3), 174–180 (2017)
8. Ali, A.A.A., et al.: Moderating effect of strategic planning on the relationship between career path planning and job performance. Sustainability **15**(11), 8490 (2023)
9. Al-refaei, A.A.-A., et al.: An integrated mediating and moderating model to improve service quality through job involvement, job satisfaction, and organizational commitment. Sustainability **15**(10), 7978 (2023)
10. Alzoraiki, M., et al.: Impact of teachers' commitment to the relationship between transformational leadership and sustainable teaching performance. Sustainability **15**(5), 4620 (2023)
11. Ali, A.A.A.: The Relationship Between Islamic Work Ethics and Employee Performance Among Medical Staff in the Public Health Sector in Bahrain: Employee Commitment as Mediating Variable. Universiti Sains Islam Malaysia (2022)
12. Ali, A.A.A., Alzoraiki, M., Al Aghbas, N.S.M.: Working remotely, social isolation and access to support and communication during Covid-19 its influence on job performance. Int. J. Intell. Human Resour. Manag. (IJIHRM) **3**(01), 23–33 (2022)
13. Ali, A.A.A., et al.: Association between Islamic work ethics and employee commitment among medical staff in the public health sector in Bahrain. Int. J. Intell. Human Resour. Manag. (IJIHRM) (2023)
14. Ateeq, A.: Knowledge Management's Effect on Learning Quality in Bahrain's Private Universities (2023)
15. Alkadash, T.M., et al.: The effects of leadership style on employee sustainable behaviour: a theoretical perspective. In: Artificial Intelligence and Transforming Digital Marketing, pp. 205–213. Springer (2023)
16. Barakat, B., et al.: Assessing the impact of green training on sustainable business advantage: exploring the mediating role of green supply chain practices. Sustainability **15**(19), 14144 (2023)
17. Ateeq, A.: The mediating role of engagement in the relationship between. Management **21**(3), 435–447 (2023)
18. Al Balushi, F.S.M., Ali, A.A.A., Alzoraiki, M.: The impact of forecasting strategy and training and development strategy on preparing leadership competencies administrative support in the sultanate of Oman. Int. J. Intell. Human Resour. Manag. (IJIHRM) **3**(01), 34–41 (2022)
19. Rahardjo, K.P., Kurniawati, K.: Caroll pyramid implementation of environmental responsibility programs of PT Pertamina EP Subang field, Tambun field, and Jatibarang field in organic agriculture. J. Soc. Res. **2**(9), 3331–3340 (2023)

Does Information Communication Technology Has Influence on Quality University Education on Gulf University: Literature Review

Ali Ateeq, Mohammed Alzoraiki, Marwan Milhem, Baligh Ali Hasan Beshr, and Sara Al Meer

Abstract Information Communication Technology (ICT) plays a vital role in shaping the quality of university education in Gulf universities. This literature review explores the influence of ICT on university education in the Gulf region, highlighting its benefits, challenges, and future trends. The benefits of ICT in university education include improved access to educational resources, enhanced communication and collaboration, personalized learning experiences, and increased student engagement. However, the adoption and implementation of ICT also present challenges such as infrastructure readiness, faculty training, digital divide, and data privacy concerns. Looking ahead, emerging trends in ICT offer opportunities to further enhance university education. These trends include mobile learning, artificial intelligence, virtual and augmented reality, open educational resources, and data privacy and ethics. To harness the benefits of ICT and address its challenges, Gulf universities must focus on developing robust infrastructure, investing in faculty professional development, ensuring equitable access to digital resources, and establishing clear policies and guidelines. Collaboration among universities, industry, and government entities is crucial in sharing best practices and resources.

Keywords Information Communication Technology · ICT · University education · Gulf universities · Benefits · Challenges · Future trends

A. Ateeq (✉) · M. Alzoraiki · M. Milhem · B. A. H. Beshr · S. Al Meer
College of Administrative and Financial Sciences, Gulf University, Building 1964, Road 4363, Sanad 743, Kingdom of Bahrain
e-mail: dr.ali.ateeq@gulfuniversity.edu.bh

M. Milhem
e-mail: dr.marwan.milhem@gulfuniversity.edu.bh

B. A. H. Beshr
e-mail: baligh.beshr@gulfuniversity.edu.bh

S. Al Meer
e-mail: sara.almeer@gulfuniversity.edu.bh

B. Awwad (ed.), *The AI Revolution: Driving Business Innovation and Research*, Studies in Systems, Decision and Control 524, https://doi.org/10.1007/978-3-031-54379-1_50

1 Introduction

Information Communication Technology (ICT) has emerged as a transformative force in the field of education, including university education [1]. The integration of ICT tools and platforms has revolutionized teaching and learning practices, offering new opportunities to enhance the quality of education. This literature review aims to explore and synthesize existing research on the impact of ICT on the quality of university education [2].

According to Serrano et al. [3], ICT has the potential to overcome traditional barriers to access educational resources. Digital technologies enable students to access vast amounts of information, scholarly articles, multimedia resources, and online libraries, thereby enriching their learning experience. Moreover, ICT provides opportunities for interactive and student-centered learning, transforming pedagogical practices [4]. Online learning management systems [1] and virtual classrooms facilitate collaborative learning and engagement among students [5]. Furthermore, ICT enables personalized learning and assessment, catering to individual student needs and preferences [6]. Adaptive learning technologies and intelligent tutoring systems have the potential to tailor educational content and instructional approaches, enhancing student engagement, motivation, and achievement [7, 8].

However, despite the potential benefits, it is important to address challenges associated with ICT integration in university education. Issues such as the digital divide, faculty training, and data privacy need to be carefully considered [9–11]. Ensuring equitable access to technology, providing adequate faculty development programs, and implementing robust privacy measures are crucial for maximizing the positive impact of ICT on the quality of university education [12].

In conclusion, this literature review will delve into the existing research to provide insights into the impact of ICT on the quality of university education. By examining various dimensions such as access to resources, pedagogical practices, student engagement, and personalized learning, this review aims to contribute to the understanding of the potential benefits and challenges associated with ICT integration in higher education.

1.1 Research Objectives (RO)

This conceptual review article aims to accomplish the following research objectives:

1. To examine the extent to which ICT enhances access to educational resources in Gulf universities.
2. To investigate the influence of ICT on pedagogical practices and teaching strategies in Gulf university classrooms.
3. To explore the impact of ICT on student engagement and collaboration in Gulf university education.

2 Literature Review

Introduction

Information Communication Technology (ICT) has become an integral part of education, revolutionizing teaching and learning practices in universities worldwide [13, 14]. This literature review aims to explore and synthesize existing research on the impact of ICT on the quality of university education. It examines the various dimensions through which ICT influences educational outcomes, including access to resources, pedagogical practices, student engagement, and learning outcomes [15].

Access to Educational Resources

Numerous studies have highlighted the positive impact of ICT on access to educational resources in university settings [16]. Digital platforms, online libraries, and repositories have made a vast amount of information readily available to students, regardless of time and geographical constraints [17]. The ability to access diverse learning materials, such as e-books, scholarly articles, and multimedia resources, has enriched the learning experience, promoted independent research, and deepened students' understanding of complex topics [18].

Pedagogical Practices and Teaching Strategies: The integration of ICT tools in university education has transformed pedagogical practices and teaching strategies. Online learning management systems [1] provide opportunities for interactive and student-centered learning [19]. Virtual classrooms, multimedia presentations, and simulation software enable instructors to create engaging and interactive learning environments [20, 21]. Studies have shown that these innovative pedagogical approaches enhance student motivation, critical thinking skills, and overall learning outcomes [22].

Student Engagement and Collaboration

ICT has revolutionized student engagement and collaboration in university education. Online discussion forums, social media platforms, and collaborative learning spaces have facilitated peer interaction, knowledge sharing, and collaborative problem-solving [23]. The use of virtual group projects, wikis, and shared documents, students can actively participate in collaborative activities, fostering a sense of community and promoting deeper learning experiences [24].

Personalized Learning and Assessment

ICT has opened new possibilities for personalized learning and assessment in university education. Adaptive learning technologies and intelligent tutoring systems can tailor educational content and instructional approaches to individual student needs [25]. Online assessment tools, such as quizzes and simulations, provide immediate feedback, allowing students to track their progress and make informed decisions about their learning strategies [26]. Research has indicated that personalized learning and assessment strategies positively impact student engagement, achievement, and motivation [26].

Challenges and Considerations

While the benefits of ICT in university education are significant, several challenges and considerations need to be addressed [27]. Digital divide issues, such as unequal access to technology and reliable internet connectivity, can hinder the effective integration of ICT tools [16, 28, 29]. Faculty training and professional development programs are crucial to support instructors in adopting innovative teaching methods and effectively utilizing ICT tools [30]. Privacy and security concerns regarding student data and online learning platforms necessitate robust policies and safeguards [31].

Conclusion: The literature review demonstrates that Information Communication Technology has a substantial impact on the quality of university education. Through improved access to resources, innovative pedagogical practices, enhanced student engagement, personalized learning and assessment, ICT has the potential to transform traditional educational models. However, addressing challenges related to access, faculty development, and data privacy is crucial to fully harness the benefits of ICT in higher education. Further research is needed to explore emerging technologies and best practices for effectively integrating ICT into university teaching and learning processes.

3 Benefits of ICT in University Education

Enhanced Access to Educational Resources: ICT provides students and educators with access to a wide range of digital resources, including online libraries, research databases, e-books, and scholarly journals. This enables students to explore diverse learning materials and stay updated with the latest developments in their fields of study [32].

Improved Pedagogical Practices

ICT tools and platforms facilitate innovative and interactive teaching methods, allowing instructors to employ multimedia presentations, simulations, and virtual laboratories. These resources can enhance students' understanding and engagement, making learning more dynamic and effective [26].

Personalized Learning Experiences

ICT enables personalized learning approaches tailored to individual student needs. Adaptive learning technologies can assess students' strengths and weaknesses and provide customized learning pathways, fostering self-paced learning and personalized feedback [25].

Collaborative Learning and Communication

ICT platforms offer opportunities for collaborative learning, enabling students to engage in online discussions, group projects, and peer feedback. This fosters teamwork, communication skills, and the ability to work effectively in virtual teams [33].

4 Challenges of ICT in University Education

While Information Communication Technology (ICT) offers numerous benefits for university education, its integration also presents certain challenges that need to be addressed. This section will discuss some of the key challenges associated with the use of ICT in Gulf universities.

1. Digital Divide: One of the main challenges is the existence of a digital divide among students, which refers to the unequal access to technology and internet connectivity [34]. Research by Zhao et al. [35] emphasizes that disparities in technology access can create inequities in students' ability to fully participate in digital learning activities and access online resources.
2. Faculty Training and Support: Integrating ICT effectively in teaching practices requires faculty members to acquire new skills and competencies [36]. Li and Yu [2, 37] highlight the importance of providing adequate training and ongoing support for faculty to enhance their digital literacy and pedagogical skills in using ICT tools for teaching.
3. Infrastructure and Technical Issues: Implementing ICT in university settings relies on a robust technological infrastructure. However, challenges related to insufficient infrastructure, limited internet bandwidth, and technical issues can hinder the smooth integration of ICT. Stanojevic [38] stress the need for universities to invest in reliable network infrastructure and address technical problems promptly.
4. Privacy and Security Concerns: The use of ICT involves the collection and storage of sensitive student data. Protecting student privacy and ensuring data security are critical challenges in the digital age. Mishra et al. [39] emphasize the importance of implementing stringent privacy measures and cybersecurity protocols to safeguard student information.

These challenges require proactive measures and strategic planning to ensure the effective implementation of ICT in university education. Collaboration between university administrators, faculty, IT departments, and policymakers is crucial in addressing these challenges and establishing a conducive environment for the successful integration of ICT in Gulf universities.

Method

This literature study used a systematic approach to source selection, with an emphasis on current and applicable studies on ICT's effects on higher education in the Gulf university. Data was extracted and then analyzed critically for source quality before being divided into advantages, difficulties, and future trends. A brief narrative described the study's main conclusions and offered some suggestions for moving forward with ICT integration at institutions, private companies, and the government.

Future Trends and Directions

As Information Communication Technology (ICT) continues to evolve, it is important to consider the future trends and directions that will shape its impact on quality university education in Gulf universities. This section explores some potential future developments in ICT and their implications for higher education.

1. Mobile Learning: The widespread use of smartphones and mobile devices opens up new possibilities for mobile learning. With the increasing availability of high-speed internet and mobile applications, students can access educational resources anytime and anywhere [40]. Mobile learning offers flexibility and personalized learning experiences, enabling students to engage with course materials at their own pace [41].
2. Artificial Intelligence (AI) and Learning Analytics: AI technologies and learning analytics have the potential to transform the way universities analyze and utilize data to enhance teaching and learning processes. AI-powered chatbots can provide personalized support to students, while learning analytics can track student progress and provide insights for instructional improvements [1].
3. Virtual and Augmented Reality: Virtual and augmented reality technologies offer immersive and interactive learning experiences. These technologies can simulate real-world scenarios, allowing students to practice skills and engage in experiential learning. Virtual and augmented reality can be particularly beneficial for fields that require hands-on training, such as medicine or engineering.
4. Open Educational Resources: The availability of open educational resources, such as open textbooks, online courses, and educational videos, has the potential to reduce costs and increase access to quality educational materials. OER promotes collaboration and knowledge sharing among educators and learners, fostering a culture of open education [42].
5. Data Privacy and Ethics: As universities increasingly rely on digital technologies and collect vast amounts of student data, ensuring data privacy and addressing ethical considerations become paramount.

   Universities must establish clear policies and guidelines for data collection, storage, and usage, while also considering the ethical implications of using student data for research or commercial purposes [43].

These future trends and directions highlight the continuous evolution and potential of ICT in shaping the future of university education. Gulf universities should stay abreast of these developments and proactively explore ways to integrate emerging technologies and pedagogical approaches to enhance teaching and learning outcomes.

5 Conclusions

In conclusion, Information Communication Technology (ICT) has a significant influence on the quality of university education in Gulf universities. Through its various applications and tools, ICT has brought numerous benefits and addressed several challenges in the educational landscape.

The literature review has highlighted the benefits of ICT in university education, including improved access to educational resources, enhanced communication and collaboration, personalized learning experiences, and increased student engagement. The integration of ICT in teaching and learning practices has the potential to foster critical thinking, creativity, and problem-solving skills among students.

However, the adoption and effective implementation of ICT in Gulf universities also come with several challenges. These challenges include issues related to infrastructure and technological readiness, faculty training and support, digital divide among students, and the need for clear policies and guidelines regarding data privacy and ethical considerations. Looking towards the future, there are several emerging trends and directions in ICT that can further enhance the quality of university education in the Gulf region. These trends include the utilization of mobile learning, the integration of artificial intelligence and learning analytics, the exploration of virtual and augmented reality technologies, the promotion of open educational resources, and the emphasis on data privacy and ethics.

To leverage the benefits of ICT and overcome its challenges, Gulf universities must prioritize the development of robust ICT infrastructure, invest in faculty professional development programs, ensure equitable access to digital resources, and establish clear policies and guidelines for responsible use of technology. Additionally, fostering collaboration and partnerships among universities, industry, and government entities can facilitate the sharing of best practices and resources in implementing ICT in university education.

In conclusion, ICT has the potential to transform and enhance the quality of university education in Gulf universities. By embracing emerging technologies, addressing challenges, and promoting a culture of innovation and collaboration, Gulf universities can create a dynamic and engaging learning environment that prepares students for the demands of the digital age.

Note: The conclusions provided above summarize the main findings and recommendations based on the literature reviewed. They should be tailored and expanded upon based on the specific research conducted and the depth of analysis in the study.

References

1. Abdulai, A.R.: Causes and effects of job stress on employees performance: a case study of banks in the tamale metropolis (2022)
2. Fluck, A.E.: Integration or transformation: a cross-national study of information and communication technology in school education. University of Tasmania (2003)
3. Serrano, D.R., et al.: Technology-enhanced learning in higher education: how to enhance student engagement through blended learning. Eur. J. Educ. **54**(2), 273–286 (2019)
4. Khalil, M.K., Abdel Meguid, E.M., Elkhider, I.A.: Teaching of anatomical sciences: a blended learning approach. Clin. Anat. **31**(3), 323–329 (2018)
5. Muhisn, Z., et al.: The impact of socialization on collaborative learning method in e-Learning Management System (eLMS). Int. J. Emerg. Technol. Learn. (iJET) **14**(20), 137–148 (2019)
6. Bhutoria, A.: Personalized education and artificial intelligence in United States, China, and India: a systematic review using a human-in-the-loop model. Comput. Educ. Artif. Intell. 100068 (2022)
7. Alam, A.: Employing adaptive learning and intelligent tutoring robots for virtual classrooms and smart campuses: reforming education in the age of artificial intelligence. In: Advanced Computing and Intelligent Technologies: Proceedings of ICACIT 2022, pp. 395–406. Springer (2022)
8. Barua, P.D., et al.: Artificial intelligence enabled personalised assistive tools to enhance education of children with neurodevelopmental disorders—a review. Int. J. Environ. Res. Public Health **19**(3), 1192 (2022)
9. Akour, M., Alenezi, M.: Higher education future in the era of digital transformation. Educ. Sci. **12**(11), 784 (2022)
10. Alzoraiki, M., et al.: Impact of teachers' commitment to the relationship between transformational leadership and sustainable teaching performance. Sustainability **15**(5), 4620 (2023)
11. Al-refaei, A.A.-A., et al.: An integrated mediating and moderating model to improve service quality through job involvement, job satisfaction, and organizational commitment. Sustainability **15**(10), 7978 (2023)
12. AlAjmi, M.K.: The impact of digital leadership on teachers' technology integration during the COVID-19 pandemic in Kuwait. Int. J. Educ. Res. **112**, 101928 (2022)
13. Snasel, V., Platos, J., El-Qawasmeh, E.: In: Proceedings of International Conference on Digital Information Processing and Communications (ICDIPC 2011), Ostrava, Czech Republic, 7–9 July 2011, vol. 188. Springer Science & Business Media (2011)
14. Aljameel, I.H.: Computer-assisted language learning in Saudi Arabia: past, present, and future. Int. Educ. Stud. **15**(4), 95–107 (2022)
15. Zhan, X., et al.: Investigating students' engagement in mobile technology-supported science learning through video-based classroom observation. J. Sci. Educ. Technol. **31**(4), 514–527 (2022)
16. Ben Youssef, A., Dahmani, M., Ragni, L.: ICT use, digital skills and students' academic performance: exploring the digital divide. Information **13**(3), 129 (2022)
17. Christopoulos, A., et al.: The effects of augmented reality-supported instruction in tertiary-level medical education. Br. J. Edu. Technol. **53**(2), 307–325 (2022)
18. Avinash, M.: Enhancing literary appreciation and critical thinking: exploring the potential of digital initiatives in teaching Paulo Coelho's novels
19. Kelton, K.P.: Instructor-generated interactions and course outcomes in online history courses. Grand Canyon University (2022)
20. Hedrick, E., et al.: Teaching & learning in virtual reality: metaverse classroom exploration. In: 2022 Intermountain Engineering, Technology and Computing (IETC). IEEE (2022)
21. Ali, A.A.A., Alzoraiki, M., Al Aghbas, N.S.M.: Working remotely, social isolation and access to support and communication during covid-19 its influence on job performance. Int. J. Intellect. Human Resour. Manage. (IJIHRM) **3**(01), 23–33 (2022)

22. Evendi, E., et al.: Assessing students' critical thinking skills viewed from cognitive style: study on implementation of problem-based e-learning model in mathematics courses. Eurasia J. Math. Sci. Technol. Educ. **18**(7) (2022)
23. Koay, K.Y., Poon, W.C.: Understanding students' cyberslacking behaviour in e-learning environments: is student engagement the key? Int. J. Human–Comput. Interact. 1–16 (2022)
24. Lin, Y.-L., Wang, W.-T.: Enhancing students' online collaborative PBL learning performance in the context of coauthoring-based technologies: a case of wiki technologies. Educ. Inform. Technol. 1–26 (2023)
25. Kem, D.: Personalised and adaptive learning: emerging learning platforms in the era of digital and smart learning. Int. J. Soc. Sci. Human Res. **5**(2), 385–391 (2022)
26. DeCoito, I., Estaiteyeh, M.: Online teaching during the COVID-19 pandemic: exploring science/STEM teachers' curriculum and assessment practices in Canada. Disc. Interdisc. Sci. Educ. Res. **4**(3), 8 (2022)
27. Ferede, B., et al.: Determinants of instructors' educational ICT use in Ethiopian higher education. Educ. Inform. Technol. 1–20 (2022)
28. Guo, C., Wan, B.: The digital divide in online learning in China during the COVID-19 pandemic. Technol. Soc. **71**, 102122 (2022)
29. Alfarras, M.: Early detection of adult valve disease-mitral stenosis using the Elman artificial neural network. Int. J. Comput. Sci. Netw. Sec. (IJCSNS) **13**(11), 101 (2013)
30. Aidoo, B., et al.: Transforming teaching with ICT using the flipped classroom approach: dealing with COVID-19 pandemic. Educ. Sci. **12**(6), 421 (2022)
31. Alam, A.: Platform utilising blockchain technology for elearning and online education for open sharing of academic proficiency and progress records. In: Smart Data Intelligence: Proceedings of ICSMDI 2022, pp. 307–320. Springer (2022)
32. Yakubu, H., Abdullah, N., Kaur, K.: Sustainability of digital collections for Nigerian academic libraries: an exploration of conception, indicators for fulfillment and accrued benefits. Malays. J. Libr. Inf. Sci. **27**(3), 73–91 (2022)
33. Ndubuisi, A., et al.: Developing global competence in global virtual team projects: a qualitative exploration of engineering students' experiences. J. Stud. Int. Educ. **26**(2), 259–278 (2022)
34. Mathrani, A., Sarvesh, T., Umer, R.: Digital divide framework: online learning in developing countries during the COVID-19 lockdown. Glob. Soc. Educ. **20**(5), 625–640 (2022)
35. Zhao, L., et al.: Determinants of the digital outcome divide in E-learning between rural and urban students: empirical evidence from the COVID-19 pandemic based on capital theory. Comput. Hum. Behav. **130**, 107177 (2022)
36. Andreasen, J.K., et al.: Professional digital competence in initial teacher education: an examination of differences in two cohorts of pre-service teachers. Nordic J. Dig. Literacy (3), 61–74 (2022)
37. Li, M., Yu, Z.: Teachers' satisfaction, role, and digital literacy during the COVID-19 pandemic. Sustainability **14**(3), 1121 (2022)
38. Stanojevic, C.: Attitudes of practitioners who provide services to individuals with autism spectrum disorder towards use of socially assistive robots. Indiana University (2022)
39. Mishra, A., et al.: Cybersecurity enterprises policies: a comparative study. Sensors **22**(2), 538 (2022)
40. Amsyar, I., Latifah, H., Wuisan, D.: Design of mobile learning content with a mobile-based live multimedia system. Int. J. Cyber IT Serv. Manage. **2**(2), 146–153 (2022)
41. Gumbheer, C.P., Khedo, K.K., Bungaleea, A.: Personalized and adaptive context-aware mobile learning: review, challenges and future directions. Educ. Inf. Technol. **27**(6), 7491–7517 (2022)
42. Purnama, S., et al.: The effect of digitalization on culinary MSMEs on increasing sales turnover during covid 19 pandemic. Aptisi Trans. Technopreneurship (ATT) **4**(3), 58–67 (2022)
43. Buchanan, E.A., Hvizdak, E.E.: Online survey tools: ethical and methodological concerns of human research ethics committees. J. Empir. Res. Hum. Res. Ethics **4**(2), 37–48 (2009)
44. Almusaed, A., et al.: Enhancing student engagement: harnessing "AIED"'s power in hybrid education—a review analysis. Educ. Sci. **13**(7), 632 (2023)

45. Ateeq, A.: Knowledge management's effect on learning quality in Bahrain's private universities (2023)
46. Islam, M.T., Tamzid, P.M.: Artificial intelligence in human resource management

Assessing Operational and Administrative Challenges Confronting Budget Planning: Analysing Al Marai Case Study

Marwan Mohamed Abdeldayem, Saeed Hameed Aldulaimi, Abdulhameed Baqi, Mohammed Al Skekhly, and Ahmed Kh. Muttar

Abstract This study analyzes the administrative and operational obstacles associated with budget planning, specifically focusing on the case study of Al Marai Saudi Company. The study highlights the inflexibility of budgeting and the potential negative outcomes that arise from overly strict budgets that struggle to adapt to unpredictable shifts in business strategies. The significance of consistent budget monitoring in response to ever-evolving market conditions is emphasized, as well as the need for key corporate decision-makers to possess specialized budgeting expertise. The analysis explores the concept of "budgetary slack" and its implications regarding data asymmetry and ambiguity. However, the study also emphasizes the importance of governance in effective budgeting and maximizing the utility of budgeting practices within the strategic planning framework. With this objective in mind, the study primarily examines various aspects of the Al Marai master budget, utilizing it as an illustrative case. The budgeting aspects studied include production, expenses, sales, and annual financial statements. Therefore, the study advocates for the significance of

M. M. Abdeldayem · S. H. Aldulaimi (✉)
Collage of Administrative Sciences, Applied Science University (ASU), Eker, Kingdom of Bahrain
e-mail: Saeed.aldulaimi@asu.edu.bh

M. M. Abdeldayem
e-mail: Marwan.abdeldayem@asu.edu.bh

A. Baqi
Directorate of Marketing and Public Relations, Applied Science University, Eker, Kingdom of Bahrain
e-mail: abdulhameed.baqi@asu.edu.bh

M. Al Skekhly
College of Engineering, Gulf University (GU), Sanad, Kingdom of Bahrain
e-mail: dr.mohammed.alshekhly@gulfuniversity.edu.bh

A. Kh. Muttar
Department of Business Administration, College of Administration and Economics, University of Anbar, Ramadi, Iraq
e-mail: ahmed.khalaf.muttar@uoanbar.edu.iq

B. Awwad (ed.), *The AI Revolution: Driving Business Innovation and Research*, Studies in Systems, Decision and Control 524, https://doi.org/10.1007/978-3-031-54379-1_51

accurate cost analysis, stakeholder engagement, budget flexibility, and a shift towards a more strategic approach to budgeting. These measures are essential for enhancing the resilience of an organization through informed financial decision-making.

Keywords Master budgets · Budget flexibility · Financial forecasting · Budgetary slack · Corporate strategy

1 Introduction

Historically, budgeting has played a prominent role in management control procedures within corporations [1]. Despite objections and criticisms, most organizations still consider budgeting as an integral part of their financial control systems [2]. However, budgeting has faced significant challenges over the years [3]. These concerns are related to how budgeting is conducted and the roles it is expected to fulfill [4]. One of the key issues with budgeting is its lack of flexibility. It is nearly impossible to accurately forecast every cost, expense, or expenditure in the corporate strategy. Many organizations struggle with setting overly tight budgets, which can lead to imbalances when unexpected changes occur. Consequently, strict budget limitations often result in negative outcomes that adversely affect overall performance in the long run [5, 6].

2 Literature Review

2.1 The Significance of Budgeting and Interconnectedness of the Master Budget

Responsiveness to transformations has become a critical success factor for today's corporations [7]. It is important to recognize that market conditions are constantly evolving, and many organizations are hesitant to regularly scrutinize their budgets. In some cases, the market itself can undergo significant changes, prompting business decision-makers to make radical adjustments to their expense strategies. However, a budget should be capable of adapting to all events, whether anticipated or unanticipated; otherwise, it can lead to a serious breakdown [8]. Corporate decision-makers and leaders must carefully review financial analysis procedures, ensuring their legal validity and accuracy. Monthly monitoring of expenditures is crucial in financial management, and it should be done with a clear understanding of how the budgetary cycle is linked to the overall strategy, as well as how budget review meetings and operations are conducted [9]. The budgeting process presents various challenges for an increasing number of businesses and key decision-makers [10]. Crafting a budget is a complex and multifaceted practice. Gathering financial data can be challenging, and budget planning requires interpretation skills to allocate resources and make forecasts that are often overwhelming. Decision-makers may lack the practical and technical financial competencies needed, which can hinder their ability to effectively fulfill budgeting obligations [11]. In some cases, decision-makers may set the organization's budget based on "budgetary slack," deliberately underestimating expected earnings or overstating costs to achieve favorable evaluations. This practice makes it easier for them to meet their targets consistently. Data asymmetry also plays a role in the relationship between budgetary assistance and budgetary slack, making it less suitable for effective planning. However, good governance is associated with higher data accuracy and analysis [12–15].

Gordey [16] emphasized the time-consuming nature of drafting and implementing strategic plans, necessitating optimization of all budgeting aspects. However, the formulation of a budget plan may involve multiple stakeholders, especially in an environment that requires consideration of numerous financial factors. This can make it challenging for decision-makers to allocate sufficient time or delay other obligations to thoroughly analyze future-oriented plans in conjunction with their existing responsibilities [17].

3 Methodology and Procedures

The purpose of this study is to investigate the renewable energy sector, utilizing a case study methodology. The case study approach allows for a comprehensive examination of a specific context, providing detailed insights into the research problem [18]. The selection of the case study will be based on predetermined criteria that align with the research objectives and the renewable energy industry. A renewable energy company with an established sustainability reporting framework and a demonstrated commitment to stakeholder engagement will be selected as the primary case for this study. The selection criteria may consider factors such as company size, reputation, industry leadership, and documented sustainability practices. Secondary data will be collected from various sources, including corporate sustainability reports, annual reports, company websites, industry publications, and academic literature. These sources will provide background information on the company's sustainability practices, reporting framework, and the broader industry context. Almarai is a joint-stock company operating in the production and distribution of agricultural, livestock, and food products in the Kingdom of Saudi Arabia and Gulf countries. Since 2009, Almarai has had a significant partnership with PepsiCo, acquiring several of the company's food manufacturing projects in the region. Almarai's manufacturing operations began in 1977, with its main facility located in Riyadh. Starting with 350 cows, the company has expanded its operations over the past three decades, increasing its herd to 170,000 cows, making it the world's largest producer of milk and dairy products. In 2005, Almarai underwent a transformation from a private company to a publicly traded company, with shares held by over 125,000 individuals.

4 Discussion and Analysis

In line with the aforementioned example, this sheds light on the budget plan of Almarai Master Company [19, 20]. Figure 1 and Table 1 present a fictitious master budget of Almarai, showcasing a comprehensive financial planning record that integrates all the financial and operating sub-budgets. Senior management and key decision-makers at Almarai can utilize the master budget for various important objectives, including expansion strategy, financial performance assessment, and competitive analysis.

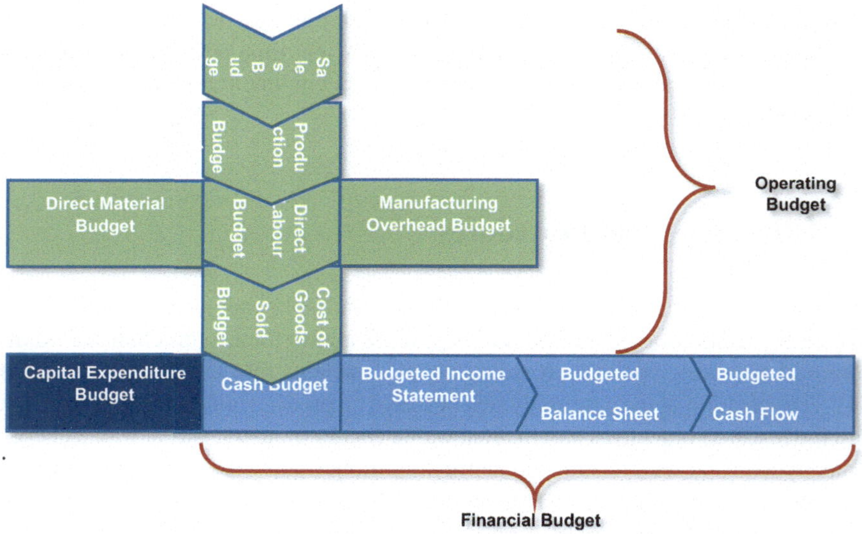

Fig. 1 Almarai master budget components

Table 1 Master budget for Almarai Company for year ended Dec 31, 2019

Master Budget for Almarai Company Dec 31, 2019					
Operating budgets					
	Q1	Q2	Q3	Q4	Total
Sales Budget for year Ended Dec 31, 2019					
Budgeted units to be sold	479,155	538,203	510,465	522,359	2,050,182
Assumed average sales price per unit	SAR 7	SAR 7	SAR 7	SAR 7	SAR 7
Total sales	3,354,083	3,767,424	3,573,257	3,656,513	14,351,277
Note 1 Assumed that Almarai desires to have an ending inventory each quarter equal to 47% of the next quarter's sales: • First quarter: Second quarte's sale 538,203 units × 47% = 252,955.41 desired ending inventories • Second quarter: Third quarte's sale 510,465 units × 47% = 245,508.6 desired ending inventories • Third quarter: Fourth quarte's sale 522,359 units × 47% = 241,207 desired ending inventories • Fourth quarter: First quarter's sales (2020) 53,674.2 × 47% = 241,207					
Production Budget for the Ended Dec 31, 2019					
Budgeted units to be sold	479,155	538,203	510,465	522,359	2,050,182
plus: Assumed desired units in ending inventory Note 1	252,955.41	239,918.55	245,508.73	241,207	241,207
Total units needed	732,110.41	778,121.55	755,973.73	763,566	2,291,389
Units in the beginning inventory	225,203	252,955	239,918	245,508	225,203
Budgeted units to be produced	506,907.41	525,166.55	516,055.73	518,058	2,066,186
Note 2 Assumed that Almarai desires to have an ending balance for the fourth quarter of SAR 94,585 which is calculated as follow: • First quarter: Second quarte's production 525,166 units x SAR 1.4 × 30% = 216,743 desired ending inventories • Second quarter: Third quarte's production 239,918 units x SAR 1.4 × 30% = 217,584 desired ending inventories • Third quarter: Fourth quarte's production 245,508 units x SAR 1.4 × 30% = 867,798 desired ending inventories • Fourth quarter: First quarter's production (2020) 225,203 × SAR 1.4 × 30% = 94,585 desired ending inventories					

(continued)

Table 1 (continued)

Master Budget for Almarai Company Dec 31, 2019					
Operating budgets					
	Q1	Q2	Q3	Q4	Total
Direct Material Budget for year Ended Dec 31, 2019					
Budgeted units to be produced	506,907.41	525,166.55	516,055.73	518,058	2,066,186
Assumed direct material cost per unit Note 2	1.43	1.43	1.43	1.43	1.43
Plus: Direct material needed for production	724,877.5963	750,988.167	737,959.694	740,822.94	2,954,645.98
Assumed desired direct materials in ending inventory Note 2	225,296.45	221,387.908	222,246.882	886,393.794	94,585.26
Total direct material needed	950,174.0463	972,376.075	960,206.576	1,627,216.734	3,049,231.24
Less: Direct material in beginning inventory	3,036,534	693,219.12	681,192.6	683,836.6	3,036,534
Budgeted purchases of direct materials	−2,086,359.954	279,156.955	279,013.976	943,380.134	12,697.24
Direct Labour Budget for year Ended Dec 31, 2019					
Budgeted Units to be produced	506,907	525,166	516,055	518,058	2,066,186
Assumed direct labour hours per unit	0.2	0.2	0.2	0.2	0.2
Direct labour hours needed for production	101,381.4	105,033.2	103,211	103,611.6	413,237.2

(continued)

Table 1 (continued)

Master Budget for Almarai Company Dec 31, 2019					
Operating budgets					
	Q1	Q2	Q3	Q4	Total
Assumed direct labour cost per hour	1.2	1.2	1.2	1.2	1.2
Budgeted direct labour cost	121,657.68	126,039.84	123,853.2	124,333.92	495,884.64
Manufacturing Overhead Budget for year Ended Dec 31,2019					
Budgeted units to be produced	506,907	525,166	516,055	518,058	2,066,186
VOH* cost per unit	1.2	1.2	1.2	1.2	1.2
Budgeted VOH	608,288.4	630,199.2	619,266	621,669.6	2,479,423.2
Budgeted FOH	270,415	270,415	270,415	270,415	1,081,660
Budgeted manufacturing overhead costs	878,703.4	900,614.2	889,681	892,084.6	3,561,083.2
Direct labour hours (DLHr)	1,013,814	1,050,332	1,032,110	1,036,116	4,132,372
Predetermined overhead allocation rate (3,561,083.2/ 4,132,372)					0.861752814
Note 3 The total projected manufacturing cost of each unit in 2019 is calculated as follows: • Direct materials cost per unit = 1.43 • Plus, direct labour cost per unit (0.20DLHr per unit x SAR 1.2) = 0.24 • Plus, Manufacturing overhead cost per unit (2DHLr per unit x SAR 0.86 per DLHr) = 1.72 Total projected manufacturing cost per unit for 2019: 1.43 + 0.24 + 1.72 = SAR 3,39					
Cost of goods sold for the Ended Dec 31, 2019					
Beginning inventory	3,874,193	3,874,193	3,874,193	3,874,193	3,874,193
Units produced and sold in 2019 @ SAR 3.39 Each (479,155, 538,203, 510,465, 522,359 units per quarter) Note 3	1,624,335.45	1,824,508.17	1,730,476.35	1,770,797.01	6,950,116.98

(continued)

Table 1 (continued)

Master Budget for Almarai Company Dec 31, 2019					
Operating budgets					
	Q1	Q2	Q3	Q4	Total
Total budgeted cost of goods sold	5,498,528.45	5,698,701.17	5,604,669.35	5,644,990.01	10,824,309.98
Selling and Administrative Expense for year Ended Dec 31,2019					
Selling expenses	609,729	585,913	594,103	586,583	2,376,328
General and Administration Expenses	94,745	99,826	79,516	126,186	400,273
Total budgeted S&A expenses	704,474	685,739	673,619	712,769	2,776,601
Financial Budgets					
Cash Budget for the year Ended Dec 31 ,2019					
Beginning cash balance	1,182,902	1,571,309.6	1,397,008.27	2,210,295.36	1,182,902
Cash receipts	3,354,083	3,767,424	3,573,257	3,656,513	14,351,277
Cash available	4,536,985	5,338,733.6	4,970,265.27	5,866,808.36	15,534,179
Cash payments					
Purchase of direct materials	12,697.24	938,640.726	930,342.912	2,047,156.52	2,084,458.56
Direct labour	1,216,576.8	1,260,398.4	123,832	1,243,339.2	4,958,846
Manufacturing overhead	878,703.4	900,614.2	889,681	892,084.6	3,561,083.2
Selling expenses	609,729	585,913	594,103	586,583	2,376,328
general and administration expenses	94,745	99,826	79,516	126,186	400,273
Zakat and income Tax	22,943	20,189	11,291	17,003	71,426
Interest expenses	130,281	136,144	131,204	153,110	550,739
Total cash payments	2,965,675.44	3,941,725.33	2,759,969.91	5,065,462.32	14,003,153.76
Ending cash balance before financing	1,571,309.56	1,397,008.27	2,210,295.36	801,346.04	1,531,025.24

(continued)

Table 1 (continued)

Master Budget for Almarai Company Dec 31, 2019					
Operating budgets					
	Q1	Q2	Q3	Q4	Total
Budgeted Income Statement for year Ended Dec 31,2019					
Sales revenue					14,351,277
Cost of goods sold					10,824,309.98
Gross profit					3,526,967.02
Selling and Administrative Expense					2,776,601
Operating income					750,366.02
Interest expenses					550,739
Income before Taxes					199,627.02
Zakat and income Tax					71,426
Net Income					128,201.02
Note 4 • Account Receivable collected in 1st Qtr. 2020 • Finished goods inventory: Production budget (241,207 units x Cost of goods sold budget 3.39) • Account Payable collected in 1st Qtr. 2020					
Budgeted Balance Sheet for year Ended Dec 31,2019					
Assets					
Current assets					
	Cash			1,531,025.24	
	Account Receivable	**Note 4:**		2,516,041	
	Raw Materials Inventory	**Note 4:**		2,954,645.98	
	Finished Goods Inventory	Note 4:		817,691.73	
	Total current Assets				7,819,403.95
Property, plant and equipment	Equipment			22,606,542	
	Less: Depreciation			1,638,626	

(continued)

Table 1 (continued)

Master Budget for Almarai Company Dec 31, 2019					
Operating budgets					
	Q1	Q2	Q3	Q4	Total
					20,967,916
Total assets					28,787,320
Liabilities					
Current liabilities					
	Account Payable	Note 4:			2,937,503
	Income tax Payable				31,578
	Derivative Financial Instruments				9166
None-current liabilities					
					12,164,269
Stockholders' equity					
Common stock				10,000,000	
Retain earnings				3,644,804	
	Total Stockholders' Equity				13,644,804
Total liabilities and stockholders' equity					28,787,320
Budgeted Cash flows for the year Ended Dec 31,2019					
Operating activities	Cash receipts from customers			14,351,277	
	Cash payments for operating expenses*			– 13,380,988.8	
	Cash payments for Zakat and income Tax			– 71,426	
	Cash payments for Interest expenses			– 550,739	
Net cash provided by operating activities					348,123.2

(continued)

Table 1 (continued)

Master Budget for Almarai Company Dec 31, 2019					
Operating budgets					
	Q1	Q2	Q3	Q4	Total
Investing activities	Appreciation of Biological Assets			– 153,167	
Net cash used for investing activities					– 153,167
Financing activities	Proceeds from Loans and Borrowings			– 816,983	
Net cash used in financing activities					– 816,983
Net increase in cash					– 622,026.8
Cash balance, 1 Jan 2019					1,182,902
Cash balance, 31 Dec 2019					560,875.2

5 Conclusion

The budget planning process is influenced by various factors beyond corporate monitoring procedures and systems. Corporate decision-makers must ensure that cost assessments are realistic, allowing flexibility for unforeseen or sudden deviations from key performance indicators (KPIs) to accommodate expenditures and incorporate input from all key stakeholders. As emphasized by [21], corporations can learn from events or challenges and utilize their experiences to enhance strategic planning through simulations of different budgeting scenarios, leading to a more financially resilient foundation.

There is considerable interest within the financial community to improve and streamline the arduous process of budget formulation [22]. Decision-makers can employ systematic and planned approaches to assess effectiveness and enhance forecast accuracy [22]. By evaluating budget performance, decision-makers can identify areas where targets were not achieved and take steps to rectify them, highlighting the importance of strategic methods and decisions.

It is important to acknowledge the potential limitations of the case study approach. As a single-case design, the findings of this study may not be generalizable to the entire renewable energy sector. However, the purpose of this research is to provide an in-depth exploration of the selected case, offering valuable insights into the stakeholder-centric approach in shaping sustainability reporting within this specific context.

References

1. Otley, D.: Management control in contemporary organizations: towards a wider framework. Manag. Account. Res. **5**(3–4), 289–299 (1994). https://doi.org/10.1006/mare.1994.1018
2. Libby, T., Lindsay, R.M.: Beyond budgeting or budgeting reconsidered? A survey of North-American budgeting practice. Manag. Account. Res. **21**(1), 56–75 (2010). https://doi.org/10.1016/j.mar.2009.10.003
3. Hansen, S.C., Otley, D.T., Van Der Stede, W.A.: Practice developments in budgeting: an overview and research perspective. SSRN Electron. J. (2003). https://doi.org/10.2139/ssrn.410544
4. Lindsay, T.L.A.R.M.: Beyond budgeting or budgeting reconsidered? A survey of North-American budgeting practice. Manage. Account. Res. Mar **21**(1), 56–75 (2010)
5. Abdeldayem, M., Aldulaimi, S.: Investment decisions determinants in the GCC cryptocurrency market: A behavioural finance perspective. Int. J. Organ. Anal. (2023)
6. Huiying Hou, D.C.H.J.K.C.A.K.K.W.L.: Tight budget constraints and the absence of strategic planning: an exploratory analysis of FM service outsourcing. Facil. Anal. FM Serv. Outsourcing 3 May, **34**(7/8), 380–393 (2016)
7. Christian Homburg, M.G.A.M.K.: Responsiveness to customers and competitors: the role of affective and cognitive organizational systems. J. Mark. **71**(3), 18–38 (2007)
8. Popovici, D.M.: Execution, Internal Monitoring, Budget Valuation and Financial Control—Elements of Dministration of Legal Persons Without Patrimonial Purpose. Timisoara, Tibiscus University in Timisoara, Faculty of Economics (2012)
9. Mathew, D., Brintha, N.C., Jappes, J.T.W.: Artificial intelligence powered automation for industry 4.0. In: Nayyar, A., Naved, M., Rameshwar, R. (eds) NewHorizons for Industry 4.0 in Modern Business. Contributions to Environmental Sciences & Innovative Business Technology. Springer, Cham (2023). https://doi.org/10.1007/978-3-031-20443-2_1
10. Lautour, J.D.: Strategic Management Accounting, Volume I [internet resource]: Aligning Strategy, Operations and Finance. 1 ed. Springer International Publishing, Cham, Switzerland (2018)
11. Nasution, M.D.T.P., Sari, P.B., Aspan, H., Rossanty, Y.: Unpacking the potential of crowdsourcing via social media to foster new product development among small and medium-sized enterprises. In: Rafiki, A., Dana, LP., Nasution, M.D.T.P. (eds) Open Innovation in Small Business. Contributions to Environmental Sciences & Innovative Business Technology. Springer, Singapore (2023). https://doi.org/10.1007/978-981-99-5142-0_9
12. Abdeldayem, M.M., Al Dulaimi, S.H.: The dynamics of crowdfunding campaigns in the middle East: Does social capital matter? Zbornik radova Ekonomskog fakulteta u Rijeci: časopis za ekonomsku teoriju i praksu **40**(1), 63–78 (2022)
13. Abdeldayem, M.M., Al Dulaimi, S.H.: Predicting crowdfunding economic success in the Gulf Cooperation Council. Int. J. Eng. Bus. Manag. **14**, 1–12 (2022b)
14. Donal Byard, Y.L.A.J.W.: Corporate governance and the quality of financial analysts' information. J. Acc. Pub. Policy **25**(5), 609–625 (2006)
15. Muhamed, N.A., Kamaruddin, M.I.H., Nazarie, W.N.F.W.M., Kamarubahrin, A.F., Radzi, R.M.: Shariah governance framework for islamic donation-based crowdfunding platforms in

Malaysia. In: Rafiki, A., Dana, LP., Nasution, M.D.T.P. (eds) Open Innovation in Small Business. Contributions to Environmental Sciences and Innovative Business Technology. Springer, Singapore (2023). https://doi.org/10.1007/978-981-99-5142-0_18
16. Gordey, O.D.: Budgeting as the basis of strategic planning in enterprises of small and medium-sized businesses in Ukraine. J. Zhytomyr State Technological Univ. **1**(1), 190–196 (2019)
17. Muttar, A.K., Alansari, S., Aldulaimi, S., Abdeldayem, M., AlKubaisi, M.: New Paradigm of behavioral finance in islamic banks: the role of leadership to facilitate creative behavior. In: International Conference on Sustainable Islamic Business and Finance, pp. 1–7, (2021). https://doi.org/10.1109/IEEECONF53626.2021.9686314
18. Feagin, J.R., Orum, A.M., Sjoberg, G. (Eds.).: A Case for the Case Study. UNC Press Books (2016)
19. Abdeldayem, M., Aldulaimi, S.: Entrepreneurial finance and crowdfunding in the Middle East. Int. J. Organ. Anal. **31**(4), 927–944 (2023)
20. Almarai (2021). www.almarai.com. [Online] Available at: https://www.almarai.com/en/corporate/investors/overview/ [Accessed 10 Jan 2021]
21. Craig, S.M., T.H.A.A.H.: Fiscal Responses to COVID-19: evidence from local governments and nonprofits. Public Adm. Rev. **80**(4), 644–650 (2020)
22. Andy Neely, M.B, C.A.: Better budgeting or beyond budgeting?. Measuring Bus. Excellence **7**(3), 22–28 (2003)

Fintech Disruption in Traditional Financial Services: Analyzing the Impact of Fintech Startups on Traditional Banking and Financial Institutions

Divya, Ashish Mathur, Abhilasha Mathur, and Vandana Gupta

Abstract This study investigates how traditional banking and financial institutions are affected by fintech innovation. Fintech firms have become important participants in the financial services sector, upending established paradigms and altering consumer expectations. The paper examines the evolving fintech disruption landscape and identifies key areas where traditional institutions are experiencing disruption. The study highlights the opportunities and obstacles for traditional financial institutions in accepting fintech disruption through a thorough examination of the literature. Additionally, it offers tactics that these organisations can use to deal with and benefit from the changes brought about by fintech businesses. The research also emphasises the importance of regulatory considerations in the fintech sector in order to foster innovation and protect consumer interests. The results of this study have significant ramifications for researchers, policymakers, and industry stakeholders. They offer insightful information about the developing fintech ecosystem and give traditional financial institutions the knowledge and tactics they need to survive disruption.

Keywords Fintech · Dispruption · Technology · Finance

1 Introduction

The financial sector is hardly an exception to how quickly technology has revolutionised other businesses. Fintech companies' development has upended incumbent banking and financial institutions, altering their long-standing procedures and

Divya · A. Mathur (✉)
Department of Management Studies, Central University of Haryana, Mahendragarh, Haryana, India
e-mail: drashish@cuh.ac.in

A. Mathur · V. Gupta
Faculty of Commerce and Management Studies, L M College of Science and Technology, Jodhpur, India

B. Awwad (ed.), *The AI Revolution: Driving Business Innovation and Research*, Studies in Systems, Decision and Control 524, https://doi.org/10.1007/978-3-031-54379-1_52

changing the competitive landscape. Fintech companies use cutting-edge technologies and business strategies to deliver financial services that are frequently more effective, client-focused, and available than those offered by conventional institutions. As a result, they are now very popular with investors and customers, posing a serious challenge to the established financial industry [13].

This paper's goal is to examine how fintech companies affect established financial and banking organisations. Researcher seek to understand the issues and opportunities that result from the emergence of fintech in the financial sector by looking at the various facets of this disruption. The changes in consumer behaviour, company models, and regulatory frameworks will be clarified by this examination.

In this paper, Researcher will first give an overview of traditional financial services and fintech, stressing their main distinctions. The disruptive nature of fintech firms and how they are changing the financial sector will then be discussed. The competitive environment, evolving client expectations, and effects on the business models of traditional banking and financial institutions will all be examined in this process. Researcher will also explore the regulatory issues surrounding the disruption caused by fintech and the function of regulators in guiding the sector [19].

Researchers hope to provide light on the relationships between fintech startups and established financial institutions through this research. Policymakers, business leaders, and investors can manage this quickly changing landscape by having a thorough grasp of the potential and difficulties posed by the fintech upheaval. In the end, our analysis will add to the ongoing discussion about the evolution of financial services and the role of fintech.

The framework of this paper is meant to conduct an exhaustive study of how fintech companies affect conventional banking and finance. The study's aims and significance are explained in the beginning. The literature review dives into the differentiation between fintech and traditional financial services, investigates the shifting environment of fintech disruption, and focuses on important areas of disruption. The study also covers methods for adapting to the disruption posed by fintech and the opportunity it presents for conventional financial institutions. The implications of regulations on the fintech industry are discussed. The final section of the paper summarises the results and makes recommendations for further research.

2 Research Objectives

1. To examine the extent of the impact of fintech startups on traditional banking and financial institutions.
2. To identify strategies for traditional financial institutions to effectively navigate and embrace fintech disruption.

3 Research Significance

Understanding how fintech disruption affects traditional banking and financial institutions is the key takeaway from this study. This report can help industry stakeholders, politicians, and scholars by assessing the potential and difficulties presented by fintech businesses. The findings can help traditional financial institutions understand the approaches required to succeed in the rapidly changing fintech sector, ultimately boosting their competitiveness and capacity to satisfy clients' shifting expectations. Additionally, by shedding light on the regulatory considerations required to promote innovation and consumer protection in the fintech sector, this research advances our knowledge of the transformational implications of technology on traditional sectors.

4 Research Methodology

The study employed a comprehensive literature review and secondary data analysis as the methodology. The research involved an extensive review of existing academic literature, industry reports, and publicly available data sources related to fintech disruption and its impact on traditional financial services. Through the systematic analysis of secondary data, this study aimed to derive insights into the key trends, challenges, opportunities, and strategies in the context of fintech's influence on traditional financial institutions.

5 Review of Literature

5.1 Understanding Fintech and Traditional Financial Services

Understanding both traditional financial services and fintech is crucial for analysing how fintech disruption will affect traditional banking and financial institutions.

Fintech describes the provision of financial goods and services through the use of technology and innovation. Digital payments, lending platforms, robo-advisory services, crowdfunding platforms, and other financial solutions are all provided by fintech businesses by using technological breakthroughs in fields like mobile applications, artificial intelligence, blockchain, and data analytics. These startups frequently operate with flexibility, agility, and a focus on the consumer, providing improved user experiences, streamlined procedures, and affordable solutions [12].

Traditional financial services, on the other hand, include the long-standing procedures and products provided by well-known banks, insurance companies, asset management organisations, and other financial institutions. These institutions have been conducting business in accordance with established regulatory frameworks and accepted business practises. They frequently offer a wide range of services, including loans, financial advice, insurance plans, savings and checking accounts, and wealth management [18].

Fintech startups have emerged as disruptive forces, upending the industry's long-standing dominance of traditional financial services. Fintech businesses have used technology to address issues in the financial sector such delayed innovation, high costs, complicated processes, and limited access to financial services. They have launched ground-breaking solutions with the goal of enhancing consumer convenience, transparency, speed, and accessibility [3].

Traditional financial institutions must adjust and react to the shifting environment as the fintech sector continues to expand and change. Developing their own digital platforms and offers, establishing partnerships or collaborations with fintech firms, and embracing technology improvements to boost their competitiveness are some examples of how they might do this [15].

5.2 Fintech Disruption: Exploring the Changing Landscape

The landscape of traditional financial services has been significantly disrupted by the rise of financial technology (fintech) firms. This section explores the shifting landscape shaped by fintech disruption and identifies the primary forces behind this change.

According to Kumari and Devi, the financial sector now has access to creative solutions thanks to the technology sector's rapid progress. Innovative business models are developed by fintech businesses using cutting-edge technology like blockchain, big data analytics, artificial intelligence, and cloud computing. In order to challenge the accepted standards of the traditional financial sector, these firms frequently work with lean structures, agile techniques, and a heavy focus on user experience [17].

Gomber et al. stated that the growing customer desire for convenient and individualised financial services is one of the major forces behind the disruption of the fintech industry. By providing user-friendly smartphone applications, digital wallets, and online platforms that offer seamless and accessible financial solutions, fintech businesses have tapped into this desire. These solutions make it easier and more convenient for people to manage their accounts, make payments, get loans, and invest [13].

The need for financial inclusion is yet another force behind fintech disruption. In order to give the unbanked and underbanked people access to financial services, fintech businesses have focused on underserved communities. These firms have been able to get past traditional banking's obstacles including physical branch infrastructure and onerous documentation requirements by utilising technology.

5.3 Impact of Fintech Startups on Traditional Banking and Financial Institutions

Traditional banking and financial institutions have been profoundly influenced by the advent of fintech firms, which have disrupted long-standing business models and changed how financial services are provided. The primary areas where fintech startups have significantly impacted the conventional financial industry are examined in this section [6].

1. Customer Experience and Convenience: By providing user-friendly interfaces, individualised services, and 24/7 accessibility, fintech firms have revolutionised the customer experience.
2. Digital Payments and Transfers: The development of electronic payments and money transfers has been significantly aided by fintech businesses. Now that peer-to-peer payment systems, mobile wallets, and blockchain-based solutions have been developed, clients can conduct quick, secure, and inexpensive domestic and international transactions.
3. Lending and Financing: By introducing cutting-edge lending models like peer-to-peer lending, crowdfunding, and alternative credit scoring, fintech businesses have rocked the financial industry.
4. Wealth Management and Investments: By providing automated investing platforms, robo-advisors, and online trading platforms, fintech businesses have democratised wealth management and investment.
5. Regulatory Compliance and Risk Management: Fintech firms have caused the financial sector to reevaluate its regulatory frameworks and risk management procedures.

5.4 Review of Literature on Fintech Disruption

The studies listed below emphasise the numerous facets of the disruption caused by fintech and what it means for conventional banking and financial institutions.

John Smith's literature review provides a comprehensive examination of the main themes and trends surrounding fintech disruption. The study explores the driving forces behind fintech innovation and investigates its impact on established financial institutions. Smith also highlights the challenges associated with regulating fintech. The central focus of the report is on the need for established institutions to adapt and collaborate with fintech startups in order to remain relevant in an ever-changing environment.

The study conducted by Bouri et al. delves into the relationship between fintech disruption and the financial health of traditional banks. By analyzing data from European banks, the researchers uncover a correlation between the increasing activity in the fintech sector and the decline in profitability and market value of traditional banks. The study emphasizes the criticality for banks to adopt fintech advances to maintain competitiveness in the industry.

Barberis and Turcotte's study investigates the influence of fintech on the digitalization of financial services. It specifically focuses on sectors such as lending, wealth management, and digital payments. The researchers underscore the necessity for established banks to make technology investments and adopt customer-centric practices to mitigate the risks posed by fintech firms. The study highlights the need for traditional banks to embrace technological advancements to stay competitive.

Jiang, Shan, and Wang's research explores the relationship between fintech disruption and financial stability in China. The study examines the impact of fintech platforms on the resilience of conventional banks and uncovers conflicting findings. While fintech disruption may enhance financial inclusion, it also introduces risks to the stability of the financial system. The study emphasizes the importance of effective regulation and risk management in controlling the potential disruptions caused by fintech.

Lenz and Heinrichs analyze the effects of fintech disruption on the structure and competition within financial markets. Their study examines the emergence of new firms, changes in market rivalry dynamics, and the potential advantages and hazards for consumers. The research places a strong emphasis on the role of regulators in ensuring fair competition, fostering innovation, and maintaining market stability amidst the disruption caused by fintech.

Agarwal and Mittal [1] seek to investigate how fintech is reshaping the financial services sector. They investigate the effects of fintech on numerous facets of the market, including banking, lending, payments, and customer experience, through a literature study. The authors draw the conclusion that fintech has the ability to revolutionise the sector and place emphasis on the necessity for incumbent institutions to change in order to stay competitive.

A thorough assessment of the literature is conducted by Carvalho [8] to investigate the phenomena of fintech disruption in the financial services industry. The study's main objectives are to present a summary of the literature, highlight significant topics, theories, and methodology used in analysing fintech disruption, and the paper continues by outlining potential topics for further investigation and the significance of continuing to watch the fintech sector.

Demeester and Fornasier [10] examine how fintech affects conventional banks with the goal of evaluating the adjustments and difficulties that these institutions have to deal with as a result of the development of fintech. The chapter's conclusion emphasises the need for traditional banks to adopt strategic solutions in order to maintain their competitiveness in the face of the changing fintech landscape.

Halaj and Kok use simulated networks to analyse interbank contagion in their study. According to their research, network-based simulations may successfully capture the dynamics of contagion and offer insightful information about the risks and vulnerabilities present in the interbank market. The results emphasise the significance of interconnection and the possibility of shock amplifying, emphasising the necessity for strong risk management practises and regulatory frameworks to improve financial stability.

From the viewpoint of fintech entrepreneurs, Heinonen et al. [14] offer a framework for disruptive innovation in financial services. According to the report, a number of elements are necessary for successful fintech disruption, including client centricity, technology-driven solutions, agility, and relationships with current financial institutions.

The effects of fintech disruption on the structure and competition of the financial markets are examined by Lenz and Heinrichs. Their study comes to the conclusion that advances in fintech have the potential to improve market performance, lower expenses, and widen access to financial services. However, they also draw attention to the difficulties associated with preserving financial stability, cybersecurity, and regulatory flexibility.

The emergence of fintech and its effects on conventional financial intermediaries are examined by Li and Ling [20]. According to their research, fintech innovations are potentially replacing or complementing the function of established intermediaries. According to the research, fintech can increase productivity, lessen information asymmetry, and boost financial inclusion. However, conventional intermediaries continue to be important in fields like regulatory compliance, relationship banking, and risk assessment.

O'Hara [24] looks at the motivations behind financial innovation and fintech in depth. The study comes to the conclusion that the main forces driving fintech innovation are changes in consumer behaviour, legislative developments, and technology advancements. The study emphasises how important it is for market actors and policymakers to comprehend how the environment is changing and how to adjust to it.

Poposki and Poposka [26] look into how fintech firms are affecting the traditional banking sector. According to the survey, fintech companies have disrupted traditional banking by providing cutting-edge goods and services, enhancing consumer satisfaction, and boosting competition. According to the findings, traditional banks must change to stay competitive in the changing market by embracing fintech collaborations, expanding digital capabilities, and rethinking their value proposition.

In his 2016 book, Schueffel examines the economic consequences of distributed ledgers, particularly blockchain technology. The study offers a thorough research plan for examining the economic implications of blockchain technology. The study stresses the importance of interdisciplinary research to comprehend the economic, legal, and social ramifications of blockchain technology and its potential to transform a number of industries.

The relationship between fintech, regulatory arbitrage, and systemic risks is examined by Tang et al. [28]. According to the study's findings, fintech innovations have the potential to take advantage of regulatory loopholes and engage in regulatory arbitrage, which could endanger the financial system's stability. The study emphasises how crucial it is to keep an eye on and manage any risks related to fintech activity in order to maintain financial stability.

In their article from 2018, Ante and Schüffel give a summary of the current research on banking and fintech. The research provides an overview of the major advancements, problems, and prospects in the fintech industry. Peer-to-peer lending, robo-advisors, and blockchain technology are discussed, with an emphasis on how they affect the traditional banking industry. To fully comprehend the effects of fintech on the banking system, the study highlights the need for additional study and collaboration between academia, business, and policymakers.

A thorough analysis of the relationship between fintech and inclusive growth is done by Beck et al. [5]. The study finds that fintech, especially in marginalised communities and developing nations, has the potential to enhance financial inclusion, loan availability, and economic growth. The paper recommends a research agenda to delve deeper into the ways that fintech might support inclusive growth and the possible hazards involved.

Using data from Lending Club, Cocco et al. [9] present evidence on fintech lending to families. The study comes to the conclusion that platforms using fintech lending, like Lending Club, have increased credit availability for underprivileged borrowers and decreased borrowing prices. According to the report, fintech lending can enhance conventional banking avenues and promote financial inclusion.

Du and Rachinsky [11] look at the relationship between bank liquidity and fintech credit. The study comes to the conclusion that the expansion of fintech credit platforms may diminish the liquidity of conventional banks. According to this, fintech credit platforms act as borrowers' secondary funding sources, potentially causing a reallocation of liquidity away from banks. The study highlights the requirement for banks to innovate and adapt in order to maintain their competitiveness and guarantee enough liquidity in the financial system.

Li et al. [21] look into the connection between systemic risk, bank lending, and fintech. The study comes to the conclusion that fintech lending platforms can affect systemic risk in both positive and negative ways. The study emphasises the importance of strong risk management, oversight from the government, and monitoring to reduce any potential systemic concerns related to fintech lending.

5.5 Key Areas of Fintech Disruption

1. Payments and Remittances: Fintech startups have launched cutting-edge payment solutions, such as peer-to-peer payment platforms and mobile wallets, challenging established payment and remittance infrastructures.
2. Lending and Financing: By offering alternative lending models including peer-to-peer lending, crowdfunding, and online lending marketplaces, fintech platforms have revolutionised lending and financing and challenged the dominance of traditional banks in this industry.
3. Wealth Management and Robo-Advisory: To challenge established wealth management companies, fintech companies have created automated investing platforms and robo-advisory services.
4. Digital Banking and Smartphone Apps: By providing convenient access to financial services and user-friendly smartphone apps, fintech firms have changed

the banking experience and put traditional brick-and-mortar banking branches to the test.

5. Insurance Technology (Insurtech): Insurtech businesses are using technology to expedite insurance processes, provide personalised plans, and improve customer experience, further disrupting the insurance market.
6. Blockchain and Cryptocurrencies: The rise of blockchain technology and cryptocurrencies has upended established financial systems, allowing for safe and decentralised transactions as well as the development of novel applications like smart contracts and virtual currencies.
7. Regtech and Compliance: The growth of regulatory technology (regtech) solutions, which automate compliance procedures and improve regulatory efficiency and transparency, has also had an impact on regulatory compliance.
8. Financial Inclusion: By giving underserved people, such as the unbanked and underbanked, access to financial services through digital and mobile-based solutions, fintech has the potential to improve financial inclusion.
9. Open Banking and API Integration: The disruption of the financial services industry has cleared the way for open banking, which permits the sharing of financial data through application programming interfaces (APIs), facilitating more cooperation between fintech startups and established financial institutions.
10. Cybersecurity and Fraud Protection: With a greater reliance on digital financial services, the fintech disruption has sparked improvements in fraud prevention and cybersecurity measures to safeguard client information and financial activities.

These focal points show the profound influence of fintech firms on numerous facets of conventional financial services, spurring the sector's innovation, effectiveness, and customer-centricity [25].

5.6 Challenges and Opportunities for Traditional Financial Institutions

1. Competition from Fintech Firms: Fintech firms that provide practical and customer-focused solutions are a fierce competitor to traditional financial institutions.
2. Digital Transformation: Traditional financial institutions must undergo digital transformation as a result of the growth of fintech.
3. Customer Expectations: The disruption brought about by fintech has increased customers' demands for seamless and individualised financial services.
4. Regulatory Compliance: Traditional financial institutions have difficulty navigating the changing regulatory landscape since fintech innovation frequently outpaces regulatory frameworks.
5. Partnership Opportunities: Traditional financial institutions can view fintech startups as potential partners rather than just competitors.

6. Enhanced Product Offerings: The disruption caused by fintech offers traditional institutions the chance to broaden their product lines and look into alternative revenue streams.
7. Data and Analytics: To gain insights and provide individualised experiences, fintech significantly relies on data analytics.
8. Trust and Security: Customers have a history of trusting traditional financial institutions, who also have a solid reputation.
9. Talent Acquisition and Retention: The disruption caused by fintech calls for a new set of abilities.
10. Customer Education: Traditional institutions may help educate clients about the advantages, dangers, and responsible use of fintech services as these solutions continue to develop.

Traditional financial institutions can survive and thrive in the quickly changing financial world by navigating these difficulties and seizing the opportunities given by fintech disruption [22].

5.7 Strategies for Traditional Financial Institutions to Embrace Fintech Disruption

1. Encourage an Innovation Culture: Develop an organisational culture that values innovation, promotes experimentation, and rewards original thought.
2. Work with Fintech companies: Promote alliances and partnerships with fintech companies to take advantage of their technological know-how and creative solutions.
3. Develop Digital Skills: Spend money on establishing strong digital platforms and updating your IT infrastructure to give your customers seamless, user-friendly experiences.
4. Adopt Open Banking: Adopt the idea of open banking by securely exchanging client information and working with other providers.
5. Build Strong Data Analytics Capabilities: To gain meaningful insights from consumer data, build strong data analytics capabilities.
6. Invest in cybersecurity: With a greater reliance on digital platforms, cybersecurity measures should be given top priority to safeguard customer information and uphold credibility.
7. Consumer-Centric Approach: Put the wants and preferences of the consumer at the forefront of corporate strategy and build products and services accordingly.
8. Agile Regulatory Compliance: Keep abreast of regulatory developments and maintain a tight working relationship with authorities to ensure compliance and promote innovation.
9. Upskill the Workforce: Invest in the workforce's education and training to provide them the abilities and information they need to successfully navigate the changing fintech landscape.

10. Continuous Monitoring and Adaptation: Keep an eye out for changing consumer expectations, new technology, and market trends.

Traditional financial institutions may embrace fintech disruption, stay competitive, and take advantage of the opportunities provided by technological breakthroughs by putting these methods into practise [23].

5.8 Regulatory Considerations in the Fintech Landscape

1. Regulatory Compliance: Businesses in the fintech sector, both startups and established financial institutions, are required to adhere to local regulations and legislation.
2. Regulatory Sandboxes: Some regulatory bodies have put in place regulatory sandboxes that permit fintech companies to test their creative solutions in a controlled setting.
3. Licencing and Authorization: In order to conduct business in the financial services industry, fintech firms frequently need to acquire licences or authorizations.
4. Consumer Protection: Fintech companies must prioritise protecting consumers and ensuring honest and open business practises.
5. Cross-Border Regulations: Due to varied legal frameworks in various jurisdictions, fintech companies operating internationally face significant obstacles.
6. Regulatory Innovation: To keep up with the rapidly changing fintech industry, regulatory authorities are increasingly looking into novel strategies.
7. Regulatory Cooperation: In order to build a supportive regulatory environment, cooperation between regulators, industry associations, and fintech companies is crucial.
8. Regulatory Agility: Because the fintech industry is continually changing, regulatory frameworks must be flexible and nimble.
9. Compliance Monitoring and Supervision: Regulatory bodies must set up efficient systems for keeping an eye on and overseeing fintech operations.
10. International Cooperation: Given the worldwide reach of fintech, regulatory agencies must work together internationally.

The fintech sector is significantly shaped by regulatory considerations, which also provide consumer protection, uphold financial stability, and promote innovation. To establish credibility, achieve compliance with regulations, and realise the full potential of their creative solutions, fintech companies must successfully traverse these regulatory considerations [2].

6 Findings

This literature study reveals several significant conclusions concerning the displacement of traditional financial services by fintech as shown in Fig. 1.

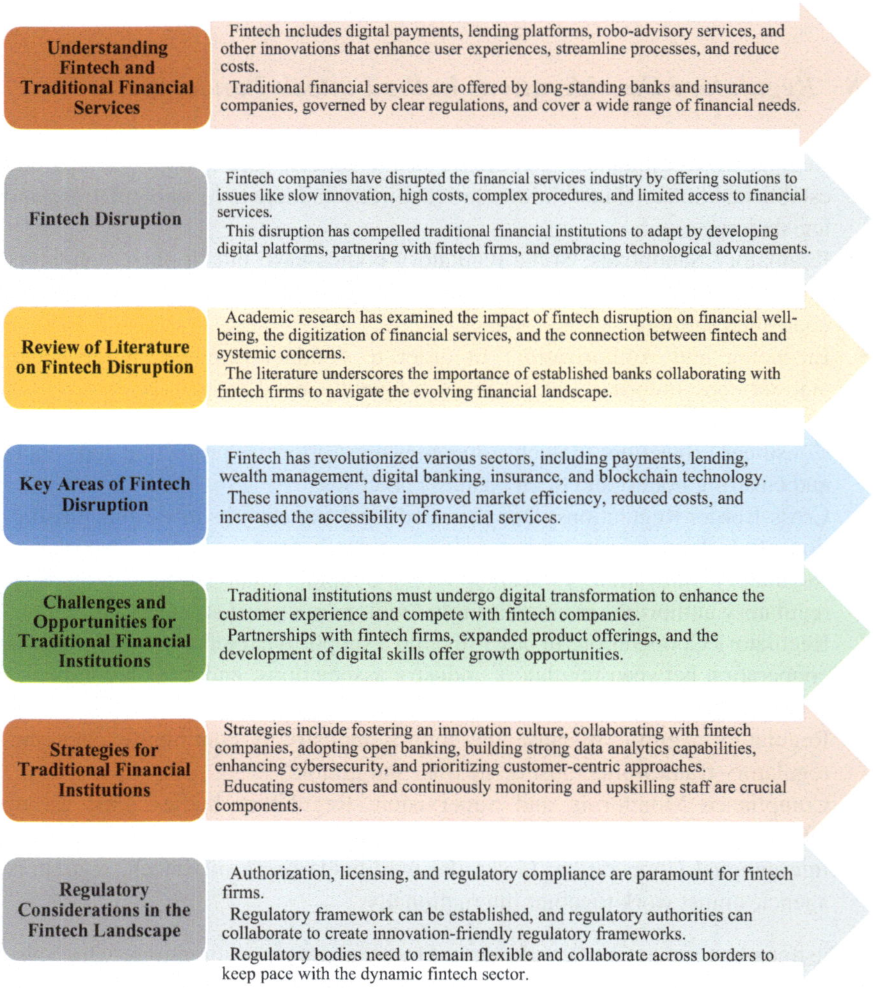

Fig. 1 Summary of findings

7 Conclusion

The development of fintech businesses has significantly disrupted the traditional financial services industry, forcing established institutions to adapt and adjust. Smith [27], Bouri et al. [7], Barberis and Turcotte [4], Jiang et al. [16], and Lenz and Heinrichs all conducted thorough literature reviews that support the idea that the fintech disruption has affected a number of aspects of financial services, including payments, lending, investment management, and client experience. According to Smith [27] and Barberis and Turcotte [4], the introduction of cutting-edge technologies and inventive business models by fintech enterprises has given consumers greater convenience, efficiency, and accessibility, creating new possibilities for financial industry growth and change.

Traditional financial institutions must embrace digital transformation, make investments in a strong technology infrastructure, and promote a creative culture if they are to manage this altering landscape. In order for banks to stay competitive, Bouri et al. [7] emphasize how important it is for them to integrate fintech innovations. As proposed by Smith [27] and Barberis and Turcotte [4], collaboration and joint ventures with fintech companies enable established organizations to work with cutting-edge startups while utilizing their knowledge and agility. Regulatory considerations are also essential for striking a balance between innovation and consumer protection, as highlighted by Jiang et al. [16]. Agarwal and Mittal [1] and Carvalho [8] establish the transformative potential of fintech, emphasizing the need for traditional financial institutions to adapt in order to remain competitive. As stated in Lenz and Heinrichs, regulators must take a proactive and adaptable stance to ensure fair playing conditions and encourage moral financial innovation. The results of the literature research are consistent with the idea that the fintech disruption heralds a change in the financial industry towards a focus on technology and customer-centricity.

By accepting these changes and improving their services and customer experiences to match clients' changing expectations, traditional financial institutions can position themselves as active participants in the fintech ecosystem. Smith [27], Bouri et al. [7], Barberis and Turcotte [4], Jiang et al. [16], and Lenz and Heinrichs have conducted studies that support the necessity of adaptability, cooperation, and regulatory flexibility in dealing with the disruptive forces of fintech.

This research provides valuable insights for traditional financial institutions aiming to navigate the fintech disruption successfully. It suggests strategies for embracing innovation, forming partnerships, enhancing customer-centricity, investing in digital infrastructure, and complying with evolving regulations. The implications for these institutions are clear—by adapting to changing consumer expectations and technological advancements, they can remain competitive in the evolving financial landscape.

However, there are limitations to this study. The reliance on secondary data sources may introduce potential biases and incomplete information. Future research could benefit from incorporating primary data collection methods for a more comprehensive understanding of the fintech landscape. Additionally, considering the dynamic nature of the fintech industry, longitudinal research is essential to capture evolving challenges and opportunities.

In conclusion, the impact of fintech disruption on traditional financial institutions is undeniable. The future direction for these institutions involves a proactive embrace of technological advancements, collaboration with fintech startups, and ongoing adaptation to regulatory changes. By doing so, they can thrive in the changing financial landscape and continue to meet the evolving demands of their customers.

References

1. Agarwal, A., Mittal, R.: Fintech: the disruptive force in the financial services industry. J. Appl. Manag. Invest. **6**(4), 251–259 (2017)
2. Anagnostopoulos, I.: Fintech and regtech: impact on regulators and banks. J. Econ. Bus. **100**, 7–25 (2018)
3. Arnaut, D., Bećirović, D.: FinTech innovations as disruptor of the traditional financial industry. In: Digital Transformation of the Financial Industry: Approaches and Applications, pp. 233–254. Springer International Publishing, Cham (2023)
4. Barberis, J.N., Turcotte, M.F.: The digital transformation of banking services: investigating the impact of fintech on traditional banks. J. Financ. Serv. Res. **53**(2–3), 329–355 (2018). https://doi.org/10.1007/s10693-017-0286-7
5. Beck, T., Chen, T.T., Lin, C., Song, F.M., Xu, L.C.: Fintech and inclusive growth: a review and research agenda. J. Econ. Surv. **33**(3), 883–902 (2019)
6. Boot, A., Hoffmann, P., Laeven, L., Ratnovski, L.: Fintech: what's old, what's new? J. Financ. Stab. **53**, 100836 (2021)
7. Bouri, A., Molnár, P., Azzi, G., Roubaud, D., Hagfors, L.I.: Fintech disruption and bank performance. J. Bank. Finance **98**, 292–305 (2019). https://doi.org/10.1016/j.jbankfin.2018.11.011
8. Carvalho, H.: Fintech disruption in financial services: a literature review. J. Innov. Entrep. **9**(1), 1–21 (2020)
9. Cocco, L., Pinheiro, R., Saita, L.: Fintech lending to households: evidence from lending club. Rev. Financ. Stud. **33**(11), 5113–5154 (2020)
10. Demeester, S., Fornasier, A.: The impact of fintech on traditional banks. In: Handbook of Blockchain, Digital Finance, and Inclusion, pp. 289–303. Springer (2020)
11. Du, Y., Rachinsky, A.: Fintech credit and bank liquidity. J. Corp. Finan. **56**, 184–203 (2019)
12. Giglio, F.: Fintech: a literature review. Int. Bus. Res. **15**(1), 80–85 (2022)
13. Gomber, P., Kauffman, R.J., Parker, C., Weber, B.W.: On the fintech revolution: interpreting the forces of innovation, disruption, and transformation in financial services. J. Manag. Inf. Syst. **35**(1), 220–265 (2018)
14. Heinonen, J., Kess, P., Ranta, P.: Disruptive innovation in financial services: a framework for fintech entrepreneurs. Int. J. Innov. Manag. **22**(04), 1–28 (2018)
15. Iwashita, N.: Why fintech is not changing Japanese banking. Asian Econ. Policy Rev. **17**(2), 297–310 (2022)
16. Jiang, L., Shan, J., Wang, X.: Fintech disruption and financial stability: evidence from China. J. Financ. Stab. **51**, 100812 (2020). https://doi.org/10.1016/j.jfs.2020.100812
17. Kumari, A., Devi, N.C.: The impact of fintech and blockchain technologies on banking and financial services. Technol. Innov. Manag. Rev. **12**(1/2) (2022)
18. Lai, K.P.: Financial advisors, financial ecologies and the variegated financialisation of everyday investors. Trans. Inst. Br. Geogr. **41**(1), 27–40 (2016)
19. Lee, I., Shin, Y.J.: Fintech: ecosystem, business models, investment decisions, and challenges. Bus. Horiz. **61**(1), 35–46 (2018)

20. Li, X., Ling, S.: The rise of fintech and its implications for the role of traditional financial intermediaries. J. Financial Intermediation **41**, 100845 (2020)
21. Li, X., Liu, H., Zhao, J.: Fintech, bank lending, and systemic risk. J. Financ. Stab. **49**, 100762 (2020)
22. Magnuson, W.: Regulating fintech. Vand. L. Rev. **71**, 1167 (2018)
23. Omarini, A.: The digital transformation in banking and the role of FinTechs in the new financial intermediation scenario (2017)
24. O'Hara, M.: Fintech and financial innovation: drivers and depth. Rev. Financ. Stud. **32**(5), 1754–1787 (2017)
25. Pillay, T.: The influence of insurtech on the existing insurance business model. Doctoral dissertation, University of Pretoria (2018)
26. Poposki, E.M., Poposka, A.: The impact of fintech startups on traditional banking industry. Eur. J. Soc. Sci. Educ. Res. **14**(1), 220–230 (2018)
27. Smith, J.: Fintech disruption: a review of the literature. Int. J. Financ. Stud. **5**(2), 12 (2017). https://doi.org/10.3390/ijfs5020012
28. Tang, Y., Wang, Q., Chen, L., Li, X.: Fintech, regulatory arbitrage, and systemic risks. J. Corp. Finan. **58**, 311–330 (2019)

Ethical Paradigms at Work: A Comparative Analysis of Consequentialism, Deontology, and Islamic Perspectives

Ali Ateeq

Abstract This study provides an in-depth review of three prominent ethical frameworks—Consequentialism, Deontology, and Islamic Ethics—and their practical implementation within the context of the workplace. Consequentialism and Deontology are contrasted with the comprehensive approach of Islamic ethics, which combines outcome-based and duty-based concepts. The examination explores the intricacies and pragmatic obstacles that each framework faces when used in professional settings. This analysis critically evaluates the tendency of Consequentialism to prioritize organizational gain above individual rights, and the possible drawback of Deontology's inflexible adherence to principles, which may impede practical decision-making. The nuanced approach of Islamic ethics, which places importance on a balanced assessment of obligations and consequences, is particularly relevant in multicultural business environments. The research explores the incorporation of different ethical frameworks, taking into account their consequences in contemporary, heterogeneous work settings. This synthesis not only emphasizes the advantages and drawbacks of each approach but also emphasizes the need for a more extensive, culturally perceptive understanding of workplace ethics. The study closes by recommending more research, with a focus on investigating ethical frameworks in a worldwide context and examining the influence of technology on ethical decision-making in professional domains.

Keywords Consequentialism · Deontology · Islamic ethics · Workplace ethics · Ethical decision-making · Multicultural work environments · Moral philosophy · Professional conduct · Cross-cultural ethics · Organizational behavior

A. Ateeq (✉)
Administrative Science Department, College of Administrative and Financial Science, Gulf University, Sanad 26489, Bahrain
e-mail: dr.ali.ateeq@gulfuniversity.edu.bh

B. Awwad (ed.), *The AI Revolution: Driving Business Innovation and Research*, Studies in Systems, Decision and Control 524, https://doi.org/10.1007/978-3-031-54379-1_53

1 Introduction

The area of workplace ethics is a complicated and diverse field of research, requiring a comprehensive comprehension of different ethical theories and their practical consequences. This research seeks to analyze and compare three well-known ethical theories—Consequentialism, Deontology, and Islamic ethical perspectives—in relation to professional behaviour and decision-making. Consequentialism, derived from Utilitarian philosophy, asserts that the moral value of an action is completely decided by its consequences [1]. Deontology, a moral theory based on Kantian ethics, promotes the strict observance of duty and laws, regardless of the outcomes or consequences [2, 3]. The ethical principles of Islam, based on the teachings of the Quran and Hadith, provide a complete moral framework that integrates religious guidance with everyday behaviour [4–6]. This comparison research aims to clarify the subtle distinctions between each paradigm and their significance in directing ethical conduct in the workplace.

The examination starts by scrutinizing consequentialism's focus on morality based on outcomes, which often gives priority to the common good in a professional context, sometimes disregarding individual rights [7]. Deontological ethics emphasizes the significance of moral obligations and principles, promoting activities that are intrinsically correct, irrespective of their consequences [8]. Islamic ethical viewpoints include a distinctive combination of moral ideals and religious beliefs, placing particular emphasis on concepts such as equity, openness, and societal accountability [9].

Ultimately, this research offers a thorough examination of how different ethical frameworks influence the process of making moral decisions in professional settings. The results of this research not only add to the scholarly discussion on workplace ethics but also provide practical knowledge for firms aiming to foster ethical cultures. Potential future investigations may delve into the implementation of these ethical theories across various corporate contexts and their influence on practical decision-making procedures.

2 Literature Review

2.1 Introduction

The study of Consequentialism, Deontology, and Islamic ethical views in the context of workplace ethics involves a wide range of philosophical and theological beliefs, each presenting unique perspectives on moral reasoning and decision-making.

2.2 Consequentialism

Consequentialism, specifically Utilitarianism, posits that the moral value of an action depends on its consequences, with the goal of maximizing overall [10]. In work environments, this method often gives precedence to judgments that favour the majority, leading to discussions concerning the possible disregard for individual rights [11]. The potential of this approach to rationalize immoral methods in order to achieve desired outcomes has been emphasized by critics [12].

Islamic ethics, based on religious teachings, promote justice, community welfare, individual rights, and moral responsibility [13]. Islamic ethics use a comprehensive approach by striking a balance between the communal good and individual rights, in contrast to the outcome-centric perspective of Consequentialism.

Consequentialism, however effective in attaining organizational objectives, poses the danger of compromising ethical principles. Islamic ethics provide a complete moral structure that combines individual and communal ethical principles. This juxtaposition emphasizes the need of using sophisticated ethical strategies while making decisions in the workplace, taking into account both the larger consequences and individual moral considerations.

2.3 Deontology

The discrepancies and crossings between Consequentialism, Deontology, and Islamic perspectives become clear in a comparative examination. Deontology, according to Kant's (1785) exposition, places emphasis on duty and obedience to principles, valuing honesty and justice as inherently correct, regardless of the consequences [14]. Conflicts may arise in professional settings when ethical rules and organizational objectives are at odds [15]. Consequentialism, on the other hand, assesses acts by their consequences, prioritizing the greatest benefit for the majority. This results-driven approach may at times rationalize morally dubious methods, as it places more importance on achieving desired outcomes rather than adhering to ethical norms [16].

Islamic ethics, which incorporates elements of both deontological concepts such as honesty and justice, also takes into account the repercussions of acts [17]. This approach adopts a more equitable stance, recognizing the significance of ethical obligations while considering the consequences for the greater welfare.

This examination compares and contrasts the distinctiveness and compatibility of different ethical systems. Deontology and Consequentialism provide explicit albeit sometimes contradictory principles, while Islamic ethics adopt a more comprehensive perspective. The presence of this variety highlights the need for a sophisticated comprehension of ethical concepts while making decisions, particularly in professional environments, where both moral uprightness and practical results are of utmost importance.

2.4 *Islamic Ethical Perspectives*

The ethical values of Islam, which are based on the Quran and Hadith, provide a moral framework that combines elements of Consequentialism and Deontology, while also presenting different points of view [18]. Islamic ethics prioritize fairness, honesty, and social responsibility, while also including spiritual accountability, as opposed to Consequentialism's emphasis on results and Deontology's focus on obligation [19].

Within Islamic societies, this amalgamation permits the making of ethical judgments that uphold moral obligations (resembling Deontology) and take into account the consequences (resembling Consequentialism), but also including an additional element of religious duty. By adopting this comprehensive approach, it is possible to achieve a greater equilibrium and promote choices that prioritize the well-being of the society. This method takes into account individual acts within a wider social and spiritual framework.

Nevertheless, incorporating religious concepts may provide difficulties in multicultural communities, where both secular and varied religious views exist simultaneously. Consequentialism and Deontology provide more secular perspectives, but Islamic ethics firmly anchor moral reasoning in a religious context. This may be advantageous for promoting communal unity, but it may also have limitations in varied environments.

3 Method

The research "Ethical Paradigms at Work: A Comparative Analysis of Consequentialism, Deontology, and Islamic Perspectives" carried out a systematic literature analysis to examine trends and insights in different ethical frameworks as they relate to workplace ethics [5, 20–22]. The process included conducting an extensive search of scholarly sources, with a particular emphasis on peer-reviewed journals and books [5, 23–26]. The keywords "Consequentialism," "Deontology," "Islamic Ethics," and "Workplace Ethics" were used to collect useful information. Subsequently, a comprehensive examination was conducted to ascertain prevalent patterns and variations in the implementation of these paradigms within professional environments [27, 28]. The research focused on using empirical information from case studies and surveys to support theoretical results. Its goal was to establish techniques for ethical decision-making in the workplace that are supported by empirical data. The brief methodology used facilitated a concentrated and equitable examination of the topic, establishing the foundation for further investigations and real-world implementations in the realm of workplace ethics.

4 Discussion

The comparative research of ethical paradigms within workplace ethics reveals the varied approaches of Consequentialism, Deontology, and Islamic ethics, each offering distinctive viewpoints and posing unique issues. Consequentialism, which emphasizes the results of acts, often clashes with Deontological ethics, which places importance on following moral obligations regardless of the outcomes [19]. This distinction becomes apparent in intricate workplace scenarios, where Consequentialism may support measures that benefit the organization, perhaps at the expense of individual rights, while Deontology adheres to moral ideals, sometimes at the cost of practical results [29, 30]. The works of scholars such as Ali Ateeq explain that Islamic ethics provide a balanced perspective that combines Deontological principles, which emphasize the importance of duties like honesty and justice, with an understanding of the consequences of actions, similar to Consequentialism [24]. This synthesis facilitates a comprehensive approach to ethical decision-making in many professional settings. Nevertheless, reconciling these ideas with practical implementations poses difficulties. Deontology offers a distinct moral framework, yet it may be too inflexible when confronted with the intricacies of contemporary commercial scenarios. On the other hand, the practical adaptability of Consequentialism may sometimes result in ethical uncertainties. Islamic ethics combine principle-based morality with concerns of outcomes, offering a realistic and balanced approach. This comparative research, including perspectives from academics such as Ali Ateeq, emphasizes the intricate nature of ethical decision-making in professional environments. The study proposes that adopting an inclusive strategy, namely by adding Islamic ethics, might provide a successful structure for resolving ethical challenges in the workplace. This method would guarantee both moral uprightness and pragmatic business concerns [5].

4.1 Potential Areas for Future Research

Subsequent investigations in the domain of ethical paradigms in the workplace might delve into the implementation of these ideas in ever varied and international work environments. Research endeavours may centre on conducting empirical evaluations of decision-making processes, examining the incorporation of artificial intelligence and technology in ethical deliberations, and conducting cross-cultural analyses to comprehend how diverse cultural backgrounds impact the understanding and application of these ethical frameworks.

5 Conclusion

In short, our investigation of Consequentialism, Deontology, and Islamic ethics in business environments emphasizes the intricate and varied nature of ethical decision-making. Every paradigm provides distinct perspectives, but also has difficulties when used in real-life situations. The amalgamation of these systems, particularly including the comprehensive approach of Islamic ethics, provides a strong basis for tackling ethical quandaries. Future endeavours in this topic should prioritize the practical implementation and cross-cultural analysis to enhance our comprehension of ethics in various professional contexts.

References

1. Herbert, R.: The moral deliberation roadmap: the US naval academy's moral reasoning framework. In: Ethics at War, pp. 37–58. Routledge
2. Vignini, S., Rusconi, G.: Business ethics: moral or amoral? An analysis from the perspective of Kantian ethics. Australas. Acc. Bus. Finan. J. **17**(3), 23–39 (2023)
3. Parthsarthi, R.G.: Liberal individualism and public health: case study of coronavirus pandemic in Gandhian duties' context. In: Relevance of Duties in the Contemporary World: With Special Emphasis on Gandhian Thought, pp. 165–186. Springer (2023)
4. Rohmatulloh, R., et al.: Energy-saving triangle: internalizing Islamic ethical values on energy saving in integrative learning. Religions **14**(10), 1284 (2023)
5. Ali, A.A.A.: The relationship between Islamic work ethics and employee performance among medical staff in the public health sector in Bahrain: employee commitment as mediating variable. Universiti Sains Islam Malaysia (2022)
6. Ali, A.A.A., et al.: Enhancing contextual performance through Islamic work ethics with mediating role of normative commitment. Math. Stat. Eng. Appl. **71**(4), 8668–8683 (2022)
7. Teays, W.: Ethical Foundations for Institutional Integrity, p. 25. The Ethical University: Transforming Higher Education (2022)
8. White, M.D., Hicks, J., Broome, J.: 17. A Kantian perspective on teaching ethics to economists. In: Handbook of Teaching Ethics to Economists: A Plurality of Perspectives, p. 278 (2023)
9. Haron, H., Jamil, N.N., Ramli, N.M.: Western and Islamic values and ethics: are they different? J. Govern. Integrity **4**(1), 12–28 (2020)
10. Sola, A.: Utilitarianism and consequentialist ethics: framing the greater good. In: Ethics and Pandemics: Interdisciplinary Perspectives on COVID-19 and Future Pandemics, pp. 61–83. Springer (2023)
11. Macdonald, S., Birdi, B.: The concept of neutrality: a new approach. J. Documentation **76**(1), 333–353 (2020)
12. Gray, C.M., et al.: Dark patterns and the legal requirements of consent banners: an interaction criticism perspective. In: Proceedings of the 2021 CHI Conference on Human Factors in Computing Systems (2021)
13. Bensaid, B., Machouche, S.: Muslim morality as foundation for social harmony. J. Al-Tamaddun (2019)
14. Maurer, D.: Sovereign, employer, community: a theory of military justice beyond discipline, obedience, and efficiency. Marquette Law Rev. **107** (2023)
15. Siakas, E., et al.: Managing ethical requirements elicitation. In: European Conference on Software Process Improvement. Springer (2023)
16. Darr, R.: Teleology and consequentialism in Christian ethics: goods, ends, outcomes. Stud. Christ. Ethics **36**(4), 906–925 (2023)

17. Elmahjub, E.: Islamic jurisprudence as an ethical discourse: an enquiry into the nature of moral reasoning in Islamic legal theory. Oxford J. Law Religion **10**(1), 16–42 (2021)
18. Suleman, M.: The balancing of virtues—Muslim perspectives on palliative and end of life care: empirical research analysing the perspectives of service users and providers. Bioethics **37**(1), 57–68 (2023)
19. Wood, N.: Virtue Rediscovered: Deontology, Consequentialism, and Virtue Ethics in the Contemporary Moral Landscape. Lexington Books (2019)
20. Ali, A.A.A., et al.: Moderating effect of strategic planning on the relationship between career path planning and job performance. Sustainability **15**(11), 8490 (2023)
21. Al-refaei, A.A.-A., et al.: An integrated mediating and moderating model to improve service quality through job involvement, job satisfaction, and organizational commitment. Sustainability **15**(10), 7978 (2023)
22. Alzoraiki, M., et al.: Impact of teachers' commitment to the relationship between transformational leadership and sustainable teaching performance. Sustainability **15**(5), 4620 (2023)
23. Ali, A.A.A., Alzoraiki, M., Al Aghbas, N.S.M.: Working remotely, social isolation and access to support and communication during Covid-19 its influence on job performance. Int. J. Intell. Human Resour. Manag. (IJIHRM) **3**(01), 23–33 (2022)
24. Ali, A.A.A., et al.: Association between Islamic work ethics and employee commitment among medical staff in the public health sector in Bahrain. Int. J. Intell. Human Resour. Manage. (IJIHRM) (2023)
25. Ateeq, A.: Knowledge management's effect on learning quality in Bahrain's private universities (2023)
26. Alkadash, T.M., et al.: The effects of leadership style on employee sustainable behaviour: a theoretical perspective. In: Artificial Intelligence and Transforming Digital Marketing, pp. 205–213. Springer (2023)
27. Ateeq, A.: The mediating role of engagement in the relationship between. Management **21**(3), 435–447 (2023)
28. Al Balushi, F.S.M., Ali, A.A.A., Alzoraiki, M.: The impact of forecasting strategy and training and development strategy on preparing leadership competencies administrative support in the sultanate of Oman. Int. J. Intell. Human Resour. Manag. (IJIHRM) **3**(01), 34–41 (2022)
29. Slee, R., Tait, G.: Ethics and Inclusive Education
30. Cox, D.J.: Descriptive and normative ethical behavior appear to be functionally distinct. J. Appl. Behav. Anal. **54**(1), 168–191 (2021)

Mediating Role of Islamic Financial Performance Index in the Islamic Corporate Governance's Impact on Sustainability Performance of Islamic Commercial Banks

Mutoharoh and Lisa Kartikasari

Abstract The increased competition in the Sharia banking industry requires Islamic Commercial Banks (ICB) to strengthen their Sustainability performance. Previous research tested the Islamic Corporate Governance (ICG) factor as an antecedent that is relevant to the Sustainability Performance (SP) of Islamic entities. In practice, the effect of ICG on SP still opens research opportunities due to the finding's inconsistency. The implementation of a sustainability strategy, however, requires adequate resources, so that the Islamicity Financial Performance Index (IFPI) factor is also involved as a mediating variable in this study. This factor is part of the novelty in the issue of the relationship between ICG and SP today. Explanatory techniques with quantitative methods are used in this study to achieve the research objectives, namely, to analyze the role of ICG on the SP of Islamic Commercial Banks in Indonesia by the IFPI mediation. This study is expected to contribute to the enrichment of the literature on sustainability issues and to become a reference for stakeholders in improving the SP of ICB.

Keywords Sustainability · Indonesian Islamic Commercial Banks · Islamic Corporate Governance · Islamicity Financial Performance Index

---

Mutoharoh (✉)
Department of Accounting, Master of Accounting, Faculty of Economics, Universitas Islam Sultan Agung, Semarang, Indonesia
e-mail: mutoharoh@unissula.ac.id

L. Kartikasari
Department of Accounting, Faculty of Economics, Universitas Islam Sultan Agung, Semarang, Indonesia

B. Awwad (ed.), *The AI Revolution: Driving Business Innovation and Research*, Studies in Systems, Decision and Control 524, https://doi.org/10.1007/978-3-031-54379-1_54

1 Introduction

Increasing awareness of the Indonesian, who are predominantly Muslim, has made a significant contribution to the development of the sharia banking industry in Indonesia. Therefore, as this phenomenon, the competition among Islamic Commercial Banks (ICB) become more competitive. Strategy is needed to maintain the existence of ICB such us improving the SP of ICB. The sustainability factor involves balances between people, planet, and profit or noted as the concept of triple bottom line (TBL). Information regarding sustainability performance increases stakeholders' confidence in business performance [1, 2]. The concept of sustainability presents a broader and more flexible paradigm for businesses in applying their responsibility practices. The findings of [3] claim that financial entities implicitly contribute to environmental degradation. Although not as extreme as the manufacturing industry, banks and other financial institutions are also obliged to contribute to saving the environment. Apart from that, various previous studies proven that social responsibility practices are very important to ensure business credibility and maintain social growth in longer term [4–6]. Collaboration in social practices and environmental awareness can open opportunities for banks to become closer to stakeholders.

Previous research examined factors that influence sustainability performance related to corporate governance [7–10]. Sharia banking with its concept of offering financial services based on sharia principles has a responsibility to meet the expectations of its stakeholders in carrying out halal and prosperous business activities. In maintaining its sustainability, [10] stated that one of the strategies that sharia banks must consider is related to the implementation of Good Corporate Government (GCG). Corporate governance in an Islamic perspective is always associated with concepts related to behavior in business governance, along with other matters that are transcendental between management and investors [11]. In line with the implementation of Sharia Enterprise Theory (SET), this is a direct consequence of a Muslim to his Creator, Allah. Good corporate governance can help create a conducive environment for efficient and sustainable growth in all corporate sectors including Islamic banking and the Islamic financial services industry. ICG is a governance concept that is more relevant for sharia entities with the issue of compliance with sharia principles. Through the Sharia Supervisory Board (SSB) aspect, ICG guarantees halal fatwas for every sharia entity activity. ICG's role in SP is a reflection that the sharia principles of *rahmatan lil alamin* as put forward in Sharia Enterprise Theory are managed seriously as sharia banks do not only focus on profit but also on social welfare and environmental preservation. Previous research [7, 12–14] states that corporate governance has a significant positive influence on business sustainability. This is different from research [8, 10] which states that corporate governance has no effect on SP. In another study, [15] found that Islamic financial institutions that concentrate more on implementing ICG have an impact on increasing TBL performance. Sustainability performance that is consistently pursued with the implementation of ICG is able to satisfy various stakeholders, including its impact on the environment. In the study, [15] also utilize insider and institutional ownership factors as variables that are proven to moderate the relationship between ICG and Sustainability.

In general, sustainability performance requires financial capability factors to support the implementation of the TBL strategy [16, 17]. Effective ICG may have a strong influence if it is supported by the company's ability to fulfill its overall responsibilities. Previous research believes that the Islamicity Financial Performance Index (IFPI) can support the sustainability of an organization as found by [18–20]. Previous research on the moderating role of financial stability index support is still very limited. The phenomenon of sustainability which continues to develop still opens up opportunities for new studies to reveal various antecedents in strengthening it as a consideration for industry best practice in the future. Sharia banks must periodically carry out evaluations to ensure that performance is achieved as planned. Understanding in assessing financial performance can rely on the IFPI index as [21] measured the performance index of Islamic banks in seven indicators which include ratio of profit-sharing ratio (PSR), zakat performance (ZPR), equitable distribution (EDR), directors-employees welfare (DEWR), Islamic Investment (IInvR), Islamic Income (IIncR), and item ratio of AAOIFI. Using the case of ICB in Indonesia, this research aims to analyze the influence of Islamic Corporate Governance on Sustainability Performance which is mediated by the Islamic Financial Performance Index by referring to [15].

As a financial industry that promises services based on Islamic values, Sharia Banks must be able to reflect Islamic ethical identity. By doing this, customers will feel comfortable and confident in making transactions. The novelty in this research modifies the research of [15] using the Islamicity Financial Performance Index (IFPI) as a moderating variable. The sharia performance index used in this research refers to the sharia index for banking developed by [21], however, in this study, IFPI is measured based on these measurement indices except for the Islamic Investment ratio and AAOIFI index. The elimination of these measurements is because the use of the AAOIFI index is generally to assess compliance with the principles in AAOIFI relating to accounting and auditing aspects. Second, regarding the sharia investment ratio compared to the non-sharia investment ratio, the existence of the SSB in Indonesia causes this ratio irrelevant since the SSB guarantees that sharia banks do not make non-sharia investments so that it cannot be traced in the financial reports of sharia banks. This research is expected to confirm the application of Sharia Enterprise Theory by ICB's study on ICG and Sustainability which are moderated by IFPI. It is also hoped that this research can meet the limitations of previous research and become a reference in the development of further research. Apart from that, the results of this research are expected to contribute to the development of the sharia industry in Indonesia so that the Muslim community obtains better facilitation in carrying out sharia transactions. This also indirectly facilitates the Muslim community in increasing its worship by minimizing transactions that are forbidden in Islam.

2 Hypothesis

2.1 The Islamic Corporate Governance's Impact on Sustainability Performance

Businesses that are run with good governance will tend to have a longer sustainability cycle. Bearing in mind that corporate governance guarantees accountability, transparency, and business efficiency, where the board of directors' functions as a check and balance tool. The SSB became a typical ICG tool, consisting of specialized legal experts in Islamic law and financing who were responsible for ensuring banking operations within Sharia rules. Lidyah [16] explained that in encouraging sharia banking practices that are financially strong and consistently refer to sharia principles, sharia banks are expected to apply GCG principles which refer to sharia provisions. This statement is also supported by previous researchers [3, 7, 8, 12] regarding the positive impact of ICG on financial performance in Islamic banks. Thus, the first hypothesis in this research is concluded:

H1: ICG has a significant positive effect on SP.

2.2 The Impact of Islamic Corporate Governance Toward Islamicity Financial Performance Index

Good governance will make it easier for management to obtain important information that supports decision making. ICG facilitates sharia financial services institutions in evaluating the achievement of sharia performance targets. Previous findings explain that the better the implementation of ICG, the better the company's performance, both in relation to its financial performance and compliance with sharia principles [22–25]. Therefore, the second hypothesis in this study is concluded:

H2: ICG has a significant positive effect on Islamicity Financial Performance Index.

2.3 The Impact of Islamicity Financial Performance Index Toward Sustainability Performance

Good performance has an impact on the sustainability of sharia financial service institutions. However, the life cycle of an entity requires resources that are well managed to produce profits and increase the welfare of the people involved in an organization. Various previous studies believe that the Islamic performance index which comprehensively considers the performance of sharia institutions based on Islamic principles of *rahmatan lil alamin* will support the sustainability of an organization as found by [18–20], so the third hypothesis in this study is concluded:

H3: Islamicity Financial Performance Index has a significant positive effect on SP.

2.4 Islamicity Financial Performance Index as Mediator Between Islamic Corporate Governance and Sustainability Performance

The sustainability of sharia banks can be achieved indirectly through the influence of sharia financial performance and sharia corporate governance. Research by [16] explains that sharia corporate governance that has been implemented by sharia entities does not guarantee the sustainability of sharia banks. However, sharia banks must maintain and improve sharia corporate governance because the implementation of sharia corporate governance will have a positive impact on improving sharia financial performance [17, 26], which in turn will have a positive impact on increasing the sustainability of sharia banks. This finding is supported by the results of other research including the findings of [11, 27] who also utilize Islamic performance with various proxies as a mediator of business sustainability. Based on previous findings, the fourth hypothesis in this research is concluded:

H4: Islamicity Financial Performance Index mediates the effect of ICG on SP.

3 Method

3.1 Research Technique

This research uses explanatory techniques to prove the truth of the proposed hypothesis. Quantitative research focuses on testing theories that are measured through relationships between variables and analyzed using statistical procedures. This quantitative approach comes from secondary data that obtained from annual reports and sustainability reports of official sources. Research data was obtained from the research object population, namely all 14 ICB registered in Indonesian Financial Services Authority (OJK) which were analyzed for 5 years, namely 2018–2022.

The research sample was determined using a purposive sampling technique for data that met the criteria according to the research objectives. These criteria include ICB is registered with OJK during the research period. Moreover, the ICB should publishes financial reports during the research period. Lastly, the ICB that published sustainability reports consecutively during the research period.

3.2 *Variable Measurement*

There are three variables in this research which are ICG (Independent Variable), SP (Dependent Variable), and Islamicity Performance Index (Mediation Variable). These three variables are defined and measured in Table 1.

3.3 *Hypothesis Testing*

Analysis and testing in this research utilize Partial Least Square (PLS) with the SmartPLS 3.0 statistical tool. Specifically, this research uses the Structural Equation Model (SEM) for several testing stages which are explained as follows:

Designing the Inner Model

The inner model design is useful for describing the relationship between latent variables. The inner model in PLS is evaluated using R-square (R^2) for the dependent construct. The R-square (R^2) value is used to show how large a percentage of the variance of the endogenous construct can be explained by the exogenous construct in the model. The higher the R-square (R^2) value, the better the prediction model of the proposed research model.

Designing the Outer Model

The outer model design is useful for defining a model of the relationship between each indicator block and the latent variable. The outer model shows the specifications of the relationship between the indicators. The reflective indicators are tested using a reliability test. Reliability testing is carried out with the aim of ensuring that the research instruments used can present concept measurements consistently without any bias. The criteria are seen from two things, namely composite reliability and Cronbach's alpha with each value > 0.70.

Hypothesis Testing

Hypothesis testing relies on direct effect tests and indirect effect tests. The direct effect test is useful for testing the hypothesis of a direct impact between the independent variable on the dependent variable. In the direct influence test, the test results are assessed based on the output of path coefficient and *p*-values. Tests in analyzing indirect effects are applied to the indirect relationship of the independent variable to the dependent variable which is mediated by the mediating variable. In this research, the effect of mediation is analyzed using the Sobel test to determine whether the relationship through a mediating variable is significantly capable of acting as a mediator in the relationship.

Table 1 Variable measurement

Variable	Operational definition	Measurement
Islamic Corporate Governance (ICG)	The concept of governance that uses elements of corporate governance in general and sharia principles [15]	1. SSB independence (SBI) = number of independent members on board 2. SSB size (SBS) = number of SSB member in the ICB 3. SSB competence (SBC) = percentage of SSB member having a higher degree and adequate experience (≥ 1 year) 4. SSB process (SBP) = number of SSB meetings in 1 year
Sustainability disclosure (SP)	Economic, social, environmental activities disclosure [15]	Content analysis (0, 1) Formula: $\frac{dj}{N}$ dj = number of disclosed item per section N = total number of item per section
Islamicity Financial Performance Index (IFPI)	The performance of sharia entities based on Islamic business objectives that is measured by: (1) profit sharing ratio, (2) zakat performance ratio, (3) equitable distribution ratio, (4) directors-employees welfare ratio, (5) Islamic investment ratio, (6) Islamic income ratio, (7) AAOIFI ratio [21]	1. PSR = (*Mudhorobah* + *Musarokah*)/total of financing 2. ZPR = zakat/net asset 3. EDR = distributed revenue/(total revenue − (zakat + tax)) 4. DEWR = average directors' remuneration: average employees' welfare 5. IIncR = Islamic income/total income

4 Summary

This paper proposes a research concept in the field of sharia banking by raising the topic of sustainability. Currently, the topic of sustainability is a very important topic for us to discuss. In Indonesia itself, research that has raised this issue is still very limited, especially those related to the sharia banking industry. This is something that is still new to examine in cases in Indonesia. Apart from that, the use of Islamic governance proxies and the IFPI makes this research even more relevant to the conditions that exist in Islamic banking.

References

1. Islami, Q.N.: Earning management of corporate social responsibility mediation and corporate governance on financial performance (an empirical study on Idx mining corporates 2016–2020). Int. J. Curr. Sci. Res. Rev. **05**(05), 1722–1740 (2022). https://doi.org/10.47191/ijcsrr/v5-i5-41
2. Khan, P.A., Johl, S.K., Johl, S.K.: Does adoption of ISO 56002-2019 and green innovation reporting enhance the firm sustainable development goal performance? An emerging paradigm. Bus. Strateg. Environ. **30**(7), 2922–2936 (2021). https://doi.org/10.1002/bse.2779
3. Jan, A., Marimuthu, M., Hassan, R., Mehreen: Sustainable business practices and firm's financial performance in Islamic banking: under the moderating role of Islamic corporate governance. Sustainability (Switzerland) **11**(23), 1–25 (2019). https://doi.org/10.3390/su11236606
4. El Gammal, W., Yassine, N., Fakih, K., El-Kassar, A.N.: The relationship between CSR and corporate governance moderated by performance and board of directors' characteristics. J. Manage. Gov. **24**(2), 411–430 (2020). https://doi.org/10.1007/s10997-018-9417-9
5. Menne, F., Winata, L., Hossain, M.: The influence of CSR practices on financial performance: evidence from Islamic financial institutions in Indonesia. J. Mod. Account. Audit. **12**(2) (2016). https://doi.org/10.17265/1548-6583/2016.02.002
6. Harun, M.S., Hussainey, K., Mohd Kharuddin, K.A., Al Farooque, O.: CSR disclosure, corporate governance and firm value: a study on GCC Islamic banks. Int. J. Account. Inf. Manag. **28**(4), 607–638 (2020). https://doi.org/10.1108/IJAIM-08-2019-0103
7. Awais, M., Ullah, N., Sulehri, N.A., Thas Thaker, M.A.M., Mohsin, M.: Monitoring and efficiency in governance: a measure for sustainability in the Islamic banking industry. Front Psychol. **13** (2022). https://doi.org/10.3389/fpsyg.2022.884532
8. Haron, H., So, I.G., Gui, A., Sari, S.A., Ramli, N.M., Jamil, N.N.: The relationship between Islamic corporate governance, human governance, usage of information technology and sustainability reporting: comparison of Shariah compliant companies in Malaysia and Indonesia. Int. J. Bus. Soc. **23**(3), 1443–1461 (2022). https://doi.org/10.33736/ijbs.5174.2022
9. Jan, A.A.A., et al.: Islamic corporate governance and Islamic banking financial performance. Int. J. Islam. Middle East. Financ. Manag. **4**(2), 164–179 (2020). https://doi.org/10.1108/JIABR-05-2016-0057
10. Siswanti, I., Salim, U., Sukoharsono, E.G., Aisjah, S.: Sustainable business of Islamic bank through on the Islamic corporate governance and Islamic financial performance. GATR J. Fin. Bank. Rev. **2**(2), 15–20 (2017). https://doi.org/10.35609/jfbr.2017.2.2(3)
11. Siswanti, I., Salim, U., Sukoharsono, G., Aisjah, S.: The impact of Islamic corporate governance, Islamic intellectual capital and Islamic financial performance on sustainable business Islamic banks. Int. J. Econ. Financ. Issues **7**(4), 316–323 (2017). [Online]. Available: http://www.econjournals.com

12. Badshah, W., Bulut, M.: A Turkish perspective of Islamic corporate governance for a sustainable financial performance. Int. J. Bus. Mark. Manag. **5**(12), 32–41 (2020). [Online]. Available: www.ijbmm.com
13. Hussain, N., Rigoni, U., Orij, R.P.: Corporate governance and sustainability performance: analysis of triple bottom line performance. J. Bus. Ethics **149**(2), 411–432 (2018). https://doi.org/10.1007/s10551-016-3099-5
14. Setyahadi, R.R., Narsa, I.M.: Corporate governance and sustainability in Indonesia. J. Asian Finance Econ. Bus. **7**(12), 885–894 (2020). https://doi.org/10.13106/JAFEB.2020.VOL7.NO12.885
15. Jan, A.A., Lai, F.W., Tahir, M.: Developing an Islamic corporate governance framework to examine sustainability performance in Islamic banks and financial institutions. J. Clean. Prod. **315**, 128099 (2021). https://doi.org/10.1016/j.jclepro.2021.128099
16. Lidyah, R.: Islamic corporate governance, Islamicity financial performance index and fraud at Islamic bank. J. Akunt. **22**(3), 437 (2018). https://doi.org/10.24912/ja.v22i3.398
17. Rahmatullah, N.Z., Tripuspitorini, F.A., Akuntansi, J., Bandung, P.N.: Analysis of the influence of Islamicity performance index on Islamic bank profitability kin Indonesia in 2014–2018. J. Appl. Islam. Econ. Financ. **1**(1), 85–96 (2020)
18. Ammar, R., Rebai, S., Saidane, D.: Toward the development of an Islamic banking sustainability performance index. Int. J. Islam. Middle East. Financ. Manag. **16**(4), 734–755 (2023). https://doi.org/10.1108/IMEFM-12-2021-0479
19. Brescia, V., Sa'ad, A.A., Hassan, R.B., Alhabshi, S.M.B.S.J., Lanzalonga, F.: Exploring sustainability from the Islamic finance perspective. Eur. J. Islam. Financ. **19**, 45–53 (2021). https://doi.org/10.13135/2421-2172/6107
20. Zakaria, S.Z.S., Majid, N.A., Arifin, K., Ramli, Z.: Analysis of Islamic performance index on Sharia business unit in Indonesia towards sustainable development. Envirobiotechjournals.Com **26**(3), 1128–1138 (2020). [Online]. Available: https://www.envirobiotechjournals.com/EEC/vol26i32020/EEC-28.pdf
21. Hameed, S., Wirman, A., Alrazi, B., Nazli, M., Pramono, S.: Alternative disclosure and performance measures for Islamic banks. In: Second Conference on Administrative Sciences: Meeting the Challenges of the Globalization Age, pp. 19–21. King Fahd University of Petroleum & Minerals, Dhahran, Saudi Arabia (2004)
22. Aslam, E., Haron, R.: Does corporate governance affect the performance of Islamic banks? New insight into Islamic countries. Corp. Gov. (Bingley) **20**(6), 1073–1090 (2020). https://doi.org/10.1108/CG-11-2019-0350
23. Buallay, A.: Corporate governance, Sharia'ah governance and performance: a cross-country comparison in MENA region. Int. J. Islam. Middle East. Financ. Manag. **12**(2), 216–235 (2019). https://doi.org/10.1108/IMEFM-07-2017-0172
24. Khan, I., Zahid, S.N.: The impact of Shari'ah and corporate governance on Islamic banks performance: evidence from Asia. Int. J. Islam. Middle East. Financ. Manag. **13**(3), 483–501 (2020). https://doi.org/10.1108/IMEFM-01-2019-0003
25. Kusuma, M., Rosadi, S.: Islamic corporate governance and Islamic banking financial performance. J. Financ. Islam. Bank. **1**(2), 164–179 (2018)
26. Umiyati, L.M.M.K.: Islamic corporate governance and Sharia compliance on financial performance Sharia bank in Indonesia. Al-IQTISHAD J. Ilmu Ekon. Syariah (J. Islam. Econ.) **12**(1), 33–50 (2020). https://doi.org/10.15408/aiq.v12i1.15053
27. Budi, I.S., Rahmawati, R., Falikhatun, F., Muthmainah, M., Gunardi, A.: Financial performance mediation in the influence of Islamic corporate governance disclosure on the Islamic social reporting. Indones. J. Sustain. Account. Manag. **3**(1), 75 (2019). https://doi.org/10.28992/ijsam.v3i1.77

Business Sustainability Model of MSMEs in the Fisheries Sector in Central Java Province

Zani Suparlan, Kiryanto, and Sri Anik

Abstract This study aims to determine the effect of financial inclusion, financial literacy and government support (marine and fisheries business assistants) on the performance and business sustainability of MSMEs in the fisheries sector in Central Java Province. The population of this study is MSME actors in the fisheries sector in Central Java with research samples taken from twelve districts/cities in the area of mentoring work carried out by TPU KP. Descriptive quantitative research methods. Type of primary data obtained through the distribution of questionnaires, interviews and observations of respondents. With the purposive sampling method, as many as one hundred respondents were taken for data processing using smart PLS version 4.0. From the results of data processing, it can be seen that the results of the study show that financial inclusion, financial literacy and government support have a positive and significant effect on business performance and sustainability and show a positive and significant influence on business performance on business sustainability.

Keywords Financial inclusion · Financial literacy · Government support · Business Performance · Business sustainability

Z. Suparlan (✉) · Kiryanto · S. Anik
Sultan Agung Islamic University, Semarang, Indonesia
e-mail: 21402200042@std.unissula.ac.id

Kiryanto
e-mail: kiryanto@unissula.ac.id

S. Anik
e-mail: srianik@unissula.ac.id

B. Awwad (ed.), *The AI Revolution: Driving Business Innovation and Research*, Studies in Systems, Decision and Control 524, https://doi.org/10.1007/978-3-031-54379-1_55

1 Introduction

1.1 Background

This research is based on the problems that exist in the fisheries sector business according to the simulation from the Directorate General of Strengthening the Competitiveness of Marine and Fisheries Products (PDSPKP) that the problems faced by fisheries and marine sector business actors are the existence of information about the lack of information on access to financing from the business actor side and the lack of information on prospective debtors on the financing institution side and most fisheries business actors have low financial literacy which is characterized. With low information on financial access, business planning and financial management that is not good.

From several previous studies as a reference for the author in conducting research. No one has conducted research on MSMEs specifically for the fisheries sector. So this is something interesting to research and encourage the author to study and examine how MSMEs in the fisheries and marine sectors can improve performance and maintain business continuity so that they will move up in class.

With the government's role in this case, the Ministry of Marine Affairs and Fisheries by appointing Marine and Fisheries Business Pendampig Workers (TPU KP). Whether the existence of business assistance personnel is influential and can solve or reduce problems that exist in MSME actors in the fisheries sector.

1.2 Problem Statement

The formulation of the problem as a formulation of important research questions must be made clearly, not ambiguously. The formulation of the problem that can be raised in this study is whether financial inclusion affects business performance, does financial literacy affect business performance, does government support affect business performance, does financial inclusion affect business sustainability. Does financial literacy affect business sustainability, does government support affect business continuity and does performance affect business continuity.

1.3 Research Objectives

This study aims to determine and analyze (1) The effect of financial inclusion on business performance. (2) The effect of financial literacy on business performance. (3) The effect of government support on business performance. (4) The effect of financial inclusion on business sustainability. (5) The effect of financial literacy on business continuity. (6) The effect of government support on business continuity. (7) The effect of performance on business continuity.

1.4 Research Benefits

The benefits of this research are providing academic contributions for the authors themselves, practical contributions for government, theoretical contributions for future research. The main thing is a practical contribution, namely as input for the Government in making policies and making mentoring programs for MSME business actors.

2 Literature Review

2.1 Grand Theory

Resource Based View Theory is a theory that describes a company can achieve competitive advantage by relying on resources so as to direct the company to be sustainable continuously. RBV theory as the basic theoretical foundation in this study which explains that financial inclusion and financial literacy are internal resources of companies or organizations that have value or potential that can support the operations of a business to achieve a competitive advantage and continuous improvement of performance and business.

The theoretical basis of research that explains government support is role theory. According to Bates and Harvey states that role theory is a set of certain norms or rules that are regulated for a particular function. Role theory is used as a basis because this theory explains how the role of the government in making regulations in carrying out the function of empowerment to the community, in this case is empowering MSMEs in the fisheries sector through the provision of credit access facilities, guidance, assistance and reinforcement assistance to improve performance and the ability to maintain business sustainability.

Expectancy theory states that an individual's motivation to perform a particular task is based on their belief that their efforts will result in high performance and that high performance will produce desired results. Expectancy Theory is the basic basis for this study explaining that improving business performance will support business sustainability. The performance measurement used in this study uses the Balanced Scorecard method commonly used for performance measurement by Government agencies. According to Dally, Balanced Scorecard is a management concept that tries to measure organizational performance in a balanced manner from various perspectives that focus on the success of implementing an organization's strategy.

Financial inclusion is the availability of access to services to various formal financial institutions regarding services and products for all levels of society with easy reach, safe and comfortable services tailored to the capabilities and needs of the community in creating welfare.

Financial literacy is the knowledge, ability and understanding and skills of a person about finance for financial products and services to make the right decisions so as to minimize financial risk.

External factors that can encourage business actors to be able to improve performance for the sustainability of their business are the Government. The role of the government in the development and empowerment of MSMEs is stated in Law Number 20 of 2008 concerning MSMEs. The law mandates the government to develop MSMEs. In article 1 paragraph 12 of the MSME Law which states that development is an effort made by the Government to empower MSMEs through the provision of credit access facilities, guidance, assistance and strengthening assistance to grow and improve the ability and competitiveness of MSMEs. Business development carried out through the fields of processing, marketing, production, human resources, design innovation and application of technology [1].

The definition of MSME performance is the achievement of overall work results compared to previously agreed planning regarding targets, targets, and criteria in a business organization with asset and turnover criteria determined by applicable regulations. The company's financial commitment to performance improvement and environmental disclosure, will give them stronger support from the community [2]. The disclosure of high-quality information as a signal enables external stakeholders including the market to assess and scrutinize the extent of the company's effort in preserving the environment, hereafter referred to as environmental performance [3].

Business sustainability is a condition where a company that is still carrying out its operations or that is going to run for an indefinite time since its establishment until now and can maintain the company's operations with the products produced.

2.2 Previous Research

Some previous studies have shown that financial inclusion has a positive impact on performance carried out by Permata Sari et al. [4], Silitonga et al. [5], Thathsarani and Jianguo [6], Liu et al. [7], Mujiatun et al. [8] and Ratnawati [9]. Furthermore, other studies have shown that financial inclusion has a positive impact on business sustainability carried out by Le et al. [10].

Financial literacy has a positive impact on research performance from Mujiatun et al. [8], Mabula and Ping [11], Agyei [12], Ye and Kulathunga [13], Farida et al. [14], Kasendah and Wijayangka [15], Buchdadi et al. [16], Wahyono and Hutahayan [17], Srisusilawati et al. [18], Frimpong et al. [19], Alshebami and Murad [20], Tumba et al. [21], Seraj et al. [22], Mawad et al. [23], Lontchi et al. [24], Silitonga et al. [5], Bidasari et al. [25], Widyastuti et al. [26], Timuneno et al. [27], Saragih et al. [28] and Wiagustini et al. [29].

Research conducted by Sabilla and Wijayangka [30], Yanto et al. [31], Wiagustini et al. [29], and Obi-Anike et al. [32] shows the results of research that financial literacy has a positive effect on business sustainability.

Research that produces a positive influence between government support on performance is carried out, among others Jin et al. [33], Hendrawaty et al. [34], Ardiyani et al. [1], Zulkiffli et al. [35], Aryanto et al. [36] and Suryani and Rita [37]. Government support does not have a positive effect on business performance as a result of research conducted by Pramestiningrum and Iramani [38]. Government support has a positive effect on business sustainability is a research conducted by Lee and Kim [39], Yemelyanov et al. [40], Jawa et al. [41], Trihastuti and Rachmawati [42], Hrivnák et al. [43], and Picas et al. [44]. Research that results in a positive influence of performance on business continuity is carried out Al Mamun et al. [45], Sobaih et al. [46], Naufal and Purwanto [47] and Maulana and Bachtiar [48].

2.3 Research Hypothesis

The hypothesis that will be put forward in this study will test the correctness of the data from the alleged influence between independent variables on dependent variables which are planned into seven hypotheses, namely (1) H1: the positive influence of financial inclusion on business performance, (2) H2: the positive influence of financial inclusion on business sustainability, (3) H3: the positive influence of financial literacy on performance, (4) H4: the positive influence of financial literacy on business sustainability, (5) H5: the positive influence of government support on performance, (6) H6: the positive influence of government support on business continuity and (7) H7: the positive influence of performance on business sustainability.

2.4 Theoretical Framework Model

Figure 1 shows the theoretical framework.

3 Research Methods

3.1 Types of Research

This research uses quantitative methods with a type of descriptive research, namely the type of scientific research that is clearly arranged about the type of data needed, who the data object is, and how large the sample needed has been compiled before data collection is carried out.

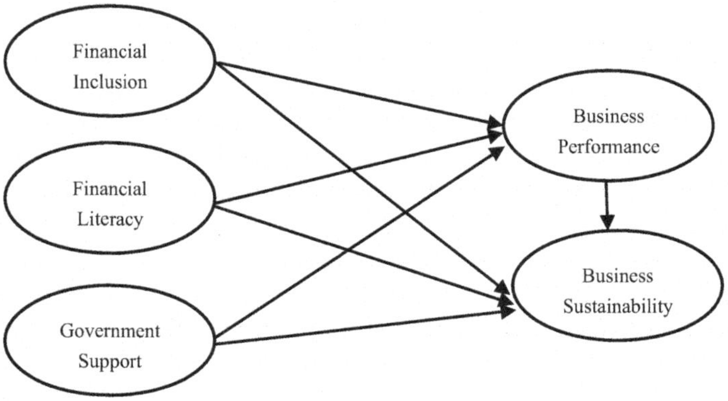

Fig. 1 Theoretical framework

3.2 Population and Sample

The population in this study is MSME business actors in the fisheries and marine sectors of Central Java province, totaling 2874 business actors. Meanwhile, the research sample was taken from the working area of assistance personnel assigned to the Regency/City area in Central Java Province as many as six marine and fisheries business assistance workers (TPU KP) who served in twelve Regencies/Cities covering Boyolali Regency, Karanganyar Regency, Sukoharjo Regency, Purworejo Regency, Kebumen Regency, Demak Regency, Pati Regency, Jepara Regency, Tegal Regency, Kudus Regency, Surakarta City and Semarang City.

The sampling technique used in this study is purposive sampling, which is taking a non-random sample from a predetermined population. Purposive sampling according to Sugiyono is sampling using certain considerations in accordance with the desired criteria to be able to determine the number of samples to be studied. The sample in this study is a group of MSMEs and individuals who run fisheries businesses based on population areas.

The sample criteria in this study are (1) MSME business actors who have received assistance from business assistance staff, (2) MSME business actors who have accessed financial institution services for the period 2021–2022.

3.3 Source and Type of Data

The data used in this study is primary data. According to Willy and Jogiyanto, primary data is raw data or data that has not been processed by anyone for certain purposes.

3.4 Data Collection Methods

To obtain this data used questionnaire method. In order to collect the data needed to meet research needs, data is obtained from the distribution of questionnaires to respondents where questions are first provided by researchers to support information data through questionnaires. To convert data from existing variables into data calculations, the author uses the Likert scale which is often used in social research. To facilitate the assessment of answers, the author will give a value from each question answer choice, namely a value of five for positive answers and a value of one for very negative answers.

3.5 Operational Definition of Variables

Research variables are the object of research, or what is the point in a study. In this study, the author's independent variables give the symbol x are financial inclusion (x1), financial literacy (x2) and government support (x3). While the dependent variables that the author gives the symbol y are performance (y1) and business continuity (y2).

Operationalization of variables is to break down the variables contained in the problem into the smallest parts so that the size classification can be known, this is done to make it easier to obtain the data needed in research. To provide understanding and interpretation of research variables, the following operational definitions are proposed:

3.6 Data Analysis Methods

This research was conducted using a type of quantitative research using a path analysis model. Then data analysis was carried out using the Partial Least Square (PLS) method. However, before conducting PLS analysis, it is necessary to conduct descriptive analysis, inferential statistical analysis, outer model and inner model.

4 Result of Research and Discussion

4.1 Sample Description

The data in this study comes from primary data in the form of questionnaires addressed to MSME business actors in the fisheries sector in twelve regencies/cities as sample areas including Boyolali Regency, Surakarta City, Karanganyar Regency,

Table 1 Characteristic of respondens

Characteristic	Category	Number of respondents	Prosentase (%)
Gender	Male	44	44
	Female	56	56
Field Business	Fish farming	21	21
	Fish processor	51	51
	Fish trader	18	18
	Fisherman	6	6
	Other	4	4
Age	20–30 year	6	6
	31–40 year	26	26
	41–50 year	33	33
	51–60 year	25	25
	61 year and above	10	10
Education	Primary school/equivalent	12	12
	Junior high school/equivalent	51	51
	High school/equivalent	12	12
	Bachelor	25	25

Sukoharjo Regency, Purworejo Regency, Kebumen Regency, Kendal Regency, Semarang City, Demak Regency, Kudus Regency, Pati Regency and Jepara Regency.

Data collection will be carried out from September 1, 2023 to October, 2023. Data collection is done online through google form. In accordance with the theory used, the questionnaire data processed in this study was as many as 100 respondents.

4.2 Descriptive Characteristic of Respondens

Based on the incoming data, it can be known the demographics of respondents as a reference in seeing the characteristics of respondents, a description of the data is made as shown in Table 1.

4.3 Descriptive Analysis of Research Variable

From the results of descriptive analysis of research variables, Table 2 shows that the total average financial inclusion variable is high with 3.87, financial literacy is high with 3.66, government support is very high with 4.28, high yield performance is 3.62 and business continuity is also high with 3.81 as shown in Table 2.

Table 2 Descriptive analysis of variables

Variable	Average	Criterion
Financial inclusion	3.87	Tall
Financial literacy	3.66	Tall
Government support	4.28	Very high
Business performance	3.81	Tall
Business sustainability	3.81	Tall

4.4 Inferential Statistical Analysis

As shown in Table 3, the descriptive statistical analysis, validity tests of research indicators were carried out. The indicators in this study are valid, all this is evidenced from the results of processing validity data, it is known that the loading factor number shows results above 0.7 as required that convergent validity testing can be known from the loading factor value for each construct indicator must be greater than 0.7.

As for testing the reliability of research variables, Table 4 shows that the variability is reliable because each variable from the data processing results shows Cronbach Alpha and Composite Reliability numbers above 0.7 and AVE values of more than 0.5 as required.

4.5 R-Square

As shown in Table 5, the measurement results show that the R square in this study is 41.8%, meaning that the variables of financial inclusion (x1), financial literacy (x2), government support (x3) on performance performance (y1), while the influence on business sustainability (y2) is 65.7%.

4.6 T-Test

As shown in Table 6, the results of data processing with Smart PLS version 4.0 testing the influence of endogenous variables on exogenous variables can be seen that t-statistics of the effect of financial inclusion variables on business performance are 3.092 > 1.96 with *p*-values of 0.001 < 0.050. Financial inclusion variables have an effect on business sustainability obtained t-statistics results of 2.772 > 1.96 and *p*-values of 0.006 < 0.050. Financial literacy variables affect performance, obtained t-statistical test results of 2.890 > 1.96 with *p*-values of 0.004 < 0.050. The effect of financial literacy variables on business sustainability was obtained by t-statistics results of 2.751 < 1.69 with *p*-values of 0.006 > 0.050. The variable of government support for the performance of t-statistics results 2.270 > 1.69 with *p*-values 0.023 >

Table 3 Test data validity

Indicators	x1 (financial inclusion)	x2 (financial literacy)	x3 (government support)	y1 (performance)	y2 (business sutainability)
x1.1	0.793				
x1.2	0.734				
x1.3	0.753				
x1.4	0.707				
x2.1		0.871			
x2.2		0.809			
x2.3		0.760			
x2.4		0.785			
x3.1			0.836		
x3.2			0.789		
x3.3			0.796		
x3.4			0.812		
x3.5			0.809		
x3.6			0.834		
x3.7			0.830		
y1.1				0.889	
y1.2				0.798	
y1.3				0.755	
y2.1					0.794
y2.2					0.715
y2.3					0.758
y2.4					0.705
y2.5					0.844

Table 4 Cronbach Alpha, Composite Reliability and AVE value

	Cronbach Alpha	Composite Reliability	AVE value	Hasil
Financial inclusion	0.736	0.835	0.599	Reliabel
Financial literacy	0.822	0.882	0.652	Reliabel
Government support	0.917	0.933	0.665	Reliabel
Business performance	0.747	0.856	0.666	Reliabel
Business sustainability	0.821	0.823	0.586	Reliabel

Table 5 R-square

	R square	R square adjust
Business performance	0.418	0.400
Business sustainability	0.657	0.643

Table 6 Test results between variables

Relationships between variables	Original sample (O)	Sample mean (M)	Standard deviation (STDEV)	T statistics (O/STDEV)	*P* values
Financial inclusion → performance	0.325	0.333	0.102	3.192	0.001
Financial inclusion → business sustainability	0.272	0.271	0.098	2.772	0.006
Financial literacy → performance	0.220	0.225	0.076	2.890	0.004
Financial literacy → business sustainability	0.212	0.217	0.077	2.751	0.006
Government support → performance	0.241	0.240	0.106	2.270	0.023
Government support → business sustainability	0.238	0.247	0.091	2.619	0.009
Performance → business sustainability	0.291	0.286	0.085	3.445	0.001

0.050 and the effect of government support variables on business sustainability with t-statistics 2.169 > 1.69 and *p*-values 0.009 < 0.050. Meanwhile, the performance variable affects business sustainability with t-statistics results of 3.445 > 1.69 and *p*-values 0.001.

5 Conclusions

5.1 Conclusion of the Hypothesis

5.1.1 The Positive Influence of Financial Inclusion on Business Performance

The higher the ability to access financial services provided by bank or non-bank financial institutions, the performance of MSMEs will increase. Financial inclusion can directly improve the performance of MSME business actors, which means all efforts to increase access and utilization of financial services so that it will improve the welfare of financial service users can improve the performance of MSMEs in the fisheries sector in Central Java Province. This is reinforced by the results of the loading factor smart PLS test version 4.0 which is reflected in the indicator of the utilization of financial services providing welfare benefits for MSME actors so that they can improve the performance of the business being run.

5.1.2 The Positive Influence of Financial Inclusion on Business Sustainability

There is a positive and significant influence of financial inclusion on business performance. This indicates that with the increasing financial inclusion of MSME business actors in the fisheries sector, easy access to financial services provided by financial institutions will be able to improve their business sustainability. Financial inclusion shows that it allows MSME business actors to access affordable financial services, the use of available financial services both savings and financing products and the quality of services provided by financial institutions. Sufficient funding to finance the company's long-term assets to facilitate business growth.

This is also evidenced by the results of loading factors on financial inclusion variables, indicators of access and use of savings and financing service products, as well as the causality of services provided by financial institutions so as to improve and maintain business continuity as evidenced by indicators of changes in assets, changes in turnover and changes in the number of customers each average period of more than 80%.

5.1.3 The Positive Influence of Financial Literacy on Business Performance

With the higher financial literacy owned by business actors, it will increase and improve business performance. The results of the study can also show that trustworthy financial literacy will be able to play an active role in improving MSME business performance. This is strongly evidenced by the results of loading factor research indicators that business actors have the ability and skills in analyzing and preparing financial statements.

5.1.4 The Positive Influence of Financial Literacy on Business Sustainability

With good literacy possessed by business actors as in questionnaire indicators, including the ability of business actors to make financial statements, financial management, making the right decisions in determining access to business financing is proven to affect the sustainability of the business being run. Financial literacy owned by fishery business actors in Central Java is very good, this is evidenced by the results of loading factors obtained by an average result of 80%.

5.1.5 The Positive Influence of Government Support on Business Performance

The success of MSMEs in Indonesia cannot be separated from government support. Government support in this case is carried out by the Ministry of Marine Affairs and Fisheries by appointing business assistance personnel in the fisheries and marine sector with one of the tasks is to assist MSMEs such as assistance and facilitation of access to financing. The financing needed by business actors is in the form of credit with cheap services such as KUR and other microloans that offer low-margin and unsecured loans.

With the assistance carried out by TPU KP has a positive impact on MSMEs, at least they feel cared for by the government in running their business. The assistance carried out by TPU KP provides enthusiasm and motivation for business actors.

For many years running a business, the government's attention through related agencies has been minimal, so the presence of these assistants is very helpful in improving their performance.

The performance of TPU KP is very good and the existence of TPU KP is very useful for business actors. This is evidenced by the results of the loading factor processing questionnaire data that the coaching carried out by TPU KP business actors to improve business performance by 83%.

5.1.6 The Positive Influence of Government Support on Business Sustainability

As is known that financial inculcation and financial literacy have a positive effect on the business sustainability of fishery MSMEs in Central Java. This is inseparable from the role of business assistants. Business continuity is strongly influenced by good financial management, decision making in accessing credit and the ability to make business books and financial statements. The role and duties of TPU KP in mentoring are to provide guidance on making books and financial statements even though they are only simple books and financial statements, the introduction of simple financial transaction bookkeeping such as SIAPIK from Bank Indonesia or other simple bookkeeping systems that can be done through the android system, but it is very useful for business actors.

MSMEs can understand and understand how to make financial transactions and know business development through financial statements so that the development of the business being carried out will be known.

Information about various kinds of financial service access facilities can open business actors' insights about financial institutions both banks and non-banks so that business actors can determine loan access options according to their needs and can avoid loans with high services, especially those that can avoid loan sharks or loan sharks.

The role of business assistants in facilitating access to capital and bookkeeping guidance has proven useful and is needed by fisheries business actors. This can be seen from the results of data processing with the results of variable loading factors of government support in questionnaire predictors on average above 80%.

5.1.7 The Positive Influence of Business Performance on Business Sustainability

With the support of three variables in this study, namely financial inclusion, financial literacy and government support, it can be concluded that with an increase in the business performance of fishery MSMEs, business continuity will certainly be maintained.

5.2 Recommendation

Based on the results of the research I described above, it can be concluded that the GISELA program at the Directorate General of Strengthening the Competitiveness of Marine and Fisheries Products (PDSPKP) by appointing marine and fisheries business assistance personnel (TPU KP) still needs to be continued. The existence of TPU KP as the government's long hand is very necessary, the function and role of assistance personnel can at least help the government and related agencies in carrying out their duties. The limited role of local agencies can be helped by the existence of TPU KP. Business actors also feel cared for by the government, although not directly, the existence of TPU KP is very helpful for business actors.

The lack of information and limited access received by fishery business actors so far regarding anything related to their business can be resolved with the existence of pendaming personnel. Complaints from the business they run they devote to assistance personnel so that it will help business actors to find a way out.

In future research that will be carried out, other variables can be added that can affect business performance and sustainability. Based on the results of the r-square influence, the dependent variable in the study affects the first independent variable, namely business performance is only 41.8%, meaning that there are still 58.2% of other variables that can affect it. The magnitude of influence on the second independent variable, namely business continuity of 65.7%, means that there are still 34.3% of other variables that can influence it.

References

1. Ardiyani, K., Syafnita, S., Aadilla: Peningkatan Kinerja UMKM Ditinjau Dari Model Pengelolaan Keuangan, Sumber Daya Manusia, Strategi Pemasaran, Dukungan Pemerintah Dan Umur Usaha. J. Ekon. Bisnis **8**, 56 (2021)

2. Ifada, L.M., Jaffar, R.: Does environmental cost expenditure matter? Evidence from selected countries in the Asia-Pacific region. Sustainability **15**, 1–16 (2023)
3. Ifada, L.M., Saleh, N.M.: Environmental performance and environmental disclosure relationship: the moderating effects of environmental cost disclosure in emerging Asian countries. Manag. Environ. Qual. Int. J. **33**, 1553–1571 (2022)
4. Permata Sari, B., Rimbano, D., Marselino, B., Aprilia Sandy, C., Ria Hairum, R.: Determinasi Literasi Keuangan dan Inklusi Keuangan Terhadap Kinerja dan Keberlangsungan Usaha UMKM. Owner **6**, 2865–2874 (2022)
5. Silitonga, H., Sianipar, R.T., Putri, J.A., Siregar, R.T.: Pengadopsian Standar Laporan Keuangan Sebagai Pemediasi Hubungan Antara Literasi Keuangan Dan Inklusi Keuangan Terhadap Kinerja UMKM Di Kota Pematangsiantar. Owner **7**, 1624–1634 (2023)
6. Thathsarani, U.S., Jianguo, W.: Do digital finance and the technology acceptance model strengthen financial inclusion and SME performance? Information **13** (2022)
7. Liu, Z., et al.: Financial inclusion and green economic performance for energy efficiency finance. In: Economic Change and Restructuring, vol. 55. Springer US (2022)
8. Mujiatun, S., Trianto, B., Cahyono, E.F., Rahmayati: The impact of marketing communication and Islamic financial literacy on Islamic financial inclusion and MSMEs performance: evidence from Halal tourism in Indonesia. Sustainability **15**, 9868 (2023)
9. Ratnawati, K.: The influence of financial inclusion on MSMEs' performance through financial intermediation and access to capital. J. Asian Financ. Econ. Bus. **7**, 205–218 (2020)
10. Le, T.H., Chuc, A.T., Taghizadeh-Hesary, F.: Financial inclusion and its impact on financial efficiency and sustainability: empirical evidence from Asia. Borsa Istanbul Rev. **19**, 310–322 (2019)
11. Mabula, J.B., Ping, H.D.: Financial literacy of SME managers' on access to finance and performance: the mediating role of financial service utilization. Int. J. Adv. Comput. Sci. Appl. **9**, 32–41 (2018)
12. Agyei, S.K.: Culture, financial literacy, and SME performance in Ghana. Cogent Econ. Financ. **6** (2018)
13. Ye, J., Kulathunga, K.M.M.C.B.: How does financial literacy promote sustainability in SMEs? A developing country perspective. Sustainability **11**, 1–21 (2019)
14. Farida, L., Afandi, M.F., Sularso, R.A., Suroso, I., Putri, N.A.: How financial literacy, innovation capability, and human capital affect competitive advantage and performance: evidence from creative MSMEs. Int. J. Sci. Technol. Res. **8**, 2300–2310 (2019)
15. Kasendah, B.S., Wijayangka, C.: Pengaruh Literasi Keuangan terhadap Kinerja UMKM. Almana J. Manaj. Bisnis **3**, 153–160 (2019)
16. Buchdadi, A.D., Sholeha, A., Ahmad, G.N., Mukson: The influence of financial literacy on SMEs performance through access to finance and financial risk attitude as mediation variables. Acad. Account. Financ. Stud. J. **24**, 1–16 (2020)
17. Wahyono, Hutahayan, B.: The relationships between market orientation, learning orientation, financial literacy, on the knowledge competence, innovation, and performance of small and medium textile industries in Java and Bali. Asia Pac. Manag. Rev. **26**, 39–46 (2021)
18. Srisusilawati, P., Malik, Z.A., Silviany, I.Y., Eprianti, N.: The roles of self efficacy and sharia financial literacy to SMES performance: business model as intermediate variable. F1000Research **10**, 1–14 (2022)
19. Frimpong, S.E., Agyapong, G., Agyapong, D.: Financial literacy, access to digital finance and performance of SMEs: evidence from central region of Ghana. Cogent Econ. Financ. **10** (2022)
20. Alshebami, A.S., Murad, M.: The moderation effect of entrepreneurial resilience on the relationship between financial literacy and sustainable performance. Front. Psychol. **13**, 1–12 (2022)
21. Tumba, N.J., Onodugo, V.A., Akpan, E.E., Babarinde, G.F.: Financial literacy and business performance among female micro-entrepreneurs. Invest. Manag. Financ. Innov. **19**, 156–167 (2022)
22. Seraj, A.H.A., Fazal, S.A., Alshebami, A.S.: Entrepreneurial competency, financial literacy, and sustainable performance—examining the mediating role of entrepreneurial resilience among Saudi entrepreneurs. Sustainability **14** (2022)

23. Mawad, J.L., Athari, S.A., Khalife, D., Mawad, N.: Examining the impact of financial literacy, financial self-control, and demographic determinants on individual financial performance and behavior: an insight from the Lebanese crisis period. Sustainability **14** (2022)
24. Lontchi, C.B., Yang, B., Shuaib, K.M.: Effect of financial technology on SMEs performance in Cameroon amid COVID-19 recovery: the mediating effect of financial literacy. Sustainability **15** (2023)
25. Bidasari, B., Sahrir, S., Goso, G., Hamid, R.S.: Peran Literasi Keuangan dan Literasi Digital dalam Meningkatkan Kinerja UMKM. Owner **7**, 1635–1645 (2023)
26. Widyastuti, M., Ferdinand, D.Y.Y., Hermanto, Y.B.: Strengthening formal credit access and performance through financial literacy and credit terms in micro, small and medium businesses. J. Risk Financ. Manag. **16** (2023)
27. Timuneno, A.Y.W., Malut, M.G., Dara, R.R., Latuheru, G.R.: Analisis Kontribusi Literasi Keuangan dan Inklusi Keuangan Terhadap Kinerja Keuangan Sektor UMKM Di Kota Kupang. Owner **7**, 1540–1552 (2023)
28. Saragih, F.M., et al.: Pengaruh financial literacy Terhadap bank's performance di Indonesia. J. Ecogen 165–176 (2023)
29. Wiagustini, N.L.P., Ramantha, I.W., Putra, I.M.W.: Financial literacy and financial behavior encouraging business sustainability by mediation of financial performance. Qual. Access Success **24**, 226–234 (2023)
30. Sabilla, S.O., Wijayangka, C.: Pengaruh Literasi Keuangan Terhadap Pertumbuhan Usaha Pada UMKM. J. Manaj. Bisnis **3** (2019)
31. Yanto, H., Kiswanto, Baroroh, N., Hajawiyah, A., Rahim, N.M.: The roles of entrepreneurial skills, financial literacy, and digital literacy in maintaining MSMEs during the COVID-19 pandemic. Asian Econ. Financ. Rev. **12**, 504–517 (2022)
32. Obi-Anike, H.O., Daniel, O.C., Onodugo, I.J., Attamah, I.J., Imhanrenialena, B.O.: The role of financial information literacy in strategic decision-making effectiveness and sustainable performance among agribusiness entrepreneurs in Nigeria. Sustainability **15**, 10416 (2023)
33. Jin, Z., Shang, Y., Xu, J.: The impact of government subsidies on private R&D and firm performance: does ownership matter in China's manufacturing industry? Sustainability **10** (2018)
34. Hendrawaty, E., Widiyanti, M., Sadalia, I.: CEO financial literacy and corporate financial performance in Indonesia: mediating role of sources of investment decisions. J. Secur. Sustain. Issues **6**, 627–636 (2018)
35. Zulkiffli, S.N.A., Zaidi, N.F.Z., Padlee, S.F., Sukri, N.K.A.: Eco-innovation capabilities and sustainable business performance during the COVID-19 pandemic. Sustainability **14** (2022)
36. Aryanto, A., Hanum, N., Syaefudin, R.: Faktor Teknologi, Organisasi, dan Lingkungan pada Penerapan Akuntansi Digital serta Dampaknya Terhadap Kinerja UMKM. Owner **7**, 632–643 (2023)
37. Suryani, I., Rita, M.R.: Pengaruh Program KUR dan BLT terhadap Kinerja UMKM dengan Strategi Diferensiasi sebagai Variabel Mediasi. J. Manaj. Keuang. **12**, 195–213 (2023)
38. Pramestiningrum, D.R., Iramani, I.: Pengaruh literasi keuangan, financial capital, kebijakan pemerintah terhadap kinerja usaha pada usaha kecil dan menengah di Jawa Timur. J. Bus. Bank. **9**, 279 (2020)
39. Lee, W., Kim, B.: Business sustainability of start-ups based on government support: an empirical study of Korean start-ups. Sustainability **11**, 1–20 (2019)
40. Yemelyanov, O., et al.: Microcredits for sustainable development of small Ukrainian enterprises: efficiency, accessibility, and government contribution. Sustainability (Switzerland) **12** (2020)
41. Jawa, S., Risnaningsih, R., Sulistiyowati, Y.: Dampak COVID-19 Dan Pemanfaatan Insentif Pajak Terhadap Keberlangsungan Usaha UMKM Di Kota Malang. Dr. Diss. Fak. Ekon. Univ. Tribhuwana Tunggadewi 10–27 (2021)
42. Trihastuti, D., Rachmawati, N.A.: Efektivitas Insentif Pajak dan Pemanfaatan Teknologi Informasi Terhadap Kelangsungan UMKM di Tengah COVID-19. Equity **24**, 223–242 (2021)

43. Hrivnák, M., Moritz, P., Chreneková, M.: What kept the boat afloat? Sustainability of employment in knowledge-intensive sectors due to government measures during COVID-19 pandemic. Sustainability **13** (2021)
44. Picas, S., Reis, P., Pinto, A., Abrantes, J.L.: Does tax, financial, and government incentives impact long-term Portuguese SMEs' sustainable company performance? Sustainability **13** (2021)
45. Al Mamun, A., Ibrahim, M.D., Bin Yusoff, M.N.H., Fazal, S.A.: Entrepreneurial leadership, performance, and sustainability of micro-enterprises in Malaysia. Sustainability **10**, 1–23 (2018)
46. Sobaih, A.E.E., Elshaer, I., Hasanein, A.M., Abdelaziz, A.S.: Responses to COVID-19: the role of performance in the relationship between small hospitality enterprises' resilience and sustainable tourism development. Int. J. Hosp. Manag. **94**, 102824 (2021)
47. Naufal, M.I., Purwanto, E.: Dampak Literasi Keuangan terhadap Kinerja Keberlanjutan UMKM (Studi Kasus Industri F&B Kecamatan Sumbersari Jember). Financ. Literacy Bus. Perform. Bus. Sustain. **16**, 209–215 (2022)
48. Maulana, B.R., Bachtiar, N.K.: Pengaruh Ketahanan Usaha, Kinerja Bisnis, Karakter Wirausaha, dan Pertumbuhan Usaha terhadap Keberlangsungan Usaha pada UMKM di Jawa Tengah. Bus. Econ. Conf. Util. Mod. Technol. **1**, 819–832 (2022)

Accountability Disclosure for Village Fund Management in Villages with Very Disadvantaged Status in North Aceh Regency

Cut Hadiatul Husna, Busra Bursa, Sri Anik, and Dista Amalia Arifah

Abstract Village Fund has been running for nine years to support the national development from the rural area. Most villages in Indonesia have grown into developing, developed and independent village. However, there are still many villages with disadvantaged and very disadvantaged status. The purpose of this study is to reveal the accountability of village fund management in villages with very disadvantaged status. The research data used in this study comes from the village development index (IDM) the ministry of villages, development of disadvantaged regions and transmigration (PDTT) in 2022 and 2023 for North Aceh Regency, Aceh province. The methodology used is explanatory research. The results showed that accountability of village fund management in villages with very disadvantaged status is very unaccountable for their rate of IKS, IKE and IKL grow very slowly and even down. The recommendation for the next study is to pay attention to factors that affect the accountability of village fund management which include village government competence, community participation, transparency and the use of information technology to have a positive impact on village development planning which is updated annually in the Village Index Development (IDM).

Keywords Accountability · Village funds · Village index development · Very disadvantaged villages

1 Introduction

The allocation of village funds to each village varies greatly, influenced by the number of villagers, number of poor people, the area of the village, IDM status, the financial capability of the village itself and others. The village fund ceiling of tens of trillions of rupiah every year since 2015 should bring significant changes to village development status, the progress of village government and villagers. Law No. 6 of 2014 on

C. H. Husna (✉) · B. Bursa · S. Anik · D. A. Arifah
Sultan Agung Islamic University, Semarang, Indonesia
e-mail: cut.hadiatul.husna@gmail.com

B. Awwad (ed.), *The AI Revolution: Driving Business Innovation and Research*, Studies in Systems, Decision and Control 524, https://doi.org/10.1007/978-3-031-54379-1_56

villages has provided opportunities for villages to independently manage village funds and develop their potential to improve the quality of life and welfare of village communities [1]. As the function of sharia governance that applies in Aceh province is for the benefit of all nature, both humans and nature so that they continue to grow in a better direction [2].

Village funds that have been running for nine years should have brought many changes to the direction of development and the level of welfare of rural communities. As the micro, small and medium enterprises (MSMEs) sector is the largest economic support sector [3]. However, much of the village fund does not manage greatly. This can be seen from the existence of villages with very disadvantaged status in 2023, especially in North Aceh Regency, Aceh Province. The number of independent villages can still be counted by fingers from a total of hundreds of villages. Here are the data in Table 1.

North Aceh as one of the regency in Aceh Province with a total of 852 villages has only 5 villages in independent status in 2022 (0.59%), the developed village are 34 or 3.99% villages, the developing village are 396 or 46.48%, the disadvantage village are 398 or 46.71% and the very disadvantaged village are 19 villages or 2.23%.

This data increases very slowly in 2023, the results of the IDM update in 2023 show an increase in the number of villages with independent status from 5 to 6 villages (0.70%). This means there is an increase of 1 independent village from the previous year or 0.11%. Developed villages are 75 villages or 8.80%. It means that there was an increase in the status of developed villages by 41 villages or 4.81% from the previous year. Developing villages are 491 or 57.63%. It means that there was an increase in the status of developing villages by 95 villages or 11.15% from the previous year. Disadvantaged villages decreased slowly from 398 to 265 villages, means 31.10%. This means that there is a decrease in the status of developing villages by 133 villages or 15.61%. Finally, the very disadvantaged or extremely backward villages also decreased from 19 to 15 villages or 1.76%. This means that there is a decrease in the status of 4 villages or 0.47%.

Table 1 Recapitulation of the 2023 update of the IDM in north Aceh regency

Village status	Year 2022		Year 2023		Range between years	
	Number of village	Percentage (%)	Number of village	Percentage (%)	Number of village (%)	Percentage
Independent	5	0.59	6	0.70	1	0.11
Developed	34	3.99	75	8.80	41	4.81
Developing	396	46.48	491	57.63	95	11.15
Disadvantaged	398	46.71	265	31.10	133	15.61
Very disadvantaged	19	2.23	15	1.76	4	0.47

Sources Ministry of Villages, 2023

The village index development was developed based on the conception that to achieve a developed and independent village. The sustainable development frameworks are social, economic, and ecological. These aspects become complementary forces, maintaining the potential and ability of villages to prosper village life [4]. This is a kind of government accounting standard. It will positively affect the quality of financial statements to increase accountability and transparency [5]. In addition, it is a kind of auditing tool that concludes the performance of the village government. So, this tool actually can change clients after being audited three times by the same auditor [6].

2 Literature Review

2.1 Accountability

The key success of improving the living standard of rural communities cannot be separated from the accountability of village fund management [7]. Accountability is an important issue for communities and organizations [8]. Accountability can improve the quality and performance of the institutions that are oriented towards the public interest because they are closely related to the welfare of the wider community [9, 10]. Therefore, a form of accountability and evaluation of the management and use of village funds is needed [11]. So, the goal of national development can be achieved, namely the acceleration of national development through villages, peripheral areas and borders [12]. This is also stated in PP (Government Regulation) No.6 of 2014 article 19 and article 2 of the Village Law, where village funds aim to support the development and empowerment of rural communities, covering the fields of governance, empowerment and development of rural communities [1].

Accountability in village government is very important because it is a form of accountability of the village government as the entity that manages village funds [13]. Apart from that, accountability village financial management is very helpful in making economic or financial decisions [14]. One of the efforts to restore economic, social and ecological conditions is to restore people's trust in the government by trying to create a clean and authoritative government or what is known as good governance which can be achieved with accountability [15]. Besides village development must meet the interests of the community [16].

Accountability has a significant effect on the effectiveness of financial management or budget performance [9, 17–20]. Therefore accountability is a key point in village development. Moreover, village funds that have been launched since 2015 in which the implementation of which refers to Minister of Home Affairs Regulation No. 20 of 2018 concerning village financial management is implemented according to principles of transparency, accountability, participation and implemented in an orderly and budgetary manner.

2.2 Stakeholder Theory

Stakeholder theory is a theory that says that companies must benefit parties who influence and are influenced, in relation to village government as village fund managers. Village funds must be able to influence internally to the prosperity of village communities and external village development must be able to contribute to the progress of the state. This theory is states about the companies must provide benefits to all parties involved, both internal and external [21]. In this case, the village government is required not only to be responsible to the government that disburses village funds, but also to prioritize the interests of the community which can influence each other's village government actions. Stakeholders in village government include the central government, village government, and village communities, Even more than that, in this theory the government system is also expected to consider environmental welfare [17].

2.3 Village Index Development (IDM)

Villages have a considerable role in efforts to determine the direction of vision to form better Indonesian human resources in the future. These efforts can be started by driving development, improving the quality of life, and welfare of rural communities. In terms of village development, the instruments that need to be known are how the actual problems experienced by the village and how big and strong the potential of the village is. The Ministry of Villages, Development of Disadvantaged Regions, and Transmigration through Permendesa No. 2 of 2016, has established several indicators that make it easier for villages to find out the extent of their village's level of independence. Just like the method of a bank's financial performance evaluation is by using ROA (Return on Assets) [23], the village fund management is evaluated by IDM besides another 7 tools more.

With this instrument, villages can recognize and explore information, the extent of the problems experienced and the potential they have to push villages out of the trap of the problem [24]. The instrument is known as the Building Village Index (IDM). PDTT is a ministry within the Indonesian government in charge of village and rural development, empowerment of rural communities, accelerated development of developing areas, and transmigration. Village ministry, PDTT classifies villages into five statuses, namely: independent, developed, developing, disadvantage and very disadvantaged villages. The classification is useful for sharpening the determination of village development status as well as a reference for development policy interventions [25]. Here are the index and the definition of each village status:

Independent Village (index > 0.8155) is the village that has the ability to carry out village development to improve the quality of life as much as possible for the welfare of the village community with social resilience, economic resilience, and ecological resilience in a sustainable manner.

Developed Village (index > 0.7072 to ≤ 0.8155) is the village that has the potential of social, economic and ecological resources, as well as the ability to manage them to improve the welfare of the village community, the quality of human life, and overcome poverty.

Developing Village (index > 0.5989 to ≤ 0.7072) is the village that has the potential to become a developing village, which has the potential of social, economic, and ecological resources but has not managed them optimally for improving the welfare of the village community, the quality of human life and overcoming poverty.

Disadvantaged Village (index > 0.4907 to ≤ 0.5989) is the village that has the potential of social, economic, and ecological resources but has not, or lack of management in an effort to improve the welfare of the village community, the quality of human life and experiencing poverty in its various forms.

Very disadvantaged Village (≤ index 0.4907) is the village that experience vulnerability due to natural disasters, economic shocks, and social conflicts so that they are unable to manage the potential of social, economic, and ecological resources, and experience poverty in its various forms.

IDM is a composite index from social resilience index, economic resilience index and environmental/ecological resilience index [24]. So, these three indexes should have an average index in 0.8 of each to get into independent village index category.

First, Social resilience index (IKS) consist of *social capital dimension* (indicators: social solidarity, tolerance, population security, social welfare), *health dimension* (indicators: health services, community empowerment, and health insurance); *education dimension* (indicators: access to primary and secondary education, access to non-formal education and access to knowledge); *settlement dimensions* (indicators: access to clean water, access to sanitation, access to electricity, and access to information and communication).

Second, the economic resilience index (IKE) consists of *economic dimensions* (indicators: diversity production of rural communities, available trade service centers, access to distribution/logistics, access to financial and credit institutions, economic institutions, and regional openness).

The last or the third, the environmental/ecological resilience index (IKL) consists of *ecological dimensions* (indicators: environmental quality and disaster-prone potential and disaster response).

IDM is updated annually on July. It is verified, validated and assigned village status in tiers. First, at the village level, the status of the village is determined by the village head and verified by the village's local assistant (PLD). Second, at the district level, the status of the village is determined by the district and verified by the village assistant (PD). Third, at the regency/city level, the village status is determined by the head of regent/city (Bappeda) and verified by the regency/city specialist village assistant (TAPM regency/City). Finally, at the provincial level the status of the village is determined by the head of the province (Bappeda) and verified by the province specialist village assistant (TAPM Province). On the deadline of July 15, the provincial IDM report have been entered into the Ministry of Finance, as one of the indicators in determining the amount of village funds for each village for the following year.

3 Research Methodology

The design of this study is explanatory research with the aim of providing a complete and detailed explanation about the study of alleviation in disadvantaged or very disadvantaged villages in North Aceh regency. This research is a kind of audit by the third party that can improve accountability of village fund management [6]. The data collection technique in this study used secondary data obtained from the village ministry. The analytical technique used in this study is qualitative descriptive analysis. The population of this study is all villages in North Aceh regency. The samples of this study were 15 villages with very disadvantaged status in North Aceh regency in 2023.

4 Findings

4.1 IKS, IKE and IKL

As shown in Table 2, from the observation of the Village Development Index (IDM) data 2023, then also compared with the update of IDM data in 2022, showed an insignificant increase in the village index. The details of the findings are as follows:

The problems of their social resilience index (IKS) are that: they do not have the availability of health workers (such as: midwives, doctors and other health workers), do not have access to village health center (poskesdes), village maternity center (polindes or posyandu), have not organized early childhood education (PAUD) activities, public activity center (PKBM/Package A-B-C) and other course activities, do not have community reading gardens or village libraries, do not have open public spaces, do not have sports groups and activities, still mass fights in the village, and finally do not have internet access at the village office.

The problems of the economic index (IKE) are the unavailability of public banking institutions and rural banks, the lack of access to credit, the unavailability of public economic institutions (cooperatives)/BUMDES, not available of food stalls, restaurants, hotels and inns, and there are no modes (public transportation, regular routes and operating hours).

Finally, the problems of the environmental/ecological index (IKL) are because there has been no planning or effort/action on potential natural disasters. Especially for disaster-prone villages, such as villages in watersheds (DAS) that have the potential to flood, flood management with the construction of river dikes, krip and dredging, is not the authority of the village, but the authority of the provincial River Basin Agency (BWS), in this case the village must build coordination with the Irrigation Public Works in realizing the construction of dikes so that IDM recommendations at this point can be implemented and the index will automatically increase.

Table 2 Recapitulation of the IKS, IKE and IKL update from 15 villages with very disadvantaged status

Village	IKS	IKE	IKL	IDM value 2022	IDM status 2022	IKS	IKE	IKL	IDM value 2023	IDM status 2023
Jurong	0.680	0.350	0.666	0.565	Disadvantaged	0.554	0.366	0.5333	0.4848	Very disadvantaged
Gunci	0.714	0.416	0.400	0.510	Disadvantaged	0.702	0.400	0.2	0.4343	Very disadvantaged
Seuneubok aceh	0.594	0.383	0.666	0.548	Disadvantaged	0.502	0.300	0.6667	0.4898	Very disadvantaged
Simpang tiga	0.485	0.366	0.600	0.484	Very disadvantaged	0.468	0.366	0.6	0.4784	Very disadvantaged
Buket linteung	0.4629	0.383	0.466	0.437	Very disadvantaged	0.451	0.383	0.4667	0.4338	Very disadvantaged
Tualang	0.577	0.266	0.533	0.459	Very disadvantaged	0.554	0.266	0.533	0.4514	Very disadvantaged
Baree blang	0.657	0.250	0.333	0.413	Very disadvantaged	0.645	0.250	0.333	0.4097	Very disadvantaged
Teupin rusep	0.674	0.383	0.400	0.485	Very disadvantaged	0.662	0.383	0.400	0.4821	Very disadvantaged
Cot lambideng	0.588	0.283	0.600	0.490	Very disadvantaged	0.594	0.250	0.6	0.4814	Very disadvantaged
Lancok	0.628	0.416	0.400	0.481	Very disadvantaged	0.645	0.416	0.4	0.4875	Very disadvantaged
Riseh tunong	0.622	0.366	0.333	0.441	Very disadvantaged	0.674	0.366	0.3333	0.4581	Very disadvantaged
Krueng lingka	0.514	0.350	0.600	0.488	Very disadvantaged	0.520	0.350	0.6	0.49	Very disadvantaged
Meunasah tuha	0.588	0.333	0.533	0.485	Very disadvantaged	0.645	0.350	0.466	0.4875	Very disadvantaged
Lubok pusaka	0.554	0.333	0.200	0.362	Very disadvantaged	0.611	0.333	0.2	0.3816	Very disadvantaged
Seureuke	0.531	0.416	0.333	0.427	Very disadvantaged	0.542	0.416	0.3333	0.431	Very disadvantaged

Sources Ministry of Villages, 2023

4.2 *Decreased Village Status and Seemingly Fixed Status*

There were three villages whose status decreased from disadvantaged to very disadvantaged status. They are Jurong village, Gunci village and Seuneubok Aceh village. Those villages were supposed to be in an increase in status to the developing status but instead, it declined. Besides the villages whose status decreased in the village index there were also villages whose status were permanent but actually decreased and permanent but increased very slowly. Here are the data in Table 3.

Table 3 shows that there are six villages with seemingly fixed status, but actually there was a decrease in the index besides it does not cause a change in the village status. They are Simpang Tiga village, Buket Linteung, Tualang, Baree Blang, Teupin Rusep and Cot Lambideng. The remaining six villages have hardly had a significant increase in the status progress. In fact, in 2022 only 12 villages have very disadvantaged status whereas in 2023 the number of villages in very disadvantaged status is sadly increased to 15 villages. It is because the total score is under 0.4907 in 2023 whereas in 2022 the total score is above 0.4907 to 0.5989.

5 Conclusions

The effectiveness of village fund management is still far from expectations where those village statuses have been increasing then decreasing. This can also be seen from the existence of villages with very disadvantaged status in 2023 in North Aceh Regency. Actually they have even had nine years of greater opportunity to take care of their own governance in accordance with the strength and potential of their villages. At the end, the number of independent villages can still be counted by fingers from a total of hundreds of villages in North Aceh Regency, Aceh Province.

In fact, when we refer to stakeholder theory, the village officials should be dedicated to advancing their village as a form of image of themselves who are authorized as leaders in the region. Moreover, Aceh is an area that focused on Islamic law that is *rahmatall lil aalamin* in the form of goodness (blessings), improvement and progress. This can be realized if a work plan is prepared to meet the principles of fairness and comply with recommendations from stakeholders [26], in this case, IDM recommendations. Since the factor that affects the success of an entity is the use of accounting information and business capital management [3].

Factors that cause an insignificant increase in the village index in villages with very disadvantaged status can be concluded as follows:

Table 3 Range of the IDM 2022 and 2023 update of the 15 villages with very disadvantaged status

Village	IDM value 2022	IDM status 2022	IDM value 2023	IDM status 2023	IDM range	Comment
Jurong	0.565	Disadvantaged	0.4848	Very disadvantaged	– 0.0808	Decreased
Gunci	0.510	Disadvantaged	0.4343	Very disadvantaged	– 0.076	Decreased
Seuneubok aceh	0.548	Disadvantaged	0.4898	Very disadvantaged	– 0.0583	Decreased
Simpang tiga	0.484	Very disadvantaged	0.4784	Very disadvantaged	– 0.0057	Permanent, but actually decrease very slowly
Buket linteung	0.437	Very disadvantaged	0.4338	Very disadvantaged	– 0.0038	Permanent, but actually decrease very slowly
Tualang	0.459	Very disadvantaged	0.4514	Very disadvantaged	– 0.0076	Permanent, but actually decrease very slowly
Baree blang	0.413	Very disadvantaged	0.4097	Very disadvantaged	– 0.0038	Permanent, but actually decrease very slowly
Teupin rusep	0.485	Very disadvantaged	0.4821	Very disadvantaged	– 0.0038	Permanent, but actually decrease very slowly
Cot lambideng	0.490	Very disadvantaged	0.4814	Very disadvantaged	– 0.0092	Permanent, but actually decrease very slowly
Lancok	0.481	Very disadvantaged	0.4875	Very disadvantaged	0.0058	Permanent but actually increase very slowly
Riseh tunong	0.441	Very disadvantaged	0.4581	Very disadvantaged	0.0171	Permanent but actually increase very slowly
Krueng lingka	0.488	Very disadvantaged	0.49	Very disadvantaged	0.0019	Permanent but actually increase very slowly

(continued)

Table 3 (continued)

Village	IDM value 2022	IDM status 2022	IDM value 2023	IDM status 2023	IDM range	Comment
Meunasah tuha	0.485	Very disadvantaged	0.4875	Very disadvantaged	0.0024	Permanent but actually increase very slowly
Lubok pusaka	0.362	Very disadvantaged	0.3816	Very disadvantaged	0.0191	Permanent but actually increase very slowly
Seureuke	0.427	Very disadvantaged	0.431	Very disadvantaged	0.0039	Permanent but actually increase very slowly

Sources Ministry of Villages, 2023

Villages do not make the results of IDM recommendations as guidelines in village development planning agreement (musrembangdes), the results of IDM updates provide recommendations on the direction of village development policies in an effort to improve village status. The recommendations consist of super priorities and priorities. It can be analyzed from the data that the recommendations are not implemented by the village government, where the recommendations are still appear in 2023 even though these recommendations have also been issued in 2022.

The increase in the development village index is not only influenced by internal village policies in the form of budgeting for activities that are recommended by IDM, but there are recommendations that cannot be followed up directly by the village, because it is not part of the village's authority, such as the procurement of health workers, and teachers, in this case the village government must coordinate with the health and education services.

6 Limitations and Future Research

Despite its contributions to accountability, this research has several limitations that should be noted. First, this study adopted a self-reported method to collect data, which may bias the results. Future studies should use both primary and secondary data due to the lack of understanding of village officials related to filling out IDM questionnaires that will have an impact on the rise or fall of village status. Second, the sample of this study represents a group of North Aceh community. Thus, the results cannot be generalized to other community.

Recommendation for the next study is to pay attention to factors that affect the accountability of village fund management which include village government competence, community participation, transparency and the use of information technology, so as to have a positive impact on village development planning which is updated annually in the Village Index Development (IDM). Villages that increase significantly in the index will get a performance allocation of 4% from the central government. It is because the government had already believed in the village government competence to handle the village fund.

References

1. Wikantosa, B.: Pemberdayaan Masyarakat Desa 2022, November, pp. 1–42 (2022)
2. Muhamed, N.A., Kamaruddin, M.I.H., Nazarie, W.N.F.W.M., Kamarubahrin, A.F., Radzi, R.M.: Shariah governance framework for Islamic donation-based crowdfunding platforms in Malaysia. In: Rafiki, A., Dana, L.P., Nasution, M.D.T.P. (eds.) Open Innovation in Small Business. Contributions to Environmental Sciences & Innovative Business Technology. Springer, Singapore (2023). https://doi.org/10.1007/978-981-99-5142-0_18
3. Rosalina, R., Ifada, L.M.: Improving Business Success Through the Use of Accounting Information and Business Capital Management, vol. 176 (2023). https://doi.org/10.1007/978-3-031-35734-3_28
4. Kemendesa Homepage, vol. 24(11) (2023). [Online]. Available: https://idm.kemendesa.go.id
5. Suprianto, E., Rusdi, D., Salim, A.: The models of improving the quality of government financial reporting. Lect. Notes Data Eng. Commun. Technol. **161**, 44–51 (2023). https://doi.org/10.1007/978-3-031-26281-4_5
6. Suprianto, E., Safitri, M., Mufarokah: The determinant factors of auditor switching in the company. In Edunomika **07** (2023)
7. Savitri, E., Andreas, A., Diyanto, V., Gumanti, T.A.: Accountability of village fund management in Riau Province. J. Ris. Akunt. Kontemporer **14**(2) (2022). https://doi.org/10.23969/jrak.v14i2.5550
8. Ridwan, M., Santosa, R.E.W.A., Suharto, S., Putri, A.P.R.Z.: Peran Moderasi Partisipasi Masyarakat Pada Hubungan Kompetensi Pengelola, Sistem Pengendalian Internal Dan Pemanfaatan Teknologi Informasi Terhadap Akuntabilitas Pengelolaan Dana Desa. Equilib. J. Ilm. Ekon. Manaj. dan Akunt. **12**(1), 92 (2023). https://doi.org/10.35906/equili.v12i1.1429
9. Garung, C.Y., Ga, L.L.: Pengaruh Akuntabilitas Dan Transparansi Terhadap Pengelolaan Alokasi Dana Desa (Add) Dalam Pencapaian Good Governance Pada Desa Manulea, Kecamatan Sasitamean, Kabupaten Malaka. J. Akunt. Transparansi Dan Akuntabilitas **8**(1), 19–27 (2020). https://doi.org/10.35508/jak.v8i1.2363
10. Taha, M.H., Eldalash, S.A.E., Mohamed, H.Z.: The role of accounting disclosure of sustainable development activities using metaverse in the field of education and training. In: El Khoury, R., Alareeni, B. (eds.) How the Metaverse Will Reshape Business and Sustainability. Contributions to Environmental Sciences & Innovative Business Technology. Springer, Singapore (2023). https://doi.org/10.1007/978-981-99-5126-0_7
11. Pangayow, B., Patma, K.: Planning, accountability and reporting of village financial management in Indonesia. Univ. J. Acc. Finan. **9**(2), 197–203 (2021). https://doi.org/10.13189/UJAF.2021.090208
12. Bawono, I.R., Kinasih, A.D.M., Rahayu, A.K.: Factors affecting accountability of village fund management through implementation of the village financial system (SISKEUDES). J. Acc. Invest. **21**(3) (2020). https://doi.org/10.18196/jai.2103160

13. Hoekstra, A., Huberts, L., van Montfort, A.: Content and design of integrity systems: evaluating integrity systems in local government. Public Integr. **25**(2), 137–149 (2022). https://doi.org/10.1080/10999922.2021.2014204
14. Khamainy, A.H., Ali, M., Setiawan, M.A.: Detecting financial statement fraud through new fraud diamond model: the case of Indonesia. J. Financ. Crime **29**(3), 925–941 (2022). https://doi.org/10.1108/JFC-06-2021-0118
15. Rahmah, Y., Rahmawati, V., Natariasari, R.: Pengaruh Pengungkapan Sustainability Report Terhadap Kinerja Keuangan Pada Perusahaan Go Public Di Indonesia dan Thailand. JOM **1**(1), 1–15 (2018)
16. Dewi, E.P.: Good Governance Dan Transparansi Rencana Strategi Terwujudnya Akuntabilitas Kinerja Pemerintah. J. Lentera Bisnis **7**(2), 81 (2018). https://doi.org/10.34127/jrlab.v7i2.236
17. Riberu, M.A.R., Fadjrih Asyik, N.: Pengaruh Akuntabilitas, Transparansi, dan Pengawasan terhadap Pengelolaan Dana Desa. J. Ris. Manaj. dan Bisnis **10**(10), 1–15 (2021)
18. Aliamin, L., Priantana, R.D.: Pengaruh Kompetensi, Akuntabilitas, dan Peran Pendamping. **11**(1), 151–164 (2023). https://doi.org/10.17509/jrak.v11i1.50434
19. Anas, D.E.A.F., Setiyowati, S.W.: Partisipasi Membentuk Hubungan Antara Aspek Akuntabilitas Dan Transparansi Terhadap Pengelolaan Dana Desa. J. Ilm. Akunt. Perad. **7**(2), 119–128 (2021). https://doi.org/10.24252/jiap.v7i2.26833
20. April, P., Akuntansi, J.R., Purnomo, B.S., Putri, C.: Akuntabilitas, Transparansi, Pengawasan dan Kinerja Anggaran Berkonsep Value For Money. Akuntabilitas, Transparansi, Pengawas. dan Kinerja Anggar. Berkonsep Value Money **6**(3), 467–480 (2018). https://doi.org/10.17509/jrak.v6i3.14886
21. de F. Langrafe, T., Barakat, S.R., Stocker, F., Boaventura, J.M.G.: A stakeholder theory approach to creating value in higher education institutions. Bottom Line **33**(4), 297–313 (2020). https://doi.org/10.1108/BL-03-2020-0021
22. Freeman, R.E.: A stakeholder theory of the modern corporation. Bus. Soc. **59**(2) (2020). https://doi.org/10.1177/0007650318773750
23. Assyarofi, A.A., Ifada, L.M.: The influence of financial ratios and Qardhul Hasan financing on financial performance in Islamic banks. Int. J. Curr. Sci. Res. Rev. **06**(08), 5549–5556 (2023). https://doi.org/10.47191/ijcsrr/v6-i8-20
24. Setyobakti, H.M.: Identifikasi Masalah dan Potensi Desa Berbasis Indek Desa Membangun (IDM) di Desa Gondowangi Kecamatan Wagir Kabupaten Malang. Wiga J. Penelit. Ilmu Ekon. **7**(1), 1–14 (2017)
25. Pranoto, G.T., Hadikristanto, W., Religia, Y.: Grouping of village status in West Java province using the Manhattan, Euclidean and Chebyshev methods on the K-mean algorithm. JISA (Jurnal Inform. dan Sains) **5**(1), 28–34 (2022). https://doi.org/10.31326/jisa.v5i1.1097
26. Ridho, M.A., Ifada, L.M.: Assessment of financial performance of RSI Sultan Agung Semarang through the Maqashid Sharia Concordance (MSC) approach. In: Mansour, N., Bujosa Vadell, L.M. (eds.) Islamic Sustainable Finance, Law and Innovation. Contributions to Management Science. Springer, Cham (2023). https://doi.org/10.1007/978-3-031-27860-0_38

A Review of Syamsul Yusof's Malaysian Box Office Hit, Mat Kilau: Kebangkitan Pahlawan (2022)

Fatimah Muhd Shukri, Mohd Saipuddin Suliman, Nur Asfarina Ab Aziz, Hidayat Hamid, Siti Nur Anis Muhammad Apandi, and Nik Zulkarnaen Khidzir

Abstract Malaysian film Mat Kilau: Kebangkitan Pahlawan (MKKP) directed by Syamsul Yusof tells the story of the Malay warrior Mat Kilau and his father Tok Gajah as they battle the British Empire to rescue Malaya from colonial rule. The film has become one of the highest-grossing Malaysian films of all time, earning RM97 million at cinemas in Malaysia, including Brunei, and Singapore in just 50 days. The film's success highlights the potential of films to support national economic expansion and cultural diplomacy. This paper uses qualitative research design and content analysis to analyze the film's portrayal of unity in Malay society in that colonial era. The five guidelines for social and psychological needs from "Uses and Gratification Theory" were adopted to validate the portrayal of unity application in the film and cinematic masterpiece production. Analysis indicates 60% of the narrative used is based on true stories, visuals, and dialogue to initiate emotional and aesthetic experiences to the audience, reflect the Personal Integrative Needs through heroics Mat Kilau character, Integrative Social Needs by highlighting the various Malaysian ethnicities act in the film and the Tension Release Needs for the audience when the climax of the film brings joy and relieve as results of the efforts and initiative to unite and fight together and gain a victory from British Colonizers. Therefore, these content analyses provide sufficient and significant evidence to conclude that MKKP entertains and effectively conveys the historical values, nationalism, and unity of Malaysian society.

Keywords Mat Kilau · Unity · Uses and Gratification Theory

F. Muhd Shukri (✉) · M. S. Suliman · N. A. Ab Aziz · H. Hamid · S. N. A. Muhammad Apandi · N. Z. Khidzir
Faculty of Creative Technology and Heritage, Universiti Malaysia Kelantan, Kota Bharu, Malaysia
e-mail: fatimah.s@umk.edu.my

B. Awwad (ed.), *The AI Revolution: Driving Business Innovation and Research*, Studies in Systems, Decision and Control 524, https://doi.org/10.1007/978-3-031-54379-1_57

1 Introduction

Films can depict Malaysians' identity, society, culture, and legacy. Like other factors, films support national economic expansion. On June 23rd, 2022, Studio Kembara and Skop Productions collaborated to create one of the most well-known movies, Mat Kilau: Kebangkitan Pahlawan, or MKKP for short. MKKP chronicles the history of the British occupation of Malaya. The film, which was directed by Syamsul Yusof, tells the story of the Malay warrior Mat Kilau and his father Tok Gajah as they battle the British Empire in a valiant effort to rescue Malaya from colonial rule. The film MKKP features several well-known actors, including Datuk Adi Putra, Beto Kusyairy, Fattah Amin, Wan Hanafi Su, Datuk Jalaluddin Hassan, Datuk Rahim Razali, Mubarak Majid, Shaharuddin Thamby, Namron, A. Galak, Ellie Suriaty, Khir Rahman Additionally, the film includes Indonesian actor Yayan Ruhian, who has made a name for himself in Hollywood [1, 2].

The film MKKP has successfully elevated the reputation of the Malaysian film industry. It has achieved remarkable milestones, including earning a ticket collection of RM97 million and becoming the highest-grossing film in the local and international categories after Munafik 2 and Avengers: End Game (with RM87 million collection in Malaysia). Additionally, MKKP has held the top position on the Malaysian box office chart for five consecutive weeks and has been listed in the Top 10 charts in nearly 20 countries, including Singapore, South Korea, Sweden, Romania, Saudi Arabia, Oman, the Philippines, Finland, and Greece. Whit that, MKKP was one of the top-grossing films in Malaysia [3].

This scenario opened possibilities for the local film industry to explore and utilize film aid and support from the government. It subsequently prompted the government to assist film producers in producing more national films through the National Film Production Fund (DEKAN) with a total allocation of RM50 million [4].

Malaysian films that are indicative of the country are still very rare, nevertheless. This raises the question of how well the film MKKP conveys the qualities of unity. This paper was conducted using qualitative research design and library research method, while content analysis was used to analyze film MKKP. This paper applied Katz, Guarevitch, and Haas's Uses and Gratification Theory. This article aims to analyze MKKP films based on the Uses and Gratification theory. To achieve the purpose of this article, the analysis is broken down into several subtopics, namely literature review, methodology, uses and gratification in MKKP, and conclusion.

2 Literature Review

Literary studies are divided into two subtopics: the influence of unity elements in the development of identity and the unification of the ummah in the context of films. Before discussing these two subtopics further, the meaning of "unity of the ummah" will be explained here. According to Leha @ Zaleha [5], Ummah refers to "…all citizens of a country." On the other hand, unity of the ummah in Islam refers to "the cooperation of people in efforts to collectively build and maintain the security, well-being, and prosperity of the country" [5]. Therefore, we would like to define "unity of the ummah" in this paper as the community of Malaysian society, regardless of race, ethnicity, or religion, coming together to preserve harmony, security, well-being, and the prosperity of the nation. Based on this definition of "unity of the ummah" in this paper, the literature review for the two aforementioned subtopics is detailed below. Lee [6] explains that knowledge and education are the main pulse to the maturity, and knowledge skills of every human being in carrying out a task and continuing the survival of life.

Nevertheless, society should be aware that all knowledge is not acquired and learned completely in the classroom but can also be learned through the process of experience they experience or through watching movies. Therefore, using film as a teaching aid is very appropriate because research has proven that film is considered an effective entertainment medium to encourage critical thinking and students' ability to see and evaluate each message presented visually. The presence of epic films such as Mat Kilau is likely to provide great opportunities for the community to acquire various knowledge and learn various things to further improve the soft skills of each individual. Hisham [7] explains that the formulation of policies through the National Creative Industries Policy (Dasar Industri Kreatif Negara or DIKN) is seen to have succeeded in re-injecting a new spirit into the film industry in Malaysia. The DIKN approved in November 2010 has also helped push the country towards a high-income country due to the rapid growth of the country's creative industry as a source of employment and income for the country. This situation positively affects many parties because the success of local films, including the film Mat Kilau, is seen not only to provide entertainment and education but also to help improve the national economy and provide employment opportunities for the community.

2.1 The Element of Unity

The issue of division and ethnic tensions in Malaysia is not new; various efforts have been made to reduce the gaps and conflicts between different communities. Razid [8] argues that Malaysia is a heterogeneous country consisting of diverse ethnicities, religions, and origins, but through unity, it can foster national consciousness.

Filmmakers in Malaysia now use films as a platform to convey patriotic sentiments and promote unity in the multicultural society. This aligns with the suggestion of filmmakers emphasizing the importance of harmony among different ethnic groups and the primary role of empathy in maintaining the integrity of a nation with diverse cultures and ethnicities, as seen in Malaysia [8]. Some films produced in Malaysia that focus on unity include "Paloh" (2003), "Muallaf" (2008), "Talentime" (2009), "Setem" (2009), "Estet" (2010), and "Ola Bola" (2016).

Films that integrate elements of unity aim to educate Malaysians to live harmoniously as one nation, complementing each other's differences. According to Shariff et al. [9], audiences are educated and motivated to accept and adapt to differences in ethnicity, religion, political sentiments, and ethnicity when such issues are explored in films. The element of unity should seek common ground among individuals from diverse backgrounds and beliefs. From there, a united ummah can emerge, capable of living in harmony despite differences in various aspects. Therefore, this paper specifically discusses the aspect of the unity of the ummah in the content of the film MKKP, based on the Uses and Gratification Theory.

2.2 The Portray of Unity in Malaysian Film Content

Films are a symbol of unity that has the potential to unite audiences [9]. Films that feature actors from different ethnicities are seen as an effort to foster unity in the multicultural society in Malaysia. The famous Malaysian actor and director, the late Tan Sri P. Ramlee, who was known for his heartwarming films, was a pioneer in bringing together actors from different ethnicities such as Malays, Chinese, and Indians to work together in films under his direction, such as "Pendekar Bujang Lapok" (1959).

This effort is a promotion of national unity, especially in Malaysia. According to Adnan (1988), the Malaysian government began to see films as a platform to foster national identity and unity. This is evident in the production of films that revolve around the themes of struggle and unity, which generate substantial returns and leave a profound impact on audiences, such as the film MKKP.

The film director, the late Mamat Khalid, also suggested that filmmakers in Malaysia produce more films that focus on ethnic issues [10]. This is because the message of unity appears to be more critical, and audiences will appreciate the importance of unity among the ummah. He saw that films that solely focused on unity were less successful in attracting the interest of some ethnic groups, such as the films directed by the late Yasmin Ahmad. The Malay community protested against the content of her films, such as "Sepet" (2005) and "Gubra" (2006), but they managed to capture the attention of the Chinese and Indian communities.

Studies on the unity of the ummah have been conducted by scholars in the past. Furthermore, the concept of unity has been presented in several Malaysian films and has been shown to the public. However, the film MKKP, which was produced in 2022 and portrays the unity of the ummah (the Malay community), has not been discussed in the context of unity in films.

3 Methodology

This paper is based on a basic research design with a qualitative study design. A literature review is also applied to collect secondary data in the form of print materials such as books, theses, magazines, and newspapers, as well as digital materials such as digital articles, digital newspapers, digital theses, and authoritative portals. Primary data is in digital form, namely through the content of the film MKKP. Content analysis is applied as a research instrument to analyze the content of the film MKKP. The content analysis conducted through the film MKKP is personal. The author, who is one of the viewers of the film MKKP, becomes an individual who evaluates and analyzes the narrative of the film MKKP. Therefore, the Uses and Gratification Theory is applied based on the author's personal experience after watching the film MKKP. In addition, secondary sources support the author's personal views in evaluating the film MKKP.

This paper applies the Uses and Gratification Theory introduced by Katz, Gurevitch, and Haas in 1973–1974 [11]. This theory is closely related to media and aims to explain the satisfaction of the audience. This theory introduces five guidelines for social and psychological needs that media users need to fulfill. The five guidelines for social and psychological needs that media users need to fulfill are detailed below:

1. Cognitive needs: the need for information, knowledge, and understanding.
2. Affective needs: the need for emotional experiences, such as pleasure, arousal, and relaxation.
3. Personal integrative needs: the need for personal identity, stability, and reinforcement.
4. Social integrative needs: the need for interaction with others, social identity, and companionship.
5. Tension release needs the need for escape and diversion from daily routines and problems.

3.1 Uses and Gratification Theory by Katz, Gurevitch, and Haas (1973–74)

Cognitive Needs

Cognitive needs refer to the need for information and knowledge to improve understanding. Cognitive needs will be examined in the overall content of the film MKKP. Questions such as Who is Mat Kilau in the film MKKP? Who are the other characters involved in the film? And how does the narrative of the film MKKP retell these characters in the form of film content to be understood by viewers who choose to watch this film?

Affective Needs

Affective needs refer to the need for aesthetic or emotional experiences. For affective needs, the question is whether the narrative of the film MKKP is successful in providing emotional and aesthetic experiences to the audience. This question is examined based on the writer's personal experience while watching the film MKKP. In addition, the visuals and dialogue in the film MKKP can also evoke emotions in the writer. Therefore, affective needs are more focused on the emotions of the writer.

Personal Integrative Needs

Integrative needs refer to the need to strengthen beliefs, status, or credibility, which has cognitive and affective components. To examine the integrative needs, it is measured how the narrative of the film MKKP reflects the characters to the extent that it can create beliefs, status, and credibility of the characters to the audience who watch the film MKKP.

Social Integrative Needs

Social integrative needs refer to the need to strengthen relationships with friends and family. For this need, firstly, it is examined and evaluated by the production crew who worked together to make the film MKKP a success. Secondly, it is also evaluated and examined based on who the audience is that comes to watch this film MKKP. These two things become aspects to examine and discuss social integrative needs in the film MKKP.

Tension Release Needs

The need to reduce tension or the need to rest is a need sought by the audience when choosing a film to watch. Therefore, to examine the need to reduce tension, it is related to social integrative needs. The question asked in the need to reduce tension is whether watching this film reduces tension for the audience. Does the film MKKP have a positive impact after watching the film MKKP? The assessment of the need to reduce tension is also more towards the personal assessment of the author who watched this film MKKP.

4 Uses and Gratification in MKKP (2022)

The discussion of the content of the film MKKP is divided into five subtopics based on the five guidelines for social and psychological needs that media users need to fulfill. Therefore, the following is a further discussion of the content of the film MKKP based on the Uses and Gratification Theory, the five guidelines for social and psychological needs that media users need to fulfill.

4.1 The Cognitive Needs Portrayed in MKKP

The film MKKP is inspired by the character and heroic story of Mat Kilau or his real name Mat Kilau bin Imam Rasu, played by Dato Adi Putra. This film is purely fictional. The scene of the film MKKP begins with a village atmosphere in Kuala Tembeling, and the residents are seen working in the rice fields. The appearance of the British disrupted the security of the village residents, and the British came to arrest three young men who were said to be rebels. Then, a verbal altercation occurred, causing the British to kill all the village residents in the rice fields. This event became a turning point in the narrative of the film MKKP and caused the Pahang leaders to agree to oppose the British. This is where Mat Kilau and his friends play the role of heroes who oppose the cruelty of the British. The film ends with the victory of the Malays with the killing of the British leaders in the MKKP narrative.

Based on the visuals and dialogue revealed in the narrative of the film MKKP, the audience is introduced to several heroic figures in Pahang. Among the characters introduced in the film's narrative are Dato Raja Haji Muhamad, Mufti Botok Haji Uthman, Tok Gajah, Dato' Bahaman, and Haji Sulong Semantan, who are the leaders of Pahang. According to the original scriptwriter of the film MKKP, 60% of the narrative is based on true stories. Although the film MKKP is a work of fiction, some parts of it are based on facts. The narrative developed based on real figures does create a sense of believability for the audience as they process the information while immersed in the dramatic storyline of the film.

4.2 The Affective Needs to Be Portrayed in MKKP

The visuals and dialogue in MKKP, evoke emotions in the audience. For example, from minute 00:03:40 to minute 00:03:54, Salmah (played by Ellie Suriaty Omar) expresses an emotional dialogue to the British soldiers after the British soldiers hired killers to stab her husband. The dialogue goes:

> There is no such thing as Malays rebelling on their land, we are defending our rights. You come to our land to rob our resources (Salmah, MKKP, 2022).

Furthermore, Salmah added by saying:

> For Allah's sake, for the sake of the Prophet, I stand on this land of Pahang under the sky of Allah that I uphold, you…, you…, you…, will witness the rise of the Malays against the cruelty of the White colonizers (Salmah, 00:04:18–00:04:41, MKKP, 2022).

The above dialogue is followed by the British officials shooting and killing Salmah. This incident caused the villagers at the scene to rise against them. A mass murder occurred, carried out by the British. Everyone at the scene, including the elderly, women, and children, was not spared and were all killed by the British. Viewers also had the opportunity to witness the emotional struggle of Mat Kilau and

Fig. 1 Rokiah who was nine months pregnant was assaulted by one of the British soldiers. *Source* from MKKP film streaming platform, Netflix

his friends to protect the villagers as well as their village, which was full of tragedy and emotions, even though it happened in one scene.

Among other tragic scenes was the death of Rokiah, the wife of Wahid, a comrade-in-arms of Mat Kilau who also fought against the British. Figure 1 depicted from the scene where the soldiers force themselves into Wahid's house and pull Rokiah who is nine months pregnant and shove her onto the ground. Apart from that, one of the soldiers kicked and assaulted Rokiah who was on the ground and this action led to Rokiah and the unborn baby's death.

This assault was witnessed by Wahid, Rokiah's husband, and Mat Kilau who was rushing home after hearing the news that the British send their soldiers to search for Wahid. Following up on that, both saw how the soldier treat Rokiah. However, Mat Kilau restrains Wahid from helping Rokiah because they are out numbers compare to the British soldiers in that scene. Figure 2 explains Wahid's emotion while seeing his wife being kicked and murdered. These two examples from visuals and dialogues captured the audience's feelings and attention and serve as the affective needs that have been portrayed in MKKP.

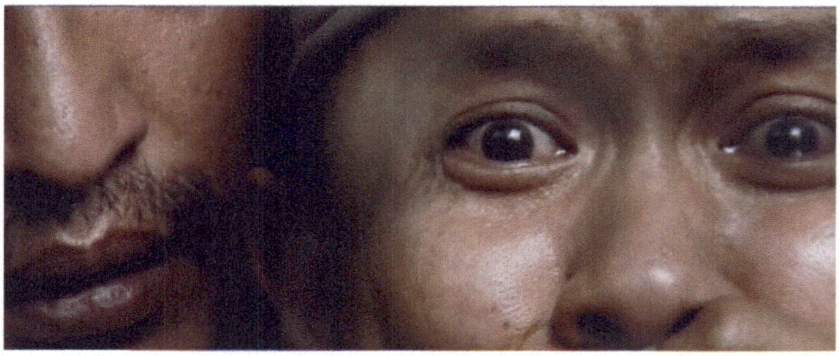

Fig. 2 Wahid saw the assault done on Rokiah and was blocked by Mat Kilau. *Source* from MKKP film streaming platform, Netflix

The aesthetic experience can nurture the spirit of love for the homeland and promote humanism and unity among the people. The audience also has the opportunity to explore the emotional experience of the struggles of the characters in Pahang, to the extent that some wear traditional Malay attire such as "kebaya" and men wear "silat" attire and "tanjak" when going to the cinema, in the capacity of unity and appreciation of aesthetic and humanistic values. Furthermore, The British Colonizers were brutal and unforgivable to the Malays, as viewers developed hatred and feel the struggle of Mat Kilau and his friends through the narrative. Overall, it is an action-packed narrative filled with up-and-down emotions that will bring joy and satisfaction throughout the film.

4.3 The Personal Integrative Needs Portrayed in MKKP

For integrative needs, the narrative of the film MKKP tries to highlight and reflect the heroic story and greatness of Mat Kilau in fighting against British colonization. From the perspective of heroism, the film MKKP is filled with silat art scenes presented by Mat Kilau and his comrades. Meanwhile, the British colonizers used advanced equipment such as guns and cannons to fight against Mat Kilau and his fellow fighters. This reflection of heroism can instill a sense of confidence in the status and credibility of Mat Kilau as a hero who opposes British power.

At this stage, the production crew tried their best to share their skills and help each other regardless of race to strengthen the belief that the Mat Kilau film could change the landscape of the local film industry. The confidence of the audience grew when the government and NGOs also supported the screening of this film. Undoubtedly, the credibility and impact of this film are difficult to question, even though not all creations are perfect. Indeed, differences in opinion are a blessing and should be celebrated.

4.4 The Integrative Social Needs Portrayed in MKKP

The making of the film MKKP involves various ethnicities in Malaysia. Malay, Chinese, and Indian ethnicities were found together in developing the production of the film MKKP. The Malay ethnicity was appointed as officers in the film MKKP, while the Chinese ethnicity was chosen as supporting actors who worked in gold mines, and the Indian ethnicity, Punjabi, as British soldiers. The unity of the production crew is very important in the success of film production. In the context of the film MKKP, the daytime and nighttime scenes are seen as balanced and equal. Therefore, the production crew worked together tirelessly to achieve the goal of producing this historical masterpiece. Certainly, gratification will be obtained equally with each person's effort.

In addition, the audience also consists of families, friends, and various ethnicities in Malaysia. They come together in a cinema to watch the film MKKP. The author himself witnessed this. The author saw several families bringing their entire family to watch the film MKKP. There were also audiences from various ethnicities in Malaysia who came to watch the film MKKP. Such moments are the best time to gather friends and family to watch this historical story. As a patriotic film, it will strengthen the relationship among Malaysian as they feel the need to protect their home, family, friends, and nation as opposed and presented in the film. This unity should be continued in the future to create unity and the theme of a harmonious and patriotic Malaysian family.

4.5 The Tension Release Needs

Viewers can take a deep breath and smile at the experience they had after viewing MKKP. The film's climax and resolution bring joy and relief as Mat Kilau and his friends gain victory over British Colonizers. All the efforts and initiative to unite and fight together were effective and they finally deserve their reward which is their land and soil, finally free without intruders.

Indeed, films are a reflection of society. If we want to see the milieu of society, then watch a film. However, we should not be too rigid in arranging the facts in a film product, as it is not like a Ph.D. thesis that requires references, validity, and so on. It is enough to appreciate the storytelling of a film and at the same time relax the mind after a tiring day of life.

5 Conclusion

This paper concludes that the MKKP films were successful in the Malaysian film industry to support economic growth. On the other hand, MKKP not only entertains but also effectively conveys the historical values, nationalism, and unity of Malaysian society.

The presentation of community content through Malaysian films is a concept that needs to be embraced by Malaysian society. Based on an analysis of the film MKKP, five guidelines for social and psychological needs that media users need to fulfill have been presented to the audience and production crew. All five of these needs, namely cognitive, affective, integrative, social integrative, and stress reduction, fulfill the audience's needs.

Based on the author's assessment, these five needs, cognitive, affective, and social integrative needs are studied more than integrative and stress reduction needs. This is because these three needs are more prominent and have an impact on the audience, as seen during the screening in the cinema. The author heard whispers and cheers from the audience, and the cinema was filled with family members.

The author also felt the cognitive, affective, and social integrative needs in choosing the film MKKP as a source of entertainment and knowledge. A film that revolves around patriotic spirit is seen as important in promoting the spirit of unity before it is lost in the tide of modernization. In this regard, filling the film MKKP with historical values and the struggle of Malay heroes during British colonization in Pahang is seen as an effective effort in uniting the Malaysian people of various races.

In conclusion, the Uses and Gratification theory can be applied to the analysis of film content by considering how individuals use films to fulfill specific needs and gratify certain desires. It also helps us understand the emotional impact of films on the audience. By examining these aspects, we can gain insights into the motivations behind film consumption and the role films play in satisfying individual needs and emotions.

References

1. Bernama: Filem Mat Kilau sarat dengan mesej patriotisme. Astro Awani. https://www.astroawani.com/berita-hiburan/filem-mat-kilau-sarat-dengan-mesej-patriotisme-367608. Last accessed 22 June 2022
2. Bernama: Kutipan filem Mat Kilau cecah RM90 juta selepas 33 hari tayangan. Sinar Harian. https://www.sinarharian.com.my/article/212939/berita/nasional/kutipan-filem-mat-kilau-cecah-rm90-juta-selepas-33-hari-tayangan. Last accessed 25 July 2022
3. Begum, M.: Malaysian film 'Mat Kilau Kebangkitan Pahlawan earns RM12.2mil at box office in 4 days. The Star Online. https://www.thestar.com.my/lifestyle/entertainment/2022/06/27/malaysian-film-mat-kilau-kebangkitan-pahlawan-earns-rm122mil-at-box-office-in-4-days. Last accessed 27 June 2022
4. Bernama: Kejayaan luar biasa, Mat Kilau disebut dalam bajet 2023. https://www.buletintv3.my/santai/kejayaan-luar-biasa-mat-kilau-disebut-dalam-bajet-2023/. Last accessed 2022/10/07
5. Leha @ Zaleha Muhamat: perpaduan Ummah Menurut Kacamata Islam. Kertas kerja dalam seminar Yayasan Dakwah Islamiah Malaysia (YADIM) dan Kolej Universiti Kejuruteraan Utara Malaysia (KUKUM), 27–29 Februari 2004, di Eagle Bay Resort Resort, Pulau Langkawi, Kedah (2004)
6. Lee, S.C.: Integrating entertainment and critical pedagogy for multicultural pre-service teachers: film watching during the lecture hours of higher education. Intercult. Educ. **30**(2), 107–125 (2019)
7. Hisham, H.T.: Cabaran industri filem Melayu, pandangan penggiat industri dan penonton. Doctoral dissertation, University of Malaya (2015)
8. Razid, M.A.: Patriotisme Nilai Jati Diri Dorong Rakyat Gigih Bangunkan Negara. Dlm. Ariah Judah. Mentari. Bil 2. Putrajaya: Bahagian Penerbitan Dasar Negara, Jabatan Penerangan Malaysia (2017)
9. Shariff, S.Z., Sualman, I., AdiIrawan, E.D.: Fostering unity among Malaysians: a case study on the local film industry. Int. J. Humanities Soc. Sci. **3**(10) (2013)
10. Umi Atikah Mohd Kamil: Kritikan Sosial Filem Talentime (2009). Kertas Penyelidikan Ijazah Sarjana Pengajian Melayu, Akademi Pengajian Melayu Universiti Malaya. Universiti Malaya, Kuala Lumpur (2017)
11. Rosengren, K.E.: Uses and gratifications: a paradigm outlined. In: Blumler, J.G., Katz, E. (eds.) The Uses of Mass Communications: Current Perspectives on Gratifications Research, pp. 269–286. Sage, Beverly Hills, CA (1974)

Correctional Through Skills: Vocational Training for Prison Inmates in Malaysia

Siti Atikah Binti Rusli, Rita Lo, Norhafizatul Akma Hamzah, and Philip Wong Pong Weng

Abstract The high rate of recidivism has been a persistent problem in the criminal justice system, with ex-inmates facing significant challenges upon their release, including a lack of employment opportunities, social stigma, and limited access to resources. In order to integrate prison inmates more effectively into society, vocational training programs were introduced in Malaysian prisons to provide inmates with practical skills and increase their employability post-release. However, the effectiveness of these programs in reducing recidivism rates and preparing inmates for employment was yet to be established. This study aimed to evaluate the effectiveness of vocational training programs in three Malaysian prisons (Sungai Udang Prison Melaka, Kajang Prison, and Bentong Prison). Data were obtained via 12 semi-structured interviews with correctional officers from the respective prisons. Findings of this research contributed to a better understanding of the role of vocational training programs in reducing recidivism and facilitating successful rehabilitation of inmates into productive members of society.

Keywords Vocational training · Prison inmates · Recidivism

1 Introduction

Correctional institutions were frequently viewed as punishment facilities, but it was important to remember that they also provided opportunities for education and rehabilitation. The public generally support correctional institutions providing education and training to inmates in order to instil useful knowledge and skills that could give them a new life [13]. In 2011, there were 688,384 male and female inmates in state and federal prisons in the United States, with an average of 1885 people released each day [4]. The statistics, however, painted a bleak picture of recidivism rates, with nearly two-thirds of ex-inmates re-arrested within three years of their release, and nearly half of them reincarcerated. There were 13,896 criminal repeat offenders apprehended in

S. A. B. Rusli (✉) · R. Lo · N. A. Hamzah · P. W. P. Weng
School of Hospitality and Service Management, Sunway University, Kuala Lumpur, Malaysia
e-mail: sitiatikahrusli27@yahoo.co

B. Awwad (ed.), *The AI Revolution: Driving Business Innovation and Research*, Studies in Systems, Decision and Control 524, https://doi.org/10.1007/978-3-031-54379-1_58

Malaysia in 2018 [11]. Investigating the efficacy of vocational training programmes was critical to promoting successful reintegration of ex-offenders into society and lowering recidivism rates. Examining how these programmes worked and identifying any barriers to completion or securing employment after release could provide valuable insights into how to improve the impact of these programmes and increase their effectiveness in reducing reoffending.

One-third of ex-inmates have difficulty finding work due to a lack of educational background and work experience [20]. Nonetheless, it was discovered that ex-inmates who had received vocational training while incarcerated had a better chance of finding work after their release. Ex-inmates who participated in employment programmes while incarcerated had a higher likelihood of finding work after their release [16]. Inmates who participated in educational programmes, including vocational training, were 43% less likely than those who did not [7]. In fact, unemployment was a significant predictor of recidivism, with many inmates returning to prison due to a lack of job opportunities. Vocational training provided individuals with the skills and knowledge necessary to secure and maintain employment, and it could significantly reduce recidivism rates [25].

Overall, these studies highlighted the importance of vocational training programs in promoting successful reintegration of ex-inmates into society and reducing recidivism rates. By equipping inmates with the skills necessary to obtain employment upon their release, these programs could help break the cycle of recidivism and improve the chances of a successful re-entry into society.

2 Literature Review

The literature review for this study was structured into three main bodies of scholarship: Vocational Training, Previous Studies related to Vocational Training for Ex-Inmates, and Theoretical Underpinnings. This funnel approach begins with a general overview of literature related to vocational training and gradually narrows down to literature specifically relevant to vocational training for ex-convicts and their employability. Furthermore, the theoretical underpinning was presented to guide the conceptual foundations of this study.

2.1 Vocational Training

Vocational education, as defined by UNESCO Institute for Statistics, is a type of education that aims to "equip learners with knowledge, skills, and competencies required for specific occupations, trades or classes of occupations or trades" [6]. Vocational education and training (VET) have been a method of education that highlights the acquisition of practical skills and the application of learning outcomes in the workplace [14]. VET differs from general education in that it is not typically structured uniformly and can be provided by a variety of training institutions, including state, non-governmental, and private providers, each with their own set of interests, administrative structures, and traditions.

Furthermore, VET has become more focused on pre-service education and training, preparing students to meet specific job market requirements. However, VET policy frequently gets shared by Ministries of Education and Ministries of Labour and/or Employment. The decline of vocational training systems in liberal market economies has resulted in a shortage of vocationally qualified labour [2]. Governments in Australia, the United Kingdom, and Canada are attempting to elevate vocational training and create new apprenticeship systems. School-based vocational training has additionally been expanded, and universities in countries with weak vocational training systems are increasing their offering of courses with strong vocational content. Vocational education and training play a vital role in bridging the gap between theoretical education and practical workplace skills. By offering direct entry into the labor market, VET enhances individual professional careers, increases earnings, improves economic productivity, and fosters social cohesion [1]. Thus, vocational training holds great promise in facilitating a successful transition from school to work, mitigating the negative consequences of early youth unemployment and idleness, and bolstering economic and social outcomes [19].

2.2 *Previous Studies Related to Vocational Training in Ex-inmates*

Recidivism rates among offenders remain elevated, raising questions about the efficacy of correctional interventions and behavioural modifications. Research has shown that ex-offenders are at high risk of recidivism due to factors such as lack of education and job skills [22]. Vocational training programs can help address these factors by providing marketable skills and opportunities for stable employment [16, 20]. However, a better understanding of the Malaysian effectiveness and challenges of prison vocational training programmes are required. A study in Kenya to investigate the impact of vocational training programs on the self-sustainability of ex-convicts [12]. The study found that although 73.6% of ex-convicts underwent vocational training while in prison, only 15.7% believed that vocational training had a positive impact on their self-sustainability. This may be due to many lacking practical experience and exposure to employability or starting their own businesses, despite passing government trade tests. Additionally, 74% of ex-convicts disapproved of the rehabilitation programs offered in prison, and 78.4% had their courses chosen for them by prison authorities, which may have affected their enthusiasm for learning.

The perceptions of correctional officers regarding vocational and trade training programs for ex-convicts found that limited access to training programs has resulted in fewer employment opportunities for ex-inmates after release [18]. Furthermore, the lack of motivation in inmates was evident in their preference to spend time in their cells, rather than participating in programs. This lack of motivation hindered their engagement in vocational and trade training programs, which are crucial for securing employment upon release. Correctional officers perceived that the programs provided greater opportunities for employment and had advantages such as less strain and attachment through social bonds with coworkers. However, the study found that the programs' impact on inmates after release was not successful, as many who had participated in them returned to prison. These studies provide valuable insights into the various factors that affect the effectiveness of these programs, including access to practical experience, freedom of choice, motivation, and the impact of economic conditions on post-release success.

3 Theoretical Foundation

The study in question draws upon Hubert and Dreyfus's Theory of Skill Acquisition (1980), which outlines a five-stage progression from novice to expert in the development of psychomotor, technical, and artistic skills. This theory serves as the foundation for the study's methodology, guiding the creation of training programs geared towards employment.

In a related study on vocational training, the theory of skill acquisition was applied specifically to the acquisition of practical skills in vocational electronic technology [15]. Their research utilized a modified version of the Dreyfus model, focusing primarily on the initial four stages. The training sessions followed the novice, advanced beginner, competent, and proficient stages of the modified model, complemented by introductory career discussions led by trainers and mentors at the start of each phase. The trainers and lecturers provided ongoing mentorship throughout the training stages. To assess the students' practical skill development, observational measures with task-related elements were employed at each stage of the Dreyfus model. Together, these studies highlight the relevance and applicability of Hubert and Dreyfus's Theory of Skill Acquisition to vocational training, showcasing how the theory can guide the design and assessment of training programs aimed at cultivating practical expertise in specific fields.

4 Methodology

Interpretive research was concerned with understanding the subjective experiences and perspectives of individuals or groups, and with generating knowledge that was grounded in these perspectives [5]. This type of research was particularly useful when little was known about a phenomenon, as was the case with the effectiveness of the vocational programme designed for inmates. Given this lack of information, we employed a qualitative research method, through semi-structured interviews, which are commonly used in interpretive research to gather participants' views and opinions. These interviews provided a deeper understanding of the participants' experiences and perspectives, as well as knowledge based on their subjective perspectives. Due to the sensitive nature of the study, the research used an individual approach to data collection rather than a focus group method.

This study utilized purposive sampling; a non-random sampling technique that involved the deliberate selection of participants based on specific qualities. Purposive sampling was chosen because it allowed for the selection of participants who possessed the necessary qualities to provide richer insights and detailed information about the research questions [8]. As the purpose of this study was to address the limited understanding of the effectiveness of the vocational programme, it was critical to choose individuals who were willing to share their experiences and insights in order to gain a deeper understanding of the effectiveness of the practises.

4.1 Data Collection

The data collection involved conducting in-depth interviews with 12 correctional officers from the Progress and Development Department of inmates (Jabatan Perkembangan dan Pembangunan Banduan) in Malaysian prisons. The participants were selected based on the following characteristics: (1) correctional officers from prisons that provided vocational training programmes and (2) those who were conducting the vocational training programmes with inmates. The selected prisons for the interviews were Penjara Kajang, Penjara Bentong, and Penjara Sungai Udang, Melaka. A formal letter was submitted to the relevant authority (Pengarah Kanan Dasar Kepenjaraan Malaysia) to request permission to conduct the interviews with the correctional officers.

Each interview lasted between 45 min up to 2 h, depending on the setting and the participant's responsiveness. The interviews were recorded during the meeting and were subsequently transcribed between March and April 2023. The interviews were conducted in Bahasa Malaysia as all the participants were Malays, and the transcripts were later translated by language lecturers from Sunway university who are fluent in both Bahasa Malaysia and English.

4.2 Data Analysis

To manage the interview transcripts effectively, NVivo was employed to assist in the initial classification and management of the data, thereby enabling a systematic and more in-depth approach to be used for the analytical process. The collected interview data were organized and categorized based on common themes to facilitate the analysis and for comparison with existing scholarly literature. To ensure the reliability and validity of the data and findings, three principles were followed, including attention to detail in both data generation and analysis, careful documentation of the thematic analysis using narrative accounts, and rigorous data maintenance and supervision to guarantee the privacy of the study participants [17]. The anonymity of the participants was deemed crucial to ethical research practices, and pseudonyms were used to report their experiences when presenting and discussing the findings.

The research conducted interviews with a total of 12 correctional officers, with 5 from Penjara Bentong, 2 from Penjara Sungai Udang Melaka, and another 5 from Penjara Kajang. Among these officers, as shown in Table 1, there were 4 males and 8 females. The majority of the officers, 9 in total, were between the ages of 30 and 39, while the remaining 2 were in the age range of 40–49. The female officers were primarily responsible for teaching sewing, pastry, and spa-related activities, whereas the male officers were instructing inmates in carpeting, ferrocement, and epoxy. All of the correctional officers possessed expertise and knowledge in their respective fields, with 11 of them having obtained SKM 3, equivalent to a diploma, and one officer already possessing a diploma certification. Among the 12 officers, 10 had accumulated 5 years of work experience, while the remaining officers had less than 5 years but were still considered knowledgeable in their field.

Table 1 Characteristics of qualitative study participants

Characteristics	*n*	%
Gender		
Male	4	33
Female	8	66
Age		
20–29		
30–39	9	75
40–49	3	25
50–59		
Education		
High school		
Sijil Kemahiran Malaysia (SKM)	11	91.6
Diploma	1	8.3
Working experience		
No experience		
< 5 years	2	16.6
> 5 years	10	83.3

5 Findings and Discussion

The sections that follow describe the findings connected with the themes and their relevance to the vocational training program in Malaysia Prison. The findings revealed the important contributions and challenges to the training program.

Skill acquisition

Skill acquisition is a vital aspect of personal development, and it plays a significant role in correctional settings. Providing inmates with opportunities to acquire new skills can contribute to their rehabilitation and improve their chances of successful reintegration into society upon release. During a discussion with a correctional officer from Kajang Prison, it was noted that inmates who successfully complete the Sistem Latihan Dual Nasional (SLDN) program provided to selected inmates are eligible for skill certification.

> Previously, inmates had no skills certificate to apply for jobs. But now, it is easier for inmates to find jobs with the skills they have and the certificates they have obtained.

This skill certification provides recognition of their competence and serves as a tangible asset that they can use to secure employment upon release. The certification also serves as evidence of the effectiveness of the vocational training programs offered within the correctional institution, which can be used to attract potential employers.

Dedicate regular time for training

Dedicating regular time for training is crucial for skill acquisition, and in some cases, certain inmates may be given longer hours for training based on various factors. While the specific allocation of training hours can vary depending on the prison system and individual circumstances, the objective is to provide inmates with ample opportunities to develop their skills [13]. An officer from Bentong Prison mentioned the schedule of training and certificate.

> Inmates involved in SLDN training have a schedule. There are specific hours in a day that they have to complete, and they have to complete the training within a specific time frame before we can give them a certificate.

The participant also added:

> For SLDN, if they meet the required hours, they will get SLDN. It is tough. But SLDN is not only for prisons. It is also available outside, in factories. But in prison, it is easier. It is easy because inmates will definitely come to the workshop and undergo this program, and meet the required hours for SLDN. Outside, it is tough. So, in prison, it is easy. In any prison, there must be a 100% chance of getting SLDN.

Another participant from the Melaka Prison mentioned:

> Usually from Monday to Friday. From 8.30am to 2pm they will be in the workshop for vocational training. Usually, they take a shower at 1pm, then return to their block at 2pm. Because there are recreational activities in the afternoon.

This ease of access to the SLDN program in prisons is due to the strict routines and schedules that govern the daily activities of inmates. The routine ensures that participants can complete the required hours for the certification, which is not always the case outside of prisons. In any prison, it is essential to provide inmates with the opportunity to participate in the SLDN program and obtain certification. This measure increases their chances of securing employment upon release and reduces the likelihood of recidivism. The program provides a tangible way for inmates to gain practical skills that can help them reintegrate into society and lead productive lives.

Apply skills

The trained convicts are expected to apply the knowledge and skills they have acquired by producing the products that have been ordered. This practice is part of the SLDN program, where inmates are engaged in internal work within the prison. By working on these orders, the convicts have the opportunity to not only enhance their skills further but also earn a wage for their work. One of the participants from Bentong Prison highlighted:

> This is all the products of Bentong Prison. These are cookies ordered from the Queen. We want to send a large number of cookies to the Palace this week or next week for Hari Raya.

By working on customer orders, the convicts gain practical experience and apply their acquired skills in a real-world setting. Additionally, by being compensated for their work, the convicts are provided with a sense of financial independence and responsibility. Another participant from the Kajang Prison mentioned on the regular purchase of frozen products:

> We have frozen products. We have suppliers who purchase our products. We also have regular wholesalers. We have our regular customers.

Due to the regular wholesale arrangements with suppliers, the focus is primarily on catering to larger-scale orders rather than individual orders of frozen food products in Female Prison.

Receive certification as a formal recognition of their abilities

Receiving certification is a formal recognition of an individual's abilities and serves as tangible proof of their skills and expertise [9]. Receiving certification through the Sistem Latihan Dual Nasional (SLDN) program serves as a formal recognition of the abilities and skills acquired by convicts who have successfully passed the SLDN test. This certification holds significant value and benefits for the individuals involved. An officer from the Kajang Prison also mentioned the uniqueness of the certification.

> When inmates obtain a skills certificate, the name of the prison is not written on the certificate. Therefore, outsiders do not know that they are former inmates. The certificate states 'Pusat Professional Industri Wanita Kajang, not Kajang Prison.

The issuance of skills certificates to inmates upon the completion of vocational training programs within correctional institutions is an important step towards improving their employability prospects upon release. However, the stigma associated with having a criminal record can hinder their ability to secure employment despite possessing the necessary skills and certification.

To address this issue, certificates issued to inmates do not bear the name of the prison where the training took place. This measure ensures that the certificates do not reveal the individual's history as an inmate and avoids potential discrimination in the hiring process. Instead, the certificate states the name of the institution providing the training, such as the 'Pusat Professional Industri Wanita Kajang', to further protect the privacy and identity of the individual. It is also consistent with the principles of restorative justice, which aims to rehabilitate and reintegrate offenders back into society as productive and law-abiding citizens. By concealing the name of the prison on the skills certificate, inmates are given a fair chance to compete in the job market and make a positive contribution to society.

Challenges

Prisons, as with any institutional setting, pose a distinct set of challenges that necessitate correctional officers to be adaptable and persistent in delivering efficacious vocational training to inmates.

Lack essential educational resources

Lack of essential educational resources in prison refers to the limited availability or inadequate provision of educational materials, tools, and resources within correctional facilities. From the interview with Bentong Prison, one of the correctional officers mentioned:

> The provision of vocational training to all inmates in Bentong Prison is challenging due to the limited capacity of the workshops. While the prison population exceeds two thousand, only a fraction of the inmates can be accommodated in the workshops, with approximately 300 inmates. The selection process for inmates is based on several criteria, such as having completed the necessary requirements, reached the working stage, and had a longer sentence without eligibility for parole or release. Interested inmates are interviewed and asked about their field of interest, following which they are recommended to the workshop officer for consideration. The final selection of participants is based on the number of spaces available, and the skills required in the workshops.

Due to the limited space available, only a fraction of inmates can be enrolled in vocational training programs, which necessitates a selective process. The available spaces in the workshops can be allocated to those inmates who can benefit the most from the training programs. Despite the challenges posed by limited resources, providing vocational training programs to inmates remains an essential aspect of the rehabilitation process in prisons. Furthermore, one of the participants expressed the desire for further skill acquisition, particularly among prison officers who have been assigned additional responsibilities to assist correctional officers in the provision of vocational training.

> We have many constraints, for example, we have our own skills. I am not talking about others, but myself. Even though I have attended 2–3 courses, it is not enough. I want to take more courses. Furthermore, we received an order for a product that we had never seen before. How can we produce a product that we have never made before? That is one of the challenges. Additionally, we have constraints in terms of equipment and environment. Our workshop is not like those outside, and it is difficult for us to go out. For example, if we want to ask anything on the spot, we have to go out and search the internet ourselves. In our workshop, we do not have a phone to contact others.

The constraints and challenges faced in the vocational training programs within prisons require correctional officers to be resourceful and resilient in their approach. Despite limitations in terms of their own skills, equipment, and environment, officers still strive to provide effective training to inmates. For instance, when faced with an order for a product that they have never produced before, officers must think outside the box to figure out how to produce it. Additionally, in a workshop that lacks access to resources like the internet and phones, officers must find alternative methods to obtain information and communicate with others. These challenges require officers to be adaptable and creative in their approach to vocational training, ultimately benefiting the inmates who are eager to learn new skills. These constraints suggest the need for more resources, training, and support for officers in providing effective vocational training programs in prisons.

Providing appropriate scaffolding to support learning

Providing appropriate scaffolding to support learning in prisons is essential due to the diverse range of inmates and their unique educational needs. Scaffolding refers to the structured support and guidance provided to learners as they acquire new knowledge and skills [23].

> When I teach different age groups, I need to explain more to older students who have a harder time understanding.

One of the correctional officers at Kajang Prison highlighted the diverse backgrounds and abilities of inmates, which presents a distinct challenge for officers who are tasked with providing vocational training. The officer also noted that some inmates struggle with writing, which is a requirement for certification.

> Here, there are various age groups. Some are unable to write. However, to obtain SLDN, one must be 16 years old and above, and be able to speak and write. If the prisoner is able to write even a little, we will provide the opportunity.

Therefore, it is crucial to implement scaffolding techniques that cater to this diversity and address the unique needs of each learner [3]. The Kajang Officers highlighted that by offering scaffolding support, they can bridge the gaps in knowledge and skills that may exist among the diverse audience.

6 Discussion

One of the participants highlighted the importance of acquiring fundamental knowledge and skills to lead a normal life, emphasizing that society now demands a higher level of knowledge and skills [21]. Failing to elevate their knowledge could result in inmates remaining marginalized and at risk of relapsing into criminal behavior. Interestingly, the study also revealed that inmates who complete vocational education have a greater likelihood of securing employment compared to those with general academic qualifications which can also be seen in other studies [10].

Reintegrating into a conventional life can be exceptionally challenging for individuals transitioning from such a background [13]. Moreover, the study found that formal qualifications alone are not the sole factor in helping ex-inmates obtain employment. The participant asserts that actively engaging in various rehabilitation activities, even without acquiring specific certifications or becoming proficient in a particular trade, is viewed positively as it demonstrates a commitment to making positive life changes.

During the interview, it became evident that correctional officers who were truly committed to their role had to be adaptable and versatile in their teaching methods. This was crucial because they had to cater to inmates from diverse backgrounds and effectively capture their attention during the instructional process. By being flexible in their approach, these dedicated officers were able to engage with the inmates and impart valuable skills. The officers' guidance emerged as a pivotal factor in the inmates' success. Through their careful mentoring and support, the officers played a crucial role in empowering the inmates to effectively complete tasks and acquire new skills [24]. This guidance served as a scaffold, providing a structured framework that facilitated the inmates' learning journey. The concept of guidance or scaffolding, proved to be of utmost importance [22, 23]. It ensured that the available resources and contextual factors were fully utilized and comprehended by the inmates.

During the course of this study, various measures were taken to minimize limitations. However, due to the sensitive nature of the research and the need to adhere to strict regulations for interview access, the study was limited to visiting three prisons in Malaysia. To enhance the quality and representativeness of future research, it is suggested that alternative methods for participant recruitment be explored, along with a larger sample size. Another constraint of this study was the narrow selection of variables, failing to consider other potential moderating variables. The focus was solely on evaluating the effectiveness of vocational training programs through interviews with correctional staff, disregarding the valuable perspective of the inmates. To address this limitation, it is recommended that future research incorporates the viewpoint of inmates to gain a more comprehensive understanding.

7 Conclusion

The study highlights the significance of the SLDN program within the prison setting. This program plays a crucial role in recognizing and certifying the training efforts of inmates. By participating in the SLDN program, inmates have the opportunity to undergo structured vocational training and skill development programs. These programs are designed to equip them with industry-relevant skills and knowledge that can enhance their employability upon release. Furthermore, through appropriate scaffolding techniques, educators can address the diverse needs of the incarcerated population, bridging knowledge gaps, and promoting inclusive learning environments.

It is important to recognize that while correctional facilities strive to provide skills training, industry players and the community also have a crucial role to play in facilitating the successful reintegration of incarcerated individuals. Industry representatives can contribute by providing input on the skills and qualifications needed in the job market and offering internship or job placement opportunities for released inmates. Besides that, community-based organizations, nonprofits, and support groups can provide resources such as job readiness training, mentoring programs, and networking opportunities. They can also assist in connecting released inmates with potential employers who are open to hiring individuals with criminal records. There is a need to create an environment that supports the successful reintegration of formerly incarcerated individuals into the workforce. It requires a collective effort from correctional facilities, industry players, the community, and policymakers to provide opportunities for change and to give individuals with criminal records a chance to rebuild their lives and contribute positively to society.

References

1. Biavaschi, C., Eichhorst, W., Giulietti, C., Kendzia, M.J., Muravyev, A., Pieters, J., Rodríguez-Planas, N., Schmidl, R., Zimmermann, K.F.: Youth unemployment and vocational training (2012)
2. Bosch, G., Charest, J.: Vocational training and the labour market in liberal and coordinated economies. Ind. Relat. J. **39**(5), 428–447 (2008)
3. Brandt, B.L., Farmer Jr, J.A., Buckmaster, A.: Cognitive apprenticeship approach to helping adults learn. In: New Directions for Adult and Continuing Education, vol. 1993(59), pp. 69–78 (1993)
4. Carson, E.A., Sobel, W.J.: Prisoners in 2011. BJS Bulletin. Bureau of Justice Statistics, Washington, D.C. (2012)
5. Creswell, J.W.: A Concise Introduction to Mixed Methods Research. SAGE Publications (2014)
6. Chamorro, T., Frankiewicz, B.: Does higher education still prepare people for jobs. Harvard Bus. Rev. (2019)
7. Duke, B.: A meta-analysis comparing educational attainment prior to incarceration and recidivism rates in relation to correctional education. J. Correct. Educ. **69**(1), 44–59 (2018)
8. Etikan, I., Musa, S.A., Alkassim, R.S.: Comparison of convenience sampling and purposive sampling. Am. J. Theor. Appl. Stat. **5**(1), 1–4 (2016)

9. Leasure, P., Andersen, T.S.: Race, criminal records, and certificates of relief: an experimental pilot study. Deviant Behav. **41**(9), 1065–1083 (2020)
10. Löfgren, S., Ilomäki, L., Toom, A.: Employer views on upper-secondary vocational graduate competences. J. Vocat. Educ. Training **72**(3), 435–460 (2020)
11. Malaysian Prison Department: Number of Prisoners with Repeated Crimes, 2014–2019 (2019). Retrieved from http://www.prison.gov.my/portal/page/portal/biru/kepenjaraan
12. Muasya, J.K.: Impact of prison rehabilitation programmes on self sustainability of ex-convicts: a case of Thika sub-county, Kenya. Doctoral dissertation, University of Nairobi (2013)
13. Mun, C.W., Ahmad: Learning support and academic achievement among Juvenile Delinquents at Malaysia's correctional institutions. Scope and Objectives, vol. 19 (2019)
14. Mustapa, M.A.S., Ibrahim, M., Yusoff, A.: Engaging vocational college students through blended learning: improving class attendance and participation. Procedia Soc. Behav. Sci. **204**, 127–135 (2015)
15. Ogbuanya, T.C., Chukwuedo, S.O.: Career-training mentorship intervention via the Dreyfus model: implication for career behaviors and practical skills acquisition in vocational electronic technology. J. Vocat. Behav. **103**, 88–105 (2017)
16. Orrick, E., Vieraitis, L.: The cost of incarceration in Texas: estimating the benefits of reducing the prison population. Am. J. Crim. Justice **40**(2), 399–415 (2015). https://doi.org/10.1007/s12103-014-9265-3
17. Patricia Benner, R.N., Christine Tanner, R.N., Catherine Chesla, R.N. (eds.): Expertise in Nursing Practice: Caring, Clinical Judgment, and Ethics. Springer Publishing Company (2009)
18. Robinson, L.V.: A look at opportunities within correctional facilities in preparation of ex-convicts attaining employment: a qualitative case study. Doctoral dissertation, Northcentral University (2022)
19. Scarpetta, S., Sonnet, A., Manfredi, T.: Rising youth unemployment during the crisis: how to prevent negative long-term consequences on a generation? (2010)
20. Sliva, S.M., Samimi, C.: Social work and prison labor: a restorative model. Soc. Work **63**(2), 153–160 (2018)
21. Tønseth, C., Bergsland, R.: Prison education in Norway—the importance for work and life after release. Cogent Educ. **6**(1), 1628408 (2019)
22. Vacca, J.: Educated prisoners are less likely to return to prison. J. Correct. Educ. **55**(4), 297–305 (2004)
23. Van de Pol, J., Volman, M., Oort, F., Beishuizen, J.: Teacher scaffolding in small-group work: an intervention study. J. Learn. Sci. **23**(4), 600–650 (2014)
24. Wright, K.A.: Time well spent: misery, meaning, and the opportunity of incarceration. Howard J. Crime Justice **59**(1), 44–64 (2020)
25. Zoukis, C.: Vocational training in prison (2015). Retrieved from https://federalcriminaldefenseattorney.com/vocational-prison-education-preps-for-life-after-prison/

"Viral Food" Obsession: Influence of Product Attributes, Personal Attitudes, and Social Group on Consumer's Purchasing Behaviour

Nur Hafizah Muhammad, Nor Maizana Mat Nawi, Nurzehan Abu Bakar, Nor Amira Mohd Razali, Nurul Fardila Abd Razak, and Nur Azimah Othman

Abstract The term 'viral food' now is a common social media vocabulary where is describe entire marketing campaigns benefited by word-of-mouth among internet community. Viral food defines as a food item produced locally that obtained fame over social media, even it does not need to have a pleasant taste to begin with. Apparently, this 'viral food' has achieved its goals in Malaysia where consumer would always seek for it once it widely spread on the internet. Henceforth, this quantitative study was conducted to investigate the consumer behaviour on purchasing 'viral food', particularly focusing on product attributes, personal attitudes and social influence. Understanding the drive to buy 'viral food' would help derive the most effective ways to improve food marketing and the product as well. Implementing convenience sampling, 400 responses collected using an online questionnaire, Google form, which had been distributed nationwide through emails, telecommunication applications and social media platforms. With OLS regression and ANOVA analysis by using STATA MP 17, results showed that product attributes ($\beta = 0.59$) were the most significant factor that urge consumer to purchase, while it also indicated that youngster and middle-income groups had higher possibility to buy 'viral food' more, if compared to other market segmentation groups. The findings concluded that consumers perceived those 'viral food' marketing through the online community creates an impulse or indicator to them to buy the food and helping to ease their decision-making processes, mainly in food procurement circumstances.

Keywords Viral food · Social media · Viral marketing

N. H. Muhammad (✉) · N. M. M. Nawi · N. A. Bakar · N. A. M. Razali · N. F. A. Razak · N. A. Othman
Faculty of Hospitality, Tourism and Wellness, Universiti Malaysia Kelantan, 16100 Kota Bharu, Kelantan, Malaysia
e-mail: hafizahm@umk.edu.my

B. Awwad (ed.), *The AI Revolution: Driving Business Innovation and Research*, Studies in Systems, Decision and Control 524, https://doi.org/10.1007/978-3-031-54379-1_59

1 Introduction

E-commerce has begun since 2000s with the advancement of information and technology infrastructure in the modern millennium. Traditional business-to-business (B2B) predominantly evolved from physical buy and sell activity to entirely electronic transaction. For the business to succeed in the sector, using the right marketing tools in virtual space is especially important to promote their products [1]. With creative minds and technology, business owners continue strategizing ways to market and convince potential customers through online digital marketing. Unintentionally, they are forced to develop attractive content and cooperate diverse channels to reach the consumers. Until recently, the latest trend of marketing called viral marketing drastically re-appears among online community where social media sites, mobile applications and mobile website used to promote products and services to the public [2].

This recent decade has witnessed the development of online communities through social media, in particular. Almost every person owns a social media account to meet their social need. They communicate with each other even without knowing them on social networking sites such as Facebook, Instagram, Twitter, TikTok, Snap Chat and many more. Besides communicating with friends and family, social media is used to share information, make connections, find inspiration, learning, to be entertained and even merely finding products to purchase. 2023 Global Digital Report reported that nearly 78.5% of Malaysia's population uses social media, and they log onto one of these sites for roughly three hours every day. The majority of them are members of generation Z and Millennials, who rely on social media to keep up with daily news and the newest trends in every area [3].

In marketing, "viral marketing" is not a new terminology. It considers an alternative marketing strategy that uses persuasion message to accelerate unpaid peer-to-peer sharing of marketing information from unidentified sources, also known as word-of-mouth (WOM) strategy [4]. Besides the growth of information and technology, the evolution of mobile industry and its related applications provides most benefit to all businesses to advertise their products without spending large resources by purely applying electronic word-of-mouth (e-WOM) [5]. Social networks are increasingly being used as the main channel for viral marketing by practically all businesses. For example, small businesses with limited budget would use these marketing tactics to welcome more crowds for them. Undeniably, viral marketing via social media is significantly effective to boost sales, particularly in food businesses. Social media is used as mediator in e-commerce nowadays [6]. A type of food is known for having an "Instagrammable" presentation, together with advance editing application that sugar-coated the actual product would invite more customers to buy and spread more news to others. The more shared the news, the more popular it would be, and it will be sold in a limited time. In social media, this 'viral food' becomes exclusively a new vocabulary that naturally known by the online communities. Customers who are always seeking fresh and interesting culinary experiences will start seeking for the 'viral food' and definitely may intend to purchase it.

Tempting 'viral food' videos or photos are spreading fast through this e-WOM strategy and trying to lure the public to purchase it. For example, Malaysians are passionate with cheese and love to mix any food with cheese, where banana fritters with mountain of shredded cheese, bubble tea with cheese, cheesy fried chicken, cheesy dessert as well as *kunafa* cheese had become trending awhile back as the result of social media viral marketing. Along with marketing to digital society, consumer's attitudes on how they perceive and effect by those viral food would be another challenge to investigate. It believes that viral food sensations have altered how consumers choose and consume food day by day. Many Malaysians have started to enjoy the dish at restaurant, seek the dish and make the dish at home, purely to follow the trend and being influenced by viral marketing. Therefore, it is worthwhile to understand consumers' behaviour on the most significant factors that contribute to their purchase intention of viral food. Despite the potential of viral marketing to attract customers, this study is believed to generate useful information on both quality and quantity of advertising materials of businesses to target for the right market. And further, the research on social media viral marketing is limited in Malaysia, even more on viral food study.

2 Literature Review

2.1 Personal Attitude

Someone attitudes towards viral marketing are expected to change continuously over time and are heavily influenced by both internal and external circumstances. Previous experience, relationships with friends and family members, advertising, reviews and comments are among the external factors that shape cognitive learning of a person [7]. The processes of a person's perception and attitudes on viral marketing are built through memory, external information, cognitive based and affect based information processing [8]. Thus, personal attitude towards viral marketing can be defined as someone's propensity to react favourably or unfavourably to a specific viral advertising stimulus within a specific exposure condition. Eventually, this will relate to the feelings and beliefs towards someone's behavioural intention towards products in a marketing context.

In Vietnam, a study shows that there is significant relationship of someone attitude on the information received. The higher informativeness on viral food marketing, the higher attitude the customer revolves [8]. The updated information on certain viral food is supposed to help make decisions and distinguish from products of other rivals. It reduces the burden of cognitive processing and develops a context of which viral food to be chosen. Moreover, viral food phenomenon among social community is not meant to end with only certain food. The food must be entertaining enough with fascinating concept to attract attention of consumer, it includes both healthy and unhealthy food. An Indonesian study proved that viral food seekers also are divided in different segment. There are healthy viral food seekers and also those that would purchase unfamiliar food such as *Indomie* ice cream which becoming popular in Indonesia [9]. Thus, someone's personal attitude notably affects their behavioural intention on viral food consumption.

2.2 Social Influence

Research of preceding decades perceived that only family and friends have always been the most influential people around someone, and obtaining an approval from them is an achievement to someone. The emerging issues of viral food marketing led to someone attention to the social media comments, reviews, videos and also "influencers" opinions when to make decision to buy viral food [10] and it might cause someone obsess to try to eat those dishes. Social media becomes a very reliable source for latest information and becomes supporters to, not only the businesses, but to the potential buyers or followers compared to the traditional marketing techniques that demand to be costly and sometimes ineffective approach [11]. A study indicates that vloggers in Tik Tok that reviews sponsorship products had play an essential role in shaping the purchasing decision as well as attitudes of their followers [10]. Convincing coverage of the food products serves as the persuasion tools for the followers or social media users and approve the purchasing decision made by them.

In fact, [12] stated that viral food on social media is classified as non-basic foods with excessive portion and youngster posted those food images on social media to inform their friends and family [13]. It included Starbucks' latest drinks, McDonald seasonal menu or local delicacies such as donut susu, local coffee menu or street-style burger. This eventually would make those non-basic food becomes more known on social media (including on the television news) and it is also called as the food that is 'in trend' at the moment. More and more people would purchase and those who are fascinated with the newest viral food would definitely seek and buy it, merely to taste and be well-versed in the current situation. For that matter discussed, not only family and friends, but social communities would positively impact someone buying viral food intention.

2.3 Attributes of Viral Food

Characteristics of a product usually act as one of the influential factors that determine someone's behavioural intention. Attributes like price, taste, shape, label, and portion size are among the tangible features typically a person is looking into on viral food [14]. In Malaysia, the well-narrated video of the food product attributes posted in social media had influenced online community to purchase it, particularly among younger generations [15]. Insight information on the famous places and cost of the product would make the viral marketing affective among the community. The videos make the audiences fascinated and make decision to buy and share to others.

Nevertheless, in economics, rational choice theory indicated that most people are motivated to engage in certain behaviour after rationally calculating the cost and benefits of their behaviour [16]. A product attributes are necessary for someone to evaluate prior deliberately make decision to consume it. But, even in viral marketing setting, an Iranian study implies the correlation between product's usefulness and purchase intention [5] which showed that a consumer did not make any procurement without cautiously evaluate it. Preferences, prices, household budget and sociodemographic factors also are important considerations in understanding demand for viral food [17]. With most preferred viral food by someone, but with budget constraints, it particularly will impact a purchasing decision. This indicates that product attributes do determine someone's intention to purchase viral food, with or without external factors.

3 Research Methods

3.1 Sample

This research attains an in-depth understanding of the purchasing intention of viral food among Malaysian. This descriptive study aims to obtain information on the obsession towards food products, in particular, resulted from viral marketing. The data were collected through an online platform and whoever owned at least one social media account such as Facebook, Instagram, WhatsApp or Tik Tok to be selected as participant. In total, after excluding invalid responses, 400 participants were finally chosen for data analysis (Table 1).

3.2 Research Instrument

A questionnaire via Google Form used as survey instrument for this study. The questionnaire is divided into two parts. The first part, consent statement, purpose of the study and demographic profile questions. While the second part, the measurement items of the variables are asked. All the three explanatory variables and an outcome variable are measured on a five-point Likert scale ranging from strongly disagree to strongly agree. Personal attitude construct measured the perception and beliefs of a person towards viral food. Meanwhile, social influence items are measuring a person affection towards the viral marketing food and product attribute items are focusing on the tangible characteristics of food item impacted someone behaviour. The items are developed based on previous study on the viral food marketing studies.

Table 1 Demographic statistics

Profile (n = 400)		%
Gender	Female	50.1
	Male	49.9
Age group	18–25	30.9
	26–33	24.4
	34–41	23.2
	42–49	12.6
	≤ 50	8.9
Race	Malay	30.6
	Chinese	43.2
	Indian	15.8
	Others	10.4
Occupation	Employed	29.9
	Self-employed	28.1
	Unemployed	6.2
	Student	31.1
	Retired	3.5
	Not seeking for work	1.2
Annual salary (RM)	Not earning	30.1
	< 12,000	16.8
	12,001–24,000	18.8
	24,001–48,000	21.2
	> 48,001	13.1

All items are modified considering the Malaysian context. Table 2 shows the mean and reliability of each item.

3.3 Data Analysis

The data analysis is divided into two segments. Ordinary least square (OLS) is a technique of estimating linear relations between a dependent variable on one hand, and a set of explanatory variables on the other. STATA MP 17 used to analyse Model 1 as shown below. The OLS command also included the robust option for estimating the standard errors using the Huber-White sandwich estimators—which can be used to estimate the variance of multiple linear regression if the underlying model is incorrect. With robust option, the estimated coefficients are exactly the same as in standard OLS, but it can deal with the standard errors (ε) issues, minor concerns of failure to meet assumptions, heteroscedasticity as well as lack of normality [18].

$$\text{Intention}(Y_n) = x_0'\beta_0 + x_1'\beta(A_n) + x_2'\beta(SI_n) + x_2'\beta(VF_n) + \varepsilon \tag{1}$$

Table 2 Items mean, reliability and standard deviation

Items	Mean	SD
Personal attitude (α = 0.70)		
Taste judgement will affect my desire to buy viral food	4.47	0.71
Emotion beliefs will affect my desire to buy viral food	4.02	1.01
I want to be among the first people to try a new viral food	3.22	1.15
I'm willing to take a risk when it comes to investing for new viral food	3.12	1.19
Before purchasing viral foods, I considered the number of calories my body needs	3.12	1.21
Before consuming viral foods, I considered the number of calories my body needs	3.13	1.19
Social influence (α = 0.70)		
I always read consumer evaluations and comments before purchasing viral food	4.14	0.83
I always read consumer evaluations and comments before eating viral food	4.22	0.87
I would not interested in purchasing meals that have received negative reviews on viral foods	3.91	1.00
I always watch video reviews on current food trends	3.87	1.05
I always keeping up with current food trends on social media	3.90	1.09
I always influenced to try viral foods after seeing advertising on social media	3.14	1.10
I prefer to eat meals that are high in nutrients for myself and others over foods that are low in nutrients and can harm the body's health	4.01	0.81
I always rate a food through the reviews given in the comments section on the authentic source of viral food	3.36	1.19
Viral food attribute (α = 0.71)		
The offered viral food has a more appealing colour combination, particularly on the packaging, which makes me want to buy it	4.10	0.89
I am tempted to try viral food because of the colour of the food exhibited	3.49	1.14
I am tempted to try viral food because of the form of the food exhibited	3.76	1.04
I am quickly attracted to portion size of viral foods	3.79	1.10
When making a purchase, I would be more concerned with the physical appearance of a viral food	3.88	0.85
I never actually consider a seller's pricing for a viral food	3.07	1.28
Intention (α = 0.88)		
I like purchasing viral food in the future	3.83	0.97

(continued)

Table 2 (continued)

Items	Mean	SD
I like eating viral food in the future	3.89	0.98
I will be consuming viral food on a regular basis	3.23	1.13
Purchasing viral food will demonstrate my distinct taste and personality	2.99	1.22
Eating viral food will demonstrate my distinct taste and personality	3.15	1.24
Purchasing viral food has the potential to enhance my quality of life	2.50	1.32
Eating viral food has the potential to enhance my quality of life	2.62	1.31

where: Y = Predicted purchase intention of viral food; A = Personal attitude of person n; SI = Social influence of person n; VF = Viral food attribute; β = Estimate coefficient of the construct.

Second segment of data analysis, ANOVA and independent T-test are also employed to explore the association between the socio-demographic and explanatory and outcome variables. One-way ANOVA is an analysis of variance used to determine whether the mean of a dependent variable is the same in two or more unrelated, independent groups.

4 Results and Discussions

Model 2 below shows the outcome of OLS estimation of Model 1 above and Table 3 indicates the model estimated coefficients. The model is statistically significant (Prob $> F = 0.00$) with $R2 = 0.429$ of the proportion of variance in the dependent variables which can be predicted by those three explanatory variables. The analysis implies that viral food attributes the most influential factors among the other two explanatory variables. So, for every unit increase in viral food attributes, 0.588 unit increase in the person intention to buy viral food, holding the other variables constant. Similarly, with other variable constants, a unit higher in social influence of viral food, approximately 0.291 higher of someone intention to buy it and a unit increase in the someone attitude, 0.27 unit increase in their intention.

$$\text{Intention}(Y_n) = -1.088 + 0.270 * A_n + 0.291 * SI_n + 0.588 * VF_n \quad (2)$$

Even though social influence and someone attitude positively impacted the consumer's behavioural intention, viral food product induces the most on the determination to purchase. It signifies that videos and photos advertising are appealing enough to exhibit the product attributes to the viewers to attract them. Similarly,

Table 3 Estimated coefficients of OLS regression

Variables	β	Robust SE	t	*p*-value
$Attitude_n$	0.270	0.072	3.89	0.00
$Social\ influence_n$	0.291	0.075	3.86	0.00
$Viral\ food\ attribute_n$	0.588	0.052	11.25	0.00
N = 400	**F (3, 396)** = 136.62	R^2 = 0.429	**Prob > F** = 0.00	

Korean-style fried chicken has become a trend due to the scrumptious look of its crispy exterior with sweet savoury sauces coated with melted cheese, which widely posted in social network [9]. Even more, entrepreneurs in Malaysia utilized social network as marketing tool to promote their product. With appetizing pictures or videos spreading among social community, it may influence consumers at the same time build relationship with them [19]. The results also show that attitude and social influence had the moderate relationship on purchasing intention of the viral food. Viral marketing not only affects consumers' choice and purchase decision but the reviews, comments and posts on social network too facilitates and distributes informative data of the viral food products [20]. It makes consumers not only understand, but also able to assess any positive or negative statement on regards of the product made available by vlogger or influencer online.

A further analysis of the differences of viral food purchase intention and three explanatory variables with regard participants' socio-demographic has been completed with the one-way ANOVA and independent t-test. The results found that only social influence variable had significant differences with the three socio-demographic profiles—age, occupation and annual income of the participants. Table 4 shows the result of the analysis.

The ANOVA (Table 4) results revealed that there is significant difference between age group as determined by $F = 3.06$ ($p = 0.00$). A Tukey post-hoc test found that senior citizens (50 and above) had less influence if compared to young age group (18–25 and 26–33) at − 0.38 ($p = 0.01$) and − 0.32 ($p = 0.05$), respectively. This seems the youngsters are prone to have the intention to buy viral food after browsing through the social network. This could be because they usually get connected with friends, role models and family on social media, compared to elders [21]. More, brands such as McDonald's and Coca-Cola posts mainly post their latest advertisement through social media and adolescents are their main target of the product [13]. For example, adolescents are keener to repost those advertisements and eventually, those products will become known to the entire online population.

Similarly, the ANOVA test of occupation group also significantly differenced at $F = 2.33$ ($p = 0.00$). Post-hoc test indicates that only participant's student has significant difference at 0.38 ($p = 0.05$) compared to the unemployed group. It appears that students have higher chances to buy viral food, while unemployed groups might face some budget constraint to even desire to have the viral food. In contrast, ANOVA results of participant's annual income displays that participants who are not earning have higher chances to be socially influenced compared to group below

Table 4 ANOVA for social influence with participants' age, occupation and salary

Demographic group: Age				
Source	SS	df	MS	F
Between groups	4.48	4	1.12	3.06
Within groups	146.01	399	0.37	
Total	150.49	403	0.37	
Post-hoc	Contrast	SE	t	$P > \|t\|$
≥ 50 * 18–25	− 0.38	0.11	− 3.30	0.01
≥ 50 * 26–33	− 0.32	0.12	− 2.73	0.05
Demographic group: Occupation				
Source	SS	df	MS	F
Between groups	4.28	5	0.86	2.33
Within groups	146.21	398	0.37	
Total	150.49	403	0.37	
Post-hoc	Contrast	SE	t	$P > \|t\|$
Student * Unemployed	0.38	0.13	2.85	0.05
Demographic group: Annual salary				
Source	SS	df	MS	F
Between groups	3.93	4	0.98	2.67
Within groups	146.56	399	0.37	
Total	150.49	403	0.37	
Post-hoc	Contrast	SE	t	$P > \|t\|$
< RM12,000 * Not earning	− 0.28	0.09	− 3.00	0.02
RM12,001 − RM24,000 * < RM12,000	− 0.32	0.12	− 2.73	0.05

Noted SE = Standard error; df = Degree of freedom; MS = Mean square; SS = Sum of squares

RM12,000. Participants who are not earning any income, would probably students, are prone to buy viral food. However, middle income participants (earned RM12,000 to RM24,000) have a higher possibility to be influenced to buy viral food in comparison to group earned below RM12,000. These novel findings indicate that a person's propensity to purchase a food item is directly tied to the level of information they found about it online that influenced their perception and thoughts.

5 Conclusion

The study has generally achieved the stated aims and identified the most influential factors on consumer buying intention. Nevertheless, the expanding review, comments together with information of the product would help consumers to make decision on their behaviour. In this new norm setting, well-informed product through online would naturally advertise the product and giving guidance to the consumers' mind to make decision. Social media is changing how businesses and consumers interact nowadays and it is an opportunity to broadcast the products effectively and to keep competitive in the market. Lastly, based on the study findings, further research could include different attributes such as the importance of information for online advertised products or services. An economic choice model may provide unique perspective when actual monetary imply to explore how much consumer willing to spend for viral food.

References

1. Labanauskaitė, D., Fiore, M., Stašys, R.: Use of E-marketing tools as communication management in the tourism. Tourism Manag. Perspectives **34**, 100652 (2020)
2. Horner, S., Swarbrooke, J.: Consumer Behaviour in Tourism. Routledge (2016)
3. Howe, S.: Social Media Statistics for Malaysia. Retrieved from Meltwater https://www.meltwater.com/en/blog/social-media-statistics-malaysia. Last accessed 27 Apr. 2023
4. Porter, L., Golan, G.J.: From subservient chickens to brawny men: a comparison of viral advertising to television advertising. J. Interact. Advert. **6**(2), 30–38 (2006)
5. Fard, M.H., Marvi, R.: Viral marketing and purchase intentions of mobile applications users. Int. J. Emerg. Mark. **15**(2), 287–301 (2020)
6. Menon, R.V., Sigurdsson, V., Larsen, N.M., Fagerstrøm, A., Foxall, G.R.: Consumer attention to price in social commerce: eye tracking patterns in retail clothing. J. Bus. Res. **69**(11), 5008–5013 (2016)
7. Ahmad, M., Lasi, A.: The attitudes on consumer perception towards viral marketing: a study on Foodpanda food delivery in Malaysia. Int. J. Res. Sci. Innov. **7**(9), 251–255 (2020)
8. Wang, S.L., Lan, N.T.: A study on the attitude of customer towards viral video advertising on social media: a case study in Vietnam. Int. J. Eng. Sci. **7**(6), 54–60 (2018)
9. Susilo, D., Chen, L.H.: Digital marketing model of Indomie ice cream's viral marketing. SIBATIK JOURNAL: Jurnal Ilmiah Bidang Sosial, Ekonomi, Budaya, Teknologi & Pendidikan **2**(5), 1397–1410 (2023)
10. Kurdi, B., Alshurideh, M., Akour, I., Tariq, E., Alhamad, A., Alzoubi, H.: The effect of social media influencers' characteristics on consumer intention and attitude toward Keto products purchase intention. Int. J. Data Netw. Sci. **6**(4), 1135–1146 (2022)
11. Rybaczewska, M., Jebet Chesire, B., Sparks, L.: YouTube vloggers as brand influencers on consumer purchase behaviour. J. Intercultural Manag. **12**(3), 117–140 (2020)
12. Qutteina, Y., Hallez, L., Mennes, N., De Backer, C., Smits, T.: What do adolescents see on social media? A diary study of food marketing images on social media. Front. Psychol. **10**, 2637 (2019)
13. Holmberg, C., Chaplin, J., Hillman, T., Berg, C.: Adolescents' presentation of food in social media: an explorative study. Appetite **99**, 121–129 (2016)
14. Hosseinikhah Choshaly, S., Mirabolghasemi, M.: The role of viral marketing strategies in predicting purchasing intention of eco-labelled products. J. Islamic Mark. **13**(5), 997–1015 (2022)
15. Ahmad, N., Rose, N.N., Ishak, A., Shakila, Chettiar, R.P.: The impacts of viral video through Whatsapp among adolecense in Penang, Malaysia. J. Phys. Conf. Ser. **1529**(3), 032014 (2020)
16. Scott, J.: Rational choice theory. In: Understanding Contemporary Society. Theories of the Present, vol. 129, pp. 126–138 (2000)
17. Cranfield, J.A.: Framing consumer food demand responses in a viral pandemic. Can. J. Agric. Econ. **68**(2), 151–156 (2020)

18. Cameron, A.C., Trivedi, P.K.: Micoeconometrics Using Stata, Revised 2nd ed. Stata Press, Texas, USA (2010)
19. Abu Bakar, N., Ahmad Zaini, A.: Literatiure review on the factors influencing the usage of social media among entrepreneurs in Malaysia. Open J. Soc. Sci. **10**, 409–419 (2022)
20. Lis, B., Neßler, C.: Electronic word of mouth. Bus. Inf. Syst. Eng. **6**(1), 63–65 (2014)
21. Boyd, D.M., Ellison, N.B.: Social network sites: definition, history, and scholarship. IEEE Eng. Manage. Rev. **38**(3), 16–31 (2010)
22. Prasanna Kumar, U., Arthi, R.: A study on utility and feasibility of digital marketing tools with lead acquisition, lead nurturing and client engagement. In: Aloysius Edward, J., Jaheer Mukthar, K.P., Asis, E.R., Sivasubramanian, K. (eds.) Current Trends in Economics, Business and Sustainability. ICEBS 2023. Contributions to Environmental Sciences & Innovative Business Technology. Springer, Singapore (2023)

Exploring the Profound Influence of Memorable Religious Tourism Experiences on Subjective Well-Being Among Malaysian Pilgrims

Nurzehan Abu Bakar, Marlisa Abd Rahim, Nur Hafizah Binti Muhammad, Muhamad Nasyat Muhamad Nasir, Mohd Hafzal Abdul Halim, and Derweanna Bah Simpong

Abstract In recent years, religious tourism has gained significant attention for its capacity to offer memorable experiences to pilgrims. While previous research has explored the impacts of religious tourism on various aspects of pilgrims' lives, there is a gap in understanding how memorable experiences within religious tourism context influence subjective well-being (SWB). Hence, this study delves into the profound impact of memorable religious tourism experiences on the SWB of Malaysian Umra and Haj pilgrims by investigating the influence of knowledge, refreshment, meaningfulness, novelty, involvement, hedonism and local culture, on pilgrims' SWB. Google Forms were used to disseminate 211 questionnaires to Malaysian Umra and Haj pilgrims, and the collected data were analyzed using the SmartPLS. The findings revealed that among the examined variables, only novelty, involvement, and hedonism demonstrated a positive correlation with SWB. This pioneering research sheds light on the concept of Memorable Tourism Experience (MTE) in religious tourism, offering valuable insights for host countries, tour operators, and travel agents to develop great marketing strategies, improving memorable pilgrimage experiences and destination image.

Keywords Memorable Tourism Experience (MTE) · Memorable religious tourism experience · Subjective Well-being (SWB)

N. A. Bakar (✉) · M. A. Rahim · N. H. B. Muhammad · M. N. M. Nasir · M. H. A. Halim · D. B. Simpong
Faculty of Hospitality, Tourism and Wellness, Universiti Malaysia Kelantan, 08544 Kota Bharu, Kelantan, Malaysia
e-mail: nurzehan.ab@umk.edu.my

B. Awwad (ed.), *The AI Revolution: Driving Business Innovation and Research*, Studies in Systems, Decision and Control 524, https://doi.org/10.1007/978-3-031-54379-1_60

1 Introduction

In general, religious tourism is a multidimensional sector that encompasses a wide range of activities and experiences. Religious tourism normally comprises of pilgrimage, religious festivals, religious conferences, and so forth [1]. Pilgrimages, which involve visiting sacred places or sites of religious significance, are one of its main components. These excursions frequently have great spiritual connotations and are considered pivotal in the practice of many faiths. Mecca and Medina in Saudi Arabia, Jerusalem in Israel, the Vatican, Varanasi in India, Lumbini in Nepal, the Golden Temple in Amritsar, Mount Sinai in Egypt, the Kashi Vishwanath Temple in Varanasi, Mount Kailash in Tibet, and Santiago de Compostela in Spain are among well-known pilgrimage destinations [2]. These sites have immense religious significance, attracting pilgrims from all over the world who seek spiritual fulfilment, participate in rituals, experience the cultural and historical wealth associated with these sacred locations.

Among the various forms of pilgrimage, the Umra and Haj pilgrimage stands out among others due to its size and diversity [3]. Saudi Arabia's Ministry Haj and Umra reported that nearly 1.8 million pilgrims from 150 countries participated in the 2023 Haj pilgrimage and nearly 6 million people have performed Umra pilgrimage until May 2023 [4, 5]. Haj, is an obligatory pilgrimage to Mecca, occurs during the concluding month of the Islamic calendar (Dhul-Hijjah) and holds great significance for Muslims worldwide [6]. On the other hand, Umra, a non-required pilgrimage with fewer rituals, offers Muslims the flexibility to undertake it throughout the year based on their convenience and resources [6]. Umra, though not mandatory, carries immense importance and can be performed multiple times, making it a spiritually enriching journey for those with the means to partake in it [6]. Both the Haj and Umra are revered by Muslims for their religious and spiritual benefits, providing an awe-inspiring and humbling experience that fosters spiritual growth [7].

Scholars have recently shown a rising interest in examining many facets of the Umra and Haj pilgrimages. This increased focus reflects the recognition of the significance of these religious journeys, prompting researchers to investigate their impact and implications. By understanding pilgrims' perspectives and experiences, experts and stakeholders can gain valuable insights essential for effectively managing both Umra and Haj, thereby enhancing pilgrims' overall well-being [1]. These insights can pave the way for more profound and life-changing pilgrimage experiences, allowing pilgrims to more thoroughly connect with their faith and experience greater fulfilment, well-being and happiness [6].

Despite the acknowledged importance of subjective well-being in pilgrims' experience, there is still a notable knowledge gap regarding the factors influencing the subjective well-being (SWB) of Umra and Haj pilgrims. Previous studies on MTE have predominantly focused on domains such as cultural tourism, urban exploration, museums, and heritage sites, while neglecting the inclusion of religious tourism [8–10]. Furthermore, prior studies on religious tourism have mostly focused on the spiritual and theological dimensions of pilgrimage encounters, with limited attention given to the direct examination of the influence of significant experiences on individuals' SWB. Thus, this study aims to bridge this gap by investigating how memorable religious tourism experiences impact the SWB of Malaysian pilgrims. The paper is structured as follows: firstly, it provides a brief introduction to Umra and Haj; secondly, it reviews the literature on MTE and SWB; thirdly, it discusses the methodology and presents the study's results; and finally, it delves into the paper's discussion and conclusion.

2 Literature Review

2.1 Subjective Well-Being

SWB can be defined as how individuals judge their own well-being [11]. SWB is essential for enhancing both one's physical and mental health as well as one's quality of life as a whole [12]. In the context of tourism, SWB has a dual significance: it not only symbolizes the social outcome of tourism development, but it also gives tourism operators, policymakers, and tourists with vital insights that help them understand the impacts of the tourism industry [13]. SWB has received a lot of attention in psychology, social sciences and tourism study. In the literature concerning tourism study, the research has shown that memories brought back from a recent vacation not only increase overall leisure time satisfaction but also other life domains (life satisfaction) over the long term [10].

2.2 Memorable Religious Tourism Experience

MTE can be defined as "positively remembered and recalled after the event has occurred" [14]. Existing research shows that fostering memorable experiences has a variety of advantages for tourism service providers [15]. For instance, visitors who have a satisfying memorable experience are more likely to return to the location [16], develop a connection to the location, and feel subjectively well-off. In addition, MTE also help tourism supplier to improve its sustainable competitive advantage [10] positioning their businesses and have a good destination image [17].

Prior research on MTE has concentrated on various forms of tourism, including eco-tourism [18] cultural and heritage tourism [19] rural tourism [20], and food consumption by tourists [21]. However, there is limited study focus on religious tourism and not all MTE characteristics can be used in a range of destination-specific tourist contexts [22]. This necessitates further investigation of MTE in a different study setting, such as religious tourism.

In the literature concerning the tourism study, the research has recently started to focus on the link between tourist experience and SWB. For instance, Saayman et al. [23], investigated the impact of foreign tourists' experience in South Africa and how it further influences their SWB. Moreover, Sthapit and Coudounaris [24] also revealed that MTE affects subjective well-being of tourists who visited Rovaniemi, Finland. For that matter discussed, the following hypothesis is proposed:

H1: Novelty associates positively with pilgrims' subjective well-being
H2: Hedonism associates positively with pilgrims' subjective well-being
H3: Involvement associates positively with pilgrims' subjective well-being
H4: Knowledge associates positively with pilgrims' subjective well-being
H5: Local culture associates positively with pilgrims' subjective well-being
H6: Meaningfulness associates positively with pilgrims' subjective well-being
H7: Novelty associates positively with pilgrims' subjective well-being
H8: Refreshment associates positively with pilgrims' subjective well-being.

3 Methodology and Result

Utilizing a quantitative research design, this study investigates the impact of MTE variables such as hedonism, involvement, knowledge, local culture meaningfulness, novelty, and refreshment on the subjective well-being of Malaysian pilgrims who have undertaken the sacred journeys of Umra and Haj in Makkah. According to News Strait Times [25], there are about 31,300 Haj pilgrims and 300,000 Umra pilgrims in the year 2023. An online questionnaire was used to collect data from pilgrims that had performed Umra and Haj, engaging a total of 211 participants. The selection of participants followed a purposive sampling approach, enabling the researchers to intentionally choose individuals with prior Umra and Haj experiences. Drawing inspiration from the work of Kim et al. [22] on MTE, the research instrument employs a 5-point Likert scale to gauge responses. There are two sections in the research instrument. The first section includes a consent declaration, the study's purpose, and demographic profile questions. While the second part measure the variables' items. Subsequently, the collected data underwent comprehensive analysis Partial Least Squares-Structural Equation Modelling (SmartPLS) to answer the research questions which encompass an array of statistical techniques including descriptive statistics, structural and measurement model to answer the research questions.

3.1 Demographic

350 Malaysian pilgrims were approached by the researchers to get their opinions. Only 211 respondents out of 350 completed the questionnaire. Based on Table 1, nearly two-thirds (57.8%) of the 211 respondents were females, while one-third (42.2%) of them were males. The majority of the respondents (n = 72, 34.1%) age

Table 1 Demographic statistics

Respondent profile	Classification	Frequency N = 211	Percentage (%)
Gender	Male	89	42.2
	Female	122	57.8
Age	< 20	15	7.1
	21–30 years old	46	21.8
	31–40 years old	72	34.1
	41–50 years old	47	22.3
	51 years old and above	31	14.7
Marital status	Married	104	49.3
	Single	98	46.4
	Other	9	4.2
Education	Secondary	34	16.1
	Form 6	11	5.2
	Diploma	72	34.1
	Bachelor of degree	73	34.6
	Master of degree	17	8.1
	PHD	4	1.9
Frequency travelling to Mecca	Once	178	84.3
	Twice	24	11.4
	More than two times	9	4.2

31–40 years old, married (n = 104, 49.3%) and a degree holder (n = 73, 34.6%). Majority have at least visited Holly Mecca and Madinah once (n = 178, 84.3%).

3.2 Measurement Model and Structural Model

Table 2 shows the mean and reliability of each item and measurement model of the current study. Overall, the mean scores ranged from 4.512 to 4.948, while the standard deviation ranged from 0.222 to 0.667. As per measurement model, the researcher assessed the loadings, average variance extracted (AVE) and the composite reliability (CR). The values of loadings should be $\geq$ 0.5, the AVE should be $\geq$ 0.5 and the CR should be $\geq$ 0.7. As shown in Table 2, the AVEs are all higher than 0.5 and the CRs are all higher than 0.7. The loadings were also acceptable with all loadings were more than 0.7 [26] which indicated that convergent validity was established.

Then the researchers assessed the discriminant validity using the HTMT criterion suggested by Henseler et al. [28] and updated by Franke and Sarstedt [27]. The HTMT values should be $\leq$ 0.85 for the stricter criterion and the mode lenient criterion is it should be $\leq$ 0.90. As shown in Table 3, the values of HTMT were all lower than the stricter criterion of $\leq$ 0.85, as such the researchers conclude that the respondents understood that the 8 constructs are distinct. Taken together both these validity test has shown that the measurement items are both valid and reliable.

Table 2 Measurement model

Code	MTE dimension	Mean	SD	Loadings	CR	AVE
	Hedonism			0.859	0.864	0.683
HE01	I am thrilled to have a new experience in Mecca and Madinah	4.777	0.470			
HE02	I took part in activities during my Umra/Haj trip	4.673	0.553			
HE03	I really enjoyed my Umra/Haj trip	4.445	0.568			
	Involvement			0.868	0.919	0.792
IN01	I visited any tourist spot that I wanted to visit when visiting Mecca and Madinah	4.787	0.464			
IN02	I enjoyed doing the activities that I really wanted to do on my Umra/Haj trip	4.853	0.392			
IN03	I was interested in the main activities offered to me	4.791	0.471			
	Knowledge			0.848	0.907	0.764
KN01	I gained a lot of information during Umra/Haj trip	4.848	0.359			
KN02	I learned new culture during my Umra/Haj trip	4.801	0.399			
KN03	I gained a new skill from this Umra/Haj trip	4.825	0.404			
	Local culture			0.726	0.847	0.649
LC01	I had a good impression of the local culture during my Umra/Haj trip	4.858	0.423			
LC02	I had the chance to experience the local culture	4.787	0.494			
LC03	The local people were friendly to me	4.479	0.677			
	Meaningfulness			0.828	0.853	0.661

(continued)

Table 2 (continued)

Code	MTE dimension	Mean	SD	Loadings	CR	AVE
ME01	I felt like I did something meaningful during my Umra/Haj trip	4.900	0.299			
ME02	I felt like I did something important during my Umra/Haj trip	4.919	0.272			
ME03	I learned something about myself from my Umra/Haj trip	4.910	0.318			
	Novelty			0.899	0.914	0.726
NO01	I had a once-in-a-lifetime experience	4.791	0.519			
NO02	I had a unique experience	4.645	0.577			
NO03	Compared to my previous travel experiences elsewhere, this one was unique	4.716	0.563			
NO04	I experienced something new	4.787	0.454			
	Refreshment			0.845	0.865	0.518
RE01	I relieve stress when performing Umra/Haj trip	4.877	0.356			
RE02	I had a refreshing experience	4.905	0.293			
RE03	I had an incredible sense of calm	4.948	0.222			
RE04	I had a peaceful soul during my Umra/Haj trip	4.948	0.222			
RE05	I felt revitalized from my Umra/Haj trip	4.929	0.257			
RE06	I felt better after performing Umra/Haj trip	4.919	0.289			
	Subjective well-being			0.899	0.919	0.591
SB01	I lead a purposeful and meaningful life	4.825	0.416			

(continued)

Table 2 (continued)

Code	MTE dimension	Mean	SD	Loadings	CR	AVE
SB02	I am engaged and interested in my daily activities	4.825	0.459			
SB03	I am competent and capable in the activities that are important to me	4.810	0.489			
SB04	I am a good person and live a good life	4.853	0.404			
SB05	My social relationship is supportive and rewarding	4.768	0.513			
SB06	I actively contribute to the happiness and wellbeing of other	4.512	0.677			

The structural model in Table 4 depicts the connections (paths) between the constructs on the suggested research model. The model's quality is measured by the strength of each structural path, which should be equal to or greater than 0.1 [29]. The result in Table 4 demonstrates that the R^2 value is more than 0.1. Hence, the predictive capability is established. Additionally, Q^2 confirms the predictive relevance of the endogenous construct. A Q^2 value of 0.334 in Table 4 that is greater than zero indicates that the model has predictive relevance.

Furthermore, hypotheses were verified to determine the significance of the relationship. Out of seven variables, only three (3) variables (hedonism: $\beta = 0.167$, $t = 2.463$, p-value $= 0.014$, involvement: $\beta = 0.274$, $t = 3.101$, p-value $= 0.002$ and novelty: $\beta = 0.260$, $t = 2.879$, p-value $= 0.004$) was found to have a positive relationship with subjective well-being, while the rest did not have significant relationship with subjective well-being (knowledge: $\beta = -0.026$, $t = 0.252$, p-value $= 0.801$, local culture: 0.143, $t = 1.601$, p-value $= 0.11$, meaningfulness: $\beta = 0.194$, $t = 1.672$, p-value $= 0.095$, and refreshment: $\beta = -0.021$, $t = 0.216$, p-value $= 0.829$).

4 Discussion

Based on Table 4, this study identifies three significant variables—hedonism, involvement, and novelty—out of the seven investigated, showing a positive link to subjective well-being within the context of religious tourism like Umra and Haj. Conversely, knowledge, local culture, meaningfulness, and refreshment do not demonstrate a significant correlation with well-being. Rituals such as circling the Kaaba, prayer, and seeking forgiveness during Umra and Haj evoke spiritual joy, fostering hedonistic feelings [30], while successful completion of these rituals leads to a sense of

Table 3 HTMT

	Hedonism	Involve	Knowledge	Local culture	Meaning	Novelty	Refreshment	Wellbeing
Hedonism								
Involvement	0.642							
Knowledge	0.423	0.522						
Local culture	0.434	0.654	0.65					
Meaningful	0.385	0.552	0.627	0.755				
Novelty	0.363	0.588	0.638	0.666	0.602			
Refreshment	0.290	0.419	0.593	0.463	0.62	0.453		
Wellbeing	0.566	0.727	0.53	0.716	0.655	0.668	0.399	

Table 4 Structural model

Variables	Original sample (O)	T-value	*P*-values	Supported/not supported
Hedonism → Wellbeing	0.167	2.463	0.014	Supported
Involvement → Wellbeing	0.274	3.101	0.002	Supported
Knowledge → Wellbeing	− 0.026	0.252	0.801	Not supported
Local culture → Wellbeing	0.143	1.601	0.11	Not supported
Meaningfulness → Wellbeing	0.194	1.672	0.095	Not supported
Novelty → Wellbeing	0.260	2.879	0.004	Supported
Refreshment → Wellbeing	− 0.021	0.216	0.829	Not supported
	R^2	Q^2		
	0.596	0.334		

achievement [31]. Active engagement in specified rituals, such as Tawaf and Sa'i, deepens the connection between pilgrims and their spiritual journey [32]. Moreover, the novelty of exploring historical sites and unique rituals enhances well-being, particularly for those engaging in Umra and Haj for the first time, evoking curiosity and enriching the pilgrimage experience [33].

5 Conclusion

This study aims to investigate the influence of memorable experiences on the subjective well-being of Malaysian pilgrims participating in the Umra and Haj. Engaging in these sacred pilgrimages, particularly to esteemed destinations such as Mecca, provides a distinctive combination of spiritual gratification, cultural enhancement, and individual accomplishment. The results of this study have practical implications for stakeholders involved in religious tourism. The findings emphasize the importance of novelty, engagement, and hedonism in improving the overall well-being of pilgrims. This observation has the potential to provide valuable guidance for destination managers and travel professionals in designing experiences that effectively enhance satisfaction and foster loyalty among pilgrims. This study also makes a valuable contribution to the academic field by emphasizing the significance of meaningful religious tourism experiences in influencing the overall well-being of Malaysian pilgrims. Due to time limitation, this particular study only focuses on Malaysian pilgrims. Hence, future study has the potential to broaden its focus by incorporating pilgrims from a range of nations and carrying out comparative evaluations of other religious pilgrimage destinations. This would contribute to a more comprehensive knowledge of the factors that influence subjective well-being within the context of religious tourism.

Acknowledgements The authors would like to thank Ministry of Higher Education Malaysia for providing FRGS grant (FRGS/1/2020/WAB01/UMK/03/2) and Universiti Malaysia Kelantan for supporting this work. The authors also would like to thank Nur Athirah Hazman, Nur Athirah Hussaini, Nur Athirah Mohamad Nordin, and Nur Atiqah Amran for helping the authors in data collection.

References

1. Quaium, A., Al-Nabhan, N.A., Rahaman, M., Salim, S.I., Toha, T.R., Hossain, J.N.M., Islam, N., Mostak, N., Islam, M.I., Mushfiq, M.M., Jahan, I., Alim, A.B.M.: Towards associating negative experiences and recommendations reported by Haj pilgrims in a mass-scale survey. Heliyon **9**(5) (2023)
2. Singh, L., Dhankhar, D.: Use of Social Media by Tourists at Religious Tourism Destinations in India, Global Development of Religious Tourism, pp. 250–267. IGI Global, Pennsylvania (2021)
3. Almuhrzi, H.M., Alsawafi, A.M.: Muslim perspectives on spiritual and religious travel beyond Haj: toward understanding motivations for Umra travel in Oman. Tourism Manag. Perspectives **24**, 235–242 (2017)
4. Bernama: Saudi Arabia says over 1.8mil pilgrims perform haj in 2023. Retrieved from News Strait Times: https://www.nst.com.my/world/world/2023/06/925377/saudi-arabia-says-over-18mil-pilgrims-perform-haj-2023. Last accessed 27 July 2023
5. Saudi Moments: More than 6 million people performed Umra during the current season. Retrieved from Saudi Moments: https://www.saudimoments.com/more-than-6-million-people-performed-Umra-during-the-current-season-583750.html. Last accessed 27 July 2023
6. Atikah, S., Tan, A.L., Trupp, A., Chong, D.K.L., Arni, A.G., Malar, V.: The emergence of a new religious travel segment: umra do it yourself travellers (DIY). GeoJ. Tourism Geosites **40**(1), 37–48 (2022)
7. Nassar, M.A., Mostafa, M.M., Reisinger, Y.: Factors influencing travel to Islamic destinations: an empirical analysis of Kuwaiti nationals. Int. J. Cult. Tourism Hospitality Res. **9**(1), 36–53 (2015)
8. Zhou, Q., Pu, Y., Su, C.: The mediating roles of memorable tourism experiences and destination image in the correlation between cultural heritage rejuvenation experience quality and revisiting intention. Asia Pac. J. Mark. Logist.Logist. **35**(6), 1313–1329 (2023)
9. Seyitoğlu, F., Alphan, E.: Gastronomy tourism through tea and coffee: travellers' museum experience. Int. J. Cult. Tourism Hospitality Res. **15**(3), 413–427 (2021)
10. Rasoolimanesh, S.M., Seyfi, S., Hall, C.M., Hatamifar, P.: Understanding memorable tourism experiences and behavioural intentions of heritage tourists. J. Destination Mark. Manag. **21** (2021)
11. Su, L., Swanson, S.R., Chen, X.: The effects of perceived service quality on repurchase intentions and subjective well-being of Chinese tourists: the mediating role of relationship quality. Tour. Manag.Manag. **52**, 82–95 (2016)
12. Kim, M.J., Lee, C.K., Preis, M.W.: The impact of innovation and gratification on authentic experience, subjective well-being, and behavioral intention in tourism virtual reality: the moderating role of technology readiness. Telematics Inform. **49**, 101349 (2020)
13. Li, Y., Song, H., Guo, R.: A study on the causal process of virtual reality tourism and its attributes in terms of their effects on subjective well-being during COVID-19. Int. J. Environ. Res. Public Health **18**(3), 1019 (2021)
14. Kim, J.H., Ritchie, J.B., McCormick, B.: Development of a scale to measure memorable tourism experiences. J. Travel Res. **51**(1), 12–25 (2012)

15. Stone, M.J., Soulard, J., Migacz, S., Wolf, E.: Elements of memorable food, drink, and culinary tourism experiences. J. Travel Res. **57**(8), 1121–1132 (2018)
16. Coudounaris, D.N., Sthapit, E.: Antecedents of memorable tourism experience related to behavioral intentions. Psychol. Mark. **34**(12), 1084–1093 (2017)
17. Kutlu, D., Ayyıldız, H.: The role of the destination image in creating memorable tourism experience. J. Tourism Serv. **12**(23), 199–216 (2021)
18. Mahdzar, M., Zaidi, N.M.M., Shuib, A.: Memorable tourism experiences in mulu national park: scales development. In: Heritage, Culture and Society, Research Agenda and Best Practices in the Hospitality and Tourism Industry, pp. 742–755 (2016)
19. Chen, H., Rahman, I.: Cultural tourism: an analysis of engagement, cultural contact, memorable tourism experience and destination loyalty. Tourism Manag. Perspectives **26**, 153–163 (2018)
20. Loureiro, S.M.C.: The role of the rural tourism experience economy in place attachment and behavioral intentions. Int. J. Hosp. Manag.Manag. **40**, 1–9 (2014)
21. Tsai, C.T.: Memorable tourist experiences and place attachment when consuming local food. Int. J. Tour. Res. **18**(6), 536–548 (2016)
22. Kim, J.H.: Determining the factors affecting the memorable nature of travel experiences. J. Travel Tour. Mark. **27**(8), 780–796 (2010)
23. Saayman, M., Li, G., Uysal, M., Song, H.: Tourist satisfaction and subjective well-being: an index approach. Int. J. Tour. Res. **20**(3), 388–399 (2018)
24. Sthapit, E., Coudounaris, D.N.: Memorable tourism experiences: antecedents and outcomes. Scand. J. Hosp. Tour. **18**(1), 72–94 (2018)
25. News Strait Times: Saudi Arabia records arrival of over 300,000 Malaysian Umra pilgrims in just 8 months. Retrieved from News Strait Times: https://api.nst.com.my/news/nation/2023/03/885188/saudi-arabia-records-arrival-over-300000-malaysian-Umra-pilgrims-just-8. Last accessed 27 July 2023
26. Hair, J.F., Risher, J.J., Sarstedt, M., Ringle, C.M.: When to use and how to report the results of PLS-SEM. Eur. Bus. Rev. **31**(1), 2–24 (2019)
27. Franke, G., Sarstedt, M.: Heuristics versus statistics in discriminant validity testing: a comparison of four procedures. Internet Res. **29**(3), 430–447 (2019)
28. Henseler, J., Ringle, C.M., Sarstedt, M.: A new criterion for assessing discriminant validity in variance-based structural equation modeling. J. Acad. Mark. Sci. **43**, 115–135 (2015)
29. Falk, R.F., Miller, N.B.: A Primer for Soft Modeling. University of Akron Press (1992)
30. Qurashi, J.: Commodification of Islamic religious tourism: from spiritual to touristic experience. Int. J. Religious Tourism Pilgrimage **5**(1), 89104 (2017)
31. Jambi, N.H.R.: Pilgrims shed tears of joy on completing Haj. Retrieved from Arab News: https://www.arabnews.com/node/2120446/saudi-arabia. Last accessed 27 July 2023
32. Muneeza, A., Zakariya, M.: COVID-19: it's impact in Haj and Umra and a future direction. J. Islamic Acc. Bus. Res. **12**(5), 661–679 (2021)
33. Nanggong, A., Hiola, P.R., Pakaya, P.: The experience and religiosity toward tourist satisfaction: the case of Umra Pilgrimage. J. Indonesian Tourism Hospitality Recreation **5**(2), 209–218 (2022)

The Antecedents of Tourist Satisfaction on Heritage Tourism Authenticity: A Systematic Literature Review

Yana Respati Dewi, Hamizah Abd Hamid, Nor Asiah Omar, and Sharizal Bin Hashim

Abstract Authenticity, which historically has an exclusive and rare value at this time is starting to be influenced by modernization. This is mainly driven by changes in market demand on tourism and hospitality industry. Many studies have emerged to analyze the effect of authenticity on tourist satisfaction due to this change. However, there has been no comprehensive analysis of the existing literature that examines the knowledge of authenticity of heritage tourism on tourist satisfaction at this time. The purpose of this review is to provide academics and tourism stakeholders with an up-to-date picture of tourist satisfaction research in the context of heritage tourism. This article uses Preferred Reporting Items for Systematic review and Meta-Analysis (PRISMA) analysis to review 35 articles in academic journals. The study showed that in addition to the value of authenticity in heritage tourism, experience and perceived authenticity are also factors that cause tourist satisfaction with heritage tourism.

Keywords Authenticity · Satisfaction · Heritage tourism · Cultural tourism

1 Introduction

Unlike other tourist attractions, heritage tourism or cultural tourism has its own aspects and value. Heritage tourism areas have historical value, that historical value contained is a uniqueness and distinctive character. The National Trust for Historic Preservation defines heritage tourism as a journey to enjoy historical places, artifacts or activities that authentically represent an earlier story or history. It can be concluded that heritage tourism provides access to enjoy various cultural customs, cultural and natural heritage objects and their contents for the purpose of providing knowledge and understanding of cultural and natural diversity for visitors. Based on this perspective, heritage tourism may influence the creation of destination images for visitors. According to the United Nations Educational, Scientific and Cultural Organization

Y. R. Dewi (✉) · H. A. Hamid · N. A. Omar · S. B. Hashim
Faculty of Economics and Management, The National University of Malaysia, 43600 Bangi, Selangor, Malaysia
e-mail: P115383@siswa.ukm.edu.my

B. Awwad (ed.), *The AI Revolution: Driving Business Innovation and Research*, Studies in Systems, Decision and Control 524, https://doi.org/10.1007/978-3-031-54379-1_61

(UNESCO), there are three types of heritage tourism attractions, namely cultural sites, natural sites and mixed sites. Examples of tourist attractions based on heritage tourism includes The Great Wall of China, Australia Great Barrier Reef and The Borobudur Temple Compounds in Indonesia.

Authenticity of heritage tourism has been discussed in many studies, especially related to the creation of a destination image [1–3], this is because the destination image will affect visit intention, tourist satisfaction and even tourist loyalty [4, 5]. The sustainability of a tourism depends on the number of visitor because the benefits of this destination image are not only in the tourism sector but also other benefits, including to the development of the local economy and also the growth of independent entrepreneurs in the surrounding [6]. Due to its cultural value, heritage tourism has important policy implications for the government, stakeholders and tourism organizations in which they must ensure all management processes to achieve sustainability goals. Therefore, there needs to be a more in-depth study of how authenticity in heritage tourism can affect visitor satisfaction in a tourism.

Among the main factors that can affect tourist satisfaction in heritage tourism is the concept of authenticity [7]. Tourism's quality will be created if the concept of authenticity applied. This concept of authenticity is considered relevant because of the history and culture contained in heritage tourism. During their visit, tourists can feel the experience of traveling with three elements of authenticity, namely objective authenticity, constructive authenticity and existential authenticity. As examples, the uniqueness and character of heritage areas such as Borobudur Temple in Indonesia or Imperial Palace in China have historical value. Both are tourist attractions low tourist involvement for the purpose of maintaining authenticity. The authenticity of the building must be maintained and preserved. The Borobudur temple restoration project carried out in 1975 is the largest construction project ever carried out by UNESCO [8]. The amenity of the two places must be maintained to be able to meet the needs and desires of tourists while in the tourist area.

The discussion on authenticity in heritage tourism is still shallow, thus slowing down our understanding of the importance of the topic. Therefore, this article aims investigate the variables involved in analyzing authenticity of heritage tourism especially in the perspective of tourist satisfaction. We do so by conducting a systematic review (SLR) to explore previous research and the depth of existing research on tourist satisfaction with authenticity in terms of heritage tourism. As an expected contribution, this study refines the definition of tourist satisfaction, linking tourist satisfaction with authenticity, discover trends in tourist satisfaction in the context of heritage tourism based on the discourse over the past nineteen years.

This paper begins with the literature review, then followed by the methods used in this SLR. Then we present the results and discuss the findings and concluded with suggestions for future research.

2 Literature Review

Heritage, or cultural heritage tourism is defined as going to experience the places, artefacts, and activities that truly represent the stories and people of the past and present [9]. Heritage is composed of the natural and cultural environments' tangible assets. Visitors' experiences to these sites are unique because it is shaped by landscapes, visual art and performances as well as special lifestyles, values, traditions and events or beliefs [10, 11]. According to Chhabra et al. [13] heritage tourism is a form of tourism that includes historical asset and socio-cultural values as a tourist attraction, which is highly related to authenticity [14].

2.1 Heritage Tourism Authenticity

Authenticity has become a concept of heritage protection which is closely related to intangible cultural heritage [15]. At a fundamental level, authenticity is the specific 'trait' of something whereas in the context of heritage tourism, authenticity is considered as a condition that must be met to be able to validate a statement of value or cultural significance [16]. For example, an old city building can be studied from its architectural style, archaeological sites can be investigated from artifacts or past objects, sanctuaries are investigated from carvings or reliefs that are on the walls of buildings and tropical forests that are observed based on the large number of living vegetative [17]. The concept of authenticity is introduced by MacCannell in, conceptualizes, who argues that heritage attraction referring only to the objective authenticity [18]. This objective perspective was developed on the grounds that authenticity can be measured by absolute, objective, pure, original and authentic criteria [19]. This concept reinforces the absence of modification as a distinguishing value. By imitating existing standards from the outside, there will be no system that can achieve authenticity, besides, authenticity is a constant criterion [20].

2.2 Tourist Satisfaction

Tourist satisfaction is one of the important concepts to achieve the success of a destination because it improves the chances of their loyalty as a customer [21]. Tourist satisfaction is defined as the level of visitor satisfaction resulting from the travel experience regarding the characteristics of the product or service that meets their wishes and expectations in relation to the trip. Tourist satisfaction can be achieved when experience exceeds or meets tourists' expectations. In the context of tourism, satisfaction is usually measured as a result of expectations and experiences before and after the trip [22]. Understanding how successful a particular destination is, understanding and meeting tourist demand, tourists' spending is the primary goal of measuring satisfaction in most of the studies on the topic [23].

The satisfaction depends on the extent of the overall enjoyment felt by tourists whereby the tourist experience is able to meet the wishes, expectations, needs of a destination [24]. The level of tourist satisfaction includes various dimensions that facilitate the psychological process of evaluating the experience of a product or service, in this case, for tourism is t destination [25]. It should be noted that tourist experiences varies according to the situational context and the subject's assessment of the activity [26]. Many of the heritage tours provide an emotional experience because of the historical value that they offer; compared to activities such as theme park tours or shopping malls that usually emphasizes tourists' involvement.

In this context, heritage tourism presents as a powerful means to build tourists' emotions of the object (architecture, relief, sanctuaries etc.). Authenticity will trigger emotions on aspects related to originality and its current conditions [27]. A tourist may feel moved or touched by a simple presence of a historic building that reminds them of an important event or an important figure, another factor is historical building's aesthetic appeal. It is this emotional aspect that is taken into account in the value and meaning attributed to a historical site, in which tourists are able to imagine a history within the site [28]. Creating an attractive destination for tourists that can meet tourist expectations is just as important as achieving tourist satisfaction and loyalty. Resources and attractions are recognized as determinants that measure the competitiveness of tourism destinations and planned marketing and efficient attribute management [29]. To be able to achieve the goal of sustainable competitive advantage in the travel and tourism sector, a destination must be able to find and define differentiating values that are developed according to tourist preferences [30]. However, there are limitations in the context of heritage tourism. On the other hand, indigenous values and cultures must be maintained so that they are not affected and eroded by post-modern authenticity. This post-modern is a threat that makes tourism lose the value of authenticity due to acculturation of shifting trends that affect the needs and desires of tourists in achieving tourist destinations.

3 Materials and Methods

The electronic databases from Scopus and Web of Science (WoS) were chosen for this SLR because they are the most extensive databases in various scientific fields, and are often used for literature search [31]. The selected terms were looked up in the paper's title, topic and abstract.

This SLR uses content analysis, as they are frequently preferred in social sciences [32], in alignment with the aim of this study, which is to understand the trends of the topic. Content analysis is an approach where we summarize specific words (codes) in a document into groups [33]. To help narrow the article search, we applied inclusive criteria of database using Preferred Reporting Items for Systematic literature reviews and Meta-Analyses, or PRISMA. The aim of PRISMA to help researchers to improve the reporting of systematic literature reviews and meta-analyses. Additionally, PRISMA can also be used as a basis for reporting systematic literature reviews of recent research [34]. We examine the papers based on the predetermined themes of this study, namely heritage tourism, authenticity and tourism satisfaction.

There are five stages used to conduct a SLR [35]: (a) defining the criteria by creating list for inclusion and exclusion, (b) defining sources of information, (c) selecting literature (this stage is used to define the keywords, exploring from the title, abstract, and also keywords), (d) collecting data, (including data extraction formulation, examining the type of article, title of journal, year published, topic, title, keywords, country, research methodology, and also the results), and (e) selecting data.

The first stage of elimination is to filter the title and abstract of the article and excluding conceptual studies. The next stage of elimination involves thorough review of the papers, ensuring that the objectives and materials of the papers and are relevant to the SLR's purposes. The extraction results were reviewed to identify emerging patterns and ascertain whether there were similarities or differences in their discussion on authenticity, heritage tourism and tourist satisfaction. The strategy of this study is described in Fig. 1.

Articles that do not meet the criteria were eliminated, resulting in 65 articles after reviewing titles and abstracts, while removing some duplicate articles contained in both databases. The filtration process led to 35 articles with a timeline spanning from 2003 to 2022. Based on research objectives of this SLR, we extracted the antecedent of tourist satisfaction and the consequences of authenticity in the context of heritage tourism.

4 Result and Discussion

4.1 Study Setting and Major Research Purposes

The following Table 1 *gives a summary of all heading levels.* There are 17 countries used as study setting for the topic, over seven continents. Overall, research on heritage tourism has been carried out throughout Asia, Europe, Africa and also America. However, Eastern Asia has dominated the research study setting (59%). A total of 14 studies were conducted in China, which was then followed by South Korea and Taiwan. This is followed by studies on European continent (12% across Italy, Spain and the United Kingdom). Southeast Asian ranks next with a proportion of 10% value spread across Malaysia and Singapore.

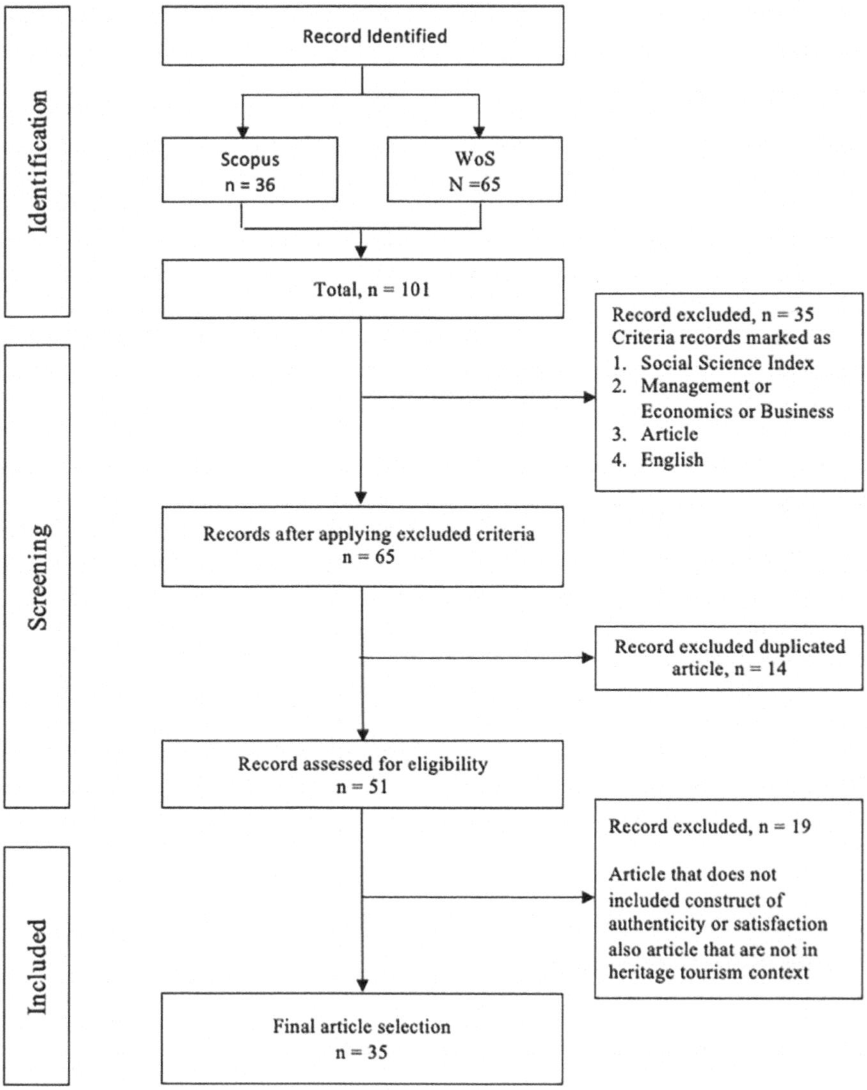

Fig. 1 PRISMA flowchart

After reviewing and analyzing the research objectives of selected published articles, we grouped the research objectives into the following categories: (1) investigating and evaluating tourist satisfaction with authenticity, (2) investigate the factors that can affect tourist satisfaction, (3) examine the concept of authenticity in heritage tourism.

For the first category of papers, which investigated factors that could affect tourist satisfaction with the authenticity of heritage tourism, is most commonly represented

Table 1 Variable used in articles

Categories	Variable / Author	[44]	[45]	[42]	[46]	[47]	[48]	[49]	[50]	[51]	[52]	[53]	[54]	[41]	[55]	[43]	[56]	[57]
Destination factor	Destination image				•							•				•		•
	Culture								•									
	Objective authenticity	•	•		•	•	•	•	•	•	•		•	•			•	
	Constructive authenticity	•	•		•	•	•	•	•	•	•		•	•			•	
	Existential authenticity	•	•		•	•	•	•	•		•		•	•	•		•	
	Place identity						•											•
Personal factor	Perceived authenticity		•	•				•		•					•	•	•	
	Motivation	•						•										
	Experiences			•								•						•
	Engagement			•							•	•						
	Involvement				•							•						
	Nostalgia					•												
	Behavioural intention						•							•				
	Satisfaction		•		•	•		•			•	•	•	•		•	•	•
	Loyalty	•								•			•		•	•		

Categories	Variable / Author	[58]	[59]	[60]	[61]	[38]	[62]	[63]	[64]	[37]	[65]	[36]	[66]	[67]	[7]	[68]	[69]	[70]	[71]
Destination factor	Destination image																		
	Culture			•						•		•							

(continued)

Table 1 (continued)

Categories	Variable / Author	[58]	[59]	[60]	[61]	[38]	[62]	[63]	[64]	[37]	[65]	[36]	[66]	[67]	[7]	[68]	[69]	[70]	[71]
	Objective authenticity	•	•		•	•	•	•	•	•	•	•		•	•			•	
	Constructive authenticity	•			•	•	•	•	•	•	•	•			•			•	
	Existential authenticity	•	•		•	•	•	•	•	•	•	•		•	•			•	
	Place identity	•																	
Personal factor	Perceived authenticity			•					•		•		•						•
	Motivation		•			•				•									•
	Experiences		•				•		•			•		•		•	•	•	•
	Engagement																•		•
	Involvement				•	•													
	Nostalgia																		
	Behavioural intention							•									•		•
	Satisfaction	•		•		•	•		•	•	•	•	•	•	•	•		•	
	Loyalty	•		•	•	•					•		•		•				

(51%, or n = 18). For the second category, based on the articles studied in this research, tourist satisfaction is evaluated from various factors. Most articles mainly analyze factors that moderate to the value of authenticity in heritage tourism. For example, an article by Zhang et al. [36] examines the relationships of cross-culture awareness, experience, authenticity, satisfaction and acculturation in South Korea [36]. A research conducted by Chhabra [37] in the US aims to identify push and pull factors that influence students to visit heritage tourism sites, examine patterns of reciprocal relationships and analyze their effect on their satisfaction. A study by Maaiah and Wouhoush [38] in Jordan analyzed the role of traditional Jordanian culture in contributing to tourist satisfaction.

The third category of articles aim to investigate and evaluate tourist satisfaction with authenticity (37% or n = 13). For example, a study conducted by Nguyen [45] aims to analyze the effect of authenticity on tourist satisfaction in China. Research in the same country was also conducted by Dai et al. [39] with the aim to examine relationship between tourists' understanding of the tourism authenticity and their satisfaction. Additionally, a research with the aim of examining the influence of authenticity on satisfaction and its relationship with engagement on social media was conducted by Kim and Kim [40]. There are also articles discussing technology, as examples an article by Atzeni et al. [41] that aims to investigate the role of authenticity in non-immersive virtual heritage tourism and Tsai [42] which aims to examine why and how location-based augmented reality applications can increase tourist satisfaction in heritage tourism.

4.2 *Identification Variable Used in Articles*

The analysis done on the articles revealed 15 variables leading to tourist satisfaction in heritage tourism. Here, we group 15 variables into two categories, destination factors and personal factors. Destination factors are factors that contribute to the attractiveness of the destination or factors that can affect the personal factors in visitors [2] and the consequences will be seen in the success or failure of the destination. Meanwhile, personal factors are factors related to the perception, social experience and emotions of tourists [43]. In other words, destination factors are factors that external from the tourists, while personal factors are factors that originate from tourist itself. The variable that we categorize as destination factors include destination image, culture value, authenticity consisting of objective, constructive and existential as well as place identity. Perceived authenticity, motivation, experiences, involvement, nostalgia, behavioral intention such as visit intention or revisit intention, satisfaction and loyalty fall into the personal factor category. This systemic analysis was carried out by filtering the dimensions of authenticity and satisfaction as the main dimensions used in previous studies. Other antecedents found were experience, perceived authenticity and loyalty. Other dimensions such as nostalgia and place identity are the two dimensions that are least discussed in the articles.

4.3 Destination Factor

We group destination image, cultural value, authenticity and place identity into destination factors. Destination image is whole visitor perception about tourist location. This perception is a key factor and stimulus for tourists when choosing a destination [69, 70]. Destination image is a result from tourist evaluation and perception which is a key factor for travel destination. Perception refers to how a person can perceive, process mentally and act in the face of information received from the environment [71]. This perception can also be formed from experiences, tourist's recommendation and attitude toward tourist destination. Previous research also explained that the experience of traveling has an influence on tourist satisfaction, through repetition of experiences with the same learning process in one destination through revisit intention. Darnell and Johnson [72] state that a degree of satisfaction can cause revisit intention.

Destination image is also defined as an individual impression of the tourist about reality [73]. It refers to the impression that tourists have about a place, which largely based on their perception of all familiar places or those that they have visited. Although it is difficult to define destination image because of its vague, subjective and immaterial nature, the development of destination image is based on rational factors and emotional factors of tourists. Affective and cognitive are the main attributes, it can be said that destination images are assessed based on resource attributes and attractiveness which then motivate tourists to visit these destinations [74]. The attribute, affective refers to the feelings and emotions evoked by tourist destinations, more like emotional component. The second attribute, cognitive attributes, refers to the natural and cultural environment, as well as unique experiences related to these destinations [75].

The essence of a culture is shaped by values where these values will be the basis for the attractiveness of tourist destinations [76]. The cultural level will be the rituals and symbols of the given culture which later also becomes the basis of the tourist destination. Culture refers to a way of life but is not limited to language, arts and sciences, thought, spirituality, social activity and interaction [77]. Therefore, it can be concluded that culture is part of the lifestyle. Similarities in language, writing, behavior, customs, heritage, ideology and even technology connect individuals of a culture. Therefore, cultural similarities or cultural nearness can form an attraction that can motivate the interest in visiting.

The study of cultural identity in the context of tourism has drawn more interest as the tourism sector has continued to grow. Authenticity has been defined as the main qualification factor regarding the value of cultural wealth while place identity is a characteristic or marker of destinations so that it has a different value from other destinations. The articles in this study explained to better understand and distinguish the tourist experience and structural comparison between modern and primitive, sacred and superficial, reality and performance; whereby authenticity is necessary. In relation to their cultural heritage and environment, cultural identity emphasis on the affirmation of common culture among individuals and society. Recognition and a sense of value serve as the foundation of cultural identity [78].

4.4 Personal Factor

Variables under the personal factor category refers to customer behavior or factors formed from within the tourist. In this article we group perceived authenticity, motivation, experiences, place engagement, involvement, nostalgia, behavioral intention, satisfaction and loyalty into personal factors. Perceived authenticity refers to how authentic someone feels and also be interpreted as authenticity that is accepted, seen or felt by tourists. This complexity of consumer perception is also defined as a cognitive exchange for a consumption experience [79]. In short, this perceived value is considered as a combined utilitarian and hedonistic response in which entertainment and emotional value must be included in the conceptualization of value [80]. The experience is unique to bond visitors in a certain way so that it is memorable. The recreational experience is described as a subjective mental state perceived by participants [81]. Experience can also be created from social interaction, in the context of tourism, local communities and the uniqueness of attractions are the main keys [82, 83].

A positive relationship between an individual and a place is either physical or social described as a place attachment. Place attachment is also known as the bond and emotional connection between the person and a place [84]. Place attachment also has a positive relationship with the image of the destination as well as the experience [85–87]. Involvement in the context of tourism can be defined as the degree to which tourists are interested in an activity and their affective responses arise based on those activities [46]. A memorable tourist experience is an experience that can be described [88]. Personal involvement in experience is an element of value creation from the experience itself. In comparison, involvement is defined as personal involvement and real-time consumption of the experience provided.

Involvement is only situational that does not last long [89]. Involvement is also related to personal values, motivation and self-awareness that influence a person in enjoying a destination [90]. From the brief definition of these variables, it can be concluded that these variables are personal variables. This category (personal factors) refers to the subjective assessment of tourist attributes. This is considered important because tourists are not only looking for a memorable experience through something modern and of high quality but also looking for an authentic experience [87].

Based on the results of the literature review analysis, in addition to satisfaction and loyalty, the variables that are widely discussed are perceived authenticity and experience. Many studies have raised the importance of assessing the quality of the tourist experience, this is because it will provide a better understanding of tourism activities [22, 92]. The study conducted by Chen and Chen [22] applies the concept of experience quality to heritage tourism and proposes three dimensions, namely engagement, peace of mind and educational experience. It has been explained at the beginning that heritage tourism is different from tourism in general because it requires the value of authenticity and knowledge of history to be able to understand the value of authenticity.

5 Conclusion

This systematic literature review analyzes the distribution of dimensions and study settings of heritage tourism research with relevant topics. This article concludes that current research on the topic of tourist satisfaction with authenticity in heritage tourism can generally be found to be related to three other variables namely perceived authenticity, experience and loyalty. The three variables are interconnected and influenced by each other. Destination factors Furthermore, this article also proposes that generally research on heritage tourism leads to resource planning and sustainability.

Because in recent years the context of intangible heritage tourism has attracted attention, traditional events, such as performances or puppet shows and even festivals have been included in the heritage tourism category although some of them cannot be assessed as object-based authenticity aspects. In this aspect, sustainability and environmental sustainability are potential research directions in the future. The results of this literature study suggest to subsequent researchers to take several variables that have not been analyzed in depth, especially such as nostalgia and place identity. These two variables are closely related to history which makes the value of the authenticity of a tourist attraction and as branding of a destination. In the preservation of this heritage attraction, we emphasize three issues. First, the planning and sustainability process starts with protecting the heritage and maintaining the authenticity of the destination. Second, planning needs to involve stakeholders. And finally, the most effective strategy to be able to create an image of the destination is to know and predict the tourist's behavior and the tourist market in the future.

It is important to preserve the cultural heritage of the world for the sake of the education of present and future generations. Tourism is highly dependent on the location, historical significance, culture and ecology underlying its products [94]. People involved in the tourism sector need to be sensitive to cultural groups that have a special interest in it to respect the cultural value of the destination. The physical structure that affects their production must also be maintained because cultural heritage resources will be essential for the sustainable cultural, social and economic growth of society. Thus, heritage tourism planning will encourage the sustainability of tourist areas. Finding suitable solutions to the problems and difficulties that arise when managing and promoting heritage tourism is essential. Because it can produce the development of a heritage tourism model that has a positive impact on overall heritage tourism.

Therefore, we hope that this research on literature studies will create an interest in academics to reveal other factors that can affect tourist satisfaction in heritage tourism. In addition, we hope that it can also be a reference for stakeholders in taking actions so that heritage tourism can continue to be preserved.

References

1. Munar, A.M.: Tourist-created content: rethinking destination branding. Int. J. Cult. Tour. Hosp. Res. **5**(3), 291–305 (2011)
2. Moreira, P., Iao, C.: Destination loyalty and destination fatigue: factors of destination image and destination attraction (2014)
3. Qu, H., Kim, L.H., Im, H.H.: A model of destination branding: integrating the concepts of the branding and destination image. Tour. Manag. **32**(3) (2011). https://doi.org/10.1016/j.tourman.2010.03.014
4. Yi, X., Fu, X., Yu, L., Jiang, L.: Authenticity and loyalty at heritage sites: the moderation effect of postmodern authenticity. Tour. Manag. **67**, 411–424 (2018). https://doi.org/10.1016/j.tourman.2018.01.013
5. Fu, X.: Existential authenticity and destination loyalty: evidence from heritage tourists. J. Destin. Mark. Manag. **12**, 84–94 (2019). https://doi.org/10.1016/j.jdmm.2019.03.008
6. Buonincontri, P., Micera, R., Murillo-Romero, M., Pianese, T.: Where does sustainability stand in underground tourism? A literature review. Sustainability **13**(22) (2021). https://doi.org/10.3390/su132212745
7. Eunkyung, P., Choi, B.-K., Lee, T.J.: The role and dimensions of authenticity in heritage tourism. Tour. Manag. **74**, 99–109 (2019). https://doi.org/10.1016/j.tourman.2019.03.001
8. Voûte, C.: The restoration and conservation project of Borobudur Temple, Indonesia. Planning: research: design. Stud. Conserv. **18**(3), 113 (1973). https://doi.org/10.2307/1505654
9. Yu, X., Xu, H.: Cultural heritage elements in tourism: a tier structure from a tripartite analytical framework. J. Destin. Mark. Manag. **13**, 39–50 (2019). https://doi.org/10.1016/j.jdmm.2019.05.003
10. Qiu, Q., Zuo, Y., Zhang, M.: Intangible cultural heritage in tourism: research review and investigation of future agenda. Land **11**(1) (2022). https://doi.org/10.3390/land11010139
11. Bagader, M.: Tourism-based approach to conserving the built heritage sites in the holy city of Makkah, Saudi Arabia. WIT Trans. Built Environ. **197**, 139–151 (2020). https://doi.org/10.2495/IHA200121
12. Park, H.Y.: Shared national memory as intangible heritage. Ann. Tour. Res. **38**(2), 520–539 (2011). https://doi.org/10.1016/j.annals.2010.11.013
13. Chhabra, D., Healy, R., Sills, E.: Staged authenticity and heritage tourism. Ann. Tour. Res. **30**(3), 702–719 (2003). https://doi.org/10.1016/S0160-7383(03)00044-6
14. Kim, H., Jamal, T.: Touristic quest for existential authenticity. Ann. Tour. Res. **34**(1), 181–201 (2007). https://doi.org/10.1016/j.annals.2006.07.009
15. Ma, W.: Recognition and introspection: a review of intangible cultural heritage protection from the perspective of cultural consciousness. J. Guangxi Teach. Educ. Univ **37**, 25–29 (2016)
16. Holtorf, C., Kono, T.: Forum on Nara +20: an introduction. Herit. Soc. **8**(2), 139–143 (2015). https://doi.org/10.1080/2159032X.2015.1126117
17. Albert, M.T.: Perspectives of World Heritage: towards future-oriented strategies with the five 'Cs' (2012)
18. MacCannell, D.: Staged authenticity: arrangements of social space in tourist settings. Am. J. Sociol. **79**(3), 589–603 (1973). [Online]. Available: http://www.jstor.org/stable/2776259
19. Reisinger, Y., Steiner, C.J.: Reconceptualizing object authenticity. Ann. Tour. Res. **33**(1), 65–86 (2006). https://doi.org/10.1016/j.annals.2005.04.003
20. Naghizadeh, M.: Identity of Iranian architecture and modernism. Honar-Ha-Ye-Ziba (7), 79–91 (2000). [Online]. Available: https://www.sid.ir/en/Journal/ViewPaper.aspx?ID=58062
21. Saayman, M., Li, G., Uysal, M., Song, H.: Tourist satisfaction and subjective well-being: an index approach. Int. J. Tour. Res. **20**(3), 388–399 (2018). https://doi.org/10.1002/jtr.2190
22. Chen, C.F., Chen, F.S.: Experience quality, perceived value, satisfaction and behavioral intentions for heritage tourists. Tour. Manag. **31**(1), 29–35 (2010). https://doi.org/10.1016/j.tourman.2009.02.008

23. Aliman, N.K., Hashim, S.M., Wahid, S.D.M., Harudin, S.: Tourists' satisfaction with a destination: an investigation on visitors to Langkawi Island. Int. J. Mark. Stud. **8**(3), 173 (2016). https://doi.org/10.5539/ijms.v8n3p173
24. Chen, C.-F., Tsai, D.: How destination image and evaluative factors affect behavioral intentions? Tour. Manag. **28**(4), 1115–1122 (2007). https://doi.org/10.1016/j.tourman.2006.07.007
25. Armario, E.: Tourist satisfaction: an analysis of its antecedents, pp. 367–382. Universidad, Soc. y Mercados Glob., 2008-01-01 (Jan. 2008). ISBN 978-84-691-5667-4
26. Gountas, J., Gountas, S.: Personality orientations, emotional states, customer satisfaction, and intention to repurchase. J. Bus. Res. **60**(1), 72–75 (2007). https://doi.org/10.1016/j.jbusres.2006.08.007
27. Vivant, E.: Nathalie Heinich, La fabrique du patrimoine. De la cathédrale à la petite cuillère. Lectures (Nov. 2009). https://doi.org/10.4000/lectures.837
28. Fortuna, C.: Heritage, tourism and emotion*. RCCS Annu. Rev. (5) (2013). https://doi.org/10.4000/rccsar.498
29. Cracolici, M.F., Nijkamp, P., Rietveld, P.: Assessment of tourism competitiveness by analysing destination efficiency. Tour. Econ. **14**(2), 325–342 (2008). https://doi.org/10.5367/000000008784460427
30. Glińska, E., Florek, M.: In searching for town brand distinguishing features—local leaders' inner perspective. Int. J. Acad. Res. **5**(4), 200–205 (2013). https://doi.org/10.7813/2075-4124.2013/5-4/B.30
31. Chadegani, A.A., et al.: A comparison between two main academic literature collections: web of science and Scopus databases. Asian Soc. Sci. **9**(5) (2013). https://doi.org/10.5539/ass.v9n5p18
32. Tranfield, D., Denyer, D., Smart, P.: Towards a methodology for developing evidence-informed management knowledge by means of systematic review. Br. J. Manag. **14**(3), 207–222 (2003). https://doi.org/10.1111/1467-8551.00375
33. Hsieh, H.-F., Shannon, S.E.: Three approaches to qualitative content analysis. Qual. Health Res. **15**(9), 1277–1288 (2005). https://doi.org/10.1177/1049732305276687
34. Liberati, A., et al.: The PRISMA statement for reporting systematic reviews and meta-analyses of studies that evaluate healthcare interventions: explanation and elaboration. BMJ **339**, b2700–b2700 (2009). https://doi.org/10.1136/bmj.b2700
35. Handayani, P.W.: Systematic review dengan PRISMA (preferred reporting items for systematic reviews and meta-analysis) (2017). https://dosen.perbanas.id/wp-content/uploads/2017/08/Sesi2-SYSTEMATIC-REVIEW-DENGAN-PRISMA.pdf. Accessed 10 Jan. 2022
36. Zhang, H., Cho, T., Wang, H., Ge, Q.: The influence of cross-cultural awareness and tourist experience on authenticity, tourist satisfaction and acculturation in world cultural heritage sites of Korea. Sustainability **10**(4), 927 (2018). https://doi.org/10.3390/su10040927
37. Chhabra, D.: Student motivations: a heritage tourism perspective. Anatolia **21**(2), 249–269 (2010). https://doi.org/10.1080/13032917.2010.9687102
38. Maaiah, B.S., Wouhoush, O.N.: Jordanian culture and traditions as a driving force for attracting Dutch tourists. Geoj. Tour. Geosites **33**(4), 1450–1455 (2020). https://doi.org/10.30892/gtg.334spl01-592
39. Dai, T., Zheng, X., Yan, J.: Contradictory or aligned? The nexus between authenticity in heritage conservation and heritage tourism, and its impact on satisfaction. Habitat Int. **107**, 102307 (2021). https://doi.org/10.1016/j.habitatint.2020.102307
40. Kim, M., Kim, J.: Destination authenticity as a trigger of tourists' online engagement on social media. J. Travel Res. **59**(7), 1238–1252 (2020). https://doi.org/10.1177/0047287519878510
41. Atzeni, M., Del Chiappa, G., Pung, J.M.: Enhancing visit intention in heritage tourism: the role of object-based and existential authenticity in non-immersive virtual reality heritage experiences. Int. J. Tour. Res. **24**(2), 240–255 (2022). https://doi.org/10.1002/jtr.2497
42. Tsai, S.: Augmented reality enhancing place satisfaction for heritage tourism marketing. Curr. Issue Tour. **23**(9), 1078–1083 (2020). https://doi.org/10.1080/13683500.2019.1598950
43. Lisa, T., Noor, M., Azila, N.: Factors influencing destination loyalty of tourists at historic cities in Malaysia—a proposed model. Soc. Sci. **11**(11) (2016)

44. Kolar, T., Zabkar, V.: A consumer-based model of authenticity: an oxymoron or the foundation of cultural heritage marketing? Tour. Manag. **31**(5), 652–664 (2010). https://doi.org/10.1016/j.tourman.2009.07.010
45. Nguyen, T.H.H.: A reflective-formative hierarchical component model of perceived authenticity. J. Hosp. Tour. Res. **44**(8), 1211–1234 (2020). https://doi.org/10.1177/1096348020944460
46. Lu, L., Chi, C.G., Liu, Y.: Authenticity, involvement, and image: evaluating tourist experiences at historic districts. Tour. Manag. **50**, 85–96 (2015). https://doi.org/10.1016/j.tourman.2015.01.026
47. Gao, J., Lin, S.T., Zhang, C.Z.: Authenticity, involvement, and nostalgia: understanding visitor satisfaction with an adaptive reuse heritage site in urban China. J. Destin. Mark. Manag. **15** (2020). https://doi.org/10.1016/j.jdmm.2019.100404
48. Ramkissoon, H.: Authenticity, satisfaction, and place attachment: a conceptual framework for cultural tourism in African island economies. Dev. South. Afr. **32**(3), 292–302 (2015). https://doi.org/10.1080/0376835X.2015.1010711
49. Nguyen, T.H.H., Cheung, C.: Chinese heritage tourists to heritage sites: what are the effects of heritage motivation and perceived authenticity on satisfaction? Asia Pacific J. Tour. Res. **21**(11), 1155–1168 (2016). https://doi.org/10.1080/10941665.2015.1125377
50. Dai, T.C., Zheng, X., Yan, J.: Contradictory or aligned? The nexus between authenticity in heritage conservation and heritage tourism, and its impact on satisfaction. Habitat Int. **107** (2021). https://doi.org/10.1016/j.habitatint.2020.102307
51. Lin, Y.C., Liu, Y.C.: Deconstructing the internal structure of perceived authenticity for heritage tourism. J. Sustain. Tour. **26**(12), 2134–2152 (2018). https://doi.org/10.1080/09669582.2018.1545022
52. Otoo, F.E., Kim, S.S., Stylidis, D.: Diaspora tourists' emotional experience. Int. J. Tour. Res. **23**(6), 1042–1058 (2021). https://doi.org/10.1002/jtr.2461
53. Zhang, T.H., Yin, P., Peng, Y.X.: Effect of commercialization on tourists' perceived authenticity and satisfaction in the cultural heritage tourism context: case study of Langzhong Ancient City. Sustainability **13**(12) (2021). https://doi.org/10.3390/su13126847
54. Lee, S., Phau, I., Hughes, M., Li, Y.F., Quintal, V.: Heritage tourism in Singapore Chinatown: a perceived value approach to authenticity and satisfaction. J. Travel Tour. Mark. **33**(7), 981–998 (2016). https://doi.org/10.1080/10548408.2015.1075459
55. Lin, Y.H., Lee, T.H.: How the authentic experience of a traditional cultural festival affects the attendee's perception of festival identity and place identity. Int. J. Event Festiv. Manag. **11**(3), 357–373 (2020). https://doi.org/10.1108/IJEFM-12-2019-0061
56. Wu, D., Shen, C.Y., Wang, E.X., Hou, Y.Y., Yang, J.: Impact of the perceived authenticity of heritage sites on subjective well-being: a study of the mediating role of place attachment and satisfaction. Sustainability **11**(21) (2019). https://doi.org/10.3390/su11216148
57. Lee, G., Lin, X., Choe, Y., Li, W.Y.: In the eyes of the beholder: the effect of the perceived authenticity of Sanfang Qixiang in Fuzhou, China, among locals and domestic tourists. Sustainability **13**(22) (2021). https://doi.org/10.3390/su132212353
58. Tian, D., Wang, Q.Y., Law, R., Zhang, M.: Influence of cultural identity on tourists' authenticity perception, tourist satisfaction, and traveler loyalty. Sustainability **12**(16) (2020). https://doi.org/10.3390/su12166344
59. Horng, J.-S., Liu, C.-H., Chou, S.-F., Yin, Y.-S., Tsai, C.-Y.: Developing a novel hybrid model for industrial environment analysis: a study of the gourmet and tourism industry in Taiwan. Asia Pacific J. Tour. Res. **19**(9), 1044–1069 (2014). https://doi.org/10.1080/10941665.2013.837399
60. Lu, W.W., Su, Y.J., Su, S.T., Zhao, J., Zhang, L.: Perceived authenticity and experience quality in intangible cultural heritage tourism: the case of Kunqu Opera in China. Sustainability **14**(5) (2022). https://doi.org/10.3390/su14052940
61. Wu, T.C., Xie, P.F., Tsai, M.C.: Perceptions of attractiveness for salt heritage tourism: a tourist perspective. Tour. Manag. **51**, 201–209 (2015). https://doi.org/10.1016/j.tourman.2015.05.026

62. Naqvi, M.H.A., Jiang, Y.S., Naqvi, M.H., Miao, M., Liang, C.Y., Mehmood, S.: The effect of cultural heritage tourism on tourist word of mouth: the case of Lok Versa Festival, Pakistan. Sustainability **10**(7) (2018). https://doi.org/10.3390/su10072391
63. Novello, S., Fernandez, P.M.: The influence of event authenticity and quality attributes on behavioral intentions. J. Hosp. Tour. Res. **40**(6), 685–714 (2016). https://doi.org/10.1177/1096348013515914
64. Dominguez-Quintero, A.M., Gonzalez-Rodriguez, M.R., Paddison, B.: The mediating role of experience quality on authenticity and satisfaction in the context of cultural-heritage tourism. Curr. Issues Tour. **23**(2), 248–260 (2020). https://doi.org/10.1080/13683500.2018.1502261
65. Lee, K.Y., Lee, H.: Traditional costume experience at a cultural heritage festival. Tour. Manag. Perspect. **32** (2019). https://doi.org/10.1016/j.tmp.2019.100555
66. Rasoolimanesh, S.M., Seyfi, S., Hall, C.M., Hatamifar, P.: Understanding memorable tourism experiences and behavioural intentions of heritage tourists. J. Destin. Mark. Manag. **21** (2021). https://doi.org/10.1016/j.jdmm.2021.100621
67. Io, M.-U.: Understanding the core attractiveness of performing arts heritage to international tourists. Tour. Geogr. **21**(4), 687–705 (2019). https://doi.org/10.1080/14616688.2019.1571096
68. Bryce, D., Curran, R., O'Gorman, K., Taheri, B.: Visitors' engagement and authenticity: Japanese heritage consumption. Tour. Manag. **46**, 571–581 (2015). https://doi.org/10.1016/j.tourman.2014.08.012
69. Lee, T.H.: A structural model to examine how destination image, attitude, and motivation affect the future behavior of tourists. Leis. Sci. **31**(3), 215–236 (2009). https://doi.org/10.1080/01490400902837787
70. Hosany, S., Prayag, G., Deesilatham, S., Caušević, S., Odeh, K.: Measuring tourists' emotional experiences. J. Travel Res. **54**(4), 482–495 (2015). https://doi.org/10.1177/0047287514522878
71. Rossi, S.D., Byrne, J.A., Pickering, C.M., Reser, J.: 'Seeing red' in national parks: how visitors' values affect perceptions and park experiences. Geoforum **66**, 41–52 (2015). https://doi.org/10.1016/j.geoforum.2015.09.009
72. Darnell, A.C., Johnson, P.S.: Repeat visits to attractions: a preliminary economic analysis. Tour. Manag. **22**(2), 119–126 (2001). https://doi.org/10.1016/S0261-5177(00)00036-4
73. Bigné, J.E., Sánchez, M.I., Sánchez, J.: Tourism image, evaluation variables and after purchase behaviour: inter-relationship. Tour. Manag. **22**(6), 607–616 (2001). https://doi.org/10.1016/S0261-5177(01)00035-8
74. Govers, R., Go, F.M.: Project destination image online: website content analysis of picture and text. Inf. Technol. Tour. **7**(2), 73–89 (2005). https://doi.org/10.3727/1098305054517327
75. Wang, C., Hsu, M.K.: The relationships of destination image, satisfaction, and behavioral intentions: an integrated model. J. Travel Tour. Mark. **27**(8), 829–843 (2010). https://doi.org/10.1080/10548408.2010.527249
76. Hofstede, G.: Cultures and Organizations. Software of the Mind. McGraw-Hill, New York (1997)
77. Csapo, J.: The role and importance of cultural tourism in modern tourism industry. In: Strategies for Tourism Industry—Micro and Macro Perspectives. InTech (2012)
78. Cui, X.: Cultural identity and its roots. J. Beijing Norm. Univ. **184**, 102–104 (2004)
79. Mathwick, C., Malhotra, N.K., Rigdon, E.: The effect of dynamic retail experiences on experiential perceptions of value: an internet and catalog comparison. J. Retail. **78**(1), 51–60 (2002). https://doi.org/10.1016/S0022-4359(01)00066-5
80. Lemmink, J., de Ruyter, K., Wetzels, M.: The role of value in the delivery process of hospitality services. J. Econ. Psychol. **19**(2), 159–177 (1998). [Online]. Available: https://econpapers.repec.org/RePEc:eee:joepsy:v:19:y:1998:i:2:p:159-177
81. Otto, J.E., Ritchie, J.R.B.: The service experience in tourism. Tour. Manag. **17**(3), 165–174 (1996). https://doi.org/10.1016/0261-5177(96)00003-9
82. Marzi, G., Marsocci, M., Zollo, L., Boccardi, A.: Understanding the intangible elements of the tourist heritage: a netnographic search. Capitale Cult. Value Cult. Herit. **13**, 267–297 (2016)
83. Romao, J.: Tourism, a place-based activity. In: Tourism, Territory and Sustainable Development: Theoretical Foundations and Empirical Applications in Japan and Europe, pp. 37–64 (2018)

84. Tsai, S.: Place attachment and tourism marketing: investigating international tourists in Singapore. Int. J. Tour. Res. **14**(2), 139–152 (2012). https://doi.org/10.1002/jtr.842
85. Jin, M., Choi, Y., Lee, C.-K., Ahmad, M.S.: Effects of place attachment and image on revisit intention in an ecotourism destination: using an extended model of goal-directed behavior. Sustainability **12**(18), 7831 (2020). https://doi.org/10.3390/su12187831
86. Low, S.M., Altman, I.: Place attachment. In: Place Attachment, pp. 1–12. Springer US Boston, MA (1992)
87. Ram, Y., Björk, P., Weidenfeld, A.: Authenticity and place attachment of major visitor attractions. Tour. Manag. **52**, 110–122 (2016). https://doi.org/10.1016/j.tourman.2015.06.010
88. Seyfi, S., Hall, C.M., Rasoolimanesh, S.M.: Exploring memorable cultural tourism experiences. J. Herit. Tour. **15**(3), 341–357 (2020). https://doi.org/10.1080/1743873X.2019.1639717
89. Zatori, A., Smith, M.K., Puczko, L.: Experience-involvement, memorability and authenticity: the service provider's effect on tourist experience. Tour. Manag. **67**, 111–126 (2018). https://doi.org/10.1016/j.tourman.2017.12.013
90. Lee, T.H., Chang, Y.S.: The influence of experiential marketing and activity involvement on the loyalty intentions of wine tourists in Taiwan. Leis. Stud. **31**(1), 103–121 (2012). https://doi.org/10.1080/02614367.2011.568067
91. Dominguez-Quintero, A.M., Gonzalez-Rodriguez, M.R., Paddison, B.: The mediating role of experience quality on authenticity and satisfaction in the context of cultural-heritage tourism. Curr. Issue Tour. **23**(2), 248–260 (2020). https://doi.org/10.1080/13683500.2018.1502261
92. Mansour, J.S.A., Ariffin, A.A.M.: The effects of local hospitality, commercial hospitality and experience quality on behavioral intention in cultural heritage tourism. J. Qual. Assur. Hosp. Tour. **18**(2), 149–172 (2017). https://doi.org/10.1080/1528008X.2016.1169474
93. Sánchez-Rebull, M.-V., Rudchenko, V., Martín, J.-C.: The antecedents and consequences of customer satisfaction in tourism: a systematic literature review. Tour. Hosp. Manag. **24**(1), 151–183 (2018). https://doi.org/10.20867/thm.24.1.3
94. Ismail, N., Masron, T., Ahmad, A.: Cultural heritage tourism in Malaysia: issues and challenges. SHS Web Conf. **12**, 01059 (2014). https://doi.org/10.1051/shsconf/20141201059

Discovering the Tourism Sensory Experience in Melaka UNESCO WHS

Nur Hidayah Abd Rahman, Noel Scott, and Hairul Nizam Ismail

Abstract The 'Tourism Experience' studies have highlighted the importance of sensory experience since the tourists engage with a destination and understand that particular place based on their senses: sight, taste, hearing, smell, and touch. Previously, most of the studies focused on the visual sense, gustatory sense and auditory sense, especially in agriculture tourism, gastronomy tourism and nature-based tourism. However, the tourism sensory experience of urban heritage destinations has not been addressed adequately. This paper explores the elements of sense-making in Melaka UNESCO World Heritage Site (WHS), Malaysia. The paper uses a combination of systematic review of the relevant literature and content analysis of brochures and other publications to identify sensory elements used in describing cultural heritage experiences in Melaka UNESCO WHS, Malaysia. The contributions of this study are (i) the sensory experiences are influenced by the destination 'theme', (ii) both positive and negative sensory impressions are associated with a heritage experience, and (iii) a systematic review of past publications and subsequent content analysis method used here provides a useful technique for understanding the tourism sensory experience of an attraction.

Keywords Melaka UNESCO world heritage site (WHS) · Tourism experience · Tourism sense-making · Tourism sensory experience · Urban heritage tourism

---

N. H. A. Rahman (✉)
Department of Tourism, Kulliyyah of Sustainable Tourism and Contemporary Languages, International Islamic University Malaysia, Pagoh Campus, Johor Darul Takzim, Muar, Malaysia
e-mail: hidayah_rahman@iium.edu.my

N. Scott
Sustainability Research Centre, University of the Sunshine Coast, Sippy Downs, Australia

H. N. Ismail
Department of Urban and Regional Planning, Faculty of Built Environment and Surveying, Universiti Teknologi Malaysia, Johor Darul Takzim, Johor Bahru, Malaysia

B. Awwad (ed.), *The AI Revolution: Driving Business Innovation and Research*, Studies in Systems, Decision and Control 524, https://doi.org/10.1007/978-3-031-54379-1_62

1 Introduction

In general, tourists engage with the tourism experiences of a destination through their senses in a holistic manner [1–7]. Their experience involves perceiving and understanding through their corporeal sensations of seeing, hearing, tasting, smelling and touching. This sensory experience has the capacity to create a 'tourist space' [8] and can be used to establish a destination profile [2, 9, 10]. Several studies have examined holistic sense-making in tourism [1, 2, 4, 5, 7, 8, 11]. Others have examined particular senses; visual [12, 13], gustatory [14–17], auditory [18, 19], and olfactory [20, 21]. Previous study has also emphasised that a tourist's sensory experience is also influenced by the destination theme [1]. In addition, the tourists' sensory experience will result to the positive or negative impressions towards their experience [22–25]. Previous tourism experience studies such as [1] have focused on a rural-based tourism context and there is a lack of research on tourist sensory experience of heritage tourism [24, 26, 27]. The aim of this paper is to discover the cultural heritage elements and items of sense-making in Melaka UNESCO WHS, Malaysia. This study combines a systematic review of the academic literature and content analysis of other materials such as non-academic books that refer to the history of Melaka, the 'grey' literature including reports such as the Conservation and Management Plan of Melaka Historic City (CMP) in 2008 and 2011 as well as the Special Area Plan (SAP) from the local authorities, and travel materials in printed form and electronic medium. The combination of systematic review and content analysis approaches complement each other [28]. The content analysis extends the previous reviews to identify the holistic sensory experience of cultural heritage elements and items [2, 29–31] in relation to the local context of Melaka UNESCO WHS.

2 Literature Review

2.1 Sense-Making in Tourism Experience

The concept of sense-making in tourism studies can be defined as 'the current restatement of the body, of sense-making and narratives of self and space, combines the senses with the cognitive and affective, thereby creating personal understanding and the storytelling that results from travel' [2] (p. 627). In short, sense-making in tourism studies is the process of giving meaning to the experience within the space. Furthermore, the two major areas in tourism studies highlight the concept of sense-making in managing and enhancing the tourist experience: There is little discussion of sensory experiences of cultural heritage beyond sight [26]. Previous studies highlight that the social construct of cultural heritage is not merely limited to the sensation of sight [26, 27]. Essentially, cultural heritage sites are typically well-known for the authenticity of the 'spirit of the place' and the positive visual image formed by the authenticity of the place. Nonetheless, the tourist experience resulting from cultural heritage activities can produce negative impressions and these negative impressions have not been sufficiently analysed by researchers [22]. For example, the touching behaviour such as vandalism, overwhelming noise produced by the residents or tourist activities, and the sight of crowding or littering might trigger disturbance and annoyance towards the attractions [32].

2.2 *Previous Studies Emphasised Sensory Experience in Malaysia*

The cultural spaces can improve the quality of the visitor experience through their sensory experiences via active learning and engagement with the local heritage attractions, such as architecture, local food, arts, music, customs, and history. In fact, satisfaction levels with heritage elements exceeds tourist expectations [33]. Previous research also identifies three main types of elements in a heritage destination: (1) heritage elements, (2) infrastructure elements, and (3) supporting infrastructure [33]. These heritage elements of the urban heritage area provide a diversity of sensory experiences based on the locals' daily, religious, cultural, and economic activities [33]. The vibrant cultural landscape of street activities such as street markets provide tourists with a unique sensory experience [33–36]. Despite the lack of studies on the multisensory experience in Malaysia, a number of studies have emphasised the need for sensory experience from the tourist perspective [3, 37, 38].

3 Methodology

This study employed a combination of a systematic review of both academic and grey/promotional literature followed by content analysis to identify the elements and items associated with each type of sensory experience: visual, gustatory, auditory, olfactory and tactile. A tri-method approach of bibliometric analysis, content analysis and a systematic literature review was conducted to discover the theoretical foundations and key themes of the tourism destination [28]. The content analysis and systematic review were performed to a total of 51 international academic journals and books, 20 academic journals which highlighted sensory experiences in Malaysia, 3 non-academic books that referred to the history of Melaka as the reference, including the "grey literature' of 3 reports of Conservation and Management Plan (CMP) in 2008 and 2011 as well as the Special Area Plan (SAP) from the local authorities, and 18 travel materials in printed form and electronic medium. Tourism papers were retrieved from Science Direct and Google Scholar using the keywords "sensory experience", "tourism", "tourist gaze", "visual" and "gastronomic tourism" [28]. Other grey literature included printed travel materials such as brochures, travel magazines and travel guides. The content analysis of the academic journals, grey literature and the travel materials was analysed line by line to discover the presence of the cultural heritage elements (i.e., the promotional activities of local people activities, local music, climate/weather) and items that will influence the senses. These elements and items will be categorised under visual, gustatory, olfactory, auditory, and tactile experience.

4 Findings

4.1 Identification of Elements and Items

Visual Experience

The visual elements of the cultural heritage urban sites that influence the tourist sensory experience involves tangible cultural heritage assets, such as the architectural building/built heritage and museum [12, 13, 26] and intangible heritage such as cultural landscape and local culture crafts [39]. Historically, these heritage buildings in Melaka were painted in white during the Dutch colonisation and then, the exterior of Stadthuys and Christ Church was painted in salmon-red colour during the British occupation [40]. Hence, the red colour of Stadthuys and Christ Church has been dominantly as the image for Melaka in travel materials. Temple Street has been dubbed the 'Street of Harmony' as it houses three different religious temples on the same side of the street and all close by. These religious communities have coexisted peacefully here for over three centuries. This article from the travel magazine has reflected on what had been discussed by [13] about the appreciation of tourism attractions through photography, by means of gazing through the lenses.

Gustatory Experience

The gustatory experience involves the local food experience that enhances the quality of tourist gastronomic experience [17, 42]. The local food experience in Melaka has highlighted the spiciness of the local dish in the travel website as follows [43]:

> Melaka has the ultimate bragging right for having the best asam pedas (sour and spicy stew dish) in Malaysia. It is thicker in consistency and uses far more spices than the average asam pedas made in other states.

Besides, the gustatory experience involves other elements such as the settings, services, prices, and the food quality [44]. For instance, the local gastronomic experience by eating with hands and the food quality were emphasised [45] (p. 11):

> When eating with your hands, wash them first and try to use your right hand; use utensils to take food from a communal plate, never your fingers.

The price also will influence in the gustatory experience, in which "...The best-seller at the moment is the creamy Musang King that goes well with their waffle cones. A single scoop of ice-cream here costs RM8.50." [46]. It also can be seen the gustatory experience in terms of the ambience or atmosphere, the services, values of money (the price), and the food quality are the main assessment in TripAdvisor for the gustatory experience.

Auditory Experience

The findings revealed the elements of local people activities, whereby cultural performance and local music have influenced the embodiment of auditory experience in urban heritage destinations [2, 8, 38, 39]. Besides, the auditory experience has been influenced by the locals' daily activities, such as vehicle noise, loud interaction among local people during retailing, and religious activities [37, 39].

Olfactory Experience

On another note, the olfactory experience involves the smell from the local food, the local people activities, and the environmental hygiene [2, 20]. The olfactory aspect resulted from local people's daily activities that reflect the identity and attachment of a place [2, 38]. For instance, vehicles or smoke smell from retail landscapes and religious activities will influence the olfactory experience [47, 48].

Tactile Experience

Interestingly, the feeling of comfort towards the climate or weather, touching the built heritage, and local culture crafts produce a significant response to the elements of tactile experience [9, 13]. The finding of tactile sense is related to the feeling of comfort from the surrounding [38]. The travel website has informed the climate of Melaka will stimulate the tactile experience by mentioning that "A gentle reminder for visitors; be prepared to embrace Melaka's tropical weather by wearing proper attire. Don't forget to bring along your shades, umbrella and apply some sunscreen on to protect yourself from the blazing sunlight." [45]. Besides, other cultural heritage elements such as touching souvenirs or local craft items and monuments or built heritage could influence tactile perception [13].

4.2 Positive and Negative Impressions

The findings show that the sensory experience will result into either positive or negative impressions of cultural heritage elements and items. The CMP [48] (p. 69) notes:

> The Conservation Area has many heritage buildings and intricate architectural elements. Various types, positions, constructions, and designs of buildings have produced an attractive visual and town appearance in the Conservation area. As such, any changes or additions conducted in the name of development or necessity must take these aspects into account to avoid any eyesore.

Therefore, the CMP [48] emphasised the mismanagement of heritage buildings and its architectural elements will affect the negative impressions towards the visual experience. Additionally, an article has emphasised on the negative impression by seeing the souvenirs, in which [41] (p. 25):

> While modern-day commercialism dominates the retailing scene, there are some stores selling local items that are remotely authentic. Sadly, most of the sarongs, fridge magnets, tea towels and other assorted items made in China, Indonesia, or Thailand and have little to do with the local culture.

This has raised the concern on the current situation on the authenticity of the local crafts. In respect to the auditory experience, the travel website described the local economic activity of trishaw as follows [49]:

> First you hear a distant blare of honking, hip-hop or techno. Then, suddenly, a convoy of three-wheeled vehicles is careening your way in a blur of fairy lights and soft toys…Music is selected according to nationality. Local opinion is divided whether these blinged-up trishaws help save the historic mode of transport or hideously distort it.

The description from this travel website provides the opinion on the influence of auditory from the trishaw, however, it brings the positive or negative impressions of the music selection and the authenticity of trishaw. Meanwhile, the vehicles contribute to environmental pollution in Melaka at an average of 69.2% [50]. This data implies that pleasant and unpleasant elements of the cultural heritage attractions could influence the olfactory aspect of the tourist sensory experience. Hence, the sensory experience is indeed the result of an individual's experience of their surroundings and the events that occur when they perform certain activities, which leads to emotional evaluations and positive or negative impressions towards the attractions.

4.3 Inference of Cultural Heritage Elements and Items for Melaka UNESCO WHS

In summary, Table 1 presents the elements of cultural heritage attractions that have been examined in through the content analysis and integrate with the systematic review, regarding urban heritage destinations of Melaka UNESCO WHS, Malaysia.

5 Conclusion

5.1 Theoretical Contributions

This paper has made three contributions, (i) the sensory experiences are influenced by the destination 'theme', (ii) both positive and negative sensory impressions of an experience are associated with a heritage experience, and (iii) the systematic review of

Table 1 Elements and items for cultural heritage elements in appreciating the urban heritage destinations in Melaka

Sensory	Cultural heritage elements	Items for Melaka	Positive/Negative impressions
Visual	Architectural building/Built heritage	The architectural trace of Melaka history for local museum	(+) visuality of the historical buildings, religious buildings, the colour of the historical buildings
		The red colour of colonial building at this historical area	
		The historical sites and monuments	
		The heritage style of *shophouses*	
		The built heritage of religious buildings	
	Cultural landscape/ Local people	The local culture that reflects the characteristics of this place	(+) visuality of the local cultural landscape and their daily activities
		The local people daily activities and lifestyles	
		The local people religious activities	
	Local culture craft/Item	The local people with traditional costume	(±) visuality of the shape and texture of the local souvenirs and handicraft
		The unique shape/texture of local souvenirs and handicraft product	
	Museum	The history and local culture by visiting local museum	(+) visuality of the history and the exhibit items in local museum
		The unique shape/texture of the exhibit item in local museum	
Gustatory	Local food experience/ Custom	Eating experience by using chopstick/fingers	(+) gastronomic experience from the local food experience
		The traditional food preparations	
	Local food/ Setting	The local food from night market	(+) gastronomic experience of the local food from the night market and local restaurant
		The local food in local restaurant	
	Service/Price	The reasonable price of local food	(+) gastronomic experience due to the service or price (−) gastronomic experience due to unreasonable price
		The service quality of local food	

(continued)

Table 1 (continued)

Sensory	Cultural heritage elements	Items for Melaka	Positive/Negative impressions
	Food quality	The quality of local food	(−) gastronomic experience due to the hygiene and food quality
Auditory	Local people activities	The trishaw as one of the local economic activities	(+) auditory experience from the local activities (−) auditory experience from trishaw
		The local daily activities and lifestyles	
		The local people religious activities	
		The local economic activities of retailing	
	Cultural performance	The cultural events/festivals	(+) auditory experience from the cultural performance
	Local music	The local music	(±) auditory experience from the local music/non-local music
Olfactory	Local food	The local food (herbs, spices, fruits)	(+) olfactory experience of local food
	Local people activities	The local people religious activities	(+) olfactory experience from the local activities
		The local economic activities	
	Environment hygiene	The fresh air of surrounding area	(−) olfactory experience from the surrounding area
		The fresh air of surrounding at Melaka River	
Tactile	Climate/ Weather	The heat and temperature	(−) tactile experience from the climate or weather
		The shades from surrounding	
	Built heritage	The unique shape/texture of the exhibit item and built heritage	(+) tactile experience from touching the built heritage
	Local craft items	The unique shape/texture of local/ non-local souvenirs and handicraft product	(±) tactile experience from touching the local/ non- local craft souvenirs

past publications and subsequent content analysis methods used here provide a useful technique for understanding the tourism sensory experience of a local attraction.

Urban Heritage Tourism Destination Theme

This study extends prior work by examining the sensory experiences in a heritage themed tourism context. The results show some similarity to Agapito's [1] studies such as the items for sensory impressions for sight will be influenced by the 'Landscape' (the landscape of the destination) and 'Green' (the colour of the environment), touch will be influenced by the 'Heat' and 'Coolness' (the climate/weather) and even smell will be influenced by 'Fresh air' (environmental hygiene). However, in comparison to the rural-based tourism context, this study highlighted the importance of the interaction between the tourist and the local people. For example, local people activities' through 'The cultural events/festivals' provide an auditory experience and 'The local food (herbs, spices, fruits)' stimulate the olfactory sense.

Positive and Negative Sensory Impressions

In tourism, the sensory experience may lead to positive or negative impressions of an experience [22, 23, 25]. Agapito et al. [1] found that the level of positive experience in respect to each sensory was higher during the visit in comparison with the post-visit. The elements noted in Table 1 are both negative and positive sensory experiences.

A Combination of Systematic Review and Content Analysis

This study employed a novel systematic review and content analysis method. Previous authors have used content analysis of brochures [29, 30], travel writing in New Zealand [2] and promotional materials [31] in order to identify sensory experiences of a destination. This study develops this method by adopting a systemic review of all available content (the academic journals, non-academic books and promotional materials as well as travel materials).

References

1. Agapito, D., Pinto, P., Mendes, J.: Tourists' memories, sensory impressions and loyalty: in loco and post-visit study in Southwest Portugal. Tour. Manage. **58**, 108–118 (2017)
2. Pan, S, Ryan, C.: Tourism sense-making: the role of the senses and travel journalism. J. Travel Tourism Mark. **26**(7), 625–639 (2009)
3. Rahman, N.H.A, Khalifah, Z, Ismail, H.N.: Addressing the importance of the sensory aspect in tourism studies—A literature review. Adv. Sci. Lett. **23**(4), 3167–3169 (2017)
4. Edensor, T.: The more-than-visual experiences of tourism. Tour. Geogr. **20**(5), 913–915 (2018)
5. Edensor, T., Falconer, E.: Sensuous geographies of tourism. In: Wilson, J. (ed.) The Routledge handbook of tourism geographies, pp. 74–81. Routledge, New York (2012)
6. Shao- M., Lin- D.: A study on how the five senses are affected when tourists experience towns with forest characteristics: an empirical analysis based on the data of Fujian, Guangdong and Sichuan in China. Sustainability **13**, 8283 (2021)
7. Fakfare, P., Cho, G., Hwang, H., Manosuthi, N.: Examining the sensory impressions, value perception, and behavioral responses of tourists: the case of floating markets in Thailand. J. Travel Tour. Mark. **38**(7), 666–681 (2021)
8. Gibson, C.: Geographies of tourism: (un)ethical encounters. Prog. Hum. Geogr. **34**(4), 521–527 (2010)
9. Diţoiu, M.-C., Căruntu, A.-L.: Sensory experiences regarding five-dimensional brand destination. Procedia-Soc. Behav. Sci. **109**, 301–306 (2014)
10. Meacci, L., Liberatore, G.A.: Senses-based model for the experiential tourism. Tourism Manage. Stud. **14**(4), 7–14 (2018)
11. Crouch, D., Desforges, L.: The sensuous in the tourist encounter: Introduction: the power of the body in tourist studies. Tour. Stud. **3**(1), 5–22 (2003)
12. Urry, J.: The Tourist Gaze. SAGE Publications (2002)

13. Urry, J., Larsen, J.: The Tourist Gaze 3.0. SAGE Publications (2011)
14. Berg, P.O., Sev´on, G.: Food-branding places—A sensory perspective. Place Branding Public Diplomacy **10**(4), 289–304 (2014)
15. Everett, S.: Beyond the visual gaze?: The pursuit of an embodied experience through food tourism. Tour. Stud. **8**(3), 337–358 (2008)
16. Hashimoto, A., Telfer, D.J.: Selling Canadian culinary tourism: Branding the global and the regional product. Tour. Geogr. **8**(1), 31–55 (2006)
17. Kivela, J., Crotts, J.C.: Tourism and gastronomy: Gastronomy's influence on how tourists experience a destination. J. Hospitality Tourism Res. **30**(3), 354–377 (2006)
18. Farsani, N.T., Shafiei, Z., Adilinasab, A., Taheri, S.: An investigation of tourists' attitudes towards promoting music niche tourism (case study: Isfahan, Iran). Tourism Manag. Perspect. **24**, 1–6 (2017)
19. Gibson, C., Connell, J.: Music, tourism and the transformation of Memphis. Tour. Geogr. **9**(2), 160–190 (2007)
20. Dann, G., Jacobsen, J.K.S.: Tourism smellscapes. Tour. Geogr. **5**(1), 3–25 (2003)
21. Low, K.E.: Presenting the self, the social body, and the olfactory: managing smells in everyday life experiences. Sociol. Perspect. **49**(4), 607–631 (2006)
22. Nawijn, J.: Determinants of daily happiness on vacation. J. Travel Res. **50**(5), 559–566 (2011)
23. Lee, I.S., Lee, C.K., Choi, Y.J.: Examining the role of emotional and functional values in festival evaluation. J. Travel Res. **50**(6), 685–696 (2011)
24. Manfredini, J.: The tourist's emotions, tourism of the senses. Histoire Urbaine **1**(1), 101–119 (2019)
25. Matteucci, X.: Tourists' accounts of learning and positive emotions through sensory experiences. Routledge Adv. Tourism Positive Tourism 54–67 (2017)
26. Poria, Y., Butler, R., Airey, D.: Tourist perceptions of heritage exhibits: a comparative study from Israel. J. Herit. Tour. **1**(1), 51–72 (2006)
27. Zhang, J.: Rethinking 'heritage' in post-conflict tourism. Ann. Tourism Res. **66**(C), 194–196 (2017)
28. Cheng, M., Edwards, D., Darcy, S., Redfern, K.: A tri-method approach to a review of adventure tourism literature: bibliometric analysis, content analysis, and a quantitative systematic literature review. J. Hospitality Tourism Res. **42**(6), 997–1020 (2018)
29. Kljenak, M., Kurdija, S., Polič, M., Golobič, M.: Experiencing Dalmatia: what constitutes the sensory landscape identity of the Dalmatia Region? Drustvena Istrazivanja **22**(2), 277–302 (2013)
30. Hassan, H.: The representation of Malaysian cultures in tourism brochures. In: Procedia—Social and Behavioral Sciences International Conference on Knowledge-Innovation-Excellence: Synergy in Language Research and Practice, vol. 118, pp. 140–151 (2014)
31. Jalis, M.H., Che, D., Markwell, K.: Utilising local cuisine to market Malaysia as a tourist destination. In: Procedia—Social and Behavioral Sciences 5th Asia Euro Conference, vol. 144, pp. 102–110 (2014)
32. Garrod, B., Leask, A., Fyall, A.: Managing visitor impacts at visitor attractions: an international assessment. Curr. Issue Tour. **9**(2), 125–151 (2006)
33. Jusoh, J., Masron, T., Hamid, N.F.A., Shahrin, N.: Tourist Expectation and satisfaction towards physical infrastructure and heritage elements in Melaka World Heritage Site. Acad. J. Interdisciplinary Stud. **2**(8), 733–739 (2013)
34. Ja'afar, N.H., Sulaiman, A.B., Shamsuddin, S.: The contribution of landscape features on traditional streets in Malaysia. In: Procedia—Social and Behavioral Sciences ASEAN Conference on Environment-Behaviour Studies, Vol. 50, pp. 643–656 (2012)
35. Kamarudin, A.M., Wahab, N.A.A., Zakariya, K.: Discovering the qualities of Ferringhi night market as an urban cultural space. UMRAN2015 Vision Establishing Green Built Environ., 1–16 (2015)
36. Zakariah, Z, Mohamad, R.K, Mohamad, W.: Boosting Malaysian tourism the food way!. In: International Proceedings of Economics Development and Research, pp. 178–183 (2012)

37. Ahmad, F., Zainol, R., Nordin, N., Ahmad, M.I.: Appreciating built heritage through urban sensory elements. J. Surveying, Constr. Property **4**(1), 1–12 (2013)
38. Zainol, R.: Appreciating a world heritage site using multisensory elements: a case study in Kinabalu Park, Sabah, Malaysia. SHS Web Conf. **12**(1080), 1–9 (2014)
39. Shaw, S.J.: Marketing ethnoscapes as spaces of consumption: Banglatown London's curry capital. J. Town City Manage. **1**(4), 381–395 (2011)
40. Witt, D.D.: Melaka from the Top. Nutmeg Publishing, Selangor, Malaysia (2010)
41. Bowden, D.: Heart of Heritage. Senses of Malaysia, May-June, 22–27 (2015)
42. Wijaya, S., Morrison, A., Nguyen, T., King, B.: Exploration of culinary tourism in Indonesia: what do the international visitors expect? In: Asia Tourism Forum 2016-The 12th Biennial Conference of Hospitality and Tourism Industry in Asia (ATF-16), pp. 374–379 (2016)
43. Lonelyplanet.: Malacca and George Town. Retrieved 4 June 2016 from https://www.lonelyplanet.com/malaysia (n.d)
44. Lin, Y.-C., Pearson, T.E., Cai, L.A.: Food as a form of destination identity: a tourism destination brand perspective. Tour. Hosp. Res. **11**(1), 30–48 (2011)
45. Tourism Malaysia Melaka Map and Guide [Brochure] (2014)
46. Tripadvisor Melaka.: Retrieved 14 Dec 2015 from https://www.tripadvisor.com.my/(n.d.)
47. Chin, L.H., Jorge, F.: Malacca: Voices from the Street. Lim Huck Chin, Malaysia (2005)
48. Melaka Historical City Council, MBMB Conservation Management Plan for Historic City of Melaka (2008)
49. TheLonelyplanet.: Melaka. Retrieved 14 Dec 2015 from https://www.tripadvisor.com.my/(n.d)
50. Mohammad, N., Hasani, Z., Jady, R.: Amin State of environmental development in Melaka: a case study. Adv. Environ. Biol. **7**(1), 207–211 (2013)

Increasing Productivity in the Malay Midwifery Entrepreneurial Sector in Kelantan Through Kaizen's Theory

Siti Nur Anis Muhammad Apandi, Farrah Atikah Saari, Suraya Sukri, Nordiana Ab. Jabar, Nurulahda Sulaiman, and Afidatul Syazwani Afandi

Abstract The performance of Bumiputera entrepreneurs within the midwifery sector attracts considerable public attention, particularly in the context of initiatives aimed at improving the economic prospects of rural communities. Various endeavors have been undertaken by all stakeholders to uphold service-oriented business domains rooted in heritage and tradition, with a specific focus on sectors like midwifery. Village midwife entrepreneurs are notably dedicated to ensuring the continuous relevance of their enterprises, thereby garnering community recognition over time. Their steadfast commitment to maintaining productivity and delivering services, while upholding traditional practices in the modern age, serves as a commendable aspiration. Consequently, an in-depth inquiry has been executed on midwife entrepreneurs situated in the Bachok district, aimed at discerning their embedded work culture and guaranteeing the enduring resonance of their services within the public sphere. This investigation establishes two principal objectives: (1) the identification of service categories offered by village midwives in the vicinity of Bachok, and (2) the analysis of the components encapsulated within the Kaizen theory (1986) within the purview of midwifery practices. The research progression was facilitated through qualitative methodologies, utilizing interview tools and focused group discussions. The application of the Kaizen theory (1986), pioneered by Masaki Imai, has been harnessed to further elucidate the work culture embraced by the collective of entrepreneurs, steering them toward desired levels of proficiency. The findings indicate that midwifery entrepreneurs within the Bachok district exhibit a notable level of commitment to safeguard the quality of services and productivity preferred to clientele, maintaining their pinnacle status. The incorporation of Kaizen theory components has demonstrated its potential in perpetuating their standing within the community, even amidst the multifaceted competition presented by contemporary and pioneering service providers.

S. N. A. M. Apandi (✉) · F. A. Saari · S. Sukri · N. Ab. Jabar · N. Sulaiman · A. S. Afandi
Fakulti Teknologi Kreatif and Warisan, Universiti Malaysia Kelanta, 16300 Bachok, Kelantan, Malaysia
e-mail: nuranis.m@umk.edu.my

S. N. A. M. Apandi
Jabatan Kesusasteraan Melayu, Akademi Pengajian Melayu, Universiti Malaya, Wilayah Persekutuan, 50603 Kuala Lumpur, Malaysia

B. Awwad (ed.), *The AI Revolution: Driving Business Innovation and Research*, Studies in Systems, Decision and Control 524, https://doi.org/10.1007/978-3-031-54379-1_63

Keywords Bachok · Confinement · Kaizen · Kelantan · Malay Women

1 Introduction

Malay women are intrinsically associated with the cultural tradition of engaging in postpartum confinement. The underpinning beliefs of this practice have deeply permeated the Malay community's heritage across generations. A significant majority of new mothers in Malaysia actively partake in this confinement custom, which encompasses a range of rituals including dietary adjustments, heat therapy, and restricted physical activities [1]. The confinement practices observed within the Malay community are intricately tied to the domain of internal health care for Malay women over a span of 40 days. During this confinement period, Malay women are tasked with the responsibilities of self-treatment, nurturing, and the restoration of their physiological well-being. This confinement phase, situated post childbirth, is closely linked to profound physiological, psychological, and sociocultural transformations. Its significance is underscored by the substantial changes a woman's body undergoes during pregnancy and childbirth, ranging from shifts in body weight to the emergence of temporary afflictions like diabetes and hypertension. The process of confinement functions as a means to restore a woman's body to its pre-pregnancy state, undoing the physical toll of childbirth. After the birthing experience, a woman's body might be rendered unclean and debilitated, marked by swollen reproductive organs, compromised veins, engorged breasts, wounds, residual bleeding, and potential exposure to spiritual and medical disturbances. In this postpartum phase, women experience a period of respite from their societal (family), financial, and religious obligations, including practices such as prayer, fasting, and Quranic reading. A notable aspect of this practice is the prohibition of sexual intercourse with spouses for a span of 40–100 days, during which women diligently prioritize self-care to attain comprehensive well-being and mitigate future health vulnerabilities.

Confinement practices that ensue childbirth hold distinctive characteristics, particularly among Malay women residing in Kelantan. This distinctiveness arises from a series of multifaceted processes intrinsic to this tradition. These processes encompass a range of rituals such as "mandi teresak" (ritual bath), "mandi air suam" (warm water bath), "berdiang" (heat application), "bertungku" (application of heated compress), "barut perut" (belly wrapping), consumption of "jamu" or "makjun" (traditional herbal medicines), and the careful regulation of dietary intake. Additionally, Malay women are advised to maintain limited mobility, engaging in leisurely walking and significant periods of bed rest. These practices are implemented with the aim of averting potential risks, such as falls, sutures becoming undone, toe injuries, and bleeding/haemorrhaging, which could culminate in fatal consequences. The decision to forgo confinement practices among Malay women might jeopardize their well-being, yielding an array of conceivable issues including obstructed birth canals, compromised fertility, and future challenges with sexual activity.

To comprehend the paradigm of postpartum customs prevalent among the Malay population, the application of the Kaizen theory has been employed. This theory, originating from Japan, is rooted in the concept of continuous enhancement and finds common implementation within organizational management. Notably, the incorporation of this theory into organizational structures was facilitated through the "Dasar Pandang ke Timur", introduced by Tun Mahathir Mohamad in 1982. This strategic utilization holds significance in light of the steadfast determination of the Japanese people in reshaping their developmental model post the aftermath of the Hiroshima and Nagasaki bombings during World War II. The corporate ethos should be informed by the commendable practices observed in Japanese and Korean enterprises. An organization, by definition, encompasses a systematically interlinked assembly of individuals, orchestrated activities, guiding principles, and objectives with a longitudinal trajectory. Proficient management of such entities holds pivotal sway in engendering novel prospects, fostering innovation, and confronting competition marked by transparency and equilibrium. The Kaizen theory comprises ten cardinal principles: questioning established norms, viewing challenges as avenues for growth, cultivating sagacity during adversities, nurturing a positive outlook, abstaining from rationalizing shortcomings, privileging creativity over capital, assimilating data and operational principles, learning through active engagement, opting for simplicity in solutions, and rectifying errors promptly. It is the comprehensive application of these tenets that begets the triumph of the organizational endeavour.

Conclusively, the confinement practices adhered to by Malay women postpartum exhibit a symbiotic relationship with the burgeoning productivity within Kelantan's entrepreneurial landscape. This congruence stems from the prevalent female dominance in the corporate realm of Kelantan. The holistic well-being of women, encompassing mental and physical dimensions, bears far-reaching implications for the advancement of entrepreneurial ventures. The stewardship of enterprises characterized by the health and vitality of its personnel can yield positive outcomes, particularly in amplifying revenue generation within the entrepreneurial domain.

2 Literature Review

The domain of midwifery entrepreneurship assumes a pivotal role in delivering essential healthcare services to both maternal and neonatal populations. Recent times have witnessed an escalating emphasis on the adoption of methodologies aimed at augmenting efficiency across diverse sectors, healthcare included. This drive seeks to elevate the calibre of services rendered while optimizing resource utilization. An eminent strategy in this endeavour is encapsulated by the Kaizen theory, which pivots on the philosophy of continuous advancement and active employee involvement. The primary objective of this scholarly investigation is to scrutinize the feasibility of instilling the principles of the Kaizen philosophy as a catalyst for amplifying productivity within the entrepreneurial landscape of Malay midwifery, specifically situated within the precincts of Kelantan.

3 Benefits of Kaizen Implementation in Midwifery

The sphere of entrepreneurial midwifery in Kelantan confronts distinctive hurdles, marked by limitations in resources, inadequate training, and adherence to customary methodologies. As highlighted by Rahman, rural midwives possess a wealth of access to contemporary techniques and technologies [2]. These challenges impede operational efficiency and compromise the delivery of high-calibre healthcare services. The application of the Kaizen philosophy within midwifery confers a spectrum of advantages. Foremost among these is Kaizen's efficacy in addressing the predicaments encountered by midwifery practices by fostering a culture that champions continual learning and advancement. The dynamic involvement and empowerment of personnel hold the potential to engender innovative solutions custom-tailored to the distinctive contours of the local milieu. The incorporation of Kaizen within a healthcare framework yielded favourable outcomes encompassing heightened collaboration among healthcare practitioners, reduction in error incidence, and elevated levels of patient contentment [3]. The triumphant integration of Kaizen is profoundly influenced by cultural nuances and their sway on workforce engagement. Particularly in the Malaysian context, characterized by veneration for hierarchical structures and deference to authority, an astute approach to employee involvement becomes indispensable. The pivotal role of cultural alignment in ensuring the efficacy of continuous amelioration endeavours was underscored by Al-Hakim [4].

4 Localizing Kaizen for the Midwifery Sector

Effectively integrating the Kaizen methodology demands a process of localization tailored to the unique landscape of the midwifery entrepreneurial realm in Kelantan. The adaptation of Kaizen principles to align with the cultural and operational intricacies holds the potential to enhance their pertinence and effectiveness. This entails a comprehensive grasp of the prevailing cultural values, beliefs, and healthcare conventions intrinsic to this specific milieu. A compendium of case studies substantiates the potency of Kaizen within the healthcare domain. As exemplified by Hussain investigation, which delved into the ramifications of Kaizen deployment in a Pakistani hospital, the discernible outcomes showcased a reduction in patient waiting durations and a discernible upliftment in employee morale [5]. These aforestated instances of triumph furnish invaluable insights into the plausible applicability of Kaizen tenets within Kelantan's midwifery industry.

5 Kaizen's Theory

Kaizen is a method of improving the quality of human life that covers aspects of personality, social, family and work. Kaizen emphasizes the comfort aspect of the workplace as well as the production of productivity by focusing on time management, and this makes the principle of kaizen that rejects delayed work matters by constantly improving the quality of service over time. Emphasis on the aspect of cooperation in a team or cooperation between both sides in doing something has made kaizen a concept that successfully shapes the success of individuals, communities, and countries [6]. Through the principles practiced in this kaizen concept, Japan has become a developed and competitive country at the global level. Kaizen is a culture that has been practiced for a long time in Japan and is constantly improving over time [7]. The implementation of the concept of kaizen has succeeded in changing the negative perception into a positive one of society towards bad practices in some job sectors, especially the business field. Although the concept of kaizen was founded and popularized by the Japanese community, it was found to be in line with Islamic values. Thus, practicing Islamic work culture behaviour and kaizen are found to always go hand in hand to improve entrepreneurial performance, especially for the Bumiputera group. This situation has encouraged this concept to always be highlighted and improved so that it is always relevant to be implemented by all parties including for individuals who are Muslims. Suhana Saad and Kartini Aboo Talib @ Khalid explained that the country strives to produce entrepreneurs who are able to compete at the global level by practicing kaizen entrepreneurial culture and at the same time not marginalizing Islamic values [8]. The combination of kaizen and Islamic entrepreneurial culture can help entrepreneurs, especially native, to increase work productivity over time. This is because the characteristics of kaizen are seen to be in line with Islamic teachings that apply positive values in every entrepreneur. Among the cultures emphasized in kaizen are persistence, helping each other and appreciating other parties who cooperate with the business owned by the entrepreneur. The theory of Kaizen presents a potentially effective approach to enhancing productivity within the entrepreneurial sector of Malay midwifery in the region of Kelantan. Through the development of an environment that prioritises ongoing enhancement, active involvement of employees, and the development of new competencies, the practice of Kaizen has the potential to effectively tackle obstacles and elevate the calibre of mother and newborn healthcare provisions. Nevertheless, the achievement of effective implementation necessitates meticulous examination of cultural elements and the distinctive operating environment within the midwifery field in Kelantan.

6 Methodology

This study employs qualitative research methods, utilizing interviews and focus group discussions as the primary data collection mechanisms. A comprehensive exploration was undertaken through in-depth interviews to glean insights into the work culture of traditional midwifery entrepreneurs situated around Bachok. The interviewee selection encompassed 9 individuals, meticulously chosen according to pre-established criteria: (I) exclusive inclusion of female midwives, (II) an experience threshold exceeding ten years as independent midwives, (III) ongoing engagement within the midwifery entrepreneurial domain, (IV) willingness to share insights on the midwifery culture integral to consumer service, and (V) a commitment to providing precise, authentic, and transparent information. Commencing with preliminary interviews, these selected respondents contributed to the delineation of themes, questions, and focal concerns earmarked for subsequent sessions. Subsequently, the second phase ensued employing the focus group discussion modality, wherein the amalgamated expertise of the 9 entrepreneurs converged to deliberate upon themes, questions, and issues rooted in the insights extracted from the initial interview phase. The subsequent interview stage primarily revolved around probing the work culture intricacies embedded within the entrepreneurs' modus operandi for sustaining their service-oriented enterprises. Within the focus group discussions, participants were encouraged to openly share their perspectives, knowledge, and expertise. However, the discourse was governed by guidelines mandating the utilization of respectful language, fostering an atmosphere of mutual regard, and active engagement without passivity. Upon the culmination of the interview process, the amassed findings underwent analysis through the prism of the Kaizen work culture theory. This theoretical framework was chosen due to its congruence with the work culture exhibited by the cohort of entrepreneurs who constituted the informants.

7 Kaizen' S Theory (1986)

The Kaizen theory stems from the Kaizen philosophy, originating in Japan. This term is deeply ingrained in the daily lives of Japanese individuals, particularly within the realms of employment and entrepreneurship. In Japanese linguistics, "kai" signifies change, and "zen" denotes good. Similarly, in Chinese, "kaizen" is known as "ghai shan," translating to "change for the better" or "improvement," with the connotation of greater profitability [9]. Kaizen as an ongoing endeavour to enhance performance, thereby augmenting both quality and productivity within various domains, including entrepreneurship [10]. The crux of this effort is to ensure continuous elevation in quality standards. Notably, the foray of Kaizen into the business sphere was primarily catalysed by Masaki Imai in 1986, with the overarching objective of instilling a culture of ceaseless quality improvement [11]. Evidently, the underlying concept encapsulated within the Kaizen culture bears paramount importance in terms of effective management, quality enhancement in work, the dissemination of knowledge, and the cultivation of adept entrepreneurs and personnel, equipped with the acumen to navigate competitors adeptly [7]. The triumphant execution of Kaizen is contingent upon the bedrock of commitment, underpinned by the right attitude and adept management leadership, ensuring the fruition of its application [12]. Six definitive attributes characterizing a Kaizen work culture: industriousness, humility, fidelity, cooperation, reverence, and a disposition to value and acknowledge others [8].

8 Analysis

The result of this study unveils a synthesis of discoveries that navigate the intricate terrain of postpartum practices within the specific backdrop of Kelantan, intricately intertwined with the tenets of the Kaizen theory, expounded in the subsequent sections.

A. **Industriousness**

Exerting diligent effort constitutes a manifestation of disciplined and conforming conduct in various legal and regulatory frameworks, commonly referred to as industriousness or to be more precise, industrious endeavours undertaken with seriousness [9]. This industrious ethos finds expression in the practice of postpartum confinement, an extensively emphasized tradition within the Malay community aimed at safeguarding maternal well-being. In this context, the role of village midwives assumes paramount significance, marked by their unflagging dedication and conscientiousness in fulfilling their responsibilities. Their rigorous commitment extends to ensuring the health and safety of mothers, often supplemented by the production of traditional remedies. This entrepreneurial work ethic permeates when these midwives make home visits to maternity mothers in need of their services. Following a massage session, midwives proffer dietary guidance, delineating permissible and prohibited foods and beverages.

Figure 1 is a "Kacip Fatimah" product produced by traditional midwives. Subsequent to the massage, midwives commonly proffer medicinal products, crafted predominantly from indigenous materials like roots and "Kacip Fatimah", embodying the essence of time-intensive craftsmanship. These medicines, aside from contributing to maternal health, also constitute a source of income for midwives. The introduction of these medicinal products is invariably accompanied by nutritional advice and recommendations for optimal use.

> Jange minum air batu kalu tok sapa seratus hari, dale pate buleh minum air kopi denge tea suam jah tapi air tea buleh jadi angin, kalu milo juga tak molep jadi sakit urat ngan peranake akan berair. Susu pon boleh minum susu tepung dan hok paling molep minum air akar kayu, ore panggil sepey kedere. Air manjakani pun boleh tapi pahit, kalu bilas ngan air biasa pun rasa pahit lagi tapi khasiat dio, dio buleh kecutkan rahim dan semboh luko bersalin.
>
> (Kak Yah 2021).

Fig. 1 Traditional medicine and traditional herbal beverages (Farrah Atikah Saari 2021)

Interviews reveal a discourse on suitable and unsuitable beverages, crucially centering on the medicinal attributes of herbal concoctions or herbs-based infusions (tonic) with respect to the mother's physique. Armed with this knowledge, mothers gravitate towards these products, acknowledging their benefits and engendering the success of the midwives' assiduous efforts in promoting these remedies, thus contributing to comprehensive maternal care. This sentiment resonates with the viewpoints expressed by Mok Ku, Mok Moh, and Kak Ani, affirming Kak Yah's assertions.

B. **Shyness**

Shyness assumes a dual connotation, encapsulating both positive and negative facets. The negative aspect of reticence often serves as a contributing factor to societal passivity, subdued demeanour, feelings of inferiority, and a dearth of self-assurance, particularly among Malays. Conversely, the positive manifestation of shyness is intertwined with decorous conduct, fostering the upholding of human dignity and national honour [13]. This interplay of shyness finds resonance in the context of the confinement process. Evident in this process is the profound attention accorded to modesty, an embodiment of shyness, particularly prevalent within the work culture of entrepreneurs, including midwives. These professionals exercise meticulous care to safeguard the "aurat," or areas of the body that require concealment.

Figure 2 is the "bertungku' equipment. The intricate "bertungku" procedure serves as an illustrative example. This multifaceted process entails massaging followed by heating stones, requiring a considerable duration to attain optimal temperature. The positioning of the mother's body on her back within an enclosed environment ensures her "aurat" remains veiled from external view. Notably, distinct charges are levied for massaging and "bertungku," with the latter commanding a higher fee due to its intricacy and time-intensive nature.

Fig. 2 River stones used for "Bertungku" and stone wrapped using Batik Cloth (Farrah Atikah Saari 2021)

> Proses diye ni keno duduk atas lata dalam keadaan ngiring dan belake tu ada bara api dan skalo ore buat atah rumoh. Diye ni jugok mesti keno pada bahagian belake tubuh sajo. Bezo dengan nyala, kalu nyala, bara api mesti keno bahagian belake, punggung, perut dan biasa ore buat dapur di bawoh rumah yang dilapik dengan lata jare. Masa proses nyala, wap panah akan masuk celoh kekeng dan kito buleh tuku atas kemaluan kito.
>
> (Mok Ku 2021).

Beyond the purview of massage and comportment, the "berdiang" and "menyalai" processes also form integral components of the confinement practice. According to Mok Ku's account, "berdiang" involves sitting sideways on the floor while experiencing applied heat along the posterior. This heat application spans the back, buttocks, and abdomen, commonly enacted beneath the shelter of a house, its floors often perforated. The meticulous execution of these practices within designated spaces is geared toward fostering maternal comfort and seclusion from external observers. This sentiment finds concurrence in Mok Cik Ani's alignment with Mok Ku's observations.

C. **Loyalty**

Loyalty embodies a genuine commitment marked by upholding promises, refraining from betrayal, and embracing sincerity. It exists as another facet of relational prowess, akin to qualities such as empathy, communication, or compromise [14]. In the context of adhering to traditional postpartum practices, midwives are entrusted with the responsibility of upholding loyalty towards the taboos, treatments, and ingredients integral to the process. This commitment is palpable through the obligatory usage of elements such as "asam" slices and salts, which are imperative components during the "pelepas" ritual. This ritual necessitates the active participation of the maternity mother's family, responsible for preparing the requisite ingredients.

> Pelepas ni sangat penting untuk dibuat sebab ibu duk dalam darah kotor selamo 40 hari. Pelepas biasanya ore buat pado hari ganjil macam hari ke 41, 43 atau pon 45. Pelepas ni penting sebab tokse buwi ibu meroye atau keno ganggu ko makluk halus. Kalu ikut ore tua dulu, bahan hok ore guno untuk Pelepas macam bene, duit ngerah 30 sen dan kalu tok buat Pelepas jadi gilo ngiye.
>
> (Mok Cik Ani 2021).

Fig. 3 *Asam* slices and Salt used in the "pelepas" Process (Farrah Atikah Saari 2021)

Figure 3 is the "pelepas" equipment. According to insights gleaned from an interview with Mok Cik Ani, this practice of "pelepas" is rooted in the belief that the mother remains impure for a period of 40 days. Therefore, the "pelepas" is conducted on odd days, such as the 41st, 43rd, or 45th day, with the overarching intention of preserving the mother's well-being and shielding her from spiritual afflictions. In her role as a midwife, Mok Cik Ani diligently adheres to ancestral wisdom regarding ingredient usage for the "pelepas" process. The materials employed, including yarn and a sum of 30 cents, align with this ancestral guidance. This perspective garners concurrence from Kak Yah, affirming the assertions made by Mok Cik Ani.

D. **Cooperation**

As explained in earlier research, cooperation entails the harmonious pursuit of a shared objective within a system, predicated upon a perspective that serves the requisites of the system itself [15]. This collaborative ethos is discernible within the context of the postpartum confinement process, evident through the demeanour of the family members in proximity to the mother. In order to ensure the mother's vitality and well-being, familial engagement, particularly from the husband or elderly and experienced individuals cohabiting with the mother, becomes paramount in furnishing a "teresap" water bath. This collaborative effort stems from the necessity for the mother to partake in a three-day regimen of "teresap" water showers during mornings, facilitating the restoration of neural pathways and invigoration of the body.

Figure 4 is the "teresap" equipment. The components utilized in concocting "teresap" water are sourced from herbal leaves typically abundant in the vicinity of the household. Additionally, midwives, during the provision of their massage services to mothers, often include "teresap" water as an accompaniment. This practice confers an incremental advantage upon midwives, bolstering their expertise in the eyes of the local community. This entrepreneurial ethos significantly reverberates in terms of the burgeoning demand for midwifery services.

Fig. 4 "Teresap" (Farrah Atikah Saari 2021)

Mandi teresap ni ore buat selamo tigo hari, ore guno air rebuse akar kayu ngan herba-herba. Mandi teresap ni ore buat selepas ngurut, Bahan yang diguna macam daun semeru, daun pine, daun sera wongi, daun kuwah, daun kunyit dan daun pande. Mandi teresap nie molep untuk kecut urat,dan jarak untuk beranok.

(Mok Moh 2021).

Mok Moh's account, she places considerable emphasis on the practice of the "teresap" bath. According to her, this bathing ritual typically spans three days and involves the utilization of boiled wood roots and carefully selected herbal components. These "teresap" baths are customarily administered subsequent to massage sessions. The concoction comprises an assortment of ingredients, including "semeru" leaves, betel leaves, lemongrass leaves, "lengkuas" leaves, turmeric leaves, and "pandan" leaves. The "teresap" bath serves to alleviate uterine discomfort and facilitate the postpartum recovery process. A midwife's unwavering commitment extends to ensuring the mother's postnatal well-being, an endeavour underscored by the significance of collaborative involvement from family members. This collective endeavour aims to guarantee optimal care and services during the confinement period. Mok Ku and Kak Ani lend their agreement to Mok Moh's observations.

E. **Gratefulness**

Demonstrating appreciation entails treating individuals with respect and propriety, whether through spoken or enacted actions. This conduct precludes any form of harm, whether verbal or physical, and necessitates gratitude and empathy towards others [16]. This practice of appreciating others finds tangible expression within the attitude exhibited by family members of postpartum mothers who esteem the midwife's proficiency in delivering services. Individuals who possess a genuine appreciation for others refrain from casting blame or causing embarrassment in public settings. When actions do not align with expectations, prudent discourse avoids immediate

Fig. 5 Raw thread, turmeric rice and hardener money (Farrah Atikah Saari 2021)

condemnation of their actions. Analogously, within the realm of midwifery, family members are inclined to heed the midwife's instructions, particularly evident during the "pelepas" process.

Figure 5 is the "pelepas" equipment. During the "pelepas" procedure, the midwife requisitions family members to assemble untreated yarn, turmeric-infused rice, and indurated currency. This collaborative response is rooted in the family's profound respect and gratitude for the midwife's skillset, prompting them to comply with the requisition. This compliance stems from a conviction that the historical heritage of the Malay community confers elevated significance upon local wisdom and traditional experts.

> Behe-behe pelepas hok Kak Yah guno nie mace bene, daun nyor, berah, duit ngeras tujoh puloh limo sen. Kito wat hari ketigo atau hari ganjil mace hari ke duo puloh satu atau duo puloh tigo. Bene hok guno tu keno sipe sapa habis pate mace hari ke pat puloh satu atau pat puloh tigo. Pitis hok kita buwi tu ore panggil keras bide dan meme wojib buwi, tok leh kalu tak dop keras bide.
>
> (Kak Yah 2021).

Based on the interview, Kak Yah elucidated the constituent elements employed, which encompass thread, coconut leaves, rice, and seventy-five cents indurated coin, for the execution of the "Pelepas" ritual. She executes this procedure on the third day or on select odd days, such as the twenty-first or twenty-third day. The threads employed are meant to be retained until the culmination of the abstinence period, extending to the forty-first or forty-third day. A crucial component of this process is the offering of "pengeras" money, a customary gesture of appreciation to midwives, imperative to the practice. The act of providing such money underscores the mother's gratitude towards the midwife's exceptional service, underscoring the profound significance of the confinement practice. The significance of confinement practices, encompassing massage, "bertungku", "berdiang" and "menyalai", "tere-sap" bathing, "barut perut" or "berbengkung", dietary aspects, dressing, abstinence positioning, and "Pelepas", holds paramount importance. This confinement regimen spans at least forty days and imparts a plethora of benefits to both maternal and infant health through time-honoured methods. The shared affirmation of Kak Yah's assertions by Mok Ku, Mok Su Ani, and Mok Moh further emphasizes their validity.

9 Conclusion

In closing, this study highlights the interconnectedness of postpartum practices within Kelantan's Malay community and the Kaizen theory's principles of continuous improvement. Traditional rituals, from massage to "Pelepas", are underpinned by values of loyalty, cooperation, and appreciation, fostering a collective approach to maternal care. This intertwining of tradition and innovation is evident in family members' involvement and midwives' expertise, reflected in practices like "teresap" bathing and giving "pengeras". The study underscores how the Kaizen theory's principles resonate with the essence of these practices, offering a framework for their refinement. This synergy showcases the enduring relevance of cultural heritage amid evolving healthcare landscapes. The fusion of tradition and modernity, exemplified by these practices, provides a promising avenue for enhancing maternal and newborn well-being.

10 Appreciation

This study was carried out with funding from Skim Geran Jangka Pendek dari Universiti Malaysia Kelantan (SGJP) bertajuk Analisis Budaya Berpantang Masyarakat Melayu Bachok: Satu Kajian Koleksi Tradisi Lisan. (No Rujukan: UMK.PC/B01.08.11/600–5/2/3 (45,553). The researchers would like to thank the support and cooperation of all parties involved, the Malaysian Ministry of Education (KPM) and the Research, Innovation and Commercialization Management Center (RMIC) in supporting this research project.

References

1. Dennis, C.L., Fung, K., Grigoriadis, S., Robinson, G.E., Romans, S., Ross, L.: Traditional postpartum practices and rituals: a qualitative systematic review. Women's Health **3**(4), 487–502 (2007)
2. Rahman, R.A., Jamaluddin, R., Othman, N.: Challenges faced by traditional midwives in rural areas. Int. J. Acad. Res. Bus. Soc. Sci. **9**(6), 775–787 (2019)
3. Radzi, A.A.M., Abdullah, R.N., Abdullah, L.: Kaizen and employees' attitudes in healthcare services. Manag. Sci. Lett. **10**(6), 1285–1292 (2020)

4. Al-Hakim, L.: Critical success factors for implementing business process reengineering. Bus. Process. Manag. J. **6**(5), 378–385 (2000)
5. Hussain, M., Al-Neaimi, A., Khan, M., Al-Qamzi, A.: Applying lean principles and kaizen to improve a healthcare service process. Health Serv. Manage. Res. **30**(1), 18–26 (2017)
6. Channon, D.: Blackwell encyclopedia dictionary of strategic management. In: Strategic Management. Blackwell Publishing Ltd. (2005)
7. Nai'mah, N., Sheera Sahid, S.: Penerapan Tingkah Laku Budaya Keusahawanan Islam Dan Budaya Kaizen Jepun Dalam Meningkatkan Prestasi Perniagaan Usahawan Bumiputera. In: Conference of Future Education and Advances (ICOFEA) 2020, p 167 (2020)
8. S. Saad, Talib@Khalid, K.A.: Implimentasi budaya kerja Jepun dalam organisasi di Malaysia (The implementation of the Japanese work ethics in a Malaysian organization). Geografia **11**(6), (2015)
9. Marzuki, I., Hakim, L.: Strategi Pembelajaran Karakter Kerja Keras. Rausyan Fikr: J. Pemikiran dan Pencerahan, 15(1) (2019)
10. Imai, M.: Comment: solving quality problems using common sense. Int. J. Qual. Reliab. Manage. **9**(5) (1992)
11. Kassim, Z.: Kajian Rekonstruksi Pelajar-Pelajar Rancangan Persediaan Khas Ke Jepun (Rpkj) Dalam Aspek Pengurusan Masa: A study on reconstruction of time management aspects of Rpkj students. J. Bahasa dan Budaya Jepun **6**, 27–51 (2016)
12. Maddox, J.: The influence of top management support on TQM indicators of defect reduction and profitability. Ph.D., Capella University, United States—Minnesota (2009)
13. Muhammad, Z., Abdullah, A., Ab Razak, R.R.: Sifat Malu Dalam Kerangka Akhlak Melayu. J. Hadhari **11**(2), 231–244 (2019)
14. Septiani, D., Azzahra, P.N., Wulandari, S.N., Manuardi, A.R.: Self disclosure dalam komunikasi interpersonal: Kesetiaan, Cinta, Dan Kasih Sayang. FOKUS (Kajian Bimbingan Konseling Dalam Pendidikan) **2**(6), 265–271 (2019)
15. Setiabudhi, D.O.: Pengelolaan aset Pemerintah Daerah Dalam Perspektif good governance. Stud. Soc. Sci. **1**(1), 7–21 (2019)
16. Susilawati, W.O., Novitasari, A., Prananda, G., Apreasta, L.: Pengaruh Pemahaman Hak Asasi Manusia (HAM) Terhadap Sikap Menghargai Pendapat Orang Lain Pada Mahasiswa program Studi PPKN FKIP UAD. Inspiratif Pendidikan **9**(2), 91–109 (2020)

Exploring Local Heritage, Virtual Reality, and English Speaking Needs to Develop World Heritage Tourism: The Case of Ombilin Coal Mining Heritage of Sawahlunto (OCMHS), Indonesia

Elfiondri, Jonny Wongso, and Riska Amelia

Abstract The study examines local heritage, virtual reality, and English speaking needs to develop World Heritage tourism with the case study of OCMHS. Recent research shows that tourist visits are low and tourists do not recognize the world heritage tourism. Local heritages around OCMHS and English speaking needs are importantly explored, and Virtual Reality is importantly designed to attract and promote the OCMHS tourism, but the study is still a gap in the literature of tourism study. The objectives are to find out: (a) local heritages around OCMHS which can support the OCMHS tourism, (b) English speaking needs needed by tourism workforce, and (c) a design of Virtual Reality as promotion media. The research applied the approach of ethnography, heritage contestation, and language needs. The study found that there are three types of local heritage which play an important role as the supporting tourist attractions of OCMHS. The local heritage-based tourism attract tourists so much that their existence in the area of OCMHS contributes to the increase of tourists' visit to OCMHS tourism. The study also yields a map of the heritages with Virtual Reality (VR) which has had the effect in the form of access online by people from different countries. English speaking skill is needed in several levels of needs to interpret the heritages, to provide information on the heritages, local people, and social norms, and to build and maintain social relationship.

Keywords Heritage · Virtual reality · English · Tourism

Elfiondri (✉)
Faculty of Humanities, Universitas Bung Hatta, Padang, Indonesia
e-mail: elfiondri@bunghatta.ac.id

J. Wongso
Faculty of Civil Engineering and Planning, Universitas Bung Hatta, Padang, Indonesia

R. Amelia
Faculty of Education, Universitas Bung Hatta, Padang, Indonesia

B. Awwad (ed.), *The AI Revolution: Driving Business Innovation and Research*, Studies in Systems, Decision and Control 524, https://doi.org/10.1007/978-3-031-54379-1_64

1 Introduction

Tourism provides benefits to improving the community's economy [1] in developing countries [2], because tourism generates a lot of jobs [3]. In many developing countries, UNESCO World Heritage sites are a tourist attraction [1]. Indonesia is a developing country that has a number of world cultural heritages. One of them is the Ombilin Sawahlunto Coal Mine World Cultural Heritage (OCMHS) in West Sumatra province, Indonesia which was newly designated by UNESCO in 2019. The OCMHS determination places tourism in the district/city government master plan in West Sumatra, and plays an important role in the development of West Sumatra tourism.

In light of closing in coal mining of Sawahlunto as social income, OCMHS tourism is a new alternative source of social income which used to rely on the Ombilin coal mine. However, OCMHS tourism has not yet developed—the level of tourist visits is low and tourists do not yet recognize OCMHS tourist attractions [4], so it is necessary to develop OCMHS tourist attractions. OCMHS has nominated areas as supporting elements of world cultural heritage [5]. Local heritages and English speaking skill based locality in the nominated areas needs to be explored and promoted to the wider community in order to develop OCMHS heritage tourism. Unfortunately, recent study shows that tourist visits to OCMHS tourism are low and tourists do not recognize the world heritage tourism [4]. As an alternative effort to solve the social problem, this study explores local heritages, virtual reality, and English speaking skill needs for the development of OCMHS tourism. The objectives are to find out English speaking skill needs, tourists' interest in OCMHS heritage tourism, and local heritages as the supporting tourist attractions for OCMHS tourism, and to design an online Virtual Reality (VR) of OCMHS and local heritages.

The study is important to develop the world heritage tourism, but still a gap in the literature of tourism study. Therefore, the study is worthy examining. The study examines local heritages, virtual reality, and English speaking skill needs in the area of the world heritage of OCMHS. It applies the ethnographic approach [6], heritage approach [7], English skill needs [8], and Virtual Reality (VR for tourism [9, 10]. The study is focused on the examination. The objectives are to find out: (a) local heritages around OCMHS which can support the OCMHS tourism, (b) English speaking needs needed by tourism workforce, and (c) a design of Virtual Reality as promotion media.

2 Literary Review

2.1 Heritage in Tourism

Yang [7] shows that heritage consists of cultural heritage and natural heritage. Natural heritage is heritage from nature such as national parks, beaches, and the like. Cultural heritage includes material and intangible cultural heritage. Intangible culture is like rituals and activities, language, and traditional arts and culture. Cultural heritage objects that are not buildings are such as heirlooms, documents, paintings, and others. Cultural heritage objects consist of buildings that include buildings in urban and rural areas. Buildings in the countryside are like village buildings, farms, mines, and museums. The legacy of urban buildings is like colonial buildings, wars, totalitarianism, places of worship, torture and death, slavery and others.

These types of inheritance exist in various forms according to the culture of local people. Mentawai people, for example, have intangible cultural heritage such as *takep, kirekat, panasai, susuru*, dances, songs, etc., various rituals [11], and traditional social norms [12, 13], cultural objects such as heirloom accessories, Uma buildings and cultural objects [14]. Cultural heritage has become a tourist attraction for tourists [15]. Cultural tourists enjoy cultural contact with various types of cultural heritage in the form of attractions and events which provide important motivations for tourism [16]. The heritage based-attractions and events alone are not enough for the tourism motivation. They should be supported by environmental quality and beauty of the destination tourism. According to Aswathy [17], environmental quality and beauty of the destination are important for the tourism development. They should be managed carefully and sustainably in order to satisfy tourists in experiencing tourism.

2.2 *Virtual Reality in Tourism*

Besides requiring language, developing tourism also needs a medium called virtual reality (VR). Virtual Reality (VR) is a technology that creates a three-dimensional virtual environment that can interact in real terms [10]. These digital technologies preserve the world's cultural heritage in ways that are attractive to future generations [18]. VR technology has been successfully applied in various sub-sectors of the tourism industry to increase tourist visits [10]. It is widely used to promote tourism, improve and create memorable tourism experiences both at tourist sites and outside tourist sites [9]. It acts as a medium for marketing tourism and creating tourist experiences at tourist sites and outside tourist sites. It is also useful as an effective instrument for heritage management and preservation by generating experiences in lieu of physical visits, thereby minimizing disruption to heritage sites [18]. In business, Karthick and Gopalsamy [19] explain that virtual reality is a technology used for smart business. It is important to build smart business based on network technologies. They add that " there are still many technological fields and information field needed for overcoming the challenges that are resolved in order to speed up the adoption of business sustainability". Technological fields and information field included virtual reality are needed to overcome challenges in order speed up the implementation of business sustainability. Based on the scholars, in the context of tourism business the technological and information fields are significant in the perspective of tourism development.

2.3 Language Needs in Tourism

Language is a very important element in tourism. It should be essentially paid attention in satisfying tourists. It plays an important role in serving foreign tourists. To satisfy foreign tourists in experiencing the heritage tourism (either cultural or natural heritage tourism), information on the heritages using a language which can be understood by the global community such as English is greatly needed in every tourism domain [20]. Language such as English is an important medium for tourism development [12]. Language as an information medium for tourism is language for specific uses, which in English is called English for Specific Purposes (ESP). In the context of English language for OCMHS tourism, it is English for tourism (tourism English). Tourism English refers to English for Specific Purposes (ESP). ESP emphasizes special needs in certain fields, jobs, or goals [21]. The language of tourism is characterized by the characteristics of cultural heritage, where the vocabulary and expressions are related to cultural heritage [15].

3 Methodology

The study uses primary and secondary data. Secondary data are collected with a library approach. Primary data was collected through a survey based on the ethnographic approach of [6] and cultural heritage approach by Yang [7]. Data collection uses interview, observation, and documentation techniques. The interview uses semi-structured and open-ended questions. Interviews were conducted using the written interview guide questions and oral questions related to the research objectives. The research respondents are local people, tourists, experts, government, NGOs, and institutions.

Data were analyzed qualitatively using Spradley's domain analysis [6]. Domain analysis is used to analyze supporting cultural heritage data and its types. Domain analysis begins with an analysis of supporting cultural heritage domains. The domain structure is analyzed using semantic relations. Domain analysis produces an overview of the supporting cultural heritage. The next step, the structure of the domain is classified using taxonomic analysis, which produces the types of supporting cultural heritage. Each type is then analyzed by component analysis (component analysis) so as to produce the smallest parts of these types. This domain analysis is supported by the concept of Yang's cultural heritage [7].

Data on level needs of English speaking skill were analyzed in accordance with [8] used by Ahmmed et al. [22]. Data of English speaking needs which were collected from survey using questionnaire were analyzed by applying descriptive analysis in Microsoft Excel to generate mean and standard deviation to determine the need of the English speaking skill for OCMHS tourism. The research used a five-point Likert scale based on Ahmmed et al. [22]. The scale ranging from highly-needed to not-needed was applied to measure the need level of English speaking sub-theme. Highly needed level was scored 4.51–5.0, Needed 3.51–4.50, Fairly needed 2.51–3.50, Less needed 1.51–2.50, and not needed 0–1.50.

4 Results and Discussion

The findings of the study are that English speaking skill in OCMHS tourism is needed to communicate a number of themes and their sub-themes related to heritage tourism (Table 1). The sub-theme is needed the levels which range from needed to highly needed level. There are no fairly, less, and not-needed levels (presented in the bar chart of level need). As indicated by Table 1, English speaking skill is needed in several functions such as providing information on the heritages, local people, etc. The next finding is that OCMHS world heritage tourism does not attract tourists so much. Emely [4] in the research entitled "Sawahlunto Towards a Sustainable and Attractive Place to Live, Work and Recreate" found that tourist visit to OCMHS tourism is low. OCMHS is not tourists' major destination, but as a side goal of their visit. In the area of OCMHS, there are three types of local heritages which support tourist attractions of OCMHS. It is also found that online Virtual Reality (VR) of OCMHS and local heritages has introduced the OCMHS to people from different countries.

4.1 Local Heritages as the Supporting Tourist Attractions for OCMHS Tourism

4.1.1 Culinary, Arts, and Dance Tourism Attraction

Culinary both in the form of food and beverages in the concept of competed heritage by Yang [7] is classifies as an intangible cultural heritage. The local cultural heritage of in Sawahlunto has a uniqueness. Culinary foods consist of food eaten with rice and with snacks. Foods categorized as heavy foods are usually eaten separately from heavy foods, but sometimes some are eaten with the staple food of rice. Heavy foods are those to eat rice such as Silungkang soup, jerked meat, Rendang, green pepper jerked Chicken, etc. Local beverages for example are egged-tea, iced sandalwood, iced durian, Silungkang coffee, leaf-coffee, buffalo milk-chips, etc. Tourists hunt the local foods and beverages to places at which they can enjoy the foods and beverages. The places are spread out in the OCMHS area that they lead tourists to visit OCMHS.

Yang [7] categorized art and dance as part of an intangible cultural heritage. There are a number of dances having developed in the area of OCMHS. They develop based on various regional cultures deriving from various regions in Indonesia. The

Table 1 Themes and sub-themes of English speaking skill

Themes	Sub-themes as functions of heritage tourism professions
Tangible heritage of OCMHS	Providing details about heritage of Museum Goedang Ransoem
	Providing details about heritage of railway system
	Providing information on heritage of Mbah Suro tunnel
	Providing information about heritage of Kayu Tanam Train Station
	Providing information on the heritage of Sawahlunto train station museum
	Providing information regarding heritage of Ombilin railway station
	Providing information regarding heritage of Tinggi Bridge
	Providing information regarding Silo Gunung Coal Storage Heritage
	Providing information regarding heritage of Soengai Durian Mining Site
	Providing information regarding mining school heritge
	Providing information regarding heritage of Coal Processing Plan Compound
	Providing information regarding heritge of headquarter of the Ombilin coal mining
	Providing information on heritage of Elite Europeans meeting hall
	Providing details about heritage of chained men prison
	Providing information regarding Emanhaven Port heritage
	Providing information regarding heritage of mining-worker meeting-hall
	Providing information regarding heritage of Kalam railway tunnel
	Providing information regarding Heritage of Mak Itam (Europe-made-E1060 locomotive)
	Providing information regarding heritage of Lapangan Segitiga
	Providing information on heritage of Batu Tabal Train Station

(continued)

Table 1 (continued)

Themes	Sub-themes as functions of heritage tourism professions
	Providing information on the heritage of Padang Panjang Train Station
	Providing information regarding heritage of old company town
	Providing information on heritage of Salak Power Plant & Ratih Power Water Pumping
Local natural heritage	Providing information regarding Satwa Kandi Park (zoo), world fantasy, water fall, monument, and other natural heritages
Local tangible heritage	Providing information regarding weaving loom, museums, mosques, etc
	Providing information regarding coal-made handycrafts
	Providing information regarding Silungkang Songket
Local intangible heritage	Providing details about regarding history of Ombilin Coal Mining of Sawahlunto heritage
	Providing information regarding various traditional arts and dances, food, drink
Building and maintaining social relationship	Introducing one self
	Greeting and bidding farewell
	Making apologies or explain the reason
Culture and social norms	Providing information regarding the culture and social norms
Financial packages	Explaining the financial packages
Attraction	Providing required information on attraction
	Answering tourists' queries of attraction
Tourists' needs	Asking about tourists' needs
Feedback and suggestions	Requesting tourists to share tourists' feedback and suggestions
Direction and guide	Providing details about direction and guide

arts are such as the art of shadow puppets, *Randai, Kuda Lumping, Ketoprak, Ronggeng, Ludruk, Keroncong, Kecapi, Brongsai, Indang Lubuak Nago Dance, Taratak Lasuang Manangih Dance, Tor-Tor, Plate Dance, Tonil, Tanun dance, layak batoboh dance, randai, Minangkabau dance, Campursari, Wayang,* and others. Sadzali [23] explained that these arts came from various cultures in the archipelago as a result of the Dutch policy which brought in artists from outside of OCMHS area. Some of the mining workers who come from various regions bring art from their respective regions. The Dutch brought these arts to attract and to make the workers feel at home in the mining area. The Dutch thought that performing arts would be able to cure the coal mine workers' fatigue and lack of entertainment so that they would feel at home living in the area.

4.1.2 Tourist Attraction of Rituals, Festivals or Carnivals, and Traditional Self-Defence

Rituals, festivals or carnivals in the area of OCMHS exist in various forms because the area is a destination with a "melting pot" culture because the OCMHS destination is inhabited by various tribes with different ethnic backgrounds such as Javanese with various rituals, Sawahlunto Randai Festival, Tanah Lapang Art Village Festival, and others. These rituals originating from the various cultures include the Grebek Suro Ritual, Bakauah Tradition, and the Ratok Silungkang Tuo Death Ritual. As released by the Spirit of West Sumatra, even though OCMHS is located in the Minangkabau culture area, the color of Javanese culture is a dominating one in the first month of Muharram welcoming rituals.

OCMHS tourism is also supported by traditional martial arts which still exist in OCMHS area. The local intangible cultural heritage of martial arts (traditional self defence). Nurnjan [24] mentioned that there are several martial arts that can become tourist attractions such as Silek Tuo (Silat Tua), Silek Harimau (Silat Harimau), Silek Lintau (Silat Lintau), Silek Sitaralak (this martial arts spread and developed in Malaysia and continues to America), Silek Pauah, Silek Sungai Patai, Silek Luncua which was developed by Pakiah Rabun, Silek Gulo-Gulo Tareh, Silek Baruah (Silat Baruh), and Silek Kumango (Kumango Silat).

4.1.3 Natural Tourist Attraction

Referring to the concept of heritage (heritage) by Yang [7], natural scenery is included as part of heritage. Survey of this study shows that Sawahlunto has a number of natural heritages as the supporting tourist attractions of OCMHS. The natural heritages have now become attractive destination. These natural heritages include Kandih Lake, Sharp Batu Peak, Bikan River Waterfall, Padang Savan Kolok, Park, Kandih Animals, Polan Peak, Microwave Peak, Blue Lake, former coal quarry, Water boom, Teletubbies Hill, Rock Lake, Kandih Hills, Dutch Cemetery, Bikan River Waterfall (Rantih Tourism Village), Kolok Savana Field, Sawahlunto Square, Triangle Square, Kandih Animal Park, Batumananggau Weaving Village, and Cemara Peak. The environmental sustainablity of the natural tourist attraction of OCMHS should be preserved due to that environment contributes to the natural attractions. Improvement in regulation plays important role. [17]) pointed that improvements in regulatory quality support environmental sustainability. The natural tourist attractions of OCMHS depends very much on natural environment. The current study survey shows that tourists visit the tourist attractions because of the natural beauty.

4.2 *Virtual Reality (VR) of Local and OCMHS Heritages*

Virtual Reality (VR) can be used to promote OCMHS tourism by online. Local heritage and OCMHS heritage can be introduced to tourists digitally using Virtual Reality (VR). VR is a technology that creates a three-dimensional virtual environment that can interact in real terms [10]. These digital technologies preserve the world's cultural heritage in ways that are attractive to future generations [18]. VR technology has been successfully applied in various sub-sectors of the tourism industry to increase tourist visits [25]. VR is widely used to promote tourism, improve and create memorable tourism experiences both at tourist sites and outside tourist sites [9]. Based on the scholars, this current study has resulted in an initial design of the map and Virtual Reality of the heritages by online. The digital map and Virtual Reality which can be accessed online at https://warisanduniaocmhs.com.

4.3 *English Speaking Skill Needs for OCMHS Heritage Tourism Development*

English speaking skill is needed in several functions to solve government and social problem on OCMHS tourism. As found by [15], tourism English is locally-colored, that is, its forms are colored by local culture. The English language has a uniqueness of locality. The English speaking skill is needed as a means of information on local people, their social norms, OCMHS world heritages, local heritages, financial packages of tourism, tourist attraction, tourists' needs, and feed back and suggestions. English skill is also needed to build and maintain social relationship with foreign tourists. Besides that, English speaking skill is a tool of heritage interpretation which is greatly essential to attract tourists to visit OCMHS tourism. Tourist attraction of heritage attracts tourists because of heritage interpretation.

Heritage interpretation is highly needed for heritage-based tourism. Providing a good interpretation of heritage which can bring tourists psychologically and emotionally into the heritage will be able to attract tourists because the interpretation can make tourists to come into an informed atmosphere. The concept of inheritance interpretation stated by father of heritage interpretation, [26] in his book "Interpreting Our Heritage". As indicated by Table 1, English speaking skill is needed to perform a number of functions of heritage tourism professions such as providing details in the form of interpretation about heritage of Museum, heritage of railway system, etc. The functions are tasks needed to work for OCMHS heritage tourism. The functions consists of themes some of which are tangible heritage, intangible heritage, natural heritage, culture and norms, and others. The tourism workers need speaking skill to provide information, explain, apologize, greet, request, and to answer those in the themes and their sub-themes.

English speaking skill in this current study is a skill needed in tourism. It is important skill in tourism industry. Jung [10] also found that English speaking skill is important skill in Indonesia tourism industry. In other tourism regions such as in Saudi tourism [27], in Jordan [28], in UK [29], in Japan [30], and [31] in Europe, speaking also becomes the important skill. Based on participants' responses in this current study, speaking becomes the most important and needed skill in tourism because information about tourism attractions can be more easily provided using spoken English rather than written English. Tourism workers from lower position (such as tour guides, front office personnel, and room boys) to higher position (managers, directors) need English speaking skill. Communication between tourism workforce and international tourists is largely spoken in nature rather than written.

Figure 1 shows the level of English speaking skill needs. The functions (providing information regarding Satwa Kandi Park and coal-made handicraft) have scores higher than 3.50, but lower than 4.51. They are in the score range of 3.51–4.50, meaning that the needs level of the tasks are "needed". The rests are "highly needed" because their scores are in the range of 4.51 to 5.00. There is no function of which the score is lower than 3.51. The needs level pictures that all of the functions are needed, and almost of them are highly needed, meaning that tourism workforce needs English speaking skill to work for OCMHS heritage tourism. In order to develop OCMHS tourism, tourism workforce particularly tour guide should have the English speaking skill of heritage tourism professions. The English skill is very essential in serving foreign tourists.

5 Conclusion

The study is limited to the examination of local heritages in the area of the world heritage of OCMHS, the design of tourism Virtual Reality, and the needs of English speaking skill to develop the world heritage tourism. The results show that a method of increasing tourists' visit to OCMHS is by way of developing local heritage-based tourism located in and around the area of OCMHS. OCMHS is surrounded by a number of local heritages which potentially attract domestic and foreign tourists. The way of developing the world heritage tourism is by introducing the local and OCMHS heritages to tourists virtually and globally by online. English speaking skill plays as important method of developing the OCMHS heritages-based tourism. In heritage-based tourism, language plays a very important role due to that it acts as a tool of heritage interpretation. The interpretation can attract tourists to visit destinations at which they can experience the heritage psychologically. A good interpretation can make tourists drift into the past. The future study will be directed to the interpretation of all heritages of OCMHS. OCMHS tourism is a tourism based on the cultural heritage of coal mining. All tangible heritages related to the mining are tourist attractions of OCMHS tourism. To attract tourists, OCMHS heritages should be interpreted orally, in written form, or physically (paintings or statues). The interpretation either in oral, written, or physical form is communicated in online virtual reality through which cultural tourists can experience OCMHS tourist attractions and the attractions are globally promoted all over the world.

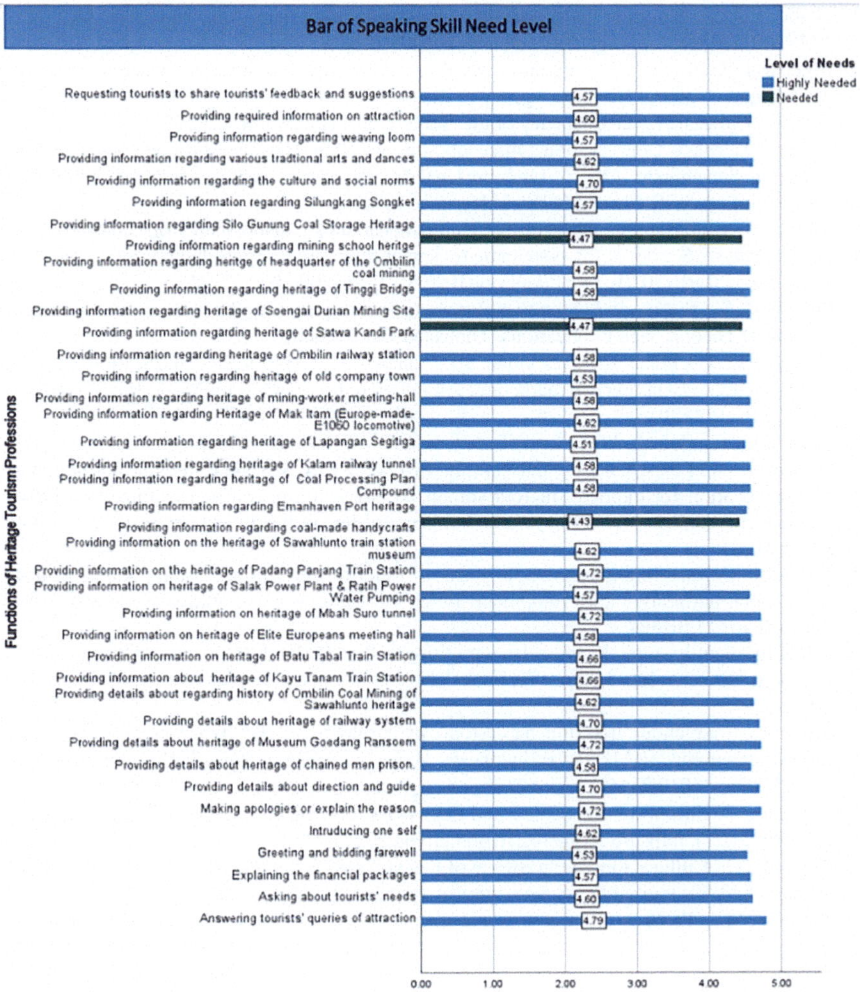

Fig. 1 Bar of speaking skill need level

Acknowledgements The paper is a partial result of the research funded by The Ministry of Research, Technology and Higher Education, Republic of Indonesia, based on a contract number 080/E5/PG.02.00.PL/2023 and 001/LL10/PG.L.JT/2023.

References

1. Hosseini, K.: World Heritage Sites in developing countries: assessing impacts and handling complexities toward sustainable tourism, (2021)
2. Assaf, A.G.: Bench-marking the Asia pacific tourism industry: A Bayesian combination of DEA and stochastic frontier. Tourism Manage. **33**(5), 1122–1127 (2012). https://doi.org/10.1016/j.tourman.2011.11.021

3. Hjalager, A.M.: Stages in the economic globalization of tourism. Ann. Tour. Res. **34**(2), 437–457 (2007). https://doi.org/10.1016/j.annals.2006.10.006
4. Emely, G., Gebert, V.: Sawahlunto towards a sustainable and attractive place to live, work and recreate. Cultural Heritage Agency of the Netherlands Ministry of Education Culture and Science. Cultural Heritage Agency of the Netherlands, Amersfoort (2020)
5. Cultural Heritage Conservation Office of Sumatra Barat, Historic Property and Museums Office of Sawahlunto. (2016). Ombilin Coal Mining Heritage of Sawahlunto: Nomination for Inscription on The UNESO World Heritage List.
6. Spradley, J.P.: Participant Observation. Holt Rinehart and Winson, New York (1980)
7. Liu, Y., Dupre, K., Jin, X.: A systematic review of literature on contested heritage. Curr. Issue Tour.. Issue Tour. (2020). https://doi.org/10.1080/13683500.2020.1774516
8. Corbin, A.S.: Basics of Qualitative Research Techniques and Procedures for Developing Grounded Theory. Sage Publications, Thousand Oaks (2008)
9. Huang, Y.C., Backman, K.F., Backman, S.J., Chang, L.L.: Exploring the implications of virtual reality technology in tourism marketing: An integrated research framework. Int. J. Tour. Res. **18**(2), 116–128 (2016)
10. Jung, T.H., tom Dieck, M.C., Lee, H., & Chung, N.: Effects of virtual reality and augmented reality on visitor experiences in museum. In: Inversini, A. & Schegg, R. (eds.) Information and Communication Technologies in Tourism, pp. 621–635. Springer, Cham, Switzerland (2016)
11. Elfiondri.: Tradisi dan Tabu Masyarakat Indigenous Mentawai. LPPM Universitas Bung Hatta, Padang (2018a)
12. Elfiondri.: Defining indigenous social norms of water-related taboo: Contemporary social attitude, social harmony and development. Int. J. Humanit. Soc. Sci. https://doi.org/10.30845/ijhss.v8n4p17 (2018b)
13. Elfiondri.: Family story on land-related tradition as base for land-use management and sustainable development. Int. J. Eng. Technol. Website: www.sciencepubco.com/index.php/IJET. (2018c)
14. Elfiondri.: Indigenous tradition based-tourism development: Foreign Tourist'S memorable tourism experience in Mentawai, Indonesia. Humanit. Soc. Sci. Rev. (2019a). https://doi.org/10.18510/hssr.2019.7665
15. Elfiondri.: Tradition, cultural contact and English for tourism: the case of Mentawai, Indonesia. Heliyon (2021). https://doi.org/10.1016/j.heliyon.2021.e07322
16. Richards, G.: Cultural tourism: a review of recent research and trends. J. Hospitality Tourism Manage. (2018), Journal homepage: http://www.journals.elsevier.com
17. Aswathy, M.J.: Problem of environmental sustainability: with the special references on houseboat tourism industry in Kerala. in Aloysius Edward, J., Jaheer Mukthar, K.P., Asis, E.R., Sivasubramanian, K. (eds.) Current Trends in Economics, Business and Sustainability. ICEBS 2023. Contributions to Environmental Sciences & Innovative Business Technology. Springer, Singapore (2023)
18. Guttentag, D.A.: Virtual reality: applications and implications for tourism. Tour. Manage. **31**(5), 637–651 (2010)
19. Karthick, A.V., Gopalsamy, S.: Role of IoT in business sustainability. In: Aloysius Edward, J., Jaheer Mukthar, K.P., Asis, E.R., Sivasubramanian, K. (eds.) Current Trends in Economics, Business and Sustainability. ICEBS 2023. Contributions to Environmental Sciences & Innovative Business Technology. Springer, Singapore (2023)
20. Stainton, H.: The commodifification of English language teaching in tourism: a sustainable solution? Tourism Manag. Perspect. **25**, 123–130 (2018). www.elsevier.com/locate/tmp
21. Basturkmen, H.: Developing Courses in English for Specifific Purposes. Palgrave Macmillan, UK (2010)
22. Ahmmed, R., et al.: A needs analysis of maritime English language skills for Bangladeshi seafarers to work on-board ships. Marine Policy. Journal homepage: http://www.elsevier.com/locate/marpol (2020).
23. Sadzali.: Cultural treasures of Sawahlunto dalam The Pearl of Sawahlunto (2019). Kementrian Kebudayaan, Olahraga dan Pariwisata Korea Selatan (2021)

24. Nurjanah, S.: 10 Aliran Silek Minangkabau (2017). https://aturanpermainan.blogspot.com/
25. Pakpahan, R.O.A.: Improving speaking skill of tourism vocational high school students through project based learning. Retain. **6**(2), (2018)
26. Tilden, F.: Interpreting Our Heritage. Fourth Edition (1957)
27. Al-Malki, E.A.A.R., Javid, C.Z., Farooq, M.U., Algethami, G.F., Al-Harthi A.A.: Analysis of the English Language needs of the Saudi Tourism workforce: a first step into designing ESP teaching materials. Int. J. Learn. Teach. Educ. Res. (2022)
28. Al-Khatib, M.: English in the workplace: An analysis of the communication needs of tourism and banking personnel. Asian EFL J. **7**(2), 174–194 (2005)
29. Davies, J.: A study of language skills in the leisure and tourism industry. Lang. Learn. J. **21**(1), 66–71 (2000). https://doi.org/10.1080/09571730085200131
30. Fujita, R.: English for tourism and hospitality. In: Terauchi, H., Noguchi, J., Tajino, A. (eds.) Towards a New Paradigm for English Language Teaching: English for Specific Purposes in Asia and Beyond, pp. 172–180. Routledge (2020)
31. Bobanovic, M.K., Grzinic, J.: The importance of English language skills in the tourism sector: A comparative study of students'/employees' perceptions in Croatia. Almatourism J. Tourism, Culture Territorial Dev. 2, 10–23. https://almatourism.unibo.it/article/view/2476/1849

Archaeological Tourism Perspectives for Oliveira De Azeméis (PT)

Adriaan De Man and João Tiago Tavares

Abstract In a small territory heavily marked by an industrial identity, based on components for the automotive industry and the footwear sector, can heritage and archeology lead to a sustainable development of cultural tourism? This is a question shared with many other regions, not only in southwestern Europe. While focusing on a case study, this paper seeks to address wider concerns about the public use of local cultural resources requiring long-term commitment and continued allocation of public monies. Enhancement of supply may lead to changes in demand and ultimately prompt induced effects on the local economy.

Keywords Rural tourism · Material culture · Tangible heritage

1 Heritage Resources

The municipality of Oliveira de Azeméis is located halfway between Porto and Aveiro, and relatively close to Coimbra, all cities with considerable heritage tourism attraction capacity. Local strategic options are therefore either based on freeriding dynamics, attempting to capture marginal segments from larger regional players, or promoting an idea of local uniqueness. A combination of both approaches has been put into place [1] in a territory between the coast and the inland mountain territories, rather distant from the Atlantic beaches and with an irregular yet not properly mountainous topography and several water courses crisscrossing the landscape. Scattered historical and archaeological remains both illustrate and suggest complex settlement since Pre-history. The first comprehensive attempt covering heritage and tourism through archaeology dates from the years 2015 to 2018, in the scope of a research project (POVOAZ, currently restarted as POVOAZ 2.1). Data are still incipient but

A. De Man (✉)
United Arab Emirates University, PO Box 15551, Al Ain, UAE
e-mail: adriaandeman@uaeu.ac.ae

J. T. Tavares
Municipality of Oliveira de Azeméis, 3720-292 Oliveira de Azeméis, Portugal
e-mail: joao.tavares@cm-oaz.pt

B. Awwad (ed.), *The AI Revolution: Driving Business Innovation and Research*, Studies in Systems, Decision and Control 524, https://doi.org/10.1007/978-3-031-54379-1_65

allow for some working hypotheses for understanding tourist flows related to heritage tangibles and ultimately for presenting a tourism offer distinguishable from that of neighboring territories.

As mentioned above, the local economy is heavily based on the manufacturing industry, which directly employs more than half of the population in the non-public sector [2] and allows for moderately low unemployment rates. The industrial productive capacity has therefore always been looked upon as the key economic driver [3], at the expense of strong commitment to other sectors, namely in the creation of tourist propositions. The general awareness of an uncharacteristic territory without notable heritage elements, containing only one classified national monument and ten monuments and sites of public interest, surely contributed to this condition. As far as archeology is concerned, merely sixty-five sites are registered in the national database Endovélico. While these seem rather modest numbers in contrast with certain other Portuguese municipalities, they are in line with surrounding municipalities of similar extension. Those with more registered sites are all located east of Oliveira de Azeméis, on a rocky area crossing Arouca, Vale de Cambra and Sever do Vouga. The degree of knowledge is a decisive factor, though, as decades of research concentrated on precisely these municipalities [4].

A fundamental question consists of whether the few sites in Oliveira de Azeméis can, by themselves, become a meaningful factor of tourist attraction for the territory. From a practical comparative standpoint, this does not seem straightforward at all, even from the marketing perspective alone. In the case of the official Oliveira de Azeméis website, anyone searching for information on places to visit will not find any reference to archaeological sites but has access to the Memórias de OAZ portal where information is available on the architectural, archaeological, and intangible heritage of the municipality. As is the case for regional competitors, archaeological heritage is not central to a municipal brand image and instead random cultural tangibles have been selected [5]. Heritage tourism offering in Oliveira de Azeméis has focused almost exclusively on three main points of interest: a watermill park connected to the traditional Ul bread production, the La Salette park, and the Ferreira de Castro house museum.

The thematic watermill park in the parish of Ul aims to rehabilitate and enhance a set of structures that symbolize the importance of milling activity in the municipality [6]. This is achieved, above all, through small elements of vernacular architecture directly related to a network of small waterways. Correspondingly, the manufacturing process of bread, with a special focus on Ul bread, a local specialty, can be experienced in the same village [7]. The park was built to create a tourist offer that promoted heritage and landscape resources, taking advantage of the relationship between the population and the milling activity, which extended beyond the traditionally cultivated cereals in the region. Local innovation led, for instance, to the creation of a rice husking cluster in a county with no tradition of rice production. This activity carried out in the rotation mills ended up gaining a dimension that led to the industrialization of this sector, giving rise to some of the companies that hustle and distribute rice. In turn, the La Salette park associates religious worship with leisure. As the result of a promise by the local population following a year of drought, a small chapel was erected on Monte dos Castros in 1870. At the beginning of the twentieth century, a request was made to Jerónimo Monteiro da Costa, Director of the Municipal Gardens of Porto, who presented a project capable of creating a leisure space on Monte that would allow the population to enjoy the space, in contrast to the inhospitable hill where the chapel had been built. The construction of the park was followed by the construction of a new chapel in neo-Manueline style, during the late 1920s [8]. The combination of these two elements, with the annual organization of a religious festival, created a dynamic attraction that produced a significant number of visitors, both during the festivities and throughout the year. Thirdly, the writer Ferreira de Castro, born in the parish of Ossela, bequeathed his family house to the municipality, along with a library that he had set up in the early 1970s, after being awarded the Grand Aigle d'Or prize at the Nice book festival. Of the three key heritage attractions, this is the least popular one in visitor numbers, although it could theoretically be included in regional cultural tourism routes [9] and gather conditions for reaching a wider museum audience.

2 Methodology

Based on these quite dissimilar heritage resources and their uneven impact on the local tourism industry, the added value of archaeology was considered as a working hypothesis. Data was collected using a thematic analysis approach, through literature review, qualitative observation, and archaeological fieldwork analysis carried out by the authors. The methodology involves the identification of key themes in the data and constructing a conceptual framework addressing the use of archaeological potential in local and regional development. The trustworthiness of this study is based on multiple data sources and reflexivity, given the researchers' own bias as archaeologists.

3 A Research Project Based on Archaeology

Scattered and few yet important Roman elements have been long known. Some were found during the demolition of the church of Santa Maria de Ul after the large earthquake that shook Portugal in 1755. The church did not collapse but suffered damage that led to its demolition for the construction of the building currently in use. During this process, three granite blocks were collected from the foundations of the primitive church, which, due to their configuration and inscriptions, were identified as Roman. Two of them turned out to be milestones, one corresponding to mile XII of the *Aeminium-Cale* road, while the other contains no text. A third piece collected in the same context is a *terminus Augustalis* [10] that plausibly marks a separation between the territories of the *Turduli Veteres* and the *Paesuri*, two pre-Roman tribes. The archaeological importance of the nowadays rural village of Ul is further visible at the hillfort site adjacent to the church. Another well-known finding is a set of golden bracelets collected during the construction of a wall at the place of Baralhas [11], about 2 km from the chapel of Senhora do Castro, which in turn can relate to the hilltop site of Ossela. Despite its provenance at this short distance from the archaeological site (Fig. 1), the findings attest to the important hoarding of goods in precious metal. Anecdotally, Ossela was mentioned in the sixteenth century by Friar Bernardo de Brito as the location of a legionary fort, yet based on a forged inscription, which led to a longstanding amateur interest in the topic.

Little fieldwork was carried out at both hilltop sites until very recently. Old and inconclusive excavations at Ossela were followed by no subsequent advances. The first campaigns at Ul happened in the mid-1980s only [12], following some occasional amateur earth turning at different locations across the hill. One other site, the Mamoa do Castelo was excavated in the 1980s and the results were first published in a local magazine and mentioned in later literature [13]. In such an overall unassertive scenario, relying on archeology as a tourist attraction seemed a rather unreasonable strategy. Without any commodified sites, what could be of interest to potential tourism demand? Despite these open questions and the lack of extensive resources, a decision was made in 2015 to carry out a research project on the municipality's archaeological settlement. This undertaking had the following objectives:

- Identifying the chronological brackets and sequences of site occupations;
- Evaluating if the larger settlements reveal synchronous activity;
- Establishing an occupation grid for the entire territory;

Fig. 1 Fieldwork in progress at Ossela (image by the authors)

- Assessing, through the acquired data, the concrete site potential, for both research and tourism.

In previous years, a regional research project (PROBA) had already included the hilltop site of Ul and served as a framework for a few minor interventions such as cutting vegetation and redefining the areas affected by old undocumented trenches and pits. This was followed by collaborations with universities, through the engagement of students and faculty, in archaeological and geophysical fieldwork. One of the driving assumptions for this project was the possibility of creating site valorization, musealization of structures, and ultimately a network of dispersed points of interest for tourists and residents. It was, however, clear that archaeological heritage attractiveness would not depend only on the sites, most of which do not possess any degree of monumentality. With potential tourism integrations in mind, four sites were selected to integrate the more ambitious POVOAZ project, namely the hilltop sites of Ul, Ossela, Monte Calbo, and Recarei, located in separate villages that cover the territory of Oliveira de Azeméis. They underwent survey and excavation to an extent where preliminary conclusions and working hypotheses can be drawn.

At Ul, a multiyear set of excavations and geophysical surveys [14] confirmed the details of several structures along the hilltop slope that had been traditionally identified as part of the set of lines of walls surrounding the village. One of these features is a small gate giving access to an inner platform, while other elements correspond to reinforcements from a later period. Materials collected at the base of these structures, although few, seem to point to a Roman imperial chronology. Surveys at Monte Calbo revealed a considerable dispersion of surface materials. None of the round huts, referred to in local folklore in association with Roman legions and perpetuated in random non-academic sources, were found. On the other hand, the oldest credible report on the site, the *Annaes do Município* (Carqueja 1908), only refers to the existence of two concentric lines of walls, which are currently also not discernible. The vast amount of scattered pottery is quite coherent, though, and proves consistent Late Bronze Age occupations. A tiny corner of the hilltop might be connected to a medieval observation point, sporadically used during military operations between Christians and Muslims, including during the 1045 battle of Villa Cesari, precisely the current parish of Cesar where Monte Calbo is located. In the case of Ossela, laconic information in the *Annaes* mentions orthogonal and circular structures from both the Iron Age and the Roman period, in addition to late medieval and modern tombs associated with the chapel of the Senhora do Castro. Despite some interesting and potentially late antique surface findings, only the medieval necropolis was confirmed. Finally, at Recarei some Roman pottery was known but excavation defined above all a set of post holes, some of them with the top still filled with pieces of clay or stones that would have been wedging wooden posts. Stratigraphically, the Middle and Modern Ages, on the one hand, and the very late Bronze and Iron Ages, on the other, are well identified, and any intermediate layers corresponding to the classical period contain very little material culture, pointing perhaps at an abandonment of sections of the hilltop (see [15] for a detailed overview and references).

This very short outline of recent fieldwork shows that additional data certainly improves the quality of archaeological information and allows for the postulating of further academic hypotheses. Current investment is, however, not directly transferrable to the advancement of local heritage tourism. Commodification and differentiation require minimum scaling levels at local and regional levels, in addition to small site conservation efforts [16]. At the current development stage of the archaeological sites in the municipality of Oliveira de Azeméis, the concept of heritage valorization is more appropriate than that of musealization, from a conceptual standpoint. None of the examples is attached to a proper site museum or has been musealized in the sense the interpretation of tangible heritage was not enhanced as a standalone attraction using, for instance, audiovisual elements or virtual reconstructions. The communication of an appropriate museum packaging does not depend on strict heritage features, though [17], and needs to be tailored according to future resource development and integrated community strategy.

What can be adequately planned for a site such as Ul is a local extension of a larger municipal entity, counting on its own facilities and resources. This includes permanent dedicated staff, opening the site to the public regularly, and fulfilling minimum and regular museum functions, namely interpretation, exhibition, and even dissemination. A slightly different solution would consist of an interpretative center with a strong educational component for the many types of public, ranging from residents to international tourists. In this picture, sustainable local development [18, 19] needs to become a premise for an exhibition-based experience, upstream from any expectations of direct return on investment. Furthermore, a practical cornerstone in any of the possible settings is the existence of an archaeological reserve, ensuring the inventory, incorporation, study, and conservation of collections. This complementary mission is vital to the tourism commodification of a territory as an umbrella concept for all forms of associated heritage that create local uniqueness for Oliveira de Azeméis. A municipal museum is not essential but would serve as an anchor for a narrative of discovery and interpretation. It could also become a platform that uses a selection of exhibited objects as an encouragement for understanding the regional cultural landscape. One immediate caveat for any museum structures installed on-site is that all archaeological remains are currently located on private land. The concept of a museum-territory, despite having been conceived with a tourist vocation, requires mature identity elements for wealth creation and economic development. On the opposite side, temporary and mobile exhibitions touring surrounding villages and the city itself can become a low-cost / high-return solution. The inherent limitations can be overcome but this would tackle some current logistical challenges of a territory still in the process of scaling up cultural tourism resources.

4 A Content Proposal for a Temporary or Small-Scale Interpretation Center

The creation of an interpretation center is to be based on a mission and a set of goals providing purpose and guaranteeing articulation with the territory and with other existing structures. The former was formulated in the following way:

- Improve the conservation conditions of historical materials, in the region between the Douro and Vouga rivers, value them as heritage, and promote their interpretation by local inhabitants and tourists.

while the latter are itemized to comprehensively incorporate future challenges:

- Aggregating knowledge of the period between the tenth century BC and the fifth century AD, with the necessary diachronic breadth for what precedes and succeeds this period;
- Valuing the municipality's archaeological and historical heritage, acting as an anchor for the movement of visitors and tourists, and supporting cultural tourism;
- Developing communication strategies that foster a relationship between the territory and archaeological heritage;
- Ensuring educational activities that make different audiences aware of the importance of safeguarding the archaeological heritage;
- Increasing visits and stays in the municipality with the explicit aim of discovering local archaeological heritage;
- Promoting knowledge of archaeological methods, through participation in practical fieldwork, including excavation, inventorying, and conservation of archaeological materials, and participation in experimental archeology activities.

The interpretation center will be dedicated to archeology and history, essentially centered on material remains collected in the municipality, but resorting to others, whenever this allows for a better presentation of the contents. Related thematic areas, such as the regional economy, environment, and geography will be considered to make the information more coherent and provide visitors with an enhanced experience.

Contents can be structured in a wide-angle presentation in which archaeology and general tangible heritage are central but serve as an attractive framework for historical storytelling.

A. Communities
  - Where and how they lived
  - What artifacts they left us
  - Culture shock—Romanization
  - The end of the Empire and the occupation of the territory.

B. The Roman milestone
  - What is it for?
  - Historical Background
  - Current location

C. Roads and transport
- How were they built and maintained?
- Roman roads in Oliveira de Azeméis
- Migrations
- The Roman army
- Roads and administration.

D. Trade and currency
- Goods and products
- Trade routes
- Long haul trade.

E. From archaeology to present times
- Urban growth
- From agriculture to industry
- The architectural heritage.

These contents produce a consistency between individual sites, objects, and the wider territory up to the present day, as recommended in previous work [20]. Visitors are to be contextualized with the sites, the corresponding material, and their different possible meanings. Following an induction at the interpretation center, itineraries can be selected according to individual preferences, namely with the use of GPS-guided smartphone apps. However, technical solutions matter less than the intuitive, emotional response to heritage [21], which constitutes the only way of securing visitor engagement in the first place.

5 Limitations and Further Research

Several directions will need to be further explored. The correlation between investment in archaeological research, on the one hand, and tourism robustness, on the other, is often taken for granted but is not spontaneous. Many academically interesting sites produce only marginal economic infiltration in the local economy, for reasons ranging from private entrepreneurship to the nature of the tangible itself. Adequately interpreting this reality will require long sequences of econometric data. Other limitations relate to the archaeological record itself, its conservation, and public use in connection with permanent leisure infrastructures. They face strict financial and operational challenges. Most archaeological remains can be successfully commodified, even when material culture lacks quantity and density. Future propositions must therefore find solutions that do not disproportionately insist on archaeology but rather optimize its expression as a constituent of larger heritage products.

6 Conclusion

This short proposal exemplifies an alternative capable of generating a cultural visitor flow in the municipality of Oliveira de Azeméis, with no substantial investment in costly additional resources. A large part of the population, both at the level of decision-makers and average residents, do not recognize the economic potential of archaeological sites. Indirect and induced effects will remain relatively modest but may yield interesting niche opportunities if gradual advances are made within in the scope of resilient and sustainable municipal planning.

In the case of Oliveira de Azeméis, an offer based on archaeological resources appears feasible and the results need to spill over to a social dimension without which heritage cannot become operational. This element is of interest to many other small communities worldwide, as it encourages a balanced, bottom-up, long-term strategy whilst discouraging excessive expectations of the positive socioeconomic results of tourism.

References

1. De Man, A., Tavares, J.T.: Local development policies for heritage tourism. Rev. Econ. Finan. **21**(1), 330–335 (2023a). https://doi.org/10.55365/1923.x2023.21.32
2. PORDATA.: Oliveira de Azeméis (2023). https://www.pordata.pt/municipios/quadro+resumo/oliveira+de+azemeis-822339. Last accessed 31 Aug 2023
3. Alces, T.: O contributo das iniciativas comunitárias para o desenvolvimento do território: o caso do Município de Oliveira de Azeméis. Universidade de Coimbra, Coimbra (2011)
4. Silva, A.M.: Ocupação proto-histórica e romana no entre Douro e Vouga litoral: breve balanço de uma investigação em curso. Actas do 1.º Congresso de Arqueologia Peninsular, **2**(Porto), 429–439 (1995)
5. Guimarães, C.S.R.F.: Competitividade territorial: as dinâmicas do concelho de Oliveira de Azeméis. Universidade de Aveiro, Aveiro (2009)
6. Paiva, D.R.C.: Turismo, Desenvolvimento e Espaços de Lazer - O Parque Temático Molinológico de Ul. Universidade de Coimbra, Coimbra (2013)
7. Galvão, A., Dias, I, Oliveira, M.: Caracterização do Pão Tradicional de Ul, Characterization of Traditional Bread from Ul. Rev. Da Ui_Ipsantarém, **2017**, 61–70 (2017)
8. Anacleto, R.: Capela de Nossa Senhora de la Salette. O Neomanuelino ou a reinvenção da arquitectura dos Descobrimentos, pp. 167–169. Comissão Nacional para as Comemorações dos Descobrimentos Portugueses, Lisboa (1994)
9. Maia, S.V.: Rotas museológicas na região de Aveiro: um estudo empírico. Universidade de Aveiro, Aveiro (2010)
10. Almeida, F.: Terminus augustalis entre Talábriga e Lancóbrica. O Arqueólogo Português (2.ª série) **2**, 209–212 (1953)
11. Vasconcelos, J.L.: Novo achado de braceletes preromanos. O Archeologo Português **2**, 86–88 (1896)
12. Marques, J.A.M.: Escavações no castro de Ul (Oliveira de Azeméis) Primeira notícia. Rev. de Ciências Históricas **4**, 65–89 (1989)
13. Carvalho, P S.: Pré-História. A milhares de anos de distância, pp. 17–28. Memórias de OAZ. MOA, Oliveira de Azeméis (2017)

14. De Man, A., Tavares, J.T., Carvalho, J.: GPR and electromagnetic induction surveys at the hilltop site of Ul (Oliveira de Azeméis, Portugal). Mediter. Archaeol. Archaeom. **17**(1), 179–188 (2017). https://doi.org/10.5281/zenodo.292622
15. De Man, A., Tavares, J.T.: Modeling municipal heritage management. AP Online J. Publ. Archaeol. **11**, 17–37 (2021). https://doi.org/10.23914/ap.v11i0.266
16. De Man, A., Tavares, J.T.: Small site conservation in the scope of territorial cohesion. In: 3rd International Conference on the Built Environment and Engineering (IConBEE) 2022: Fostering a Transpicuous Construction Industry, AIP Conference Proceedings (2023b)
17. De Man, A.: Tangential heritage values in museum communication. Sustainability **15**(6), 5169 (2023). https://doi.org/10.3390/su15065169
18. Aswathy, M.J.: Problem of environmental sustainability: with the special references on houseboat tourism industry in Kerala. In: Aloysius Edward, J., Jaheer Mukthar, K.P., Asis, E.R., Sivasubramanian, K. (eds) Current Trends in Economics, Business and Sustainability. ICEBS 2023. Contributions to Environmental Sciences & Innovative Business Technology. Springer, Singapore (2023). https://doi.org/10.1007/978-981-99-3366-2_33
19. Brown, K.: Museums and local development: an introduction to museums, sustainability and well-being. Museum Int. **71**(3–4), 1–13 (2019). https://doi.org/10.1080/13500775.2019.1702257
20. Tavares, J.T.: Políticas Culturais Municipais—Um Balanço e Algumas Propostas. Universidade de Coimbra, Coimbra (2022)
21. Smith, L.: Emotional heritage: Visitor engagement at museums and heritage sites. Routledge, Oxon (2021)

Factors Affecting Malaysian Consumers' Perceptions of Chinese Smartphone Advertisements and Their Implications for Other Sectors

Julia Tan Yin Yin and Alexander Stark

Abstract China is currently playing a significant role in the world economy. However, the majority of buyers worldwide still lack sufficient confidence in goods produced by Chinese firms. This study intends to identify the advertising appeals utilized in Chinese brand smartphone advertisements (both online and offline). It also looks at the way Malaysian consumers perceive these appeals. In a qualitative study, ten participants were interviewed in-depth about their perceptions regarding Chinese advertisements. The participants watched Chinese smartphone advertisements and were asked about their impressions. It was found that these commercials elicited specific feelings and increased their familiarity with the details surrounding the products. It was shown that many buyers focused primarily on these details, which affected their purchasing decisions. The researchers also provided an assessment of the potential impact of these findings on other industries.

Keywords Advertising appeals · Chinese smartphones · Malaysian consumers

1 Introduction

1.1 Introductory Words

Advertising is a prominent topic in communication studies. It focuses on the best method for persuading customers to buy a given product. Advertisers have employed a range of tactics to attain their objectives. Some of these tactics were of major

This study comprises a brief version of a part of Julia Tan Yin Yin's unpublished Ph.D. Thesis with the title, The Impact of Advertising Appeals on Malaysian Consumers' Preference towards Chinese Smartphone Brands. The thesis was submitted in 2021 at Xiamen University, China. the supervisor was Professor Dr. Yan Li Feng.

J. T. Yin Yin (✉) · A. Stark
Universiti Malaysia Kelantan, Kota Bharu, Malaysia
e-mail: julia@umk.edu.my

B. Awwad (ed.), *The AI Revolution: Driving Business Innovation and Research*, Studies in Systems, Decision and Control 524, https://doi.org/10.1007/978-3-031-54379-1_66

interest to the scientific community. They looked at how far advertising influences consumer behavior. Advertising must rely on appeals, sometimes intellectual, sometimes emotional, to be successful in this persuasive drive [1].

Almost everywhere in the world, people use smartphones. They play a crucial role in social interactions nowadays. These interactions include communication, advertising, and market discourse. As earnings rise, the need for more individualized communication and a wider range of phone applications grows, and so does the number of smartphone users. Because of this, businesses are now using the smartphone market to advertise their goods. For the Chinese economy, the smartphone market is of growing importance. Currently, the two main smartphone companies, Apple and Samsung, dominate the market. On the other hand, companies like Huawei are becoming more important and popular.

For Chinese companies, the Asian market is very promising. Malaysia is an attractive market. Consumers in Malaysia frequently use smartphones made in China. Both countries have established strong trading links over the years.

In this research, the researchers want to examine how consumers perceive Chinese smartphone advertisements and what factors influence their purchase behavior. Due to its vast population of smartphone owners and its participation in the production and supply chains of this crucial device, China has emerged as a major participant in the global smartphone market. Thanks to intensive marketing and advertising, Chinese-brand smartphones have continued to grow in popularity across the globe, especially in Malaysia.

For a number of parties, including academics and active Chinese brand marketers, this research has some theoretical and practical value. It will contribute significantly to scholarly knowledge and provide evidence-based answers to marketing, in this case, Chinese smartphone commercials in Malaysia. This study provides deeper insights into understanding potential clients in pluralistic, multiethnic, multicultural countries such as Malaysia for international Chinese enterprises, particularly top Chinese brands.

The findings of this study can potentially be applied to other fields. The researchers will use the tourism industry as a brief example. It might be put into practice if the tourism industry is aware of the variables that are significant to potential visitors in the sphere of advertising.

1.2 Research Questions

The following are the study's research questions:

(1) How do Malaysian consumers perceive the use of commercial appeals in Chinese smartphone brand advertisements?
(2) Can the research results be implemented in other sectors?

1.3 Structure of the Research Paper

The article consists of five major parts: After the introductory section, in part 2, there will be a discussion of relevant literature. The researchers engage in a comprehensive examination of the existing body of literature pertaining to advertising appeals, the country-of-origin effect, and the theories surrounding soft-sell and hard-sell appeals.

In part 3, the research methodology will be outlined. Subsequently, Chap. 4 will encompass a comprehensive analysis and examination of the collected data. The advertising strategies of Chinese smartphone companies will be illustrated. It will be described how these strategies influence consumer behavior. It will also be discussed whether the results can be implemented in other business sectors.

Chapter 5 will discuss the limitations of this study and provide some recommendations before making some concluding remarks.

2 Literature

2.1 Advertising Appeals

Dix and Marchegiani offer more insight into why scholars have paid so much attention to advertising appeals [2]: "Advertising appeals sit at the heart of advertising creative to connect the brand to the consumer wants and needs. Advertising appeals is the sticky glue that hooks the reader or viewer to the advertising message." Global brands have common international acceptance, demand, impact, and recognition. A global brand is described as a brand that is considered international and is promoted not only locally but also internationally. This term refers to the fact that as a brand's perceived multimarket presence grows, so does its perceived worldwide presence [3]. Kotler stated that advertising appeal is the main subject of an advertisement [4]. In order for an audience to receive a message effectively, advertisers have to put some driving power into the message. Appealing is the driving power for advertisers. Every advertising appeal represents an attraction that arouses consumers' desires. All in all, advertising appeals have an impact on attitudes towards advertisements and purchase intentions. Different appeals chosen from advertisements will have different influences on different product categories. For low-involvement and hedonic products, studies have proven that emotional appeals are believed to be effective. On the other hand, for high-involvement and utilitarian products, rational appeals are suggested to be effective. High-involvement products here refer to higher-priced goods that are irregularly bought, while utilitarian goods are generally instrumental and functional.

2.2 Country-of-Origin (COO) Effect

Country-of-origin (COO) is an essential factor in influencing consumers' brand evaluation, perceptions, and purchasing behavior, offering brands a significant competitive advantage. The effects of the Country of Origin (COO) refer to positive or negative product quality evaluations, attitudes towards products of brands, or purchase intentions of the consumers towards the products originating in a certain country or as perceived to originate from that country [5]. There are three significant ways of product labelling with COO to the company, as mentioned by Insch and Florek [6]. Firstly, it serves as the product quality indicator. Secondly, it serves as a geographic referential for customers who prefer products from specific places, probably influenced by consumer ethnocentrism, pride, and prestige. Thirdly, it preserves the positivity that a good brand country has for the product. It may refer to where a brand is based, a product designed or manufactured, or other value-creation forms aligned to a country.

2.3 Hard-Sell and Soft-Sell Appeals Theory

Okazaki et al. identified three main dimensions of soft-sell appeals: feeling (creative, instinctive, imaginative, and abstract), implicitness (insinuation, appealing, subjective, and expressive), and image (entertaining, interpretive, playful, and impression-based) [7]. In the same way, hard-sell appeals comprise three dimensions: thinking (rational, logical, analytic, factual, and concrete), explicitness (precise, explanation, convincing, persuasion, and instructive), and fact (educational, descriptive, realistic, informative, and evidence-based).

According to Laskey, Day, and Crask, hard-sell appeal is usually direct, but it is traditionally a type of fair, truthful, and informative ad. Informative attractions aim to concentrate on the product's characteristics or advantages [8].

According to Mueller, a hard-sell appeal seems straightforward, with a strong sales emphasis and frequent mentions of the company name and product reviews [9]. According to preliminary results, hard-sell appeal is more prevalent in individualistic markets, while a soft-sell request is more common in collectivistic societies [9–11]. A recent study by Okazaki, Mueller, and Diehl discovered that soft-sell attraction is more successful and causes less frustration [12].

Hard-sell versus soft-sell appeals in advertising have remained relevant in society for over 100 years as advertising appeals to society, affecting persuading attitudes and beliefs towards a brand [7]. "Hard-sell ads encompass a more direct and fact-based approach with a focus on a quick sale, while soft-sell ads encompass a more subtle and indirect approach. Studying hard-sell and soft-sell ads under transportation will provide a further understanding of the persuasive mechanisms [13].

3 Research Method

3.1 A Qualitative Research

A qualitative approach, which includes in-depth interviews and the analysis of these data, was used. In order to gain more profound knowledge and avoid misinterpretation, in-depth interviews are essential. The appropriate number of participants is debatable among scientists: Some cite the numbers 1–10 [14], while others give the numbers 5–25 [15]. Ten (10) respondents were chosen to take part in the in-depth interviews for this research, as advised by the literature. The ten people who participated in the in-depth interviews were chosen from advertising and marketing experts, friends of the researchers, and friends of friends who are presently using or have used Chinese brand smartphones within the past three to five years. All of them freely consented to participate in the interviews. There were 10 participants, 5 of whom were Chinese, 4 Malay, and 1 Indian.

3.2 Procedures and Interviewing Techniques

In the qualitative stage of the in-depth interviews, respondents were asked several questions before and after showing two YouTube advertisements of Chinese smartphone brands. Individual interviews were conducted for each candidate. Telephone appointments were scheduled with respondents to arrange a meeting time, date, and location. All in-depth interviews were conducted face-to-face since the researchers believed it was essential to learn the respondents' instant reactions and perceptions "on the spot" following the YouTube advertisements. This procedure was in accordance with Shuy's advice to promote naturalness and build rapport [16]. Each interview lasted roughly 70–120 min, with two respondents taking nearly 120 min. All of the questions had an "open-ended design," allowing respondents to freely express their perceptions and opinions about Chinese smartphone brand advertisements. According to Groenewald's suggestion, questions pertaining to respondents' user experiences, opinions, and sentiments were put forth with the theme in mind [17].

4 Data and Discussion

4.1 Advertising Strategies of Chinese Smartphone Companies

A variety of Chinese smartphone advertising that could be viewed on YouTube was shown to the participants. Furthermore, the participants also looked at posters that advertised Chinese smartphones. In this paper, two smartphone advertisements will be exemplary.

Vivo 5 Commercial. Vivo continued to utilize the same persuading arguments in its outdoor marketing as it did in its inside ones. For instance, four young women striking a sensual pose while promoting the Vivo V5 product may be seen in Fig. 1. The utilization of young women is consistent with the employment of women to promote products in many other ads. Young women, who make up the majority of the target audience, urban, metropolitan smartphone users, as well as admirers of beauty and attractive women, are persuaded by this strategy. Advertising experts contend that sex appeals aim to attract attention, pique interest, and create an imprinted memory.

Sex in advertising, according to Hultin and Lundh, is the use of sexual desires as a persuasive tactic to highlight a certain product for consumption [19]. It is regarded as one of the most potent tools available to marketers, especially advertisers. The models strike selfie poses as they demonstrate the functional appeal. As this outdoor advertisement plainly tries to portray the Vivo V5 as the best device to do that leisure activity, high-quality selfies have been a big selling point for many smartphones.

Huawei Nova 5T. In Fig. 2, the major sequences of a commercial for the Huawei Nova 5T smartphone can be seen

.

The motto "pushing beauty to the age" is shown to the audience as the advertisement begins, set against a purple background and a swarm of lights. The main character in the advertisement then arrives in what appears to be a lake at night with light reflections below; the setting is naturally beautiful. Once more focusing on status appeal and utilizing the attractive and youthful model, the advertisement primarily targets young smartphone users. Given that the majority of smartphone users belong to that demographic, this is not surprising. As the advertisement aims to make a big impression on a wealthy audience, the technical attraction is highlighted by the "Huawei AI Quad" camera, as a side-mounted fingerprint sensor is disclosed to the viewers even as the lead female character snaps a "selfie." As the

Fig. 1 An outdoor advertisement of Vivo (*Source* Tan [18])

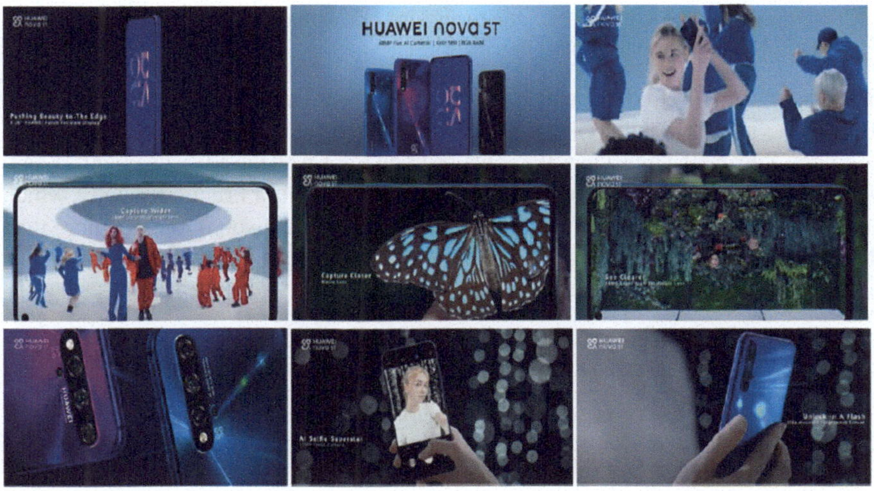

Fig. 2 Sequences of a Huawei Nova 5T Malaysia TV commercial video (retrieved from https://www.youtube.com/watch?v=7V7yo0AMDgg) (*Source* Tan (2021))

smartphone's marketers attempt to persuade the potential customer that this product is the best for capturing images and having fun in a relaxed environment such as a garden and in the company of friends during parties or other celebrations, "See better" is flashed across the screen, and quality and functional appeals in the advertisement are supported. In an effort to inspire consumers' confidence that the battery is built to last long—possibly as long as the party or the fun moment that deserves to be captured—3750 mAh is highlighted on the screen at the conclusion of the advertisement. MAh is an abbreviation for milliampere hour, which is used to express the electricity capacity of smaller batteries, such as those used by mobile phones. "Nova 5T—Inspire your moment" is written on the screen as the commercial comes to an end.

4.2 Consumers' Perceptions of Chinese Advertisements: Emerging Themes

In the chapter before, the way of advertising Chinese smartphones was outlined. In this chapter, it will be illustrated what types of topics emerged.

Emotions (soft-sell appeals). The investigation revealed that consumers' choice of a certain brand is significantly influenced by emotion. This is particularly evident when respondents discuss their prior brand experiences, which ultimately led them to choose that particular brand when they went to purchase the phone. The following excerpt from the interviews shows that the advertisements evoke emotions:

> It portrays many types of emotion…the happy kind of emotion. Because it portrays such a happy emotion, we also feel happy too.

Spokesperson (soft-sell appeals). Almost all of the interviewees were able to relate to how the phone brand spokesman helped them form a strong impression of the company. During the interview, the spokesperson improved the advertisement's recall value and relatability. This is especially important in light of Fruchter's study, which suggested that the manner in which an advertisement is delivered matters more than the content of the advertisement [20]. The following excerpts show the impressions of the customers:

> I prefer well-known international artists. That tells me how much the stakeholders are willing to invest in the products, and that usually shows how good the product is.

> I think the spokesperson in the advert also plays an important role. An artist like Siti Nurhaliza and Neelofa is a good spokesperson for the local Malaysian market, where such a public figure has a big influence in the local market.

A good example of a prominent spokesman is Dato Siti Nurhaliza. She is a famous Malaysian singer and well-known to many people. Studies in psychology and society on consumerism have revealed how influential people influence and evoke emotions by projecting positive images into the minds of their audiences. Therefore, the spokesperson should be carefully picked based on the traits of the target audience in order to have a significant influence on them.

Product Specification (hard-sell appeals). The interviews revealed that the majority of modern smartphone buyers are well-educated and have vast technological knowledge. Companies need to be aware of the swarming requests for the newest phone features and specifications. The commercial needs to include features that the average consumer wants in a smartphone in order to stay current with this trend. The following excerpts of interviews show that many consumers are well-versed in technological features.

> For me (the most important phone function) is the RAM and Memory because I do online gaming like Mobile Legend, PUBG, COD (Call of Duty).

> Normal cameras use a camera lens and only have one lens; this is their limitation…when you zoom in, it is actually an optical zoom. They are moving the length of the lens, which is like DSLR …they are moving the distance of the length and the aperture…aperture is where it opens its shuttle….phone cameras basically just have one lens; now some have two lenses, three lenses, Huawei has four lenses. So basically, what it does is it has different lenses.

Pertinent/Relevant (hard-sell appeals). Modern technology is unquestionably well-liked by the younger generation. With its distinctive features, user-friendly interface, reasonable price, and, of course, the thirst for fresh, original ideas that facilitate interaction, it quickly captures the attention of this generation.

> The product's quality, the second would be the brand. Brand is more difficult if you're not used to using the smartphone; everything that you do with the smartphone is difficult. The next thing would be the smartphone's durability. The next would be the price. And finally, promotion. Oh, sorry, the last one after the promotion is the postal service.

This excerpt gives a very good overview of the different elements that influence consumer behavior.

Advanced Features (hard-sell appeals). The need for the most recent features in smartphones is fueled by the necessity to communicate better through social networking and other instant communication technology, which sometimes require the latest model or version of the application or the smartphone itself. Other advanced features mentioned by the respondents are as follows:

> Watching beautiful things at night or in dark places, but when trying to capture (them) using a phone, not able to capture. Frustrated because I cannot share the beauty I am seeing at that moment (the need for good lighting in the dark).

4.3 Impact on Other Business Sectors

In this part, the researchers want to show in brief that the results regarding the Chinese smartphone advertisements have an impact on other business sectors. It was decided to take the tourism industry as an example. The customers were influenced by some factors like hard-sell appeals, famous product 'ambassadors', etc. These factors have to be considered if a local government wants to promote a specific area. For potential tourists, such an area is attractive if the following components are fulfilled:

- Nowadays, many tourists are well-informed about a special region due to the accessibility of the Internet. The potential tourist can compare the prices and attractions of each place. Somehow, the situation is similar to the situation of advertising a smartphone. The customers know well what they want. Customers are well aware of what they want. In the case of a camera, they might search for a high resolution, and in the case of a particular location, they might search for amenities like clean air and water.
- Attractive people and landscapes might help in the decision-finding process of the potential traveller.

5 Conclusion

5.1 Limitations

The present study is subject to some limitations. Initially, the focus was limited to the Malaysian market, namely targeting the smartphone industry. Furthermore, the scope of the study was limited to the analysis of YouTube clips and billboards.

5.2 *Recommendations*

The preceding paragraph provides a clear indication of the factors that necessitate more investigation, including the impact of artificial intelligence (AI) [21] and a more comprehensive examination of the digital market [22]. By doing so, certain overarching themes can be identified that have the potential to enhance firms' marketing efforts.

6 Summary

In this study, the researchers examined Chinese smartphone commercials to see how much of an impact they had on Malaysian consumers purchasing decisions. It was shown that many customers are knowledgeable about the things they buy. For many customers, the so-called hard-selling appeals are crucial in helping them choose the proper product.

These attractions make the goods appear more appealing, which may encourage potential customers to purchase the product. For instance, if they are undecided, seeing a renowned person can persuade them to purchase it.

The results of this research might be implemented in other sectors. The field of tourism was chosen as exemplary. Nowadays, many potential customers are well-informed, and the respective sector has to provide them with information. Furthermore, some soft appeals might be helpful in the decisive-finding process.

References

1. O'Guinn, Allen, C. Scheinbaum, A: Advertising and Integrated Brand Promotion, 8th Edition. Cengage Learning, Singapore (1997)
2. Dix, S., Marchegiani, C.: Advertising appeals. J. Promot. Manag. **19**(4), 393–396 (2013). https://doi.org/10.1080/10496491.2013.817218
3. Steenkamp, J.-B., Batra, R., Alden, D.: How perceived brand globalness creates brand value. J. Int. Bus. Stud. **34**, 53–65 (2003)
4. Kotler, P.: Marketing Management: Analysis, Planning, Implementation and Control, 9th edn. Prentice-Hall, New Jersey (1997)
5. Magnusson, P., Westjohn, S.A., Zdravkovic, S.: What? i thought samsung was Japanese: Accurate or not, perceived country of origin matters. Int. Mark. Rev. **28**(5), 454–472 (2011)
6. Insch, A., Florek, M.: Prevalence of country of origin associations on the supermarket shelf. Int. J. Retail Distrib. Manage. **37**(5), 453–471 (2009)
7. Okazaki, S., Mueller, B., Taylor, C.R.: Global consumer culture positioning: testing perceptions of soft-sell and hard-sell advertising appeals between US and Japanese consumers. J. Int. Mark. **18**(2), 20–34 (2010)
8. Laskey, H.A., Day, E., Crask, M.R.: Typology of main message: Strategies for television commercials. J. Advert. **18**, 36–41 (1989)
9. Mueller, B.: Reflections of culture: An analysis of Japanese and American advertising appeals. J. Advert. Res. **27**(3), 51–59 (1987)

10. Lin, L.Y.: The impact of advertising appeals and advertising spokespersons on advertising attitudes and purchase intentions. Afr. J. Bus. Manage. **5**(21), 8446–8457 (2011)
11. Maynard, M., Taylor, C.: Girlish images across cultures: Analyzing Japanese versus U.S. Seventeen magazine ads. J. Advertising **28**(1), 39–48 (1999)
12. Okazaki, S., Mueller, B., Diehl, S.: A multi-country examination of hard-sell and soft-sell advertising comparing global consumer positioning in holistic-and analytic-thinking cultures. J. Advert. Res. **53**(3), 258–272 (2013). https://doi.org/10.2501/jar-53-3-258-272
13. Daniels, S.: Exploring advertisements: The effects of hard-sell and soft-sell advertisement appeals on transportation. Being a thesis president in atrial fulfilment of Master degree. Cleveland State University (2018)
14. Starks, H., Brown, S.B.: Choose your method: A comparison of phenomenology, discourse analysis, and grounded theory. Qual. Health Res. **17**(10), 1372–1380 (2007)
15. Creswell, J.W.: Qualitative Inquiry and Research Design: Choosing Among Five Traditions. Sage Publications, Thousand Oaks (2016)
16. Shuy, R.W.: In-person versus telephone interviewing. In Gubrium, J.F., Holstein, J.A. (eds.), Handbook of Interview Research: Context and Method. Sage, Thousand Oaks, CA (2002)
17. Groenewald, T.: A phenomenological research design illustrated. Int J Qual Methods **3**, 1–26 (2004)
18. Tan, J.Y.Y.: The Impact of Advertising Appeals on Malaysian Consumers' Preference towards Chinese Smartphone Brands. Ph.D. Thesis, Unpublished. Xiamen University, China (2021)
19. Hultin, M., Lundh, E.: Sex in advertising: How it influences young men and women. Psychology (2004)
20. Fruchter, G.E., Van den Bulte, C.: Why the Generalized bass model leads to odd optimal advertising policies. Int. J. Res. Mark. **28**(3), 218–230 (2011)
21. More, A.B.: Implementing digital age experience marketing to make customer relations more sustainable. In: Nayyar, A., Naved, M., Rameshwar, R. (eds) New Horizons for Industry 4.0 in Modern Business Technology. Springer, Berlin (2023) https://doi.org/10.1007/978-3-031-20443-2_5
22. Prasanna Kumar, U., Arthi, R.: A study on utility and feasibility of digital marketing tools with lead acquisition, lead nurturing and client engagement. In: Aloysius, Edward, J., Jaheer Mukhtar, K.P.A., Sivasubrmaniam, E.R. (eds.) Current Trends in Economics, Business and Sustainabilty. ICEBS2023. Contributions to Environmental Business Technology. Springer, Singapore (2023). https://doi.org/10.1007/978-981-99-3366-2_24

Empowerment in Mosque Tourism Sustainability: Facilities Management (FM) Strategy

Zafirah Ab. Muin, Maimunah Sapri, Maryanti Mohd Raid, Aminah Mohsin, and Nur Khairiyah Mohammad

Abstract A mosque is a sacred place of worship for Muslims. Nevertheless, it also become a one stop centre that is always visited by the local community and outside visitors. However, disagreements over the mosque's condition would frequently arise. The cleanliness, size, design, layout, and other aspects of the mosque's facilities frequently cause users to express dissatisfaction. In the tourism industry, loyalty is crucial since travelers are the ones who return to a location or suggest it, which is a deciding element in tourist site management. Facility management (FM) is the process of ensuring that an organization's physical environment and services meet its essential operations and objectives. FM is critical to generating and sustaining a pleasant customer experience, which can lead to loyalty and retention. Thus, this study aims to look into the efficacy and applicability of the FM approach to mosque tourism sustainability. The study's purpose is to highlight that implementing a management platform based on FM principles can give a better foundation for facilitating the process of mosque tourism sustainability. This qualitative research encompassed a systematic study involving philosophical and academic perspectives with past works of literature. Due to the fact that this study involves library research from various primary data sources, library studies and critical analyses were used to evaluate relevant resources, data, and knowledge on mosque tourism and FM, particularly tourist attraction and loyalty concerns. This study concludes that function of mosque can be optimized not only as a main place of worship but also as a tourist destination through an effective FM strategy. According to the study, an excellent FM could promote mosque tourism by creating a pleasant environment for users to flourish the mosque and attracting the curiosity of tourists, particularly non-Muslims, to obtain experience in comprehending Muslim worship. It is hoped that efficient mosque tourism management will provide tourists with information and great experiences at mosques.

Z. Ab. Muin (✉) · M. M. Raid · A. Mohsin · N. K. Mohammad
Real Estate Department, Faculty of Built Environment and Surveying, Universiti Teknologi Malaysia, Skudai, Malaysia
e-mail: zafirah2@utm.my

M. Sapri
Centre for Real Estate Studies, Universiti Teknologi Malaysia, Skudai, Malaysia

B. Awwad (ed.), *The AI Revolution: Driving Business Innovation and Research*, Studies in Systems, Decision and Control 524, https://doi.org/10.1007/978-3-031-54379-1_67

Keywords Mosque · Mosque tourism · Facilities management · Tourist attraction · Customer loyalty · Malaysia

1 Introduction

1.1 Tourism Sector in Malaysia

Nowadays, tourism is one of the world's fastest expanding service industries. As a result, the Malaysian government has designated tourism as a crucial sector for driving Malaysia's long-term economic growth [1–3]. According to World Bank [4], tourism-based development has a critical role in contributing to Gross Domestic Product (GDP) growth in both developed and developing countries.

Tourism contributes to socioeconomic development in many ways include creating jobs, decreasing unemployment rates and developing innovative businesses [5, 6]. Therefore, tourism industries in Malaysia are becoming increasingly important in order to boost the Malaysian economy while also increasing popularity and awareness through publicity.

Malaysia, which has a majority Muslim population, is one of the countries that dominate Muslim tourism. For that reason, Malaysia gained in terms of domestic and international Muslim tourist flows because it became a preferred destination that easily fits the Muslims' special needs in customs and religious practice.

Because of the prominence of Islam in Malaysia, the country is aggressively seeking to become a popular destination for Muslims all over the world. In 2023, thus, Malaysia has been awarded the 'Best Muslim-Friendly Destination' by the Organisation of Islamic Cooperation (OIC) [7]. This accomplishment has strengthened Malaysia's position and branding as a regional and international leader in Muslim-Friendly Tourism and Hospitality (MFTH).

1.2 Mosque as a Tourist Attraction Place

In particular, Islamic tourism presents a novel strategy for expanding Malaysia's tourism sector while upholding Islam's morals, culture, and values. According to some scholars, mosque tourism is a significant aspect of rising tourism as well as an important component of Islamic tourism [8].

Thus, the mosque is one of the attractions for tourists visiting the Muslim country [9]. Recent years have seen an increase in Muslim and non-Muslim tourists visiting mosques. Mosque tourism is booming as many mosques feature unique architectural and spectacular designs, as well as tradition and culture that attract tourists.

However, Islamic tourism is not just described as a visit to the mosque; it also includes visit that are directly tied to nature, culture, or other forms of creativity that incorporate Islamic principles. This made mosque tourism a vital and strategic tool for sustaining Islamic tourism [8].

On the other hand, mosque tourism also showed a potential source of income and a cutting-edge travel trend. For instance, mosque tourism could raise demand for goods and services (such tour guides, food and drink, hotel, transportation, entertainment, and shopping) and lead to the creation of jobs. In this regard, the policymakers and stakeholders should emphasize the mosque tourism market [10].

Refer to study conducted by Ganesan et al. [11], the settings and specialties of various locations do influence Muslims' decision to visit a mosque. They also claimed that the sustainability of the mosque depends on the quality of the services provided. Low quality of service provided by the mosque management led to the mosque being unsustainable and receiving less visits from people.

2 Empowerment in Mosque Tourism Management and Development

2.1 Enhancing Tourist Experiences

According to Budeanu [12] and Moscardo and Murphy [13], tourism is a product that attempts to meet tourist demands,thus, it has been researched from the perspectives of providers and customers (tourists). Many countries hope to boost their economy by improving tourist experiences in order to attract more visitors and repeat visits [14].

Therefore, enhancing visitors' religious experiences is receiving more attention than before. Based on Chen et al. [15] and Al-Ghamdi et al. [16], enriching the visitor experience is associated with increased satisfaction and intent to return. In this regard, enriching the tourist experience will benefit all stakeholders.

According to Mahazan, a mosque is highly depending on the human factor to prosper the mosque. It can be said that a mosque that meet the needs and requirement of a Muslim is able to encourage or attract them to visit the mosque. Conceptualisation of human needs has been proposed by Maslow's theory. Based on the theory, people who grew in an environment in which their needs are met, they will likely to function as healthy individuals. It shows that motivational factors play an important role in influencing people behaviours or attitudes.

Najafi and Mustafa [17] stated that a place is a space that constructed by the physical features, activities, and meanings. They claimed that user's experience in a place has a significant impact on their attachment to the place. Based on the understanding, place attachment is defined as the emotional and affecting bond established between people and places.

Facilities have a great impact on organisation performance. Thus, inadequate facilities translate to poor performance in which affected the whole operations of an organisation. This is supported by Khalit [18] who found that inadequate facilities become the constraint to the mosque to effectively function.

According to the literature on the theory of commitment or loyalty, the expected quality [19], perceived quality, satisfaction [20], and motivation [21] are the most crucial elements in the relationship between a Muslim tourist and a place or business.

One of the key elements in determining whether a visitor will return to a location and, consequently, their loyalty, is the comfort they feel while there, especially at the monument they have just visited [22, 23]. Meanwhile, perceived quality is frequently regarded as one of the key determinants of loyalty.

According to Lichtenstein et al. [24], perceived quality is the sum total of all of a visitor's feelings regarding their time spent at a particular place. Regarding its connection to customer satisfaction, perceived quality is significant because of how it affects customer happiness [25].

Based on the characteristics of the monument or site, tourists evaluate their overall experience as favourable or bad, positive or negative. Therefore, perceived quality and customer satisfaction are positively correlated, and visitors are more likely to return or recommend a place if they give it a favourable or positive review.

Tourists' overall evaluation of the service they received in comparison to their expectations might be described as their level of satisfaction. As a result, improving the visitor experience while at a mosque is essential for empowering mosque tourism.

2.2 Potential of FM Approach to Promote Sustainable Mosque Tourism

According to a study conducted by Pirali Khairabadi et al. [26], there are various variables that encourage Muslim and non-Muslim tourists to visit mosques in Malaysia. The findings reveal that tourists are motivated by religious and spiritual motivations, as well as a desire to satisfy their curiosity and snap photographs. They went to mosques to learn about the culture, legacy, and history of the buildings, as well as its iconic architectural and creative design. Because of their unique designs and gorgeous architecture, mosques can be a destination for both Muslim and non-Muslim tourists.

However, the study reveal that tourist were not satisfied with food and beverages, public toilets, and road signage [26]. In this context, the customer's satisfaction and loyalty were reflected in the Jemaah's sense of satisfaction and loyalty among those who visited the mosque. The situation has inspired the researcher to study the idea of developing FM in order to achieve mosque sustainability.

Mosques serve many important purposes for Muslims, hence it is important for the management of a mosque to maintain the building and provide the basic facilities that are needed by the people. Yet, the Malaysian mosque management and administration are still viewed as being inefficient, which results in poor management or inefficiency in the tourism industry [27].

Refer to Navajas et al. [28], the correct management of the heritage site can influence the feelings of the Islamic tourist, which increases their loyalty to the site and affects the recommendations they make. This indicates that the tourism industry should consider these variables in order to promote tourism loyalty. Policymakers should explore new management methods to give this type of tourist a memorable experience when they visit heritage sites [28].

Therefore, Facilities Management (FM) is a viable alternative for mosque management to consider, and it is considered to be a strategy that will aid in the continued operation of the mosque [29]. FM is a fast-expanding field nowadays. The industry has several incentives to use this notion in each of their projects in order to ensure that the fundamental business goal is met. FM, without a doubt, adds a new dimension to the market by improving organisational performance and customer service. FM is projected to become a significant aspect of organisational decision-making as a result of the need to increase client loyalty in using their services.

3 Materials and Methods

Data scanning based on primary keywords was deemed insufficient because the study's goal is to look into the efficacy and applicability of the FM approach to mosque tourism sustainability. The study's purpose is to highlight that implementing a management platform based on FM principles can give a better foundation for facilitating the process of mosque tourism sustainability. Fink [30], Piper [31], and Paul et al. [32] argue that the key to a successful independent literature review is a well-developed methodology and structure. The approach used should be simple to comprehend, thorough in its treatment of the subject at hand, and reproducible by other researchers while permitting the identification of the constraints, significance, and viewpoint of the existing literature.

As a result, scientometrics and content analysis were used to evaluate the literature in-depth and analyse the data acquired for the study. Okoli and Schabram [33] identified eight steps for conducting a systematic literature review, which are followed in this study (refer Fig. 1):

4 Results and Discussion

Customer loyalty is a crucial factor to take into account as it has a direct impact on how profitable products and services are. It has elements that affect whether Muslim visitors intend to visit a mosque and spread the word about it.

Since FM is an approach that focus on integrating between people, place and process, thus its involvement in the whole organisation of mosque is makes sense. The position of FM at the centre implies will enhance cooperation among the key factors in mosque. Thus, an effective FM that involved in strategic, tactical and

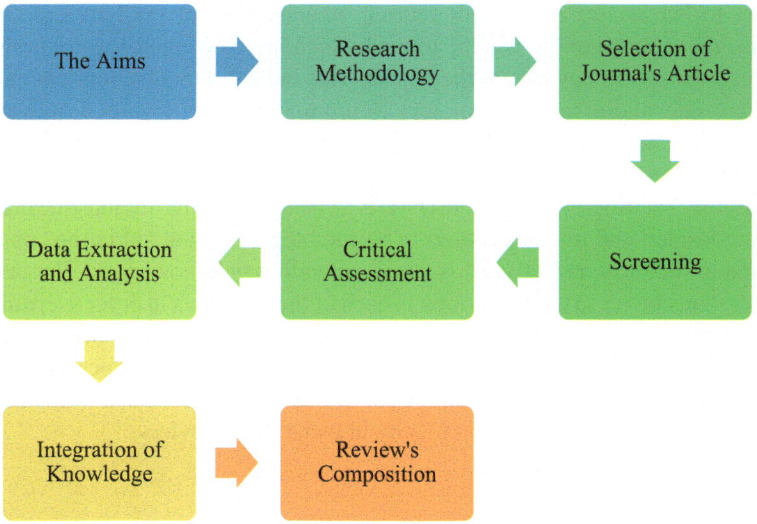

Fig. 1 Eight (8) steps for conducting systematic literature review. *Source* Okoli and Schabram [33]

operational level is required as enabler that helps to align the mosque objectives and missions, so they can be achieved successfully. Hence, it is hope that a mosque with good facilities will enhance the congregation to attend the mosque frequently. Moreover, it is hope that it enables to create a pleasant environment in the mosque in which could assist the congregation perform their worship and activities more perfectly.

FM is a core management function that necessitates the active participation of the facilities manager in the mission, business, and asset management. To support the vital role of people in the business, FM concentrates resources on satisfying user demands and tries to continuously enhance quality, decrease risk, and assure value for money in service delivery. FM, according to Amaratunga and Baldry [34], is a method that is essential in delivering and maintaining an operational environment to suit an organization's strategic needs. As a result, integrating FM in an organisation is beneficial, as FM must be evaluated at four levels: corporate, strategic, tactical, and operational.

Many researchers have lately formed the notion of facilities management, focusing on areas such as healthcare, offices, infrastructure, airports, and many more. However, there has not been scientific research conducted and gives attention to FM for mosque. Malaysia as a country with almost all of the community is Muslim, thus it indicates a need to conduct research related to FM within the mosque environment in order to sustain its function as the place to the community assemble and perform their worship.

According to Mudrak et al. [35] in order to stay in business, FM organisations should be able to provide services that exceeds the expectations of customers,this

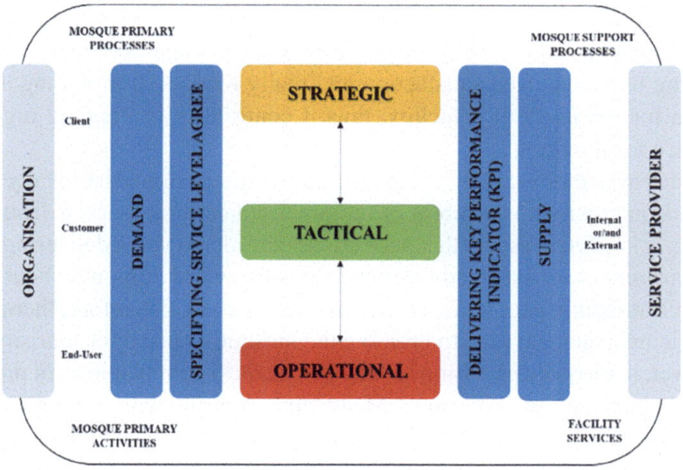

Fig. 2 FM Model (Adaptation from BSI 2007)

nature of services could be achieved by adopting new developments while continuously innovating the FM process. In this way, the core business of the host organisation could benefit from the FM value adding. Then and Tan [36] propose that FM by nature is a dynamic business activity and must be responsive to changing clients' needs in order to fulfil its role as a support service function. They added that this variation is influenced by the nature of the business, the competence of the in-house management and FM services provider(s). The capability of FM to lead the organisation change is illustrated in Fig. 2.

The model emphasises the connection between demand and supply of FM services while illustrating the importance of FM as support services to businesses' principal activities. The model also depicts the many levels of FM interactions, stating that organisations must rely on their primary processes while FM serves as the primary process's support system to help the organisation achieve its central goals. Additionally, this model illustrates how crucial it is to implement FM at all three levels of management: operational, tactical, and strategic.

5 Conclusion

The mosque is the most important institution in Islam. It is a place where Muslims go on a daily basis. Some people visit the mosque once, twice, three times, and five times in a 24-hour period. Unquestionably, the mosque is a place where Muslims can have an experience that will help them enhance their spirituality. As a result, a Muslim who frequently attends the mosque is naturally brought closer to the Creator, Allah S.W.T. This has resulted in the requirement for mosques to operate as a venue for community development over the years.

FM defined as an integrated approach that maintains, improves and adapts the built environment to support the primary business of the organisation. It is about able in managing user perceived satisfaction and really focuses in providing a pleasant experience for the users of a facility, thus it contributes to improve organization performance for long run.

This study was motivated by the growing importance of tourist loyalty to places visited. Mosque tourism is a subset of religious and cultural tourism. Thus, quality of facilities is important aspect that needs to be considered when delivering services. Satisfied customers are expected to spread a positive word of mouth about the institutions, thus attracting people with lower marketing costs. Therefore, there is a need for strategic management plan to promote the mosque as place for tourism.

Moreover, inadequate provision of facilities leads to poor planning of programme for various target groups in mosque. Meanwhile, mosque with low quality service will probably contribute to absenteeism of the congregation in visiting mosque. Thus, the mosque management should aim to enhance satisfaction of the tourist and reduce their dissatisfaction to ensure the sustainability of mosque functions that has the potential as tourist attraction place in Malaysia.

Acknowledgements The funding for this article is made possible through UTM Encouragement Research (UTMER) grant obtained from Universiti Teknologi Malaysia [Q.J130000.3852.20J03].

References

1. Umar, M.H., Ashaari, M.F.: The implementation and management of tourism mosque in Malaysia. *Journal of Islam and Contemporary Society*. ISSN 1985-7667, e-ISSN: 2289-6325 (2019). http://journal.unisza.edu.my/jimk
2. Habibi, F.: The Determinants of Inobound Tourism to Malaysia: A Panel Data Analysis. Curr. Issues Tourism **20**(9), (2017)
3. Wan Suzita, W.I., Habibah Ahmad, H.A., Hamzah Jusoh, H.J., Mohd Fuad, M.J., Mushrifah Idris, M.I.: Tourists' movement and impacts on regional tourism destination (2016)
4. World Bank.: Tourism and Competitiveness (2022). https://www.worldbank.org/en/topic/competitiveness/brief/tourism-and-competitiveness LNCS Homepage, http://www.springer.com/lncs, Last accessed 21 Nov 2016
5. Nor, N.A.M., Salleh, N.H.M., Faroby Falatehan, A.: The effect of tourism expenditure on the economy: a new evidence. J. Ekon. Malays. **55**(3), 23–34 (2021). https://doi.org/10.17576/JEM-2021-5503-02
6. Shaliza, A., Syafiqah, W.N., Sharida, N.: The impact of tourism industry on local economy: case of Malaysia. J. Islamic, Soc. Econ. Dev. (JISED), **8**(53), 361–374 (2023)
7. Global Muslim Travel Index Report.: 8th edition of the Mastercard-Crescentrating Global Muslim Travel Index (GMTI) 2023, June 2023. Retrieved on 24 Sept 2023 (2023). https://www.crescentrating.com/reports/global-muslim-travel-index-2023.html
8. Kessler, K.: Conceptualizing mosque tourism: a central feature of Islamic and religious tourism. Int. J. Religious Tourism Pilgrimage **3**(2), 11–32 (2015)
9. Utaberta, N., Leong, Y.C., Tahir, M.M.: Evaluating the potentials of mosque as a tourist attraction place in Malaysian urban context. J. Design Built **14**(1), 25–35 (2021). ISSN 1985-6881

10. Moghavemi, S., Sulaiman, A., Taha, A.Z., Abd Shukor, M.E., Mustapha, M.Z., Mohtar, M., Jaafar, N.I., Zakaria, S.B.: Factors contributing to mosque tourism in 6(23) (June 2021), 71–78. 10/35631/JTHEM.623008 Copyright © GLOBAL ACADEMIC EXCELLENCE (M) SDN BHD—All rights reserved 78 Malaysia. A research paper funded by Konsortium Institut Halal IPT Malaysia, University of Malaya MOHE grant MO001–2018 (2020)
11. Ganesan, Y., Pitchay, A.A., Mydin, A.M.: Does service quality matters to ensure the loyalty of the Jemaah of a mosque? Int. J. Islamic Mark. Branding **5**(2), 77–98 (2020)
12. Budeanu, A.: Sustainable tourist behaviour ? A discussion of opportunities for change. Int. J. Consum. Stud. **31**(5), 499–508 (2007). https://doi.org/10.1111/j.1470-6431.2007.00606.x
13. Moscardo, G., Murphy, L.: There is no such thing as sustainable tourism: Re-conceptualizing tourism as a tool for sustainability. Sustainability (Switzerland) **6**(5), 2538–2561 (2014). https://doi.org/10.3390/su6052538
14. Amer, M.S., Karban, A.S., Majrashi, A.A., Majid, M.R., Saleh, A.: Identifying the influential factors enriching the experiences of visitors of touristic religious sites: the case study of Al-Khandaq battle site. Int. J. Built Environ. Sustain. IJBES **8**(2), 1–9 (2021)
15. Chen, X., Cheng, Z., Kim, G.-B.: Make it memorable: Tourism experience, fun, recommendation and revisit intentions of Chinese outbound tourists. Sustainability **12**(5), 1904 (2020)
16. Al-Ghamdi, K.A., Karban, A.S., Said, I., Rasidi, M.H., Al-Ghamdi, S.A.: Enrich pilgrims experience by improving the wayfinding system in (Meeqat Zul Halifa) Abiar Ali. In: The 20th Scientific Forum for the Hajj, Umrah, and Visit Research. pp. 427–436. Umm Al-Qura University, Makkah (2021)
17. Najafi, M., Mustafa Kamal, M.S.: Place attachment to contenporary public-funded mosques in Malaysia. Doctoral dissertation, Universiti Putra Malaysia (2012)
18. Khalit, N.A.: Revitalizing roles and Functions of mosques in Kuala Lumpur. Doctoral dissertation, Universiti Teknologi MARA (2011)
19. Mohseni, S., et al.: Attracting tourists to travel companies' websites: the structural relationship between website brand, personal value, shopping experience, perceived risk and purchase intention. Curr. Issue Tour. **21**(6), 616–645 (2018)
20. Kim, M.J., Hall, C.M.: A hedonic motivation model in virtual reality tourism: Comparing visitors and non-visitors. Int. J. Inf. Manage. **46**, 236–249 (2019)
21. Buhalis, D., Lo´pez E.P., Martinez-Gonzalez J.A.: Influence of young consumers' external and internal variables on their e-loyalty to tourism sites. J. Destination Mark. Manage. **15**, 100409 (2020)
22. Loureiro, S.M.C., González, F.J.M.: The importance of quality, satisfaction, trust, and image in relation to rural tourist loyalty. J. Travel Tourism Mark. **25**(2), 117–136 (2008)
23. Kozak, M., Rimmington, M.: Tourist satisfaction with Mallorca, Spain, as an off-season holiday destination. J. Travel Res. **38**(3), 260–269 (2000)
24. Lichtenstein, D.R., Netemeyer, R.G., Burton, S.: Distinguishing coupon proneness from value consciousness: An acquisition-transaction utility theory perspective. J. Mark. **54**(3), 54–67 (1990)
25. Gamedze-Mshayisa, D.I., Kuo, S.C., Liu, C.Y., & Lu, Y.Y.: Factors associated with women's perception of and satisfaction with quality of intrapartum care practices in Swaziland. Midwifery **57**, 32–38 (2018)
26. Pirali Khairabadi, E., Sedigheh Mousavi, S., Momtaz, H., Nikokhah, F., Hosseini Shekarabi, S.P., Raissy, M.: Prevalence and phylogenetic analysis of Listeria monocytogenes isolated from the fillets of two farmed fish in Shahrekord in 2018. J. Food Microbio. **7**(4), 81–93 (2020)
27. Sabri, M.F.M., Tonot, H.: Comparison of mosque quality management against mosque given star recognition. J. Bus. Innovation **2**(1), 25–41 (2017)
28. Navajas-Romero, V., Herna´ndez-Rojas, R.D., Hidalgo-Ferna´ndez, A., Jimber del Rio, J.A.: Tourist loyalty and mosque tourism: The case of the Mosque-Cathedral in Co´rdoba (Spain). PLoS ONE **15**(12), e0242866 (2020). https://doi.org/10.1371/journal.pone.0242866
29. Muin, Z.A., Sapri, M., Sipan, I., Hashim, H.A.: An investigation of facilities management implementation process for Malaysian state mosque. Int. J. Real Estate Stud. **15**(2), 103–113 (2021)

30. Fink, A.: Conducting research literature review: From the internet to paper (2nd ed.), Thousand Oaks. Sage Publication Ltd., London (2005)
31. Piper, R.J.: How to write a systematic literature review: A guide for medical students. National AMR, Fostering Medical Res. **1**, 1–8 (2013). Retrieved from http://cures.cardiff.ac.uk/files/2014/10/NSAMR-SystematicReview.pdf%0Acures.cardiff.ac.uk/files/2014/%E2%80%89%E2%80%A6%E2%80%89/NSAMR-Systematic-Review.pdf
32. Paul, J., Lim, W.M., O'Cass, A., Hao, A.W., Bresciani, S.: Scientific procedures and rationales for systematic literature reviews (SPAR-4-SLR). Int. J. Consum. Stud. (2021). Retrieved from https://doi.org/10.1111/ijcs.12695
33. Okoli, C., Schabram, K.: A guide to conducting a systematic literature review of information system research. Rev. Inf. Syst. 10–26 (2010). Retrieved from https://doi.org/10.2139/ssrn.1954824
34. Amaratunga, D., Baldry, D.: A conceptual framework to measure facilities management performance. Prop. Manage. **21**(2), 171–189 (2003)
35. Mudrak, T., van Wagenberg, A., Wubben, E.: FM as a scientific field of inquiry: Theory, research and practice. Facilities. **23**(3/4) (2005)
36. Then, D.S.S., Tan, T.H.: Aligning facilities management performance to business needs—an exploratory model linking FM performance to business performance. In: Haugen, T.I., Mourn, A., Brochner, J. (eds), In: Proceedings of trondheim CIBW70 international symposium—Changing users demands in buildings, pp 340–349. Trondheim, Norway, 12–14 June (2006)

Exploring the Innovation-Based Capabilities on Innovation Performance: A Case Study of Small and Medium Tourism Enterprises (TSMEs) in Malaysia

Fadhilahanim Aryani Abdullah, Nurhaiza Nordin, Nurnaddia Nordin, Dzulkifli Mukhtar, Nik Maheran Nik Muhammad, and Intan Nurliana Abdullah

Abstract The continuous expansion and competitiveness of Malaysia's Small and Medium Tourism Enterprises (TSMEs) are greatly influenced by innovation. We explore the complex dynamics of how TSMEs create and use innovation-based capabilities to improve their innovation performance. We identify the crucial elements that shed light on this relationship using survey. Purposive sampling was used in this study's methodology, with 213 TSME chosen from the Ministry of Tourism and Culture (MOTAC) list as samples. The study's findings provide insight into how TSME view innovation-based capabilities and their top priorities. Notably, it was found that they correctly recognize the importance of factors associated to innovation capabilities and performance. These components are seen to be crucial for encouraging an innovative culture in Malaysian SMEs. The study intends to offer insightful information for TSMEs, decision-makers, and business professionals who are trying to harness innovation for long-term growth in Malaysia's vibrant business environment.

Keywords Tourism SME · Innovation capability · Innovation performance · Business sustainability

---

F. A. Abdullah (✉)
Faculty of Hospitality, Tourism and Wellness, Universiti Malaysia Kelantan, Kota Bharu, Malaysia
e-mail: aryani.a@umk.edu.my

N. Nordin · N. Nordin · D. Mukhtar · N. M. N. Muhammad
Faculty of Entrepreneurship and Business, Universiti Malaysia Kelantan, Kota Bharu, Malaysia

I. N. Abdullah
Malaysian Graduate School of Entrepreneurship and Business, Pengkalan Chepa, Malaysia

B. Awwad (ed.), *The AI Revolution: Driving Business Innovation and Research*, Studies in Systems, Decision and Control 524, https://doi.org/10.1007/978-3-031-54379-1_68

1 Introduction

According to the SME Annual Report for 2014–15, Malaysia's SMEs contribute between 30 and 53% of the country's GDP and 19% to 31% of exports. Along with creating jobs and income, SMEs play a significant role in promoting entrepreneurship. As a result, SMEs play a vital role in the economic growth and development of the member states [1]. In line with the strategic directions of the Sustainable Development Goals (SDG) of the United Nations, the National Tourism Policy enhances the value of already-existing tourist experiences by creating new products, improving enablers, and removing barriers [2] that need the participation of TSME. This promotes innovation and competitiveness. Research by [13] stated that empowering small businesses faces several problems and according to National Tourism Policy 2020, there are signs that the tourism sector in Malaysia is stuck in its comfort zone. Existing tourism offerings are growing "tired" and unappealing due to a silo mentality, an over-reliance on the government, and a lack of originality and innovation. In comparison to our ASEAN neighbors, our incapacity to cultivate a service culture has led to a real or perceived decline in service quality [14]. Figure 1 shows the factors that influence the Malaysia's tourism industry is stuck in comfort zone. One of the factors is lack of innovation.

Stated differently, the tourism industry's low official rates of technological innovation can be attributed to the numerous "hidden" innovations that take place within it. The question is whether the tourism industry faces sector-specific barriers to innovation or if measurement techniques based on scoreboards designed for the manufacturing or general services industries can skew a less creative approach by tourism companies, undervaluing the true innovation that takes place in this sector. 84% of businesses believe that their strategy depends on innovation to propel their businesses forward in the long run [1]. Theoretical and practical observations indicate that business model innovation (BMI) proves to overcome and recover the business during the

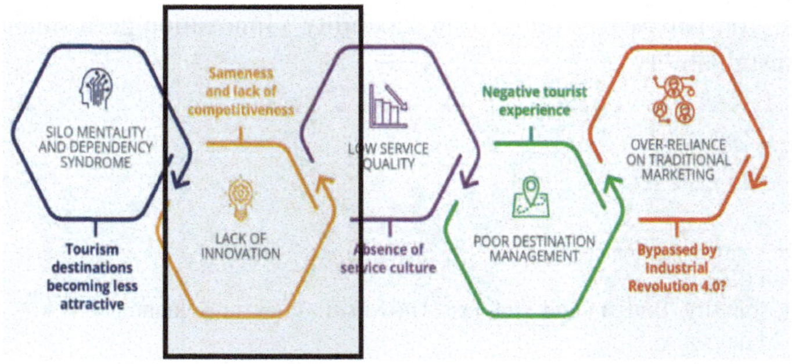

Fig. 1 The factors that influence the Malaysia's tourism industry is stuck in comfort zone (National tourism policy 2020)

pandemic whereby they have successfully transformed their business models [8]. But there is a disconnect between the company's goals and capabilities, which leads to the failure of innovation projects or challenges maintaining their effectiveness.

When studying and measuring innovation performance in tourism, it is important to define specific methodological issues with and limitations of public sources and models to distinguish between actual and measured innovation more clearly. It goes without saying that organisations must evaluate the effectiveness of their innovation activities. Establishing metrics to assess national activities and guide government policymaking, as well as business activities to direct management action and stakeholders directly involved in the company's future, has taken a great deal of time and work. On the right metrics to use, however, there doesn't appear to be much agreement.

Managers have not given much thought to the relationship between innovation performance and innovation competency and capability, and there is a dearth of thorough justification regarding the degree of association between the constructs examined. There isn't much agreement on how this measurement could guide actions to enhance performance because of these factors. However, very little research has been done to link and empirically assess entrepreneurial skills, innovation, and the sustainable performance of TSMEs. Thus, this study sets out an objective to close this important gap by thoroughly evaluating the function of innovation within TSMEs, considering their restrictions and unique characteristics. The lack of thorough studies on the adoption of innovation within organisations has been correctly noted by educational institutions [12, 15]. This gap needs to be filled right away because it is particularly obvious in the TSME sector.

2 Literature Review

The ability of firms to compete and produce value depends on innovation, which is typically hailed as a complex and multifaceted phenomenon. To develop new or improved products, services, and manufacturing techniques, innovation fundamentally entails the development, application, and sharing of information [7]. This all-encompassing perspective recognises that innovation is about more than just creating new products, which is crucial for Small and Medium Enterprises (SMEs) in all industries. SMEs have proven to be tremendous engines for innovation despite having little resources, debunking the popular misconception that innovation exclusively happens in larger businesses [3]. Schumpeter first proposed the idea that SMEs might serve as sources of innovation in 1934, and it has subsequently received empirical support. Because of this, SMEs have become a significant force in the innovation scene [2]. According to some scholars, a lack of resources may increase the effectiveness of the innovation process and motivate SMEs to adopt innovative practises. Both academic and commercial communities have recently shown a greater interest in researching the symbiotic relationship between innovation and the success of SMEs [16]. It is generally agreed that innovation serves as a potent catalyst, increasing the competitiveness of SMEs. Because of their capacity for technical innovation, SMEs are ultimately well-positioned to be adaptable and flexible organisations in dynamic. The idea that "innovation" is the "engine that drives revenue growth" has been correctly defined and a key principle of company survival [10].

In essence, effective management of the innovation process is dependent on a grasp of the intricate aspects that promote continual business progress. Innovation's process, fostering, and management are crucial for an organisation to survive and prosper [12]. Numerous scholars have used various measurements to detect innovation adoption and performance in the firm. Effective results from these measurements include innovation, competitive advantage, the firm's organizational culture, strategy, structure, systems, behaviours, and processes, as well as better financial performance and increased organizational learning. Priority metrics for measurement included firm-specific components, competitive advantage in a market economy, and capability.

Scholarly discourse surrounding this phenomenon has primarily focused on the notion of innovation capability. The capacity for innovation can be defined as the ability to produce new and valuable goods or knowledge. "The ability to continuously transform knowledge and ideas into new products, processes, and systems for the benefit of the firm and its stakeholders" is how they define innovation capability. Thus, the ability to innovate is essential for companies looking to outperform their rivals. This insight has caused a rising emphasis on creating and managing innovation capabilities that span a larger spectrum of inventive aspects than only technical innovation. A crucial subject in the literature on modern innovation management is the definition of innovation capability (IC) at the company level. Many attempts have been made to operationalize and conceptualize this capability. However, they either do not differentiate the components that define IC or do not provide elements that are consistent and similar across different models or frameworks.

Studies have demonstrated the importance of institutional support, which is a flexible and organic structure, in fostering innovation. Power dynamics and decision-making agility are also key components of this support. Prior studies discovered a positive correlation between IC and intraorganizational determinants (size, formal organizational structures), and they concluded that the existence of innovation-focused personnel was important. A culture that is conducive to innovation is a result of shared norms, beliefs, and values (mission and strategy) that are equally held by members of groups and the organization. This is known as the "culture antecedent." A firm's resource endowments and the way these resources are used through a clear strategic orientation can be positioned as influencing innovation performance. Innovation in service industries, such as the tourism industry, is a multifaceted phenomenon that demands expertise in various fields that leads to growth in all branches of society [4]. The basis for talent management, retention, and development is provided by competency models. Competency models are used to identify, define, and select the behaviors and abilities necessary for success in a specific field or discipline. The ability to innovate has, therefore, become a required and significant

Table 1 Hypothesis postulated

H1:	There is a significant relationship between innovation capability and innovation performance in Malaysia tourism industry
H1a:	There is a significant relationship between institutional support and innovation capability towards innovation performance in Malaysia tourism industry
H1b:	There is a significant relationship between Intraorganization determinant and innovation capability towards innovation performance in Malaysia tourism industry
H1c:	There is a significant relationship between Mission and Strategy and innovation capability towards innovation performance in Malaysia tourism industry
H2:	There is a significant relationship between innovation competency and innovation performance in Malaysia tourism industry
H2a:	There is a significant relationship between managerial abilities and innovation competency towards innovation performance in Malaysia tourism industry
H2b:	There is a significant relationship between skill in product or process innovation and innovation competency towards innovation performance in Malaysia tourism industry
H2c:	There is a significant relationship between Knowledge and innovation competency towards innovation performance in Malaysia tourism industry

competency for success [5]. The role of managerial skills, or aspects of strategic orientation, in utilizing company resources to boost performance has also been studied in the past. These include performance in the areas of market, learning, technology, and entrepreneurship, as well as innovation. The ability to perform and attain organizational goals in an effective and efficient manner is referred to as managerial ability. Therefore, we can postulate hypothesis from this research as according to Table 1:

The analysis of the literature reveals that a growing body of research is highlighting the importance of innovation-based competences for encouraging innovative performance within TSMEs, including those in Malaysia. Even though this relationship has been the focus of many studies, more context-specific research is necessary to fully comprehend the unique opportunities and challenges faced by Malaysian SMEs. Based on the previous study related to innovation and TSMEs, the framework and hypotheses for this study has been formed. Figure 2 shows a research model which has been applied for this study.

3 Methodology

To fill the research gap, this study proposes three constructs which are innovation capability, innovation competency and innovation performance and develops a research framework to further their relationships based on Resource Based View (RBV) and Diffusion of Innovation (DOI) theories. The model was tested using the data collected from 213 respondents who held managerial positions in the Malaysia SME tourism companies. The related literature was reviewed to generate a theoretical model of the factors above. This study used a correlational cross-sectional research

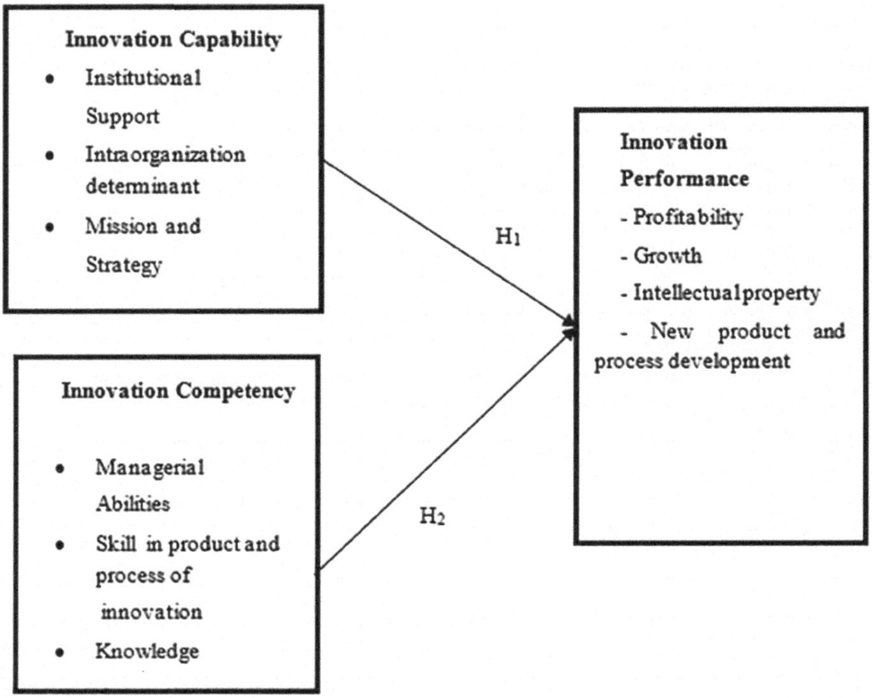

Fig. 2 Research model

design, and the findings partially support the study's goal. A survey approach was used in this study, and quantitative research methodology was specifically used. Cross-sectional surveys were thought to be a suitable way to collect primary data, especially when working with a population that was too large to be directly served. The study used a purposive sample approach, in which the sample of organizations and responders was drawn from a variety of websites and targeted agencies. As a result, 213 businesses in all, representing a valid response rate of 58.8% among Malaysia's TSMEs, were chosen as the study's sample. The Ministry of Tourism and Culture (MOTAC) of Malaysia provided a detailed list from which these businesses were selected. Next, PLS-SEM was used to analyze the gathered data. Prior to testing the research hypotheses, the reflective and formative models of the measurement model for this study were evaluated. Convergent validity in the evaluation of the reflective measurement model met all suggested values. Convergent validity of the formative measurement model was evaluated by means of the significance of weight and loading for each item, as well as collinearity (VIF) [11]. Next, discriminant validity was proven for the HTMT and Fornell-Larcker criteria. Furthermore, CR with loadings greater than the cut-off value of 0.5 and above was used to determine reliability analysis. After establishing the measurement model's validity and reliability, the structural model was assessed to see if the study's hypothesised relationship held true.

4 Results and Discussion

Based on the results of the descriptive analysis, it is shown that the respondents agree that the innovation capability and innovation competency (institutional support, intraorganizational determinants, mission and strategy, managerial abilities, skill, knowledge) are moderately essential for them. The results show that among the six factors rated, the respondents preferred to intraorganizational determinants (mean = 3.895; std. deviation = 0.814) followed by institutional support (mean = 3.883; std. deviation = 0.814), mission and strategy (mean = 3.846; std. deviation = 0.814), knowledge (mean = 3.831; std. deviation = 0.789), skill (mean = 3.809; std. deviation = 0.815) and managerial activities (mean = 3.701; std. deviation = 0.822) based on their level of innovation capability and competency. Lastly for innovation performance, the descriptive analysis has a mean value of 3.692 with a standard deviation of 0.773. The result of descriptive analysis among variables of this study is good and at a satisfactory level. From these values, respondents slightly agreed that the innovation capability and competency are important for them to achieve successful innovation performance, and this can be seen in the current study in Table 2.

The results of the direct relationships between innovation performance, innovation competency, and innovation capability are shown in the section below. The relationships in this study are represented by the standard beta values, and the structural model relationship was tested using a significance level of $P < 0.1$ (greater than or equal to 1.282) [9]. There are two hundred and thirteen (213) usable questionnaires collected. The data of Malaysian Tourism Industry were analysed. From the test of goodness of data, it was found that the data was reasonably good and provided a useful basis for the analysis and inferences. Convergent validity was verified by means of

Table 2 The results of descriptive analysis

Construct (N = 213)	Minimum	Maximum	Mean	Std. Deviation
Institutional support (IS)	1.00	5.00	3.8826	0.79812
Intraorganizational determinants (ID)	1.00	5.00	3.8949	0.81432
Mission and strategy (MS)	1.00	5.00	3.8460	0.81380
Managerial abilities (MA)	1.00	5.00	3.7014	0.82238
Skill (SK)	1.00	5.00	3.8088	0.81500
Knowledge (KN)	1.70	5.00	3.8313	0.78956
Innovation performance (IP)	1.63	5.00	3.6925	0.77317

Note Scale 1 = Strongly Disagree, 5 = Strongly Agree

item loading, average variance extracted (AVE) value, and composite reliability after a range of analytic techniques were employed in this investigation. Concurrently, a test for discriminant validity was carried out utilizing the Fornel and Larcker, cross-loading, and HTMT criteria. The results verified the attainment of discriminant validity. Using SmartPLS 3, this study examined the research hypotheses of the direct and indirect relationships, further evaluating the measurement and the structural models. The results shown in Table 3 agree that all direct relationships are supported and significant.

Based on Table 3, the path coefficient on the relationship of innovation capabilities and innovation performance is significant at 0.000 with $\beta = 0.406$ and $t = 3.512$, institutional support and innovation performance is significant at 0.000 with $\beta = 0.367$ and $t = 33.114$, intraorganizational determinants and innovation performance

Table 3 Result of path coefficient (bootstrapping result)

	Std. beta	Sample mean	Standard deviation	T-Values	*P*-values	Relationship
Innovation capabilities → Innovation performance	00.406	00.411	00.116	30.512	00.000[a]	Significant
Institutional support → Innovation performance	00.367	0.366	00.011	33.114	00.000[a]	Significant
Intraorganizational determinants → Innovation performance	00.351	0.350	0.013	27.800	00.000[a]	Significant
Mission and strategy → Innovation performance	00.330	0.331	00.013	25.394	00.000[a]	Significant
Innovation competency → Innovation performance	00.261	00.244	00.143	1.824	00.034[a]	Significant
Managerial abilities → nnovation performance	00.240	00.240	00.012	20.608	00.000[a]	Significant
Skill → Innovation performance	00.392	00.391	00.009	43.157	00.000[a]	Significant
Knowledge → Innovation performance	00.406	00.407	00.013	31.951	00.000[a]	Significant

Note [a] $P < 0.01$, indicates the item is significant at 1% significant level. Nine (9) direct relationships were supported based on their t-values and [b] $P < 0.10$, indicates the item is significant at 10% significant level, [c] $P < 0.05$, indicates the item is significant at 5% significant level

Table 4 Summary of Hypotheses Testing

Hypotheses	Statement of Hypotheses	Decision
H1	There is a significant relationship between innovation capability and innovation performance in Malaysia tourism industry	Supported
H1a	There is a significant relationship between institutional support and innovation capability towards innovation performance in Malaysia tourism industry	Supported
H1b	There is a significant relationship between Intraorganizational determinant and innovation capability towards innovation performance in Malaysia tourism industry	Supported
H1c	There is a significant relationship between Mission and Strategy and innovation capability towards innovation performance in Malaysia tourism industry	Supported
H2	There is a significant relationship between innovation competency and innovation performance in Malaysia tourism industry	Supported
H2a	There is a significant relationship between managerial abilities and innovation competency towards innovation performance in Malaysia tourism industry	Supported
H2b	There is a significant relationship between skill in product or process innovation and innovation competency towards innovation performance in Malaysia tourism industry	Supported
H2c	There is a significant relationship between Knowledge and innovation competency towards innovation performance in Malaysia tourism industry	Supported

Note [a] $P < 0.10$ (indicates the item is significant at 10%), [b] $P < 0.5$ (indicates the item is significant at 5%), [c] $P < 0.10$ (Indicates the item is significant at 1%)

is significant at 0.000 with $\beta = 0.351$ and $t = 27.800$, mission and strategy and innovation performance is significant at 0.000 with $\beta = 0.330$ and $t = 25.394$, innovation competency and innovation performance is significant at 0.034 with $\beta = 0.261$ and $t = 1.824$, managerial abilities and innovation performance is significant at 0.000 ($\beta = -0.240$, and $t = 20.608$), skill and innovation performance is significant at 0.000 with $\beta = 0.392$ and $t = 43.157$, while knowledge and innovation performance is significant at 0.000 with $\beta = 0.406$ and $t = 31.951$. Therefore, Table 4 represents a summary of the hypotheses testing of the relationship between factors in this study.

5 Discussion and Conclusion

Innovation is an essential part of competitiveness and long-term viability for organizations of all sizes. Its base is the acquisition of new knowledge and research. This study extends the current understanding of the various determinant factors of innovation performance. *It provides a fresh conceptual framework that connects both motivating elements* (innovation competency and innovation capability) and innovation performance. By testing the framework, this study fills the literature gap by providing empirical evidence of relationships between the driving factors and performance. The practical implications and recommendations derived from this research underscore the importance of building and nurturing innovation capabilities for TSMEs, however it is only covering the firms related to tourism SME in Malaysia. Due to this research suggested several practical implications to the TSME in Malaysia. TSMEs in Malaysia should allocate resources strategically to build and enhance innovation-based capabilities. This may include investing in employee training, R&D, and technology infrastructure.

Strong leadership and management support for innovation are key components of the resources and guidance that leaders in TSME should provide to cultivate a culture of innovation. Establishing strategic alliances with government agencies, academic institutions, and other SMEs can facilitate access to capital, knowledge, and experience. Putting money into talent development or employee skills and training programs will encourage a more innovative workforce. Both internal training programs and outside educational opportunities can accomplish this. To support and promote innovation in SMEs, TSME also encourages members to look for grants and funding that are available for innovation projects and are offered by organizations and governments. Investigate new market niches both domestically and globally to increase the clientele, as market diversification can benefit TSME.

In conclusion, the research provides valuable insights into the determinants of innovation-based capabilities and their impact on innovation performance in TSMEs. Because of the unique and disruptive nature of the emerging innovation performance, it is necessary to maximize the relevance of concepts to the context, which makes contextualizing these terms crucial. For future research, we would suggest exploring how innovative capabilities can help businesses to adapt to market changes, technological advancements, and shifting customer preferences. SMEs with strong innovation capabilities are better equipped to withstand economic downturns and disruptions, as they can pivot and adapt more effectively. It is essential for TSME to understand the comprehensive significance of all elements within the innovation capabilities and innovation performance frameworks to truly drive innovation and ensure sustainable company prospects. Therefore, it is the responsibility of TSME to develop successful strategies that encourage firms to prioritize and allocate resources to each of these components. A complete set of policy recommendations should be considered to enable Small and Medium Enterprises (SMEs) to harness innovation as a catalyst for sustainable growth. Priority one should be given to implementing initiatives aimed at increasing awareness and providing training in innovation management.

References

1. Abdullah, F.A., Nik Muhammad, N.M., Redzuan, R., Hong, L.M.: Malaysia health Tourism SME's innovation performance in the relationship with innovation capability and competency.

In: Yusoff, M.N.H. (eds) Industry Forward and Technology Transformation in Business and Entrepreneurship. InCEBT 2022. Springer, Singapore (2023)
2. Abereijo, I., Ilori, M., Taiwo, J., Adegbite, S.: Innovation and firm performance: a study of the Nigerian manufacturing industry. Afr. J. Bus. Manage. **1**(6), 146–154 (2007)
3. Andries, P., Czarnitzki, D.: Small firm innovation performance and employee involvement. Small Bus. Econ. **39**(2), 419–434 (2012)
4. Barak, M., Usher, M.: The innovation profile of nanotechnology team projects among face-to-face and online learners. Comput. Educ. **137**, 1–11 (2019)
5. Barak, M., Watted, A., Haick, H.: Motivation to learn in massive open online courses: examining aspects of language and social engagement. Comput. Educ. **94**, 49–60 (2016)
6. Barak, M., Watted, A., Haick, H.: Establishing the validity and reliability of a modified tool for assessing innovative thinking of engineering students. Assess. Eval. High. Educ. (2019). https://doi.org/10.1080/02602938.2019.1620680.Barak,2017
7. Bigliardi, B.: Innovation trends in the food industry: The case of functional foods. Trends Sci. Technol. **31**(2), 118–129
8. Faisal, Y.A., Kanti, R.A., Alamanda, A.R., Apriliadi, A.: The role of business model innovation (BMI) in social enterprises during pandemic COVID-19 in Indonesia: a case of islamic boarding schools' business units. In: Rafiki, A., Dana, LP., Nasution, M.D.T.P. (eds) Open Innovation in Small Business. Contributions to Environmental Sciences & Innovative Business Technology. Springer, Singapore. https://doi.org/10.1007/978-981-99-5142-0_1 (2023)
9. Farjam, F., Shojaei, P., Askarifar, K.: A conceptual model for open innovation risk management based on the capabilities of SMEs: a multi-level fuzzy MADM approach. Technovation, Volume 127, ISSN 0166-4972, https://doi.org/10.1016/j.technovation.2023.102844 (2023)
10. Galvez, J.L., Camargo, M., Rodriguez, L., Morel, L.: The importance of innovation as a competitive factor in a developing economy: The case of Paraguay. World J. Entrepreneurship Manag. Sustain. Dev. **9**(1), 35–48 (2013)
11. Hair, J.F., Jr., Hult, G.T.M., Ringle, C., Sarstedt, M.: A Primer on Partial Least Squares Structural Equation Modeling. Sage Publications, PLS-SEM (2016)
12. Humphreys, P., McAdam, R., Leckey, J.: An exploration of the nature of operational innovation in the UK. Int. J. Oper. Prod. Manag. **25**(1), 41–63 (2005)
13. Kusumastuti, R., Virgianita, A., Juwono, V.: Co-creation strategy in empowering Indigenous Women's innovation in Indonesia: empirical evidence of weaver community entrepreneurs in West Nusa Tenggara. In: Rafiki, A., Dana, L.P., Nasution, M.D.T.P. (eds) Open Innovation in Small Business. Contributions to Environmental Sciences & Innovative Business Technology. Springer, Singapore (2023). https://doi.org/10.1007/978-981-99-5142
14. National tourism policy (2023). https://www.pmo.gov.my/2020/12/national-tourism-policy-2020-2030, Last accessed 21 Nov 2023
15. Pantano, E.: Innovation factors in retail industry: The case of SMEs. J. Innov. Knowl. **1**(2), 69–76 (2014)
16. Van Auken, H., Madrid-Guijarro, A., Garcia-Pérez-de-Lema, D.: Innovation in entrepreneurial firms: the role of intention and industry. Int. J. Entrepreneurship Innovation Manage **8**(1), 95–110 (2008)

Empowering Women's Participation in Malaysia's Tourism Industry: A Comprehensive Analysis

Nur Amiera Zaidi, Fadhilahanim Aryani Abdullah, Nur Dalila Mat Yusoff, Munirah Mahshar, Raja Rosnah Raja Daud, and Asanah Radhi

Abstract The tourism industry in Malaysia has experienced a significant transformation, with women taking on increasingly pivotal roles and catalysing growth across various economic sectors. Various initiatives and activities have been implemented to empower women within the travel and tourism sector. This study endeavours to explore the motivating factors that drive women's engagement in the field of tourism, aligning with Malaysia's broader strategy to encourage entrepreneurship among its citizens. A nationwide survey was conducted, drawing responses from participants across Malaysia, with 384 individuals successfully completing an online questionnaire. The data gathered were analysed using the Statistical Package for Social Science (SPSS) version 29.0, which provides a comprehensive platform for handling the entire statistical analysis process, including data preparation, analysis, and reporting. This study specifically delved into the relationship between leadership, well-being, digitization, and women's participation in the tourism industry using the Pearson correlation coefficient. The results underscore a statistically significant association between these factors and women's active involvement in the tourism sector. Through the findings of this study, we gain a deeper understanding of the intricate interplay between skills and training, leadership, well-being, and digitization in shaping women's participation and contribution to the dynamic landscape of the tourism industry in Malaysia. The insights generated from this research can guide policy makers, businesses, and stakeholders in the tourism sector to design targeted strategies and support mechanisms that further empower women, fostering sustainable growth and innovation in the industry.

Keywords Tourism industry · Women participation · Digitalization · Leadership

---

N. A. Zaidi (✉) · F. A. Abdullah · N. D. Mat Yusoff
Faculty of Hospitality, Tourism and Wellness, Universiti Malaysia Kelantan, Kota Bharu, Malaysia
e-mail: nuramierazaidi@gmail.com

M. Mahshar · R. R. Raja Daud
Faculty of Entrepreneurship and Business, Universiti Malaysia Kelantan, Kota Bharu, Malaysia

A. Radhi
Faculty of Bioengineering and Technology, Universiti Malaysia Kelantan, Kota Bharu, Malaysia

B. Awwad (ed.), *The AI Revolution: Driving Business Innovation and Research*, Studies in Systems, Decision and Control 524, https://doi.org/10.1007/978-3-031-54379-1_69

1 Introduction

The history of the study and the issues with the research topic are covered in this chapter. In response to the problems mentioned in the research problem, research objectives are created. The significance of the study is discussed in the chapter's final section. This study is done practically so that it can contribute to the improvement of women's participation and tourism in Malaysia. Women are regarded as cultural caretakers across the world. They are the chefs, weavers, farmers, and housewives, the keepers of traditions and stories. According to academics, women engaging in tourism can increase their expertise. For example, they can acquire a variety of languages based on the tourists that visit their locations, leading to a deeper understanding of themselves. According to research, a high-quality education can have an influence on entrepreneurial growth by increasing an individual's self-efficacy and self-confidence (GEM 2012).

When considering the industry, the involvement of women in tourism will strengthen the local economy. The current study found a significant relationship between gender and perceived business performance, indicating that gender positively impacts business success. A better work-life balance for society may be indicated by a higher gender balance in entrepreneurship (GEM 2013). Persuading intelligent women to consider initiating a business and overcoming obstacles that might be difficult for others could be beneficial for economic growth. Women who exhibit boldness and success as entrepreneurs might play a crucial role as mentors or role models in that regard [1].

When seen as a whole, this underlines the critical relationship between macroeconomic and social welfare policies for women to enjoy any real empowering effects from tourism engagement [2]. A schism often exists between women working in tourism, occasionally labelled "money ladies" because they earn money or "stand out," and other community members. This emphasizes the significance of taking 'intersectionality' into account when creating gender equality measures in tourism. This includes considering the various identities and privileges that women hold, encompassing factors such as geography, race, nationality, socioeconomic background, disability, and other identity markers, as well as the intersections of these identities.

There are four objectives of this research:

a. To examine the relationship between skills and training towards women participation in tourism.
b. To examine the relationship between leadership towards women participation in tourism.
c. To examine the relationship between well-being towards women participation in tourism.
d. To examine the relationship between digitalization and participation towards women participation in tourism.

1.1 *Significance of the Study*

This study focuses on the importance of women's participation in the tourism sector in Malaysia. Women are regarded as cultural caretakers and can increase their expertise by learning different languages. A high-quality education can have an influence on entrepreneurial growth by increasing self-efficacy and self-confidence. Women's engagement in tourism can also boost the local economy, and there is a strong association between gender and perceived business performance. To ensure gender equality in tourism, it is important to consider women's identities and privileges in terms of geography, race, nationality, socioeconomic background, disability, and other identity markers.

2 Literature Review

This chapter reviews research in which skill and training, leadership, well-being, and digitalization are independent variables, and female tourism engagement is a dependent variable. It also goes into hypothesis development and the research framework.

Women make about 70% of the over 200 million workers in the tourist sector, which employs over 200 million people overall [3]. Two of the major contributors to this segmentation are gender norms and preconceived notions about gender. For the growth of entrepreneurship and employment, the MSME sector is crucial in the Philippines and India [4]. Inspiring and guiding their staff, women leaders serve as role models [5].

The tourism industry's workforce in Malaysia, like many other nations, is undergoing a significant transformation. One of the key factors influencing women's participation in this industry is their skills and training. Research suggests that women often face barriers related to skill development and access to quality training. Empirical evidence demonstrates that investing in women's education and skill development is integral to enhancing their participation in tourism. The development of workplace skills is the goal of programs offered by businesses. This is particularly pertinent in segments like hospitality, where specific skills are vital. Skills and training programs that cater to the unique needs and aspirations of women can lead to increased employability and upward mobility within the sector [6].

WLT encourages and promotes the participation of women in tourism's highest-level corporate governance organizations [7]. Women should be appointed to key positions in corporate leadership, as only 37% of managers are women, 28% of board members are women, and fewer than 1/5% hold top executive positions [8]. Well-being encompasses both personal and subjective, while safety protects against unintended harm and intentional damage [9]. The well-being of women in the tourism industry is a multifaceted concern. The sector often entails long and irregular working hours, which can negatively impact the well-being of female employees. Studies have shown a strong connection between job satisfaction, work-life balance, and overall well-being. Therefore, addressing well-being-related issues is essential to retaining and motivating women in the tourism workforce. Initiatives aimed at improving working conditions, promoting mental and physical health, and enhancing work-life balance are crucial in nurturing women's long-term commitment to the sector [10].

Digitalization uses digital technologies to change a business model and generate new revenue and value [11]. These digital technologies improve access to markets, customers, manufacturing operations, technologies, marketing, and business information, knowledge, and resources. Women micro-entrepreneurs use mobile technology to facilitate client payments and banking [12]. Digitalization is increasingly shaping the tourism industry globally, with technology playing a pivotal role in all aspects, from marketing to customer experience. Women's participation in this digital transformation is crucial, as it presents both opportunities and challenges. On one hand, digital platforms offer flexibility that can benefit women who often juggle multiple roles. On the other hand, the digital divide remains a concern, with disparities in access to technology and digital skills. Bridging this divide is vital to ensure women can harness the full potential of digitalization and contribute effectively to the sector. Moreover, digital skills and innovation can lead to entrepreneurial opportunities, enabling women to create their businesses and assert greater control in the industry [13].

The United Nations World Tourism Organization (UNWTO) has chosen "Tourism Opening Doors for Women" as the topic for the 2007 Tourism Day activities. Women in the tourism industry contribute to the informal economy by setting up stalls, selling trinkets and food, and interacting with hawkers [14]. Family income is boosted in part by women, but their influence in society is still limited. Women comprise 46% of the industry's workforce, which is greater than the global average of 34–40% [15]. The percentage of women working in the industry varies widely by nation, from 2% to more than 80%. Despite the lack of clear regional trends, women make up around half of the workforce in nations with more developed tourism industries [16]. Leadership opportunities for women in the Malaysian tourism industry have traditionally been limited. Nevertheless, recent initiatives, such as the Women in Hospitality and Tourism (WiHT) program, indicate a shift towards encouraging women to assume leadership roles. The underrepresentation of women in leadership positions is a well-documented issue, and research underscores the importance of mentorship and leadership development programs in breaking these barriers. Leadership skills and mentorship programs can be vital in inspiring and preparing women to take on leadership roles, promoting gender diversity and inclusive decision-making processes in the tourism sector [17].

In conclusion, the factors of skills and training, leadership, well-being, and digitalization play a significant role in shaping women's participation in Malaysia's tourism industry. Empirical studies emphasize the importance of targeted initiatives and policies that address these factors, fostering an inclusive and diverse workforce that can contribute to the growth and innovation of the tourism sector. The combined efforts of government bodies, businesses, and industry stakeholders are essential to creating an environment where women can thrive and realize their potential within the tourism industry in Malaysia. The hypothesis in the study is to find out whether there any correlations or relationship between dependent variables and independent variables:

H1: Skills and training have a significant relationship with the participation of women in tourism.
H2: Leadership has a significant relationship with participation of women in tourism.
H3: Well- being has a significant relationship with participation of women in tourism.
H4: Digitalization has a significant relationship with the participation of women in tourism.

2.1 Research Framework

Figure 1 shows the research framework used for this study.

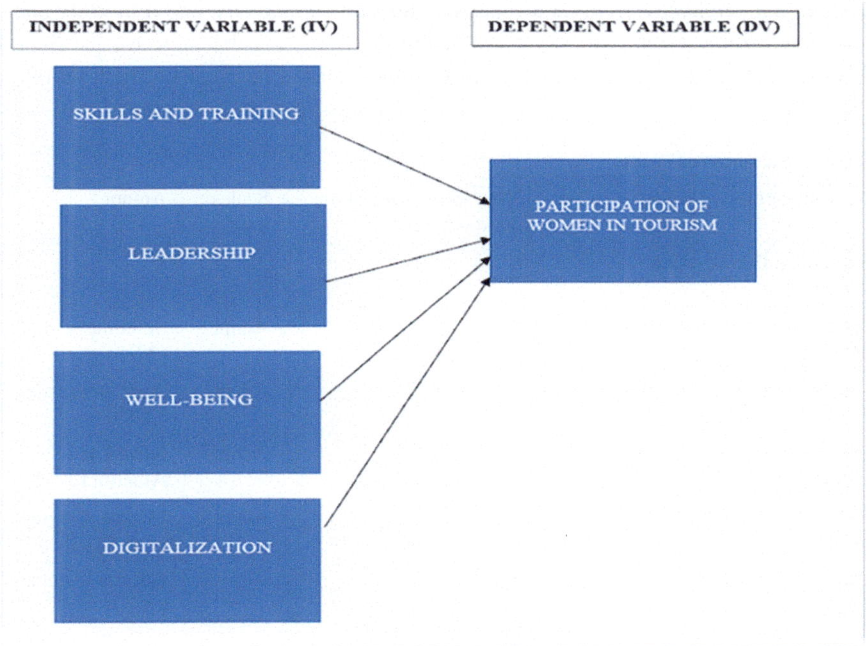

Fig. 1 Research framework

3 Methodology

3.1 Research Design

This study employs a descriptive research approach and uses a quantitative method, collecting data through a questionnaire sent to Malaysian citizens, requiring data, logical reasoning, and an unbiased perspective for the research. The focus of quantitative research is on precise, convergent thinking as opposed to divergent reasoning [18]. For this study, questionnaires will be used to collect data. It is adopted and adapted from [1, 3, 9, 19, 20]. Based on the population of residents in Malaysia, the population of residents in Malaysia exceeds 1 million. To support our population, our researchers put the justification to support our research. The amount of sample size for the population is 384.

This study used questionnaires distributed through an online survey, which typically includes both closed- and open-ended questions. The survey was widely disseminated. The study's purpose was clearly communicated through a covering letter distributed to respondents, enabling them to provide accurate information. After completion, questionnaires were collected and analyzed. This study employed three types of data analyses: frequency analysis, descriptive analysis, and reliability analysis. The acquired data were analyzed using the Statistical Package for the Social Sciences (SPSS) to demonstrate the links. Table 1 shows the measurement items.

Table 1 Measurement items

Measurements	Number of items in the questionnaires	Scales used	Author
Section A: Demographic Information	6	Adopt	
(Dependent Variables) Section B: Women Participation in Tourism	5	Adopt	Serafimoya et al. (2018)
(Independent Variables) Section C: Skills and Training	5	Adopt	Andrades et al. (2019)
(Independent Variables) Section D: Leadership	5	Adopt	Samad and Alharthi (2022)
Independent Variables) Section E: Well-being	5	Adopt	Moreno Gonzales et al. (2020)
Independent Variables) Section E: Digitalization	5		Khurramov et al. (2020)
Total	31		

4 Findings

Table 2 shows the result of reliability Coefficient Alpha for the independent variables and dependent variable. Cronbach's Alpha Coefficients was shown in Table 2. From the table, all variables were over 0.7 and overall variables were 0.979. Malaysian women's tourism participation was measured using five skills and training questions. Table 2 shows a very good Cronbach's Alpha of 0.856 for this section's question. Thus, the skills and training variable coefficients were accurate. Five questions measured leadership's impact on Malaysian women's tourism. This section's Cronbach's alpha coefficient, 0.932, is excellent. Leadership variable coefficients were reliable. Finally, five questions were utilized to measure the well-being variable that influenced Malaysian women's tourism. This section's Cronbach's Alpha was excellent at 0.942.

Thus, wellbeing variable question coefficients were reliable. Five questionnaires measured the digitalization variable that affected Malaysian women's tourism industry engagement. The query's Cronbach's alpha score was 0.929, which is excellent. Thus, the coefficients for digitalization variable inquiries were reliable. The Cronbach's Alpha for this section's question was 0.936, indicating outstanding. Thus, these questions' coefficients on Malaysian women's tourism involvement were credible. Since the variables Cronbach's Alpha charge exceeded 0.7, surveys are credible, and the study can continue. Overall, reliability showed that respondents understood the questions, approving the questionnaires for this study.

The finding of this study showed that to examine the relationship between skills and training, leadership, well-being, and digitalization among women participation in tourism. Table 3 shows Pearson's correlation result between hypotheses.

Table 3 confirmed that all hypotheses on skills and training, leadership, well-being, and digitalization among women in tourism were accepted through Pearson's correlation analysis at a 0.01 significance level. The results demonstrated a strong and positive correlation between the independent variables and women participation.

Table 2 Result of reliability coefficient alpha for the independent variables and dependent variable

Variable	Number of items	Cronbach's alpha	Strength of association
Skills and training	5	0.856	Very good
Leadership	5	0.932	Excellent
Well-Being	5	0.942	Excellent
Digitalization	5		Excellent
Women Participation in Tourism	5	0.936	Excellent
Overall variables	25	0.979	Excellent

Table 3 Pearson's correlation results

Hypothesis	Pearson's correlation results
H1: There are relationship between skills and training among women participation in tourism	r = 0.767, $p < 0.01$ Strong
H2: There are relationship between leadership among women participation in tourism	r = 0.793, $p < 0.01$ Strong
H3: There are relationship between well-being among women participation in tourism	r = 0.808, $p < 0.01$ Strong
H4: There are relationship between digitalization among women participation in tourism	r = 0.794, $p < 0.01$ Strong

5 Discussion and Recommendations

The findings of this study reveal a promising and significant relationship between skills and training, leadership, well-being, and digitalization, and women's participation in the tourism industry in Malaysia. The confirmation of all hypotheses through Pearson's correlation analysis at a 0.01 significance level underscores the importance of these factors in shaping and promoting women's engagement in the sector.

The positive correlation between skills and training and women's participation is consistent with previous research that highlights the pivotal role of education and skill development in enhancing employability and career advancement. To build on this finding, tourism stakeholders should focus on tailored training programs that address the unique needs and aspirations of women in the industry. Recommendations include creating scholarships, mentorship programs, and vocational training that not only impart technical skills but also foster leadership and entrepreneurship capabilities.

The significant correlation between leadership and women's participation reinforces the im-portance of empowering women to assume leadership roles within the tourism sector. To capitalize on this relationship, organizations and government bodies should invest in leadership development programs, mentorship opportunities, and gender-inclusive leadership structures. Encouraging women to take on decision-making positions and supporting their career progression will contribute to a more balanced and dynamic industry.

The positive correlation between well-being and women's participation highlights the intrinsic connection between employee satisfaction and commitment. To further empower women in tourism, there is a need to address the well-being challenges that female workers often face. Strategies such as flexible working hours, wellness programs, and support for work-life balance can enhance the overall well-being of women in the sector, improving job satisfaction and retention rates.

The findings also suggest that digitalization can be a significant enabler for women in tourism. The positive correlation underscores the potential of technology to offer flexible work arrangements and entrepreneurial opportunities. However, it is essential to address the digital divide by providing access to digital resources and training. Furthermore, promoting digital literacy and encouraging women to embrace technology-driven innovation will open new avenues for their participation and leadership within the industry.

As recommendation, we would suggest a few like Develop and implement training programs specifically designed for women in the tourism sector. These programs should not only focus on technical skills but also include leadership, entrepreneurship, and digital literacy components. An investment in leadership development initiatives for women, including mentorship programs and opportunities for women to take on decision-making roles within organizations also should be considered. Other than that, Promotion on well-being and work-life balance by introducing flexible working hours, wellness programs, and support mechanisms for female employees in the tour-ism industry also suggested. Digital inclusion also can address the digital divide by providing access to technology and digital resources. Offer training and mentorship to enhance digital skills among women working in tourism. Policy Support from government bodies should work in collaboration with the private sector to implement policies that promote gender diversity and inclusion within the tourism industry. This includes setting diversity targets and incentivizing organizations that actively support women's participation.

6 Conclusion

In conclusion, the significant relationships between skills and training, leadership, well-being, and digitalization and women's participation in the tourism industry highlight a path for sustainable growth and innovation. By actively addressing these factors, stakeholders in the Malaysian tourism sector can not only empower women but also tap into a more diverse and talented workforce, ultimately enhancing the industry's competitiveness and success. The current study can help researchers better understand the long-term career sustainability of women in tourism, including the problems they experience at various life phases and how policies and support systems can be adapted to address them. It's also possible that technology's role in shaping women's participation in industry is changing quickly. Women working in the tourism industry should investigate how technological advances, such as remote work and online marketing, affect women's opportunities and challenges in the industry. While one of the limitations is generalization, it is important to recognize that what encourages women's participation may vary depending on the specific subsectors within the tourism industry. Furthermore, this study contributes to future study improvement by encouraging greater women participation in the tourism industry, which can have significant economic benefits such as increased innovation and improved service quality, making the sector more competitive globally. Beyond the tourism industry, the study's conclusions will affect society. They serve as a reminder of the advantages a diversified workforce may have for society and highlight the significance of gender equality in all fields. As a result, it is hoped that all the knowledge gathered throughout this research will be useful to the next researcher.

References

1. Khurramov, O.: The role of the tourism sector in the digitalization of the service economy. Econ. Innovative Technol. 2020(1) (2020)
2. Suess, C., Woosnam, K.M., Erul, E.: Stranger-danger? Understanding the moderating effects of children in the household on non-hosting residents' emotional solidarity with Airbnb visitors, feeling safe, and support for Airbnb. Tour. Manage. **77**, 103952 (2020)
3. Andrades, L., Dimanche, F.: Destination competitiveness in Russia: tourism professionals' skills and competences. Int. J. Contemp. Hospitality Manag. (2019)
4. Solina, M.F.L.: Assessment of business environment of Women involved in micro, small, and medium enterprises (MSMEs) in the philippines: a comparative study with select ASEAN countries. JAS (J. ASEAN Stud.) **8**(2), 191–211 (2020)
5. Voumik, Liton Chandra, M.S., Rahman, M.H.: Effects of tourism on African women employment: using generalized methods of moments and quantile regression (2022)
6. Gursoy, D., et al.: Industry's expectations from hospitality schools: what has changed? J. Hosp. Tour. Educ. **24**, 32–42 (2013)
7. Rajapaksha, S.N., Wijesinghe, W.P.M., De Silva, K.H.A.L., Wijethunga, W.M.K.R., Weligodapola, M., Rathnayake, R.M.N.M.: The impact of organizational culture on the use of business analytics in Sri Lankan Companies in Sri Lanka. J. Manage. Stud. **4**(1), 49–69 (2022)
8. Akomea-Frimpong, I., Asogwa, I.E., Tenakwah, E.J.: Systematic review of sustainable corporate governance of SMEs: conceptualisation and propositions. Corp. Ownersh. Control. **19**(3), 74–91 (2022)
9. Moreno-González, A.-A., et al.: Health destination image: the influence of public health management and well-being conditions. J. Destin. Mark. Manag. **16**, 100430 (2020)
10. Gupta, P., S. Srivastava: Work–life conflict and burnout among working women: a mediated moderated model of support and resilience. Int. J. Organ. Anal. Ahead-of-print (2020)
11. Vrana, J., Singh, R. Digitization, digitalization, and digital transformation (2021)
12. Alhajri, A., Aloud, M.: Female digital entrepreneurship: a structured literature review. Int J Entrepreneurial Behav. Res. Ahead Print (2023)
13. Hammad, R., El Naggar, R.: The role of digital platforms in Women's Entrepreneurial opportunity process: does online social capital matter? Hum. Behav. Emerg. Tech. **2023**, 5357335 (2023)
14. Cheng, X., Li, J.: The impact of virtual tourism experience on stress and emotion under the stay-at-home restrictions due to COVID-19 epidemic. Tour. Trib. **35**, 13–23 (2020)
15. Abdul Ghani Azmi, I., Che Hashim, R., Mohamed, H.AB., Sahol Hamid, N.: A Study of Muslim Women Entrepreneurs' SMEs Challenges and Motivation in the Asia Pacific Region. Contributions to Environmental Sciences and Innovative Business Technology. Springer, Singapore (2023)
16. Zhu, H., Deng, F.: How to influence rural tourism intention by risk knowledge during COVID-19 containment in China: mediating role of risk perception and attitude. Int. J. Environ. Res. Public Health **17**(10), 3514 (2020)
17. Mooney, S.: Gender research in hospitality and tourism management: time to change the guard. Int. J. Contemp. Hospitality Manage. ahead-of-print (2020)
18. Bhandari, P.: What is quantitative research? | Definition, Uses & Methods. Scribbr (2020)
19. Samad, Samad, S., Alharthi, A.: Untangling factors influencing women entrepreneurs' involvement in tourism and its impact on sustainable tourism development. Adm. Sci. **12**(2), 1–19
20. Serafimova, M., Petrevska, B.: Female entrepreneurship in tourism: a strategic management perspective in Macedonia. J. Appl. Econ. Bus. **6**(1), 21–32 (2018)

Changes in Recreational Tourism: A Thematic Analysis During the COVID-19 Pandemic and Its Aftermath

Maryanti Mohd Raid, Nur Khairiyah Mohammad, Aminah Mohsin, Zafirah AB. Muin, and Wan Ibrisam Fikry Wan Ismail

Abstract In contemporary society, marked by a rapid and dynamic lifestyle, individuals demonstrate an increased propensity for choosing vacations involving leisurely activities and immersing themselves in natural surroundings. The adoption of modern lifestyles has also been on the rise due to the diversity of global trends in recreational tourism. The COVID-19 pandemic has introduced a new set of societal norms, including practices such as social distancing, lockdowns, and guidelines for outdoor activities. The emergence of this novel pandemic has brought about changes in social engagements, particularly in the domains of recreation and tourism. This study aimed to explore the changes in recreational activities within the tourism industry during the COVID-19 pandemic and its aftermath. Using thematic analysis, a total of 67 publications between 2020 and 2023 on the impacts of COVID-19 on recreational tourism were synthesized from three journal databases, namely Emerald Insights, Scopus, and Science Direct. The findings derived from the analysis of code-to-document using ATLAS.ti 8 indicated that prior studies had placed significant emphasis on socio-psychological, economic, environmental, and governmental concerns. This study serves as the foundation for future comprehensive research focused on enhancing recreational activities within the tourism sector, particularly in the aftermath of the COVID-19 pandemic.

M. Mohd Raid (✉) · N. K. Mohammad · A. Mohsin · Z. AB. Muin · W. I. F. Wan Ismail
Department of Real Estate, Faculty of Built Environment and Surveying, Universiti Teknologi Malaysia, Skudai, Malaysia
e-mail: maryanti@utm.my

N. K. Mohammad
e-mail: nur.khairiyah@graduate.utm.my

A. Mohsin
e-mail: aminahmohsin@utm.my

Z. AB. Muin
e-mail: zafirah2@utm.my

W. I. F. Wan Ismail
e-mail: wanibrisam@utm.my

B. Awwad (ed.), *The AI Revolution: Driving Business Innovation and Research*, Studies in Systems, Decision and Control 524, https://doi.org/10.1007/978-3-031-54379-1_70

Keywords Recreational activity · Tourism · COVID-19 pandemic

1 Purpose and Background

In contemporary times, the debate on leisure tourism has gained significant traction due to multiple reasons, including the various recreational activities pursued by individuals or collectives of tourists. The trend of recreational tourism has experienced a significant surge in popularity in the current twenty-first century and may shape future trends within the tourism industry. The United Nations World Tourism Organization [1] proposes that recreational resources present distinct opportunities for enhancing the advancement of tourism across various nations.

The term 'tourism' was initially defined in 1942 by Krapf and Hunziker as encompassing 'the collection of phenomena and interactions that arise from the travel and temporary stay of individuals who are not residents, provided that such activities do not result in permanent residence and are not associated with any income-generating activities' [2, 3]. Nevertheless, [4] criticised the definition of tourism, arguing that it is overly broad. In response, he proposed a revised definition that characterises tourism as a collection of interconnected activities and occurrences associated with travel and consumption outside of one's usual place of residence to fulfil recreational and cultural desires. Including the term 'recreational' in the definition is warranted as it represents the primary motivational needs of tourists when engaging in travel activities. The World Tourism Organisation also issued an official definition of tourism. It defined tourism as the engagement in activities by individuals who travel to and reside in locations outside their customary surroundings for a duration not exceeding one uninterrupted year, intending to pursue leisure, recreation, business, or other objectives [5].

The significance of recreation in the context of tourism has been acknowledged as a fundamental component of the tourism industry. Scholars such as [3, 6, 7] have explored the interconnectedness and close relationship between rest, leisure, recreation, and tourism. Recreation is defined as a form of entertainment or engagement in activities pursued for personal enjoyment or relaxation, as opposed to professional or occupational endeavours. Recreational areas refer to publicly accessible spaces designated for engaging in sports and games, frequently featuring amenities such as children's playgrounds. Hence, outdoor recreation is intricately interconnected with the field of tourism.

According to [8], recreational tourism involves activities that impose moderate physical demands and are tailored to accommodate individuals of varying ages and physical capabilities. Recreation or recreational or leisure tourism encompasses a specific segment within the broader tourism industry [9]. The primary objective of recreational tourism is to enhance the overall well-being and physical condition of the individuals participating in such activities [10]. Thus, recreational tourism pertains to the act of travelling and engaging in activities primarily for the purpose of relaxation, pleasure, and recreation. This includes outdoor expeditions, cultural immersion, wellness getaways, and entertainment options instead of professional or business-related activities.

The outdoor recreational behaviours of tourists are often influenced by various factors, including subjective assessments of safety and hygiene, social-peer influence and responsibility, the image of the destination and the behaviours exhibited by local communities [11]. However, the implementation of restrictions due to the COVID-19 pandemic has affected individuals' leisure pursuits in the tourism industry. In late 2019, the global outbreak of the novel coronavirus, known as COVID-19, resulted in implementing lockdown measures that limited people's mobility. This viral infection was first detected in Wuhan, China, in December 2019. Subsequently, numerous countries implemented lockdowns, including stay-at-home directives, mandatory quarantines, adherence to physical distancing guidelines, closure of non-essential establishments (e.g., restaurants, educational institutions, and businesses), and restrictions on large-scale public gatherings [12] due to its easy virus transmission which was through respiratory droplets during human interactions [13]. As a result, there was a significant decrease in the recreational tourism trend.

In Malaysia, there was a significant decrease in travel expenditure, amounting to a substantial loss of RM54 billion in 2020 compared to the previous year, indicating a 82% decline [14]. This is due to the pandemic significantly affecting tourists' cognitive, affective, and emotional aspects, which has led to potential alterations in their decision-making processes related to outdoor activities. This paper examined recreational tourism changes during and after the COVID-19 pandemic. The following research questions were formulated to address the issue: (i) What are the changes in recreational activities in the tourism sector during the COVID-19 pandemic? and (ii) What are the changes in recreational activities in the tourism sector after the COVID-19 pandemic? This article reviewed the changes in recreational activities in the recreational tourism sector during and after the COVID-19 pandemic, spanning publication from 2019 to September 2023. Thus, the finding of this analysis is significant to serve as a foundation for further research focusing on recreational activity demand in tourism industry that need to be emphasize by the local authority in providing recreational area.

2 Methodology

The present study employed thematic analysis, a methodological approach characterised by identifying patterns and constructing themes through an extensive review of relevant literature [15]. The literature selection process adhered to three specific criteria. Firstly, the publications must be within 2019 to 2023. Secondly, the selected literature had to include the keywords 'recreational activity,' 'tourism,' and 'COVID-19.' The purpose of selecting these keywords was to provide an outline for discussing the changes in recreational tourism that occurred both during and after the COVID-19 pandemic. The selected articles were subsequently examined using ATLAS.ti version

8 to facilitate the thematic analysis process, following the guidelines provided by [16]. The search strings from three databases are shown in Table 1.

Using the thematic review method, a comprehensive literature search was performed across three databases: Emerald Insight, Scopus and Science Direct. This approach employs various methodologies to examine a certain subject or motif comprehensively. [15] present a comprehensive six-step methodology for doing a thematic analysis. The first step involves developing a thorough understanding of the data. In the second step, initial codes are generated to identify meaningful patterns and concepts within the data. The third step entails searching for overarching themes that emerge from the coded data. These themes are then reviewed and refined in the fourth step. The themes are clearly defined and described in detail in the fifth step. Finally, in the sixth step, the findings are synthesised and presented in a written format.

In this study as illustrated in Fig. 1, the purpose of the thematic review was to locate scholarly publications pertaining to the transformations in recreational tourism during and after the COVID-19 pandemic. The search produced 217 articles across the three databases, with 70 articles found in Emerald Insight, 58 in Scopus, and 89 in Science Direct. However, the study included a total of 67 papers after applying strict inclusion and exclusion criteria. These criteria involved evaluating the articles' relevance to the issue, identifying and deleting duplicate articles, and considering their accessibility.

Following the first screening process, a total of 67 articles were selected and subsequently imported into the Mendeley software platform, where they were stored as metadata. The metadata was then imported into the ATLAS.ti version 8 software to facilitate data organisation. The articles were categorised according to six parameters: author, issue number, periodical, publisher, volume, and year of publication. The utilisation of this methodology facilitated a methodical and structured examination of the data, hence facilitating the extraction of significant information.

In the initial coding phase, a total of 36 codes were created, considering the country and year of publication as key factors. Figure 2 illustrated the geographical dispersal of the 67 articles selected for analysis. The represented countries included Argentina, Australia, Bangladesh, Brazil, China, Ecuador, Egypt, India, Indonesia,

Table 1 Search string from Emerald insight, SCOPUS, and science direct database

Databases	Keyword	Results	After exclusion
Emerald insight	Recreational activity AND tourism AND COVID-19 Date range: from 2020 to 2023	70 results	29
SCOPUS	Title-ABS-KEY ('recreational activity' AND 'tourism' AND 'COVID-19' PUBYEAR 2020–2023	58 results	14
Science direct	Recreational activity AND tourism AND COVID-19. Year(s): 2020–2023	89 results	24
Total		217	67

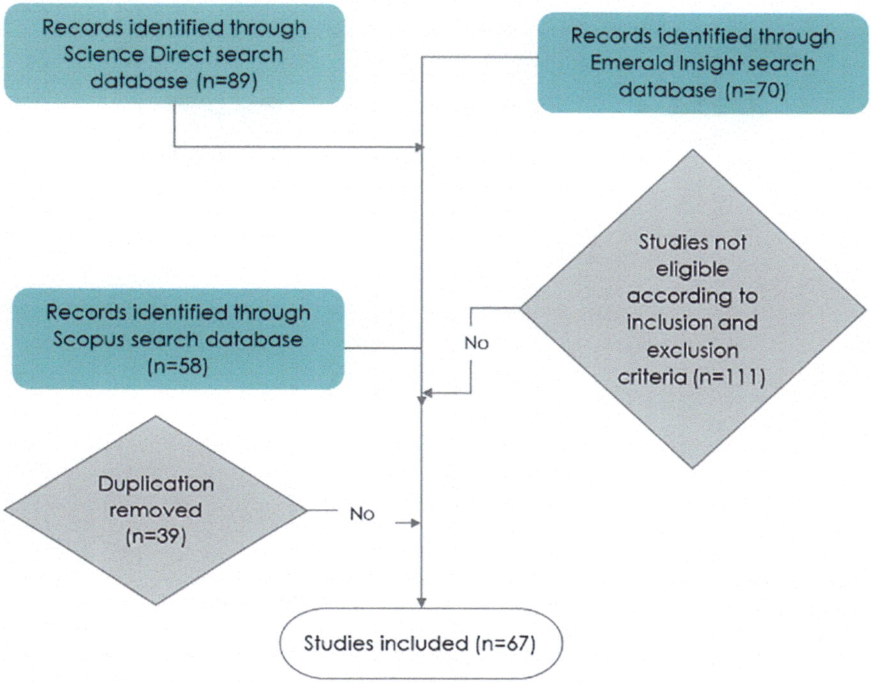

Fig. 1 Process of screening (inclusion and exclusion criteria) in the thematic review

Jamaica, Japan, Jordan, Korea, Malaysia, Mauritius, Mexico, New Zealand, Poland, Portugal, Romania, Russia, South Africa, Spain, Sweden, Taiwan, Thailand, Turkey, Ukraine, United Kingdom, United States, Uzbekistan, and Vietnam. The countries in dark green were with the highest number of publications in the field of recreational tourism, namely the United States (7), China (7), and India (6). In contrast, the countries in light green indicated the countries with the lower publication.

In addition, the coding also provided information on the number of publications based on the year of publication, as shown in Table 2. The analysis revealed a notable rise in the frequency of publications pertaining to the subject matter. The current study was conducted in September 2023, so the number of publications for 2023 could not be determined. Subsequently, the second coding phase aimed to identify the themes pertaining to the research question on the changes in recreational activities in the tourism sector during and after the COVID-19 pandemic.

After a comprehensive review of 67 selected articles, a total of 98 codes emerged. Following several rounds of re-coding and codes merging on ATLAS.ti 8, the final coding based on the patterns of discussion in the articles was grouped into four themes: socio-cultural concerns, environmental concerns, economic concerns, and governmental concerns.

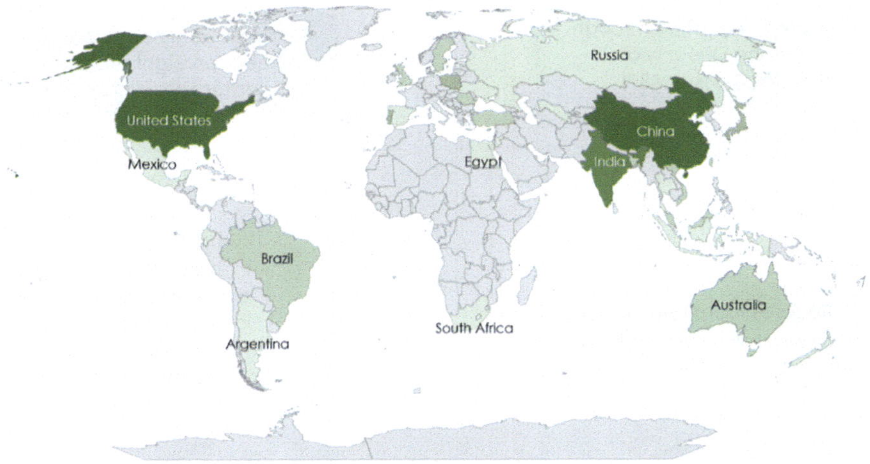

Fig. 2 Geographical dispersal of selected articles

Table 2 Paper reviewed according to year of publication

Year	Number of articles
2019	1
2020	3
2021	22
2022	25
2023	16

3 Findings and Discussion

The recreational tourism industry experienced enormous changes during and after the COVID-19 pandemic. Stringent travel restrictions imposed by most nations worldwide during the pandemic, including lockdowns, border closures, and quarantine procedures, significantly reduced internal and worldwide travel, impacting the recreational tourism industry. In addition, strict health and safety regulations were implemented, including stricter cleanliness guidelines, mask requirements, social isolation, and fewer guests allowed in hotels, restaurants and attractions.

These changes have impacted the recreational tourism industry and create new trend. For example, tourists prefer socially isolated outdoor and natural activities, increasing the popularity of outdoor recreation activities like hiking and camping. The deployment of digital technologies has change tourism norm to reduce physical touch such as online reservations, contactless check-in, and virtual tours that have become more common these days.

Concerns about future pandemics have prompted policymakers and researchers to consider how the recreational tourism industry sector should be structured after COVID-19. Quarantine, lockdown, and social isolation are crucial measures in future preparedness initiatives implemented on a global scale, encompassing both national and local levels in the wake of future pandemics [17–19]. Even when the pandemic is over, tourists might demand higher levels of hygienic practices. As a result, the following four major concerns in the recreational tourism industry during and after the COVID-19 outbreak were examined.

3.1 Socio-Phycological Concerns

Due to the COVID-19 pandemic, the operation of the recreational tourism industry underwent a substantial transformation, resulting in notable changes during and after the pandemic. For example, there is a noticeable shift in people's travel preferences. Most tourists preferred personal cars or hired vehicles as their chosen method of transportation to mitigate their perceived risks [20]. They preferred individual or small-group activities that facilitated the practice of social distancing in accordance with their safety [21, 22]. They also sought to avoid large crowds and popular tourist destinations [12, 23]. Not only that, tourists were inclined towards engaging in outdoor recreational activities that are nature-based, as well as their propensity to exhibit ecologically responsible behaviour, such as visiting national parks and mountain destinations and seeking open-air spaces [24, 25]. They sought to fulfil their need for a connection with nature, such as by hiking, cycling, fishing, and walking [26–29].

According to [30], tourists felt more secure when travelling to rural and non-dense destinations rather than cities, decreasing the demand level for urban tourism. [31] expects rural tourism to become one of the leading destinations as the COVID-19 pandemic lasts. Indeed, tourists preferred leisure tourism, utilising such occasions as an opportunity to relax while immersing in stunning natural landscapes and learning about culture and heritage [32]. Tourists could actively interact with local communities and get a deeper appreciation for their cultural heritage in the aftermath of the COVID-19 pandemic [33]. Efforts would be on preserving historical towns and structures, archaeological sites, the resuscitation of dance, theatre, singing, and ceremonial festivals, and establishing handicraft centres. Still, as tourists needed high-quality hotel accommodations and diverse dining options, they could favour seaside and spa resorts [8, 34]. Additionally, prior studies also discovered staycation programmes and short trips would have in future destination planning among tourists [20, 35].

3.2 Economic Concerns

The topic of post-COVID-19 pandemic transformation of visitor behaviours has been extensively addressed in previous research, with particular attention given to economic considerations. The shift necessitates changing and adapting both tourism demand and marketing strategies. For example, the rise in visitor demand for nature-based destinations has led to a corresponding increase in agritourism [32].

Recreational farming is a notable agricultural enterprise providing various tourism activities. Operational farms offer options for holiday experiences [36, 37]. Owners of recreational farm owners can evaluate their business model and align it with governmental programmes to promote innovation and enable the development of new firms or products.

Moreover, the emergence of new rules has led to the development of additional innovations in recreational tourism services [38]. The COVID-19 pandemic has led to a significant increase in the digitalisation of various aspects of people's lives, including the tourist sector. This includes the use of digital platforms for activities such as hotel reservations, booking recreational activities, meal reservations, and other related services [25]. The utilisation of the internet and digitisation has been favoured by tourists to compare pricing, make purchases of travel packages, and rent vehicles, enhancing the overall tourist experience even before its commencement [39, 40].

Convenience and cost-effective price are also one of the economic concerns among tourists [41]. It is because a growing number of travellers are currently seeking vacation options that are less ostentatious, more cost-effective, and have a pleasant experiential aspect. Furthermore, the prevalence of long-distance travel has also experienced a decline, and solo tourism has become a new norm due to prioritising the media value of spending time with oneself, clearing work hours, and meeting new people, among others [42].

Moreover, the luxury yacht, cruise, and boat sector emerged as a potential avenue for such investment opportunities [43]. Indeed, most Caribbean islands and numerous other small islands developing states rely heavily on tourism, particularly during the cruise season. It has been proven that much prior research revealed that more tourist desire willing to pay premiums, indicating their behavioural intention concerning green hotels [44–47]. A recent study by [48] found that community self-efficacy, particularly within indigenous communities, plays a crucial role in their ability to adapt and effectively take advantage of changing conditions. Thus, to seize possibilities in the highly competitive recreational tourism business, it is imperative for the community to enhance its cultural dynamics, values, beliefs, and interests.

3.3 Environmental Concerns

Environmental concern moderates the relationship between green trust and behavioural intention among tourists [44]. In general, tourists exhibit increased care for the environment through their willingness to reduce spending patterns to protect the natural environment, such as by booking green hotels and spending time on nature-based outdoor recreation activities, as discussed in the socio-phycological concerns section. Moreover, [24] discovered that in the aftermath of the COVID-19 pandemic, recreational activities that offer proximity to natural environments have become increasingly appealing to tourists seeking to engage with nature while simultaneously reducing their perceived risk and fulfilling their desire for travel. For instance, mountain biking and hiking are gaining in popularity, and people prefer urban parks with networks of mountain bike trails, whereas some people prefer more remote large parks to conserve nature [49]. According to the findings of [34], tourist experiences in Beijing were characterised by many attractions such as hot springs, the Great Wall, snow scenes, skiing, and delectable cuisine, which served as a means of escape for visitors. Also, the Costa Maya community in Mexico is concerned regarding the potential degradation of the ecosystem caused by the cruise industry [21]. The absence of tangible advantages from cruise tourism and the inadequate provision of public services and infrastructure such as water and power, sewage treatment, and waste management have elicited disappointment within the local community [50, 51]. Furthermore, [52] stressed the significant influence of environmental characteristics on the quality of divers' experiences underscores the notion that tourism damages the natural ecosystems. Hence, environmental issues have gained increasing significance among tourists, particularly in relation to the preservation of natural resources rather than solely focusing on leisure and enjoyment during vacations.

3.4 Government Concerns

In the wake of the COVID-19 pandemic and the implementation of border closures, many governments of the world made substantial financial commitments towards tourism programmes, including financial support allocated to organisations to foster the creation of novel products and experiences that fit the domestic market [27, 53]. For instance, the New Zealand government allocated NZD10 million for the development of website content, NZD15 million for domestic transformation, and NZD20 million to support regional tourist associations with local campaigns as part of a NZD400 million tourism recovery package that included website content pertaining to attractions, destinations, and recreational activities [54].

According to [55], the recreational tourism industry has increasingly prioritised the use of retargeting strategies. This marketing practice focuses on particularly targeting individuals who have visited websites but have not completed a purchase. [55] posited initiatives, such as 'shop local, support local' campaigns, were devised to encourage businesses to leverage the concept of good governance. In addition, 'Love Local campaigns were introduced in Wellington to explore their region, culture, and food through festivals [56]. Localism has become the new trend in tourism that is catalysed by a sense that society is in decline or that urbanisation and globalisation carry detrimental consequences for communities [57].

The government of Rio de Janeiro, Brazil, emphasised implementing regulations to establish protocols and restrictions for granting access to employees and customers within the hotel industry. These measures are intended to facilitate the reopening of the tourism sector [58]. Furthermore, a campaign called 'More Rio for Less' was strategically devised to allow visitors to avail of two-week savings. Additionally, an initiative known as the 'Party Waits for You' programme was introduced. The measures implemented aimed to facilitate the gradual resumption of tourist operations and engage all sectors of the tourism industry in adhering to health norms, thereby assuring a secure experience for visitors while simultaneously promoting the city.

Previous research also suggests that there is a necessity to construct a tourism system that is more sustainable in the aftermath of the COVID-19 pandemic, with the aim of promoting sustainable development [59–61]. [53] believed that tourism is oriented towards making a positive contribution to the attainment of Sustainable Development Goals (SDGs). In this scenario, the majority of proposed remedies necessitate a comprehensive examination of the fundamental causes and advantages for tourists and the cohesive development of sustainable tourism, mass tourism, and high-value tourism [62–66]. Hence, to endure the challenges posed by the pandemic and ensure long-term viability and expansion in the aftermath, the government within these industries must possess a capacity for originality and ingenuity and a well-defined and adaptive strategic approach. Furthermore, it is imperative that requisite knowledge and skills are applied to navigate and adapt to forthcoming transformations effectively.

4 Conclusions

This article reviewed the changes in recreational activities in the recreational tourism sector during and after the COVID-19 pandemic, spanning publication from 2019 to September 2023. The findings derived from the analysis of code-to-document using ATLAS.ti 8 indicated that prior studies had placed significant emphasis on socio-psychological, economic, environmental, and governmental concerns. Based on the reviewed articles, prior research focused more on tourist behaviours comprising those four concerns while they travel. The current scenario resembles a pandemic outbreak, substantially altering several aspects of individuals' lives, including their lifestyle and cognitive processes.

Various sociological theories exist to elucidate the mechanisms through which societal change occurs. In accordance with the tenets of change processes, it is imperative for individuals and society to synchronise their actions and values with the prevailing demands of the era. The main drivers influencing individuals' travel choices are socio-physiological concerns, preferences for convenient accommodations, and a desire for nature-based recreational activities. This aligns with the increased knowledge and appreciation for nature conservation following the outbreak of COVID-19.

Furthermore, the government's concerns have been the subject of discussion, with a few instances of its policies and programs aimed at improving the local tourism offering and prioritising the local population rather than focusing solely on international visitors. For the economic concerns, recreation providers suggested implementing social marketing strategies to enhance their support, operational, and empowering initiatives to stimulate tourist behaviour motivation in the future. For instance, an organisation can contemplate implementing a social marketing approach utilising various social media platforms such as YouTube, Facebook, X, TikTok, and others to acquire new clients and maintain current visitors' loyalty.

Moreover, previous research has thoroughly explored the preferences of tourists regarding short-duration trips, cost-efficient travel options, and the desire for enjoyable experiences elements. Therefore, pursuing further research on the changes in the recreational tourism industry after the COVID-19 pandemic will stimulate and improve tourists' travel satisfaction, necessitating the investigation of their inclination towards recreational activities while considering future trip plans.

Acknowledgements The authors would like to thank Universiti Teknologi Malaysia through the UTM Encouragement Grant (UTM-ER) VOT number Q.J130000.3852.31J24 for the study fund.

References

1. UNWTO.: International tourist arrivals could fall by 20–30% in 2020 [Internet] 2020. Available from: https://www.unwto.org/news/international-tourism-arrivals-could-fall-in-2020
2. Alfonso, C., Angel, M., Andres, A., Diaz, J., Agamez, L.J., Esteban, M., et al.: Perception of the city of Montería from an international tourism perspective. **XI**(5), 1993–2006 (2023)
3. Gjorgievski, M., Kozuharov, S., Nakovski, D.: Tipology of recreational-tourism resources as an important element of the tourist offer. Spec. Issue. UTMS J. Econ. **4**(1), 53–60 (2013)
4. Panov, N., Taleska, M., Dimeska, H.: The importance of mountain regions for tourism development in republic of Macedonia. Hilly- Mt Areas Probl. Perspect. **2017**(845), 12–15 (2013)
5. WTO: National and regional tourism planning: methodologies and case studies (English version). World Tourism Organization (1994)
6. Hadi, D.M., Naeem, M.A., Karim, S.: The exposure of the US tourism subsector stocks to global volatility and uncertainty factors. Curr. Issues Tour. **26**(4), 603–616 (2023)
7. Stephen, W.: Tourism Geography: A New Synthesis. Taylor & Francis (2009)
8. Bogomolova, E., Dovlatova, A.: Recreational tourism as a growth driver for tourist destination **94**(Jahp), 263–266 (2019)
9. Wu, T.P., Wu, H.C., Wu, C.F., Chen, C.H., Zhang, M.Q.: Tourism activities and house price nexus: a multivariate quantile-on-quantile estimation. J. Policy Res. Tour. Leis Events. 1–24 (2023)
10. Mathew, P.V., M NP.: Sustainable tourism development: discerning the impact of responsible tourism on community well-being. J. Hosp. Tour. Insights. **5**(5), 987–1001 (2022)
11. Humagain, P., Singleton, P.A.: Reprint of: exploring tourists' motivations, constraints, and negotiations regarding outdoor recreation trips during COVID-19 through a focus group study. J. Outdoor Recreat Tour. [Internet]. **41** (September 2021), 100626 (2023). Available from: https://doi.org/10.1016/j.jort.2023.100626
12. Humagain, P., Singleton, P.A.: Exploring tourists' motivations, constraints, and negotiations regarding outdoor recreation trips during COVID-19 through a focus group study. J. Outdoor Recreat Tour [Internet]. **36** (September), 100447 (2021). Available from: https://doi.org/10.1016/j.jort.2021.100447
13. Wilder-Smith, A., Freedman, D.O.: Isolation, quarantine, social distancing and community containment: Pivotal role for old-style public health measures in the novel coronavirus (2019-nCoV) outbreak. J. Travel Med. **27**(2), 1–4 (2020)

14. Abdullah, M.F., Noor, M.I., Ahmad, B.E., Yusoh, M.P., Pardi, F.: Managing the impact of COVID-19 crisis on tourism sector in protected area: a case study in Pahang National Park. In: IOP Conference Series: Earth and Environmental Science 2023 Jul 1, Vol. 1217, No. 1, p. 012024. IOP Publishing (2023)
15. Clarke, V., Braun, V.: Teaching thematic analysis: overcoming challenges and developing strategies for effective learning. Psychologist **26**(2), 120–123 (2013)
16. Zairul, M.: A thematic review on student-centred learning in the studio education. J. Crit. Rev. **7**(2), 504–511 (2020)
17. Brooks, S.K., Webster, R.K., Smith, L.E., Woodland, L., Wessely, S., Greenberg, N., et al.: The psychological impact of quarantine and how to reduce it: rapid review of the evidence. Lancet [Internet]. **395**(10227), 912–920 (2020). Available from: https://doi.org/10.1016/S0140-6736(20)30460-8
18. Torales, J., O'Higgins, M., Castaldelli-Maia, J.M., Ventriglio, A.: The outbreak of COVID-19 coronavirus and its impact on global mental health. Int. J. Soc. Psychiatry **66**(4), 317–320 (2020)
19. Xie, J., Luo, S., Furuya, K., Sun, D.: Urban parks as green buffers during the COVID-19 pandemic. Sustainability **12**(17), 1–17 (2020)
20. Das, P., Singh, A.K., Maitra, S.: Effect of COVID-19 on recreational trips to tourist destination—An Indian context. Asian Transp. Stud. [Internet] **8**(October), 100088 (2022). Available from: https://doi.org/10.1016/j.eastsj.2022.100088
21. Magio, K.O.: Sustainability-Focused Recovery Approach for Cruise Tourism After COVID-19: Key Issues and Opportunities in the Port Destination of Mahahual—Costa Maya. Emerald Handb. Destin Recover Tour. Hosp. 445–462 (2022)
22. Viñán-Merecí, C., Celi-Sánchez, K., Correa-Quezada, R., Durán-Sánchez, A.: Impacts of COVID-19 on tourism-related activities: a case study of ecuador by scenarios. Pandemics Travel COVID-19 Impacts Tour Ind. 157–172 (2021)
23. Vasko, M., Volfova, N., Zikova, A.: Factors affecting the tourists' approach to health and safety information in reviews during the COVID-19 pandemic. In: Online Reputation Management in Destination and Hospitality. Emerald Publishing Limited, p. 131–48 (2023)
24. Chi, N.T.K.: Environmentally responsible behaviour in outdoor recreation: the moderating impact of COVID-19 related risk perception. J. Tour. Futur. 1–16 (2022)
25. Gómez-Prado, R., Alvarez-Risco, A., Sánchez-Palomino, J., Anderson-Seminario, M. de las M., Del-Aguila-Arcentales, S.: Current Trends in Sustainable Tourism, Hospitality, and Recreation 65–80 (2023)
26. Andrieieva, O., Blystiv, T., Byshevets, N., Moseychuk, Y., Balatska, L., Liasota, T., et al.: Assessment of the impact of outdoor activities at leisure facilities on the physical activity of 15-year-old schoolchildren during the COVID-19 pandemic. J. Phys. Educ. Sport. **22**(8), 1839–1847 (2022)
27. Yeoman, I.S., Schänzel, H.A., Zentveld, E.: Tourist behaviour in a COVID-19 world: a New Zealand perspective. J. Tour. Futur. **8**(2), 155–176 (2022)
28. Beery, T., Olsson, M.R., Vitestam M.: COVID-19 and outdoor recreation management: Increased participation, connection to nature, and a look to climate adaptation. J. Outdoor Recreat Tour [Internet]. **36**, 100457 (2021). Available from: https://doi.org/10.1016/j.jort.2021.100457
29. Indu, B., Does S.K.: Your post-COVID-19 travel dream talk about sustainability? Insights from Potential Tourists in India. Tour. Destin Manag. Post-Pandemic Context Glob. Issues Destin Manag. Solut. 273–288 (2021)
30. Anguera-Torrell, O., Vives-Perez, J., Aznar-Alarcón, J.P.: Urban tourism performance index over the COVID-19 pandemic. Int. J. Tour. Cities. **7**(3), 622–639 (2021)
31. Zhu, H., Deng, F.: How to influence rural tourism intention by risk knowledge during COVID-19 containment in China: mediating role of risk perception and attitude. Int. J. Environ. Res. Public Health **17**(10), 1–23 (2020)
32. Kim, E.J., Park, S.M., Kang, H.W.: Changes in leisure activities of the elderly due to the COVID-19 in Korea. Front Publ. Heal. **10**(2) (2022)

33. Galvani, A., Lew, A.A., Perez, M.S.: COVID-19 is expanding global consciousness and the sustainability of travel and tourism. Tour. Geogr. [Internet]. **0**(0), 567–76 (2020). Available from: https://doi.org/10.1080/14616688.2020.1760924
34. Yi-Chen, D.B., Dai, J., Ma, E.: Winter sports mega-event destination experience: an experience economy model perspective during the post-COVID-19 pandemic. Sport Tour. 79–94 (2023)
35. Lin, Z. (CJ), Wong, I.K.A., Kou, I.T.E., Zhen, X. (Christine).: Inducing wellebing through staycation programs in the midst of the COVID-19 crisis. Tour. Manag. Perspect. [Internet] **40** (February), 100907 (2021). Available from: https://doi.org/10.1016/j.tmp.2021.100907
36. Arru, B., Furesi, R., Madau, F.A., Pulina, P.: Economic performance of agritourism: an analysis of farms located in a less favoured area in Italy. Agric. Food Econ. **9**(1) (2021)
37. Sznajder, M., Przezbórska, L., Scrimgeour, F.: The economic and social importance of agritourism. In: Agritourism. CABI Wallingford UK, pp. 15–42 (2009)
38. Sheresheva, M., Efremova, M., Valitova, L., Polukhina, A., Laptev, G.: Russian tourism enterprises' marketing innovations to meet the COVID-19 challenges. Sustainability **13**(7) (2021)
39. Agarwal, S., Kumar, S., Adichwal, N.K.: Effect of marketing strategies on the market performance and a comparative study of online travel agencies in India. J. Public Aff. **22**(1), e2381 (2022)
40. Farhangi, S., Alipour, H.: Social media as a catalyst for the enhancement of destination image: evidence from a mediterranean destination with political conflict. Sustainability **13**(13), (2021)
41. Tiwari, P., Chowdhary, N.: What makes Indian domestic tourists crowd-friendly in the post-COVID-19 phase? Int. J. Tour. Cities. (August) (2022)
42. Shende, S., Bhaduri, E., Goswami, A.K.: Analyzing changes in travel patterns due to COVID-19 using Twitter data in India. Case Stud. Transp. Policy [Internet]. **12**(February), 100992 (2023). Available from: https://doi.org/10.1016/j.cstp.2023.100992
43. Ajagunna, I., Ilori, M.O., McLean, E.: What are the implications for the post-pandemic Caribbean cruise industry? Worldw Hosp. Tour. Themes. **14**(2), 193–198 (2022)
44. Dwivedi, R.K., Pandey, M., Vashisht, A., Pandey, D.K., Kumar, D.: Assessing behavioural intention toward green hotels during COVID-19 pandemic: the moderating role of environmental concern. J. Tour. Futur. 1–17 (2022)
45. Agag, G.: Understanding the determinants of guests' behaviour to use green P2P accommodation. Int. J. Contemp. Hosp. Manag. **31**(9), 3417–3446 (2019)
46. Balaji, M.S., Jiang, Y., Jha, S.: Green hotel adoption: a personal choice or social pressure? Int. J. Contemp. Hosp. Manag. **31**(8), 3287–3305 (2019)
47. Yadav, R., Balaji, M.S., Jebarajakirthy, C.: How psychological and contextual factors contribute to travelers' propensity to choose green hotels? Int. J. Hosp. Manag. [Internet]. **77**(August 2018), 385–395 (2019). Available from: https://doi.org/10.1016/j.ijhm.2018.08.002
48. Hanafiah, M.H., Zamzuri, N.H., Sharif, T.I.S.T.: Chartrungruang B on. Reframing Indigenous tourism entrepreneurial personality, experience, sense of community and challenges in community-based tourism-related business. J. Enterprising Communities (2023)
49. Smith, I., Velasquez, E., Norman, P., Pickering, C.: Effect of the COVID-19 pandemic on the popularity of protected areas for mountain biking and hiking in Australia: Insights from volunteered geographic information. J. Outdoor Recreat Tour [Internet] **41**(October 2022), 100588 (2023). Available from: https://doi.org/10.1016/j.jort.2022.100588
50. Lopes, H.D.S., Remoaldo, P.C., Ribeiro, V., Martín-Vide, J.: Effects of the COVID-19 pandemic on tourist risk perceptions—The case study of Porto. Sustainability **13**(11), 1–29 (2021)
51. Beirman, D.: Restoring Tourism Destinations in Crisis: A Strategic Marketing Approach [Internet]. Taylor & Francis (2020). Available from: https://books.google.com.my/books?id=vEX1DwAAQBAJ
52. Marconi, M., Giglio, V.J., Pereira-Filho, G.H., Motta, F.S.: Effects of the COVID-19 pandemic on SCUBA diving experience in marine protected areas. J. Outdoor Recreat Tour. [Internet]. **41**(November 2021), 100501 (2023). Available from: https://doi.org/10.1016/j.jort.2022.100501

53. Fenitra, R.M., Abbas, A., Ekowati, D., Suhairidi, F.: Strategic intent and strategic leadership: A review perspective for post-COVID-19 tourism and hospitality industry recovery. Emerald Handb. Destin Recover Tour. Hosp. 23–44 (2022)
54. Ministry of Business, Innovation and Employment. 2020 Support - tourism recovery package [Internet] (2020). Available from: https://www.mbie.govt.nz/immigration-and-tourism/tourism/tourism-recovery/2020-tourism-recovery-package/
55. Kangas, I., Schwoerer, M., Bernardi, L.J.: Recommender systems for personalized user experience: Lessons learned at Booking.com. In: RecSys 2021—15th ACM Conference Recommended Systems, pp. 583–586, 13 Sep 2021
56. Meikle, S.: Food tourism's moment in the New Zealand sun. J. Tour. Futur. **8**(2), 234–239 (2022)
57. Oliver, H.: From localism towards localism. Policy Q. **5**(21), 14–18 (2019)
58. Trentin, F., de Almeida Moraes, C.C., de Fátima Fogaça, I., Lidizia Soares, C.A.: Impact of the COVID-19 Pandemic on Brazilian Tourism: Public Policies, Coordination and Government Functions. In: Pandemics and Travel: COVID-19 Impacts in the Tourism Industry. Emerald Publishing Limited, pp. 67–82 (2021)
59. Chen, H., Huang, X., Li, Z.: A content analysis of Chinese news coverage on COVID-19 and tourism. Curr. Issues Tour. **25**(2), 198–205 (2022)
60. Haywood, K.M.: A post COVID-19 future-tourism re-imagined and re-enabled. Tour. Geogr. **22**(3), 599–609 (2020)
61. Sharma, G.D., Thomas, A., Paul, J.: Reviving tourism industry post-COVID-19: a resilience-based framework. Tour. Manag. Perspect. **37**, 100786 (2021)
62. Goh, H.C.: Strategies for post-COVID-19 prospects of Sabah's tourist market–Reactions to shocks caused by pandemic or reflection for sustainable tourism? Res. Glob. **3**, 100056 (2021)
63. Faisal, Y.A., Kanti, R.A., Alamanda, A.R., Apriliadi, A.: The role of business model innovation (BMI) in social enterprises during pandemic COVID-19 in Indonesia: a case of Islamic boarding schools' business units. In: Open Innovation in Small Business: Creating Values for Sustainability Sep 26 Sep 2023, pp. 1–16. Springer Nature Singapore, Singapore
64. Dhanalakshmi, K., Jwalapuram, S.: A study on business resilience of MSME's with special reference to pandemics/crises. In: International Conference on Economics, Business and Sustainability, 13 Jan 2023
65. Abdul Ghani Azmi, I., Che Hashim, R., Mohamed, H.A.B., Sahol Hamid, N.: A Study of Muslim Women Entrepreneurs' SMEs Challenges and Motivation in the Asia Pacific Region. In: Open Innovation in Small Business: Creating Values for Sustainability, pp. 195–213. Springer Nature Singapore, Singapore (2023)
66. Kusumastuti, R., Virgianita, A., Juwono, V.: Co-creation strategy in empowering indigenous Women's innovation in Indonesia: empirical evidence of weaver community entrepreneurs in West Nusa Tenggara. In: Open Innovation in Small Business: Creating Values for Sustainability, 26 Sep 2023, pp. 17–37. Springer Nature Singapore, Singapore

Understanding the Significant Woodcarving Motifs of the Traditional Timber Royal Palace Panels

Siti Aisyah Muhammad

Abstract Royal sovereignty is a social acceptance between the commoners and rulers that develops the mutual social status hierarchy. The effect of this understanding is presented among the traditional wood carvers, whereby only the chosen one will be honored to become the wood carver for the royal buildings. It is essential to identify the culture that has been inherited to select the royal wood carver and the customary aspect of sovereignty that prohibits the commoners from applying the selection of wood carving designs to their houses. This study will examine the intricacy of the significant woodcarving motifs in traditional timber royal palaces. The methodology selected to achieve the study's objective was qualitative research. The research tools consisted of interviews with the targeted master craftsmen, a review of data collection and measured drawings, and the precedent study. This study focused on a traditional timber palace, namely Istana Balai Besar, Kelantan, and Istana Tengku Long, Terengganu. The findings show that the elements of the wood carvings applied in both palaces were dedicated to the Ruler during the era and were incomparable to the commoners' houses. The calligraphy motifs of Quranic verses, mixed motifs of flora and fauna, and the intricate 3D incisions of the timber panels are the superior designs of this palace's wood carving ornamentation. Thus, the Ruler's sovereignty controlled the custom, including the elements of an aesthetic of the building during the era indirectly, to show the social status between the Ruler and the ruled persons. The study's implication is the enhanced understanding of mutual acceptance of the sovereignty concept in social life among the wood carvers, which indirectly regulates the wood carving ornamentation at the palace and the house. This study also exposes the beauty of the traditional royal palaces with intricate wood carving motifs that contribute to the local heritage identity and vital landmarks for tourism.

Keywords Royal palace · Woodcarving · Ornamentation

S. A. Muhammad (✉)
Department of Interior Design, College of Architecture and Design, Prince Sultan University, Rafha Street, Riyadh 11586, Saudi Arabia
e-mail: amuhammad@psu.edu.sa

B. Awwad (ed.), *The AI Revolution: Driving Business Innovation and Research*, Studies in Systems, Decision and Control 524, https://doi.org/10.1007/978-3-031-54379-1_71

1 Introduction

The current scenario indicates a gap in the Malay woodcarving heritage advancement and analysis, especially in traditional Malay royal palaces. These include developing new motifs for contemporary use, competing with worldwide woodcarving items, and correctly documenting cultural heritage products. Woodcarving knowledge has been passed down through the centuries to live up to history. Very little work is done to document and research ornaments' forms, placement, and meanings in our heritage buildings, from houses to shops, mosques, palaces, temples, and labor houses [1]. The issues raised in this study are based on only a few timber palaces standing in Malaysia. There is a dialectic debate about preserving Malay architecture, focusing on the Malay house. Besides, scholars have been interested in the Malay Mosque's ornamentation and relics since the early 1980s, but there has been little follow-up about the timber palaces.

Similarly, academics still need to produce extensive research and documentation to record the Malays' traditional wood carving in the royal palaces. While the rapid advancement of technologies and modern architecture has imposed a risk of overshadowing the art of woodcarving [2] due to the present technological advancements also, there has been a tendency of modern Malay houses or buildings being created or existing houses being rebuilt employing non-traditional materials and building components. One of the functions of woodcarving is to reflect the social status of its owner and how this is related to royal sovereignty and the woodcarving composition at timber palaces [3]. Besides, the importance of Malaysia's sovereign interest can no longer be denied, as cited in the National Principles, which is loyalty to the king and the country, demonstrating the significance of the royal hierarchy and the king occupying the highest position and protecting the people. This study will explicitly examine the sociocultural influences of sovereignty and determine the significant intricate ornamentation of the wood carvings, especially in the traditional timber royal palaces. The structure of this writing introduces the background of the traditional Malay wood carving motif and its relation with Royalty. Then, the sample of the case studies with the most complex motif consisting of the higher motif elements such as the flora, calligraphy, and geometry composition will be discussed further about the significant intricacy of the traditional palace wood carving motifs that prohibited for commoners. This study focused on the remaining preserved traditional timber palaces, namely Istana Balai Besar and Istana Tengku Long, as the case study.

2 Literature Review

There is no disputing that the art of wood carving in Kelantan and Terengganu is influenced by countries with which they have a commercial relationship. Through this trade, foreign influence permeated the Malay population, which was accepted through adaptation to local natural conditions to establish its own identity. The Petalawati Indera Sri Kayangan has become the pride of the Kelantan. However, it was carved in neighboring Thailand during the reign of the Chulalongkorn emperor as a gift to the Kelantan government for delivering *bunga emas* yearly (personal interview with Mr. Muhaimin, 2022). Besides, the carvings found in Kelantan today link to neighboring nations such as Jawa. The carvings of flowers and leaves found in the palaces and Borobudur Temple on the East Coast contain the same carving motifs. Terengganu states also possess a few beautiful carved palaces, such as Istana Tele, Istana Raja, Istana Kolam, Istana Tengku Long, and Rumah Tengah. It is stated that most of the wood carvings are influenced by Siamese carving motifs, as Siam once influenced the state. This is based on the carving patterns of mosques, waqfs, palaces, and old residences, which share many similarities with Siamese state sculpture. Thus, it is expected that the wood carvers in this country originated in Patani, Southern Thailand. The socio-culture of Malay society comprises law, customs, order, and advice. The most vital aspect of the Malay socio-culture is the customs and traditions. It differentiates the social class and status of society. Malay rulers and royalties were held in high regard in traditional Malay society. The royals' socio-culture, which was similar to but more refined than ordinary Malays, was generally reflected in palace architecture [4]. The ornamentation was designed especially for the kings, with the most complex and intricate to reflect the owner's status. Previously, the ornamentation motifs were influenced by Hinduism, and as Islam was introduced, the application of animal and human figures as motifs was prohibited. The common motifs that can be seen are the fauna, flora, Islamic calligraphy, geometric patterns, and the universe. Table 1 shows the motif's application in the traditional woodcarvings.

3 Methodology

To achieve the study's objective, the methodology applied for this research was the qualitative method. A case study was selected at Istana Balai Besar, Kota Bharu, Kelantan, and Istana Tengku Long, Terengganu, with a review of outcomes through documentation and measured drawing by Gapura Measured Drawing [5] and Terrapuri Measured Drawing [6]. An in-depth interview with craft master Mr. Muhaimin was done to understand the woodcarver philosophy and techniques. Besides, reviewing the secondary data from the journals, books, and unpublished thesis was also implemented.

3.1 Istana Balai Besar

Istana Balai Besar (Fig. 1) has been the residence of the rulers of Kelantan since the nineteenth century. It was officially turned into the Office of the Sultan of Kelantan. Istana Balai Besar is famous for having a large front hall, a throne, and a collection

Table 1 The development of motifs before and after the presence of Islam

Before Islam (Before fourteenth century)	Description	After Islam (After fourteenth century)
Dewa/God	Appears on the hilt and believes it protects the owner invincible in a fight	Forbidden
Kala	Disobedient creature who was supposedly sentenced to eat himself Shiva later reversed the penalty and made it a Protector and was used as *acuan putu*	*Punca bunga ukir/Ibu/Masa*
Makara	The head and body of *makara* are shaped like fish, and an elephant's trunk is attached to the mouth	*Sulur bayu*
Mount Meru	Refer to God in Hinduism	*Kepala*
Stupa	Representation of the earthen stack in the Indian cemetery and a symbol of Buddha entering the Nirvana	*Buah Buton*
Pohon Beringin	Another type of animal immortalized in Malay woodcarving is the *naga* or dragon	*Pohon Budi*
Padma/Teratai	A fragrant flower	*Bunga Tunjung*
Naga	Not applied in motifs	Discourage in usage
Garuda/Jentayu	The *garuda* is renowned as Vishnu's chariot, and it was also utilized to convey the royals in any important procession in the Malay realm	*Petala Indera/Gagak Sura*

of the King of Kelantan royal instruments. Sultan Muhammad II built Istana Balai Besar to replace his palace on Saba Island, which was destroyed due to floods and riverbank erosion. The palace's location was picked based on its accessibility at the time, as it was neither too far nor too near the river of Kelantan, where traders used to do their business, and also due to the belief that the nearby Calumpang tree was magical. The Malay community at the time believed that when a Calumpang tree produced flowers, it would prevent flooding to the surrounding area. After the Istana Balai Besar was completed, the people called it 'Kota Bharu' in conjunction with 'Istana or a new city that was completed.' Over time, this name became a household and official name for the State Capital of Kelantan, Kota Bharu. Istana Balai Besar also symbolizes the sovereignty of the institution of the King of Kelantan, which is magnificent in the heart of the state capital of Kelantan. Table 2 shows the summary of the palace chronology.

Fig.1 The facade of Istana Balai Besar, Kelantan [5]

Table 2 The timeline of Istana Balai Besar, Kelantan

Year	Description
1844	Sultan Muhammad II built Istana Balai Besar to replace the previous Istana Kota Lama located at Kota Saba
1839	The Sultan Muhammad II residence
1886	The Sultan Ahmad's residence
1889	The Sultan Muhammad III residence
1890	The Sultan Mansur residence
1899	The Sultan Muhammad IV residence

3.2 Istana Tengku Long

Istana Tengku Long (Fig. 2) was built in 1904 by Tengku Long because of his marriage to Tengku Embung from the relatives of the great king of Terengganu. According to a discussion with Dr. Zukifli Abdullah, in 1920, Kampung Pengkalan Teluk, a densely populated village, was located on the river's edge and caused the soil to be eroded by the river. Therefore, the resident school there had to move to a new site for the second time, namely Balai Istana Tengku Long. The hall was the former residence of Raja Besut, owned by Tengku Long at the time. The school was also moved from the palace for the third time after a significant flood, namely the Red Flood in 1926, which submerged almost the entire city of Kuala Terengganu and several other districts. In 1929, The governance of Besut district, which borders the state of Kelantan, took over by Tengku Long from Tengku Chik Haji, and Tengku Long officially became the second Besut commissioner. The palace was built during the reign of Tengku Long with the help of two artisans, Mrs. Long and Cik Hat, who were famous skilled workers and craftsmen from Pahang. Mrs. Long and Cik Hat were appointed as a contractor for the project building, and the palace had very distinctive and unique decorations that manifested the richness of traditional Malay architecture of the period. Table 3 shows the palace's historical background.

Fig. 2 Istana Tengku Long, Kuala Terengganu [6]

Table 3 The timeline of Istana Tengku Long, Terengganu

Year	Description
1899	Palembang was made part of the State of Terengganu and became a British navy. The name of Palembang was changed to the name of Besut
1904	Istana Tengku Long was built
1919	Tengku Wok Palace was built
1920	Kampung Raja National School was transferred to Tengku Long Palace's *Balai*
1926	Hit by a red flood while under the British Adviser
1929	The governance of Besut district took over by Tengku Long from Tengku Chik Haji and Tengku Long
1984	The Malaysian Heritage Agency and PETRONAS reached an agreement to buy the palace from Tengku Zainal Abidin and relocate at the Museum of Terengganu

4 Findings and Discussion

4.1 Woodcarvings Panels in Istana Balai Besar

The intricate design of the woodcarvings inside Istana Balai Besar shows the hierarchical status between royals and commoners. There are twelve panels of woodcarving located at Istana Balai Besar, and the flora motif is the highest application here. The panels located at this palace are decorative panels with no structural or functional purposes. Figure 3 shows the 3D redevelopment of the main hall in the palace with the intricate woodcarving panels. The most intricate motif incomparable to other palaces is the combination motif of flora with Arabic words calligraphy stating the names of four Sahabi of Prophet Muhammad (peace be upon him): Abu Bakar As-Siddiq, Umar Al-Khatab, Usman Bin Affan and Ali Bin Talib located at the Front

Fig. 3 The 3D development of the woodcarving panels at main hall of Istana Balai Besar

Hall (*Balai Surung Hadapan*). Besides, the floral motifs, such as flowers, become the most prominent elements of ornamentation in Istana Balai Besar. The detail of the significant motif of the panel will be elaborated in Table 4.

The uniqueness of this panel (Table 4) intentionally presents the faith of the Royalty as Muslim and the King or Sultan as the religious leader for more decades. The intricacy of the panel combined the element of floral motif (*Bunga Bayam Peraksi*), The Arabic calligraphy with the name of the four caliphs, and composed in a symmetrical geometry frame. The dedicated royal wood carver applied the deep grove, although this panel is only for decoration and not for ventilation.

4.2 Woodcarvings Panels in Istana Tengku Long

There are 31 panels located at Istana Teng Long, and most of the panels are functional panels that work as louvers for ventilation and lighting. The 3D redevelopment of the hall is shown in Fig. 4. The intricacy of the woodcarving panel motifs in this palace is presented with the 3-dimensional incision of the elements. The most significant motif incomparable with other palaces in this world is the panel with floral motifs that illustrated the two birds representing the verse from AlQuran (61: 13): "..*help from Allah and an imminent victory*..". These panels are located at the right elevation of the palace and significantly face the primary access towards the palace from the river. This verse gives the meaning of hope for gaining the victory of a person in his life and any effort and is probably used as a prayer. In addition, calligraphy ornamentation is acceptable as a culture only but rejected as a belief. Istana Tengku Long was a presidential palace for Royalty and the Ruler of Besut. Therefore, it was believed

Table 4 Details of the sample wall panel

Woodcarving with a combination floral motif (*Bunga Bayam Peraksi)* and Arabic words calligraphy	Detail
A wall carving panel with the names of the four Rulers after Prophet Muhammad (saw) with circle-shaped flower motifs (*Bunga Bayam Peraksi*). Four named are Abu Bakar As-Siddiq (1), Umar Al-Khatab (2), Usman Bin Affan (3), and Ali Bin Talib (4). It is made of Chengal wood and painted in gold. The perforated carving comes from a piece of board or several pieces of board that are drilled using a saw drill (drilling saw) so that it penetrates flower shapes or patterns	Style: *Ukiran Bunga Timbul* Types of Flower: *Bunga Bayam Peraksi* Floral Element: Flower, Leaves Composition Pattern: Complete Pattern Principle Design: Symmetrical Principle Form: Single Rectangular

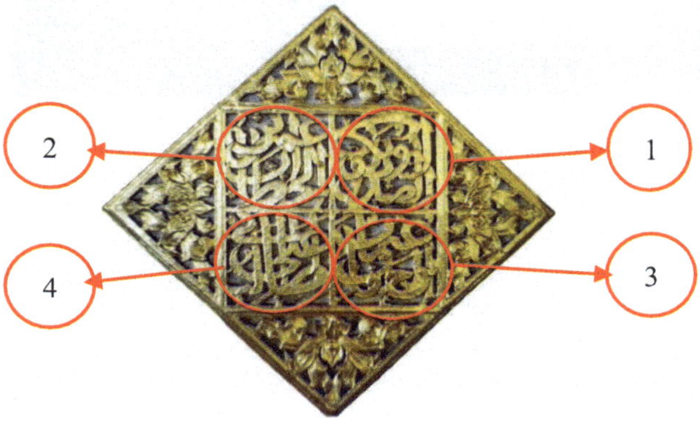

that the importance of this verse in their daily life was a prayer to gain victory in ruling the state and any intention these Royalties wished. Table 5 shows an example of the most complex intricacy of a panel in this palace.

The panel in this hall (Table 5) is a functional panel with ventilation and natural lighting purposes. The intricacy of the wood carving represented the belief in Royalty and also the cultural obedience in forbidding the fauna motif to be visualized as decoration. The image of the bird claimed on his panel can only be seen with focused eyes.

Reviewing the panels of woodcarving located at both palaces, the percentage of motif application is summarized in Table 6. The palace, as the royal monarchy system applied the hierarchical concept between rulers and ruled persons, influenced the ornamentation of the palace as well. The woodcarving on both palaces' panels symbolized the vital position of the royalties that no commoners' houses could decorate with a similar design. As the ornamentation can be the structural, functional, or decorative elements, both palaces show different purposes of the woodcarving panels. Istana Balai Besar implemented the woodcarving panels as the decorative

Fig. 4 The 3D image of the hall showing the woodcarving at Istana Tengku Long

Table 5 Details of the wall panel

Woodcarving with a combination floral motif (*Sulur Kacang*) and Quranic verse calligraphy	Detail
This ornament was called *Sisir Angin*. The motif of calligraphy has a bird-like shape that is talking to each other. Consists of the verses 13 Surah As-Saff:	Motif: Calligraphy and beanstalk Pattern: *Awan Larat/Pola lengkap pemidang* Technique: Relief carving without incision Source: Source from the seeds Composition Pattern: Complete Pattern Principle Design: Symmetrical

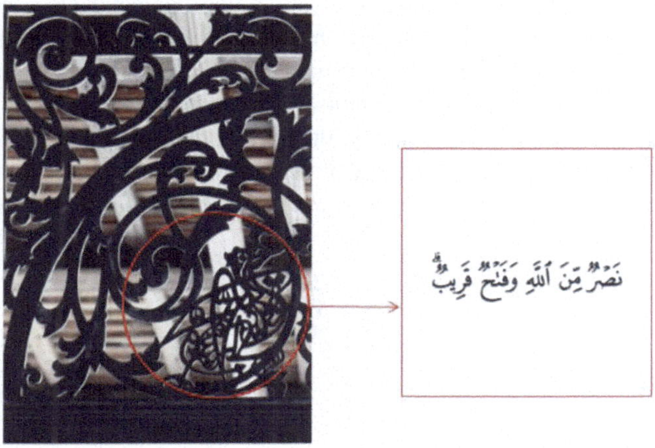

Table 6 Comparison of the elements of woodcarving analysis at both palaces

Motif	Istana Balai Besar	Istana Tengku Long
Flora	/(80%)	/(80%)
Fauna	–	/(5%)
Geometry	/(10%)	/(5%)
Islamic calligraphy	/(10%)	/(10%)
Universe	–	–

elements, while Istana Tengku Long applied the woodcarving panels as the fenestration. Therefore, the woodcarving panels in Istana Tengku Long are mostly relieved with solid and void design.

5 Conclusion

Malay woodcarvings are an ornament that is a symbol of the identity and creativity of the Malay community. Besides that, it is vital to safeguard our legacy and keep it in the best condition possible to prevent its potential from being lost. The carving is more than just a decoration; it requires eloquence in its construction, and every choice of motif, pattern, source, incision, technique, and philosophy arrangement to form a woodcarving composition conveys many emotions and experiences. The most vital part is that the woodcarving at the palace is undoubtedly a symbol of royal status and dignity, as the royal works as the civilization's identity. The significance of the selected motifs represents the royal symbol of the Ruler, and the intricacy was forbidden for the commoners to duplicate. This research will contribute to understanding the vital values of the appreciation of traditional royal palaces and woodcarving as the identity of Royalty. Unfortunately, both traditional palaces need proper preservation and conservation as both were made of timber. The ingenious ingenuity of the wood carvers for both palaces need to be recognized, and the potential documentation with current preservation technology can contribute to prolonging the buildings for future generations. Convincedly, the current technology in wood carving alone can only produce the most intricacy of the wood carving motifs, such as in both palaces, with understanding the meaning and philosophy of the Ruler and ruled persons.

References

1. Utaberta, N., Sojak, S.D.M., Surat, M., Che-Ani, A.I., Tahir, M.M.: The typology study of traditional mosque ornamentation: an analysis of mosque elements. Int J Civil Architectural Sci Eng **6**(7), 10–17. Dover, USA (2012)
2. Silah, S., Ponimin, Taswadi, Ambarwati, D.R.S.: The beauty of Malay woodcarving motif ornamentation. In: International conference on art, design, education and cultural studies (ICADECS), pp. 81–89. (2021). https://doi.org/10.18502/kss.v5i6.9181
3. Muhammad, S.A.: Potential Heritage tourism of an old palace, Istana Sepahchendera: understanding the influence of socio-culture of royalty. In: Alareeni, B., Hamdan, A. (eds) Impact of artificial intelligence, and the fourth industrial revolution on business success. ICBT 2021. Lecture Notes in Networks and Systems, vol 485. Springer, Cham (2023). https://doi.org/10.1007/978-3-031-08093-7_68
4. Muhammad, S.A., Rosdi, N.M.: IOP Conf Ser.: Earth Environ Sci **1102** 012069 (2022). https://doi.org/10.1088/1755-1315/1102/1/012069
5. Report, G.M.D.: Istana Balai Besar. Universiti Malaysia Kelantan, Unpublished Report (2021)
6. Report, T.M.D.: Istana Tengku Long. Universiti Malaysia Kelantan, Unpublished Report (2022)

Knowledge, Attitude, Practice (KAP) of Food Handlers and Food Quality on Customer Retention

Akua Serwah, Derweanna Bah Simpong, and Marlisa Abdul Rahim

Abstract The objective of this paper is to determine the influence of attitude, knowledge, and practice and food quality on customer retention of food handlers in Ghana. The knowledge-attitude-practice (KAP) paradigm served as the theoretical underpinning. The cross-sectional data were gathered from 500 food handlers via a questionnaire survey. The partial least square structural equation modeling (PLS-SEM) second generation statistical technique was used to analyse the data. The study results revealed that knowledge, attitude, practice and food quality exerted a positive and significant influence on customer retention. This article will assist researchers and policymakers in deeper comprehending the elements impacting customer retention in Ghana's food and beverage business. The discussion, recommendations, and limits concluded the study.

Keywords Attitude · Knowledge · Practice · Food quality · Customer retention

1 Introduction

Academicians have called attention to the potentials of the foods and beverages sectors within the hospitality industry, studies in Indonesia indicates how the food and beverage sector is the second largest contributor to the economy [1]. The Food service industry accounts for over twenty percent of service earnings and has an average growth rate per year of more than 3%. It is the second greatest earnings generator in the hospitality sector [2]. According to Edward [3] this sector supplies food and beverages that are purchase and eaten both within and outside the home. The food handler's knowledge, attitude and practice are crucial in determining the end product of the cuisine [4]. Food and beverage service is an area of interest for research on customer behaviour. According to [5, 6], the past decades has demonstrated how the food and beverage service industry has made significant impact in terms of

A. Serwah (✉) · D. Bah Simpong · M. Abdul Rahim
Faculty of Hospitality, Tourism and Wellness, Universiti Malaysia Kelantan, 16100 Kota Bharu, Malaysia
e-mail: serwahlady426@gmail.com

B. Awwad (ed.), *The AI Revolution: Driving Business Innovation and Research*, Studies in Systems, Decision and Control 524, https://doi.org/10.1007/978-3-031-54379-1_72

research, and also an improvement in their excellence food service. The study also made discoveries that there have been transforming and developing components of the food service industry restaurant operations in the areas of service quality, food service restaurant finance, food service marketing, food safety and health, and industry technology.

Irrespective of the discoveries, the search for alternatives has not progressed far enough in terms of indigenous food [7] and ties with consumer retention. This means that, despite multiple studies conducted over the period of ten years, it is entirely reasonable to argue that some critical topics have been overlooked and should be examined. In such a situation, Ghana's food and beverage sector has to present its offerings with a strategic marketing approach that focuses on identifying the key factors influencing the retention of customers in the food industry. Therefore, this paper examines the significant impact of KAP-Q (Knowledge, Attitude, Practice and Quality) on customer retention in Ghana.

2 Literature Review

2.1 Theoretical Foundation

The knowledge-attitude-practice (KAP) paradigm was used as the theoretical framework in this study. Knowledge acquisition, attitude development, and behaviour creation are the three sequential processes that comprise the "KAP theory," which was established by [8] and has since been adopted as a theory of health behaviour modification [9]. Knowledge, according to the KAP model, has a positive impact on a person's attitude, which in turn influences practices or behaviour [10]. The KAP model argues that individuals' behaviour or practices on their knowledge has significant impact and could bring change in attitude and, as a result, a transformation in behaviour or practice [11]. Furthermore, food handlers' attitudes about the preparation and cooking of indigenous cuisine, as well as how these dishes are presented play a significant role in achieving customer expectations. According to the theory, knowledge serves as the foundation for behaviour change, while attitudes and beliefs serve as the trigger [9].

2.2 Customer Retention

Customer retention (CR) is a crucial aspect of relationship-based marketing. It is defined as the percentage of initial customers that continue to be clients at the end of the period [12]. Customer retention impressions and applications are currently extremely valuable for enterprises [13]. Various studies have been conducted in order to uncover factors that influence customer retention [14].

2.3 Hypotheses Development

2.3.1 Attitude Toward Customer Retention

The notion of attitude is clearly defined as consistently positive or negative responses to a specific thing [15]. Attitude is made up of three parts: cognitive, affective, and behavioural. Attitudes can be formed in the food preparation surroundings based on behavioural beliefs and a variety of factors such as previous exposure in the area of training and current environment, and that attitude leads in progressive behaviour change. According to [16], attitudes have a favourable and considerable impact on consumer retention. Thus, the following hypothesis is formulated:

H1: Attitude is positively associated to customer retention.

2.3.2 Food Quality

There are specific indicators of quality cuisines that the food handlers must introduce to their customers through their menus, such as the nutrient content, variety on the menu, the service style adopted, the way food is garnished and presented on the plate and how the food taste [17]. These determine how perceived food quality is significantly associated with repeat business [18]. Food quality is described as "fitness for use" or, more precisely, "fitness for consumption," which is highly associated with consumer enjoyment [19]. In order for consumers to consume the food properly and be satisfied, the quality of food must meet the expectation. According to [20], food quality is defined by various factors, including freshness, flavour, and nutritional value. [21], discovered a robust and beneficial connection between customer retention and food quality. As a result, this study formulated that:

H2: Food quality is positively associated to customer retention.

2.3.3 Knowledge

Knowledge is defined as the acquisition, retention, and application of information or abilities [8]. Acquiring knowledge involves cognition, concept understanding, and the ability to discriminate acquired information from feelings experienced. Education and experience are both sources of knowledge. According to [22], customer knowledge management has a direct and considerable impact on customer retention. Thus, this study stated that:

H3: Knowledge is positively associated to customer retention.

2.3.4 Practice

Badran [8] Defines practice as "the application of knowledge and rules to take a final action." The acts or behaviours that can help improve the performance of various restaurants are referred to as "best practices." As a result, practice focuses on presenting previously acquired knowledge. Recent empirical studies found that food sustainability practices significantly influence consumers' satisfaction [23, 24]. Thus, the following hypothesis is formulated:

H4: Practice is positively associated to customer retention.

3 Research Methodology

This study used a cross-sectional research design and gathered quantitative data from food handlers in Ghana. The research population for this study was operating in in five regional capitals namely Accra, Tema, Takoradi, Kumasi and Koforidua. This study's sample size was calculated to be 172 using G-Power 3.1, with a power of 0.95, an effect size of 0.15, and 5 model predictors [25]. Also, for structural equation modeling using partial least squares, sample sizes ranging from 50 to 200 are appropriate [26], with a minimum of 100 samples necessary [27]. Furthermore, [28] stated that a sample size of more than 200 is considered high for a non-complex model. As a result, to prevent any complications caused by the limited sample size, the research collected data from 500 food handlers.

The questionnaire items were adopted from previous research validated scales. Items for attitude were obtained from [29]; Items for food quality were adopted from [30]. Items for practice were adopted from [31]. Items that measured knowledge and customer retention were retrieved from [32].

3.1 Reliability and Validity

According to Table 1, CA, DG rho, and CR all had values greater than 0.821 [26]; thus, every one of them are judged reliable. Likewise, the AVE value for all items was greater than 0.50, indicating adequate convergent validity [26]. In addition, the VIF for each construct and all values were below 3, indicating that multi-collinearity among the study variables was not an issue [26]. As a result, these research constructs met both the criterion for reliability and convergent validity.

The Foronell-Larcker criterion and the Hetrotrait and Mono-trait ratio (HTMT) are shown in Table 2. Both of these validity tests, when combined, suggest that the measuring items are both valid and reliable. Furthermore, Table 4 (Appendix), which displays the discriminant validity of cross-loading, showed that all cross-loading values for the constructs met the minimum threshold value of 0.708 [26].

Table 1 Reliability analysis

Variables	CA	Dg Rho	CR	AVE	VIF
Attitude	0.866	0.878	0.891	0.509	1.452
Food quality	0.821	0.886	0.861	0.564	2.827
Knowledge	0.901	0.929	0.925	0.645	2.511
Practice	0.909	0.926	0.925	0.58	1.242
Customer retention	0.966	0.968	0.972	0.811	

Notes CA Cronbach's alpha; *DG rho* Dillon-Goldstein's rho; *CR* composite reliability; *AVE* average variance extracted; *VIF* variance inflation factor

Table 2 Discriminant validity

	AT	CR	QT	KN	PR
Heterotrait-monotrait ratio (HTMT)					
AT					
CR	0.328				
QT	0.511	0.878			
KN	0.297	0.796	0.844		
PR	0.522	0.202	0.319	0.178	
Fornell-Larcker criterion					
AT	0.713				
CR	0.344	0.901			
QT	0.431	0.845	0.681		
KN	0.291	0.753	0.774	0.803	
PR	0.439	0.2	0.228	0.15	0.762

As a result, the outcomes reveal that discriminant validity is satisfied for the study's constructs.

3.2 Path Analysis

As shown in Table 3, the adjusted r^2 value for customer retention (CR) with exogenous constructs attitude (AT), food quality (QT), knowledge (KN), and practice (PR) signified that (73.9%) of the change in the CR explicated by the exogenous variables.

The path coefficient's values for AT ($\beta = 0.022$ and $p < 0.05$), QT ($\beta = 0.661$ and $p < 0.05$), KN ($\beta = 0.244$ and $p < 0.05$), and PR ($\beta = 0.122$ and $p < 0.05$), exerted a positive and significant influence on CR. The confidence intervals bias corrected 95% also did not show any intervals straddling a 0, thus confirming the results of the study. Therefore, H1, H2, H3, and H4 were supported. The f^2 value of 59.3%

Table 3 Hypothesis testing

Hyp.	Relationship	Beta	t-value	*p*-value	95% BCI LL	95% BCI UL	R^2	f^2	Decision
H1	AT → CR	0.022	7.470	0.027	0.071	0.126	0.739	0.021	Supported
H2	QT → CR	0.661	14.676	0.000	0.585	0.733		0.593	Supported
H3	KN → CR	0.244	5.215	0.000	0.166	0.32		0.091	Supported
H4	PR → CR	0.122	8.36	0.001	0.026	0.163		0.032	Supported

obtained indicates a large effect of QT on CR, while AT, KN and PR had almost small effect on CR.

4 Discussion, Implication and Limitations

The current study investigates if food handlers AT, KN, PR, and QT influence CR. According to the study analysis, food handlers AT, KN, PR, and QT exhibited a significant beneficial impact on CR, supporting H1, H2, H3, and H4. The outcome was consistent with [33] finding that KN had a significant impact on CR. Thus, food handlers should focus on knowledge because, according to [34], when customer satisfaction increases, it translates into customer retention, and this may be as a result of food handlers becoming knowledgeable about their recipes in terms of preparation, cooking, and serving method of native foods.

The study's findings matched those of [35], who discovered that AT had a substantial influence on CR. The survey also advised restaurant managers to cultivate an attitude of efficiently resolving customer concerns in order to keep them. When it comes to maintaining online customers, [36] believes that attitude is critical. The study finding is also compatible with that of [23], who displayed that PR exerted a significant influence on CR. According to the findings of the study, QT was the most crucial predictors to predict CR, and this is consistent with past research that QT had a significant effect on CR [37, 38].

5 Implications

The current work enhances the theoretical stance that the perceptions of food handler attitude, knowledge, practice and food quality on customer retention. Thus, the current study contributes to the knowledge-attitude-practice theory.

Again, the study findings offer empirical and practical implications for the food service industry. The attitude of food handlers is critical in defining the entire dining experience and client satisfaction. Restaurants and food enterprises should prioritise food handler training, incentive, and assistance to ensure they retain a positive attitude and give exceptional customer service, resulting in higher customer retention rates. Furthermore, food handler knowledge has a direct impact on customer retention by enhancing the quality of the indigenous cuisines, through adequate knowledge on the Ghanaian recipe. Restaurants and food sector should invest in training and educational programs to guarantee that their food handlers are aware and up to speed on industry best practices and developing customer expectations. Additionally, food handler practices have a direct impact on customer retention by impacting food quality, presentation, and efficiency, adaptability, increasing variety and using wholesome ingredients. Food businesses should prioritize training and monitoring to ensure that food handlers constantly adhere to best practices and provide great dining experiences in order to keep their valued consumers. Moreover, food quality is an important factor that has a direct impact on customer retention. Regularly providing outstanding and appetising food not only satisfies customers but also generates a strong brand reputation, fosters loyalty, and draws new customers. Sustaining good food quality should be a key priority for food and beverage sector in order to rub shoulders with competitor on the market.

6 Limitations

The present research makes valuable empirical contributions to both the literature and practice. Nevertheless, the current study has three significant drawbacks. The current study used a quantitative research methodology with restricted applicability and did not completely investigate the research topic at hand. In the future, studies should use a mixed technique or qualitative research design to properly understand consumer retention in the food and beverage industry. Future research should look into a variety of elements that influence client retention in the food industry. Finally, customer retention in the food service sector is a process that takes time to modify. Adoption of modern technology in the food business may lead to increased client retention. Future study could presume the use of technological items in the food business.

7 Conclusions

Indigenous cuisine is vital within the Food and Beverage sector and a contributing factor to the Gross Domestic Product of many nations globally. The objective of this study is to empirically examine the effect of food handler knowledge, attitude and practice, as well as food quality on customer retention. Via structural equation modelling, food handlers' factors were tested based on the sample of 500 food handlers operating in food services. The outcomes of the study displayed that food handler's knowledge, attitude and practice, as well as food quality exerted a positive and significant effect customer retention. The findings of this research lead to expansion of knowledge-attitude-practice (KAP) model while also improving our comprehension of the importance of KAP factors in enhancing customer retention in developing country like Ghana.

Appendix

See Table 4.

Table 4 Outer loadings and cross loadings

	ATT	CUR	FQT	KNW	PRA
ATT1	0.595	0.28	0.345	0.237	0.213
ATT2	0.724	0.192	0.213	0.141	0.355
ATT3	0.599	0.373	0.454	0.345	0.167
ATT4	0.656	0.153	0.232	0.164	0.33
ATT5	0.793	0.171	0.223	0.139	0.427
ATT6	0.773	0.192	0.241	0.121	0.391
ATT7	0.745	0.17	0.223	0.128	0.345
ATT8	0.789	0.189	0.256	0.16	0.397
CUR1	0.297	0.88	0.743	0.632	0.157
CUR2	0.333	0.849	0.76	0.644	0.157
CUR3	0.267	0.898	0.752	0.681	0.188
CUR4	0.348	0.805	0.691	0.607	0.158
CUR5	0.317	0.95	0.785	0.713	0.193
CUR6	0.328	0.951	0.793	0.72	0.189
CUR7	0.295	0.939	0.799	0.722	0.2
CUR8	0.297	0.924	0.763	0.695	0.192
FQT1	0.179	0.159	0.271	0.168	0.191
FQT2	0.295	0.413	0.585	0.479	0.179
FQT3	0.268	0.691	0.762	0.738	0.051
FQT4	0.405	0.143	0.311	0.138	0.323
FQT5	0.299	0.706	0.798	0.598	0.153
FQT6	0.363	0.743	0.852	0.636	0.242
FQT7	0.365	0.755	0.88	0.679	0.193
FQT8	0.338	0.557	0.695	0.432	0.157

(continued)

Table 4 (continued)

	ATT	CUR	FQT	KNW	PRA
KNW1	0.204	0.666	0.672	0.857	0.095
KNW2	0.194	0.277	0.31	0.418	0.257
KNW3	0.214	0.7	0.698	0.884	0.087
KNW4	0.218	0.716	0.723	0.914	0.115
KNW5	0.273	0.612	0.625	0.845	0.122
KNW6	0.31	0.592	0.657	0.82	0.114
KNW7	0.255	0.553	0.57	0.778	0.16
PRA1	0.293	0.077	0.106	0.038	0.543
PRA2	0.358	0.171	0.208	0.127	0.795
PRA3	0.352	0.122	0.131	0.034	0.73
PRA4	0.331	0.155	0.164	0.117	0.833
PRA5	0.37	0.174	0.213	0.162	0.804
PRA6	0.337	0.144	0.186	0.115	0.812

References

1. Bui, T.D., Aminah, H., Wang, C.H., Tseng, M.L., Iranmanesh, M., Lim, M.K.: Developing a food and beverage corporate sustainability performance structure in Indonesia: enhancing the leadership role and tenet value from an ethical perspective. Sustainability (Switz.) **14**(6), 1–24 (2022). https://doi.org/10.3390/su14063658
2. Bondzi-Simpson, A., Ayeh, J.K.: Assessing hotel readiness to offer local cuisines: a clustering approach. Int. J. Contemp. Hosp. Manag. **31**(2), 998–1020 (2019). https://doi.org/10.1108/IJCHM-12-2017-0820
3. Edwards, J.S.A.: The foodservice industry: eating out is more than just a meal. Food Qual. Prefer. **27**(2), 223–229 (2013). https://doi.org/10.1016/j.foodqual.2012.02.003
4. Gruenfeldova, J., Domijan, K., Walsh, C.: A study of food safety knowledge, practice and training among food handlers in Ireland. Food Control **105**, 131–140 (2019). https://doi.org/10.1016/j.foodcont.2019.05.023
5. Kim, W.G., Li, J.J., Brymer, R.A.: The impact of social media reviews on restaurant performance: the moderating role of excellence certificate. Int. J. Hosp. Manag. **55**, 41–51 (2016). https://doi.org/10.1016/J.IJHM.2016.03.001
6. DiPietro, R.: Restaurant and foodservice research: a critical reflection behind and an optimistic look ahead. Int. J. Contemp. Hosp. Manag. **29**, 1203–1234 (2017). https://doi.org/10.1108/IJCHM-01-2016-0046
7. Mcilvaine-newsad, H.: Together at the table: sustainability and sustenance in the American agrifood system. Gastronomica **6**(2), 103–104 (2006). https://doi.org/10.1525/gfc.2006.6.2.103
8. Badran, I.G.: Emhj_1995_1_1_8_16.Pdf. East. Mediterr. Health J. **1**(1) (1995)
9. Samosa, R., Rose, M., Camintoy, M.C., Dayo, R.J., Policarpio, S.J.O.: Crop Production Knowledge, Attitude and Practices of San Joseños in the Midst of COVID-19: Basis for Sustainable Crop Production Action Plan Philippines Which is Considered an Agricultural Country. But Due to the Pandemic, the Imposed the Primary Goal, pp. 40–55 (2022)
10. Kwol, V.S., Eluwole, K.K., Avci, T., Lasisi, T.T.: Another look into the knowledge attitude practice (KAP) model for food control: an investigation of the mediating role of food handlers' attitudes. Food Control **110**, 107025 (2020). https://doi.org/10.1016/j.foodcont.2019.107025

11. Zanin, L.M., da Cunha, D.T., de Rosso, V.V., Capriles, V.D., Stedefeldt, E.: Knowledge, attitudes and practices of food handlers in food safety: an integrative review. Food Res. Int. **100**, 53–62 (2017). https://doi.org/10.1016/j.foodres.2017.07.042
12. Ngacha, W.J., Onyango, F.E.V.: The Role of a Customer-Oriented Service Culture in Influencing Customer Retention in the Hotel Industry (2017)
13. Dastane, D.O., Fazlin, I.: Re-investigating key factors of customer satisfaction affecting customer retention for fast food industry. Int. J. Manag. Account. Econ. **4**(4), 379–400 (2017)
14. Anees, R.T., Nordin, N.A., Anjum, T., Cavaliere, L.P.L., Heidler, P., et al.: Evaluating the impact of customer relationship management (CRM) strategies on customer retention (a study of fast food chains in Pakistan). Bus. Manag. Strat. **11**(2), 117–133 (2020)
15. Krosnick, J.A., Judd, C.M., Wittenbrink, B.: The measurement of attitudes. In: The Handbook of Attitudes. Basic Principles, vol. 1, pp. 45–105. Routledge (2018)
16. Yang, B., Liu, M.-T., et al.: How consumers' attitudes towards alternative food network affect their consumption satisfaction-empirical evidence from Zhengzhou. The. Vjesn. **28**(5), 1568–1574 (2021)
17. Xiao, A., Yang, S., Iqbal, Q.: Factors affecting purchase intentions in generation Y: an empirical evidence from fast food industry in Malaysia. Adm. Sci. **9**(1), 4 (2018)
18. Yuchen, G.: Factors affecting online food quality control among delivers in delivery process in Malaysia. IOSR J. Bus. Manag. **22**(5), 26–29 (2020)
19. Omar, S.S.: Food safety knowledge of street food vendors in downtown Amman-Jordan. Eurasian J. Biosci. **14**(2), 3601–3606 (2020)
20. Ryu, K., Lee, H.-R., Kim, W.G.: The influence of the quality of the physical environment, food, and service on restaurant image, customer perceived value, customer satisfaction, and behavioral intentions. Int. J. Contemp. Hosp. Manag. **24**(2), 200–223 (2012)
21. da Costa Oliveira, Z.C., Saldanha, E.S., Vong, M.: The mediation effect of restaurant image on the relationship between service & food quality and customer satisfaction. Timor Leste J. Bus. Manag. **2**, 10–20 (2020)
22. Kabue, H.W.: Enhancing customer retention: the role of customer knowledge management. Int. J. Bus. Manag. Commer. **6**(1), 1–11 (2021)
23. Chaturvedi, P., Kulshreshtha, K., Tripathi, V., Agnihotri, D.: Investigating the impact of restaurants' sustainable practices on consumers' satisfaction and revisit intentions: a study on leading green restaurants. Asia Pac. J. Bus. Adm. (2022)
24. Zamani, N., et al.: A study on customer satisfaction towards ambiance, service and food quality in Kentucky Fried Chicken (KFC), Petaling Jaya. Malays. J. Soc. Sci. Humanit. (MJSSH) **5**(4), 84–96 (2020)
25. Fishbein, M.: An investigation of the relationships between beliefs about an object and the attitude toward that object. Human Relat. **16**(3), 233–239 (1963)
26. Hair, J.F., Risher, J.J., Sarstedt, M., Ringle, C.M.: When to use and how to report the results of PLS-SEM. Eur. Bus. Rev. **31**(1), 2–24 (2019)
27. Reinartz, W., Haenlein, M., Henseler, J.: An empirical comparison of the efficacy of covariance-based and variance-based SEM. Int. J. Res. Mark. **26**(4), 332–344 (2009)
28. Kline, R.B.: Principles and Practice of Structural Equation Modeling. Guilford Publications (2023)
29. Johnston, A.C., Warkentin, M.: Fear appeals and information security behaviors: an empirical study. MIS Quart., 549–566 (2010)
30. Qin, H., Prybutok, V.R.: Service quality, customer satisfaction, and behavioral intentions in fast-food restaurants. Int. J. Qual. Serv. Sci. **1**(1), 78–95 (2009)
31. Koo, H.-C., Poh, B.K., Ruzita, A.T.: Assessment of knowledge, attitude and practice towards whole grains among children aged 10 and 11 years in Kuala Lumpur. Int. J. Food Sci. Nutr. Diet. **4**(1), 171–177 (2015)
32. Salamah, A.A., et al.: Customer retention through service quality and satisfaction: using hybrid SEM-neural network analysis approach. Heliyon **8**(9) (2022)
33. Nabalayo, A.O., Makomere, J., Karei, R.: Moderating effect of customer product knowledge on cuisine quality and customer satisfaction in selected hotels within North Rift Region, Kenya. Afr. J. Educ. Sci. Technol. **7**(1), 302–311 (2022)

34. Cruz, C.: Regulatory Disclosure Policies and Potential Induced Changes in Behavior: An Outcome Evaluation of Santa Clara County's Enhanced Food Safety Program Elements (2019)
35. Abdul Rashid, S., Ariffin, H.F., Sanny, M., Ungku Zainal Abidin, U.F., Abdul Mutalib, N.-A., Mohd Izani, H.F.: Knowledge, attitude, and practice on food safety culture among kitchen employees of Malaysian government hospitals. ESTEEM J. Soc. Sci. Human. **4**, 180–195 (2020)
36. Al-Ghraibah, O.B.: Online consumer retention in Saudi Arabia during COVID 19: the moderating role of online trust. J. Crit. Rev **7**(9), 2464–2472 (2020)
37. Al-Tit, A.A.: The effect of service and food quality on customer satisfaction and hence customer retention. Asian Soc. Sci. **11**(23), 129 (2015)
38. Atsnawiyah, D., Rizan, M., et al.: The influence of cafe atmosphere and food quality on customer satisfaction in building customer loyalty of Masalalu Café Rawa Domba Jakarta. J. Dinamika Manajemen Dan Bisnis **5**(1), 113–138 (2022)

Conceptualizing the Decision Motivation of Tourist Behaviour Using Neuro-Tourism Analysis

Syazwani Sahwandi, Nor Syuhada Zulkefli, Nik Alif Amri Nik Hashim, and Shida Irwana Omar

Abstract A crisis can influence the behaviour of visitors in a variety of different ways, from the decisions they make about their travel plans to the things they choose to do while they are there. In the wake of these disasters, the tourism industry is confronted with several obstacles that must be overcome to recover, one of which is to comprehend the conduct of tourists in reaction to catastrophic occurrences. Although many studies have been conducted to explore the tourism dilemma, the majority focus on the rise and fall of the number of tourists that visit. However, very little is known at this time regarding how various crises have contributed to the development of distinct patterns of tourist behaviour over time. This research is designed to explore the concepts of Decision Choice Motivators (DCMs) using the Neuro-Tourism method. This research employs qualitative approaches to investigate the tourist behaviour pattern and millennial tourists' destination choice criteria during a crisis. Semi-structured interviews complemented with the brain imagine method and content analysis will be conducted with Malaysian Inbound Tourism Associations (MITA) members and millennial tourists to gather the information. This study will provide new insights for academics and practitioners, particularly tourism organisations, destination management organisations (DMOs), and local tourism suppliers, in marketing the destinations that match tourists' needs.

Keywords Decision choice motivators (DCMs) · Tourist's behaviour · Millennial tourists · Neuro-tourism · Post-crisis · Malaysia

S. Sahwandi · N. S. Zulkefli (✉) · N. A. A. N. Hashim
Faculty of Hospitality, Tourism and Wellness, Universiti Malaysia Kelantan, Kota Bharu, Malaysia
e-mail: syuhada.z@umk.edu.my

S. I. Omar
School of Housing, Building and Planning, Universiti Sains Malaysia, Gelugor, Malaysia

B. Awwad (ed.), *The AI Revolution: Driving Business Innovation and Research*, Studies in Systems, Decision and Control 524, https://doi.org/10.1007/978-3-031-54379-1_73

1 Introduction

The outbreak has brought the whole tourism industry to a stop, as countries throughout the world have taken steps to prevent the spread by closing their international borders. In the current situation, the COVID-19 pandemic issue has put the tourism industry under pressure by imposing entry bans for international tourists and travel limitations for local tourists for specified times. As a result, Malaysia's number of tourist arrivals contracted to a massive 78.6% during the first nine months of COVID-19, a decline from 20.1 million to 4.29 million in 2020 [35]. Simultaneously, as the third-largest contributor of income to the country, tourism revenues have plunged 80.9% from RM66.1 billion to RM12.6 billion in the previous corresponding period [11]. The tourism sector suffered approximately RM 3370 million in losses within the first two months of the Movement Control Order (MCO) [11]. From this point of view, the tourism sector is now returning to the domestic market to help the country's economy recover by understanding travel behaviour and choosing a destination for Malaysian tourists during and after the crisis.

According to [10], tourist behaviours is the result of the relationship between intrinsic (motivation, behaviours, and beliefs) and extrinsic (economic, security, and sociocultural) elements. The outcome of the tourist's processing of stimuli is evaluated in accordance with intrinsic and personal preferences. Additionally, external factors influence how travellers perceive and decide when they travel. Due to the long-term effects of COVID-19, how do tourists plan to travel? [27] In order to rebuild the tourist sector, the travel industry needs to understand the relationships between travel preferences, past travel history, and afterwards travel behaviour [27, 35] Previous studies have revealed that tourists showed several types of decision choices motivations for different destinations during their vacation process before this pandemic [4, 7, 9]. If before, tourists can travel anywhere, even abroad; however, the pandemic has imposed numerous travel restrictions that would affect the landscape of future travel decisions. For example, tourists can only go to nearby destinations and within the country but still need to be selective in the destination due to avoiding the crowded place and practising new norms [9, 14].

The millennial generation or 'Y generation' represents young people born between 1980 and 1995 [2], and they are a significant new market segment that could change the tourism landscape in the world market. This generation is a unique group of tourists whose considered travelling experience is a priority, and their consumption choices and preferences differ significantly from previous generations [13]. Compared to their older counterparts, millennial tourists are one of the largest generations in history, grown in a sheltered environment, more independent, and more likely to revisit some tourist destinations [3]. Responding to this, World Travel and Tourism Council expected a massive market of outbound travellers in the Asia Pacific region in the year 2030, as these new travellers depict six personalities as simplicity searcher, reward hunter, social capital seeker, cultural purist, ethical traveller, and obligation meeker. Therefore, the [34] observed millennial travel as a high-value, resilient market, direct spending to local communities, purposeful travel, flexible and attractive. Given the prominent role of millennials in driving significant changes within the travel sector, an inquiry emerges regarding the forthcoming patterns in travel subsequent to the resolution of the crisis. Basically, this inquiry pertains to the potential shifts in tourist behaviors among the millennial demographic preceding and subsequent to the epidemic. As observed in previous instances, a crisis often gives rise to novel concerns, anxieties, and visitor expectations. Hence, it is imperative to anticipate the trajectory of shifts in tourist behavior regarding decision-making incentives as a proactive approach to effectively address the situation. This study intends to apply a new technique/approach in evaluating tourist behaviour using Neurotourism, which has not been previously investigated in the literature, particularly in the tourism field.

This paper follows the following structure: The data and technique are presented and discussed in Sect. 3, while the literature is reviewed in Sect. 2. Section 4 wraps up the document.

2 Literature Reviews

2.1 Crisis in Tourism

According to research [18, 24], a "crisis" is defined as a "sudden and unexpected event related to the internal environment that causes chaos and destroys organizations that lack decisive action." The economic, social, cultural, environmental, and political effects of tourism are at the heart of the problem. [14] said that the crisis has had a major effect on the tourism industry, specifically through decreased tourist flow, canceled trips, increased unemployment, and decreased corporate investment. In addition, the literature on tourism is replete with examples of crises and their effects on visitor numbers and industry activity. Exogenous shocks to the tourism industry have been triggered by a wide variety of events, not just natural disasters, or geological upheavals. Foot-and-mouth disease, the SARS epidemic, and the Avian Flu pandemic all contributed to a drop in demand across many areas of the tourism industry [5, 7]. Nonetheless, the tourism sector has bounced back since the crisis [31].

The tourism economy is currently facing an unparalleled crisis due to the COVID-19 pandemic, which has caused a significant and sudden disruption in the sector. The global pandemic has brought about significant changes in the tourism industry, necessitating the adoption of new practices commonly referred to as the "New Norms" [12, 14]. During periods of pandemics, individuals tend to exhibit an increased perception of risk associated with various forms of travel, leading them to refrain from visiting locations that are thought to possess a moderate to high level of risk [17]. The aforementioned scenario exemplifies the challenges associated with resuming normalcy, given the previous state of normalcy itself was the underlying issue [3]. Consequently, there has been a significant transformation in the dynamics, functioning, and mobility of tourism activities. In light of the future prospects of the tourism industry, it is vital to comprehend the alterations occurring on the demand side. It is imperative to evaluate their response to these changes through external mediation and examine how immediate consequences can result in long-term alterations in travel behavior, considering both a survey and strategic standpoint [22].

2.2 Underpinning Theory—Theory Decision Making

Researchers have focused on studies about travel decision-making for over four decades to understand how tourists make their leisure choices. According to research that was conducted in the tourism industry [29], the decision-making process for tourists reflects a funnel in that it begins with a wide variety of potential locations and gradually narrows down to a single option. Dreaming, Planning, Reserving, Experience, and Loyalty are the stages that make up the decision-making process, which is considered multi-staged rather than a one-step process [29]. According to [28], decision-making is a dynamic and complex process that is not made up of fixed sequential steps. Numerous factors, both sociopsychological and nonpsychological, have an impact on it.

Theoretically, according to [29], the decision-making process is not fully evolved. The present scenario of research regarding decision-making in tourism lacks a consistent viewpoint that reflects the qualities of tourism services, which results in this lack of progress. In addition, research that focuses on comprehending and characterising the dynamic nature of the decision itself is required because decision-making is a complicated and dynamic process. Therefore, it is likely that the decision-making process will be affected differently under various circumstances, such as the COVID-19 pandemic, and various generations.

2.3 Tourist Behaviour

The literature has presented the connection both to the crisis and tourism in recent years, particularly regarding the risk around specific destinations. COVID-19 is expected to have a significant impact on the commercial behaviour of tourists. It is revealed that once the pandemic crisis ends, tourists would not feel comfortable travelling to tourist destinations as compared before this pandemic [16]. Several studies (e.g. [19, 21]); reported that some factors affected tourists' behaviour after the crises, such as fear of infections, lack of trust, hygiene, and perceived risk. The result also observed new influences when tourists made their travel decisions, such as domestic travel with the family, focusing on sanitation, infection control, and a reliable health care system [19].

Travel behaviour is generally seen as a continuous process that includes various but interrelated steps and concepts that cannot always be analysed separately. In Malaysia's context, the existing literature offers minimal insights into the travel behaviour choice in the post-crisis [20, 35]. Table 1 shows the summary of selected literature on tourist behaviour and decision motivation studies related to this paper's area of discussion.

2.4 Neuro-Tourism

The concept of neuro-tourism integrates the worlds of tourism and neuroscience. How important is neuroscience in the modern tourism context? The brain's capacity to transform and continuously update prior knowledge based on new information, as well as its active generation of internal predictions that direct adaptive behaviour and decision-making, are what enable flexible and intelligent conduct [1]. According to [24], individuals' decisions and interpretations may be influenced by the integration of a variety of disciplines, including sociology, psychology, medical, and tourism. The field of neurotourism was created through the fusion of neuroscience and travel. However, there aren't enough studies being done in this area. For the purpose of tourism study, neuro-tourism examines the brain underpinnings of visitor behaviour [30]. Tourists' reasons for going, how they behave while they're there, and how they respond to challenges are all psychologically significant factors [32]. The brain processes information, detects, frames challenging situations, and selects appropriate actions using a psychological process-oriented approach [30]. Tourists make decisions based on a variety of factors, including psychological ones such as personality, way of life, perception, motivation, and attitude, as well as demographic ones such as age, gender, income level, occupation, educational attainment, and family size, as well as sociocultural ones such as family, advisory groups, social classes, and culture [28, 36]. Finding solutions to the questions of what influences tourists' choice of destination and how their decision-making process is affected is one that should be taken into account initially. Tourists should choose their trip destination based on its amenities and appearance in order to portray it as an appropriate and sufficient location for "a good and safer vacation" in light of this pandemic. This research technique will be used to look at how tourists choose to travel while carefully adhering to the Neuro-tourism method's philosophy [25].

Research shows that focus group discussions, questionnaires, and in-person observations are the most effective ways for the hospitality and tourism industries to learn about the attitudes and actions of their customers. These age-old techniques, however, are effective at exposing tourists' conscious thought processes. A range of cultural, social, personal, and psychological factors have an impact on tourists' decisions. Additionally, as human behaviour, preferences, attitudes, and intentions are stored in the unconscious mind, prior experiences have a major impact on these traits [8, 15]. Therefore, methods that can explain and reveal this behaviour in more depth and with greater objectivity are required. The aspects of consumer behaviour and how it may impact their decisions and emotions might be better understood by collaborating across disciplines [25, 28]. The fields of tourism and hospitality research have been motivated to be accepting towards cognitive neuroscience research and more

Table 1 Summary of selected research on decision-making

Research title	Authors	Variables	Methodology and findings	Limitation
Travel decision determinants during and after COVID-19: the role of tourist trust, travel constraints, and attitudinal factors	[28]	Travel decision factors: tourist trust; travel constraint; extended theory of planned behaviour; travel decision	The results of this quantitative analysis highlight certain significant tourist market groups and the negligible influence of interpersonal considerations on travel choices explains the emergence of a new travel segment: solitary travellers	• This study concentrated on Korean travellers • Post-pandemic contexts will be more appropriate to investigate travel intention for international travel
A systematic review of cognitive biases in tourist decisions	[33]	Pre-trip experience. On-site and post-trip experiences; others; cognitive biases in each tourism stage	• Qualitative research (systematic review of supporting evidence) • The three stages of tourism management, including marketing, social media, and pricing, as well as policy implications, indicate that tourists make irrational decisions	• This study only uses three selected journals • No cognitive bias phenomenon papers from other tourism journals
Investigating tourist destination choice: effect of destination image from social network members	[26]	Social networks; choice behaviour; destination image	• Quantitative study • Tourists' destination image is an essential influence on tourist destination choice • The influence of social networks on the picture that tourists have of a place will be greater if those tourists do not have an image of the destination	• This study focuses on Dutch in the Netherlands • Only listed four dimensions of destinations • Tourists' personality traits did not support preferences towards destinations
Travel inspiration in tourist decision making	[8]	Inspiring sources; travel inspiration; travel intention; destination awareness	• In a qualitative study, the motivational state known as "travel inspiration" is what propels potential travelers to carry out their freshly acquired vacation plans • In this article, the subject of how travel inspiration influences travelers' decisions was further investigated	• This research note is purely conceptual • Only provides practical implications for tourism marketing against the backdrop of COVID-19

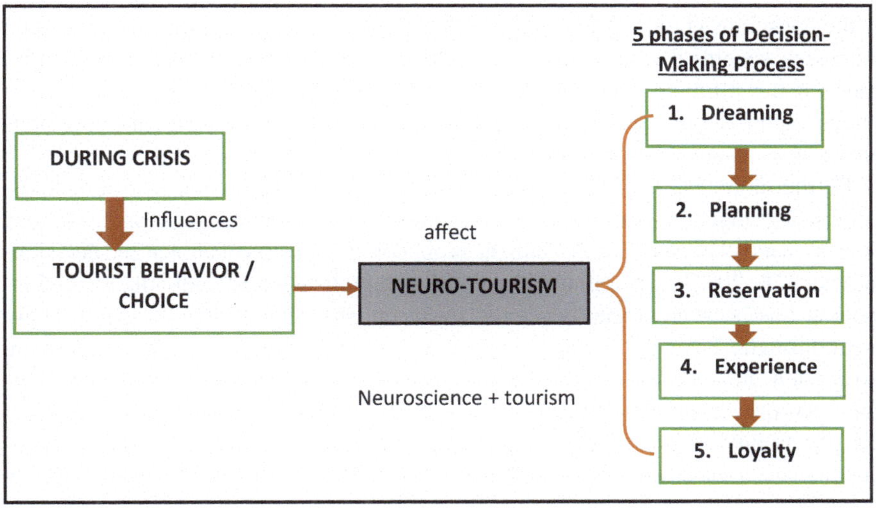

Fig. 1 The conceptual framework of decision motivation and neuro-tourism

particularly, brain imaging data by novel technology for investigating the neurological foundations of mind and behaviour. This results in a thorough comprehension of the emotional and mental processes underlying the behaviour of the tourists. Future tourism research in the field of neuro-tourism may be launched as a result of this multidisciplinary expertise [32]. Investigating the brain processes that underlie tourist behaviour is suggested by neurotourism [25]. Figure 1 illustrates the integration and fusion of neurotourism, crisis tourism, and the theory of decision-making based on the literature study discussed above.

2.5 *Bridging the Gap in Studying Decision Making Using the Neuro-Tourism Method*

In the field of cognitive and affective neuroscience, there have been notable advancements that have brought out novel methodologies and frameworks for comprehending human behavior and cognition, as well as investigating subconscious emotional and cognitive reactions [1, 24]. The emergence of several interdisciplinary endeavors, such as "neuroeconomics" and "neuromarketing," has been observed in response to this phenomenon. These methodologies have the capacity to provide an unparalleled framework for understanding and assessing the subconscious factors that impact an individual's decision-making process [24, 28]. Based on empirical evidence, the preliminary investigation indicated the absence of significant obstacles in the pursuit of this goal. According to [23], the brain exhibits activity during the process of perceiving a product and predicts the forthcoming decision several seconds prior to the conscious choice. Additionally, other studies indicate that unconscious brain processes, specifically those related to evaluating outcomes, exert a substantial influence on the ultimate decision [15, 32, 34]. Research has indicated that unconscious cognitive processes have the potential to substitute for or originate conscious preferences and decision-making.

Furthermore, the utilization of neuroscience not only furnishes resources and empirical evidence pertaining to unconscious decision-making processes, but also presents a novel perspective that enhances our comprehension of choice and decision-making [28, 29]. In several situations, the comprehension of individual decision-making has been significantly altered by recent theoretical advancements. These breakthroughs have particularly impacted our understanding of the perception, expression, and recollection of beliefs, as well as their consequential influence on the decision-making process. Multiple research studies have repeatedly demonstrated that the brain's responses exhibit a discernible behavioral influence on decision-making, even in cases where individuals assert that their choices were random [25, 32].

The utilization of neuroscience in the examination of different forms of human decision-making, specifically in emerging fields such as consumer neuroscience and neuromarketing, has led to notable advancements in the ability of individuals to assess, comprehend, and anticipate decision-making behaviors [25]. The utilization of neuroscience in the examination of many facets of human decision-making has facilitated these advancements. In a similar manner, scholarly investigation into the perception, preference, and selection of travel destinations by consumers, which can be appropriately categorized as a type of consumption behavior, ought to incorporate a blend of theoretical deliberation and empirical examination of the subconscious cognitive and emotional responses that influence destination choice [28]. Nevertheless, this has proven to be significantly inadequate. Consequently, the researchers employed hardware-enabled eye tracking and brain imaging techniques in an applied neuroscience investigation to gain further insights into the influence of direct and unconscious cognitive and emotional reactions on decision-making processes related to travel destinations [1].

3 Methodology

enlargethispage-30ptThis research will be used a qualitative approach with the intention of discovering new information. Using a procedure that used brain imaging, the purpose of the study is to discover the factors that are most important to tourists when deciding on a travel choice destination. The intention is to understand tourists' preferences in this manner without questioning them or providing them with any background information. During the process of the research, a controlled environment will be developed in which those involved were not exposed to any kind of stimulant. In this stage, an EPOC emotive 16-channel EEG instrument will be utilized. The results of the research will be interpreted with the help of the application software known as Test-bench and 3D brain map [1].

During the qualitative stage of the research, the Atlas.ti application will be analyzed the responses from the open-ended question approach by using the content analysis method. The qualitative data will be evaluated by employing the software program Atlas.ti to do a theme analysis of important episodes. The interpretation of similar data, once it has been grouped and organized within the context of specific concepts and topics, constitutes the primary step in the content analysis process. The primary procedure in the content analysis process is the analysis of similar data after it has been categorized and structured within the framework of particular concepts and subjects. The fundamental objective of conducting a content analysis is to discover ideas and connections that are capable of providing an explanation for the material that has been compiled. In this way, the data are described, and the facts that were hidden inside the data are brought to light [6].

4 Conclusion

This paper will develop a theoretical framework that is helpful for future research in this area. Besides provided a new insight view on Neuro-tourism studies. This study intends to apply a new technique/approach in evaluating tourist behaviour using Neuro-tourism with the aim in mind. This method has not been previously investigated in the literature, particularly in the tourism field; thus, work in this area is extensively important and critical. Previous studies have not yet identified the neuro-tourism approach/technique (brain imagine method) potentials to our knowledge and extensive studies.

As well, integrating behavioural and neurocognitive characteristics with the intention of predicting the behaviour of customers might be a good way for constructing a knowledge database that includes stimulus data pertinent to a number of different vacation destinations. This could be a valuable method. The establishment of a knowledge database using this strategy can turn out to be quite beneficial. The characteristics of social networks are directly related to the impact that a destination's image has on visitors' perceptions of the location. In addition, the findings of this study will contribute new information to the Theory of Decision Choice Making. Furthermore, researchers will integrate this theory with the neuroscience method in order to modify the theory so that it may be used in the period following a crisis. Since this is a conceptual paper, however, there are no facts collected that could be used to make a broad statement about other countries.

This study will define "neuroscientific tourism research" in order to identify works that use neuroscience methodology in the context of tourist research and contribute to a better understanding of this approach, which draws from a variety of domains. The study of sociocultural and genetic variability in the way people think, brain activity, and behavior over many time scales, mirror neurons, and neuroethics that can respond to tourist destination choices, is the goal of tourism neuroscience. Neuroscience influences tourist decision-making by combining medical, psychology, sociology, and tourism. The exploration of the psychological and physiological elements that contribute to visitor behavior is the most important aspect of tourism research. Neuroscientific is a tool for learning about neurophysiology as well as cognitive and emotional processes. This enables these technologies to assess certain characteristics of the human body, such as happiness, satisfaction, and reasons for returning to a specific region.

Acknowledgements This research has been financially supported by the Ministry of Higher Education (MOHE) with code research "*FRGS/1/2022/SS01/UMK/02/3*" and the title grant "*Constructing the Decision Choice Motivators (DCMs) Model among Malaysian Millennial Tourists during the Crisis*".

References

1. Alvino, L., Pavone, L., Abhishta, A., Robben, H.: Picking your brains: where and how neuroscience tools can enhance marketing research. Front. Neurosci. **14** (2020). https://doi.org/10.3389/fnins.2020.577666
2. Benckendorff, P., Moscardo, G., Pendergast, D.: Tourism and Generation Y. CABI (2010)
3. Benjamin, S., Dillette, A., Alderman, D.H.: "We can't return to normal": committing to tourism equity in the post-pandemic age. Tour. Geogr. **22**(3), 476–483 (2020). https://doi.org/10.1080/14616688.2020.1759130
4. Björk, P., Kauppinen-Räisänen, H.: Destination foodscape: a stage for travelers' food experience. Tour. Manage. **71**, 466–475 (2019). https://doi.org/10.1016/j.tourman.2018.11.005
5. Borzyszkowski, J.: Destination management organizations (DMOs) and crisis management. J. Tour. Serv. **4**, 6–17 (2013). https://www.researchgate.net/publication/318589329_Destination_management_organizations_DMOs_and_crisis_management
6. Bowen, J., Baloglu, S.: Common themes across social media research. Worldwide Hosp. Tour. Themes **7**(3), 314–319 (2015). https://doi.org/10.1108/WHATT-04-2015-0022
7. Calgaro, E., Lloyd, K., Dominey-Howes, D.: From vulnerability to transformation: a framework for assessing the vulnerability and resilience of tourism destinations. J. Sustain. Tour. **22**(3), 341–360 (2014). https://doi.org/10.1080/09669582.2013.826229
8. Dai, F., Wang, D., Kirillova, K.: Travel inspiration in tourist decision-making. Tour. Manage. **90**, 104484 (2022). https://doi.org/10.1016/j.tourman.2021.104484
9. De Vos, J.: The effect of COVID-19 and subsequent social distancing on travel behaviour. Transp. Res. Interdiscip. Perspect. **5**, 100121 (2020). https://doi.org/10.1016/j.trip.2020.100121
10. Dimanche, F., Andrades, L., Ilkevich, S.: Tourist behaviour and trends. In: Tourism in Russia: A Management Handbook. Emerald Editors (Chapter 4) (2015)
11. Dzulkifly, D.: Muhyiddin: Tourism industry hit hardest by Covid-19, faces RM3.37b loss, 13 Mar 2020. Malay Mail. https://www.malaymail.com/news/malaysia/2020/03/13/muhyiddin-tourism-industry-hithard-By-covid-19-to-loserm3.37b-while-GDP-s/1846323
12. Galvani, A., Lew, A.A., Perez, M.S.: COVID-19 is expanding global consciousness and the sustainability of travel and tourism. Tour. Geogr. **22**(3), 567–576 (2020). https://doi.org/10.1080/14616688.2020.1760924
13. Glover, P., Prideaux, B.: Using population projections to identify aspects of future tourism demand. Adv. Hosp. Leis. **4**(9), 185–209 (2008). https://doi.org/10.1016/S1745-3542(08)00009-X

14. Gössling, S., Scott, D., Hall, C.M.: Pandemics, tourism and global change: a rapid assessment of COVID-19. J. Sustain. Tour. **29**(1), 1–20 (2021). https://doi.org/10.1080/09669582.2020.1758708
15. Guidici, E., Dettori, A., Caboni, F.: Neurotourism: futuristic perspective or today's reality? In: 20th Excellence in Services International Proceedings. University of Verona, 7–8 Sept 2017
16. Gursoy, D., Chi, C.G.: Effects of COVID-19 pandemic on hospitality industry: review of the current situations and a research agenda. J. Hosp. Market. Manag. **29**(5), 527–529 (2020). https://doi.org/10.1080/19368623.2020.1788231
17. Hotle, S., Murray-Tuite, P., Singh, K.: Influenza risk perception and travel-related health protection behavior in the US: Insights for the aftermath of the COVID-19 outbreak. Transp. Res. Interdisc. Perspect. **5**, 100127 (2020). https://doi.org/10.1016/j.trip.2020.100127
18. Hyndman, D.W., Hyndman, D.W.: Natural Hazards and Disasters, 5th edn. Brooks/Cole Australia, (2016)
19. Ivanova, M., Ivanov, I.K., Ivanov, S.: Travel behaviour after the pandemic: the case of Bulgaria. Anatolia **32**(1), 1–11 (2021). https://doi.org/10.1080/13032917.2020.1818267
20. Khan, A.M.A., Hashim, H.: The effect of covid-19 on tourism and hospitality industry in Malaysia, resurgence in the post-pandemic era: a conceptual criterion. Int. J. Tour. Hosp. Rev. **7**(2), 54–62 (2020). https://doi.org/10.18510/ijthr.2020.726
21. Kim, C., Cheon, S.H., Choi, K., Joh, C.-H., Lee, H.-J.: Exposure to fear: changes in travel behavior during MERS outbreak in Seoul. KSCE J. Civ. Eng. **21**(7), 2888–2895 (2017). https://doi.org/10.1007/s12205-017-0821-5
22. Kourgiantakis, M., Apostolakis, A., Dimou, I.: COVID-19 and holiday intentions: the case of Crete, Greece. Anatolia **32**(1), 148–151 (2021). https://doi.org/10.1080/13032917.2020.1781221
23. Knutson, B., Bossaerts, P.: Neural antecedents of financial decisions. J. Neurosci. **27**(31), 8174–8177 (2007)
24. Ma, Q.G., Hu, L.F., Pei, G.X., Ren, P.Y., Ge, P.: Applying neuroscience to tourism management: a primary exploration of neurotourism. Appl. Mech. Mater. **670–671**, 1637–1640 (2014). https://doi.org/10.4028/www.scientific.net/AMM.670-671.1637
25. Michael, I., Ramsoy, T., Stephens, M., Kotsi, F.: A study of unconscious emotional and cognitive responses to tourism images using a neuroscience method. J. Islam. Mark. **10**(2), 543–564 (2019)
26. Pan, X., Rasouli, S., Timmermans, H.: Investigating tourist destination choice: effect of destination image from social network members. Tour. Manage. **83**, 104217 (2021). https://doi.org/10.1016/j.tourman.2020.104217
27. Polyzos, S., Samitas, A., Spyridou, A.E.: Tourism demand and the COVID-19 pandemic: an LSTM approach. Tour. Recreat. Res. **46**(2), 175–187 (2021). https://doi.org/10.1080/02508281.2020.1777053
28. Shin, H., Nicolau, J.L., Kang, J., Sharma, A., Lee, H.: Travel decision determinants during and after COVID-19: the role of tourist trust, travel constraints, and attitudinal factors. Tour. Manage. **88**, 104428 (2022). https://doi.org/10.1016/j.tourman.2021.104428
29. Sirakaya, E., Woodside, A.G.: Building and testing theories of decision-making by travellers. Tour. Manage. **26**(6), 815–832 (2005). https://doi.org/10.1016/j.tourman.2004.05.004
30. Sop, S.A.: Neuroscientific tourism research: current state and perspectives. Metin, T.C. (ed.), New Trends in Leisure Experience Around the Persona, pp. 103–121. LAP Lambert Academic Publishing (2021)
31. Tejan, M.A., Safaa, L.: Crises, learning, and resilience: retrospective analysis of tourist activity in Morocco from 1991 to 2016. Focus on the destination Marrakech. Caribb. Stud. **2**. https://doi.org/10.4000/etudescaribeennes.13975 (2018)
32. Tosun, C., Ozdemir, S., Cubuk, F.: Usage of neuro-tourism methods in hotel preferences of the consumers. In: The 2016 WEI International Academic Conference Proceedings, pp. 19–26. The West East Institute (2016)
33. Wattanacharoensil, W., La-ornual, D.: A systematic review of cognitive biases in tourist decisions. Tour. Manage. **75**, 353–369 (2019). https://doi.org/10.1016/j.tourman.2019.06.006

34. World Tourism Organization: UNWTO Annual Report 2015 (2016). https://doi.org/10.18111/9789284418039
35. Zainal, F., Chung, C.: Be aware of over-tourism. The Star, 6 Jan 2020. https://www.thestar.com.my/news/nation/2020/01/06/be-aware-of-overtourism
36. Zhou, Q., Sotiriadis, M., Shen, S.: Using tiktok in tourism destination choice: a young Chinese tourists' perspective. Tour. Manag. Perspect. **46**, 101101 (2023). https://doi.org/10.1016/j.tmp.2023.101101

Exploration of the Travel Motivations of Gastronomy Tourism: A Case Study in East Region Area

Nor Syuhada Zulkefli, Siti Salina Saidin, Muhammad Nasyat Muhammad Nasir, and Zaimatul Awang

Abstract Gastronomic tourism is a novel form of cultural tourism that endeavors to cater to the demands of customers in the food sector and position itself as a pivotal determinant in travellers' deliberations over their vacation destinations. Nevertheless, the extent of academic research into the field of gastronomic tourism has been relatively restricted. The objective of this study is to discover the travel reasons of gastronomy tourists, focusing specifically on the insights provided by tourist guides. This study chooses the East Coast region as a case study in accordance with UNESCO's creative cities initiative. The study employed the qualitative research method, specifically utilising a descriptive phenomenological approach as the research methodology. The participants in this study were millennial travellers. The researchers opted for a data collection tool that was appropriate for capturing the phenomenological pattern, and they employed the semi-structured interview technique. A total of fifteen tourist guides were subjected to interviews employing the procedures. The study yielded the identification of four primary themes, accompanied by ten sub-themes and a total of ninety-two codes. The findings indicate that gastronomy motive significantly influences tourists' destination selection, while their satisfaction and loyalty levels are affected by their culinary experiences.

Keywords Gastronomy tourist · Travel motivation · Millennial tourists · Qualitative · Malaysia

1 Introduction

Tourism has emerged as a crucial and rapidly expanding industry within the contemporary global landscape. Tourism exhibits a strong correlation with the natural landscape and delicious local cuisine [1, 16]. Food is no longer merely a fundamental human need; it is now one of the primary factors that motivate tourists to travel

N. S. Zulkefli (✉) · S. S. Saidin · M. N. M. Nasir · Z. Awang
Faculty of Hospitality, Tourism and Wellness, Universiti Malaysia Kelantan, Kota Bharu, Malaysia
e-mail: syuhada.z@umk.edu.my

B. Awwad (ed.), *The AI Revolution: Driving Business Innovation and Research*, Studies in Systems, Decision and Control 524, https://doi.org/10.1007/978-3-031-54379-1_74

[9]. Gastrotourism changes the environment in which it develops in addition to bringing about certain changes. Maintaining positive attitudes among all participants in gastro-tourism is crucial for a number of reasons, including the happiness of visitors to a specific destination [28]. Gastronomy is a vital component of the global tourism industry's expansion. According to [3], gastronomic tourism can be defined as an investigation and comprehension of culture and history through the lens of gastronomy, hence contributing to the creation of memorable experiences. Calvo Porral and Levy [7] argue that the establishment of consumer loyalty relies on the importance of certain culinary attributes and the lasting impression that food creates on visitors.

Gastronomic tourism is one of the most active and innovative aspects of tourism due to its economic influence and social and cultural interaction [28]. Since its creation, gastronomic tourism has been a popular form of travel and, as a subset of travel, a source of income for cantons with particular potential in this field of tourism. This encourages the expansion of small and medium-sized businesses with a gastronomic offering and their revenue. The cuisine of a place becomes an attraction of the visited tourist destination [23]. However, gastronomy is more than just eating or feeding; it implies integrating the culture of the destination in which it is located. Bondzi-Simpson and Ayeh [5] discovered that local cuisine is a significant tourist draw. Moreover, indigenous foods contribute to tourists' satisfaction and loyalty to a destination [21]. Almeida-García [2] found that tourist gratification and motivation are dependent on food quality, the significance of the destination, and the natural atmospheric environment [8]. Several studies were conducted and have indicated that the use of local cuisine can serve as a competitive strategy and a determining factor for the appeal and loyalty of tourists towards specific attractions and destinations [9, 22, 29].

According to [23], the significance of local cuisine in the selection of a holiday place is one of the elements that is increasingly taken into consideration in the tourist industry. Studies that focus on meals, wines, and gastronomy in general as determining variables of the so-called "gastronomic experience" are becoming increasingly popular [3]. In point of fact, regional cuisine has emerged as an important resource for the tourism industry [6] and it is now integrated into the cultural legacy of popular vacation spots. As a result, food and tourism have a very close relationship that can be divided into four categories [10]: (1) food as a part of the local cultural identity; (2) food as a desire for tourists; (3) food as an investment for tourists; and (4) food as an experience for tourists.

The WTO (World Tourism Association) is concerned that gastronomic tourism is not being adequately promoted, as 15% of all tourists choose destinations based on the abundance of local cuisines [12]. After an interest in the local culture and the natural environment, gastronomy is the third most common reason that people choose to travel to a particular location. The importance of local cuisines, culinary experiences, and food tourism has increased in the tourism industry [20, 24]. According to [23], gastronomy has emerged as a pivotal determinant of the competitive edge possessed by tourism destinations, serving as a way to allure a growing cohort of tourists that exhibit a heightened inclination towards exploring local elements and immersing themselves in the cultural fabric of the visited locale [15]. This study has the potential to help assess gastronomy tourism after the tourism sector has reopened (after the COVID-19 pandemic).

The paper is organized as follows: Sect. 2 reviews relevant literature, while Sect. 3 describes and explains the data and methodology. Section 4 discusses the result analysis, and Sect. 5 concludes the paper.

2 Literature Review

2.1 The Concept of Gastronomy Tourists

Gastronomy tourism encompasses the engagement of tourists in various gastronomy-related endeavours, such as attending festivals or workshops, within a specific destination. The notion of gastronomy tourism was introduced by [24] as a means of delineating the interplay between culinary experiences and broader tourist endeavours. Huertas et al. [17] definition of a gastronomy tourist is a person who has the time, inclination, and financial means to travel to a city or country other than their own, primarily for the purpose of partaking in culinary experiences and beverages.

According to [3], who stress that the gastronomy tourist is a special interest tourist, the term "gastronomy tourist" is defined as "*a special interest tourist participating in gastronomy tourism, which is considered special interest tourism, satisfying the curiosity of new food and beverage by participating in gastronomic activities and gaining experiences by participating in gastronomic activities.*" Gastronomy tourism, often known as a blend of the sense of taste and the act of traveling, has emerged as a well-liked pastime that gives travelers access to a variety of unique dining opportunities [7, 9]. Amusement for tourists and an opportunity for them to make memories that will last a lifetime, gastronomy tourism provides both.

2.2 Gastronomy Tourism and Tourism Motivation

Currently, the primary factor motivating tourist development is the gastronomic experience [25]. Gastronomic experiences are derived from the consumption of regional foods and cuisines, taking into account additional elements such as the cost of food and the general ambiance of the destination, while also ensuring the provision of exceptional customer service. The present study aims to examine the factors that drive individuals to choose a certain destination for the purpose of engaging in culinary pursuits. Gastronomic tourism encompasses the act of visiting establishments involved in food production, attending food and beverage festivals, exploring restaurants, and exploring specific locations associated with particular culinary offerings [9]. This form of tourism entails engaging in activities such as sampling distinctive dishes, seeing the intricate procedures involved in their creation and preparation, or indulging in a specially crafted meal served by a globally recognised culinary expert [26].

According to [3], motivation is defined *as "a state of need, a condition that serves as a driving force to display different types of behavior towards certain types of activities, developing preferences, and achieving some expected satisfactory outcomes.*" Numerous tourism researchers have attempted to determine what motivates different people to travel to various destinations and what engages them in tourism activities offered by various destinations. Dann [11] defined travel motivation as "a meaningful state of mind that adequately disposes an actor or group of actors to travel and that can be interpreted by others as a valid explanation for such a decision."

In the present day, gastronomy is widely acknowledged as a significant factor that motivates tourists to visit various destinations [21]. Motivations exhibit a significant influence on consumption patterns while acknowledging that they do not operate in isolation and are not the sole determinant of behavioral outcomes. Motivations are forcing that manifest in response to an unmet need, thereby generating behaviour [24, 26].

According to research conducted by [27], an increasing number of tourists are going to certain locations in order to partake in specific dining experiences. Food is a main activity and a tourist attraction [23]. Gastronomy was previously secondary activity in tourism choice destinations. This view has evolved recently. According to [14], food tourism is "the type of tourism that includes the primary or secondary motivation, such as the visit to local producers, gastronomic festivals, restaurants, or other places where food tasting and/or the gastronomic experience are key elements." In addition to gastronomy, travellers seek to sample the cuisine of a destination for a variety of reasons [21]. Visitors' motivations are categorized. Okumbus and Cetin [24] classify tourists' location choices into two key motivations: the desire to escape and the desire to seek. While, according to [25], visitors travel for one of two reasons: either they are impelled to do so by certain motives or internal variables, or they are drawn to the locations themselves as a result of the characteristics of those locations. The term "gastronomic tourism" refers to travel in which the primary focus is on the cuisine, the exploration of new flavors, and the learning of the local history or culture [24]. Thus, all tourists must consume food when traveling, and for some, good food can be a significant draw and primary motivator [3].

In addition, [21] identified nine distinct motivational factors for tourists to consume local cuisine. These factors are enumerated as an escape from routine, exciting experience, health, level of learning, togetherness, genuine experience, prestige, physical environment, and sensory satisfaction. For consumer behavior research, it is essential to understand who gastronomy tourists are and what motivates them. The assessment of a geographical location will heavily rely on the gastronomic experience, as stated by [3]. This assessment consists of many various aspects such as the methods employed in food preparation, the cultural significance of food, patterns of food consumption, the style of cuisine, and customary gastronomic practices, as highlighted by [10]. Additionally, [25] emphasize that the overall quality excellence of the food will also be taken into account.

3 Methodology

The research employed qualitative research approaches in order to discover the travel motivations of gastronomy tourists, as perceived by tourist guides. The research issue was initially recognised with the aim of examining the motivations behind culinary tourists' journeys. The primary research question is the determinants that drive gastronomy tourists to engage in travel activities. Consistent with the fundamentals of qualitative research, the study employed an exploratory approach, wherein specific inquiries were posed to the participants, while maintaining focus on the research's purpose and primary issue. A research design was established to outline the methodology and procedures that will be employed in conducting the study.

The interview is widely regarded as a very suitable method for collecting data in the context of phenomenological research design. This research employed a semi-structured interview methodology, utilizing millennial travellers as the selected sample of respondents. The millennial generation also referred to as Generation Y, encompasses those born between 1980 and 1995 [4]. This demographic represents a significant emerging market segment with the capacity to bring about transformative changes in the international tourism industry. The current generation of travelers has a unique set of characteristics, placing a high value on a meaningful and deliberate travel experience. Their preferences and consumption patterns diverge significantly from those of preceding generations [13].

In the conducted study, all interviews were conducted in person. On average, the duration of the interviews was approximately 40 min. The interviews were conducted solely based on the participants' voluntary participation. The study utilised content analysis as a method of data analysis. The study utilised frequency analysis, a form of content analysis. Consequently, the research themes were quantitatively analysed to identify their frequency of occurrence, presented in numerical, percentage, and proportional formats.

4 Result Analysis

As part of the research, interviews were conducted with Millennial groups, and the data obtained were analyzed. First, the demographic data of the tourist guides were presented in Table 1 and the respondents were labeled as R1, R2, R3, … and R15. Information on gender, age, education level, and frequency of traveling in a year was provided for each respondent.

Table 1 Demographic characteristics of the respondents

Respondents	Gender	Age	Frequency of traveling in the year
R1	Male	30	5
R2	Male	33	4
R3	Female	28	5
R4	Female	35	6
R5	Female	29	5
R6	Male	31	4
R7	Female	27	4
R8	Male	26	3
R9	Male	27	6
R10	Female	29	4
R11	Female	25	5
R12	Male	27	3
R13	Female	25	4
R14	Female	28	4
R15	Male	26	4

4.1 Respondents' Demographic Characteristics

Of the 15 respondents included in the study, 8 were female and 7 were male. The age range of the respondents varied from 25 to 30 years. The participants, most respondents have at least a college degree and at least 4 times to travel in one year.

4.2 Influencing Factors of Gastronomy Tourism Motivation

The primary themes that were identified encompassed intrinsic elements, extrinsic aspects, promotion factors, and desired locations. The intrinsic aspects encompassed several sub-themes, namely interest, hedonism, experience, and the pursuit of novelty and adventure. The extrinsic elements encompassed several sub-themes such as the local culinary culture, festivals, and fairs, as well as the style and conceptions of restaurants. The aspects contributing to promotion encompassed the utilization of social media and viral marketing. Ultimately, the favored destinations encompassed both widely recognized and less well-known sub-themes. The explanations have been shown in Table 2.

Table 2 Explanation and quotation of main themes, sub-themes, density, and quotation

Main themes	Sub-themes	Density	Selected from quotation
Intrinsic	Interest	7	"*I'm interested to explore the local food at certain places...Eating and shopping is one of my intention to travel...*" (*R1*)
	Hedonism	8	"*I think it's hedonism. Eating is one of the best pleasures in my life...*" (*R4*)
	Escape from routine	10	"*I want to have experience for traveling and tasting their local food...*" (*R8*) '*...It's possible that the food that people ate in the past was very different from what we consume today, either at home or in restaurants. It would be interesting to me to learn about the food that people in the past consumed and how it differed from the meals that are served today...*' (*R5*)
	Seeking difference/ adventure	11	"*I like to travel and try many local foods at certain destinations...*" (*R10*)
Extrinsic	Local culinary culture	8	"*I think that only the cuisine that is food and beverage culture to travel... food is the main starting point.*" (*R9*)
	Festivals and fairs	9	"*Because of food festivals event... Especially local cuisines, unique cuisines are very beneficial in this respect... street delicacies or extreme dishes*" (*R2*)
	Restaurant style and concepts	11	"*I want to taste the food at the new restaurant... certain restaurant has their own style and concept for their menu...*" (*R7*)
Promotion	Trustworthiness in media programmes	14	"*Now being shown so much on social media through Instagram phenomena so much that everyone wants to go and eat it once... this place is viral because of food...*" (*R16*)
	Popular place	9	"*I think tourists who are trying to become new gastro-tourists prefer popular destinations. I say that I think more popular places are preferred.*" (*R15*)

4.2.1 Theme 1: Intrinsic

According to [9], the most significant intrinsic motivation is seeking cultural experiences and adventure. This finding is consistent with previous research by [18] into the motivations for consuming local produce. Food and foodways have been thoroughly investigated by scholars as a kind of destination identity and a basic manifestation of place [18]. Similarly, gastronomy serves as a significant indicator of cultural heritage and tourism [19, 28]. Kim and Iwashita [19] argue that the local food legacy of a particular area can effectively serve as a reflection of the process of identity formation, as well as the characteristics of the place and its culinary traditions. The primary motivation for the participants in this research was to visit Kota Bharu in order to establish a correlation between the local cuisine and the cultural, historical, and regional identity of Kelantan. This is similar to the behavior observed among Japanese food tourists who consume noodles and associate specific regions known for noodle production with a distinct sense of place [20].

4.2.2 Theme 2: Extrinsic

This research discovered various sub-themes within the extrinsic theme, which refers to local destination appeals. These sub-themes encompass cultural activities that feature food in addition to other customs associated with a culinary destination. The examination of this external element has indicated that it aligns with the marketing research undertaken by the Travel Industry Association of America on the profiles of culinary tourists. Calvo Porral and Levy-Mangin [7] highlighted in their study the inclination of food tourists to engage in a diverse range of activities during their visit, encompassing those that are closely associated with the cultural and historical aspects of the location. The participants emphasized the significance of sensory encounters and the gastronomic aspect of local food experiences, including engaging with several senses during the local food festival and culinary demonstrations. They considered this to be their second most important reason on an individual level.

4.2.3 Theme 3: Promotion

The media, which included travel guides, pamphlets, maps, TikTok applications, as well as radio and television programmes, all played an important part in encouraging domestic travelers to sample regional food. Furthermore, it is evident that social media exerted significant influence on the younger participants (interviewees 3, 4, 6, and 16), since they all acknowledged utilizing social media platforms to make informed decisions regarding their dining choices. This discovery indicates that social media had a notable impact on the younger participants during the interviews. Based on the aforementioned findings, it is evident that younger interviewees assign significant importance to the impact of social media on their decision-making process. This discovery aligns with the outcomes of other research conducted on tourism, indicating that the utilization of social media platforms has significantly transformed the manner in which individuals strategize their trips and make decisions regarding their dining preferences [19, 22]. This observation aligns with the results of numerous studies in the field of tourism, which suggest that the utilization of social media platforms has significantly transformed the decision-making process of individuals when it comes to selecting their dining preferences. According to this discovery, the use of social media has had a substantial impact on the manner in which individuals select the foods that they will consume for their meals.

According to this research, the exposure that a place receives in the media is one of the most important factors in attracting visitors there. The previously stated impact is primarily derived from credible sources, including documentaries and esteemed national television channels, which collaborate to skilfully construct a comprehensive portrayal of the destination's gastronomic delights [8, 9]. This result aligns with and affirms the significant influence of media on regional food tourism, as previously documented in the research conducted by [21] on Japanese domestic tourists' experiences in a food tourism destination. This discovery serves as a confirmation of the substantial influence the media exerts on regional food tourism.

4.3 Discussion and Implication

The gastronomy tourism market is characterized by its high level of dynamism and ongoing growth. This type of tourism refers to the study of tourists whose primary purpose for traveling is to experience the cuisine, tradition, and, in particular, the local product of the city that they are visiting. Specifically, this type of tourism focuses on food tourism. This study demonstrates the significance of cuisine as a tourist motivation by analyzing a number of links between factors. The study's primary focus is on the concept of visitor satisfaction with the location. Within the context of food tourism, the method of gastronomy tourism motivation provides a grasp of destination features as motivational factors, as new parts of the concept of motivation. In particular, the research reveals three categories of reasons why people engage in gastronomic tourism. These are intrinsic reasons, extrinsic reasons, and promotional reasons.

In recent times, there has been a notable surge in the demand for gourmet tourism, owing to the heightened degree of interest in the field of gastronomy. Gastronomy tourism offers individuals a diverse array of experiences that are beneficial to their well-being. When considering gastronomy tourism, the primary aspect that immediately springs to mind is engagement in culinary and beverage-related pursuits. Nevertheless, gastronomy tourism extends beyond the mere act of consuming food and beverages. Gastronomy tourism encompasses a wide range of supplementary endeavors, including but not limited to, touring gastronomic production facilities, engaging in gastronomic events or activities, acquiring knowledge about local products and harvests, engaging in conversations, exploring gastronomic museums, witnessing gastronomy-oriented competitions, enrolling in gastronomy training programs. The evolving demands and expectations of tourists have led to a prevailing belief that contemporary travelers are increasingly inclined towards participating in types of tourism that prioritize experience activities. Currently, gastronomy tourists enjoy a more advantageous position compared to other categories of tourists.

This study is important for people who work in the tourism and hospitality industries. Knowing how tourists feel about the city's food and the area around it will help to provide better service, improve the food in bars and restaurants, and add new flavors to the classic ones. All of this takes into account how important food is becoming in how people travel and where they choose to visit.

5 Conclusion

The empirical findings of this study provide significant information for food destination organizations to use in the development of strategies that cater to the demands and motivations of food tourists. These findings have implications for management as well as for the practical application of these findings. Food manufacturers or suppliers at a location ought to come up with a selection of food-related goods and services so that they can provide tourists interested in food with a comprehensive set of advantages. Specifically, foodies have the opportunity to try regional cuisine in real restaurants or culinary villages, which give settings that are traditional in nature. It is a wonderful opportunity for food travelers to sample home-cooked meals and discover the culture of the local area by dining with local folks. A further aspect for companies that are destinations is the need to enhance cultural experiences through the use of cuisine and activities related to eating. Tourists can learn something new and interesting about the local culture, people, and culinary traditions at each location they visit thanks to the narratives that are available. As a consequence of this, the tactic of storytelling can be of assistance to destination marketing organizations in the promotion of exceptional travel experiences and culinary tourism, as well as in the development of a unique food destination brand and a unique selling point.

There are limitations to this study. First, the research focuses exclusively on the east coast area. In order to generalize the results, it is recommended to compile data from a variety of tourist destinations. Second, only a limited number of interviews with Millennial travelers were conducted for the study. Therefore, future researchers will be able to collect the perspectives of various gastronomy tourism stakeholders, such as tour guides, government agencies, and restaurant proprietors. Utilizing multiple case studies will improve the dependability and generalizability of the research findings. The study of travel motivation is essential for understanding travel behavior, as it serves as the fundamental basis. Furthermore, there is potential for future research to explore the connections between food travel motivation and several other factors, such as visitors' attitudes, engagement, or intention towards food tourism.

Acknowledgements This research has been financially supported by the Universiti Malaysia Kelantan (UMK) with code research "*R/COM/A1100/01868A/002/2022/01144*" and the research grant title "*Memperkasakan Potensi Gastronomi Pelancongan (Makanan Tradisional) Dalam Kalangan Pengusaha Tempatan Melalui Model Keusahawanan Pemasaran Digital*".

References

1. Agyeiwaah, E., Otoo, F.E., Suntikul, W., Huang, W.-J.: Understanding culinary tourist motivation, experience, satisfaction, and loyalty using a structural approach. J. Travel Tour. Mark. **36**, 295–313 (2019)
2. Almeida-García, F., Domigunez-Azcue, J., Mercadé-Melé, P., Pérez-Tapia, G.: Can a destination really change its image? The roles of information sources, motivations, and visits. Tour. Manag. Perspect. **34**, 100662 (2020)
3. Berbel-Pineda, J.M., Palacios-Florencio, B., Ramírez-Hurtado, J.M., Santos-Roldán, L.: Gastronomic experience as a factor of motivation in tourist movements. Int. J. Gastron. Food Sci. **18**, 100171 (2019)
4. Benckendorff, P., Moscardo, G., Pendergast, D. (eds.).: Tourism and Generation Y. Cabi (2010)
5. Bondzi-Simpson, A., Ayeh, J.K.: Assessing hotel readiness to offer local cuisines: a clustering approach. Int. J. Contemp. Hosp. Manag. **31**(2), 998–1020 (2019)
6. Björk, P., Kauppinen-Räisänen, H.: Exploring the multi-dimensionality of travellers' culinary-gastronomic experiences. Curr. Issues Tour. **19**(12), 1260–1280 (2016)
7. Calvo Porral, C., Levy-Mangin, J.P.: Food private label brands: the role of consumer trust on loyalty and purchase intention. Br. Food J. **118**(3), 679–696 (2016)
8. Carvache-Franco, W., Carvache-Franco, M., Carvache-Franco, O., Hernández-Lara, A.B.: Motivation and segmentation of the demand for coastal and marine destinations. Tour. Manag. Perspect. **34**, 100661 (2020)
9. Chang, R.C., Mak, A.H.: Understanding gastronomic image from tourists' perspective: a repertory grid approach. Tour. Manage. **68**, 89–100 (2018)
10. Chen, Q., Huang, R.: Understanding the role of local food in sustaining Chinese destinations. Curr. Issue Tour. **22**(5), 544–560 (2019). https://doi.org/10.1080/13683500.2018.1444020
11. Dann, G.M.S.: Tourism motivation: an appraisal. Ann. Tour. Res. **8**, 187–219 (1981)
12. FITUR: International Tourism Trade Fair. FITUR, Madrid, Spain (2018)
13. Glover, P., Prideaux, B.: Using population projections to identify aspects of future tourism demand. In: Advances in Hospitality and Leisure. Emerald Group Publishing Limited (2008)
14. Hall, C.M., Sharples, L., Mitchell, R., Macionis, N., Cambourne, B.: Food Tourism Around the World. Butterworth Heinemann, Oxford (2003)
15. He, L., Mu, L., Jean, J.A., Zhang, L., Wu, H., Zhou, T., Bu, H.: Contributions and challenges of public health social work practice during the initial 2020 COVID-19 outbreak in China. Br. J. Soc. Work., bcac077 (2022)
16. Hernandez-Rojas, R.D., Folgado-Fernandez, J.A., Palos-Sanchez, P.R.: Influence of the restaurant brand and gastronomy on tourist loyalty. A study in Córdoba (Spain). Int. J. Gastron. Food Sci. **23**, 100305 (2021)
17. Huertas, T., et al.: Gastronomic festivals: attitude, motivation and satisfaction of the tourist. Br. Food J. **119**, 267–283 (2016). https://doi.org/10.1108/BFJ-06-2016-0246
18. Kim, S., Ellis, A.: Noodle production and consumption: from agriculture to food tourism in Japan. Tour. Geogr. **17**(1), 151–167 (2015). https://doi.org/10.1080/14616688.2014.978812
19. Kim, S., Iwashita, C.: Cooking identity and food tourism: the case of Japanese udon noodles. Tour. Recreat. Res. **41**(1), 89–100 (2016)
20. Kim, J.H., Jang, S.: Determinants of authentic experiences: an extended Gilmore and Pine model for ethnic restaurants. Int. J. Contemp. Hosp. Manag. **28**(10), 2247–2266 (2016)
21. Kim, S., Park, E., Lamb, D.: Extraordinary or ordinary? Food tourism motivations of Japanese domestic noodle tourists. Tour. Manag. Prospect. **29**, 176–186 (2019)
22. Lin, L., Mao, P.C.: Food for memories and culture. A content analysis study of food specialities and souvenirs. J. Hosp. Tour. Manag. **22**, 19–29 (2015)
23. López-Guzmán, T., Uribe Lotero, C.P., Pérez Gálvez, J.C., Ríos Rivera, I.: Gastronomic festivals: attitude, motivation and satisfaction of the tourist. Br. Food J. **119**(2), 267–283 (2017)
24. Okumbus, B., Cetin, G.: Marketing Istanbul as a culinary destination. J. Dest. Mark. Manag. **9**, 340–346 (2018)

25. Şahin Perçin, N., Eren, D., Sezgin, A.: The effects of local food consumption motivations on customer satisfaction: a research on Turkish food. J. Soc. Res. Behav. Sci. **7**(13), 400–423 (2021)
26. Sormaz, U., Akmese, H., Gunes, E., Aras, S.: Gastronomy in tourism. Proc. Econ. Financ. **39**, 725–730 (2016)
27. Torres, L.C., Phakdee-Auksorn, P.: Understanding international tourists' attitudes towards street food in Phuket, Thailand. Tour. Manag. Prospect. **21**, 66–73 (2017)
28. Vukolic, D., Gajić, T., Penic, M.: The effect of social networks on the development of gastronomy—the way forward to the development of gastronomy tourism in Serbia. J. Tour. Futures (2022)
29. Wijaya, S., King, B., Nguyen, T.H., Morrison, A.: International visitor dining experiences: a conceptual framework. J. Hosp. Tour. Manag. **22**, 19–29 (2013)

Exploring Staycation Experiences Drivers Among Millennials' Tourists

Siti Salina Saidin, Nor Syuhada Zulkefli, Nur Aliah Mansor, Nik Alif Amri Nik Hashim, and Nguyen Duc Tri

Abstract Globally, tourists have become more anxious to travel overseas due to increased uncertainties posed by disastrous global-scale incidents, like the Covid-19 pandemic, the war in Ukraine, the earthquake in Turkey, or even the increased assertiveness of the People's Republic of China on the world's stage. The uncertainties crisis outbreak has significantly impacted on triggered important changes in tourism mobility, social behavior, consumption patterns, and leisure. As a result, people have significantly changed their long-term traveling into near-term travel decisions. Therefore, this research aims to explore the staycation motivation drivers among Malaysian vacationers in the proximity of the tourism concept. The study applies an interpretive phenomenological approach to explore post-pandemic travel behaviour. The research employs an exploratory method with initially in-depth interviews providing rich empirical data as a precursor to a survey. Data were collected via personal in-depth semi-structured interviews with Malaysian Vacationers. Thematic analysis of the transcripts was conducted to arrive at discussion themes. This study will not only shed some light on the concept of staycation, within the Malaysian context, and position this concept within the wider proximity of tourism but also frame the new dimension of domestic tourism in the country.

Keywords Staycation · Proximity tourism · Malaysian vacationers · Qualitative

S. S. Saidin (✉) · N. S. Zulkefli · N. A. A. N. Hashim
Faculty of Hospitality, Tourism and Wellness, Universiti Malaysia Kelantan, Kota Bharu, Malaysia
e-mail: salina.s@umk.edu.my

N. A. Mansor
Faculty of Business and Finance, Universiti Tunku Abdul Rahman, Kampar, Perak, Malaysia

N. D. Tri
School of Tourism, College of Business, University of Economics Ho Chi Minh, Ho Chi Minh City, Vietnam

B. Awwad (ed.), *The AI Revolution: Driving Business Innovation and Research*, Studies in Systems, Decision and Control 524, https://doi.org/10.1007/978-3-031-54379-1_75

1 Introduction

In an era characterized by heightened levels of insecurity and uncertainty, numerous prospective visitors may perceive nearby sites as comparatively less risky. This inclination is particularly pronounced as their spending power may have been impacted by the economic crisis stemming from the pandemic. Romagosa [1] asserts that the Covid-19 pandemic has had a substantial impact on the tourism industry, leading to considerable transformations in mobility, social conduct, consumer habits, and recreational activities. This assertion posits that the rise in societal and ecological apprehensions will motivate post-crisis travelers to opt for places in closer proximity to their residences and engage in localized travel, commonly referred to as a staycation [2]. The concept of staycation has emerged as a contemporary tourism trend, particularly among urban residents, owing to shifts in lifestyle patterns and a growing emphasis on achieving a harmonious work-life equilibrium [3]. The concept of staycation has evolved into a multifaceted cultural phenomenon, characterized by individuals opting to spend their holidays in close proximity to their place of residence while endeavoring to recreate the ambiance and experiences typically associated with a conventional vacation. Engaging in tourism is a deliberate choice made by individuals, primarily driven by their desire to explore unfamiliar aspects of their surroundings and their approach to uncovering such unfamiliarity [4]. Nevertheless, the existing empirical research on tourist behavior in relation to the staycation industry lacks clarity. However, the specific activities that are capable of attracting inhabitants from urban regions, as well as the profiles and motivations of persons who engage in staycation behaviors, are two unresolved matters.

In particular areas, such as Europe and China, there seems to be a sustained interest in travel despite the ongoing global health crisis, as seen by the rising demand for both domestic and international travel [5, 6]. Staycations have garnered significant interest among tourists in Western countries such as the United States and the United Kingdom, as well as in Asia, including Hong Kong [7]. This phenomenon might be considered a manifestation of the "new normal" in leisure activities. According to James et al. [8], a "staycation" refers to the practice of individuals spending their vacation time within their own country and engaging in activities that involve visiting local sites, as opposed to traveling to a different location. In response to individuals' inclination for travel and the need to sustain hotel revenue amidst the COVID-19 epidemic, the hospitality and tourism sector initiated the formulation and promotion of staycation packages [9].

According to Eloksari [10], an analysis of Google Trends data reveals an upward trend in the popularity of the search "staycation" during the mid-year of 2020. The notion of staycation garnered heightened interest among individuals seeking vacations in the previous year [11]. However, the academic community has provided limited definitions and conceptualizations of this phenomenon. Currently, a standardized definition of "Staycation" does not exist because of the varying interpretations and usage of the term by individuals seeking vacation experiences. However, there is still a significant amount of research that has to be conducted in order to determine the potential of familiar and geographically close locations to become appealing tourist destinations. Furthermore, within the realm of scholarly writing, several concepts have been explored in relation to the phenomenon of staycations. These concepts encompass notions such as "local tourism," "proximity tourism," and "near-home travel." However, it is worth noting that these concepts do not fully encapsulate the comprehensive understanding of the term "staycation."

According to Barclays [12], millennials constitute a significant consumer demographic when it comes to the consumption of staycation package products. In contrast to preceding generations, millennials exhibit notable proficiency in technology use and display a heightened awareness of prevailing trends [13]. Moreover, they demonstrate a proclivity towards prioritizing the pursuit of a high standard of living and engaging in activities that hold personal significance [14]. According to Siegel and Wang [15], millennials, who exhibit a willingness to explore novel experiences and possess a strong inclination towards travel, may contemplate opting for staycations as a substitute for their cancelled vacations amidst the ongoing coronavirus epidemic. Additionally, given their substantial financial resources, this demographic group may find staycations to be a viable alternative. Due to the distinctive attributes of the millennial generation and the prevailing circumstances of the COVID-19 pandemic, the perspectives and encounters of millennials in relation to staycations may exhibit variations when compared to individuals visiting from other regions or in comparison to pre-pandemic circumstances.

Nevertheless, the motivations behind the travel, consumption patterns, travel behavior, preferences, and destination selection of these individuals remain mostly undisclosed. There is a scarcity of related studies undertaken in Malaysia, with minimal available research conducted by Yusoff et al. [16] and Ratthinan and Selamat [17]. This study aims to investigate the fundamental elements that lead to individuals' travel style within the context of staycation, and how these aspects influence their travel decisions. This research field warrants further investigation because to its growing significance in comprehending the staycation motivational model among contemporary Malaysian vacationers. Such understanding is crucial for effectively accessing the potential of this profitable tourist industry and accurately forecasting future demand for tourism services.

The present paper is structured in the following manner: Sect. 2 provides a comprehensive assessment of pertinent literature, while Sect. 3 presents a detailed exposition of the data and technique employed in this study. The paper is concluded in Sect. 4.

2 Literature Review

2.1 Proximity Tourism Concepts

Proximity tourism can be defined as a specific type of tourism that places emphasis on nearby attractions, limited travel lengths, and the utilization of transportation methods with fewer carbon emissions [4]. In addition, proximity tourism can be described as a form of tourism that focuses on visiting sites within close proximity to one's place of residence, with a particular emphasis on locations that are not situated at a considerable distance [2]. However, it is important to note that proximity tourism extends beyond mere physical distance. It encompasses a type of tourism that revolves around the concept of perceiving nearby, familiar environments in a fresh and novel manner [18]. The notion of closeness poses a challenge to the conventional approach of defining tourism based on the notions of proximity and distance, or by examining the dynamic relationship between the ordinary and remarkable [18]. Proximity tourism has garnered heightened interest in recent times, mostly driven by the prevailing COVID-19 pandemic and the growing recognition of climate change's repercussions, as well as the pressing nature of the environmental predicament we confront. According to the conventional theory of travel, motivated by a need for novelty, visitors intentionally select destinations that are farther away from their habitual place of residence in order to fulfill their objective of seeking distinctive and fresh experiences. Nevertheless, in light of the COVID-19 pandemic, there have been limitations imposed on long-distance travel. Due to heightened levels of social and environmental consciousness, a considerable number of prospective travelers tend to regard places in close proximity as being comparatively less hazardous [19]. Therefore, proximity tourism presents a viable alternative for travelers during the ongoing epidemic.

The importance of proximity tourism in promoting local development and bridging the gap between urban areas and interior regions is gaining recognition [20]. Numerous researchers have emphasized the potential of proximity tourism to mitigate seasonality and decrease reliance on long-haul tourism markets [21]. Additionally, it is argued that proximity tourism can offer novel prospects for tourism marketing and destination branding [21]. In the context of this research, Jeuring and Haartsen [22] investigate the perspectives of inhabitants in the Province of Friesland, located in the Netherlands, regarding proximity tourism. The authors utilize a survey to examine the attitudes of these individuals towards proximity tourism and their preferences for their home province as a tourism destination. The results of their study indicate a division among participants into two distinct profiles, each characterized by contrasting sociodemographic traits and motivations.

The rise of proximity travel as a phenomenon can be seen as being intricately linked to the growth of environmental consciousness within the context of sustainable tourism [22]. This development also brings out novel prospects and possibilities [22]. Proximity tourists seek to engage in tourism experiences that are characterized by their near proximity. Proximity travel not only encourages proximity travelers to visit familiar areas but also contributes to the growth and advancement of neighboring destinations [4].

2.2 *Staycation Concept*

A staycation is defined as the practice of transforming one's residence into a vacation destination, hence avoiding the need to travel [23]. Broadly speaking, the notion of a staycation pertains to the practice of engaging in tourism activities by visiting sites that are in closer proximity to one's place of residence. Individuals may have a tendency to minimize the amount of time dedicated to the process of planning and eagerly anticipating a vacation, resulting in feelings of exhaustion and heightened anticipation [24]. According to Wong et al. [11], a staycation can be defined as a form of local travel, while the experience of engaging in tourism within one's own country is commonly referred to as proximity tourism. To clarify, the duration of a staycation is typically brief, as it primarily encompasses local outings that can be reached by car. According to Rosu [2], the concept of a staycation refers to a brief period of vacation spent within one's own residence or hometown. The purpose of this study is to examine the ways in which individuals utilize a socially natural setting to consistently fulfil their personal sense of style and the subsequent impact it has on reclamation. Given the prevailing conditions, a portion of individuals are opting to spend their vacations inside their local vicinity [22]. This trend, known as a staycation, is gaining traction as evidenced by the growing availability of staycation options. This study examines the staycation phenomenon, which involves travelers disengaging from their usual residential environment in order to engage in tourism-related activities within nearby areas, typically involving an overnight stay.

According to Terry Massey [25], the concept of a "staycation" has emerged as a viable alternative to traditional vacations during the current crisis, as discussed by James et al. [8]. According to earlier research conducted by de Bloom et al. [26] and James et al. [8], Terry Massey [25] has been recognized as the individual who first employed the term in 2003, The concept of "staycation," as discussed by Vada et al. [27] has gained significant popularity as a form of tourism activity, particularly in situations where individuals are limited in their mobility due to border lockdowns [28]. The emergence of the staycation notion might potentially be attributed to the economic crisis that occurred in the early 2000s [29]. The phenomenon of staycation gained traction during the global COVID-19 pandemic, and its prevalence surged as the virus disseminated on a global scale [30]. The emergence of staycations can be attributed to the increasing number of individuals who were unable to financially support extensive or far-reaching travel endeavors, prompting them to seek alternative means of relaxation and entertainment.

The media portrayed staycations as a state of stillness and labeled them as 'undesirable'. This representation positioned staycations as either a type of consumer mobility or as a lack of mobility, as discussed by Molz [29]. This highlights the construction of representations of mobility. Moreover, the concept of staycation has emerged as a viable and practical solution within the tourism and hospitality industry, taking into account the limitations imposed by the Covid-19 pandemic on travel and accessibility. This stands in contrast to the less developed recovery strategies, such as the "tourism bubble," which have proven to be disappointing [31]. Expanding upon the notion that staycation ostensibly imparts novel connotations to the act of traveling, a comparably growing concept known as slow tourism arises. Dickinson et al. [32] argue that travel extends beyond mere physical mobility, instead emphasizing its integral role in the overall experience. Slow tourism is a conceptual framework that encompasses a consumer-oriented approach to tourism, characterized by a harmonious integration of experiential elements within the trip experience. Conversely, the notion of staycation places a significant emphasis on the concept of distance. Nevertheless, it is worth noting that a staycation can encompass local activities such as bike rides, concerts, visits to museums, and parks [3]. Individuals frequently engage in and actively participate in these activities. In addition, the activities of cycling, walking, and local transport are also seen as components of slow tourism, as discussed by Dickinson et al. [32]. Hence, a staycation appears to be ascribing novel connotations to the concepts of travel and tourism.

2.3 Tourist Experiences

Jeuring and Diaz-Soria [33] also examine the concept of "unfamiliarity" in tourist experiences inside domestic sites in their research on proximity tourism. Given that distance plays a significant role in the travel behavior of different visitor mobilities. The model proposed by Ram et al. [34] aims to examine the relationship between visitors' experiences and their mobility, specifically in the context of pleasure in tourism. According to the source cited, there is a potential challenge in promoting behavioral change regarding sustainable tourist mobility behavior due to the possibility of reduced happiness (ibid, p. 1025). The importance of the quality of the tourist experience in the field of tourism is significant due to its strong correlation with several elements, including the sense of visitor satisfaction [35]. Furthermore, it is worth noting that a proficient website supplier possesses the ability to enhance favorable affective states, such as happiness, hence influencing the behavioral inclinations of tourists [36]. According to Pujiastuti et al. [37], there are three key elements that can influence the overall quality of the tourist experience. These aspects include the characteristics and behaviors of the tourists themselves, the offerings and services provided by the tourist destination, and the involvement and attitudes of the local population, including both tourism personnel and the surrounding community. Experience Quality refers to the assessment of the comprehensive experience at a certain destination during a defined timeframe.

From a scholarly perspective, it may be observed that the staycation, defined as a holiday where individuals choose to stay at or near their residence while simulating the ambiance of a conventional vacation [38], may exhibit distinct characteristics for millennials during the COVID-19 pandemic compared to a conventional tourist encounter. A staycation refers to a leisure activity where an individual chooses to remain within their local vicinity and engages with the physical surroundings of a nearby hotel, as well as explores local attractions, as opposed to undertaking travel to a different location [8]. The environment and culture encountered during their travels may not exhibit substantial divergence from their own habitual experiences. The concept of a staycation involves individuals engaging in recreational activities within a confined geographical area, such as a hotel or resort, that is situated within their immediate community. This phenomenon encompasses the social interactions and behaviors exhibited by individuals within their respective social networks. However, the actions of staycationers may deviate from the norm, since they assume the dual role of both residents and tourists inside the local region [39]. Hence, consumer behaviors linked to their imaginations play a significant role in shaping their overall experience, particularly in the context of the ongoing epidemic. Expanding upon the preceding discourse, the present investigation delineates the staycation encounter of millennials as "the engagement of millennials in both physical and mental pursuits during a hotel staycation in close proximity to their place of residence." The staycation experience encompasses a combination of familiarity and novelty within the immediate physical surroundings. This study aims to investigate the construction of the staycation experience in light of the global health crisis, with a particular focus on the perspectives of millennials.

2.4 Bridging the Gap in Studying Staycation Concept

The literature on staycations reveals the presence of several gaps in study. The existing body of research offers limited insights into the structure of staycation experiences, mostly focusing on categorizing various local activities associated with staycations. As an illustration, Besson [39] delineates outdoor leisure, adventure, sightseeing, social bonding, relaxation, and gustatory pleasures as the predominant activities. In the study conducted by Rosu [2], the author examines several staycation habits through the analysis of internet postings. These practices frequently revolve on the appreciation and engagement with natural landscapes and activities. According to Kalista's [40] findings, families with children have a greater propensity to engage in novel activities when opting for a staycation at a nearby resort. Typically, individuals' actions and encounters have a greater influence on their sense of relaxation and contentment compared to the duration of their stay [26, 40]. Existing literature mostly focuses on providing descriptions of staycation habits, with limited attention given to investigating situational aspects that influence staycation experiences, particularly in the context of the COVID-19 pandemic.

In their study, Kay and Wang [41] emphasize the notion of staycations, which refers to brief excursions to nearby locales, as an emerging promotional strategy for local locations. Staycations have emerged as a pertinent industrial strategy throughout the COVID-19 pandemic [28]. This is because individuals who possess the financial means to travel may find themselves pushed to contemplate staycations as a result of the pandemic-induced restrictions on their travel choices. This research addresses the aforementioned gaps in the literature by examining the staycation experiences of millennials during the pandemic, as well as their perspectives of staycations in the post-COVID-19 period. In contemporary times, there has been scholarly attention directed towards the exploration of travel patterns in the aftermath of the pandemic [42–44]. Nevertheless, the utilization of qualitative methodology in this area of research remains relatively limited [45, 46]. Miao et al. [47] argue in favor of employing a qualitative methodology to enhance comprehension of tourist activities in the post-pandemic era. They contend that such an approach is advantageous as it enables the examination of relevant circumstances and attitudes, hence facilitating a more comprehensive knowledge of the phenomenon.

3 Methodology

The present study will employ a qualitative research strategy in order to gather and analyze data with the purpose of addressing the research inquiries. This study aims to investigate the staycation trend in Malaysia and analyze the underlying themes derived from the perspectives of Malaysian vacationers regarding the concept of staycations. The purpose of the interview will be to ascertain the suitability of the measure in assessing the idea of staycation. The primary objective of this study is to examine the concept and essential criteria of staycations among Malaysian tourists within the context of Malaysia. The interview will be extended till the point of data saturation has been achieved. Data saturation occurs when a sufficient amount of information has been collected and no more novel data is acquired from participants.

The researchers employed a semi-structured interview approach during the individual interviews, allowing them to pose both pre-determined questions and subsequent inquiries to the participants [48]. The formulation of the questions was informed by the existing literature and aligned with the research aims of this study. The interview protocol utilized in this study was created in accordance with the guidelines proposed by Creswell [49]. Initially, a concise explanation of the research objective was provided, followed by the acquisition of informed consent for the audio recording of the interview. The interview comprised three distinct sections of inquiry, which centered on the following aspects: the motivating factors that prompt individuals to engage in the consumption of staycation products amidst the influence of the COVID-19 pandemic, in comparison to their previous experiences with traditional hotel stays; the participants' recent encounters with staycations, encompassing their activities and emotional responses to these experiences; and the participants' anticipated future behaviors in relation to staycations. According to Glaser and Strauss [50], interviews ceased to be done when the material being obtained did not present a significant level of novelty. Every interview session will be recorded using digital technology and transcribed afterward. The process of recording and analyzing data will be conducted with the qualitative software known as Atlas.ti. Thematic analysis was employed to analyze the data obtained from interviews.

4 Discussion and Conclusion

This study offers valuable insights into the existing body of literature concerning staycations and the behavior of millennial visitors. This study aims to provide a deeper understanding of the staycation phenomenon by presenting a clear description and constructing a thorough framework based on the perspective of millennials. The previous studies have mostly focused on categorizing the activities that occur during staycations and emphasizing the interpersonal relationships among the individuals involved, with the understanding that this analysis is contingent upon the established definition of staycation as proposed by James et al. [8]. This research has been conducted by scholars such as Besson [39] and Rosu [2]. In relation to the staycation experience experienced by consumers during the COVID-19 pandemic, there has been a limited amount of research conducted thus far that focuses on the quality of staycation reported experiences [51]. Additionally, a few studies have explored factors associated with staycations, such as stress reduction and escapism [11, 25]. The current body of research has only examined the staycation experience from a narrow perspective, focusing mostly on the perceptions of a specific demographic group, namely millennials [8]. This study contributes to the existing body of literature on staycations by examining the structure of millennials' staycation experiences in the current period of the COVID-19 pandemic. It highlights the emergence of staycations as a new normal, characterized by a combination of both ordinary and extraordinary elements. Millennials exhibit an awareness of the potential hazards associated with undertaking travel activities amidst the ongoing coronavirus epidemic. Nevertheless, they continue to harbor a strong inclination towards embarking on such journeys, driven by a desire to seek respite from the routine demands of their everyday existence.

This inquiry pertains to the positioning of the staycation concept within the broader discourse on tourism mobility. In light of this, the present study seeks significant contributions from key stakeholders in policy-making, including Tourism Malaysia, the Ministry of Tourism, Arts, and Culture (MOTAC), and business entities involved in the tourism industry. The findings of this study can be utilized as a point of reference for hotel and hospitality practitioners and destination marketers in order to enhance the planning and development of staycation products as viable strategies to mitigate the severe impacts of the Covid-19 pandemic. This novel finding will shed light on the likely mechanisms underlying staycation qualities that prioritize the parameters influencing the choice of tourism destinations.

Acknowledgements This paper has been written in the frame of the research grant with the title "*New Staycation Motivators Model for National Vacationers on Proximity Tourism*". This research has been financially supported by the Ministry of Higher Education (MOHE) with code research "*FRGS/1/2023/SS04/UMK/02/1*".

References

1. Romagosa, F.: The COVID-19 crisis: opportunities for sustainable and proximity tourism. Tour. Geogr. **22**(3), 690–694 (2020). https://doi.org/10.1080/14616688.2020.1763447
2. Rosu, A.: Making sense of distance. Mobility in staycation as a case of proximity tourism. Lund University, Lund (2020)
3. Pawłowska, A., Matoga, Ł: Staycation as a way of spending free time by city dwellers: examples of tourism products created by local action groups in Lesser Poland voivodeship in response to a new trend in tourism. World Sci. News **51**, 4–12 (2016)
4. Diaz-Soria, I.: Being a tourist as a chosen experience in a proximity destination. Tour. Geogr. **19**(1), 96–117 (2017). https://doi.org/10.1080/14616688.2016.1214976
5. ForwardKeys: Tourism Strategies to Aide the Travel Recovery—By Luis Millan. Hotel News Resource. Available at: www.hotelnewsresource.com/article117807.html (2021)
6. Lane, M.: Chinese Traveler Survey Reveals Increased Wanderlust and Higher Confidence in Travel Safety. The Moodie Davitt Report. Available at: www.moodiedavittreport.com/chinese-traveller-survey-reveals-increased-wanderlust-and-higher-confidence-in-travel-safety/deals (2021)
7. Hong, J., Lam, E.: Hong Kong Economy Gets a Boost from Staycations, Dining Deals. Bloomberg Quint. Available at: www.bloombergquint.com/global-economics/hong-kong-economy-gets-a-boost-from-staycations-dining (2020)
8. James, A., Ravichandran, S., Chuang, N.K., Bolden, E., III.: Using lifestyle analysis to develop lodging packages for staycation travelers: an exploratory study. J. Qual. Assur. Hosp. Tour. **18**(4), 387–415 (2017)
9. Chan, J., Gao, Y.L., McGinley, S.: Updates in service standards in hotels: how COVID-19 changed operations. Int. J. Contemp. Hosp. Manag. **33**(5), 1668–1687 (2021)
10. Eloksari, E.A.: Indonesian Internet Users Hit 196 Million, Still Concentrated in Java: APJII Survey. The Jakarta Post (2020)
11. Wong, I.A., Lin, Z., Kou, I.E.: Restoring hope and optimism through staycation programs: an application of psychological capital theory. J. Sustain. Tour. **1**, 1–20 (2021)
12. Barclays: Staycation Boom Driven by Millennials. Barclays. Available at: https://home.barclays/news/press-releases/2019/05/staycation-boom-driven-by-millennials (2019)
13. Sethi, R.S., Kaur, J., Wadera, D.: Purchase intention survey of millennials towards online fashion stores. Acad. Mark. Stud. J. **22**(1), 1–16 (2018)
14. Ketter, E.: Millennial travel: tourism micro-trends of European Generation Y. J. Tour. Futures 7(2) (2020). https://doi.org/10.1108/JTF-10-2019-0106
15. Siegel, L.A., Wang, D.: Keeping up with the Joneses: emergence of travel as a form of social comparison among millennials. J. Travel Tour. Mark. **36**(2), 159–175 (2019)
16. Yusoff, Z.M., Aziz, I.S., Naharudin, N., Rasam, A.R.A., Ling, O.H.L.: Mobility and proximity coefficient to high-traffic volume in daily school operations. Plan. Malays. **20** (2022)
17. Ratthinan, S.P., Selamat, N.H.: Understanding the push and pull motivational factors of technology usage for traveling among women students in USM, Penang. J. Inf. Syst. Technol. Manag. **2**(4), 52–63 (2017)

18. Rantala, O., Salmela, T., Valtonen, A., Höckert, E.: Envisioning tourism and proximity after the anthropocene. Sustainability **12**(10), 3948 (2020)
19. Rogerson, C.M., Rogerson, J.M.: COVID-19 tourism impacts in South Africa: government and industry responses. Geo J. Tour. Geosites **31**(3), 1083–1091 (2020)
20. Salvatore, R., Chiodo, E., Fantini, A.: Tourism transition in peripheral rural areas: theories, issues and strategies. Ann. Tour. Res. **68**, 41–51 (2018)
21. Canavan, B.: The extent and role of domestic tourism in a small island: the case of the Isle of Man. J. Travel Res. **52**(3), 340–352 (2013)
22. Jeuring, J.H.G., Haartsen, T.: The challenge of proximity: the (un)attractiveness of near-home tourism destinations. Tour. Geogr. **19**(1), 118–141 (2017)
23. Sharma, S.: The great American staycation and the risk of stillness. M/C J. **12**(1) (2009). Available at: https://doi.org/10.5204/mcj.122
24. Smith, M.K., Diekmann, A.: Tourism and wellbeing. Ann. Tour. Res. **66**, 1–13 (2017)
25. Jacobsen, J.K.S., Farstad, E., Higham, J., Hopkins, D., Landa-Mata, I.: Travel discontinuities, enforced holidaying-at-home and alternative leisure travel futures after COVID-19. Tour. Geogr. **1**, 1–19 (2021)
26. De Bloom, J., Nawijn, J., Geurts, S., Kinnunen, U., Korpela, K.: Holiday travel, staycations, and subjective well-being. J. Sustain. Tour. **25**(4), 573–588 (2017)
27. Vada, S., Prentice, C., Scott, N., Hsiao, A.: Positive psychology and tourist well-being: a systematic literature review. Tour. Manag. Perspect. **33**, 100631 (2020). https://doi.org/10.1016/j.tmp.2019.100631
28. Le, D., Phi, G.: Strategic responses of the hotel sector to COVID-19: toward a refined pandemic crisis management framework. Int. J. Hosp. Manag. **94**, 102808 (2021)
29. Molz, J.G.: Representing pace in tourism mobilities: staycations, slow travel and the amazing race. J. Tour. Cult. Chang. **7**(4), 270–286 (2009)
30. Novani, S.: Examining factors influencing people's intention to staycation during COVID-19: an extended model of goal-directed behaviour. Tour. Hosp. Manag. **28**(2), 361–380 (2022)
31. The Star:. Interstate Travel Only Allowed for Langkawi Tourism Bubble, Says Tourism Ministry. https://www.thestar.com.my/news/nation/2021/09/29/interstate-travel-only-allowed-for-langkawi-tourism-bubble-says-tourism-ministry (2021)
32. Dickinson, J.E., Lumsdon, L.M., Robbins, D.: Slow travel: issues for tourism and climate change. J. Sustain. Tour. **19**(3), 281–300 (2011)
33. Jeuring, J., Diaz-Soria, I.: Introduction: proximity and intraregional aspects of tourism. Tour. Geogr. **19**(1), 4–8 (2017)
34. Ram, Y., Nawijn, J., Peeters, P.M.: Happiness and limits to sustainable tourism mobility: a new conceptual model. J. Sustain. Tour. **21**(7), 1017–1035 (2013)
35. Yan, N., Halpenny, E.A.: Tourists' savoring of positive emotions and place attachment formation: a conceptual paper. Tour. Geogr. **24**(2–3), 369–389 (2022). https://doi.org/10.1080/14616688.2019.1647454
36. Ratnasari, R.T., Gunawan, S., Mawardi, I., Kirana, K.C.: Emotional experience on behavioral Intention for halal tourism. J. Islamic Mark. **12**(4), 864–881 (2020). https://doi.org/10.1108/JIMA-12-2019-0256
37. Pujiastuti, E.E., Nimran, U., Suharyono, S., Kusumawati, A.: The antecedents of behavioral Intention regarding rural tourism destination. Asia Pac. J. Tour. Res. **22**(11), 1169–1181 (2017). https://doi.org/10.1080/10941665.2017.1377270
38. Wixon, M.: The Great American Staycation: How to Make a Vacation at Home Fun for the Whole Family (and Your Wallet!). Simon and Schuster, New York, NY (2009)
39. Besson, A.: Everyday aesthetics on staycation as a pathway to restoration. Int. J. Hum. Cult. Stud. **4**(2), 34–52 (2017)
40. Kalista, E.: Family Play Experience During a Weekend Staycation. California State University. Available at: https://csusdspace.calstate.edu/bitstream/handle/10211.3/215203/FAMILY%20PLAY%20EXPERIENCE%20DURING%20A%20WEEKEND%20STAYCATION.pdf?sequence=1 (2020)

41. Kay, M.J., Wang, Y.: Marketing the staycation: the salience of the local in destinations branding. In: Proceedings of the Northeast Business and Economics Association (2010)
42. Chansuk, C., Arreeras, T., Chiangboon, C., Phonmakham, K., Chotikool, N., Buddee, R., Pumjampa, S., Yanasoi, T., Arreeras, S.: Using factor analyses to understand the post-pandemic travel behavior in domestic tourism through a questionnaire survey. Transp. Res. Interdiscip. Perspect. **16**, 100691 (2022)
43. Li, J., Nguyen, T., Coca-Stefaniak, J.A.: Coronavirus impacts on post-pandemic planned travel behaviours. Ann. Tour. Res. **86**, 102964 (2020). https://doi.org/10.1016/j.annals.2020.102964
44. Zaman, U., Raza, S.H., Abbasi, S., Aktan, M., Farıas, P.: Sustainable or a butterfly effect in global tourism? Nexus of pandemic fatigue, covid-19-branded destination safety, travel stimulus incentives, and post-pandemic revenge travel. Sustainability **13**(22), 12834 (2021)
45. Oktaviani, D., Sutisna, D.: An analytical study of the readiness of eco-tourism based transactions in the mount putri Bandung (Indonesia) tourism area. Rev. Int. Geogr. Educ. Online **11**(3), 229–234 (2021)
46. Panzer-Krause, S.: Rural tourism in and after the COVID-19 era: revenge travel" or chance for a degrowth-oriented restart? Cases from Ireland and Germany. Tour. Hosp. **3**(2), 399–415 (2022)
47. Miao, L., Im, J., Fu, X., Kim, H., Zhang, Y.E.: Proximal and distal post-COVID travel behavior. Ann. Tour. Res. **88**, 103159 (2021). https://doi.org/10.1016/j.annals.2021.103159
48. Patton, M.Q.: Qualitative Research and Evaluation Methods, 4th edn. Sage Publications, Thousand Oaks, CA (2015)
49. Creswell, J.: Research Design: Qualitative, Quantitative and Mixed Methods Approaches, 4th edn. Sage Publications, Thousand Oaks, CA (2014)
50. Glaser, B.G., Strauss, A.L.: Discovery of Grounded Theory: Strategies for Qualitative Research. Routledge, New York, NY (2017)
51. Yan, Q., Shen, H., Hu, Y.: A home away from hem: exploring and assessing hotel staycation as the new normal in the COVID-19 era. Int. J. Contemp. Hosp. Manag. **34**(4), 1607–1628 (2022)

Women Hoteliers in the COVID-19 Crisis: A Conceptual Paper

Andi Tamsang Andi Kele, Yuko Kyogoku, Arif Kamisan, Jennifer Kim Lian Chan, Sharija Che Shaari, and Wirawati Yi Xe Ngui

Abstract The COVID-19 crisis has negatively affected people's health and economies worldwide. Being subject to negative external threats, the hotel industry has been greatly hit by the COVID-19 crisis. In the global labour market, the hotel industry has long been a key employer, supplying many employees. The majority of hotel employees worldwide are women and given how hard the COVID-19 pandemic struck the hotel business, women's employment in the hotel industry has suffered significantly, which is no exception in the case of Sabah. Compared to the earlier crises, such as the 2003 SARS outbreak, the COVID-19 crisis has a greater effect and is more volatile. Thus, it is critical to explore the impact of the COVID-19 pandemic on the Sabah hotel industry in general and women hotel employees in particular. For this research, the interpretivism paradigm of the descriptive phenomenological approach will guide the research process. Semi-structured and in-depth interviews with women employed in the Sabah hotel industry will be conducted qualitatively. The data will be analysed using a thematic approach to generate a holistic perspective of the impact of COVID-19 on women working in the Sabah hotel industry. The expected findings of this research contribute theoretically to gender studies of the hotel industry in crisis and provide practical guidance to hotel managers on how to establish a more favourable work environment for women employees in Sabah.

Keywords Women · Hotelier · Covid-19 · Crisis

A. T. A. Kele (✉) · Y. Kyogoku · A. Kamisan · J. K. L. Chan · S. C. Shaari
Faculty of Business, Economics and Accountancy, Universiti Malaysia Sabah, Kota Kinabalu, Sabah, Malaysia
e-mail: andi@ums.edu.my

W. Y. X. Ngui
Faculty of Psychology and Education, Universiti Malaysia Sabah, Kota Kinabalu, Sabah, Malaysia

B. Awwad (ed.), *The AI Revolution: Driving Business Innovation and Research*, Studies in Systems, Decision and Control 524, https://doi.org/10.1007/978-3-031-54379-1_76

1 Introduction

1.1 Background

COVID-19 is a coronavirus, a type of acute disease identified in 2019 and has since evolved into a global health crisis. Since it was announced as a pandemic on 22 March 2020 by the World Health Organization, this crisis drove numerous countries to shut borders and lockdown cities as a way of containing the virus [1]. COVID-19 quickly escalated into an economic catastrophe, with devastating outcomes at every level.

The hotel business is sensitive to crisis, as external threats such as the COVID-19 pandemic affect demand for travel and tourism-related products and services [2]. Globally, it was estimated that tourism-related sectors lost 100.8 million jobs [3]. Women make up a considerable part of the hotel industry, and the United Nations World Tourism Organization (UNWTO) [3] reports that women make up 54% of those employed in tourism worldwide. Women make up most of the hospitality workforce, and the COVID-19 pandemic has affected their employment significantly in the hospitality industry. Thus, it is critical to examine the effect of the COVID-19 pandemic on women employed in the hospitality sector, particularly in the hotel sector.

Sabah is Malaysia's second-biggest state, which relies on minimally processed primary industry exports and other commodities such as agriculture and tourism for its economy [4]. Tourism contributes to Malaysia's income and economic growth, especially in Sabah [5]. Tourist arrivals in Sabah from 2013 to 2020 are shown in Fig. 1. According to the Sabah Economic Development and Investment Authority (SEDIA) Revenue from the tourist sector also grew, going from RM6.52 billion in 2015 to RM8.34 billion in 2018, demonstrating positive development in Sabah's tourism economy [5]. However, due to the COVID-19 pandemic, which storms and affects worldwide, it showed a tremendous drop in 2020.

Given the magnitude of the COVID-19 pandemic's impact on the Sabah hotel industry, changes brought about by COVID-19 are certain to influence women's livelihood. Therefore, this research attempts to assess the impact COVID-19 had on women hotel employees and the extent to which they were impacted in the Sabah context and their well-being. Thus, four research questions are proposed:

1. How does the COVID-19 crisis affect women employees of Sabah's hotel industry?
2. What are the coping strategies used by the women employees?
3. What are the key issues and challenges faced by women employees to remain in the hotel industry?
4. What is the future of work for women in the hotel industry during the COVID-19 crisis?

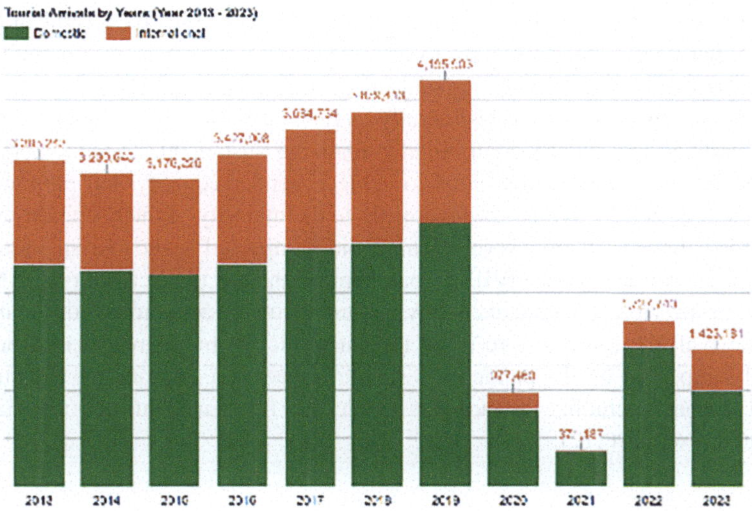

Fig. 1 Sabah tourist arrival from 2013 to 2020. *Source* Sabah Tourism Board (2023)

The structure outline of this paper begins with an introduction and the significance of the research, a literature review that reviews relevant literature in the field of crisis and gender, the methodology that will be applied, and a conclusion that emphasises implication and the future research direction.

2 Literature Review

2.1 Women and the Hotel Industry

Due to low entry barriers, flexible work schedules and job availability [2], the hotel industry may be desirable to a potential employee, especially for women. The hotel, catering, and tourist sector is a rapidly expanding service industry with a global employment rate for women of 55.5% [6]. Furthermore, the hotel nature of shift employment and part-time is frequently viewed as a way for women working while simultaneously fulfilling their household obligations; however, women need to face a new challenge on how to balance between work and family [7].

Moreover, the traditional view of the hotel industry is that many hotel professions are feminised and hence viewed as more appropriate for women [6]. Hospitality-related work is compatible with women's conventional gender roles and duties and has been stereotyped as their career [8]. On the contrary, [9] complimented the hotel industry for enabling women to utilise gender norms. The overly optimistic explanations ignore the fact that this sort of employment perpetuates women's subjugation to conventional inequities [10] and emphasises the notion that women should be devoted only to domestic life. Even though women are more successful in hotels than in other sectors, they still have lower success rates than men, as women have a lower representation in executive management roles than men [11].

The hospitality industry has historically served as a platform for women's empowerment and has consistently contributed to realising Sustainable Development Goal 5 of the 2030 Agenda, which aims to achieve gender equality and empower all women and girls [12]. Women, who constitute the majority of the hospitality workforce (54%), have been more severely and rapidly impacted by the tourism industry's economic blow due to the COVID-19 pandemic than their male counterparts. Moreover, domestic violence, unpaid caregiving, and women's economic and social insecurity have all increased as a result of the pandemic. Women have been disproportionately impacted by the pandemic's severe impacts on the tourism industry for several reasons, including the fact that females are frequently left out of social and institutional support [12].

2.2 Crisis and the Hotel Industry

The conceptualisation of crises as unforeseen events that disrupt daily life and result in death and destruction has been a prevalent feature in crisis research. Factors such as poverty, environmental degradation, and unequal power dynamics can exacerbate the severity of the crisis. Consequently, the notion of an individual's vulnerability, which encompasses their social, economic, and physical circumstances or status, is often invoked to elucidate further [13]. Crises can manifest at the national or global level, with global crises encompassing financial collapses, world wars, natural disasters such as tsunamis, and healthcare pandemics. Recent examples of global crises include the Global Financial Crisis and the Coronavirus Global Pandemic (COVID-19), both of which have had catastrophic social, economic, and financial ramifications for individuals, societies, and economies.

Due to the reliance on the tourism industry, which is highly seasonal, the hotel industry is vulnerable to a crisis. Not only do crises result in a decline in demand for hotel services, but they also result in large decreases in the guest average expenditure [14]. The hotel business was impacted differently by several crises before COVID-19. The global financial crisis resulted in a global economic recession as People were less inclined to take vacations and could not afford them, which had an indirect impact on the tourist and hospitality sectors [15]. Another crisis such as the Severe Acute Respiratory Syndrome health crisis directly affected the hotel industry. SARS caused a greater than 29% decrease in the share values of the hospitality sector, compared to other industries [16]. Another major issue that affected the hospitality sector in the 2000s was the 9/11 attacks in New York, which resulted in a quick and severe drop in hotel occupancy and rates [17].

Every time a crisis strikes, demand for hospitality services decreases, and employees, especially women, become subject to dismissal [6]. For instance, [18] showed that during the 2008 global financial crisis, most industries in the United

States of America suffered from the effects, resulting in widespread unemployment, with women's unemployment rates in several industries higher than men's, including the hotel industry [18]. As a result of the possibility of unemployment, women working in the hotel industry are susceptible during times of crisis.

2.3 Women's Role and Employment

According to social role theory, gendered traits and stereotypes emerge from the distinct function of men and women participating in society. Behavioural differences between the sexes arise from the differential social roles that women and men inhabit, particularly those related to the division of labour [19]. Throughout history, women and men have been assigned labour tasks that align with their physical characteristics due to demands from the economy, the environment, society, and technology. Consequently, Men were more likely to perform activities that needed quickness, strength, and the capacity for long periods away from home, while Women were more likely to perform duties connected to the house and family since they were largely responsible for birthing. These varied societal responsibilities have led to the development of gender roles, which set expectations for the traits and actions of men and women. Men are specifically expected to embody the masculine gender role, which represents agentic attributes, while women are meant to embody the feminine gender role, which reflects community qualities [20].

In contrast to women, males have more wealth and influence because of their superior physical qualities that have allowed them to participate in economic production [20]. Consequently, women have tended to develop communal traits such as empathy, caring, and kindness, which are essential for successful caregiving, while men have developed agentic traits such as aggression and ambition, which apply to their roles as hunters, farmers, workers, and leaders [21].

Recently, social role theory has been expanded to encompass the roots of the labour division between males and females. This development was spurred by the advent of evolutionary psychology, which offered an essentialist justification for many social behaviour variations between the sexes. However, an alternate theory contends that differences in male and female behaviour result from a complex interaction of social and biological factors, rather than solely relying on evolutionary psychology to explain ultimate origins. The underlying causes of male and female behaviour come from evolved physical variations between the sexes, particularly women's reproductive functions and men's larger size and strength, which interact with the needs of people's social and economic environments. Because they need the time- and energy-consuming processes of gestation, nursing, and infant care, women's reproductive activities during pregnancy and lactation function as strong limitations. These activities make it difficult for women to participate as fully as males in duties that call for rapid movement, continuous work, extensive training, or far-off travel. Consequently, women typically contribute little to activities like

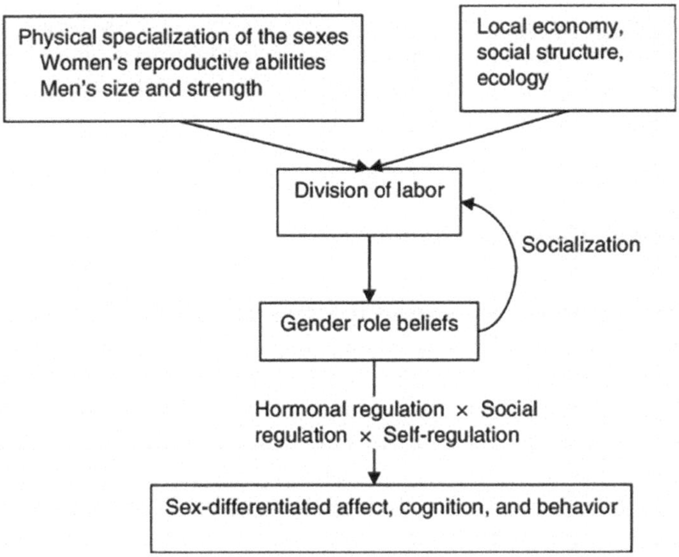

Fig. 2 Gender roles guided sex differences and similarities through the biosocial process [22]

hunting large animals, fighting, and ploughing in foraging, horticultural, and agricultural civilisations. Women prefer activities that are more conducive to childcare instead. Gender roles influence sex differences and similarities through biosocial processes, as seen in Fig. 2.

Before the onset of any health crisis, women encounter difficulties in managing multiple roles, which is further compounded by the perception that certain roles are exclusively reserved for women, such as domestic work and childcare. Despite the increased participation of women in full-time paid employment, women continue to bear the brunt of household chores and childcare responsibilities [23]. This is largely attributed to societal gender expectations, which dictate that domestic responsibilities are primarily the domain of women, thereby placing significant strain on working women to balance their work and home commitments [24] effectively.

The expectation placed on women to fulfil multiple roles, not only as mothers but also as employees and leaders in organisations, has presented additional challenges. This research aims to highlight the difficulties that women face in managing these various roles, as outlined in the Social Role Theory. The study seeks to understand how women can cope with these challenges and hopes to shift cultural acceptance and mindset regarding the role of women. Rather than solely focusing on what women can do to manage their responsibilities, this research presents an opportunity to examine how policymakers and others can support women during times of crisis, particularly regarding childcare. The COVID-19 pandemic has prompted many businesses to adopt work-from-home and telecommuting options, which may lead to increased workplace flexibility in the future. Considering that mothers now shoulder a disproportionate amount of the responsibility of juggling work and child-care responsibilities, they will benefit from these improvements more than males. Additionally, husbands and fathers will now have to shoulder additional childcare and home-schooling responsibilities. This crisis presents an opportunity to challenge gender stereotypes, particularly the persistent notion of men as the breadwinners [25].

3 Methodology

For this research, the interpretivism paradigm of the descriptive phenomenological approach will guide the research process. Phenomenology is a philosophical world-view that aims to assess experience [26]. It directs the researcher to study what it is like to be human, particularly regarding the things that matter and create [27]. Descriptive methods are one type of phenomenological inquiry that attempts to describe a lived experience without seeking to give it meaning [27]. In other words, to study phenomena as experienced, the researcher should set aside their past knowledge and beliefs during data collection and analysis.

For data collection, semi-structured and in-depth interviews with women working in the Sabah hotel industry will be conducted. A semi-structured, in-depth interview is a conversation in which participants are questioned to uncover any potential motivators, viewpoints, desires, attitudes, and sentiments regarding the subject at hand [28]. The semi-structured style can be effective if the researcher is particularly interested in conducting interviews that are relevant to the topic, particularly if the interviewer must ask a series of questions [29]. More importantly, it enables the interviewer to ask additional questions to explain the research objectives and make the interview more interactive and engaging [30].

The purpose of this study is to confirm the participation of Sabah-based hotels that are officially recognised by Malaysia's Ministry of Tourism, Arts, and Culture in 2020. All hotel organisations operating in Malaysia are required to adhere to the official hotel categories, which the Ministry of Tourism, Arts, and Culture enforces. According to the evaluations recorded as of 2020, Table 1 shows the number of lodgings in Sabah.

Table 1 Accommodation in Sabah based on star ratings in 2019

Start-rating	Number of hotels	Number of rooms
5-star	11	4231
4-star	17	2019
3-star	50	4848
2-star	33	4738
1-star	56	2111
Total	187	17,947

The research will be conducted in Kota Kinabalu, Sabah's capital city as almost 50% of the star-rated hotels in the state of Sabah are located in Kota Kinabalu. The researcher will approach the Malaysia Association of Hotels (MAH), which serves as the industry's gatekeeper, to conduct research. The Human Resources Departments of each hotel will then be contacted to get clearance and to provide access to interview participants. The researcher chooses a small number of participants who satisfy the research's inclusion criteria and asks them to recommend other individuals who satisfy them [30]. Female participants who were employed as hotel staff in Sabah during the COVID-19 outbreak were the selection criteria. These sampling techniques can potentially remove bias in sample selection and necessitate the classification of the appropriate person to be hired, improving participant selection. Saunders et al. [30] add that the technique is frequently the only option to acquire difficult-to-reach respondents.

In qualitative research, a large sample size is not required; Interviews are conducted until participant responses become repetitive and saturated [31]. In other cases, the number of interviews is controlled by the saturation point of the interview, which is the point at which data repetition occurs and the themes emerge. At this time, the researcher will call the interview closed because additional information will not change our understanding of the phenomenon [32]. To analyse data, The use of theme analysis is selected because it enables researchers and participants to uncover rich contextual material and creates a comprehensive picture of how COVID-19 has affected women working in the Sabah hotel business [29]. Figure 2 shows the style of data analysis, which adheres to [32] six-step data analysis recommendation.

4 Conclusion

There haven't been many studies on how diseases affect the economy, particularly the tourism industry, at least in the case of Sabah, Malaysia. Malaysia's tourism industry is feeling negative repercussions as a result of the recent COVID-19 pandemic [4]. On that note, it could not be accentuated more, importance of the hotel industry to Sabah. Therefore it is crucial to conduct this research in the Sabah hotel industry context. Many things are changing rapidly causing the need to find out the impact of the health crisis on its full course. This has made it crucial to remember that the situation is still developing and that factors that were critical a few weeks or months ago might not be as significant now, such as the rise and drop in the number of cases and their variable appearance in different geographical contexts [33].

There are several studies conducted on the implication of the pandemic towards the tourism and hotel industry in Malaysia and Sabah such as [34]. To create a comprehensive understanding of how women perceive gender roles during the pandemic, more studies on various female experiences should be conducted. Having a comprehensive grasp of what it was like to be a woman during the pandemic can help future efforts to lessen gender inequality [35]. There is research that is looking at the health crisis impact on women generally and women from different industries such as [36].

However, there is yet to be research conducted on the implication of the current global health crisis on women in the hotel industry. Thus, this research hopes to shed light on the impact of the current global health crisis from the perspective of women in Sabah's hotel industry, together with its contribution to gender studies. When planning and executing longerterm recovery and development initiatives, it's crucial to keep in mind that the crisis just exacerbates these enduring forms of gender inequality, not causes them [37]. This research aims to be the voice of women in the hotel industry, with the hope that this voice will be heard and considered by the civil society, private sector, government and parties that are responsible for the future development of crisis recovery plans, to put women at the front and centre of this plan, to ensure a more sustainable and equal recovery for all. To create effective interventions and inform future regulations, it is also important to recognise the various coping methods used by those groups [38].

The COVID-19 pandemic had a negative impact on the hotel industry, particularly in Sabah; however, there is currently limited information on its effects on women hotel employees. COVID-19 is a new catastrophe with unforeseeable ramifications. This research examines women's employment in the hotel industry during the COVID-19 pandemic to better grasp how the COVID-19 pandemic impacts the human experience. Additionally, the outcomes of this research contribute theoretically to the crisis in the hotel industry's gender studies and provide functional guidance to hotel management on how to establish a more favourable work environment for women employees in Sabah. In recent years, the terminology and substance of Studies on gender have evolved substantially, reflecting the growing interest in gender equality. To give new insights into gender studies in the hotel industry, this study investigates women's working lives under the COVID-19 framework. Regarding practical contributions, hotel management can begin comprehending women employees' value. This research could assist hotel employees in dealing with the COVID-19 pandemic by providing information about their employment.

Acknowledgements The Universiti Malaysia Sabah Research Grant Scheme SDK0332-2022 financially supports this project.

References

1. World Health Organization: Timeline: WHO's COVID-19 response. https://www.who.int/emergencies/diseases/novel-coronavirus-2019/interactive-timeline. Last accessed 02 Jan 2020
2. Baum, T., Mooney, S., Robinson, R., Solnet, D.: COVID-19's impact on the hospitality workforce—new crisis or amplification of the norm? Int. J. Contemp. Hosp. Manag. **32**, 2813–2829 (2020). https://doi.org/10.1108/ijchm-04-2020-0314
3. Statista: COVID-19: job loss in travel and tourism worldwide 2020–2021, by region. http://www.statista.com/statistics/1104835/coronavirus-travel-tourism-employment-loss/
4. Idris, R., Mansur, K.: Sabah economic model: an overview. Int. J. Acad. Res. Account. Financ. Manag. Sci. **10** (2020). https://doi.org/10.6007/ijarafms/v10-i3/8156
5. Sabah Economic Development and Investment Authority: SEDIA Annual Report 2018. https://sedia.com.my/annualreport/index.html

6. Baum, T.: International Perspectives on Women and Work in Hotels, Catering and Tourism. International Labour Office (2013)
7. World Travel and Tourism Council: News-article. wttc.org (2023)
8. Obadić, A., Marić, I.: The significance of tourism as an employment generator of female labour force. Ekonomska Misao I Praksa **18**, 93–114 (2009)
9. Gentry, K.M.: Belizean women and tourism work. Ann. Tour. Res. **34**, 477–496 (2007). https://doi.org/10.1016/j.annals.2006.11.003
10. Carvalho, I., Costa, C., Lykke, N., Torres, A.: Agency, structures and women managers' views of their careers in tourism. Women's Stud. Int. Forum. **71**, 1–11 (2018). https://doi.org/10.1016/j.wsif.2018.08.010
11. International Tourism Highlights, 2019 edn. UNWTO (2019). https://doi.org/10.18111/9789284421152
12. Global Report on Women in Tourism, 2nd edn. (2019). https://doi.org/10.18111/9789284420384
13. Su, Z.B., Yee, P.C.: A qualitative research on genders in crisis: young adults' perspective and experiences towards the COVID-19. Malays. J. Soc. Sci. Human. **7**, e001368 (2022). https://doi.org/10.47405/mjssh.v7i3.1368
14. Smeral, E.: Impacts of the world recession and economic crisis on tourism: forecasts and potential risks. J. Travel Res. **49**, 31–38 (2009). https://doi.org/10.1177/0047287509353192
15. Kubíčková, M., Kırımhan, D., Li, H.: The impact of crises on hotel rooms' demand in developing economies: the case of terrorist attacks of 9/11 and the global financial crisis of 2008. J. Hosp. Tour. Manag. **38**, 27–38 (2019). https://doi.org/10.1016/j.jhtm.2018.10.002
16. Chen, J., Subbarao, K.: The immunobiology of SARS. Annu. Rev. Immunol. **25**, 443–472 (2007). https://doi.org/10.1146/annurev.immunol.25.022106.141706
17. Kosová, R., Enz, C.A.: The terrorist attacks of 9/11 and the financial crisis of 2008. Cornell Hosp. Quart. **53**, 308–325 (2012). https://doi.org/10.1177/1938965512457021
18. Author_Id, N.: Full report—world of work report 2010: from one crisis to the next? World Work Rep. (2010). https://doi.org/10.1002/wow3.19
19. Eagly, A.H., Wood, W., Diekman, A.B.: Social role theory of sex differences and similarities: a current appraisal. In: The Developmental Social Psychology of Gender, pp. 123–174 (2000)
20. Eagly, A.H., Wood, W.: Social Role Theory. SAGE Publications Ltd. eBooks, pp. 458–476 (2012). https://doi.org/10.4135/9781446249222.n49
21. Schneider, M.C., Bos, A.L.: The application of social role theory to the study of gender in politics. Polit. Psychol. **40**, 173–213 (2019). https://doi.org/10.1111/pops.12573
22. Eagly, A.H., Wood, W.: Gender identity: nature and nurture working together. Evol. Stud. Imag. Cult. **1**, 59–62 (2017). https://doi.org/10.26613/esic.1.1.10
23. Lyonette, C., Crompton, R.: Sharing the load? Partners' relative earnings and the division of domestic labour. Work Employ Soc. **29**, 23–40 (2014). https://doi.org/10.1177/0950017014523661
24. Craig, L., Sawrikar, P.: Work and family: how does the (gender) balance change as children grow? Gend. Work Organ. **16**, 684–709 (2009). https://doi.org/10.1111/j.14680432.2009.00481.x
25. Alon, T., Doepke, M., Olmstead-Rumsey, J., Tertilt, M.: The impact of COVID-19 on gender equality (2020). https://doi.org/10.3386/w26947
26. Charlick, S.J., Pincombe, J., McKellar, L., Fielder, A.: Making sense of participant experiences: interpretative phenomenological analysis in midwifery research. Int. J. Doct. Stud. **11**, 205–216 (2016). https://doi.org/10.28945/3486
27. Tindall, L., Smith, J.A., Flower, P., Larkin, M.: Interpretative phenomenological analysis: theory, method and research. Qual. Res. Psychol. **6**, 346–347 (2009). https://doi.org/10.1080/14780880903340091
28. Malhotra, N.K., Nunan, D., Birks, D.F.: Marketing Research: An Applied Approach (2000)
29. Creswell, J.W.: Research Design: Qualitative, Quantitative, and Mixed Methods Approaches (2014)

30. Saunders, M., Lewis, P., Thornhill, A.: Research Methods for Business Students, 5th edn. (2009)
31. Guba, E.G., Lincoln, Y.S.: Paradigmatic controversies, contradictions, and emerging confluences. In: The SAGE Handbook of Qualitative Research (2005)
32. Braun, V., Clarke, V.: Using thematic analysis in psychology. Qual. Res. Psychol. **3**, 77–101 (2006). https://doi.org/10.1191/1478088706qp063oa
33. Kaushal, V., Srivastava, S.: Hospitality and tourism industry amid COVID-19 pandemic: perspectives on challenges and learnings from India. Int. J. Hosp. Manag. **92**, 102707 (2021). https://doi.org/10.1016/j.ijhm.2020.102707
34. Chan, J.K.L., Lily, J., Idris, S., Kele, A.T.A.: Impacts and measures covid-19 pandemic and tourism industry in Sabah. J. Sustain. Sci. Manag. **17**, 1–21 (2022). https://doi.org/10.46754/jssm.2022.08.001
35. Parlak, S., Çakıroğlu, O.Ç., Gül, F.Ö.: Gender roles during COVID-19 pandemic: the experiences of Turkish female academics. Gend. Work Organ. **28**, 461–483 (2021). https://doi.org/10.1111/gwao.12655
36. Adisa, T.A., Aiyenitaju, O., Adekoya, O.D.: The work–family balance of British working women during the COVID-19 pandemic. J. Work Appl. Manag. **13**, 241–260 (2021). https://doi.org/10.1108/jwam-07-2020-0036
37. O'Donnell, J.P., Cárdenas, D., Orazani, N., Evans, A., Reynolds, K.J.: The longitudinal effect of COVID-19 infections and lockdown on mental health and the protective effect of neighbourhood social relations. Soc. Sci. Med. **297**, 114821 (2022). https://doi.org/10.1016/j.socscimed.2022.114821
38. Rudert, S.C., Janke, S.: Call me maybe: risk factors of impaired social contact during the COVID-19 pandemic and associations with well-being. Br. J. Soc. Psychol. **62**, 281–301 (2022). https://doi.org/10.1111/bjso.12546

Indonesian Sharia Tourism Towards a Sustainable Halal Industry

Wazin, Nihayatul Maskuroh, Hadi Peristiwo, and Asep Dadan Suganda

Abstract The tourism industry has the potential to enhance job opportunities and foster economic growth, with a rising demand for halal tourism aligning with the increasing number of Muslim tourists annually. The surge in Muslim tourists presents opportunities and challenges for enhancing Indonesia's sustainable halal tourism sector. The objective of this research is to explore the challenges and opportunities within the halal tourism industry, along with identifying appropriate strategies for the development of Indonesia's halal tourism sector. Employing qualitative methods, this research adopts descriptive techniques, primarily through documentation studies. The findings reveal that Indonesia's halal tourism industry holds promising opportunities and potential as an integral part of the sustainable national tourism sector, contributing to increased government revenues. Despite its tourism potential, Indonesian halal tourism faces challenges in areas such as demand and market dynamics, technology and information integration, regulatory compliance, financing, and research and development.

Keywords Halal tourism · Opportunities · Challenges

1 Introduction

Islam is recognized as a holistic lifestyle that, through its teachings, directly or indirectly covers every facet of human life. The Qur'an specializes in tourism, considering its stance towards tourism as a significant blessing. According to the holy text of the Qur'an, Muslims are encouraged to travel, appreciate the beauty of Allah's creation, and visit friends and relatives. Additionally, they hold the duty of providing hospitality to visitors who, by Islamic law, have the entitlement to the rights granted to citizens. Historically, the connection between tourism and religion has been closely

---

Wazin · N. Maskuroh · H. Peristiwo · A. D. Suganda (✉)
Department of Islamic Economics, Faculty of Islamic Economics and Business, Universitas Islam Negeri Sultan Maulana Hasanuddin Banten, Serang, Indonesia
e-mail: asep.dadan@uinbanten.ac.id

B. Awwad (ed.), *The AI Revolution: Driving Business Innovation and Research*, Studies in Systems, Decision and Control 524, https://doi.org/10.1007/978-3-031-54379-1_77

intertwined, serving as a strong motivation for travel. From an academic perspective, the scope of tourism encompasses terminology and its theological foundation and characteristics (ontology). This includes the management of objects and the carrying capacity of destinations, along with various related aspects. Furthermore, it addresses issues related to the roles and behavior of each community involved in tourism, whether as rulers, entrepreneurs, *ulama*, or tourism users [13].

The inclination towards religious travel is not a novel occurrence, as it has long been a fundamental motive for travel and is often regarded as the earliest form of economic travel. Simultaneously, the increased awareness of halal products has contributed to the burgeoning growth of the halal industry. This heightened awareness has given rise to new phenomena, such as halal tourism. Islamic tourism, as a phenomenon emerging from the expansion of the halal industry, has gained traction [19]. Various sources in the literature highlight those Muslim tourists, when visiting destinations, prioritize the consumption of Sharia-compliant products and services. The escalating number of Muslims worldwide has led to a growing demand for halal products each year. While the appeal of halal products extends beyond the Muslim demographic, the prevailing social culture allows them to be comfortably embraced by Muslims [20]. Consequently, there is a need for exceptional service to ensure that tourists feel respected, thereby creating a positive image [17]. The scope of the halal tourism sector has expanded annually, with major tourism markets globally expressing keen interest in Islamic tourism [18]. For tourists, traveling serves as a means of refreshment, providing new experiences and insights that support personal and professional development in their daily lives. The concept of tourism is intricate, encompassing different social, behavioral, economic, political, cultural, and environmental dimensions. It involves a series of activities, services, and benefits aimed at providing a unique experience for tourists. Both conventional and religious tourism can potentially support the implementation of halal tourism, provided they align with the principles of Islamic law [21].

In the contemporary globalized era, the tourism sector constitutes a significant economic sector, encompassing both national and international movements. Many countries professionally manage their tourist attractions to maximize their economic value, recognizing tourism as a prospective business [27]. The rapid development of the national and international tourism sector has positioned it as a significant revenue generator for numerous countries in the process of development [23]. Tourism is a pivotal modern human behavior that brings about substantial changes, influencing the economy, conditions, culture, and traditions (Monsef 2017). While tourism was once considered an exclusive pursuit for a particular privileged class, it is now commonplace in the broader community, recognized as a personal, communal, and constitutional entitlement for all. Governments of numerous nations are creating tourism websites to provide information about potential destinations, facilitating accessibility for tourists [22].

Global travel expenditure for the Muslim tourism segment is projected to reach US $300 billion by 2026, encompassing hospitality services. Halal tourism, stemming from the global increase in Muslim tourists and the swift growth of the halal business sectors, defined as the engagement of Muslim tourists in activities and establishments aligned with Islamic teachings [25]. Consequently, halal tourism offerings, including hotels, restaurants, and other facilities, utilize halal materials, adhering to certification requirements. This sector presents a chance for hoteliers and other entities in tourism to enhance their competitive edge by tailoring services to the standards and preferences of Muslim travelers [30].

The entry of new entrepreneurs into the market, particularly those producing Muslim-friendly products in halal tourism, has attracted customers who were previously not aligned with the modern consumer lifestyle. This has led to the development of tourism products tailored to these emerging markets, alongside expanded media and social institutions promoting new services and products to meet evolving consumer needs [15]. Sharia tourist destinations must showcase a unique character that distinguishes them from conventional tourism, emphasizing management systems and attractions that embody religious nuances [14].

In the globalized international tourism market, sustaining competition in the global market requires substantial investments in financial and human resources for countries, companies, and governments. This includes activities such as making realistic forecasts about current tourism trends, understanding the motivations behind travel, and addressing the diverse demands, needs, and expectations of consumers determined by their income levels and utilization of technology. Performing market research, investing in research and development, encouraging innovation, handling public relations, leveraging online resources, advertising, and promotion, and keeping a vigilant eye on future trends and developments are crucial aspects of this competitive landscape [1]. Traditionally, the tourism sector has displayed favorable and unfavorable traits, covering economic, social, environmental, and cultural dimensions.

A novel tourism concept, known as halal tourism or hospitable halal food, has emerged alongside the concept of halal food. The legality of travel (tourism) is determined by its purpose, implying that if the intention is obligatory, then it becomes a religious obligation [5]. Halal tourism, a contemporary trend in tourism, has gained prominence in various countries, particularly in the Middle East, with many nations actively embracing and promoting this form of tourism [16]. The growing interest in halal tourism is directly proportional to the increasing number of Muslim tourists annually [6].

Recognizing the potential of Islamic tourism, several nations have initiated the provision of tourism-related products, facilities, and infrastructure tailored to the specific requirements of Muslim tourists. However, there remains a challenge as many stakeholders in the tourism sector lack a comprehensive understanding of halal tourism. Promoting the tourism sector as a means to position tourist destinations on the map is a lucrative segment of the industry [10]. The marketing of religious objectives plays a pivotal role in influencing consumer behavior, as it can engage visitors on spiritual and emotional levels, fostering a sense of community [28].

Tourist satisfaction can be categorized into two categories: satisfaction specific to a transaction, which pertains to contentment within a particular context or dimension, and overall satisfaction, which refers to satisfaction with the entire service [12]. This research stands out from previous studies by employing the value chain analysis of the Indonesian halal tourism industry as an analytical tool. The aim is to identify opportunities and challenges and devise specific strategies for the industry's development. The rise in the number of Muslim tourists presents both opportunities and challenges that Indonesia faces in enhancing its halal tourism sector.

2 Research Method

This research is a descriptive qualitative study that explores the challenges and opportunities within the realm of halal tourism. The employed method for this research is descriptive analysis, involving the organization of data, description into units, synthesis, compilation into patterns, identification of crucial aspects to be studied, and drawing conclusions that can be elucidated. The primary focus of this study is on the challenges and opportunities presented by halal tourism in Indonesia. Qualitative research typically involves methods such as interviews, observations, and the utilization of documents.

3 Results and Discussion

3.1 Global and National Conditions of Halal Tourism

Tourism stands out as a promising industry on a global scale. The escalating number of destinations and investments in tourism has elevated its status as a significant contributor to foreign exchange earnings, employment opportunities, and the development of businesses and infrastructure [29]. Furthermore, in comparison to other sectors, tourism is recognized as one of the largest and fastest-growing industries. Globally, the tourism sector plays a pivotal role in influencing macro variables such as economic growth, and the mitigation of unemployment. The worldwide progress of tourism and travel is closely tied to the accessibility of information and transportation for both foreign and domestic tourists during their visits [4].

The international tourism industry generates substantial tourist flow and high-volume revenues globally and comprises integral activities dynamically structured in conceptual dimensions. This industry's structural attributes are influenced by the global economic downturn, where demand shifts occur due to economic and political uncertainties, both within and between countries, and fluctuate with changes in technology, knowledge, destinations, demands, expectations, and income levels.

Due to the increasing demand and satisfaction of tourists, various activities and services in halal tourism have become well-known in the tourism industry [7]. The increasing global Muslim population contributes to the rising demand for halal tourism, which involves tourism activities for recreation and social purposes adhering to Sharia law [2]. Halal tourism encompasses various facets, including the participation and involvement of Muslims, tourist attractions, products like lodging, food, entertainment, drinks, and more, as well as dimensions such as social, economic, cultural, and religious aspects. Additionally, service management aspects like marketing, cultural considerations, and religion play a crucial role. Considering the significance of travel in Islam, there has been an attempt to position itself as a global halal center [26]. Therefore, this emerging tourism industry requires professional development to establish strong competitiveness on the global stage, ultimately contributing to foreign exchange earnings for respective countries (Damanik 2013). Ultimately, it is anticipated that the tourism industry will have an increasingly substantial economic impact on broader communities.

The development of halal tourism is imperative for Indonesia to attract foreign tourists from the Middle East, predominantly Muslim countries with the highest tourism travel expenses globally. Middle Eastern tourists are the primary consumers of halal tourism worldwide, making it crucial for Indonesia to devise strategies to attract them. Indonesia's natural wealth and diverse arts and culture, stretching from Sabang to Merauke, should be leveraged to diversify the marketing of superior tourism products. It's important to note that the concept of halal tourism doesn't exclusively target Muslim tourists but is also inclusive of non-Muslims. Halal tourism is a sub-category that aligns with the values of religious tourism and travel, adhering to Islamic principles throughout all processes.

3.2 Indonesian Halal Tourism Industry Value Chain

To confront global competition effectively, the halal tourism industry must possess and cultivate high competitiveness, a crucial aspect for sustaining superiority in any industry. The industry's ability to achieve optimal international trade performance hinges on the effectiveness of its value chain. An effective value chain serves as the cornerstone for a competitive advantage, generating added value for the industry. The value chain encompasses all activities involved in moving a product from the design stage through various production phases, delivery to the end consumer, and recycling after use, involving physical transformation and input from various service providers. The halal tourism industry is closely connected to both the primary and ancillary sectors, constituting a series of entry points. These entry points collectively form the value chain of the halal tourism industry, including transportation, destinations, restaurants and cafes, hotels and accommodation, as well as tours and travel.

Table 1 illustrates the intricate relationships between the halal tourism industry and various sectors, including both main and supporting industries. This underscores the significant multiplier impact that the growth of the halal tourism sector can exert on the economy. Particularly in halal tourism, the related industries encompass the halal, real, and financial sectors, contributing significantly to the advancement of the Islamic economy in Indonesia. The Islamic tourism industry has garnered recognition as an increasingly significant niche within the broader tourism sector. It involves tourists visiting destinations to explore past, present, and future economic activities. The concept of Islamic tourism destination attractiveness has sparked growing interest in its effective measurement within Islamic tourism literature [24].

To advance halal tourism, Indonesia needs to collaborate with various entities, such as the National Sharia Economic and Finance Committee (KNEKS), the Indonesian Ulama Council (MUI), the National Sharia Council (DSN), and Business Certification Institutions (LSU). This collaboration, as outlined in the Minister of Tourism and Creative Economy's Regulation, emphasizes the development of tourism with a focus on cultural and religious values. Additionally, human resource training, socialization, and capacity building are crucial for supporting industries, particularly the training and development sector. Collaborations with industry associations, organizations like the Association of the Indonesian Tours and Travel (ASITA) and the Indonesian Hotel and Restaurant Association (PHRI), are underway to provide halal lodging, halal food menus, and sharia tour packages.

Table 1 Indonesian halal tourism industry value chain

Value chains	Main industry	Supporting industry
Destinations	1. Media and recreation 2. Food and beverage 3. Modest fashion 4. Hospitality 5. MSME 6. Renewable energy 7. Islamic finance	1. Technology 2. Capacity building and skill enhancement 3. Construction
Airlines and transport hub	1. Media and recreation 2. Food and beverage 3. Modest fashion 4. MSME 5. Islamic finance	1. Technology 2. Capacity building and skill enhancement 3. Construction
Hotels and accommodation	1. Media and recreation 2. Food and beverage 3. Modest fashion 4. MSME 5. Islamic finance	1. Technology 2. Capacity building and skill enhancement 3. Construction
Cafes and restaurants	1. Food and beverage 2. MSME	1. Technology 2. Capacity building and skill enhancement 3. Construction
Tours and travel	1. Food and beverage 2. Hospitality 3. Airlines 4. MSME	1. Technology 2. Capacity building and skill enhancement

Islamic tourism represents a growing global trend, particularly among Muslims residing in urban areas, who exhibit high educational levels and a desire for extensive travel with family. They actively seek goods and services aligned with their Muslim identity [8]. Travel and tour service providers serving tourism destinations should concentrate on promotional materials that aid tourists in assessing halal service offerings based on Islamic travel experiences. These entities need to function not only as intermediaries but also play a role in shaping the demand for these destinations [28]. Traditionally, the tourism industry has not given sufficient consideration to the requirements of Muslim tourists when formulating tourism products [11].

Intense competition in marketing Islamic tourism destinations, coupled with a saturated market, necessitates a deep understanding of factors influencing satisfaction and loyalty of tourists. The strategic plans formulated by ASEAN member countries in the global tourism industry context for 2016 to 2025 underscore the crucial role of the tourism industry in its economic growth. Member nations have prioritized delivering quality services to tourists, investing in human resource development, engaging in marketing efforts, investing in significant projects, encouraging community participation, promoting sustainable development, and attracting more tourists to the local market.

3.3 *Opportunities and Challenges*

Indonesia, with a population of 275.7 million people, stands the position as the fourth most populous country globally and stands as the largest Muslim-majority nation globally. Moreover, it boasts the 10th largest economy globally, considering its purchasing power parity and membership in the G20. The tourism sector holds significant importance, contributing substantially to Indonesia's national GDP and employment rates. Indonesia's strategic geographical conditions, featuring a tropical climate and rich biodiversity, position it as a prime tourist destination, emphasizing efforts to develop halal tourism. The values associated with these attributes are crucial in influencing the purchasing decisions of Muslim consumers, who often prioritize functionality [3].

The Ministry of Tourism and Creative Economy of the Republic of Indonesia has recognized 13 provinces, namely Aceh, West Sumatra, Riau, Lampung, Banten, West Java, DKI Jakarta, Central Java, Yogyakarta, East Java, South Sulawesi, Bali, and, West Nusa Tenggara (NTB) as being prepared to establish themselves as halal tourism destinations. Among these, Aceh and West Nusa Tenggara have notably developed halal tourism for foreign Muslim tourists. Aceh, known as the *Serambi Mekah*, showcases a strong Islamic culture reflected in daily life, marked by the application of a Sharia-based system. Beyond Aceh, the government of NTB provincial actively collaborates with the MUI, LPPOM, Culture, and Tourism Office, and MSMEs to obtain halal certification for various establishments, resulting in the issuance of 644 halal certificates. Boasting a Muslim majority of 90%, NTB is home to 4500 mosques distributed across 598 villages and sub-districts, earning it the name: "*Island with*

Numerous Mosques". Indonesia strives to promote its halal tourism internationally, evidenced by its participation in the World Halal Tourism event in Abu Dhabi, where it secured 12 out of 16 awards.

In the ecosystem of the Indonesian halal tourism industry, a myriad of opportunities and challenges must be optimized (Table 2). Key aspects of the halal tourism industry ecosystem encompass demand and market dynamics, technology and information, regulations, financing, and research and development.

Table 2 Opportunities aspect and challenges aspect

Aspect	Opportunities	Challenges
Market and demand	1. The rise in the Muslim population in both Indonesia and globally 2. The increasing income of the burgeoning Muslim middle class 3. The emergence and progression of advocates for halal tourism 4. Growing awareness among the public regarding the adoption of a halal lifestyle	1. The progress of halal tourism in nations worldwide, encompassing both Muslim and non-Muslim countries 2. The global presence of top-notch branding for Islamic tourism is currently lacking 3. The quality of infrastructure for halal tourism in rival nations appears to be comparatively superior
Information and technology	1. The advancement of digitalization within the broader tourism sector 2. The continuous evolution and innovation in information media	1. The adoption of information technology within the halal tourism sector remains limited
The law and regulation	1. Support from the government for the growth of halal tourism 2. Implementation of local regulations about halal tourism in various provinces 3. Collaboration among multiple institutions/ministries in advancing halal tourism 4. Existing government strategies and policies focused on the advancement of halal tourism and the tourism industry	1. Indonesia lacks a central legal authority overseeing the implementation of halal tourism 2. There is no dedicated regulation mandating Islamic financial institutions to offer financing for the halal industry
Finance/fund	1. The expansion and diversification of Sharia-compliant financing 2. Prospects for partnership in securing funds from banks and issuing sukuk 3. The establishment and execution of Sharia-compliant insurance programs for mainstream tourism travel	1. Islamic financial institutions currently lack a designated funding focus for the halal industry, which encompasses halal tourism
The research and development	1. The existence of a dedicated tourism institution that incorporates halal tourism education 2. The initiation of research endeavors focused on halal tourism	1. Inadequate research has been conducted on market segmentation and preferences for halal tourism in Indonesia 2. A national-level curriculum for halal tourism is not yet in place

Indonesia has emerged as a prominent destination for Islamic tourism, offering favorable year-round weather, cost-effective travel options, widespread availability of Halal food, and a multitude of accessible mosques across the country [31]. However, unfamiliar destinations may present increased vulnerability as a result of differing norms and values in comparison to those of the tourists [9]. Indonesia's entrance into the global market, particularly in the realm of halal tourism, has been facilitated by various factors.

The integration of these factors, along with the halal tourism ecosystem, serves as a crucial component in creating opportunities for the advancement of halal tourism in Indonesia. A well-structured, cohesive, and efficient halal tourism ecosystem plays a vital role in development. This ecosystem is essentially a system that unifies the entire halal tourism value chain, connecting all key industrial components and supporting industries seamlessly from upstream to downstream.

4 Conclusion

Halal tourism in Indonesia presents promising opportunities and holds significant potential within the national tourism industry, contributing to increased government revenue. Emphasizing Sharia principles in tourism management, this approach aims to create a courteous and welcoming environment for all tourists while respecting the surroundings. The strategic development plan to position Indonesia as a global halal tourism center involves enhancing the tourism competitiveness index through the expansion of tourism destinations, transportation options, hotels, accommodation, restaurants, cafes, and travel services. The emergence of halal tourism has introduced new business potential, attracting both domestic and international tourists. This industry has a positive influence on the Indonesian economy, specifically in terms of generating foreign exchange earnings and providing employment opportunities. Islamic tourism, being a burgeoning segment, presents a substantial potential and opportunity for the broader tourism sector.

However, there are challenges in developing halal tourism in Indonesia, particularly in terms of demand and market aspects. These challenges include the global development of halal tourism in both Muslim and non-Muslim countries, weak branding of international-class halal tourism, and the relatively superior quality of halal tourism infrastructure in competing nations. In the realm of technology and information, a notable challenge is the low adoption of information technology within the halal tourism industry. Regulatory challenges involve the absence of a supreme legal body overseeing the implementation of halal tourism in Indonesia and the lack of specific regulations mandating Islamic financial institutions to provide financing to the halal industry.

Furthermore, financing-related challenges include the absence of specific targets for Islamic financial institutions in supporting the halal industry, including Islamic tourism. Research and development pose challenges due to insufficient studies on market segmentation and preferences for halal tourism in Indonesia, along with the absence of a standardized national curriculum for halal tourism. Indonesia has actively pursued becoming a developed country in the Asian region, prioritizing physical and infrastructure development in tourism projects to bolster economic prospects. The continuous rise in Muslim tourists presents a significant potential for advancing the Indonesian halal tourism sector. While other countries, both with Muslim and non-Muslim majorities, are also venturing into halal tourism, Indonesia's distinct advantage lies in its commitment to creating a genuinely Muslim-friendly atmosphere, aligning with halal tourism concepts and principles.

References

Book

1. Abdul Manan, S.K.: Contemporary Issue and Development in the Global Halal Industry. Springer Science+Bussiness Media, Singapore (2016)
2. Abror, A., Wardi, Y., Trinanda, O., Patrisia, D.: The impact of Halal tourism, customer engagement on satisfaction: moderating effect of religiosity. Asia Pac. J. Tourism Res. **24**(7), 633–643 (2019)
3. Al-Ansi, A., Han, H.: Role of halal-friendly destination performances, value, satisfaction, and trust in generating destination image and loyalty. J. Dest. Mark. Manag. **13**, 51–60 (2019). https://doi.org/10.1016/j.jdmm.2019.05.007
4. Andayani, F., Sucipto, H.: Wisata Syariah: Karakter, Potensi, Prospek, Dan Tantangan. Grafindo, Jakarta (2004)
5. Bahammam, F.S.: Panduan Wisatawan Muslim. Pustaka Al-Kautsar, Jakarta (2012)
6. Battour, M., Ismail, M.N.: Halal tourism: concepts, practises, challenges and future. Tour. Manag. Perspect. **19**, 150–154 (2016). https://doi.org/10.1016/j.tmp.2015.12.008
7. Bhuiyan, A.H., Darda, A.: Prospects and potentials of halal tourism development in Bangladesh. J. Tour. **4**(2), 93–106 (2018). https://doi.org/10.26650/jot.2018.4.2.0007
8. Biancone, P.P., Secinaro, S., Radwan, M., Kamal, M.: Halal tourism: an opportunity for the global tourism industry. Tour. Anal. **24**(3), 395–404 (2019). https://doi.org/10.3727/108354219x15511865533112
9. Brown, L., Osman, H.: The female tourist experience in Egypt as an Islamic destination. Ann. Tour. Res. **63**, 12–22 (2017). https://doi.org/10.1016/j.annals.2016.12.005
10. Calderwood, E.: The invention of Al-Andalus: discovering the past and creating the present in Granada's Islamic tourism sites. J. North Afr. Stud. **19**(1), 27–55 (2014). https://doi.org/10.1080/13629387.2013.862777
11. Carboni, M., Perelli, C., Sistu, G.: Is Islamic tourism a viable option for Tunisian tourism? Insights from Djerba. Tour. Manag. Perspect. **11**, 1–9 (2014). https://doi.org/10.1016/j.tmp.2014.02.002
12. Carboni, M., Janati, M.I.: Halal tourism de Facto: a case from Fez. Tour. Manag. Perspect. **19**, 155–159 (2016). https://doi.org/10.1016/j.tmp.2015.12.007
13. Diarta, I.K.S., Pitana, I.G.: Pengantar Ilmu Pariwisata. Andi, Yogyakarta (2010)
14. Djakfar, M.: Pariwisata Halal Perspektif Multidimensi (Peta Jalan Menuju Pengembangan Akademik & Industri Halal di Indonesia). UIN-Maliki Press, Malang (2017)
15. Elaziz, M.F., Kurt, A.: Religiosity, consumerism and Halal tourism: a study of seaside tourism organizations in Turkey. Tourism **65**(1), 115–128 (2017)
16. Fawzy, N.M., Qoura, O.: Halal tourism: evaluating opportunities and challenges in the Middle East "Jordan and Egypt". J. Fac. Tour. Hotels Fayoum Univ. **10**(2), 343–358 (2016)
17. Gayatri, P.G., Pitana, I.G.: Sosiologi Pariwisata. Andi, Yogyakarta (2015)

18. Geetanjali, R.C.: Halal tourism; a new goldmine for tourism. Int. J. Bus. Manag. Res. (IJBMR) **4**(6), 45–62 (2014). http://www.tjprc.org/view-archives.php?year=2014_48_2&id=32&jtype=2&page=4
19. Haq, F.M.: The significance of partnership as a marketing strategy for Islamic spiritual tourism. J. Islamic Mark. **5**(2), 258–272 (2014). https://doi.org/10.1108/JIMA-11-2013-0079
20. Henderson, J.C.: Managing tourism and Islam in Peninsular Malaysia. Tour. Manage. **24**(4), 447–456 (2003). https://doi.org/10.1016/S0261-5177(02)00106-1
21. Hermantoro, H.: Creative-Based Tourism Dari Wisata Rekreatif Menuju Wisata Kreatif. Aditri, Depok (2011)
22. Horng, J.S., Tsai, C.T.S.: Government websites for promoting East Asian culinary tourism: a cross-national analysis. Tour. Manag. **31**(1), 74–85 (2010). https://doi.org/10.1016/j.tourman.2009.01.009
23. Khazaei Pool, J., Khodadadi, M., Asadi, A.: The impact of congruence between self-concept and destination personality on behavioural intentions for visiting an Islamic-historical destination. Tour. Hosp. Res. **18**(3), 378–387 (2018). https://doi.org/10.1177/1467358416663820
24. Lee, C.F.: An Investigation of factors determining industrial tourism attractiveness. Tour. Hosp. Res. **16**(2), 184–197 (2016). https://doi.org/10.1177/1467358415600217
25. Nur Sa'adah, M., Sulaiman, S., Adham, K.A., Said, M.F.: Halal tourism: literature synthesis and direction for future research. Pertanika J. Soc. Sci. Humanit. **27**(1), 729–745 (2019)
26. Razzaq, S., Michael Hall, C., Prayag, G.: The capacity of New Zealand to accommodate the Halal tourism market—or not. Tour. Manag. Perspect. **18**, 92–97 (2016). https://doi.org/10.1016/j.tmp.2016.01.008
27. Sedarmayanti: Membangun Dan Mengembangkan Kebudayaan & Industri Pariwisata (Bunga Rampai Tulisan Pariwisata). PT. Refika Aditama, Bandung (2014)
28. Taheri, B.: Emotional connection, materialism, and religiosity: an Islamic tourism experience. J. Travel Tour. Mark. **33**(7), 1011–1027 (2016). https://doi.org/10.1080/10548408.2015.1078761
29. Todaro, M.P., Smith, S.S.: Pembangunan Ekonomi. Erlangga, Jakarta (2006)
30. Vargas, J. M.: Management of turfgrass diseases. CRC Press (2018)
31. Winarti, O.: Halal tourism in Indonesia: does it attract only Muslim tourists? J. Stud. Komunikasi (Indonesian J. Commun. Stud.) **1**(3), 232–239 (2017). https://doi.org/10.25139/jsk.v1i3.139

Stressors and Mental Well-Being Among Hospitality Employees During COVID-19 Pandemic

Ruzanifah Kosnin, Nur Athirah Sumardi, Noor Nabila Jafarudin, Siti Salina Saidin, and Ahmad Fahme Mohd Ali

Abstract The hospitality and tourism industry has been known for its labor-intensive and stressful work environment. The novel coronavirus (COVID-19) pandemic has added stress and mental health impacts on the hospitality workforce. Therefore, the objective of this article is to investigate stressors affecting hospitality and tourism employees during the COVID-19 pandemic and their mental well-being. This conceptual paper sets forth a framework based on the tourism and hospitality literature on stressors and mental well-being. From the review of the literature, stressors among hospitality and tourism employees during the COVID-19 pandemic can be divided into financial insecurity, occupational stress, psychological stress as well as stigma. Stressors during the pandemic may prolonged post-pandemic and will affect the mental well-being of employees in the hospitality industry. This study broadens the hospitality management literature on stressors and enables hospitality industry providers to strategize efforts to enhance the mental well-being of employees during the post-pandemic. Future research can take on the coping strategies to reduce the negative impact of stressors on employee's mental well-being.

Keywords COVID-19 · Hospitality employees · Mental well-being · Stressors

1 Introduction

Growing data predicts that depression, which is already one of the world's most serious health issues, could overtake diabetes as the second most frequent disease by 2020 [1–3]. Data showed that 15.7% of U.S. employees have had a major depressive disorder at least once and 8.6% of working adults have had a major depressive disorder in the twelve months before being questioned [4]. Aside from the apparent negative

R. Kosnin (✉) · N. N. Jafarudin · S. S. Saidin · A. F. M. Ali
Faculty of Hospitality, Tourism and Wellness, Universiti Malaysia Kelantan, 15000 Kota Bharu, Kelantan, Malaysia
e-mail: ruzanifah@umk.edu.my

N. A. Sumardi
Universiti Teknologi MARA, 42300 Bandar Puncak Alam, Selangor, Malaysia

B. Awwad (ed.), *The AI Revolution: Driving Business Innovation and Research*, Studies in Systems, Decision and Control 524, https://doi.org/10.1007/978-3-031-54379-1_78

impact on people's quality of life, depression is also recognized to be one of the contributing factors towards low productivity, increased absenteeism and disability as well as premature early retirement [5]. Moreover, more than six million people were infected and 500,000 people were killed by the COVID-19 pandemic in 188 countries. This has threatened the tourism and hospitality-related employees [6]. Most tourist and hospitality personnel, specifically flight attendants, cleaners, and front-desk employees need to work at their respective workplaces, as the nature of their work does not allow them to work from home. Airline, cruise ship, and hotel reservations have plummeted, leaving these companies battling to stay competitive. As a result, it is not unexpected that tourist and hospitality operators are dealing with high levels of stress, worry, and uncertainty [7]. During a pandemic, front-line tourism and hospitality personnel endure the added stress of interacting with consumers. For example, a flight attendant and hotel front line must wear gloves and face masks to lower the danger of infection and alleviate anxiety. On the other hand, the ground workers in the aviation business are concerned about unpaid brief furloughs and salary reductions [2].

The hospitality and tourism industry has been known for its "labor-intensive and stressful work environment" [8]. The novel coronavirus (COVID-19) pandemic has added stress and mental health impacts on the hospitality workforce. In Malaysia, workplaces were economically challenged with the cost of RM 14.46 billion due to their employees' mental health problems in 2018 [9]. While the sorrow and distress an individual feels are very much focused, the negative consequences of the feelings should be the main priority [10]. Compared to the stress encountered in daily life, the COVID-19 pandemic also results in tremendous stress that is pervasive and uncontrollable [11]. For this reason, this paper explores a review of the literature on stressors among hospitality and tourism employees during COVID-19. Occupational stress, psychological stress, and stigma are the identified stressors in this industry. Therefore, by investigating the contributing factors to employees' mental well-being, organizations in the tourism and hospitality industry will have a better understanding and trust in protecting their employees' physical and psychological well-being. This study is divided into three sections. Section 1 provides an introduction on the stressors due to the COVID-19 pandemic which will highlight the study's purpose. Section 2 explores the related previous studies and literature on the types of stressors that affect the mental well-being of hospitality and tourism industry employees. Section 2.3 describes the stressors' effects on the mental well-being of hospitality industry employees. Finally, Sect. 3 concludes the findings of this paper.

2 Literature Review

Stress can be referred to as a relative connection between an individual and his or her environment which is self-assessed as taxing or exceeding his or her resources that may endanger his or her well-being [12]. Due to the reciprocal influence of people and their settings, the interaction between a person and their environment is seen as a transactional process. World Travel and Tourism Council (WTTC) has stated over 50 million tourism hospitality industry employment was threatened during the COVID-19 pandemic [2]. Current studies on occupational stress have recognized how the individuals' behavioral responses to stressors may not only be harmful to the individuals themselves, rather they may also affect the organization [13]. Mental well-being is how to respond to the ups and downs of life [13]. In this straightforward definition of mental well-being, lies a deeper meaning and suggestion for lives. It combines the way an individual considers, handles feelings (spirited well-being), and acts. For example, hotel staff is one of the parties that have to deal with hotel guests because it is part of their job role, and guests sometimes have certain types of behavior on how they want to be treated properly but mistreat others. Especially during COVID-19, high pressure on hotel employees due to the pandemic will affect the mental well-being of employees in this industry. Especially when speaking with customers, and handling customers' belongings, hotel personnel may endure psychological loads and acute stress [14]. Yu et al. [14], also mentioned that stress brought on by the appearance of previously unidentified viruses and diseases can also be extremely harmful. Hotel personnel may experience a variety of stress-related symptoms as a result of unfavorable external conditions like epidemics.

2.1 Underpinning Theory

According to the Mental Health Foundation [15], stress is an individual body's response to pressure that can be triggered when something unexpected threatens our sense of self, feeling overwhelmed and unable to deal with unmanageable pressures. A stressor is defined as a stimulus or threat that creates stress which may be composed of challenge, threat, demands, or any structural constraints which through its existence and occurrence, the operating integrity of a company can be called into question. Due to the correlative influence of people and their settings, the interaction between a person and their environment is seen as a transactional process. Stressors affecting human behavior are frequently explained using the conservation of resources (COR) theory [16]. This theory has proposed that individuals will actively seek to create and retain new resources whenever they feel threatened by the resources that they value [16]. Resources can be in the forms of various objects, for instance, a house; personal traits such as time, knowledge, or social networks; conditions such as one's social status which cumulatively are important and valuable to any individual. In psychology, when there is a resource loss or a threat of resource loss to individuals, this environment creates stressful events for them [17]. Adverse or undesirable events may involve a loss of something that people value, such as status or money. Consequently, these individuals will be adjusting to their environment to achieve a gain in resources. Nevertheless, employees in the hospitality industry are highly susceptible to stress as the COVID-19 pandemic has challenged them with uncertainty through the global tourism crisis.

2.2 *Stressors in the Hospitality Industry During the COVID-19 Pandemic*

The COVID-19 pandemic significantly affected the hospitality and tourism industry when movement restrictions were implemented, resulting in fewer tourists and visitors who will use tourism-related services. This pandemic eventually contributed to additional stressors for the employees in this industry. Employees experienced a variety of stress-related symptoms because of the pandemic. According to Yu et al. [14], employees were under stress due to the possibility of contracting the disease; psychological anxiety that they might spread it to their family and friends; the burden of decreased hotel sales as a result of the pandemics (unstable employment, unpaid leave, overwork as a result of employee leave); and the uncomfortable social gaze because they work in hotels where direct or indirect contact with various customers occurs. Chen [18] identified pandemic-induced panic, unemployment, and weak social support as the major stressors for employees in China's tourism and hospitality industry during the COVID-19 pandemic.

Financial Insecurity

The COVID-19 pandemic has taken a toll on employment in the hospitality industry around the world. Many employees were affected as they experienced the fear of losing jobs, especially among those who were not directly involved with the COVID-19 patients. During this pandemic, feelings of job insecurity and financial anxiety were focused on, as many organizations opted for restructuring and downsizing operations to reduce costs [19]. Mensah et al. [20] found financial anxiety mediated the COVID-19 risk perception and psychological distress among hotel employees while weakening their social functioning.

Occupational Stress

Occupational or work-related stress is defined as the responses of individuals when their coping ability is challenged as they are pressured and demanded by their work responsibilities that are unmatched by their knowledge and abilities [21]. Role ambiguity, role conflict, excessive workload, and work-life balance are among the contributing factors to stress for hotel employees [22]. With a stressful workload, hotel employees are forced to change jobs with the hope that other jobs will fulfill their well-being than their previous job. It is undeniably difficult to have work-life balance but stress will be triggered when there is an imbalance between the work demands and personal life which eventually will decrease the employees' productivity [23]. Occupational stress is a negatively seen trait that results from inadequate coping mechanisms with pressures at work and has detrimental effects on both mental and physical health. This suggests that people must first adversely perceive a stressor before they start exhibiting stress symptoms, and they must subsequently show poor coping skills. Occupational stress develops when a person's capacity to meet the demands of their job is insufficient to meet those obligations [24]. Interchangeably, work stress can be identified as any particular problem individuals face in customer-oriented fields as they are commonly experiencing conflicting demands from the customers, supervisors as well and the company, which subsequently creates dissonance for them [25]. In addition, hotel service jobs are characterized by long and irregular work hours, low wages, and poor working conditions [26]. According to Faulkner and Patiar [27], these traits frequently result in high levels of occupational stress among hotel staff, and if not properly controlled, occupational stress can become dangerously severe [28].

There are two basic sources of stress at work: personal and environmental. Work schedule, workplace safety, work pace, travel time to and from work, transportation difficulties, client volume, and nature of the customers are all examples of environmental elements [29]. Excessive noise, heat, light, workload, level of responsibility, and amount of supervision are other factors that cause stress on employees in organizations [29]. Besides, difficulty, complexity, change, organizational and work structure, and human relations within the work environment are also factors [29]. These factors may have some direct and indirect impacts on the employees themselves in terms of stresses at work. Next, interpersonal conflicts and work overload are the two most typical workplace stressors in the hotel sector, and research has linked them to reduced job satisfaction, poor physical health, and increased turnover intentions [30]. It is impossible to escape stress at work thus stressed-out employees are more likely to be lack of motivation, perform poorly, and less productive [31]. In addition, employees who face job ambiguity find it difficult to accomplish their individual goals because they lack guidance, knowledge, resources, or any combination of these.

Psychological Distress

Individuals' psychological distress may be attributed to multiple life stressors during infectious disease outbreaks and quarantines. This may be caused by a lack of support, lack of empathy, emotional exhaustion, inadequate feedback, and undervalue in life. A fundamental aspect of burnout is emotional exhaustion, characterized by strain brought on by extended exposure to chronic stressors as well as a persistent state of physical and emotional tiredness [32]. Emotional exhaustion is a psychological symptom that negatively affects significant work outcomes. In the COVID-19 context, this emotional exhaustion was found to be exacerbated, which is consistent with previous research suggesting that younger people and women exhibit more severe psychological symptoms [33].

To solve a problem or manage emotions brought on by stressful events, social support can be employed as a problem and emotion-focused coping technique. Social support has a positive impact on psychological health either directly by improving mental health or indirectly by acting as a stress-relieving mechanism for those who are under more strain in their lives [34]. Although face-to-face interactions were prohibited during quarantine, people were nevertheless able to communicate virtually and support one another emotionally [35]. Supervisor support safety is the perception of safety among employees that includes communication, encouragement, and other supportive safety behaviors from their superiors [36]. If these coping strategies were taken and applied as well then psychological and emotional stress could be reduced among hospitality employees before, during, and after the COVID-19 pandemic.

Stigma

Stigma or social exclusion can be defined as a simple standardized perception of an individual's dishonor which is commonly assigned when he or she poses a threat to a group of people [37]. Due to the difficulty in determining who is infected and who is not, social stigma in the form of discrimination, harassment, and hatred is widely used in social communities [38]. The nation has directly and indirectly engaged in this social exclusion particularly when the Malaysian government has strictly imposed the rules of Movement Control Order (MCO). To break the chain of COVID-19, the nation has to maintain social distancing in which the pressure is so strong to the extent that people have to pass their time in stress, fear, loneliness, and anxiety [39]. As a matter of fact, during the era of COVID-19, the infected patients were stigmatized to the extent of some individuals dying in their houses just to ensure that they would not be seen, seeking treatment [40]. In the context of stigma or social exclusion, it can be categorized into two perspectives which are based on how the hospitality employees apply the term to themselves and how society possesses the power of excluding the employees from the society.

From the first perspective, hospitality employees may exclude themselves from society or even family and friends as the fear of being the virus spreader 'obliged' them to be socially responsible. Hence, they will distance themselves from their family and friends so as the feeling of burden as a possible virus carrier. This condition is also known as internalized self-stigma. Though hospitality employees were reported to endure mental health challenges during COVID-19 which include, loneliness, depression, and anxiety [41], not all of these employees perceived the same level of risk during the pandemic [42]. Some Peruvians are strongly concerned about their infection risks of COVID-19 while others are assured about them being contracting the virus, which indicates how individuals may differently perceive the risk of COVID-19 [43]. The individual's depression symptoms weigh heavily on how they perceive the risks of COVID-19 [44]. Some of these hospitality employees may act as the frontliners during the COVID-19 pandemic as they are part of the team who work in quarantine hotels. They need to serve the quarantine patients and, in some cases, the inbound tourists who came into Malaysia when the traveling restriction was lifted by the Malaysian government. COVID-19 has increased hospitality employees' worries about job prospects and health, threatening their mental health [45]. Therefore, it is well-understood why these employees have excluded themselves from others due to the respect of being the first party to engage with patients or potential patients. A similar analogy can be applied to the healthcare employees who serve as the frontliners as these hospitality employees. Healthcare professionals have difficulties providing healthcare services to patients as they have conflicting and complex feelings and thoughts concerning their roles as providers and parents, contradicting feelings on the need to be professionally responsible for COVID-19 patients, and the fear of associating with them [46]. Besides, it was found that internalized stigma was found among nurses with lower levels of education, fearing infections from the patients as compared to medical doctors who have higher educational levels [47].

On the other hand, the other perspective of stigmatization concerns how society involved in stigmatizing these hospitality employees as the frontliners, excluded them from the community. Though there was a scarce number of studies, investigating the stigmatization impact on the hospitality employees themselves, the analogy of healthcare providers was taken into consideration as hospitality employees were also part of frontliners in Malaysia. Thus, the basis of the situation is similar to the previous perspective which is solely based on the fear of getting infected with the virus. Society believes that the employees may be the potential carriers of the virus, forcing them to alienate or isolate these employees. Increasing prevalence and mortality, caused by COVID-19 has extremely pushed the front liners among healthcare providers to a greater risk of facing psychological problems in terms of stigma, social isolation, and discrimination [48]. As potential patients or carriers of the COVID-19 virus, the employees were seen to be threatening society's effectiveness in social living [49] and consequently considered as being 'bullied' through stigmatization [50]. Between 60 and 100% of COVID-19 patients were rejected or verbally abused in the hospitals and neighborhoods which also includes some cases related to family and friends [51]. Though it was widely known that these front liners or hospitality employees were obliged to comply with the Standard Operating Procedures (SOPs) for serving the respective patients or potential ones, fear has taken control over their awareness. Despite the symptoms of COVID-19 being informed to be similar to SARS which has occurred before, society did not feel right, and they have treated the virus to be an anomaly. This has compelled them to be very strict with those who are involved in the prevalence of COVID-19. Job-related stressor has been widely identified as a determinant of employee well-being [52]. Thus, the conceptual model of this study posits that job-related stressors may lead to mental distress.

2.3 Mental Well-Being

Well-being is defined as the extent of an individual's assessment which his or her contribution to improving the quality of life [53]. In other words, employees' well-being can be referred to as the idea of improvement in the quality of one's life through the happiness, health, comfort, and tranquility that employees experience while working. Mental well-being also refers to a person's overall mental state, encompassing aspects such as emotional and psychological health, cognitive function, and social well-being. It goes beyond the absence of mental illnesses and involves a positive and fulfilling state of mind. A mentally well individual is typically able to cope with the normal stresses of life, work productively, build and maintain meaningful relationships, and contribute to their community [13]. Positive mental health is described as "a condition of well-being in which an individual understands his or her strengths, can cope with everyday pressures, can work productively, and can contribute to his or her community" [54].

According to global statistics, incidents of psychological stress, anxiety, depression, marital violence, and self-harm have increased during COVID-19 [55]. Psychological health and psychological distress served as indicators of mental health [56]. In contrast to psychological distress, which is characterized by negative emotional states like hopelessness, loss of interest, and sadness as well as anxiety-related symptoms like feeling tense and restless, psychological well-being can be defined as a combination of positive emotional states and functioning with maximum effectiveness in one's personal and social life [57]. Positive emotional states include feelings of happiness, contentment, satisfaction, and a sense of purpose. Psychological well-being also involves a deeper sense of life satisfaction and the ability to navigate challenges with resilience.

3 Conclusion and Discussion

This research intends to provide an understanding of the independent variables which are stressors and the dependent variable which is mental wellbeing. Studies in the hospitality and tourism industry show that stressors and coping strategies play an important role. Generally, this research discussed the causes of stress that could lead to and affect mental health, and also coping strategies to handle them so hotel employees could have a good state of mental well-being. Previous studies on mental well-being in the hospitality and tourism industry concluded that it is necessary to increase the well-being awareness perceived by employees to reduce stress and maintain a good state of mental well-being. It is thus urged future research to investigate the long-term impacts of the COVID-19 crisis on tourism and hospitality employees and their mental well-being.

Acknowledgements The authors would like to thank the Ministry of Higher Education Malaysia for providing the FRGS grant (FRGS/1/2021/SS0/UMK/02/8) and (R/FRGS/A1100/01807 A/002/2021/00958). Also, this work was supported by Universiti Malaysia Kelantan.

References

1. Murray, C.J.L., Alan, D.L.: Evidence-based health policy—lessons from the global burden of disease study. Science **274**(5288), 740–743 (1996)
2. Kang, S.E., Park, C., Lee, C.K., Lee, S.: The stress-induced impact of COVID-19 on tourism and hospitality workers. Sustainability **13**(3), 1327 (2021)

3. Khuong, M., Linh, U.: Influence of work-related stress on employee motivation, job satisfaction and employee loyalty in the hospitality industry. Manag. Sci. Lett. **10**(14), 3279–3290 (2020)
4. Marcotte, D.E., Virginia, W.-G., Patrick, R.D.: Prevalence and patterns of major depressive disorder in the United States labor force. J. Mental Health Policy Econ. **2**(3), 123–131 (1999)
5. Blackmore, E.R., Stephen, A.S., Iris, W., Sarah, M., Brandon, M.Z., Donna, E.S.: Major depressive episodes and work stress: results from a national population survey. Am. J. Public Health **97**(11), 2088–2093 (2007)
6. World Travel and Tourism Council: Coronavirus Puts Up to 50 Million Travel and Tourism Jobs at Risk Says WTTC. Available online: https://wttc.org/News-Article/Coronavirus-puts-up-to-50-million-Travel-and-Tourism-jobs-at-risk-says-WTTC. Accessed on 17 June 2020
7. Hartman, G., Nickerson, N.P.: Tourism-Related Business Owners Speak Out About COVID-19 Impacts. Institute for Tourism and Recreation Research, Missoula, MT, USA, pp. 1–35 (2020)
8. Rosemberg, M.A.S., Adams, M., Polick, C., Li, W.V., Dang, J., Tsai, J.H.C.: COVID-19 and mental health of food retail, food service, and hospitality workers. J. Occup. Environ. Hyg. **18**(4–5), 169–179 (2021)
9. Zakaria, R.: Addressing direct, indirect COVID-19 impact on mental health. New Straits Times. https://www.nst.com.my/opinion/columnists/2020/10/632791/addressing-direct-indirectcovid-19-impact-mental-health. Last accessed by 16 Oct 2020
10. Scheid, T.L., Wright, E.R.: (eds.) A Handbook for the Study of Mental Health. Cambridge University Press (2017)
11. Wang, M.T., Del Toro, J., Scanlon, C.L., Schall, J.D., Zhang, A.L., Belmont, A.M., Plevniak, K.A.: The roles of stress, coping, and parental support in adolescent psychological well-being in the context of COVID-19: a daily-diary study. J. Affect. Disord. **294**, 245–253 (2021)
12. Lazarus, R.S., Folkman, S.: Stress, Appraisal, and Coping. Springer Publishing Company, New York, NY (1984)
13. Peterson, T.: What Is Mental Wellbeing? Definition and Examples, HealthyPlace. https://www.healthyplace.com/self-help/self-help-information/what-mental-wellbeing-definition-and-examples. Last accessed 26 Dec 2021
14. Yu, J., Park, J., Hyun, S.S.: Impacts of the COVID-19 pandemic on employees' work stress, well-being, mental health, organizational citizenship behavior, and employee-customer identification. J. Hosp. Market. Manag. **30**(5), 529–548 (2021)
15. Mental Health Foundation: https://www.mentalhealth.org.uk/explore-mental-health/a-z-topics/stress. Last accessed 11 June 2023
16. Hobfoll, S.E.: Conservation of resources: a new attempt at conceptualizing stress. Am. Psychol. **44**(3), 513–524 (1989)
17. Hobfoll, S.E.: Conservation of resources theory: its implication for stress, health, and resilience. In: Folkman, S. (ed.), The Oxford Handbook of Stress, Health, and Coping, pp. 127–147. Oxford University Press (2011)
18. Chen, C.: Psychological tolls of COVID-19 on industry employees. Ann. Tour. Res. (2020). https://doi.org/10.1016/j.annals.2020.103080
19. Baum, T., Mooney, S., Robinson, R., Solnet, D.: COVID-19's impact on the hospitality workforce—new crisis or amplification of the norm? Int. J. Contemp. Hosp. Manag. **32**(9), 2813–2829 (2020). https://doi.org/10.1108/IJCHM-04-2020-0314
20. Mensah, C., Edem, M.A.-G., Eunice, F.A., Evelyn, A.: COVID-19, financial anxiety and the psychological well-being of hotel workers. Int. J. Hosp. Tour. Adm. **24**(1), 142–165 (2023). https://doi.org/10.1080/15256480.2022.2055692
21. World Health Organization: Occupational health: stress at the workplace. https://www.who.int/news-room/questions-and-answers/item/ccupational-health-stress-at-the-workplace. Last accessed 18 Oct 2022
22. Wen, B., Zhou, X., Hu, Y., Zhang, X.: Role stress and turnover intention of front-line hotel employees: the roles of burnout and service climate. Front. Psychol. **11** (2020). https://doi.org/10.3389/fpsyg.2020.00036
23. Bataineh, K.A.: Impact of work-life balance, happiness at work, on employee performance. Int. Bus. Res. **12**(2), 99–112 (2019)

24. Cooper, C.L., Cooper, C.P., Dewe, P.J., O'Driscoll, M.P.: Organizational Stress: A Review and Critique of Theory, Research, and Applications. Sage Publications, Thousand Oaks, CA (2001)
25. De Ruyter, K.O., Wetzels, M., Feinberg, R.: Role stress in call centers: its effects on employee performance and satisfaction. J. Interact. Mark. **15**(2), 23–35 (2001)
26. Lo, K., Lamm, F.: Occupational stress in the hospitality industry—an employment relations perspective. N. Z. J. Employ. Relat. **30**(1), 23–47 (2005)
27. Faulkner, B., Patiar, A.: Workplace-induced stress among operational staff in the hotel industry. Int. J. Hosp. Manag. **16**(1), 99–117 (1997). https://doi.org/10.1016/S0278-4319(96)00053-9
28. Johanson, M.M., Youn, H., Woods, R.H.: A study of stress levels among hotel general managers—a comparison between 1998 and 2008. J. Hum. Resour. Hosp. Tour. **10**, 32–44 (2011). https://doi.org/10.1080/15332845.2010.500185
29. Dessler, G.: Human Resource Management. Prentice Hall, New Jersey (2000)
30. O'Neill, J.W., Davis, K.: Work stress and well-being in the hotel industry. Int. J. Hosp. Manag. **30**, 385–390 (2011). https://doi.org/10.1016/j.ijhm.2010.07.007
31. McCarthy, J.M., Trougakos, J.P., Cheng, B.H.: Are anxious workers less productive workers? It depends on the quality of social exchange. J. Appl. Psychol. **101**(2), 1–13 (2016)
32. Maslach, C., Schaufeli, W.B., Leiter, M.P.: Job burnout. Annu. Rev. Psychol. **52**, 397–422 (2001). https://doi.org/10.1146/annurev.psych.52.1.397
33. Fu, W., Wang, C., Zou, L., Guo, Y., Lu, Z., Yan, S., Mao, J.: Psychological health, sleep quality, and coping styles to stress facing the COVID-19 in Wuhan, China. Transl. Psychiatry **10**, 1–9 (2020). https://doi.org/10.1038/s41398-020-00913-3
34. Cohen, S.: Social relationships and health. Am. Psychol. **59**(8), 676 (2004)
35. Qian, Y., Hanser, A.: How did Wuhan residents cope with a 76-day lockdown? Chin. Soc. Rev., 1–32 (2020)
36. Christian, M.S., Bradley, J.C., Wallace, J.C., Burke, M.J.: Workplace safety: a meta-analysis of the roles of person and situation factors. J. Appl. Psychol. **94**(5), 1103–1127 (2009)
37. Neuberg, S.L., Smith, D.M., Asher, T.: Why people stigmatize: toward a biocultural framework. In: Todd, R.E.K., Heatherton, F., Hebl, M.R., Hull, J.G. (eds.) The Social Psychology of Stigma, pp. 31–61. Guilford Publications (2000)
38. Riyasad, N.: COVID-19 and mental health around the world. New Age. Retrieved 20 Aug 2020 from https://www.newagebd.net/article/105119/COVID-19-and-mental-health-around-the-world
39. Mahmud, A., Islam, M.R.: Social stigma as a barrier to COVID-19 responses to community well-being in Bangladesh. Int. J. Community Well Being **4**(3), 315–321 (2021). https://doi.org/10.1007/s42413-020-00071-w
40. Ting, S.-H., Shamsul, M.H.: Stigma-marking of COVID-19 patients in Facebook and Twitter of youth in Malaysia in 2020–2021. Youth **2**(4), 717–732 (2022). https://www.mdpi.com/2673-995X/2/4/51
41. Bruine de Bruin, W.: Age differences in COVID-19 risk perceptions and mental health: evidence from a National U.S. Survey conducted in March 2020. J. Gerontol. Ser. B **76**(2), e24–e29 (2020). https://doi.org/10.1093/geronb/gbaa074
42. Yan, J., Kim, S., Zhang, S.X., Foo, M.-D., Alvarez-Risco, A., Del-Aguila-Arcentales, S., Yáñez, J.A.: Hospitality workers' COVID-19 risk perception and depression: a contingent model based on the transactional theory of stress model. Int. J. Hosp. Manag. **95**, 102935 (2021). https://doi.org/10.1016/j.ijhm.2021.102935
43. Caycho-Rodríguez, T., Tomás, J.M., Barboza-Palomino, M., Ventura-León, J., Gallegos, M., Reyes-Bossio, M., Vilca, L.W.: Assessment of fear of COVID-19 in older adults: validation of the fear of COVID-19 scale. Int. J. Ment. Health Addict. **20**(2), 1231–1245. https://doi.org/10.1007/s11469-020-00438-2. Last accessed 01 Apr 2022
44. Ding, Y., Xu, J., Huang, S., Li, P., Lu, C., Xie, S.: Risk perception and depression in public health crises: evidence from the COVID-19 crisis in China. Int. J. Environ. Res. Publ. Health **17**(16), 5728 (2020). https://www.mdpi.com/1660-4601/17/16/5728
45. Alonso, A.D., Kok, S.K., Bressan, A., O'Shea, M., Sakellarios, N., Koresis, A., Solis, M.A.B., Santoni, L.J.: COVID-19, aftermath, impacts, and hospitality firms: an international perspective. Int. J. Hosp. Manag. **91**, 102654 (2020)

46. Ramaci, T., Barattucci, M., Ledda, C., Rapisarda, V.: Social stigma during COVID-19 and its impact on HCWs outcomes. Sustainability **12**(9), 3834 (2020). https://www.mdpi.com/2071-1050/12/9/3834
47. Yufika, A., Pratama, R., Anwar, S., Winardi, W., Librianty, N., Prashanti, N.A.P., Sari, T.N.W., Utomo, P.S., Dwiamelia, T., Natha, P.P.C., Salwiyadi, S., Asrizal, F.W., Ikram, I., Wulandari, I., Haryanto, S., Fenobilire, N., Wagner, A.L., Jamil, K.F., Mudatsir, M., Harapan, H.: Stigma associated with COVID-19 among health care workers in Indonesia. Disaster Med. Public Health Prep. **16**(5), 1942–1946 (2022). https://doi.org/10.1017/dmp.2021.93
48. Xiong, Y., Peng, L.: Focusing on healthcare providers' experiences in the COVID-19 crisis. Lancet Glob. Health **8**(6), e740–e741 (2020). https://doi.org/10.1016/S2214-109X(20)30214-X
49. Barreto, M.: Experiencing and coping with social stigma. In: Mikulincer, M.E., Shaver, P.R., Cooper, M., Larsen, R.J. (eds.) APA Handbook of Personality and Social Psychology. Group Processes, vol. 2, pp. 473–506. American Psychological Association (2015). https://doi.org/10.1037/14342-018
50. Leymann, H.: The content and development of mobbing at work. Eur. J. Work Organ. Psychol. **5**(2), 165–184. https://doi.org/10.1080/13594329608414853. Last accessed 01 June 2023
51. Muhidin, S., Vizheh, M., Moghadam, Z.B.: Anticipating COVID-19-related stigma in survivors and health-care workers: lessons from previous infectious diseases outbreaks—an integrative literature review. Psychiatry Clin. Neurosci. **74**(11), 617–618 (2020). https://doi.org/10.1111/pcn.13140
52. Haver, A., Olsen, E., Akerjordet, K.: Well-being among hotel managers: a study on the influence of job stressors and cognitive reappraisal. Int. J. Contemp. Hosp. Manag. **31**(4), 1819–1835 (2019)
53. Grzeskowiak, S., Sirgy, M.J.: Consumer well-being (CWB): the effects of self-image congruence, brand-community belongingness, brand loyalty, and consumption recency. Appl. Res. Qual. Life **2**(4), 289–304 (2007)
54. World Health Organization: Mental health: a state of well-being. Retrieved from http://www.who.int/features/factfiles/mental_health/en/. Last accessed 19 Oct 2022
55. Serafini, G., Parmigiani, B., Amerio, A., Aguglia, A., Sher, L., Amore, M.: The psychological impact of COVID-19 on the mental health in the general population. QJM Int. J. Med. **113**(8), 531–537 (2020)
56. Zakaria, R.: Addressing direct, indirect COVID-19 impact on mental health. New Straits Times. https://www.nst.com.my/opinion/columnists/2020/10/632791/addressing-direct-indirectcovid-19-impact-mental-health, 16 Oct 2020
57. Mirowsky, J., Ross, C.E.: Depression, parenthood, and age at first birth. Soc. Sci. Med. **54**(8), 1281–1298 (2002)

The Impact of Smart Tourism Technologies and Destination Image on Tourists' Intention to Revisit Malaysia: The Mediator Role of Memorable Tourists' Experiences

Kwee Kim Peong, Kwee Peng Peong, Sook Fern Yeo, Sakiru Adebola Solarin, Cheng Ling Tan, and Kah Boon Lim

Abstract The concept of smart tourism technologies (STTs) has emerged as the most crucial element in the tourism industry. Its emergence enables the integration of modern informational tools and tourist resources which in turns offers more valuable information to tourism industry's stakeholders, in terms of timely data and interconnection. STTs also help mitigate the effects of losses suffered by the tourism sector as a result of the 2019 coronavirus disease (COVID-19). Malaysia's retails and tourism sectors have been drastically impacted where more businesses have to close as a result of the prolonged pandemic. Therefore, STTs plays a vital role in bolstering up the foundation for all tourism-related goods and services and provides visitors a complete destination experience. However, the studies on smart technology and psychological well-being in tourism are still in its early stages. As the review of literature suggested, more comprehensive research ought to be carried on STTs, destination image (DI), memorable tourist experiences and behaviour intentions. Only a few research has examined memorable visitor experiences as a mediating variable

K. K. Peong (✉) · S. F. Yeo · K. B. Lim
Faculty of Business, Multimedia University, Melaka, Malaysia
e-mail: kkpeong@mmu.edu.my

S. F. Yeo
e-mail: yeo.sook.fern@mmu.edu.my

K. B. Lim
e-mail: kblim@mmu.edu.my

K. P. Peong
Accounting Department, MLK Management Services, Melaka, Malaysia

S. A. Solarin
University of Nottingham, Semenyih, Selangor Darul Ehsan, Malaysia
e-mail: sasolarin@nottingham.edu.my

C. L. Tan
Graduate School of Business, Universiti Sains Malaysia, Gelugor, Malaysia
e-mail: tanchengling@usm.my

B. Awwad (ed.), *The AI Revolution: Driving Business Innovation and Research*, Studies in Systems, Decision and Control 524, https://doi.org/10.1007/978-3-031-54379-1_79

thus far. The current study aims to fill the gap by exploring STTs, DI, memorable visitors' experiences, and behavioural intents through the theory of planned behaviour (TPB), in order to determine if STTs and DI affect behavioural intentions through mediating factors like memorable tourist experiences. The study employs purposive sampling approach method and a systematic questionnaire using a seven-point Likert scale in gathering quantitative data from Malacca's Generation Y (Gen Y). The measurement and structural model are evaluated using the Smart PLS 4 version.

Keywords Smart tourism technologies · Destination image · Memorable tourist' experiences

1 Introduction

Tourism is seen as a crucial industry of growth as it is deemed as one of the largest foundations of wealth creation, livelihoods, and income worldwide [1]. The conventional tourism sector has been significantly boosted by the development of information and communication technology (ICT) in the era of smart technology, or "tourism 4.0" [2]. Nowadays, the tourism business and information technology go hand in hand. This brings about the idea of smart tourism technologies (STTs). The idea of STTs has emerged as the most important component of the travel and tourism sector. The integration of cutting-edge information technologies and tourist resources provide timely data, interconnectivity and more useful experience among tourism stakeholders [3]. The emergence of smart tourism and the subsequent phenomenon of digital transformations has drawn attention to the positive impact of information technology on the travel and tourism industry.

STTs have aided in reversing the effects of the tourism industry's losses during the coronavirus disease 2019 (COVID-19) [2]. According to [4], the prolonged pandemic compelled more firms' closures particularly in badly affected industries like tourism and retail in Malaysia. In 2020, Malaysia had a significant drop in visitor arrivals of 78.6% and income losses of 80.9% in the tourism industry. There were 4.3 million tourists who arrived in Malaysia overall, which was the lowest number of visitors from 2000 to 2020 (see Fig. 1). With 131.7 million visitors as opposed to 239.1 million in 2019, domestic visitor visits were down 44.6%. This resulted in a reduction of 60.8% in overall spending, from RM103.2 billion in 2019 to RM40.4 billion in 2020. The tourism industry has seen significant losses, totaling more than RM100 billion ($24.6 billion) and negative growth in the short-haul (ASEAN), medium-haul and long-haul markets of 78.8%, 80.0% and 74.0%, respectively [5].

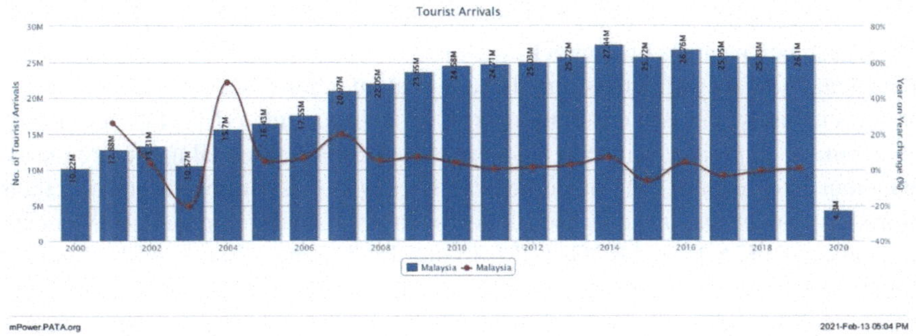

Fig. 1 The total number of tourist arrivals in Malaysia from 2000 to 2020

Numerous studies have been conducted on smart tourism destinations from the perspectives of various stakeholders due to the growing importance of technology and smartness [2, 3, 6–9]. Smartness can be seen as the appropriate use of information and communication technologies, which is the fundamental concept behind a smart destination [10]. Additionally, the idea of memorable tourism experiences has drawn a lot of attention recently and is viewed as a crucial differentiator for firms with a competitive edge. Recalling past events can have an impact on attitudes and actions in the present and the future [11]. Tourists' views of destinations for instance, are made up of their thoughts and feelings. Destinations are the key component of the travel and tourism sector and are the basis for all tourism-related goods and services as well as the entire tourist experience [7]. However, [2, 9] highlighted that the research of smart technology in tourism and psychological well-being is still at infant stage. Adding to this, a review of the literature suggested the need for more integrated research to be carried out on STTs, destination image (DI), memorable tourists' experiences [7, 12], and behaviour intentions [6]. Besides, the literatures are lacking of studies about memorable tourists' experiences as a mediating variable such as [7, 12, 13].

In the tourism industry, its stakeholders are more likely to feel that they will gain memorable tourists' experiences if they have more favourable opinion of the target country such as stunning natural scenery and friendly people, which in turn increases their likelihood of returning [13]. Clearly, the availability of research on theory of planned behaviour (TPB) to forecast travellers' overall behaviour while choosing a travel destination [1] is understudied. Therefore, this study extends TPB in ascertaining whether STTs and DI affect behavioural intentions through mediating factors like memorable tourists' experiences by taking into account the STTs, DI, memorable tourists' experiences, and behavioural intents. Since travel experience is a major antecedent of behaviour intention, hence it is crucial to comprehend how do STTs and DI influence tourists' entire travel experiences at their tourism destination. Notably, this implies that STTs and innovation have become driving and essential forces for tourist locations and firms in generating new skills growth and capacities for providing a pleasant and joyful service [14].

The present study adds to the body of research on tourism in two aspects. First, it offers fresh perspectives on how STTs, DI and memorable tourists' experiences are created and how they affect visitors' behavioural intentions. Second, its recommendations can assist tourism destination firms in incorporating STTs, DI and memorable tourists' experiences in their business strategic plans when taking future tourist behaviour into account. The outcome of this study can assist governments in their efforts to organise major events and support smart tourism development to align with the goals of KEGA 15 and SDGs 8, 12, 14 and 15. The organisation of this study is structured to three different segments. Firstly, the current study expands on existing knowledge and adds to the body of tourism literature by exploring the importance of STTs, DI and memorable tourists' experiences on tourists' behavioural intentions to revisit. Secondly, the study discusses the research methodology. Thirdly, the present study sums up the study by discussing implications of findings, limitations, and future research.

2 Literature Review

2.1 Theoretical Background

The TPB model is a psychological theory founded by [15] and has been utilised in numerous tourism-related research. The model explains how tourists observe various destinations [1]. According to [6, 15], this theory states that the mental processes underlying human behaviour intention and behavioural intentions generally precede actual behaviour. TPB can predict people's intentions to engage in a conduct at a particular time and location. It suggests that a person's behaviour is motivated by their objectives. Notably, whether the results are positive or negative, the impact of technology on people's conduct intention is quite substantial. This has driven the tourism industry to integrate technologies into popular tourist locations [5]. The paradigm is also in line with the core human psychological process, which is the model of cognitive, affective, and behavioural (CAB) that was introduced by [16]. Chang [7] mentioned attitudes are derived from CAB where individuals move from awareness to knowledge, detect emotional responses and display buying behaviours, correspondingly [16]. Thus, the proposed research framework for this study is shown in Fig. 2.

2.2 Smart Tourism Technologies on Memorable Tourists' Experiences and Behavioural Intention

As the tourism industry has been profoundly impacted by technology in several ways, tourist locations are not an exception to the use of smart technology [6]. Smart tourism

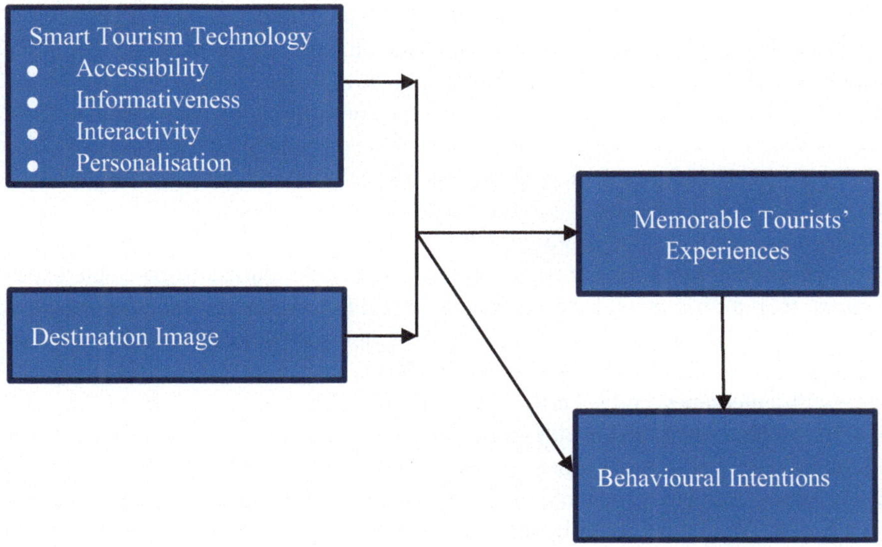

Fig. 2 The research framework

entails specific applications that could improve visitors' experiences and add value for customers [3]. According to [3, 7], smart tourism is a new, customer-oriented tourism model that employs intelligent technology to offer intelligent services and management as to satisfy the different needs of travellers and increase the tourist's value of the services and experiences.

More precisely, STTs can be defined as "*all forms of online tourism applications and information sources such as online travel agents, personal blogs, public websites, company websites, social media, smartphone apps, and so on*" [17]. In other words, STTs are particular tools, goods, and services such as artificial intelligence, the Internet of Things (IoTs), cloud computing, augmented reality, social networks sites, smart cards, radio-frequency identification, near-field communication, integrated payment method, and mobile communication that typically create value by encouraging greater connectivity, interactivity, personalisation, and co-creation and improve the entire travel experiences [3, 6].

The use of STTs technologies is important in a smart tourism setting for furnishing tourists and service providers with interactions, accurate information, better decision support, increased mobility, and high-quality travel experiences [3] and promoting tourism activities such as the dissemination of pertinent information and engagement-building [18]. STTs have changed the way tourism businesses are operated and managed, as well as tourist consumption patterns [9]. Therefore, tourists will choose a location that has a superior information infrastructure, for instance, networking and faster internet connectivity [3].

STTs have been classified into four distinct characteristics namely accessibility, informativeness, interactivity, and personalisation to quantify the usefulness of STTs in tourism destinations [3, 6, 7]. Accessibility is the first characteristic which refers to the availability, frequency, quantity, and accuracy of information available to tourists and provision of information at the tourist location employing different types of STTs [6, 7]. This has shorten the amount of time needed to make decisions and brought about the tourism experience enhancement [7]. STTs with high levels of accessibility have enabled tourists to gain more information and make their travelling trips more memorable. Hence, accessibility may enable tourists to co-create experiences, making it a pivotal indicator of memorable travel experiences and loyalty at tourism destinations [6, 9]. Mobile applications are the best example of smart tools in tourism. They help travellers make relevant decisions by providing them not only with emergency and security alerts, but also marketing and service information on weather alerts and health monitoring, coupons, discounts, tickets, banking, shopping, and reservations. These services furnish high levels of accessibility and convenience to service providers and travellers [18]. Furthermore, [17] disclosed that STT's accessibility and usability are interrelated.

The second characteristic of STTs is informativeness. It is defined as the quality and trustworthiness of information for tourists at tourist locations [9]. Since the tourism industry is intangible, credibility and quality of information are significant elements that might influence tourists' overall experience at the tourism site. One of the best examples is augmented reality. Tourists who utilise the augmented reality or virtual reality at the smart tourism destination, can easily grasp the richness of information scope for their tourism activities. As such, their travel experience at the smart tourism site could be enriched [6]. Tourists must know where to look for the most relevant information like finding the best deals (i.e., air tickets, car rentals, and hotels) and even how to begin a conversation with suppliers or other tourists online [17]. The third characteristic of STTs is interactivity as it encourages vital interaction between stakeholders through the interaction of real-time communication and information sharing [7]. This active interaction among individuals enables STTs to offer more useful and pertinent information that ease effective travel information searching. Therefore, when there is a higher level of interactivity, tourists will have more favourable opinions of information technologies and feel more convenience [6, 7]. This aids tourists in building up preliminary ideas about travel and facilitates effective travel information search [17]. Furthermore, the interactivity of STTs enables smart tourism destinations' operators to gather dynamic traveller data that could provide insights to destination marketers in creating and providing more specialised offerings [3, 6].

The last characteristic is personalisation. Personalisation relates to one's own personal expectations in meeting their specific needs, enhancing their trip experience, and increasing their satisfaction with smart tourism locations [3, 7, 9]. For example, traffic-routing apps could furnish travellers with the most effective path to minimise driving time, suffer less stress from traffic jams and eventually enhance their experience at smart tourism sites [6]. Another example is a hotel booking website where they can personally managed their bookings not only on hotel rates, locations, and dates, but also surrounding landmarks, service options and sorts of other preferences [7]. Kim [19] highlighted that overall satisfaction is a more accurate predictor of future behaviour. A cumulative satisfaction construct is more accurate in predicting future behaviours and economic performance. Besides, a consumer's memory of past events also mediates consumer behaviour by influencing the consumer's decision-making processes and eventually establishing the framework for upcoming purchase transactions. Considering the relationships discussed in the studies mentioned above, it is reasonable to conclude that STTs' attributes are the primary influences on tourists' experiences, satisfaction, and behavioural intentions [3, 6, 7, 9, 10, 17]. Accordingly, it is expected that:

H1 (a–d): STTs' (a) accessibility, (b) informativeness, (c) interactivity and (d) personalisation have a significant impact on behavioural intentions at tourist destinations.

2.3 Destination Image on Memorable Tourists' Experiences and Behavioural Intention

Destination Image (DI) refers to the tourist's knowledge, ideas, impressions, expectations, sentiments, and overall perception of a place [20, 21]. According to [7], DI describes a core tourism product image associated with visitor attractions and services that can satiate visitors' demands directly while stressing the location's image after their visit. There are three components of DI namely affective, cognitive, and conative dimensions. The affective dimension describes travellers' feelings towards a place. The cognitive component deals with tourists' assessments of the properties or features of a tourist destination. The conative aspect explains how visitors behave at a tourism destination based on the images they develop in their minds [19, 21]. Melon et al. [12] emphasised that DI affects satisfaction, which relates to quality and lasting nature of the tourist's experience. In contrast, the study by [1, 7, 12] indicated that DI influences future tourist behavioural intention at a tourism site. For instance, In Iran, tourists demonstrate that the country's reputation as a sport tourism destination has a beneficial impact on tourists' intentions to return. There have been several studies in the literature reporting the effects of memorable tourists' experiences on DI such as [7, 19, 22]. However, only few studies explore the relationship between DI and memorable tourists' experiences [12, 23]. Hence, it can be suggested that:

H2: DI has a significant impact on behavioural intentions at tourist destinations.

2.4 Memorable Tourists' Experiences and Behavioural Intention

Memorable tourists' experiences are a tourist's unique encounter that is fondly recalled and retained in his or her memory [10]. As highlighted by [19], individuals who have prior expertise with a product should utilise this knowledge in streaming their search. Information that is based on his or her prior experiences increases motivation and involvement. In addition, as past experiences are viewed as being quite reliable, they are a vital source of knowledge and are preserved in an individual's memory. [10, 11] noted that due to their influence on traveller preferences, the decision to return, and positive word-of-mouth, memorable tourists' experiences are regarded as crucial destination elements. Tourists frequently rely on their memories as reliable sources of information when making decisions. While choosing a destination, travellers think back on their previous visits. Memories can influence an individual's decision-making process and subsequent behaviour. Obviously, a memorable experience directly influences behaviour intention, albeit the patterns of influence can vary [6, 11, 13]. On the other hand, behavioural intention refers to tourists' perceptions of behaviours [7]. Behaviour intentions are the most accurate indicators of an individual's propensity to carry out a behaviour as well as their ability to forecast their future behaviour [6]. Tourists' intentions to return are highly valued for business growth and success. Regular visits from tourists lower marketing and promotion expenses, thus improve the profitability and viability of the tourism industry [1]. When a person has a good intention to execute a behaviour, they are more likely to do it. A strong desire to carry out a certain behaviour is thus highly indicative of its execution. For instance, recommendation intent is influenced by outstanding local cuisine experiences [6, 23]. Tourists who have unfavourable technical experience can have a greater impact on their contentment. They will decide whether to use the technology again. Tourists' prior experiences are strongly positively correlated with both their contentment in using technology and their future behaviour intentions [6]. Chang [7], Shin et al. [11] and Zhang et al. [13] discovered that DI has direct and indirect effects on the intention to return. The scope and importance of the mediation analysis should be evaluated for indirect consequences. Thus, this study examines how memorable tourists' experiences, acting as a mediator, can indirectly affect behavioural intention by proposing STTs and DI. Accordingly, the following hypotheses are formulated:

H3: Memorable tourists' experiences have a significant impact on behavioural intentions at tourist destinations.
H4 (a–d): Memorable tourists' experiences mediate the effect of STTs' (a) accessibility, (b) informativeness, (c) interactivity and (d) personalisation on behavioural intentions at tourist destinations.
H5: Memorable tourists' experiences mediate the effect of DI on behavioural intentions at tourist destinations.

3 Research Methodology

3.1 *A Sampling Method, Questionnaire Design and Data Collection*

It is determined that using the exploratory strategy is appropriate to precisely determine the links between the various variables of the study's components. However, it was not the case for this study due to the dearth of research on STTs and destination image toward behavioural intention in Malaysian tourism. The purposive sampling method is one of the more practical ways of collecting data for the present study. The justifications for using this technique are predicated on the premise that, given the goals and objectives of the study, certain types of people may have distinctive and significant opinions regarding the theories and problems under consideration, necessitating the inclusion of the sample of those individuals [24].

The current study employs a structured questionnaire with a seven-point Likert scale ranging from strongly agree (= 1) to strongly disagree (= 7) for the quantitative data collection from Generation Y (Gen Y) in Malacca. The United Nations Educational, Scientific, and Cultural Organisation (UNESCO) formally recognises Malacca as a World Heritage Site [25]. According to [26], Gen Y are those born between 1983 and 1996 and they are the ones who use the Internet the most compared to other age groups. As noted by [18], age reduces the amount of time spent online. The researchers assured participants that their participation in the study is fully voluntary and no confidential information would be exposed to unauthorized third parties. Demographic information will be kept separately from the data collection to guarantee their privacy. The participants were informed that they can respond as they wish and there are no right or wrong answers.

The self-administered questionnaire for this study is designed into three sections. First, two filter questions appeared as suggested by [9] (i.e.: Are you a tourist? For this travelling trip, did you utilise the smart tourism app?) The participants in the follow-up survey are invited when they responded "yes" to these filter questions. Second, the questionnaire contains items to gauge the primary key constructs. The current study is concerned with the conceptions of STTs, destination image, memorable tourists' experiences, and behavioural intention. Third, demographic data about the respondents is collected, including their age, ethnicity, gender, occupation, level of education, monthly income as an individual, number of years they have used STTs, and the most popular travel applications and websites they have explored.

3.2 *Data Analysis*

As survey research is typically not normally distributed [27], the current research employs partial least squares (PLS) modelling using the Smart PLS 4 version [28] as the statistical tool to assess the measurement and structural model. PLS is useful in estimating complex models if the sample size is comparatively small [8, 13]. The present study will run the issue of common method bias as suggested by [29, 30] to test the full collinearity. All the variables of this study will be regressed against a common variable. If the VIF ≤ 3.3, it indicates that there is no bias from using the single source data. The researchers follow the suggestions of [31] to analyse the data using a two-step approach. First, this study tests the measurement model to examine the validity and reliability of the instruments as proposed by [32] followed by the structural model to test the developed hypothesis. Second, the discriminant validity utilising the heterotrait-monotrait ratio (HTMT) criterion as suggested by [33] and modified by [34] is assessed. The HTMT values consist of the stricter criterion and the mode lenient criterion. The stricter and the mode lenient criteria should be $\leq$ 0.85 and $\leq$ 0.90, respectively.

4 Conclusion

Memorable experiences play a crucial role in the success of tourist services, not only as a measure of the competitiveness of a location or a firm, but also as influential elements pertaining to visitors' intentions to return. Consequently, tourism scholars and industry experts are starting to recognise the significance of memorable tourists' experiences [14]. Different people have various perspectives on technology [8]. In terms of marketing and industry, STTs play a vital part in the social, economic, and cultural development of a nation [2]. Similarly, [35] added that the metaverse creates enormous prospects that are extremely beneficial to the tourism industry. Customers can gain metaverse tourism benefits by experiencing a new and more immersive form of virtual travel. STTs pledge to offer travellers and service providers wider target customers, more practical information, greater freedom, sound decision assistance, and good vacation experience [14]. The degree of experience, which varies depending on the number of visitor contacts with a location, can be determined by finding a strong correlation between DI and the degree of experience [19]. Even event planners can utilise the metaverse to increase operational efficiency. The metaverse provides a fantastic platform for travel and hospitality firms to improve their marketing efforts [35]. In addition, [36] shows that through the effective transformation of their business models, business model innovation (BMI) demonstrates how to tackle and recover the business throughout the pandemic crisis. Hence, this study is expected to develop marketing tactics (i.e., STTs) for the tourism industry that emphasise on offering memorable tourists' experiences or boosting positive DI are essential for influencing future behaviour, specifically in a developing country, Malaysia. The current study has a few implications. This study will add to the body of knowledge as it encapsulates a thorough understanding of STTs in a developing nation. It enriches the little empirical study which has been conducted on STTs in the developing market like Malaysia. The present study also analyses how innovative tourism practices such as STTs, DI and memorable tourists' experiences can enhance behavioural intentions to revisit tourist destinations. The findings suggest that the tourism industry players should combine diverse informational resource systems to meet the needs of individual tourists. Furthermore, the tourism industry should use strategies to develop memorable tourists' experiences such as incentivising visitors to snap photos and share their experiences on social media by giving them presents and planning interactive events. Tourism authorities and industry players could collaborate to create a favourable DI and encourage visitors' intents to revisit a tourist destination by emphasising local cuisine, culture, and architecture. However, the study has its own limitation. The present study employs purposive sampling methods based on tourists who visited Malacca. Thus, future researchers could expand the study by covering more states and sample sizes so that the accurate generalisation could be made. Given that this study uses a quantitative approach, adopting a qualitative approach and different population samples will improve the validity of future research findings.

5 Appendices

See Figs. 1 and 2.

Acknowledgements The Malaysian Ministry of Higher Education (MOHE) provided the Fundamental Research Grant Scheme (FRGS) (Ref: FRGS/1/2022/SS01/MMU/03/1) to fund this research, for which the authors are truly grateful.

References

1. Abbasi, G.A., Kumaravelu, J., Goh, Y., Singh, K.S.D.: Understanding the intention to revisit a destination by expanding the theory of planned behaviour (TPB). Span. J. Mark. ESIC **25**(2), 280–307 (2021)
2. Gani, M.O., Roy, H., Faroque, A.R., Rahman, M.S., Munawara, M.: Smart tourism technologies for the psychological well-being of tourists: a Bangladesh perspective. J. Hosp. Tour., 2514–9792 (2023)
3. Azis, N., Amin, M., Chan, S., Aprilia, C.: How smart tourism technologies affect tourist destination loyalty. J. Hosp. Tour. Technol. **11**(4), 6603–6625 (2020)
4. OECD: OECD Economic Surveys Malaysia-Executive Summary. Shutterstock.com. OECD (2021).
5. Peong, K.K., Yeo, S.F., Peong, K.P., Tan, C.L., Solarin, S.A., Lim, K.B.: Smart tourism technology framework: a gateway to enhancing tourists' experiences and psychological behaviour. In: 2023 International Conference on Digital Applications, Transformation and Economy (ICDATE), pp. 219–225. IEEE Xplore, Sarawak (2023)
6. Jeong, M., Shin, H.H.: Tourists' experiences with smart tourism technology at smart destinations and their behavior intentions. J. Travel Res., 1–4 (2019)
7. Chang, S.: Can smart tourism technology enhance destination image? The case of the 2018 Taichung World Flora exposition. J. Hosp. Tour. Technol. **13**(4), 590–607 (2022)
8. Shin, H.H., Jeong, M., Cho, M.: The impact of smart tourism technology and domestic travelers' technology readiness on their satisfaction and behavioral intention: a cross-country comparison. Int. J. Tour. Res. **23**(5), 726–742 (2021)

9. Ng, K.S.P., Wong, J.W.C., Xie, D., Zhu, J.: From the attributes of smart tourism technologies to loyalty and WOM via user satisfaction: the moderating role of switching costs. Kybernetes **52**(8), 2868–2885 (2023)
10. Shin, H.H., Kim, J., Jeong, M.: Memorable tourism experience at smart tourism destinations: do travelers' residential tourism clusters matter? Tour. Manag. Perspect. **46**, 101103 (2023)
11. Stavrianea, A., Kamenidou, I.: Memorable tourism experiences, destination image, satisfaction, and loyalty: an empirical study of Santorini Island. EuroMed. J. Bus. J. Bus. **17**(1), 1–20 (2022)
12. Melon, M.P.A., Fandos-Herrera, C., Sarasa, R.G.: Analysis of antecedents and consequences of memorable tourist experiences (MTEs): a Spanish case study. J. Vacat. Mark., 1–15 (2021)
13. Zhang, H., Wu, Y., Buhalis, D.: A model of perceived image, memorable tourism experiences and revisit intention. J. Destin. Mark. Manag. **8**, 326–336 (2018)
14. Elshaer, A.M., Marzouk, A.M.: Memorable tourist experiences: the role of smart tourism technologies and hotel innovations. Tour. Recreat. Res., 1–13 (2022)
15. Ajzen, I.: The theory of planned behavior. Organ. Behav. Human Decis. Process. **50**(2), 179–211 (1991)
16. Lavidge, R.J., Steiner, G.A.: A model for predictive measurements of advertising effectiveness. J. Mark. 59–62 (1961)
17. Huang, C.D., Goo, J., Nam, K., Yoo, C.W.: Smart tourism technologies in travel planning: the role of exploration and exploitation. Inf. Manag. **54**, 757–770 (2017)
18. Tavitiyaman, P., Qu, H., Tsang, W.L., Lam, C.R.: The influence of smart tourism applications on perceived destination image and behavioral intention: the moderating role of information search behavior. J. Hosp. Tour. Manag. **46**, 476–487 (2021)
19. Kim, J.H.: The impact of memorable tourism experiences on loyalty behaviors: the mediating effects of destination image and satisfaction. J. Travel Res. **57**(7), 856–870 (2018)
20. Fakeye, P.C., Crompton, J.L.: Image differences between prospective, first-time, and repeat visitors to the lower Rio Grande Valley. J. Travel Res. **30**(2), 10–16 (1991)
21. Viet, B.N., Dang, H.P., Nguyen, H.H.: Revisit intention and satisfaction: The role of destination image, perceived risk, and cultural contact. Cogent Bus. Manag. **7**(1), 1796249 (2020)
22. Hu, F., Shen, H.: Memorable tourism experiences, destination image, satisfaction and revisit intention of Chinese outbound tourists to South Pacific Islands. Adv. Hosp. Leis. **17**, 103–128 (2021)
23. Rasoolimanesh, S.M., Seyfi, S., Hall, C.M., Hatamifar, P.: Understanding memorable tourism experiences and behavioural intentions of heritage tourists. J. Destin. Mark. Manag. **21**, 1–14 (2021)
24. Campbell, S., Greenwood, M., Prior, S., Shearer, T., Walkem, K., Young, S., Bywaters, D., Walker, K.: Purposive sampling: complex or simple? Research case examples. J. Res. Nurs., 1–10 (2020)
25. Teo, C.B.C., Khan, N.R.M., Rahim, F.H.A.: Understanding cultural heritage visitor behavior: the case of Melaka as World Heritage City. In: INCOMaR 2013. Proc. Soc. Behav. Sci. **130**, 1–10 (2014)
26. Ramirez-Correa, P.E., Grandon, E.E., Arenas-Gaitan, J.: Assessing differences in customers' personal disposition to e-commerce. Ind. Manag. Data Syst. **119**(4), 792–820 (2019)
27. Chin, W.W., Marcolin, B.L., Newsted, P.R.: A partial least squares latent variable modeling approach for measuring interaction effects: results from a monte carlo simulation study and an electronic-mail emotion adoption study. Inf. Syst. Res. **14**(2), 189–217 (2003)
28. Ringle, C.M., Wende, S., Becker, J.-M.: SmartPLS 4. Oststeinbek: SmartPLS. Retrieved from https://www.smartpls.com (2022)
29. Kock, N., Lynn, G.S.: Lateral collinearity and misleading results in variance-based SEM: an illustration and recommendations. J. Assoc. Inf. Syst. **13**(7), 546–580 (2012)
30. Kock, N.: Common method bias in PLS-SEM: a full collinearity assessment approach. Int. J. e-Collab. **11**(4), 1–10 (2015)
31. Anderson, J.C., Gerbing, D.W.: Structural equation modeling in practice: a review and recommended two-step approach. Psychol. Bull. **103**(3), 411–423 (1988)

32. Hair, J.F., Thomas, G., Hult, M., Ringle, C.M., Sarstedt, M.: A Primer on Partial Least Squares Structural Equation Modeling, 3rd edn. Sage, Thousand Oakes, CA (2022)
33. Henseler, J., Ringle, C., Sarstedt, M.: A new criterion for assessing discriminant validity in variance-based structural equation modeling. J. Acad. Mark. Sci. **43**(1), 115–135 (2015)
34. Franke, G., Sarstedt, M.: Heuristics versus statistics in discriminant validity testing: a comparison of four procedures. Internet Res. **29**(3), 430–447 (2019)
35. Thapa, P.: Metaverse and tourism industry: a conceptual proposition. In: El Khoury, R., Alareeni, B. (eds.) How the metaverse will reshape business and sustainability. Contributions to Environmental Sciences and Innovation Business Technology. Springer, Singapore (2023)
36. Faisal, Y.A., Kanti, R.A., Alamanda, A.R., Apriliadi, A.: The role of business model innovation (BMI) in social enterprises during pandemic COVID-19 in Indonesia: a case of Islamic boarding schools' business units. In: Rafiki, A., Dana, L.P., Nasution, M.D.T.P. (eds.) Open Innovation in Small Business. Contributions to Environmental Sciences and Innovative Business Technology. Springer, Singapore (2023)

A Purchase Factor for AirAsia Berhad e-Ticketing System Based on Customer's Perception: An EFA Analysis

Nor Alina Ismail, Nurzulaikha Abdullah, and Fakhitah Ridzuan

Abstract The use of the IR4.0 technology is now growing fast and most countries and industries now apply at least the simple IR4.0 technology in their business environment including in the Airline Industry in Malaysia. This technology has also been applied by the famous low-cost airline, AirAsia Berhad. The changes in technology usage (for example, online purchasing, check-in, and so on) have forced airline companies, into intense competition since now in Malaysia, many airline companies such as FireFly, Malindo, Malaysia Airlines, and so on. The use of e-ticketing systems in the Airline industry has been widely used for the past 10 years in selling flight tickets online. This study is trying to discover what is the feedback from customers when they are purchasing flight tickets using the AirAsia Berhad Berhad e-ticketing system. The measurement of Cronbach alpha and EFA was conducted in this study. Data was collected from 510 AirAsia Berhad customers to get their feedback and points of view regarding the current online e-ticketing system provided by the company. This study highlighted in detail the Exploratory Factor Analysis process for 'the purchase construct in the online transaction cycle. Based on the findings, the customer felt satisfied using the e-ticketing system and tended to be loyal to the company as the e-services provided by the company meet their needs.

Keywords e-ticketing system · Airline industry · AirAsia · Exploratory factor analysis · Online transaction cycle · Purchase

1 Introduction

The Fourth Industrial Revolution (IR4.0) refers to the current technology of advancement considered by the integration of digital, physical, and biological. The technology of IR4.0 has also been widely used in the airline industry particularly. It encompasses various types of technologies such as artificial intelligence, the Internet

N. A. Ismail (✉) · N. Abdullah · F. Ridzuan
Faculty of Data Science and Computing, Universiti Malaysia Kelantan, Kota Bharu, Malaysia
e-mail: alina.i@umk.edu.my

B. Awwad (ed.), *The AI Revolution: Driving Business Innovation and Research*, Studies in Systems, Decision and Control 524, https://doi.org/10.1007/978-3-031-54379-1_80

of Things, automation, and so on [1]. In the context of the airline e-ticketing system used by AirAsia Berhad company, IR4.0 had a significant impact on the way airline ticketing and travel processes are conducted. Online E-ticketing can be defined as "a digitalization of airline ticketing process, that replaces the traditional paper tickets before. With IR4.0 technology, there are several technologies been influenced such as the digitalization of booking and reservation systems, personalization and Data analytics, contactless Travel and biometrics, Automated customer support, real-time updates and notifications, eco-friendly initiatives, and blockchain technology" [1].

Recent statistics show that the competition becomes very violent to cover airline companies' operating costs and break even [1, 2]. Most airline companies in Malaysia especially AirAsia Berhad choose electronic systems as a platform to do their business with the customers. Same with other countries, our airline industry in Malaysia is also affected by the current trends in technology. This factor causes the low-cost company to improve their services in getting customers. Malaysia had 33.03 million internet users at the beginning of 2023, with a percentage of internet users of 96.8% in January 2023 [3]. As more people get access to the internet and grow more comfortable with online transactions, they are increasingly turning to e-ticketing as a quick and cost-effective option for purchasing airline tickets.

Exploratory Factor Analysis (EFA) was conducted to identify the items that affect customer satisfaction in purchasing flight tickets online. In the literature review section, the related studies will be discussed and Sect. 3 will present the results and findings from the EFA. Section 4 will provide insight into the future direction of airline e-ticketing and Sects. 4 and 5 will summarize in this paper.

2 Literature Review

In the literature review section, the discussion on online transaction cycle, purchase factor, and related studies will be discussed. The reason for reviewing is the need to get a better justification for the research idea.

2.1 Online Transaction Cycle

According to past studies, the online transaction cycle had 3 main stages such as pre-purchase, purchase, and post-purchase. In general, past studies explain that pre-purchase can be defined as "the activities of surveying, searching, and comparing the price of the product before buying". While Purchase can be defined as "an activity of placing a booking and making a payment online". Lastly, post-purchase can be defined as tracking the order or reviewing a product after making a purchase [4]. This paper will focus on the purchase factor and the discussion of the data collection result will be discussed in the next section.

3 Research Design

A pilot test was conducted before proceed with main survey. A total 150 questionnaires distributed to respondents who had experience purchase flight tickets online. The sampling is fall under random sampling. The questionnaire content was developed to carter the study area and based on the literature review. Before distribute the questionnaire, the instrument been check by expertise and industry expert in order to get a better understanding on the study nature. The questionnaire been developed using English language as the target sampling is random.

The questionnaire has a few sections including customer demography, online transaction section and recommendation or suggestion section. In the purchase feature section, there are 8 items been designed to measure the customer satisfaction and loyalty towards online e-ticketing system in Malaysia [5].

Table 1 shows the descriptive analysis result for the Purchase construct. The data were analysed using IBM SPSS.

Based on Table 1, there were three (3) constructs and eight (8) items used for this study. This result found that most of the respondents seems like happy using all the services during the purchase stage. Hence, this finding is giving a positive feedback to the company so that they can improve their online services in the future [5].

Table 1 Descriptive analysis

Construct/items	Mean	Std. dev.
e-Booking		
I can easily make a booking for flight tickets through the website	6.51	1.0
The websites always display booking charges	6.22	1.1
Payment methods		
Different payment methods to choose from are available in airline e-ticketing services in Malaysia	6.39	0.994
The airline websites offer convenient payment procedures	6.32	1.1
I accept the payment methods provided by airline websites	6.32	0.998
Privacy and security		
The websites never misuse my personal information	6.14	1.2
I feel secure giving my personal information to the website	6.20	1.1
I feel safe completing transactions on these airline websites	6.34	1.1
Average	6.31	

4 Research Methodology

A survey was conducted via online platform in order to collect data from the respondents. A total 510 collected. The reliability test been conducted in order to identify whether the factor is reliable or not. After done the reliability test, purchase construct had 0.930 of reliability value which indicates as accepted and reliable. After conducting a test for Cronbach alpha, an exploratory factor analysis (EFA) was conducted in the study in order to test the factor loading among items in the questionnaire. The decision to determine the number of factors and items are based on factor loading and the value is must > 0.50 for each item [5].

5 Exploratory Factor Analysis (EFA)

According to [5, 6] study, they mentioned EFA is one of the applied statistical techniques. As this study was adopted the validated instruments from previous research, EFA tested need to be done again because to make sure all the procedure was followed by the researcher. In addition [6], mentioned if the researcher makes any changes to the existing instruments, the researcher is required to conduct the EFA procedure to get the latest result as it has been conducted in this study.

In the EFA procedure, we concern on the factor loading value need to exceed 1.0 and also, we look into the factor extracted into different components [5, 6]. The factor loading with > 0.6 result will go for further analysis and considering the item with > 0.5 of reliability result.

5.1 EFA Data Analysis

Table 2 shows that the EFA result has extracted into one component from the 3 main components. Only the factor loadings with the value above 0.5 will be retained for this study in order to get a consistent result and finding from previous studies [5].

As shown in Table 2, all the items in the purchase construct were loaded into one factor. Result shows that the range between 0.778 and 0.86. Before EFA conducted, the three main constructs were proposed to be examined in this purchase phase which are (1) e-booking, (2) Payment Method, and (3) Privacy and Security. However, after the EFA was performed in this analysis, a study found that all eight items strongly belong to one factor. Hence, after reviewing and put the consideration of past studies, the study found that the name “Purchase Cycle” is very suitable for this study.

Table 2 EFA for purchase result

Items	Factor loading
PayMethod_02	0.860
Privacy_03	0.849
PayMethod_03	0.830
e-Booking_01	0.828
e-Booking_02	0.828
PayMethod_01	0.807
Privacy_02	0.802
Privacy_01	0.778

In the purchase factor, the ppayment methods is one of the important elements of customer satisfaction, that will lead to the customer loyalty as suggested in the proposed model. This is because an e-payment method is another way of make a payment online, such as credit card, direct debit, or PayPal [6]. In addition, online payment or e-payment is essential in e-commerce in order to meet the target in online business. Customers normally like to have many payment methods when making an online purchase.

While another factor in the purchase construct is the privacy and security factor in was presented as the level of security and data protection. Alhaiou [6] stated that this factor is one of the most important elements because the security aspect of the websites needs to be strong to give customers sufficient confidence to use their online services. Besides, this factor can protect all the information provided by customers. While [7] explained that one of the primary barriers to e-shopping is the worry of privacy and security. Security in the context of e-commerce is considered as an important factor that can affect customer satisfaction toward e-services.

6 Conclusion and Discussion

This study underwrites to the measurement of Purchase construct particularly in the context of AirAsia Berhad especially towards an e-ticketing system. The EFA result of this study produced a structure that extracted a factor of purchase feature which is named as "purchase cycle". This may reflect the items that apply to this research. In In addition, the management and Service Quality in e-ticketing system is an unusual surprise that is shaped by a wide range of aspects such as usability, web development, information security, website specializations, and consumer pleasure. By addressing these concerns is critical for an organization to create an e-ticketing architecture that first identifies consumer wants and produces a high component of satisfaction.

The rise in the number of internet, social networking, and media users has resulted in massive data management [8] which has further increased the demand for a reliable and efficient cloud computing environment to manage the massive amounts of passenger data generated from e-ticket transactions. Cloud computing provides the flexibility to scale up or down resources based on demand, thereby reducing the capital expenditure and operational costs of maintaining an IT infrastructure. In addition, it not only provides a cost-effective solution for data storage and processing, but it also ensures the security and privacy of passenger information, which is critical for e-ticketing services.

There are a few aspects can be seen for the future study such as, the cloud computing is a type of virtual computing in information technology that uses virtual data centres and requires the highest level of privacy protection [8]. Since cloud computing relies on remote servers to store, manage, and process data, privacy and security are critical concerns. To address these concerns, cloud computing providers offer robust security measures such as firewalls, data encryption, access control, and monitoring to ensure the highest level of privacy protection for the users.

Besides, Dai et al. [9] proposed an e-ticketing system using cloud computing for sports competitions. The study found that when 6000 users made ticketing inquiries simultaneously, the e-ticketing system responded within an average of 583 ms with a 100% success rate, indicating that the system had load space at that time and would respond even faster if it was built on the cloud.

Electronic payment systems are limited to a few strategically positioned institutions, but the COVID-19 pandemic has changed the way people interact with electronic payments [10]. Service providers are seeking to stay up with rising technical progress and current technology deployment in a range of industries, such as electronic payment, which allows clients to complete all transaction payments electronically [11]. The integration of cloud computing and e-ticketing has the potential to enhance the efficiency of airline operations, provide cost-effective solutions, and improve the customer experience. Besides, [12] mentioned in order to get a sustainability of each business, the IR4.0 systems and technology need to be adapted and must offered a customised product. This is also in getting a satisfaction among customer.

Thus, the rapid adoption of technology in the airline industry has led to intense competition among companies in Malaysia, which has pushed them to adopt innovative measures to improve customer satisfaction. E-ticketing and cloud computing are two technological innovations that have been increasingly used by airline companies to provide a more convenient and efficient service to customers. The study conducted has shown that e-booking, payment, privacy, and security are the main factors that influence customer satisfaction in purchasing the e-ticket. Therefore, it is essential for airline companies to ensure that these factors are carefully managed and implemented to provide a smooth and hassle-free experience for customers. Furthermore, the integration of e-ticketing and cloud computing is expected to play a significant role in the future direction of the airline industry. The use of cloud computing allows companies to store and process vast amounts of data, including passenger information and flight schedules, in a cost-effective and efficient manner. This will enable airlines to better manage customer information and improve service delivery. Other than that [13] mentioned in their studies that the digital transformation also might cause the main important thing in industrial production including the use of computer vision, internet of things, robotics and so on.

Overall, the airline industry in Malaysia has embraced technological advancements to enhance their business operations and stay competitive in the market. The integration of e-ticketing and cloud computing has been a significant step in this direction. It is crucial for airline companies to continue to explore and innovate using the latest technologies to improve customer satisfaction and drive sustainable economic and financial growth.

References

1. Subramanya, K., Kermanshachi, S., Pamidimukkala, A.: Evaluation of E-ticketing technology in construction of highway projects: a systematic review of adoption levels, benefits, limitations and strategies. Front. Built Environ. **8**, 1–14 (2022). https://doi.org/10.3389/fbuil.2022.890024
2. Mahdzar, M.: Airline E-ticket online purchase intention among university student. Adv. Bus. Res. Int. J. **7**, 53 (2021). https://doi.org/10.24191/abrij.v7i2.12068
3. Kemp, S.: Digital 2023: Malaysia. https://datareportal.com/reports/digital-2023-malaysia. Last accessed 16 Feb 2023.
4. Qteishat, M.K., Alshibly, H.H., Al-ma'aitah, M.A.: The impact of e-ticketing technique on customer satisfaction: an empirical analysis. J. Inf. Syst. Technol. Manag. **11** (2014). https://doi.org/10.4301/S1807-17752014000300001
5. Ismail, N.A., Daud, S.N.: Exploratory factor analysis of pre-purchase construct for airlines e-ticketing system in Malaysia in the context of e-transaction. Int. J. Adv. Trends Comput. Sci. Eng. **8**, 255–258 (2019). https://doi.org/10.30534/ijatcse/2019/25822019
6. Alhaiou, T.: An empirical study on electronic customer relationship management (E-CRM) implementation and E-loyalty at different adoption stages of the transaction cycle (2011)
7. Teck, C.L., Choon, L.K., Hoi, P.: Airline e-ticketing service: how e-service quality and customer satisfaction impacted purchase intention. Int. Bus. Manag. **5**, 200–208 (2013)
8. Agarkhed, J., Ramegowda, A.: Security and privacy for data storage service scheme in cloud computing. Int. J. Inf. Eng. Electron. Bus. **9**, 7–12 (2017). https://doi.org/10.5815/ijieeb.2017.04.02
9. Dai, W., Shao, J., Zhang, X.: Research on the design and application of sports competition ticketing platform based on edge computing. Complexity (2021). https://doi.org/10.1155/2021/3682165
10. Odeh, M., Yousef, M.: The effect of Covid-19 on the electronic payment system: usage level trust and competence perspectives. Indones. J. Electr. Eng. Comput. Sci. **22**, 1144–1155 (2021). https://doi.org/10.11591/ijeecs.v22.i2.pp1144-1155
11. Al Farawn, A., Rjeib, H.D., Ali, N.S., Al-Sadawi, B.: Secured e-payment system based on automated authentication data and iterated salted hash algorithm. TELKOMNIKA Telecommun. Comput. Electron. Control. **18**, 538–544 (2020). https://doi.org/10.12928/TELKOMNIKA.V18I1.15623
12. Mary Josephine, M., Muninarayanappa, M.: Industry 4.0—It's Impact and Scope Towards Sustainable Business Operations (2023)
13. Aloysius Edward, J., Jaheer Mukthar, K.P., Asis, E.R., Sivasubramanian, K. (eds.) Current Trends in Economics, Business and Sustainability. ICEBS 2023. Contributions to Environmental Sciences and Innovative Business Technology. Springer, Singapore. https://doi.org/10.1007/978-981-99-3366-2_15

SPRINGER NATURE

GPSR Compliance

The European Union's (EU) General Product Safety Regulation (GPSR) is a set of rules that requires consumer products to be safe and our obligations to ensure this.

If you have any concerns about our products, you can contact us on ProductSafety@springernature.com

In case Publisher is established outside the EU, the EU authorized representative is:

Springer Nature Customer Service Center GmbH
Europaplatz 3
69115 Heidelberg, Germany

The manufacturer's authorised representative in the EU is Springer Nature Customer Service Centre GmbH, Europaplatz 3, 69115 Heidelberg, Germany. If you have any concerns regarding our products, please contact ProductSafety@springernature.com

Printed and bound by CPI Group (UK) Ltd, Croydon, CR0 4YY

01/10/2026

02225874-0003